LOGIC AND DISCRETE MATHEMATICS

A Computer Science Perspective

WINFRIED KARL GRASSMANN

Department of Computer Science
University of Saskatchewan

JEAN-PAUL TREMBLAY

Department of Computer Science
University of Saskatchewan

Prentice Hall, Upper Saddle River, New Jersey 07458

Library of Congress Cataloging-in-Publication Data

Grassmann, Winfried K.
    Logic and discrete mathematics : a computer science perspective /
  Winfried Karl Grassmann, Jean-Paul Tremblay.
        p.   cm.
    Includes bibliographical references and index.
    ISBN 0-13-501206-6
    1. Computer science — Mathematics.   I. Tremblay, Jean-Paul
  II. Title.
  QA76.9.M35G725   1996
  005.1'01'5113—dc20   95-38351
                                                                          CIP
                                                                           AC

                Acquisitions editor: *Alan Apt*
                Editorial/production supervision: *Maes Associates*
                Copy editor: *William O. Thomas*
                Cover designer: *Bruce Kenselaar*
                Manufacturing buyer: *Donna Sullivan*

 © 1996 by Prentice-Hall, Inc.
Pearson Education
Upper Saddle River, New Jersey 07458

Printed in the United States of America

10   9   8   7   6

ISBN 0-13-501206-6

Prentice-Hall International (UK) Limited, *London*
Prentice-Hall of Australia Pty. Limited, *Sydney*
Prentice-Hall Canada Inc., *Toronto*
Prentice-Hall Hispanoamericana, S.A., *Mexico*
Prentice-Hall of India Private Limited, *New Delhi*
Prentice-Hall of Japan, Inc., *Tokyo*
Prentice-Hall (Singapore) Asia Pte. Ltd., *Singapore*
Editora Prentice-Hall do Brasil, Ltda., *Rio de Janeiro*

To my wife Louise Grassmann and
to my children Stephanie and Bettina Grassmann
—W.K.G.

To my wife Deanna Tremblay and
to my grandchildren
Robert, Leanne, Lisa, and Nicole Tremblay
—J.P.T.

Contents

Preface

Most universities require that a course in discrete mathematics be taken by every undergraduate computer science student, and rightly so. Indeed, discrete mathematics gives the appropriate theoretical foundations for computer science, foundations that are not only beneficial for doing theoretical computer science, but also for the practice of computer science as evidenced by the recent proliferation of books on formal methods. The areas covered in a course in discrete mathematics vary, but they traditionally include logic, sets, relations, functions, and graphs. All these topics are included in this book. Moreover, this book reflects several recent trends in computer science. In particular, it gives a more thorough exposure to logical reasoning than most other texts. It also shows how to use discrete mathematics and logic for specifying new computer applications and how to reason about programs in a systematic way. The book contains chapters on languages and grammars, the Z specification language, and relational databases. There is a chapter describing Prolog, a programming language based on logic, and a section discussing Miranda, a language based on functions. In all chapters, numerous examples relate the mathematical concepts to problems in computer science. We found that such examples are essential for keeping the student motivated.

The outline of the book is as follows. Chapter 1 covers propositional calculus. We also define what is to be understood by a formal derivation. In fact, formal derivations are introduced as refinements of the type of logical reasoning we use in daily life. The chapter also contains an extensive discussion on algebraic manipulation of logical expressions.

Chapter 2 discusses predicate calculus. We cover predicates and quantifiers, and we try again to relate this theory to everyday thinking. Chapter 2 also introduces unification, a topic that is needed in order to understand logical languages such as Prolog (covered in Chapter 4) and resolution theorem proving (described in Chapter 11). Chapter 2 also introduces equational logic. Equational logic is concerned with the manipulation of mathematical equations, and this is obviously of fundamental importance. As it turns out, equational logic is closely related to functions, and indeed, without functions, equational logic would not have attained the dominance in mathematical reasoning which it now has. This motivated us to introduce functions at this point.

Chapter 3 deals with induction and recursion, which is an important part of logic, mathematics, and, above all, computer science. We discuss several proof methods, including mathematical induction, strong induction, proof by recursion, and structural induction. The notions of matrices and sums are also introduced. As an application, recursive programming techniques are explored. This chapter also includes material on recursive functions and decidability. Though recursive functions are very important, they constitute advanced material which may be omitted.

Chapter 4 covers Prolog. The coverage is sufficient to allow the reader to write nontrivial Prolog programs. The connections between logic and Prolog are also explored. A section in Chapter 4 explains what kind of logical statements can be translated into Prolog. Another section shows how to use Prolog to program manipulations involving logical expressions. The study of Prolog also reinforces the concept of recursion, since almost every nontrivial Prolog program makes use of recursion.

Computer scientists deal with many different types of objects, which can in turn be combined in different ways. To understand these compound objects and their manipulation, the student has to understand the concepts of sets and relations, which are discussed in Chapter 5. The important operations on these constructs are described by numerous examples, many of which are directly related to computer applications.

Chapter 6 deals with several topics regarding functions, such as algorithmic analysis and computational complexity. These concepts are illustrated with a number of important examples. In addition to the topics mentioned, the language Miranda is introduced as an example of a functional language.

Chapter 7 gives an overview of graphs and trees, concepts that are fundamental to computer science. In addition to introducing basic terminology dealing with paths, reachability, and connectedness, the emphasis in this chapter is on computing paths, minimum paths, minimum weighted paths, spanning trees, and minimum spanning trees for weighted graphs. This chapter concludes with a discussion of the application of graphs to project planning and management.

Chapter 8 is about Z, a language applying concepts from logic and set theory for specifying and analyzing requirements in software development.

Chapter 9 shows how to use logic, and to a certain extent sets and functions, to reason about programs. As in Chapters 1 and 2, we start with commonsense reasoning, and we introduce formal correctness proofs as a refinement of normal reasoning.

Chapter 10 deals with context-free grammars as a vehicle for defining the syntax of programming languages and their use in syntax analysis. Emphasis is given to an LL(1) parser generator system.

Chapter 11 gives additional information about derivations. In particular, it shows how to use natural deduction and the techniques of resolution theorem proving.

The book concludes with a discussion of relational databases in Chapter 12. This chapter first focuses on the relational data model and its associated relational algebra. An alternative way of describing queries based on predicate calculus is also given. Finally, an overview of a structured query language for specifying queries is given.

This text would appeal to a student in computer science at the freshman (second term), sophomore, or junior level. Preliminary versions of this book were used in a one-semester course given to second-year computer science students. In this course, most of the material contained in Chapters 1 through 6 was covered. To this, we selectively added topics from Chapters 7 through 12. However, the book contains enough material for a full year course, but in discrete mathematics, such courses are not normally offered. Whatever the length of the course, it should lay the mathematical foundations for many classes, including classes in database design, syntactical analysis and parsing, artificial intelligence, programming languages, logic programming, functional languages, and computer hardware.

The application chapters allow the instructors to shift the emphasis according to their interests. In all cases, the material of Chapters 1, 2, and 5 is essential, and most instructors want to cover Section 6.1, which discusses functions. With respect to the other chapters, we would like to suggest the following four scenarios, which can be combined or altered, depending on the preferences of the instructor.

> *Emphasis on Procedural Programming* For procedural programming, recursion is very important, and this is covered in Chapter 3. The same chapter contains a discussion of recursive functions, and it shows what can and what cannot be solved by computers. Section 6.3 gives an introduction to computational complexity which should prove useful to the student. The instructor may also want to cover Chapter 7, which discusses graphs and trees, and Chapter 9, which shows how to prove that a program is correct.

> *Emphasis on Logic and Logic Programming* Since recursion is very important for all types of logic programming, the instructor should cover recursion, which is described in Chapter 3. Chapter 4 then gives an extensive treatment of Prolog. At this point, Chapter 10, grammars and languages, and Chapter 11, which shows resolution theorem proving, provide suitable topics.

> *Emphasis on Functions and Functional Programming* To provide a functional programming emphasis, all of Chapter 3 should be covered. From Chapter 6, Sections 6.1 and 6.3 are important. Of course, Section 6.5, Miranda, is central under this scenario. For the remaining topics, the instructor may select material from Chapter 8, requirement specification, Chapter 9, correctness proofs, and Chapter 10, grammars and languages.

> *Emphasis on Information Systems* The slant to systems analysis requires some knowledge of graphs and trees as it is discussed in Chapter 7, which in turn have the first two sections of Chapter 3 as a prerequisite. Other relevant material can be found in Chapters 8 (specification in Z) and 12 (relational database systems).

The text is geared to the student, and every effort has been made to explain the material as clearly as possible. In general, we motivate all concepts and derivations by means of examples, preferably examples from computer science. This approach allows for an informal introduction to the topic that can be formalized later. This gradual approach helps the students to appreciate the value of formal arguments and prepares them for their later studies in computer science.

As teaching and learning aids, the book includes more than 300 examples and more than 550 problems with detailed solutions provided for all even-numbered problems.

We would like to thank Wayne Mackrell, Christy Kenny, David Haugen, Allan Rempel, Jacob Wickland, and Michael Zaleski for their help in entering the manuscript and suggesting improvements to the text. Wayne and Christy have also assisted us in formulating solutions to the problems and in producing the computer-generated diagrams. A special thank you to Cyril Coupal who suggested the title of the book. Deanna Tremblay proofread many versions of the manuscript during its period of preparation.

We would like to thank our colleagues for their valuable help. Jim Greer read and commented on the first three chapters. Eric Neufeld and Grant Cheston made important suggestions for which we are very grateful. John Cooke read and commented on the relational database chapter.

Finally, we thank the students who used earlier versions of these notes and who made many valuable comments.

Winfried Karl Grassmann
Jean-Paul Tremblay

1

Propositional Calculus

Logic plays a central role in many areas, particularly in mathematics. Logic is also essential for constructing and testing computer programs. However, sometimes our logical reasoning is faulty and errors can result. Therefore, it is important to identify the laws underlying logical derivations. Basic to logical derivations are *propositions*, which are statements that are either true or false. Propositions can be combined and manipulated in various ways. These manipulations are the subject of propositional calculus.

This chapter first presents a number of logical arguments. All logical arguments involve *atomic propositions*, which cannot be further subdivided. The atomic propositions are then combined by various *connectives* to form *compound propositions*. The connectives are discussed in detail, and their use in constructing complex compound propositions is analyzed. There are certain propositions, called *tautologies*, that are always true. Tautologies give rise to *logical implications* and *logical equivalences*. Logical implications are basic to sound reasoning, and logical equivalences provide the means to manipulate propositions algebraically. All these issues are addressed in this chapter.

1.1 LOGICAL ARGUMENTS AND PROPOSITIONS

1.1.1 Introduction

It is important to distinguish between arguments that are logically sound and arguments that are not. In the following subsections, we will provide examples of arguments that any reasonable person would consider logical. We will then identify the basic structure of these arguments. While doing this, we will discover that logical arguments consist of certain propositions that cannot be further subdivided. These *atomic propositions* are held together by *logical connectives*.

1.1.2 Some Important Logical Arguments

As an example of a sound logical argument, consider the following three statements.

1. If the demand rises, then companies expand.

2. If companies expand, then they hire workers.

3. If the demand rises, then companies hire workers.

This logical argument has three lines, and each line contains a statement. The statements of lines 1 and 2 provide the *premises* of the argument, and line 3 contains the conclusion. One can argue against the premises and claim that they are wrong. However, as soon as the premises are accepted, the conclusion must be accepted also, because it logically follows from the premises and, therefore, the argument is sound.

Here is a second example of a sound logical argument.

1. This computer program has a bug, or the input is erroneous.

2. The input is not erroneous.

3. This computer program has a bug.

Again, there are two premises, which are given in lines 1 and 2 of the argument. These premises may be true or false in a particular case. However, if the premises are true, then we cannot avoid the conclusion that the computer program has a bug. This conclusion follows logically from its premises.

Using statements like the ones given, Aristotle developed patterns of correct and faulty arguments. First, he noticed that many of the statements used in logic are really *compound* statements; that is, they consist of several parts, each of which is a statement in its own right. The first example contains the following statement as one of its premises:

If the demand rises, then companies expand.

This statement has two parts, which are statements in their own right. They are "demand rises" and "companies expand." These two statements are connected by the "If ... then" construct. Our second argument contains the statement

This computer program has a bug, or the input is erroneous.

Again, this statement consists of two parts: "this computer program has a bug" and "the input is erroneous." Both parts are themselves statements, and they are connected by the word "or."

To see which arguments are correct and which are not, Aristotle abbreviated the essential statements of the arguments by substituting letters. In this fashion, the pattern of valid arguments can be described concisely. Following the convention of Aristotle, we will use capital letters to denote the essential statements. The letter P may express the statement that "demand rises," the letter Q may express the statement "companies expand," and the letter R may stand for the statement "companies hire workers." Using these symbols, we can express the argument involving rising demand as follows:

1. If P, then Q.

2. If Q, then R.

---

3. If P, then R.

Aristotle even gave a name to this type of argument; he called it the *hypothetical syllogism*. In the hypothetical syllogism, P, Q, and R each can stand for any statement. For instance, if P stands for "The cat sees the goldfish," Q for "The cat catches the goldfish," and R for "The cat eats the goldfish," then the hypothetical syllogism becomes

1. If the cat sees the goldfish, then the cat catches the goldfish.

2. If the cat catches the goldfish, then the cat eats the goldfish.

---

3. If the cat sees the goldfish, then the cat eats the goldfish.

The premises of this argument may be right or wrong. The cat may be well trained and never do a thing as nasty as catching a goldfish. Moreover, even if the cat catches the goldfish, the cat may decide that cat food is preferable to live fish. However, as soon as we accept the premises, we have no choice but to accept the conclusion, which is that the cat eats the goldfish if it sees the goldfish.

Another pattern for a correct argument can be demonstrated by making P and Q stand for statements in the program bug argument. Let P stand for "This computer program has a bug," and let Q stand for "the input is erroneous." Using these abbreviations, the argument can be expressed in the following way:

1. P or Q.

2. Not Q.

---

3. P.

This argument is called the *disjunctive syllogism* and is a fundamental argument in logic. Notice, again, that the premises can be either true or false, but if they are considered to be true, then the conclusion is inevitable.

There is one extremely important logical argument, called the *modus ponens*. The modus ponens can be formulated as follows:

1. If P, then Q.

2. P.

---

3. Q.

For instance, if P is "The light turns red," and Q is "Cars stop," then the premises "If the light turns red, then cars stop" and "The light turns red" allow one to conclude "Cars stop."

Determining which logical reasoning patterns are sound, and which are unsound is an essential part of logic. We return to this at the end of this chapter. First, however, we need to discuss how logical statements are constructed, how their truth values can be found, and how they can be manipulated. This is done next.

1.1.3 Propositions

The statements involved in logical arguments have a special property: they are either true or false, and they can never be anything else. In logic, such statements are called propositions.

Definition 1.1:   Any statement that is either true or false is called a *proposition*.

For instance, the statement "the program has a bug" is a proposition if it is either true or false. We have only two choices, or, to use the technical term, we have a *dichotomy*. This sometimes causes problems. Consider, for instance, the following statement: "the number 8 smells awful." One would normally say that this statement is meaningless. Yet if we admit the possibility of a meaningless statement, we admit that there is a third category besides true or false, that is, meaningless, and we are no longer dealing with propositions. Of course, it is impossible to associate truth values with commands and questions; that is, neither commands nor questions are propositions.

Many statements we make in everyday life are not completely true, yet they are not entirely false either. For instance, if a program gives an unexpected output, it is often not clear whether this is caused by a bug or merely a consequence of poor specifications. However, logic only deals in black and white or, to be more specific, with truth and falsity, and anything not fitting into this dichotomy is ruled out.

We mentioned earlier that Aristotle introduced variables, such as P, Q, and R, in order to formulate which patterns of reasoning are sound. We call these variables *propositional variables*. Propositional variables can only assume two values, true or false. There are also two *propositional constants*, T and F, that represent true and false, respectively. Any propositional variable can be assigned the value T or F. A possible *assignment* to the variable P is, for instance, T. The only other possible assignment to P is, of course, F. The term *assignment* can also be used in connection with expressions. An assignment of expression A consists of assigning each variable a truth value. For instance, a possible assignment of the proposition "P or Q" would be $P = F$, $Q = T$, but there are others. We also admit assignments to propositions containing only propositional constants. In this case, the assignment has no effect.

Propositional variables and propositional constants are *atomic* propositions; that is, they cannot be further subdivided. By combining several atomic propositions, one obtains *compound* propositions. For instance, "P or Q" is a compound proposition, and so are "P and Q" and "not P." The function of the words "or," "and," and "not" is to combine propositions, and they are therefore called *logical connectives*. Generally, one can classify propositions as atomic or compound, using the following definition:

Definition 1.2: A proposition consisting of only a single propositional vari-able or a single propositional constant is called an *atomic* proposition. All nonatomic propositions are called *compound* propositions. All compound propositions contain at least one logical connective.

Logical connectives are discussed in Section 1.2, and compound propositions are fully explained in Section 1.3.

Now that we have a definition of a proposition, the challenge is to associate with each proposition, and with each assignment, a unique truth value. If the proposition is atomic, this truth value is provided directly by the assignment. Otherwise, rules must be established showing how to calculate the truth value of the compound proposition. These rules are given by the meaning of the connectives. For instance, according to the English language, under the assignment $P = $ T and $Q = $ F, P or Q should yield T: If P is true, and Q is false, P or Q is true. Finding the truth value of a given assignment of a more complex proposition is more difficult. In all cases, however, a useful tool is the *truth table*, which is defined as follows:

Definition 1.3: A *truth table* of a proposition gives the truth values of the proposition under all possible assignments.

Truth tables will be used extensively in this chapter. In particular, all logical connectives are defined by means of truth tables.

Problems 1.1

1. Use the propositional variables P and Q to formalize the following logical arguments:
    (a) If 10 is a prime, 10 cannot be equal to 2 times 5. 10 is equal to 2 times 5. Therefore, 10 cannot be a prime.
    (b) If it rains frequently, farmers complain. If it does not rain frequently, farmers complain. Consequently, farmers complain.
2. Which of the following statements are propositions?
    (a) Is this true?
    (b) John is a name.
    (c) 8 is prime.
    (d) 8 is not prime.

3. Which of the following statements are atomic propositions and which are compound propositions?
   (a) Every cat has seven lives.
   (b) Fred is tall, and so is Jim.
   (c) Fred and Jim are tall.
   (d) The car involved in the accident was green or blue.
4. Assign the logical constants T or F to the following propositions.
   (a) 7 is even.
   (b) New York is a city.
   (c) Canada is a city.

1.2 LOGICAL CONNECTIVES

1.2.1 Introduction

Logical connectives are used to combine propositions to form new propositions. These new propositions are *compound* propositions because they consist of several components. So far, we have used words from the English language to serve as connectives. For instance, we used constructs such as "P or Q" or "if P, then Q." Unfortunately, statements formulated in natural languages are frequently ambiguous because the words used in these statements can have more than one meaning. We want to avoid this. Moreover, many natural language connectives reflect concern with such things as causality, intent, emphasis, beliefs, and time. For instance, the word "if" often implies some type of causality. Standard logic, on the other hand, only deals with truth and falsity, which means that the notion of causality cannot be expressed at all. For these reasons, it is best to introduce new mathematical symbols to take the role of connectives. These symbols can then be defined unambiguously.

There is one convention of the English language that is also used in logic. Stating a proposition in English implies that this proposition is true. There is no need to explicitly state that the proposition is true. For instance, instead of saying "it is true that cats eat fish," one simply says "cats eat fish." Similarly, if P is a proposition, then "P" means "P is true." To stress that P is indeed true, we sometimes will use the phrase "P holds."

1.2.2 Negation

The simplest connective is the negation, which is defined as follows:

Definition 1.4:   Let P be a proposition. The compound proposition $\neg P$, pronounced "not P," is the proposition that is true if P is false, and that is false otherwise. $\neg P$ is called the negation of P. The connective $\neg$ may be translated into English as "It is not the case that," or simply by the word "not."

Table 1.1 represents this definition in tabular form. Generally, the definitions of all connectives can be expressed in tabular form or, to use the technical term introduced in Definition 1.3, by truth tables.

TABLE 1.1.
Truth Table for Negation

P	$\neg P$
T	F
F	T

Example 1.1   Let P stand for the proposition "London is a city." Then $\neg P$ stands for the proposition "It is not the case that London is a city" or "London is not a city."

1.2.3 Conjunction

Frequently, we want to express the fact that two statements are both true. In this case, we use the conjunction, which is defined as follows:

> **Definition 1.5:**   Let P and Q be two propositions. Then $P \wedge Q$ is true if and only if both P and Q are true. $P \wedge Q$ is called the *conjunction* of P and Q, and the connective $\wedge$ is pronounced "and." The connective $\wedge$ may be translated into English by the word "and."

The truth table corresponding to this definition is given in Table 1.2.

TABLE 1.2.
Truth Table for Conjunction

P	Q	$P \wedge Q$
T	T	T
T	F	F
F	T	F
F	F	F

Example 1.2   Let P stand for the proposition "The program has a bug" and Q for the proposition "The input is erroneous." The conjunction of P and Q is $P \wedge Q$, which can be translated as "The program has a bug, and the input is erroneous."

In the English language, we often use shortcuts that are not allowed in logic statements. For instance, the sentence "He eats and drinks" is really understood as "He eats, and he drinks." Similarly, the statement "Jack and Jill went up the hill" is really understood as "Jack went up the hill and Jill went up the hill." In logic, every statement must have its own subject and its own predicate. To translate "Jack and Jill went up the hill," one first has to convert the statement into two complete sentences: "Jack went up the hill" and "Jill went up the hill." At this stage, we can define P to stand for "Jack went up the hill" and Q for "Jill went up the hill." Using these definitions, "Jack and Jill went up the hill" becomes $P \wedge Q$.

Sometimes, we use words other than "and" to denote a conjunction in the English language, such as "but," "in addition to," and "moreover." Also, not all instances of the word "and" denote conjunctions. For instance, the word "and" in the statement "Jack and Jill are cousins" is not a conjunction at all.

Example 1.3   Translate the following statement into logic: He earns more than \$50,000, but less than \$80,000.

> **Solution**   Let P stand for "He earns more than \$50,000," and let Q stand for "He earns less than \$80,000." Then the statement translates into $P \wedge Q$.   ■

1.2.4 Disjunction

In Section 1.1, we used the statement "The computer program has a bug, or the input is erroneous." Statements containing an "or" can usually be translated as *disjunctions*. Formally, a disjunction is defined as follows:

Definition 1.6:   Let P and Q be two propositions. Then $P \vee Q$ is false only if both P and Q are false. If either P or Q or both are true, then $P \vee Q$ is true. $P \vee Q$ is called the *disjunction* of P and Q, and the connective $\vee$ is pronounced "or." The connective $\vee$ can usually be translated into English by the word "or."

This definition leads to the truth table of a disjunction as given by Table 1.3. Note that under the assignment $P = \mathrm{T}$ and $Q = \mathrm{T}$, $P \vee Q$ yields T as indicated in the first row of the truth table.

TABLE 1.3.
Truth Table for Disjunction

P	Q	$P \vee Q$
T	T	T
T	F	T
F	T	T
F	F	F

Example 1.4   Let P be "There is a bug in the program" and Q be "The input is erroneous." Then the disjunction of P and Q is $P \vee Q$, and it means "There is a bug in the program, or the input is erroneous."

The English word "or" has two different meanings. In the sentence "You can either have soup or salad," the presumption is that you can either take soup *or* salad, but not both. The "or" is thus used in an exclusive sense. On the other hand, the statement "the computer program has a bug, or the input is erroneous" does not preclude the possibility of both a bug and erroneous input. To see how $\vee$ is defined, we consult the truth table, which indicates that $P \vee Q$ is true if P or Q or both are true. Hence, $\vee$ is used as an *inclusive or*, rather than an *exclusive or*. To avoid ambiguity, one should therefore translate $P \vee Q$ as "P or Q or both."

When forming the disjunction of two sentences (or statements), always make sure that the sentences are complete: each sentence must have its own subject and predicate. For instance, the sentence "There was an error on line 15 or 16" must first be expanded to "There was an error on line 15, or there was an error on line 16."

1.2.5 Conditional

In logical arguments the "If . . . then" construct is very important. This construct expresses the conditional.

Definition 1.7:   Let P and Q be two propositions. Then $P \Rightarrow Q$ is false if P is true and Q is false, and $P \Rightarrow Q$ is true otherwise. $P \Rightarrow Q$ is called the *conditional* of P and Q. The conditional $P \Rightarrow Q$ may be translated into English by using the "If . . . then" construct, as in "If P, then Q." In other words, $P \Rightarrow Q$ means that, whenever P is correct, so is Q. The statement P is called the *antecedent* and Q the *consequent*.

The truth table of the conditional is given by Table 1.4. The next example demonstrates the conditional, using a statement encountered earlier.

TABLE 1.4.
Truth Table for Conditional

P	Q	$P \Rightarrow Q$
T	T	T
T	F	F
F	T	T
F	F	T

Example 1.5   Let P be "The demand rises," and let Q be "companies expand." The conditional of P and Q is $P \Rightarrow Q$, which translates into "If demand rises, companies expand." The antecedent of this conditional is "The demand rises," and the consequent is "companies expand."

If the antecedent is true, then the truth value of a conditional is equal to the truth value of the consequent. For instance, if it is true that the demand rises, then the statement "If demand rises, companies expand" is true if and only if companies expand. If the antecedent is false, then the conditional is said to be trivially true. For instance, if the demand does not rise, then the statement "If demand rises, companies expand" is trivially true. This is sometimes in conflict with the use of the word "if" in the English language. For instance, suppose Jim did not have a good supper. In this case, the statement "If Jim had a good supper, then $4 + 4 = 9$" would normally be considered as false by English speakers, even though this statement is considered to be trivially true by the logician. For a logician, every conditional in which the antecedent is false is considered to be trivially true. The definition of the conditional has been attacked repeatedly because of such conflicts. However, these objections have not been successful, and practically all logicians agree with the definition of the conditional as presented here, and this is the reason why.

Generally, logic deals with the consistency of statements, and the truth values for the conditional given in Table 1.4 are consistent with the conditional of the English language. To see this, consider the following statement, which may appear on the wall of a laboratory.

If a bottle contains acid, it carries a warning label.

This statement is clearly a conditional, and it can be expressed as $P \Rightarrow Q$, where P stands for "the bottle contains acid" and Q for "the bottle carries a warning label." There is no possible contradiction to the statement on the wall if a bottle does not contain acid. In fact, if the bottle does not contain acid, but, say, a strong poison, then it, too, should carry a warning label. This is an example for which the antecedent is false and the consequent is true, but this does not make the statement posted on the wall false. As a second example, suppose that a bottle contains apple juice and carries no warning label. This is certainly consistent with the message on the wall as well. The two examples are both trivially true.

Understanding the meaning of the conditional is extremely important, and to gain further insight, we now list all possibilities that make the conditional true.

P	Q	$P \Rightarrow Q$
T	T	T
F	T	T
F	F	T

This shows that, if the conditional holds, then P is true only if Q is true. P is true in fewer cases than Q, which makes it a *stronger* condition. Q is therefore *weaker* than P. In other words, if Q is false, so is P. In this sense, one can say that $P \Rightarrow Q$ translates into "P only if Q," as in "The bottle contains acid only if it carries a warning label." If it carries no warning label, it contains no acid. The warning label is a *necessary condition* for the bottle to contain acid. On the other hand, the fact that a bottle contains acid is a *sufficient condition* for it to carry a warning label. There may be other conditions that also merit a warning label. This discussion suggests a number of ways to express the conditional in the English language. They are as follows:

1. If P, then Q.
2. Whenever P, then Q.
3. P is sufficient for Q.
4. P only if Q.
5. P implies Q.

These expressions are all logically equivalent in the sense that they all express the truth assignment given in Table 1.4.

In English, we can reverse the order of the antecedent and the consequent. For instance, instead of saying "If P then Q," we can say "Q if P," as in "The bottle carries a warning label if it contains acid." Similarly, we can write $Q \Leftarrow P$ instead of $P \Rightarrow Q$. The $\Leftarrow$ can thus be translated by the word "if." Other ways to express $Q \Leftarrow P$ are as follows:

1. Q if P.
2. Q whenever P.
3. Q is necessary for P.
4. Q is implied by P.

For instance, "The bottle carries a warning label if it contains acid" may also be expressed as "A warning label is necessary for bottles containing acid."

The beginner must be aware of one pitfall. As mentioned, the word "only if" must be translated as $\Rightarrow$. The word "if," however, corresponds to $\Leftarrow$.

Example 1.6   Clearly, the two statements "Cars stop if the light turns red" and "Cars stop only if the light turns red" are different. The first statement is consistent with cars stopping for reasons other than the change of the light, such as pedestrians in the intersection. In general, the first statement is true. The second statement, on the other hand, is generally false. If cars stop only if the light turns red, then the light must have turned whenever one observes that cars stop. Nothing else could cause the cars to stop.

1.2.6 Biconditional

The biconditional is defined as follows.

Definition 1.8:   Let P and Q be two propositions. Then $P \Leftrightarrow Q$ is true whenever P and Q have the same truth values. The proposition $P \Leftrightarrow Q$ is called *biconditional* or *equivalence*, and it is pronounced "P if and only if Q." When writing, one frequently uses "iff" as an abbreviation for "if and only if."

The truth table of the biconditional is given by Table 1.5. The following is an example of a biconditional. Let P be the proposition that "x is even" and Q be the proposition that "x is divisible by 2." In this case, $P \Leftrightarrow Q$ expresses the statement "x is even if and only if x is divisible by 2."

TABLE 1.5.
Truth Table for Biconditional

P	Q	$P \Leftrightarrow Q$
T	T	T
T	F	F
F	T	F
F	F	T

1.2.7 Further Remarks on Connectives

We have introduced the connectives $\neg$, $\wedge$, $\vee$, $\Rightarrow$, and $\Leftrightarrow$. The connective $\neg$ is the only *unary* connective; that is, $\neg$ only negates a single proposition. All other connectives are *binary connectives*, which means that they require two propositions, which are joined by the connective. For instance, the two propositions P and Q can be joined by the connective $\wedge$ to form the conjunction $P \wedge Q$. The binary connectives $\wedge$, $\vee$, and $\Leftrightarrow$ are symmetric in the sense that the order of the two propositions joined by the connective does not affect the truth value of the resulting expression. For instance, the truth value of $P \wedge Q$ is equal to the truth value of $Q \wedge P$. The connective $\Rightarrow$, on the other hand, is not symmetric: $P \Rightarrow Q$ and $Q \Rightarrow P$ have different truth values. The truth values of all connectives are summarized in Table 1.6. In Table 1.6, P and Q are placeholders, and they may be replaced with any other pair of propositional variables. Hence, if P is replaced by R, and Q is replaced with S, then one can use the table to find that $R \wedge S$ yields T for the assignment $R = $ T and $S = $ T. One can even replace one or both of the variables by constants. One can form, for instance, $\text{T} \wedge Q$. The truth value of this expression can be found in Table 1.6 by using only the rows that contain a T in the column headed by P. One easily verifies in this way that $\text{T} \wedge Q$ always has the same truth value as Q. Two important propositions are $\text{F} \Rightarrow Q$ and $Q \Rightarrow \text{T}$. Both propositions always yield T, as is easily verified.

TABLE 1.6.   Summary of Connectives

P	Q	$\neg P$	$\neg Q$	$P \wedge Q$	$P \vee Q$	$P \Rightarrow Q$	$P \Leftrightarrow Q$
T	T	F	F	T	T	T	T
T	F	F	T	F	T	F	F
F	T	T	F	F	T	T	F
F	F	T	T	F	F	T	T

Problems 1.2

1. Using the statements R and H for "Mark is rich" and "Mark is happy," respectively, write the following statements in symbolic form:
   (a) Mark is not rich.
   (b) Mark is rich and happy.
   (c) Mark is rich or happy.
   (d) If Mark is rich, then he is happy.
   (e) Mark is happy only if he is rich.

2. Identify all atomic propositions in the following sentences, and abbreviate them with symbols such as P, Q, or R. Then convert the sentences into propositional calculus.
   (a) If Jim is in the barn, then Jack must be in the barn as well.
   (b) The getaway car was red or brown.
   (c) The news is not good.
   (d) You will be on time only if you hurry.
   (e) He will come if he has time.
   (f) If she was there, then she must have heard it.

3. Prepare the truth table for the exclusive or.

4. Give the truth tables for $P \wedge P$, $P \vee P$, $P \wedge \mathrm{T}$, and $P \wedge \mathrm{F}$.

1.3 COMPOUND PROPOSITIONS

1.3.1 Introduction

By using logical connectives, one can combine propositions, whether they are atomic or are themselves compound. Unless precautions are taken, the resulting expressions may be ambiguous in the sense that they can be interpreted in several ways. To avoid this, we first introduce *fully parenthesized expressions*. The resulting expressions can then be split in a unique way into subexpressions, or, to use the technical term, they can be *parsed* in a unique way. To make it easier to read logical expressions, one often makes use of *precedence rules*, which allow one to omit certain parentheses without making the expression ambiguous. Whereas the truth value of an atomic proposition is given externally, the truth value of a compound proposition must be inferred from the truth values of its constituents, as will be shown. We also show how to prepare truth tables for compound propositions. Finally, examples involving compound propositions are presented.

1.3.2 Logical Expressions

Any proposition must somehow be expressed verbally, graphically, or by a string of characters. A proposition expressed by a string of characters is called a *logical expression* or a *formula*. For instance, $P \wedge Q$ is a logical expression and so is Q, provided that P and Q are logical variables. Logical expressions can either be *atomic* or *compound*. An atomic expression consists of a single propositional variable or a single propositional constant, and it represents an atomic proposition. Compound expressions contain at least one connective, and they represent compound propositions.

Unless precautions are taken, logical expressions may be ambiguous. For example, let P be "Mary finishes her report," let Q be "Mary is happy," and let R be "Mary goes to the movies tonight." Consider now the expression $P \Rightarrow Q \wedge R$. Without further rules, this expression can be interpreted in two ways. It can mean $(P \Rightarrow Q) \wedge R$, which translates into "If Mary finishes her report, she will be happy, but in any event, she will go to the movies." Alternatively, it could mean $P \Rightarrow (Q \wedge R)$, which translates into "If Mary finishes tonight, she will be happy and go to the movies." Hence, the expression $P \Rightarrow Q \wedge R$ is ambiguous. To avoid ambiguities, rules must be provided to show how to group the different subexpressions. Alternatively, one can use parentheses. We first explore the second alternative.

In expressions like $P \wedge Q$, $P \vee Q$, and others, the variables P and Q belong together, and to indicate this, one often writes $(P \wedge Q)$ and $(P \vee Q)$, respectively. If this is done, then no misunderstanding can arise if these expressions are later used as constituent parts of larger expressions, such as, say, $((P \wedge Q) \Rightarrow (P \vee Q))$. The resulting expressions are said to be *fully parenthesized*, or fpes as we will call them.

One can use identifiers to refer to expressions. For instance, the expression $(P \wedge Q)$ could be called A, and the expression $(P \vee Q)$ could be named B. Note that the expression A contains the propositional variables P and Q. However, A is not a propositional variable because the truth value of A must be inferred from the truth values of its propositional variables P and Q. It is important to be aware of this distinction. This is particularly important when using *schemas*.

Definition 1.9: All expressions containing identifiers that represent expressions are called *schemas*.

Example 1.7 If $A = (P \wedge Q)$ and if $B = (P \vee Q)$, then the schema $(A \Rightarrow B)$ stands for $((P \wedge Q) \Rightarrow (P \vee Q))$.

The same schema can be instantiated in different ways. For instance, the schema $A \Rightarrow B$ can mean $((P \wedge Q) \Rightarrow (P \vee Q))$ as in Example 1.7. However, if $A = (P \vee Q)$ and $B = R$, then the same schema instantiates to $((P \vee Q) \Rightarrow R)$.

An instance of the schema $(A \wedge B)$ is said to be of the form $(A \wedge B)$, and the same is true for $(A \vee B)$, and so on. If A and B denote expressions, one has the following:

1. Any expression of the form $(\neg A)$ is called a *negation*.
2. Any expression of the form $(A \wedge B)$ is called a *conjunction*.
3. Any expression of the form $(A \vee B)$ is called a *disjunction*.
4. Any expression of the form $(A \Rightarrow B)$ is called a *conditional*.
5. Any expression of the form $(A \Leftrightarrow B)$ is called an *equivalence*.

From now on, all expressions considered in this chapter must either be atomic, or they must be negations, conjunctions, disjunctions, conditionals, or equivalences.

Example 1.8   The expression $((P \wedge Q) \Rightarrow (P \vee Q))$ is a conditional, consisting of the conjunction $(P \wedge Q)$ and the disjunction $(P \vee Q)$.

Under the conditions given previously all fully parenthesized expressions (fpes) can be constructed according to the following formation rules:

1. Every atomic expression is an fpe.
2. If A is an fpe, so is $(\neg A)$.
3. If A and B are fpes, so are $(A \wedge B)$, $(A \vee B)$, $(A \Rightarrow B)$, and $(A \Leftrightarrow B)$.
4. No other expression is an fpe.

Hence, $((P \wedge Q) \Rightarrow (P \vee Q))$ is an fpe because both $(P \wedge Q)$ and $(P \vee Q)$ are fpes. $(P \wedge Q)$ is an fpe because $(P \wedge Q)$, being a properly formed conjunction, is an fpe, and so is $(P \vee Q)$. Expressions that are defined by explicit formation rules are usually called *well-formed formulas*, which is abbreviated wffs. Hence, every fully parenthesized expression is a well-formed formula; that is, every fpe is a wff.

Note that the definition of what is and what is not an fpe is *recursive*; that is, it uses the term "fpe" in its own definition. Generally, recursion is a convenient way to define an infinite number of objects. For another example of a recursive definition, consider the definition of the word "descendant." Clearly, a child is a descendant, and so is every descendant of the child. This statement completely defines the term "descendant": every descendant satisfies this definition, and only descendants satisfy it.

In the expression $(\neg A)$, A is frequently called the *scope* of the negation, and the connective $\neg$ is called the *main connective* of the expression $\neg A$. In conjunctions, that is, in expressions of the form $(A \wedge B)$, the main connective is $\wedge$, and A and B are called, respectively, the *left scope* and the *right scope* of the conjunction. A similar definition applies to expressions of the form $(A \vee B)$, $(A \Rightarrow B)$, and $(A \Leftrightarrow B)$. The scope or the scopes may be compound, in which case the connectives found in the scopes are *subconnectives* of the expression in question. For instance, the main connective of the conjunction $((P \vee Q) \wedge R)$ is $\wedge$, and the subconnective of the left scope is $\vee$. The right scope, which is R, does not have a subconnective. The situation for disjunctions, conditionals, and equivalences is similar. For instance, in the expression $((P \wedge Q) \Rightarrow (P \vee Q))$, $\Rightarrow$ is the main connective, and $\wedge$ and $\vee$ are two subconnectives. Of course, $\wedge$ is the main connective of the left scope, and $\vee$ is the main connective of the right scope.

1.3.3 Analysis of Compound Propositions

Fully parenthesized expressions are one method to represent compound propositions. However, there are others. In particular, there are all kinds of facilities to represent compound propositions in the English language. Moreover, one can represent compound propositions graphically. No matter how the proposition is expressed, one can distinguish between negations, conjunctions, disjunctions, and so on. All compound propositions have subpropositions and these subpropositions may, in turn, be identified as conjunctions, disjunctions, and so on. To see how this works, consider the following proposition:

If Michelle wins at the Olympics, everyone will admire her, and she will get rich; but if she does not win, all her effort was in vain.

This proposition is a conjunction, and the scopes of this conjunction are given by the following propositions.

If Michelle wins at the Olympics, everyone will admire her, and she will get rich.

and

If she does not win, all her effort was in vain.

Both scopes are again compound, and they can therefore be analyzed in a similar fashion. For instance, the left scope can be divided into the two statements "Michelle wins at the Olympics" and "everyone admires her, and she will get rich." The first of these two statements is atomic, and it cannot be further divided. The second, however, is again compound and can be written as the conjunction of the two atomic propositions "Everyone admires Michelle" and "Michelle will get rich." A similar analysis can be done for the right scope of the main propositions. The separation of a statement into its components is called *parsing*, and the result can be represented graphically by a *parse tree*. The parse tree of the statement in question is given in Figure 1.1.

Parse trees are constructed top-down. First the entire expression is used to obtain the top node, and the scopes of this expression are used to find the nodes of the next level. These nodes give rise to further nodes, and this continues until one reaches the atomic expressions, which form the leaves of the parse tree. An expression with a given parse tree can easily be converted into a fully parenthesized expression. To do this, we define the following propositional variables:

P: Michelle wins at the Olympics.

Q: Everyone admires Michelle.

R: Michelle will get rich.

S: Michelle's effort was in vain.

With the help of these propositions, the expression in question becomes

$$((P \Rightarrow (Q \wedge R)) \wedge ((\neg P) \Rightarrow S)) \tag{1.1}$$

Like the parse tree, the fully parenthesized expression reflects what belongs together and what does not. For instance, (1.1) contains the subexpression $(P \Rightarrow (Q \wedge R))$, which indicates that P must not be associated with Q, but with $(Q \wedge R)$. Propositions stated in English, on the other hand, may be ambiguous. In the statement "If Michelle wins at the Olympics, everyone will admire her, and she will get rich," for instance, one could interpret the consequent as either consisting of "everyone will admire her, and she will get rich" or as consisting only of "everyone will admire her." In the second interpretation, she will become rich without regard to her winning at the Olympics.

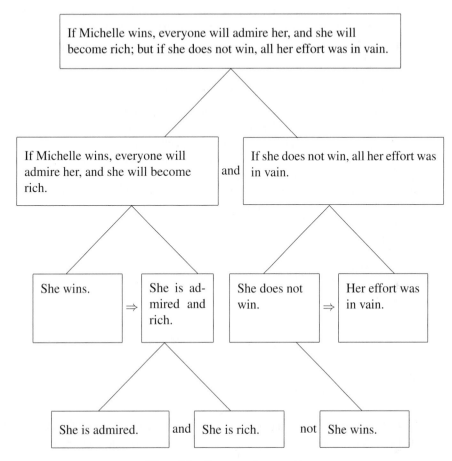

Figure 1.1. Parse tree of a proposition

If E is any compound expression, then the scopes of the main connective in E are subexpressions. For instance, if E is of the form $(A \wedge B)$, then both A and B are subexpressions. These subexpressions will be called the *immediate subexpressions* of E. Moreover, A and B may be themselves compound expressions, in which case they, too, have subexpressions. All subexpressions of both A and B are also subexpressions of E. The statement about Michelle, for instance, is of the form $(A \wedge B)$, with $A = (P \Rightarrow (Q \wedge R))$ and $B = ((\neg P) \Rightarrow S)$. Here, A has the subexpressions P and $(Q \wedge R)$, and $(Q \wedge R)$ has the subexpressions Q and R. All these expressions are subexpressions of the original proposition. Generally, the subexpressions of an expression E are defined as follows:

1. E is a subexpression of E.

2. If E is of the form $(\neg A)$, then A is a subexpression of E.

3. If E is of the form $(A \wedge B)$, $(A \vee B)$, $(A \Rightarrow B)$, or $(A \Leftrightarrow B)$, then A and B are both subexpressions of E. These subexpressions are called *immediate subexpressions*.

4. If A is a subexpression of E and if C is a subexpression of A, then C is a subexpression of E.

5. No other expression is a subexpression of E.

Note that E itself is a subexpression of E. This subexpression is called an *improper* subexpression. All other subexpressions are called *proper* subexpressions of E. The scopes of an expression E are always the immediate subexpressions of E.

Generally, parsing an expression amounts to finding all its subexpressions. The parse is represented by the parse tree in the sense that all subexpressions of an expression are present in the tree. The root of the parse tree represents the expression to be parsed, and all nodes below the root represent proper subexpressions. This was illustrated with the help of Figure 1.1. The same can also be achieved by a fully parenthesized expression and, in fact, there is an immediate correspondence between parse trees and fully parenthesized expressions, as the reader may verify.

One special type of expression, the *literal*, is of particular importance; it is defined as follows:

Definition 1.10: A proposition is called a *literal* if it is of the form Q or $\neg Q$, where Q is a propositional variable. The two expressions Q and $\neg Q$ are called *complementary literals*.

If P and Q are propositional variables, then P, Q, and $\neg Q$ are all literals, but $\neg(P \vee Q)$ is not a literal. P and $\neg P$ are two complementary literals, but P and Q are not.

1.3.4 Precedence Rules

Very few people work with fully parenthesized expressions because such expressions are lengthy and often difficult to read. In particular, the outer parentheses of an expression are almost always omitted. Hence, instead of $(P \wedge Q)$, one writes $P \wedge Q$, and instead of $((P \wedge Q) \Rightarrow (P \vee Q))$ one writes $(P \wedge Q) \Rightarrow (P \vee Q)$. When doing this, one must never forget to add parentheses back on when the expression in question is to be compounded with some other expression. Parentheses inside an expression may also be omitted. To correctly interpret the resulting expression, so called *precedence rules* are used. Generally, each connective is given a precedence, and connectives with a higher precedence introduce a stronger binding than connectives with a lower precedence.

The connective $\neg$ always has the highest precedence. Consequently, $\neg P \vee Q$ is to be understood as $(\neg P) \vee Q$, and not as $\neg(P \vee Q)$. In the case of binary connectives, the highest precedence is given to $\wedge$, followed by $\vee$, $\Rightarrow$, and $\Leftrightarrow$, in that order. In the expression $P \wedge Q \vee R$, for instance, $\wedge$ takes precedence over $\vee$ when forming subexpressions; that is, $P \wedge Q \vee R$ is to be understood as $(P \wedge Q) \vee R$. Similarly, $P \Rightarrow Q \vee R$ is to be understood as $P \Rightarrow (Q \vee R)$, because $\vee$ takes precedence over $\Rightarrow$. The connective $\Leftrightarrow$ is given the lowest precedence, which implies that $P \Leftrightarrow P \Rightarrow Q$ must be understood as $P \Leftrightarrow (P \Rightarrow Q)$. Because of the precedence rules, one can write the proposition about Michelle as follows:

$$(P \Rightarrow Q \wedge R) \wedge (\neg P \Rightarrow S)$$

Rules involving precedence are, of course, familiar from arithmetic expressions. For instance, in Pascal, $*$ has precedence over $+$, which means that $a + b * c$ must be interpreted as $a + (b * c)$.

In some expressions, the precedence rules are not sufficient to remove all ambiguities. For instance, the expression $P \Rightarrow Q \Rightarrow R$ could either be understood as $P \Rightarrow (Q \Rightarrow R)$ or $(P \Rightarrow Q) \Rightarrow R$. Which interpretation is used depends on the associativity of the connective $\Rightarrow$. Generally, $\Rightarrow$ is an operator, just as $+$ and $/$ are operators in Pascal.

> **Definition 1.11:**   A binary operator is called *left associative* if the operator on the left has precedence over the operator on the right. A binary operator is called *right associative* if the operator to the right has precedence over the operator on the left.

All binary logical connectives are left associative. Consequently, $P \Rightarrow Q \Rightarrow R$ must be understood as $(P \Rightarrow Q) \Rightarrow R$. This is consistent with programming languages such as Pascal, where $a/b/c$ is interpreted as $(a/b)/c$, rather than $a/(b/c)$. Hence, the binary arithmetic operators in Pascal are also left associative.

The following definition introduces two terms related to our discussion that are useful for describing logical operators.

> **Definition 1.12:**   Let $\circ$ be the main operator of an expression. Then, $\circ$ is said to be in *prefix* position if it precedes its operands, in *infix* position if it is inserted between the operands, and in *postfix* position if it follows its operands.

The connectives $\wedge$, $\vee$, $\Rightarrow$, and $\Leftrightarrow$ are all used in infix position. For instance, we write $P \wedge Q$, which is infix notation. In prefix notation, the same expression would be $\wedge P, Q$, and in postfix notation, it would be $P, Q \vee$. Of course, the logical connective $\neg$ precedes its scope, and it is therefore in prefix position.

1.3.5 Evaluation of Expressions and Truth Tables

The truth values of nonatomic expressions must be derived from the truth values of their constituent atomic expressions. Before this can be done, the expression in question must be parsed; that is, all subexpressions of the expression must be found. As indicated in the previous section, this can be done by means of a parse tree. The parse tree is constructed top-down, starting with the entire expression, which is recursively broken into subexpressions until one reaches the atomic expressions, which form the leaf nodes of the parse tree. Only the atomic expressions can be assigned truth values directly. The truth values of all other subexpressions must be inferred in turn from their immediate subexpressions. This means that the evaluation of the truth values must be bottom-up: the process starts with the truth values of the atomic expressions and culminates with the determination of the truth value for the entire expression. Usually, it is not necessary to draw the parse tree. Indeed, any fully parenthesized expression conveys the same information as a parse tree, and it can be used in its place. One can even omit some of the parentheses and use precedences. The procedure is essentially the same in all cases. To illustrate how to find the truth value of an expression, consider the following statement.

> If you take a class in computers, and if you do not understand recursion, you will not pass.

We want to know exactly when this statement is true and when it is false. To do this, we define

P: You take a class in computers.
Q: You understand recursion.
R: You pass.

Using these definitions, the statement in question becomes

$$(P \wedge \neg Q) \Rightarrow \neg R$$

We identify this expression by the identifier A; that is, whenever we write A, we refer to $(P \wedge \neg Q) \Rightarrow \neg R$. A is of the form $B \Rightarrow C$, where $B = (P \wedge \neg Q)$, and $C = \neg R$. B, in turn, is of the form $P \wedge D$, where P is a propositional variable, and $D = \neg Q$. The truth values of the subexpressions B, C, and D can now be found, given an assignment for P, Q, and R. This assignment determines the truth value of the entire expression. For instance, the assignment $P = $ T, $Q = $ F, and $R = $ T yields $D = \neg Q = $ T, $B = P \wedge D$, which is T, and $C = \neg R$, which yields F. Consequently, $A = B \Rightarrow C$ yields F.

A truth table gives the truth values of an expression for all possible assignments. Since A contains the three propositional variables P, Q, and R and since each variable can be assigned two values, T and F, there are $2^3 = 8$ assignments in total. Generally, a truth table with n propositional variables has 2^n assignments, which makes this method impractical for large n. This will be discussed further in Sections 6.3.3 and 6.3.7. Truth tables are very convenient for small n, however.

Table 1.7 lists all these assignments, and it shows the calculation of the truth value of A for each of them. The first row contains the assignment that makes the variables P, Q, and R all true, the second makes P and Q true, and R false, and so on. Table 1.7 also contains the identifiers A, B, C, and D, which denote the subexpressions of A. There is only one assignment that makes A false: $P = $ T, $Q = $ F, and $R = $ T.

TABLE 1.7.   Truth Table for $(P \land \neg Q) \Rightarrow \neg R$

			D	B	C	A
P	Q	R	$\neg Q$	$P \land \neg Q$	$\neg R$	$(P \land \neg Q) \Rightarrow \neg R$
T	T	T	F	F	F	T
T	T	F	F	F	T	T
T	F	T	T	T	F	F
T	F	F	T	T	T	T
F	T	T	F	F	F	T
F	T	F	F	F	T	T
F	F	T	T	F	F	T
F	F	F	T	F	T	T

1.3.6 Examples of Compound Propositions

In this section, we translate a number of statements into logic. We start with the following sentence:

> No work, no pay.

To translate this statement into a logical expression, reword it as follows: "If no work was done, then there is no pay." We use the variables P and Q to express the propositions "work was done" and "there is pay," respectively, and we obtain

$$\neg P \Rightarrow \neg Q$$

Next, we translate the following:

> Goods bought in this store can be returned only if they are in good condition, and only if the purchaser brings the receipt.

The atomic propositions of this statement are P, which represents "goods can be returned," Q, which represents "goods are in good condition," and R, which represents that "the purchaser brings the receipt." Hence, the following logic expression results:

$$P \Rightarrow (Q \land R)$$

The parentheses can be omitted because $\land$ has precedence over $\Rightarrow$.

We now provide an example from mathematics, and we translate:

> If p is a prime number, then, for even integers n, $n^p - n$ is divisible by p.

We make the following definitions: P is "p is a prime," Q is "n is an integer," R is "n is even," and S is "$n^p - n$ is divisible by p." When we combine these statements, we obtain

$$P \Rightarrow (Q \land R \Rightarrow S)$$

A statement of importance in Chapter 5 is the following:

Set C is the intersection of set A and set B if and only if every element of set C is an element of both A and B.

We define P to mean "Set C is an intersection of set A and set B," Q to mean "Every element of A is an element of C," and R to mean "Every element of B is an element of C." The result is

$$P \Leftrightarrow Q \wedge R$$

A particularly fruitful area for the application of logic is the law, as the next statement demonstrates.

Nobody is allowed to turn on a freeway. Police officers on duty are exempt from this rule.

We define P as "X is allowed to turn on a freeway," Q as "X is a police officer," and R as "X is on duty." The main connective of the proposition is "exempt," which is really an or. We therefore obtain

$$\neg P \vee (Q \wedge R)$$

The expression $P \vee Q$ always means P or Q or both. In other words, the connective $\vee$ must be understood as an *inclusive or*. If we want to use the *exclusive or*, we have to state explicitly that we mean P or Q, but not both. Since the word "but" translates into an $\wedge$, we obtain

$$(P \vee Q) \wedge \neg(P \wedge Q)$$

Problems 1.3

1. Analyze the following statements by means of a parse tree. Make sure that all leaves of the tree are truly atomic propositions that do not contain proper subexpressions.
   (a) If the hare is alert and quick, neither the fox nor the lynx will catch him.
   (b) If I am not mistaken, she drove a red car, and there was a man sitting beside her.
   (c) We can either try to get the mortgage approved and buy the house, or we can wait and see whether a better deal comes up.

2. Insert parentheses into the following expressions such that they are unambiguous without the use of precedence rules.
   (a) $P \wedge Q \wedge R \Rightarrow P$ (c) $\neg(P_1 \wedge P_2) \Rightarrow \neg Q \vee P_1$
   (b) $P \wedge R \vee Q \Leftrightarrow \neg R$ (d) $P \Rightarrow Q \Leftrightarrow \neg Q \Rightarrow \neg P$

3. Identify the atomic propositions of the following sentences and replace them by propositional symbols. Then translate the sentences into propositional calculus.
   (a) If you do not leave, I will call the police.
   (b) Two children have the same uncle(s) if and only if they have the same mother and the same father.
   (c) It is a nice day if it is sunny, but only if it is not hot.
   (d) If $i > j$, then $i - 1 > j$, else $j = 3$.

4. Which of the following are not literals?
   (a) P (c) $\neg P_2$
   (b) $\neg\neg P$ (d) $P \vee Q$

5. Convert the following sentences into propositional calculus, using propositional variables for atomic propositions only. Moreover, give the scope of all logical connectives.

(a) Mary is tall, but Jim is small and nimble.

(b) If x is greater than y, and y is greater than z, then x is greater than z.

(c) Mineral extraction is profitable if the ore concentration is high, but only if the distance to the market is short.

6. Given P and Q are true and R and S are false, find the truth values of the following expressions:

(a) $P \vee (Q \wedge R)$

(b) $(P \wedge (Q \wedge R)) \vee \neg((P \vee Q) \wedge (R \vee S))$

(c) $(\neg(P \wedge Q) \vee \neg R) \vee (((\neg P \wedge Q) \vee \neg R) \wedge S)$

7. Given P and Q are true and R and S are false, find the truth values of the following expressions:

(a) $(\neg(P \wedge Q) \vee \neg R) \vee ((Q \Leftrightarrow \neg P) \Rightarrow (R \vee \neg S))$

(b) $(P \Leftrightarrow R) \wedge (\neg Q \Rightarrow S)$

(c) $(P \vee (Q \Rightarrow (R \wedge \neg P))) \Leftrightarrow (Q \vee \neg S)$

8. Construct the truth tables for the following formulas:

(a) $\neg(\neg P \vee \neg Q)$ (d) $P \wedge (Q \wedge P)$

(b) $\neg(\neg P \wedge \neg Q)$ (e) $(\neg P \wedge (\neg Q \wedge R)) \vee (Q \wedge R) \vee (P \wedge R)$

(c) $P \wedge (P \vee Q)$ (f) $(P \wedge Q) \vee (\neg P \wedge Q) \vee (P \wedge \neg Q) \vee (\neg P \wedge \neg Q)$

9. Frequently, fully parenthesized expressions (fpes) are defined by the following formation rules.

1. Every propositional variable is an fpe.

2. If A is an fpe, so is $\neg A$.

3. If A and B are two fpes, so are $(A \wedge B)$, $(A \vee B)$, $(A \Rightarrow B)$, and $(A \Leftrightarrow B)$.

4. Nothing else is an fpe.

What is the difference between this definition and the one given in Section 1.3.2? In particular, is $\neg\neg P$ an fpe according to the definition here? Is this expression an fpe according to the definition in Section 1.3.2?

1.4 TAUTOLOGIES AND CONTRADICTIONS

1.4.1 Introduction

Truth tables allow us to classify logical expressions as follows:

Definition 1.13: A logical expression is a *tautology* if it is true under all possible assignments.

Definition 1.14: A logical expression is a *contradiction* if it is false under all possible assignments.

> **Definition 1.15:**   A logical expression that is neither a tautology nor a contradiction is called *contingent*.

The most direct way to determine if a logical expression is a tautology, a contradiction, or contingent is by means of truth tables. If all rows make the expression true, then the expression is a tautology. If all rows make the expression false, then it is a contradiction. If some rows yield T while others yield F, the expression is contingent.

We have already encountered a number of simple tautologies. In Section 1.2.7, for instance, we mentioned that $F \Rightarrow Q$ and $P \Rightarrow T$ are always true. Hence, both expressions are tautologies. Further tautologies will be discussed in Section 1.4.2. Tautologies are important because every logical argument can be reduced to a tautology. Alternatively, one can show that an argument is sound by using contradictions. In both tautologies and contradictions, one is allowed to replace any propositional variable by an (atomic or compound) expression. The following sections elaborate on these topics.

1.4.2 Tautologies

Consider the expression $\neg(P \wedge Q) \vee Q$. Is this expression a tautology? To decide this, we establish the truth table for the expression in question. This is done in Table 1.8. It can be seen that for any possible assignment, the expression $\neg(P \wedge Q) \vee Q$ is true, which establishes that it is a tautology.

TABLE 1.8.   Truth Table for a Tautology

P	Q	$P \wedge Q$	$\neg(P \wedge Q)$	$\neg(P \wedge Q) \vee Q$
T	T	T	F	T
T	F	F	T	T
F	T	F	T	T
F	F	F	T	T

Since tautologies are so important, a special symbol has been created to indicate that a logical expression is a tautology. Instead of writing A is a tautology, we write $\models A$. Thus, we proved above that

$$\models \neg(P \wedge Q) \vee Q$$

At this point, we should point out an important distinction. Propositional calculus can be thought of as a language for formulating propositions. When we speak about this language, we use English to describe the facts of interest about this language. Generally, a language used to describe another language is called a *metalanguage*. For instance, if Spanish is taught to students whose mother tongue is English, the facts of the Spanish language are explained in English, in which case English is the metalanguage used to teach

Spanish. Clearly, English terms to describe entities occurring in Spanish grammar do not belong to the Spanish language, but to the metalanguage, which is English.

When we say that a proposition is always true, then this statement says something about the language of propositional calculus, and thus it is not part of propositional calculus, just as the English words used to describe certain entities occurring in Spanish do not belong to the Spanish language. In this sense, we conclude that $\models$ is not part of propositional calculus, but it is part of our metalanguage describing the propositional calculus. We will introduce other metasymbols later.

There are a number of very simple, yet important tautologies. The first of these is the law of the excluded middle. It states that $P \vee \neg P$ is a tautology. In other words, P is either true or false, and everything else is excluded. Table 1.9 proves this law by showing that it is true for all assignments. Of course, there are only two assignments, because P is either true or false. Two additional tautologies are $F \Rightarrow P$ and $P \Rightarrow T$. We discussed them already in Section 1.2.7.

TABLE 1.9.
Proof of the Excluded Middle Law

P	$\neg P$	$P \vee \neg P$
T	F	T
F	T	T

If A is a tautology that contains the variable P, one can determine a new expression by replacing all instances of P by an arbitrary expression. The resulting expression A' is again a tautology. For instance, $P \vee \neg P$ is a tautology. We can now replace all instances of P by any expression we like, say by $P \wedge Q$. The resulting expression $A' = (P \wedge Q) \vee \neg(P \wedge Q)$ is again a tautology. Similar substitutions can be made in any other tautology. This is true because, when evaluating the truth value of any expression, only the truth values of its immediate subexpressions have any effect. On whether this truth value has been obtained directly through assignment or by some type of prior evaluation is irrelevant. Hence, if A is an expression containing P, the value of A does not change if P is replaced by an expression that has the same truth value as P. If A is a tautology, then A remains a tautology, irrespective of whether P is true or false, and replacing P by any expression does not affect this. One can even replace all propositional variables of a tautology by expressions, in this way converting the tautology into a schema. This leads to the following theorem:

Theorem 1.1.   Let A be a tautological expression, and let $P_1, P_2, \ldots, P_n$ be the propositional variables of A. Suppose that $B_1, B_2, \ldots, B_n$ are arbitrary logical expressions. In this case, the expression obtained by replacing P_1 by B_1, P_2 by $B_2, \ldots, P_n$ by B_n is a schema, and every instance of this schema is a tautology.

Example 1.9   Use the fact that $\neg(P \wedge Q) \vee Q$ is a tautology to prove that $\neg((P \vee Q) \wedge R) \vee R$ is a tautology.

Solution   Let B and C be two arbitrary expressions. Theorem 1.1 allows us to convert $\neg(P \wedge Q) \vee Q$ into the schema $\neg(B \wedge C) \vee C$. If $B = (P \vee Q)$ and if $C = R$, then this schema yields $\neg((P \vee Q) \wedge R) \vee R$, which must be a tautology as well.   ∎

1.4.3 Tautologies and Sound Reasoning

A logical argument is sound if the conclusion logically follows from the premises. If all premises are true (i.e., the conjunction of all the premises yields true), then the conclusion must also be true. Hence, if the conjunction of the premises is A and if the conclusion is C, then $A \Rightarrow C$ must be true under all assignments: it must be a tautology. For instance, the disjunctive syllogism indicates that as long as $P \vee Q$ and $\neg P$ are both true then Q must be true. Hence, $((P \vee Q) \wedge \neg P) \Rightarrow Q$ must be a tautology. This idea is used in Table 1.10 to prove that the disjunctive syllogism is sound. For the proof, one only has to establish that $((P \vee Q) \wedge \neg P) \Rightarrow Q$ is always true, and this is shown in the last column of the table. The soundness of other logical arguments can be demonstrated in a similar fashion.

TABLE 1.10.   Truth Table for the Disjunctive Syllogism

P	Q	$P \vee Q$	$\neg P$	$(P \vee Q) \wedge \neg P$	$((P \vee Q) \wedge \neg P) \Rightarrow Q$
T	T	T	F	F	T
T	F	T	F	F	T
F	T	T	T	T	T
F	F	F	T	F	T

In all tautologies, one is allowed to replace the propositional variables by expressions. Consequently, every sound argument can be converted into a schema.

Example 1.10   Show that from $P \wedge Q$ and $P \wedge Q \Rightarrow \neg R$ one may conclude $\neg R$.

Solution   According to the modus ponens, the premises P and $P \Rightarrow Q$ allow one to conclude Q. This expression can be converted into a schema, indicating that, for any expressions A and B, A and $A \Rightarrow B$ have B as a logical consequence. By setting $A = P \wedge Q$ and $B = \neg R$, one finds that the premises $P \wedge Q$ and $P \wedge Q \Rightarrow \neg R$ have $\neg R$ as a logical consequence.   ∎

1.4.4 Contradictions

An expression is a contradiction if it yields F under all assignments. One can check by using a truth table whether or not this is true. As an example, we prove that $P \wedge \neg P$ is a contradiction. This is done in Table 1.11, which shows that, for all possible assignments, $P \wedge \neg P$ is false.

Contradictions are closely related to tautologies. In fact, if A is a tautology, $\neg A$ is a contradiction, and vice versa. Like tautologies, contradictions can be converted into schemas. This means, for instance, that $(P \Rightarrow Q) \wedge \neg(P \Rightarrow Q)$ is a contradiction because it follows the schema $A \wedge \neg A$, which can in turn be derived from $P \wedge \neg P$.

Contradictions can be used to prove that logical arguments are sound. To do this, note that an argument cannot be sound if all premises are true, yet the conclusion is false.

TABLE 1.11.
Truth Table for Contradiction

P	$\neg P$	$P \wedge \neg P$
T	F	F
F	T	F

In other words, it is impossible that the negation of the conclusion and the premises are all true simultaneously. The conjunction of all premises with the negation of the conclusion must therefore always be false. It must be a contradiction. To demonstrate this, we prove the disjunctive syllogism once more. If formulated as a contradiction, it says that, for all possible assignments, $(P \vee Q) \wedge \neg P \wedge \neg Q$ must be false. Table 1.12 shows that this is indeed the case, demonstrating once more that the disjunctive syllogism is sound.

TABLE 1.12.   Proof of the Disjunctive Syllogism by Contradiction

P	Q	$P \vee Q$	$\neg P$	$(P \vee Q) \wedge \neg P$	$\neg Q$	$(P \vee Q) \wedge \neg P \wedge \neg Q$
T	T	T	F	F	F	F
T	F	T	F	F	T	F
F	T	T	T	T	F	F
F	F	F	T	F	T	F

1.4.5 Important Types of Tautologies

There are two important types of tautologies, *logical implications* and *logical equivalences*. They are defined as follows:

Definition 1.16:   If A and B are two logical expressions and if $A \Rightarrow B$ is a tautology, we say that A *logically implies* B, and we write $A \Rrightarrow B$.

Definition 1.17:   If A and B are two logical expressions and if A and B always have the same truth value, then A and B are said to be *logically equivalent*, and we write $A \equiv B$. In other words, $A \equiv B$ if and only if $A \Leftrightarrow B$ is a tautology.

It is important to distinguish between the symbol pair $\equiv$, $\Rrightarrow$ and the symbol pair $\Leftrightarrow$, $\Rightarrow$. Neither $\equiv$ nor $\Rrightarrow$ is a connective. Both $\equiv$ and $\Rrightarrow$ belong to the metalanguage. This is the case because they are defined in terms of $\models$, which is also a symbol belonging to the metalanguage. To distinguish between $A \equiv B$ and $A \Leftrightarrow B$, one frequently uses the term *material equivalence* for the equivalence $A \Leftrightarrow B$, as opposed to the *logical equivalence*, which is denoted by $\equiv$. In the same fashion, one says that A *materially* implies B for $A \Rightarrow B$ and that A *logically* implies B for $A \Rrightarrow B$.

There are important connections between general tautologies, logical equivalences, and logical implications. Obviously, A is a tautology if $A \equiv T$, and A is a contradiction if $A \equiv F$. One can also easily show that if A is a tautology then $T \Rrightarrow A$, and if A is a contradiction, $A \Rrightarrow F$. Finally, from $A \equiv B$, one can conclude $A \Rrightarrow B$ and $B \Rrightarrow A$. Hence, each logical equivalence can be used to derive two logical implications. Conversely, if $A \Rrightarrow B$ and $B \Rrightarrow A$, one can always conclude $A \equiv B$.

Problems 1.4

1. Construct the truth table for each of the following expressions. Indicate for each expression whether it is a tautology, a contradiction, or contingent.
   (a) $(P \wedge (P \Rightarrow Q)) \Rightarrow Q$
   (b) $(P \Rightarrow Q) \Leftrightarrow (\neg P \vee Q)$
   (c) $((P \Rightarrow Q) \wedge (Q \Rightarrow R)) \Rightarrow (P \Rightarrow R)$
   (d) $(P \Leftrightarrow Q) \Leftrightarrow ((P \wedge Q) \vee (\neg P \wedge \neg Q))$
   (e) $(Q \wedge (P \Rightarrow Q)) \Rightarrow P$
   (f) $\neg(P \vee (Q \wedge R)) \Leftrightarrow ((P \vee Q) \wedge (P \vee R))$

2. Use the tautology $P \vee \neg P$ to prove that the following expressions are tautologies.
   (a) $(P \Rightarrow Q) \vee \neg(P \Rightarrow Q)$
   (b) $\neg P \vee \neg\neg P$
   (c) $((P \wedge S) \vee Q) \vee \neg((P \wedge S) \vee Q)$

3. Show that $(P \Rightarrow Q) \wedge (\neg P \Rightarrow Q) \Rightarrow Q$ is a tautology. Convert this tautology into a schema with A replacing P and B replacing Q. Use this schema to prove that

$$(\neg P \Rightarrow \neg Q) \wedge (\neg\neg P \Rightarrow \neg Q) \Rrightarrow \neg Q$$

4. Discuss the difference between logical implication and material implication.

5. The dilemma is an argument that allows one to conclude R, given the premises $P \vee Q$, $P \Rightarrow R$, and $Q \Rightarrow R$.
   (a) Convert the dilemma into a tautology and prove it by using a truth table.
   (b) Convert the dilemma into a contradiction and prove it by using a truth table.

1.5 LOGICAL EQUIVALENCES AND THEIR USE

1.5.1 Introduction

Let us consider first the following two statements:

The program is well written and well documented.

The program is well documented and well written.

Clearly, these two statements always have the same truth values, and they are therefore by definition logically equivalent. To make this notion precise, let us translate them into logic. If P expresses the proposition "the program is well written" and if Q expresses the proposition "the program is well documented," then the first of the two statements translates into $P \wedge Q$, whereas the second translates in $Q \wedge P$. A glance at the definition of the connective $\wedge$ confirms that these two expressions have the same truth values for all possible assignments; that is, $(Q \wedge P) \Leftrightarrow (P \wedge Q)$ is a tautology. This proves that the two expressions are logically equivalent. Statements that are logically equivalent can be substituted for each other without affecting their truth values. Moreover, since logical equivalences are tautologies, they can be converted into schemas and used as such. We will show how these two properties can be used as the basis for an algebra for logical expressions, an algebra that is similar to the standard algebra used for manipulating expressions involving real numbers.

1.5.2 Proving Logical Equivalences by Truth Tables

Consider the following two statements.

> He is either not informed, or he is not honest.

> It is not true that he is informed and honest.

Intuitively, these two statements are logically equivalent. We prove this now. Define P and Q to represent the statements that he is honest and that he is well informed, respectively. The first statement translates into $\neg P \vee \neg Q$, whereas the second translates into $\neg(P \wedge Q)$. Our claim is that the two expressions are logically equivalent; that is, $\neg(P \wedge Q) \Leftrightarrow (\neg P \vee \neg Q)$ is a tautology. This is proved in Table 1.13. For simplicity, we omit the columns for $\neg P$ and $\neg Q$. The last column shows that the expression in question is a tautology, which means that the two statements in question are indeed logically equivalent. The same result can be obtained by comparing the previous two columns and checking if they are identical. This allows one to omit the last column altogether. The logical equivalence proved in Table 1.13 is important. It is called *De Morgan's law.*

TABLE 1.13.   Truth Table for $\neg(P \wedge Q) \Leftrightarrow (\neg P \vee \neg Q)$

P	Q	$P \wedge Q$	$\neg(P \wedge Q)$	$\neg P \vee \neg Q$	$\neg(P \wedge Q) \Leftrightarrow \neg P \vee \neg Q$
T	T	T	F	F	T
T	F	F	T	T	T
F	T	F	T	T	T
F	F	F	T	T	T

Next, consider the following pair of statements:

> If the goods were not delivered, the customer cannot have paid.

> If the customer has paid, the goods must have been delivered.

If Q and P stand for "goods were delivered" and "customer paid," respectively, then these two statements translate into $\neg Q \Rightarrow \neg P$ and $P \Rightarrow Q$. $P \Rightarrow Q$ and $\neg Q \Rightarrow \neg P$ are *contrapositives* of each other. Table 1.14 shows that contrapositives are logically equivalent; that is,

$$\neg Q \Rightarrow \neg P \equiv P \Rightarrow Q$$

The logical equivalence is established by the last two columns of the table, which are identical.

TABLE 1.14.   Truth Table for $\neg Q \Rightarrow \neg P$ and $P \Rightarrow Q$

P	Q	$\neg P$	$\neg Q$	$\neg Q \Rightarrow \neg P$	$P \Rightarrow Q$
T	T	F	F	T	T
T	F	F	T	F	F
F	T	T	F	T	T
F	F	T	T	T	T

The following is a very important logical equivalence, which has already been used in our discussion.

P and Q have the same truth value.

If P, then Q, and if Q, then P.

The second statement can be reworded "P if and only if Q." The conversion of the two statements into logic is straightforward. The first statement becomes $P \Leftrightarrow Q$ and the second $(P \Rightarrow Q) \land (Q \Rightarrow P)$. Table 1.15 shows that these two expressions are logically equivalent; the two columns corresponding to the two expressions in question have identical truth values.

TABLE 1.15.   Truth Table for $(P \Leftrightarrow Q)$ and $(P \Rightarrow Q) \land (Q \Rightarrow P)$

P	Q	$P \Leftrightarrow Q$	$P \Rightarrow Q$	$Q \Rightarrow P$	$(P \Rightarrow Q) \land (Q \Rightarrow P)$
T	T	T	T	T	T
T	F	F	F	T	F
F	T	F	T	F	F
F	F	T	T	T	T

1.5.3 Statement Algebra

In standard algebra, expressions in which the variables and constants represent numbers are manipulated. In statement algebra, logical expressions, that is, expressions in which the constants and variables represent truth values, are manipulated. Before discussing statement algebra, let us have a look at standard algebra and see how expressions are manipulated. Consider, for instance, the expression

$$(a + b) - b$$

One sees at a glance that this expression yields a. In fact, we are so accustomed to performing such algebraic manipulations that we are no longer aware of what is behind each step. Let us therefore retrace all the steps in detail. To begin with, we have the following identities:

$$(x + y) - z = x + (y - z) \tag{1.2}$$
$$y - y = 0 \tag{1.3}$$
$$x + 0 = x \tag{1.4}$$

In these three laws, one can replace the variables x, y, and z by arbitrary expressions; that is, the laws can be interpreted as schemas. In fact, any algebraic identity is a schema. We instantiate (1.2) with $x = a$, $y = b$, and $z = b$, which yields

$$(a + b) - b = a + (b - b)$$

Next, we conclude from (1.3) that $b - b = 0$. This means that $b - b$ and 0 are equivalent in the sense that the value of an expression does not change if one is replaced with the other. Hence, we can write

$$(a + b) - b = a + (b - b) = a + 0$$

The term on the right, that is, $a + 0$, can now be simplified, using (1.4), which yields

$$(a + b) - b = a$$

Consider now the following logical expression:

$$(P \wedge Q) \wedge \neg Q$$

This logical expression can be simplified in a similar way, except that logical equivalences take the place of algebraic identities. Specifically, the following schemas are used:

$$(A \wedge B) \wedge C \equiv A \wedge (B \wedge C) \tag{1.5}$$
$$A \wedge \neg A \equiv F \tag{1.6}$$
$$A \wedge F \equiv F \tag{1.7}$$

We can now apply (1.5), with $A = P$, $B = Q$, and $C = \neg Q$, to obtain

$$(P \wedge Q) \wedge \neg Q \equiv P \wedge (Q \wedge \neg Q)$$

Schema (1.6) allows us to conclude that $Q \wedge \neg Q$ is logically equivalent to F, which allows us to replace $Q \wedge \neg Q$ by F. Hence,

$$(P \wedge Q) \wedge \neg Q \equiv P \wedge (Q \wedge \neg Q) \equiv P \wedge F$$

Finally, from (1.7), we conclude that the expression on the right is F. Therefore,

$$(P \wedge Q) \wedge \neg Q \equiv F \tag{1.8}$$

Note that in algebra one does not really differentiate between an expression and a schema. The reason is that any expression can easily be converted into a schema, and vice versa. For the same reason, we will no longer distinguish between a logical equivalence and the schema it generates.

Statement algebra has many applications. Here, we use statement algebra to prove that a particular argument is sound. In this case, it is often easiest to show that it is impossible that all premises are true, yet the conclusion is false. In other words, one proves that the conjunction of the premises and the negation of the conclusion cannot be true simultaneously.

Example 1.11   Prove that, once $P \wedge Q$ is established, one may conclude Q.

> **Solution**   The argument in question is sound iff $(P \wedge Q) \wedge \neg Q$ is a contradiction. It follows from (1.8) that this is the case. Incidentally, the law $P \wedge Q \Rightarrow Q$ is known as the *law of simplification*.   ∎

1.5.4 Removing Conditionals and Biconditionals

Since the symbolic treatment of conditionals and biconditionals is relatively cumbersome, one usually removes them before performing further algebraic manipulations. To remove the conditional, one uses the following logical equivalence. The reader may prove this equivalence by the truth table method.

$$P \Rightarrow Q \equiv \neg P \vee Q \tag{1.9}$$

There are two ways to express the biconditional:

$$P \Leftrightarrow Q \equiv (P \wedge Q) \vee (\neg P \wedge \neg Q) \tag{1.10}$$
$$P \Leftrightarrow Q \equiv (P \Rightarrow Q) \wedge (Q \Rightarrow P) \tag{1.11}$$

The version given by (1.10) expresses the fact that two expressions are equivalent if they have the same truth value. The version given by (1.11) stresses the fact that two expressions are equivalent if the first implies the second and if the second implies the first. Since we want to remove all instances of $\Rightarrow$, we use (1.9) to rewrite (1.11) as follows:

$$P \Leftrightarrow Q \equiv (\neg P \vee Q) \wedge (P \vee \neg Q) \tag{1.12}$$

There are thus two equations that allow one to remove the $\Leftrightarrow$, (1.10) and (1.12). One can use whichever form is more convenient under the circumstances.

Example 1.12   Remove $\Rightarrow$ and $\Leftrightarrow$ from the following expression:

$$(P \Rightarrow Q \wedge R) \vee ((R \Leftrightarrow S) \wedge (Q \vee S))$$

Solution   The equivalence (1.9) allows us to replace $P \Rightarrow Q \wedge R$ by $\neg P \vee (Q \wedge R)$, which yields

$$(P \Rightarrow Q \wedge R) \vee ((R \Leftrightarrow S) \wedge (Q \vee S)) \equiv (\neg P \vee Q \wedge R) \vee ((R \Leftrightarrow S) \wedge (Q \vee S))$$

Next, we use (1.12) to convert this to the following logically equivalent expression:

$$(\neg P \vee Q \wedge R) \vee (((\neg R \vee S) \wedge (R \vee \neg S)) \wedge (Q \vee S))$$

This expression contains no further $\Rightarrow$ and $\Leftrightarrow$, and it is therefore the answer to our problem. ∎

1.5.5 Essential Laws for Statement Algebra

In this section, we discuss a number of important equivalences that are useful to simplify or otherwise manipulate logical expressions. We begin by discussing the laws given in Table 1.16. Some of these laws have already been proved by the truth table method. This method can be used to prove all the other laws. Note that, with the exception of the double-negation law, all the laws of Table 1.16 come in pairs, called *dual pairs*. For each expression, one finds the dual by replacing all T by F and all F by T and by replacing all $\wedge$ by $\vee$ and all $\vee$ by $\wedge$. The excluded middle law, for instance, has the contradiction law as its dual. This dual can be obtained by replacing the $\vee$ by $\wedge$, yielding $P \wedge \neg P$, which is logically equivalent to F, which corresponds to T in the excluded middle law. At first glance, it appears that the double-negation law does not have a dual. However, on closer inspection it turns out that this law is its own dual. In this sense, one can say that all laws given in Table 1.16 have a dual, and the same must obviously be true for all results derived from these laws.

The laws of Table 1.16 up to and including the double-negation law allow one to simplify an expression, and it is normally a good idea to apply them whenever possible. Consider, for instance, the expression $\neg\neg P \wedge (Q \vee \neg Q)$. The law of double negation allows us to write this as $P \wedge (Q \vee \neg Q)$, and because of the excluded middle law, this becomes $P \wedge$ T. The identity law shows that this is P, and we conclude that

$$\neg\neg P \wedge (Q \vee \neg Q) \equiv P$$

The commutative, associative, and distributive laws have their equivalents in standard algebra. In fact, the connective $\vee$ is often treated like $+$, and the connective $\wedge$ is often treated like $\times$. For instance, $P \vee Q \equiv Q \vee P$ corresponds to the commutative law $a + b = b + a$ in standard algebra. The analogy between the standard algebra and statement algebra carries quite far. In some cases, however, the analogy breaks down. In particular, $(a + b) \times (a + c) \neq a + (b \times c)$, yet $(P \vee Q) \wedge (P \vee R) \equiv P \vee (Q \wedge R)$.

From the laws given in Table 1.16, one can derive further laws. Of particular importance are the *absorption* laws, which are

$$P \vee (P \wedge Q) \equiv P \tag{1.13}$$
$$P \wedge (P \vee Q) \equiv P \tag{1.14}$$

TABLE 1.16.   Laws of Conjunction, Disjunction, and Negation

Law	Name
$P \vee \neg P \equiv \text{T}$ $P \wedge \neg P \equiv \text{F}$	Excluded middle law Contradiction law
$P \vee \text{F} \equiv P$ $P \wedge \text{T} \equiv P$	Identity laws
$P \vee \text{T} \equiv \text{T}$ $P \wedge \text{F} \equiv \text{F}$	Domination laws
$P \vee P \equiv P$ $P \wedge P \equiv P$	Idempotent laws
$\neg(\neg P) \equiv P$	Double-negation law
$P \vee Q \equiv Q \vee P$ $P \wedge Q \equiv Q \wedge P$	Commutative laws
$(P \vee Q) \vee R \equiv P \vee (Q \vee R)$ $(P \wedge Q) \wedge R \equiv P \wedge (Q \wedge R)$	Associative laws
$(P \vee Q) \wedge (P \vee R) \equiv P \vee (Q \wedge R)$ $(P \wedge Q) \vee (P \wedge R) \equiv P \wedge (Q \vee R)$	Distributive laws
$\neg(P \wedge Q) \equiv \neg P \vee \neg Q$ $\neg(P \vee Q) \equiv \neg P \wedge \neg Q$	De Morgan's laws

Equivalence (1.13) is proved as follows:

$$
\begin{aligned}
P \vee (P \wedge Q) &\equiv (P \wedge \text{T}) \vee (P \wedge Q) & &\text{Identity law} \\
&\equiv P \wedge (\text{T} \vee Q) & &\text{Distributive law} \\
&\equiv P \wedge \text{T} & &\text{Domination law} \\
&\equiv P & &\text{Identity law}
\end{aligned}
$$

The proof of (1.14) is similar, except that one uses the dual equivalences. The absorption laws are very useful when expressions need to be simplified. Two other important laws are the following:

$$(P \wedge Q) \vee (\neg P \wedge Q) \equiv Q \tag{1.15}$$

$$(P \vee Q) \wedge (\neg P \vee Q) \equiv Q \tag{1.16}$$

To prove equivalence (1.15), apply the distributive law to obtain $(P \vee \neg P) \wedge Q$, which in turn simplifies to $\text{T} \wedge Q$ because of the excluded middle law. At this point, one applies the identity law, which yields Q, and the result follows. The proof of (1.16) is similar.

1.5.6 Shortcuts for Manipulating Expressions

In the previous section, the laws of Table 1.16 were applied exactly as stated. For practical purposes, this is often cumbersome, and shortcuts are possible. To see how this works, consider the conjunction

$$((P_1 \wedge P_3) \wedge P_2) \wedge (\text{T} \wedge P_1) \tag{1.17}$$

Since $(\text{T} \wedge P_1) \equiv P_1$, one can drop the T. It is easy to see that, because of the associative law, it does not matter where the parentheses are. One may as well drop them. Moreover, the commutative law allows one to reorder the terms freely. Hence,

$$((P_1 \wedge P_3) \wedge P_2) \wedge P_1 \equiv P_1 \wedge P_1 \wedge P_2 \wedge P_3$$

It is now clear that the second P_1 may be dropped because of the idempotent law. The expression (1.17) therefore simplifies to $P_1 \wedge P_2 \wedge P_3$.

Next, consider the expression

$$P \wedge \text{F} \wedge Q$$

Two applications of the domination law show that this expression yields F. In general, any conjunction in which the term F appears yields F. The same is true if a conjunction contains two complementary literals. For example, $P \wedge Q \wedge \neg P$ yields F. To see this, rewrite $P \wedge Q \wedge \neg P$ as $P \wedge \neg P \wedge Q$ and apply the contradiction law to obtain $\text{F} \wedge Q$, which is F.

In summary, the following rules are available to simplify conjunctions containing only literals. Before applying these rules, it is best to sort the literals lexicographically, according to variable names.

1. If a conjunction contains complementary literals or if it contains the logical constant F, it always yields F; that is, it is a contradiction.
2. All instances of the logical constant T and all duplicate copies of any literal, may be dropped.

To deal with disjunctions, the duals of these two rules can be used.

1. If a disjunction contains complementary literals or if it contains the logical constant T, it always yields T; that is, it is a tautology.
2. All instances of the logical constant F and all duplicate copies of any literal may be dropped.

For example, $P \vee Q \vee \neg P$ yields T, and $P \vee Q \vee P \vee \text{F}$ yields $P \vee Q$. Note the distinction between conjunctions and disjunctions. A disjunction with complementary literals or a disjunction containing T always yields T. Hence, it is a tautology. A conjunction containing complementary literals or containing the logical constant F is always a contradiction.

There are obviously conjunctions with four, three, or two terms. For instance, $P_1 \wedge P_2 \wedge P_3 \wedge P_4$ is a conjunction with four terms, and $P \wedge Q$ is a conjunction with two terms. A single propositional variable can be interpreted as a conjunction with one term. Moreover, it turns out to be convenient to define T as the conjunction with zero terms. One can justify this, for example, by considering the following four expressions:

$$P_1 \wedge P_2 \wedge P_3 \wedge \text{T}$$
$$P_1 \wedge P_2 \wedge \text{T}$$
$$P_1 \wedge \text{T}$$
$$\text{T}$$

If the T is dropped, the first expression becomes a conjunction of three terms, the second a conjunction of two terms, and the third a conjunction of one term. This suggests that the fourth expression should become a conjunction of zero terms. Conjunctions with zero terms are useful when establishing correct patterns for proof. In this case, one has to form the conjunction of the premises, and if conjunctions with zero terms are allowed, one can investigate conclusions that are valid without having to resort to any premises. Indeed, all tautologies can be interpreted as conclusions that do not require any premises. By defining a conjunction with zero terms to yield T, all tautologies can be viewed as special cases of correct logical arguments. This will be discussed later. There are also disjunctions with zero terms. However, a disjunction with zero terms always yields F.

Example 1.13   Simplify

$$(P_3 \wedge \neg P_2 \wedge P_3 \wedge \neg P_1) \vee (P_1 \wedge P_3 \wedge \neg P_1)$$

Solution   The expression in question is the disjunction of two conjunctions. First, we simplify the conjunctions. In the first conjunction, P_3 appears twice, and one P_3 may be dropped. The second conjunction contains both P_1 and $\neg P_1$, and it therefore yields F. Consequently,

$$(P_3 \wedge \neg P_2 \wedge P_3 \wedge \neg P_1) \vee (P_1 \wedge P_3 \wedge \neg P_1) \equiv (\neg P_1 \wedge \neg P_2 \wedge P_3) \vee F$$
$$\equiv \neg P_1 \wedge \neg P_2 \wedge P_3 \qquad \blacksquare$$

1.5.7 Normal Forms

Generally, it is useful to have standard forms for expressions, because these make the identification and comparison of two expressions easier. There are also standard forms for logical expressions, and these are called *normal forms*. There are two types of normal forms, *disjunctive normal forms* and *conjunctive normal forms*. In both cases, one has *full* forms, that is, full disjunctive and full conjunctive normal forms, respectively, that are unique. However, it is often more convenient to use normal forms that are not full normal forms because they are shorter.

> **Definition 1.18:**   A logical expression is said to be in *disjunctive normal form* if it is written as a disjunction, in which all terms are conjunctions of literals. Similarly, a logical expression is said to be in *conjunctive normal form* if it is written as a conjunction of disjunctions of literals.

Examples of disjunctive normal forms include the disjunctions $(P \wedge Q) \vee (P \wedge \neg Q)$, $P \vee (Q \wedge R)$, and $\neg P \vee T$. The disjunction $\neg(P \wedge Q) \vee R$, on the other hand, is not in normal form because it contains a negated nonatomic subexpression. Only atomic subexpressions can be negated, and negated atomic expressions are literals. Examples of conjunctive normal forms include $P \wedge (Q \vee R)$ and $P \wedge F$. However, $P \wedge (R \vee (P \wedge Q))$ is not in conjunctive normal form, because the disjunction $(R \vee (P \wedge Q))$ contains a conjunction as a subexpression. To qualify as a conjunctive normal form, no disjunction must contain a conjunction as a subexpression. Literals, such as P or $\neg P$, are simultaneously conjunctive normal forms and disjunctive normal forms, and so are T and F.

In this section, we discuss how to obtain conjunctive normal forms through algebraic manipulations. Disjunctive normal forms can be obtained in a similar way. Moreover, Section 1.5.8 shows how to obtain disjunctive normal forms from truth tables.

Three steps are required to obtain a conjunctive normal form through algebraic manipulations.

1. Remove all $\Rightarrow$ and $\Leftrightarrow$.

2. If the expression in question contains any negated compound subexpressions, either remove the negation by using the double-negation law or use De Morgan's laws to reduce the scope of the negation.

3. Once an expression with no negated compound subexpression is found, use the following two laws to reduce the scope of $\vee$.

$$A \vee (B \wedge C) \equiv (A \vee B) \wedge (A \vee C) \tag{1.18}$$

$$(A \wedge B) \vee C \equiv (A \vee C) \wedge (B \vee C) \tag{1.19}$$

The laws (1.18) and (1.19) follow from the commutative and the distributive laws.

Example 1.14   Convert the following expression into normal form.

$$\neg((P \vee \neg Q) \wedge \neg R)$$

Solution   The conjunctive normal form can be found by the following derivation:

$$
\begin{aligned}
\neg((P \vee \neg Q) \wedge \neg R) &\equiv \neg(P \vee \neg Q) \vee \neg\neg R && \text{De Morgan} \\
&\equiv \neg(P \vee \neg Q) \vee R && \text{Double negation} \\
&\equiv (\neg P \wedge \neg\neg Q) \vee R && \text{De Morgan} \\
&\equiv (\neg P \wedge Q) \vee R && \text{Double negation} \\
&\equiv (\neg P \vee R) \wedge (Q \vee R) && \text{By (1.19)}
\end{aligned}
$$
∎

This method for finding a conjunctive normal form will always succeed. In fact, if any compound expression is negated, either the double-negation law or one of the De Morgan laws applies. In the end, all negations must be parts of literals. Similarly, if any subexpression contains a conjunction, either (1.18) or (1.19) applies. In the end, all disjunctions are within the scope of some conjunction, at which time the expression is in normal form.

Example 1.15   Convert the following expression into conjunctive normal form.

$$(P_1 \wedge P_2) \vee (P_3 \wedge (P_4 \vee P_5))$$

Solution

$$(P_1 \wedge P_2) \vee (P_3 \wedge (P_4 \vee P_5))$$
$$\equiv (P_1 \vee (P_3 \wedge (P_4 \vee P_5)))$$
$$\wedge (P_2 \vee (P_3 \wedge (P_4 \vee P_5))) \qquad \text{By (1.19)}$$
$$\equiv (P_1 \vee P_3) \wedge (P_1 \vee P_4 \vee P_5) \qquad \text{By (1.18)}$$
$$\wedge (P_2 \vee P_3) \wedge (P_2 \vee P_4 \vee P_5) \qquad \text{By (1.18)}$$

All occurrences of $\vee$ now have the smallest possible scope, and the expression is in conjunctive normal form. ∎

Once a conjunctive normal form is obtained, it pays to check if further simplifications are possible. Hence, one should check if any of the equivalences (1.13)–(1.16) are applicable. Another source of equations that can be used for the purpose of simplification is the first nine laws of Table 1.16. In particular, all disjunctions containing complementary literals can simply be dropped. Moreover, one is allowed to remove literals that occur more than once in the same disjunction.

Example 1.16   Simplify the following normal form:

$$(P \vee Q) \wedge P \wedge (Q \vee R) \wedge (P \vee \neg P \vee R) \wedge (\neg Q \vee R)$$

Solution   The second-to-last disjunction $P \vee \neg P \vee R$ may be dropped because it yields T. Moreover, the law of absorption allows one to drop $(P \vee Q)$, which is absorbed by P. Finally, according to (1.16), $(Q \vee R) \wedge (\neg Q \vee R)$ yields R.

$$(P \vee Q) \wedge P \wedge (Q \vee R) \wedge (P \vee \neg P \vee R) \wedge (\neg Q \vee R) \equiv P \wedge R \qquad ∎$$

The laws described in this section are often useful for simplifying expressions. Considerable simplifications can already be obtained by using De Morgan's laws, because they allow one to rewrite the equivalence such that the double-negation law becomes applicable. Also, expressions containing no negated compound subexpressions are more flexible, which increases the opportunity for simplifications.

1.5.8 Truth Tables and Disjunctive Normal Forms

So far, we have shown how to find the truth table of a logical expression. The reverse is also possible. One can convert any given truth table into an expression. The expression obtained in this way is already in disjunctive normal form. In fact, the conceptually easiest method to find the normal form of an expression is by using truth tables. Unfortunately, truth tables grow exponentially with the number of variables, which makes this method unattractive for expressions with many variables.

Generally, a truth table gives the truth values of some proposition for all possible assignments. Table 1.17 gives an example of a truth table for a certain proposition f. The truth value of f depends on the three variables P, Q, and R. This makes f a truth function, with the arguments P, Q, and R.

To convert a function given by its truth table into a logical expression, one makes use of *minterms*.

TABLE 1.17.

Truth Table for a Truth Function

P	Q	R	f
T	T	T	T
T	T	F	F
T	F	T	T
T	F	F	F
F	T	T	F
F	T	F	F
F	F	T	T
F	F	F	F

Definition 1.19:   A *minterm* is a conjunction of literals in which each variable is represented exactly once.

For example, if a truth function has the variables P, Q, and R, then $P \wedge \neg Q \wedge R$ is a minterm, but $P \wedge \neg Q$ and $P \wedge P \wedge \neg Q \wedge R$ are not. In $P \wedge \neg Q$, R is missing, and in $P \wedge P \wedge \neg Q \wedge R$, the variable P occurs more than once.

Each minterm is true for exactly one assignment. For example, $P \wedge \neg Q \wedge R$ is true if P is T, Q is F, and R is T. Any deviation from this assignment would make this particular minterm false. A disjunction of minterms is true only if at least one of its constituent minterms is true. For instance, $(P \wedge Q \wedge R) \vee (P \wedge \neg Q \wedge R) \vee (\neg P \wedge \neg Q \wedge R)$ is only true if at least one of $P \wedge Q \wedge R$, $P \wedge \neg Q \wedge R$, or $\neg P \wedge \neg Q \wedge R$ is true. If a function, such as f, is given by a truth table, we know exactly for which assignments it is true. Consequently, we can select the minterms that make the function true and form the disjunction of these minterms. The function f, for instance, is true for three assignments:

1. P, Q, R are all true.

2. P, $\neg Q$, R are all true.

3. $\neg P$, $\neg Q$, R are all true.

The disjunction of the corresponding minterms is logically equivalent to f, which means that we have the following logical expression for f:

$$f \equiv (P \wedge Q \wedge R) \vee (P \wedge \neg Q \wedge R) \vee (\neg P \wedge \neg Q \wedge R)$$

Since the minterms are conjunctions, we have expressed the function in question in disjunctive normal form. Actually, we have a special type of a normal form, the *full disjunctive normal form*.

Definition 1.20:   If a truth function is expressed as a disjunction of minterms, it is said to be in *full disjunctive normal form.*

The full disjunctive normal form is usually not the shortest disjunction of conjunctions available to express a given function. In fact, one can simplify it in the normal fashion. In our case, one can apply (1.15) twice to write f as

$$f \equiv (P \wedge R) \vee (\neg Q \wedge R)$$

1.5.9 Conjunctive Normal Forms and Complementation

In this section, a concept closely related to duality, *complementation*, will be introduced, and it will be used to obtain conjunctive normal forms from truth tables. If A is any expression, one obtains the complement of A by first forming the dual of A (see Section 1.1.5) and then replacing all literals by their complements. In fact, the complement of A is always the negation of A, and complementation is an efficient way to negate an expression.

Example 1.17   Negate the expression $A = (P \wedge Q) \vee \neg R$ by using complementation.

Solution   The dual of A is $(P \vee Q) \wedge \neg R$. To find the complement, one replaces every literal by its complement, which yields $(\neg P \vee \neg Q) \wedge R$. One concludes that

$$\neg A \equiv \neg((P \wedge Q) \vee \neg R) \equiv (\neg P \vee \neg Q) \wedge R \qquad \blacksquare$$

We now prove that complementation is negation. Hence, let comp A be the complement of A. If A consists of one literal only, say $A = P$, then comp $A = \neg P$, and this is the negation of A. A similar result applies if $A = \neg P$. Also, if $A = $ T, its complement is $A = $ F, and comp $A \equiv \neg A$, and the same holds for $A = $ T. Consider now conjunctions of literals. If $A = (P \wedge Q)$, its complement is $(\neg P \vee \neg Q)$, and this is logically equivalent to $\neg(P \wedge Q)$ because of De Morgan's law. Similarly, the complement of $P \vee Q$, which is $\neg P \wedge \neg Q$, is logically equivalent to the negation of $P \vee Q$. Moreover, conjunctions and disjunctions of terms A and B have complements, which are logically equivalent to $\neg(A \wedge B)$ and $\neg(A \vee B)$, respectively, provided that $\neg A \equiv$ comp A and $\neg B \equiv$ comp B. A similar statement applies to negations. Hence, all statements built from expressions for which complementation means negation inherit this property: their complement, too, is logically equivalent to their negation. It is not difficult to see that this implies that the complement of every expression is logically equivalent to negation. This will be discussed in further detail in Section 3.3.5.

Complementation is also the reason why all equivalences of Table 1.16 appeared in dual pairs. If two expressions A and B are logically equivalent, so are their negations and, with it, their complements. To find the dual from the complement, one only needs to replace all literals by their complements. This operation converts tautologies into tautologies. In summary, if $A \equiv B$, then comp $A \equiv$ comp B, and, as a consequence, the dual of $A \equiv B$ is a logical equivalence.

Complementation can be used to find the conjunctive normal form from a truth table of some function f. One first determines the disjunctive normal form of $\neg f$. If the resulting disjunctive normal form is A, then $A \equiv \neg f$, and the complement of A, being the negation of A, must be logically equivalent to f. The following example shows the details of this approach.

Example 1.18   Find the full conjunctive normal form for f_1 given by Table 1.18.

TABLE 1.18.
Truth Table for a Truth Function

P	Q	R	f_1
T	T	T	T
T	T	F	T
T	F	T	F
T	F	F	F
F	T	T	T
F	T	F	T
F	F	T	F
F	F	F	T

Solution $\neg f_1$ is true only for the following assignments:

$$P = \text{T}, \qquad Q = \text{F}, \qquad R = \text{T}$$
$$P = \text{T}, \qquad Q = \text{F}, \qquad R = \text{F}$$
$$P = \text{F}, \qquad Q = \text{F}, \qquad R = \text{T}$$

The disjunctive normal form of $\neg f_1$ is therefore

$$(P \wedge \neg Q \wedge R) \vee (P \wedge \neg Q \wedge \neg R) \vee (\neg P \wedge \neg Q \wedge R)$$

This expression has the complement

$$f_1 \equiv (\neg P \vee Q \vee \neg R) \wedge (\neg P \vee Q \vee R) \wedge (P \vee Q \vee \neg R)$$

This complement is the desired conjunctive normal form.   ∎

Problems 1.5

1. Use equations (1.9) and (1.10) to remove $\Rightarrow$ and $\Leftrightarrow$ from the following expressions:
   (a) $(P \Rightarrow Q) \wedge (Q \Rightarrow R)$
   (b) $(P \Rightarrow Q) \Leftrightarrow ((P \wedge Q) \Leftrightarrow Q)$
   (c) $\neg P \Rightarrow \neg Q$
2. Using the laws of Table 1.16, prove the following equivalences:
   (a) $(P \wedge Q) \vee (Q \wedge R) \equiv Q \wedge (P \vee R)$

(b) $\neg(\neg(P \wedge Q) \vee P) \equiv F$

(c) $\neg(\neg P \vee \neg(R \vee S)) \equiv (P \wedge R) \vee (P \wedge S)$

(d) $(P \vee R) \wedge (Q \vee S) \equiv (P \wedge Q) \vee (P \wedge S) \vee (R \wedge Q) \vee (R \wedge S)$

3. Simplify the following expressions:

    (a) $(P_3 \wedge T) \wedge (P_2 \wedge T)$

    (b) $(P_3 \wedge (P_2 \wedge (P_1 \wedge P_3)))$

    (c) $(P_3 \wedge T) \wedge (P_2 \wedge \neg P_3)$

4. Simplify the following expressions:

    (a) $(Q \wedge R \wedge S) \vee (Q \wedge \neg R \wedge S)$

    (b) $(P \vee R) \wedge (P \vee R \vee S)$

    (c) $(P \vee (Q \wedge S)) \vee (\neg Q \wedge S)$

5. Move all negations inward as far as possible.

    (a) $\neg(P_1 \wedge \neg(P_2 \vee \neg P_3))$

    (b) $P \vee \neg(P \wedge \neg(Q \vee R))$

6. Use statement algebra to simplify the following expressions:

    (a) $P \vee \neg Q \vee (P \wedge Q) \wedge (P \vee \neg Q) \wedge \neg P \wedge Q$

    (b) $(P \vee \neg Q) \wedge (\neg P \vee Q) \vee \neg(\neg(P \vee \neg R) \wedge Q)$

    (c) $\neg((P \vee Q) \wedge R) \vee Q$

7. Find the conjunctive normal form of the following expressions by using statement algebra.

    (a) $(P \Rightarrow Q) \Leftrightarrow (P \Rightarrow R \vee Q)$

    (b) $(P \vee Q) \wedge (P \vee (R \wedge S)) \vee (P \wedge Q \wedge S)$

8. Find the full disjunctive normal form of the logical function given in Table 1.19.

TABLE 1.19

P	Q	R	f
F	F	F	F
F	F	T	T
F	T	F	F
F	T	T	F
T	F	F	T
T	F	T	F
T	T	F	F
T	T	T	T

9. Find the full conjunctive normal form for f, where f is given in Table 1.19.

10. Prove equations (1.14), (1.16), and (1.9).

1.6 LOGICAL IMPLICATIONS AND DERIVATIONS

1.6.1 Introduction

In this section, we will discuss logical implication and how it can be used as a basis for sound reasoning. There are, of course, arguments that are not sound, and some of these will be identified in the following subsections. Unsound arguments are called *fallacies*. By using the truth table method, one can distinguish between sound arguments and fallacies.

Reasoning patterns can be expressed in a variety of ways. In the English language, the conclusion is typically stated after the premises, and it is introduced by such words as "therefore," "as a consequence," and "in conclusion." In the arguments presented at the beginning of this chapter, we listed all the premises first, one below the other. There was a horizontal line underneath the last premise, and below that line, there was the conclusion. In the case of the modus ponens, this looks as follows:

1. P

2. $P \Rightarrow Q$

3. Q

If the reasoning pattern is sound, but only then, we use the symbol $\models$ to separate the premises from the conclusion. To take again the example of the modus ponens, this looks as follows:

$$P, P \Rightarrow Q \models Q$$

We have used the symbol $\models$ already to denote tautologies. This agrees with our present use: a tautology is merely a reasoning pattern with no premises.

An argument is sound if the conclusion follows logically, provided all premises are met. This means that the conjunction of all premises logically implies the conclusion. Hence, if A is the conjunction of all premises and if B is the conclusion, one has to show that $A \Rightarrow B$ is a tautology. In other words, one has to demonstrate that $A \Rrightarrow B$ or, equivalently, $A \models B$.

Many arguments contain a sequence of steps. This is particularly true for mathematical proofs, but it is also true in other application areas of logic. In this section we discuss such sequences of arguments and how to apply them. This leads to the notion of a *derivation* or a *formal proof*. Several systems for derivations will be discussed briefly in this chapter. Additional details are given in Chapter 11.

1.6.2 Logical Implications

Any tautology of the form $A \Rightarrow B$ is called a *logical implication*. We have already encountered a number of logical implications. In Section 1.2.7, for instance, we stated that $P \Rightarrow T$ and $F \Rightarrow P$ are expressions that are true for all values of P, which makes both expressions tautologies. Hence, $P \Rrightarrow T$ and $F \Rrightarrow P$. Logical implications are tautologies, and any logical tautology can be used as the basis of a schema. Specifically, if A is any expression, $P \Rrightarrow T$ justifies the schema $A \Rrightarrow T$. The schema $F \Rrightarrow A$ can be proved in a similar fashion.

Logical equivalences create logical implications, as indicated by the following theorem.

Theorem 1.2.   If C and D are two expressions and if $C \equiv D$, then $C \Rrightarrow D$ and $D \Rrightarrow C$.

The proof of this theorem is left to the reader. As an application of the theorem, consider the equivalence $(P \lor Q) \land (\neg P \lor Q) \equiv Q$ [see (1.16)]. This equivalence leads to the following important logical implication:

$$(P \lor Q) \land (\neg P \lor Q) \Rrightarrow Q \qquad (1.20)$$

Of course, any logical implication can be proved by the truth table method. Indeed, if we want to prove $A \Rrightarrow B$, where A and B are two expressions, we prepare the truth table for $A \Rightarrow B$. For instance, Table 1.20 proves $P \Rrightarrow (P \lor Q)$, and Table 1.21 proves $(P \land Q) \Rrightarrow P$. Both of these logical implications have names: they are called, respectively, the *law of addition* and the *law of simplification*. The disjunctive syllogism was proved earlier in the same way (see Table 1.10).

TABLE 1.20.
Truth Table for $P \Rightarrow (P \lor Q)$

P	Q	$P \lor Q$	$P \Rightarrow (P \lor Q)$
T	T	T	T
T	F	T	T
F	T	T	T
F	F	F	T

TABLE 1.21.
Truth Table for $(P \land Q) \Rightarrow P$

P	Q	$P \land Q$	$(P \land Q) \Rightarrow P$
T	T	T	T
T	F	F	T
F	T	F	T
F	F	F	T

1.6.3 Soundness Proofs through Truth Tables

An argument is sound if the premises together logically imply the conclusion. Hence, if $A_1, A_2, \ldots, A_n$ denote the premises and if C denotes the conclusion, one must have

$$A_1, A_2, \ldots, A_n \models C$$

As shown previously, this can be proved through the truth table method by showing that the following expression is a tautology.

$$A_1 \land A_2 \land \cdots \land A_n \Rightarrow C$$

We will now use a shortcut. First find the truth values of all premises. If, for a given assignment, they all yield T, then write a T under a special column with the heading "Premises." Another column is reserved for the truth value of the conclusion. A last column, labeled

"Valid," indicates whether or not the premises imply the conclusion. To demonstrate how this works, consider the hypothetical syllogism, which can be expressed as

$$P \Rightarrow Q, Q \Rightarrow R \models P \Rightarrow R$$

Table 1.22 is the truth table for this expression. Since all possible assignments lead to a T in the Valid column, the hypothetical syllogism is sound.

TABLE 1.22.   Truth Table of the Hypothetical Syllogism

P	Q	R	$(P \Rightarrow Q)$	$(Q \Rightarrow R)$	Premises	$(P \Rightarrow R)$	Valid
T	T	T	T	T	T	T	T
T	T	F	T	F	F	F	T
T	F	T	F	T	F	T	T
T	F	F	F	T	F	F	T
F	T	T	T	T	T	T	T
F	T	F	T	F	F	T	T
F	F	T	T	T	T	T	T
F	F	F	T	T	T	T	T

The second logical argument is taken from "A Study in Scarlet" by Conan Doyle, and it appears in a passage where Sherlock Holmes explains to Dr. Watson how he derived the motive of the murder by logical reasoning. We will discuss the entire passage later. At this point, we only analyze the following argument from this passage.

Robbery had not been the object of the murder, for nothing was taken.

This sentence really boils down to the following argument

1. If it was robbery, something would have been taken.

2. Nothing was taken.
    _____

3. It was not robbery.

Note that Sherlock Holmes assumes that his listener, Dr. Watson, knows that if it was robbery then something would have been taken, and, indeed, this is implied by the definition of robbery. However, in logic, all these facts must be stated explicitly.

To convert the Sherlock Holmes argument into logic, let P represent the statement "Robbery was the object of the murder," and let Q be the statement "Something was taken." Hence, the argument becomes

1. $P \Rightarrow Q$

2. $\neg Q$
    _____

3. $\neg P$

Is this pattern sound? Table 1.23 shows that it is by the truth table method. If all the premises are true, and this is the case only if P and Q are both F, then the conclusion $\neg P$ is true. As indicated in the caption, the argument in question is called the *modus tollens*. Since all assignments show a T in the Valid column, the argument is sound.

TABLE 1.23.   Truth Table to Prove the Modus Tollens

P	Q	$P \Rightarrow Q$	$\neg Q$	Premises	Conclusion $(\neg P)$	Valid
T	T	T	F	F	F	T
T	F	F	T	F	F	T
F	T	T	F	F	T	T
F	F	T	T	T	T	T

The following argument is not sound; it is a fallacy.

1. $P \vee Q$
2. P

Table 1.24 shows why. There is an assignment, namely $P = $ F and $Q = $ T, for which the premise is true, yet the conclusion is false.

TABLE 1.24.   A Fallacy

P	Q	Premise $P \vee Q$	Conclusion P	Valid $(P \vee Q) \Rightarrow P$
T	T	T	T	T
T	F	T	T	T
F	T	T	F	F
F	F	F	F	T

1.6.4 Proofs

Many logical arguments are really compound arguments in the sense that the conclusion of one argument is a premise for the next. Every proof is a sequence of such arguments. Most arguments used in practice are informal. There are also formal logical arguments, which are referred to as *derivations* or *formal proofs*. The discussion is based on two examples. The first example involves the effect of an if-statement in Pascal, and the second is the complete argument used by Sherlock Holmes to find the motive of a murder.

The first example involves the following statement:

if X > Max **then** X = Max

We want to prove that, after execution, it is impossible that X be greater than Max.

For the proof, one must consider two cases, (1) the case when $X > Max$ is true before the execution of the if-statement, and (2) the case when it is false. In either case, $X > Max$ is false after execution. In the first case, the assignment statement is executed, and $X = Max$ after execution. This, in turn, implies that $X > Max$ becomes false. In the second case, no operation is done, and $X > Max$ is false before and after execution. In either case, $X > Max$ is false after execution.

Let us now express this argument in the language of logic. We define

P: $X > Max$ Before execution

Q: $X = Max$ After execution

R: $X > Max$ After execution

Note that we have to distinguish between the state of the system before and after the execution. We return to this point when discussing specifications languages and proofs for program correctness (see Section 8.4.4 and Chapter 9).

The actual proof consists of two parts. First, $P \Rightarrow Q$ is true because of the way the if-statement works: if $X > Max$ before execution, then $X = Max$ afterward. Next, $Q \Rightarrow \neg R$ is true because if two variables are equal, it is impossible that one be greater than the other. Hence, $P \Rightarrow Q$ and $Q \Rightarrow \neg R$, and we can apply the hypothetical syllogism to conclude that $P \Rightarrow \neg R$. Formally,

1. $P \Rightarrow Q$

2. $Q \Rightarrow \neg R$

   ───────────

3. $P \Rightarrow \neg R$

In any sequence of logical arguments, the premises may include expressions that have been proved. In particular, $P \Rightarrow \neg R$ can be used as premise for further argumentation. We now use the following schema, which can be derived from (1.20).

$$A \Rightarrow B, \neg A \Rightarrow B \models B \qquad\qquad (1.21)$$

Use this schema with $A = P$ and $B = \neg R$ to obtain

1. $P \Rightarrow \neg R$

2. $\neg P \Rightarrow \neg R$

   ───────────

3. $\neg R$

We now give what will later be shown to be a formal argument, or a *derivation*, which combines both arguments into one proof. This proof is given in Figure 1.2. In each line of the derivation, there is one expression. This expression is either a premise, or it is derived from previous lines by using rules from a given list of rules. In the present example, only two rules are used. One rule is the hypothetical syllogism, which is abbreviated by HS. The other rule is the *law of cases*, which is the schema given by (1.21). This rule is abbreviated

Prove: $P \Rightarrow Q,\ Q \Rightarrow \neg R,\ \neg P \Rightarrow \neg R \models \neg R.$

Formal Derivation	Rule	Comment
1. $P \Rightarrow Q$	Premise	If X $>$ Max before execution, then X $=$ Max after execution.
2. $Q \Rightarrow \neg R$	Premise	If X $=$ Max after execution, then X $>$ Max is not true at this point.
3. $\neg P \Rightarrow \neg R$	Premise	If X $>$ Max is not true before execution, then it is not true after execution.
4. $P \Rightarrow \neg R$	1, 2, HS	If X $>$ Max before execution, then X $>$ Max is not true after execution.
5. $\neg R$	3, 4, Cs	Therefore, no matter whether before execution X $>$ Max is true or false, after execution, X $>$ Max is false.

Figure 1.2.   Proof of the Pascal if-statement

by Cs. For instance, line 4 is obtained from lines 1 and 2 by using the hypothetical syllogism. The conclusion is obtained from lines 3 and 4 by using the law called Cs.

The second example deals, as promised, with the reasoning given by Sherlock Holmes in connection with a particular murder. We quote from "A Study in Scarlet."

> And now we come to the great question as to the reason why. Robbery has not been the object of the murder, for nothing was taken. Was it politics, then, or was it a woman? That is the question which confronted me. I was inclined from the first to the latter supposition. Political assassins are only too glad to do their work and fly. This murder had, on the contrary, been done most deliberately, and the perpetrator had left his tracks all over the room, showing he had been there all the time.

To express this quote, we use the following propositional variables.

P_1: It was robbery.

P_2: Something was taken.

P_3: It was politics.

P_4: It was a woman.

P_5: The assassin left immediately.

P_6: The assassin left tracks all over the room.

The proof of the Sherlock Holmes' argument, given in Figure 1.3, uses the modus ponens (MP), the modus tollens (MT), and the disjunctive syllogism (DS). This completes the formalization of the argument given by Sherlock Holmes. Of course, some of the premises

Prove: Premises as stated $\models P_4$.

Formal Derivation	Rule	Comment
1. $P_1 \Rightarrow P_2$	Premise	If it was robbery, something would have been taken.
2. $\neg P_2$	Premise	Nothing was taken.
3. $\neg P_1$	1, 2, MT	It was not robbery.
4. $\neg P_1 \Rightarrow P_3 \vee P_4$	Premise	If it was not robbery, it must have been politics or a woman.
5. $P_3 \vee P_4$	3, 4, MP	It was politics, or it was a woman.
6. $P_3 \Rightarrow P_5$	Premise	If it was politics, the assassin would have left immediately.
7. $P_6 \Rightarrow \neg P_5$	Premise	If the assassin left tracks all over the room, he cannot have left immediately.
8. P_6	Premise	The assassin left tracks all over the room.
9. $\neg P_5$	7, 8, MP	The assassin did not leave immediately.
10. $\neg P_3$	6, 9, MT	It was not politics.
11. P_4	5, 10, DS	Consequently, it was a woman.

Figure 1.3.   Proving the motivation to a crime

may be somewhat shaky. However, if we accept all the premises, we must accept the conclusion.

1.6.5 Systems for Derivations

We indicated that the formalized proofs given in the previous section were in fact derivations. In this section, we justify this claim and define exactly what is meant by the word derivation. There are actually different systems for doing derivations. All systems have the following features in common.

1. There is a given list of admissible logical arguments, called *rules of inference*. This list will be referred to as L.

2. The derivation itself is a list of logical expressions. Originally, this list is empty. Expressions can be added to this list if the added expression is a premise or if it can be obtained from previous expressions by applying one of the rules of inference. This process continues until the conclusion is reached.

If there is a derivation for the conclusion C, given that $A_1, A_2, \ldots, A_n$ are the premises and given that L is the set of admissible rules of inference, then we write

$$A_1, \; A_2, \; \ldots, \; A_n \; \vdash_L \; C$$

Sometimes, L is not pertinent to the discussion, and in this case, $\vdash$ is used in place of $\vdash_L$.

Example 1.19   Show that the two proofs given in Figures 1.2 and 1.3 are derivations, and list the rules of inference used.

> **Solution**   In the proof of Figure 1.2, the list of expressions constituting the derivation is given by the lines of the proof. All lines are either premises or they are obtained from previous lines by applying (1.21) and the hypothetical syllogism. If these two rules are denoted by Cs and HS, respectively, then $L = \{\text{Cs}, \; \text{HS}\}$. The proof becomes a derivation under L. For Figure 1.3, we define L_1 as consisting of MP, MT, and DS, which stand for modus ponens, modus tollens, and disjunctive syllogism, respectively. In this case, the proof is a derivation under L_1.   ∎

Note that in most systems for formal derivations L is fixed. No rule of inference can be used unless it is included in L as an admissible rule of inference. In this book, we are less formal. However, we will essentially restrict ourselves to the rules of inference given in Table 1.25. The rules are given in schema form, and A, B, and C are expressions.

TABLE 1.25.   Main Rules of Inference Used in Text

$A, \; B \models A \wedge B$	Law of combination
$A \wedge B \models B$	Law of simplification
$A \wedge B \models A$	Variant of law of simplification
$A \models A \vee B$	Law of addition
$B \models A \vee B$	Variant of law of addition
$A, \; A \Rightarrow B \models B$	Modus ponens
$\neg B, \; A \Rightarrow B \models \neg A$	Modus tollens
$A \Rightarrow B, \; B \Rightarrow C \models A \Rightarrow C$	Hypothetical syllogism
$A \vee B, \; \neg A \models B$	Disjunctive syllogism
$A \vee B, \; \neg B \models A$	Variant of disjunctive syllogism
$A \Rightarrow B, \; \neg A \Rightarrow B \models B$	Law of cases
$A \Leftrightarrow B \models A \Rightarrow B$	Equivalence elimination
$A \Leftrightarrow B \models B \Rightarrow A$	Variant of equivalence elimination
$A \Rightarrow B, \; B \Rightarrow A \models A \Leftrightarrow B$	Equivalence introduction
$A, \neg A \models B$	Inconsistency law

The first law, the law of combination, is trivial and needs no further explanation. The next six laws and their variants have been discussed, and they should be clear by now. The law of cases was applied in the proof in Figure 1.2. This law is often used to divide a problem into two smaller subproblems that are solved independently. Specifically, if B is a mathematical theorem, it may be easier to make two proofs, one assuming $\neg A$, the other assuming A. Here, A is an assumption chosen to make the two subproofs easier. The equivalence laws follow from (1.11) and Theorem 1.2. The final law is the inconsistency law. It follows from the fact that $A \wedge \neg A$ is always F, and $F \Rightarrow B$ is trivially true. There is a major consequence of the inconsistency law. If one can derive a single contradiction from the premises, then every imaginable logical expression B can be obtained. One must therefore check carefully that the premises do not allow one to derive any contradiction, because, otherwise, everything can be proved.

Of course, the rules of inference must be chosen in such a way that one can only derive results that are sound. This means that L must not contain any fallacy. A fallacy allows one to find a conclusion that is not implied by the premises and that is therefore not sound. In fact, the possibility of introducing fallacies is the prime reason one often restricts the rules of inference to a fixed list L. From now on, we will assume that only sound rules of inference are used. This means that if there is a derivation showing $A \vdash B$ we can conclude that $A \models B$.

A system for doing derivations should not only be sound, but also *complete*. By this, we mean that it must be possible to derive *every* conclusion that logically follows from the premises. For instance, the system given by Table 1.25 is not complete. To see this, consider the law of excluded middle, which is $P \vee \neg P$. This law holds without any premises; that is, $\models P \vee \neg P$. It is easy to see that, without any premises, none of the laws given in Table 1.25 can be used. Consequently, $P \vee \neg P$ cannot be derived if L is given by Table 1.25. The next section will show how to overcome this problem. If a system is sound and complete, then $A \vdash B$ if and only if $A \models B$.

There are several major types of systems for doing derivations. Here, we describe only *Hilbert systems* and *systems of natural derivation*. Both systems have several variations. Hilbert systems use only two connectives, $\neg$ and $\Rightarrow$. Every logical expression is written with the help of these two connectives. For instance, instead of writing $P \wedge Q$, one writes $\neg(P \Rightarrow \neg Q)$. Hilbert systems have only one rule of inference, the modus ponens. In addition to this, Hilbert systems use schemas that are tautologically true. These schemas are called *axiom schemas*. The most popular set of axiom schemas is the following:

$$(A \Rightarrow (B \Rightarrow A))$$
$$(A \Rightarrow (B \Rightarrow C)) \Rightarrow ((A \Rightarrow B) \Rightarrow (A \Rightarrow C))$$
$$(\neg B \Rightarrow \neg A) \Rightarrow ((\neg B \Rightarrow A) \Rightarrow B)$$

Systems for natural derivation use no axiom schemas, and they use the complete set of connectives. For each connective, two rules are provided, one for introducing the connective and one for eliminating it. For instance, the rule of combination (see Table 1.25) is used to introduce the connective $\wedge$, and the rule of simplification is used to eliminate it. Similar rules are provided for all other connectives. The names of the rules of inference are suggestive. For instance, the rule for eliminating the connective $\wedge$ is denoted by $\wedge E$, and the rule for introducing it is called $\wedge I$. We will discuss the system of natural derivation in greater detail in Chapter 11.

A number of terms related to derivations is now introduced. In logic, a *theory* is given by a set of premises, together with all conclusions that can be derived from the premises. The premises of a theory are often called *axioms*. The conclusions that can be derived from the axioms are called *theorems*.

Example 1.20   Suppose that P and $\neg Q$ are axioms. Some theorems that can be derived from these axioms are $P \vee Q$ and $Q \Rightarrow P$. $P \vee Q$ follows directly from the axiom P and the law of addition, while $Q \Rightarrow P$ is true since its consequent, P, is an axiom. However, $P \wedge Q$ is not a theorem of this theory because it does not logically follow from the axioms, and it therefore cannot be derived by any sound system.

1.6.6 The Deduction Theorem

To prove $A \Rightarrow B$ in mathematics, one often uses the following informal argument.

1. Assume A, and add A to the premises.
2. Prove B, using A, if necessary.
3. Discharge A, which means that A is no longer necessarily true, and write $A \Rightarrow B$.

Example 1.21   A couple has a boy, and they are expecting a second child. Prove that if the second child is a girl then the couple has a girl and a boy.

Solution   Let P be "the first child is a boy," and let Q be "the second child is a girl." We want to prove $Q \Rightarrow P \wedge Q$, given that the premise is P. According to the method under discussion, this can be done as follows:

1. P is true: the couple has a boy.
2. Assume Q; that is, assume that the second child is a girl.
3. From P and Q, conclude $P \wedge Q$ by the law of combination.
4. At this stage, one is allowed to conclude that $Q \Rightarrow P \wedge Q$. Q can now be discharged; that is, $Q \Rightarrow P \wedge Q$ is true even if Q turns out to be false: in this case, $Q \Rightarrow P \wedge Q$ is trivially true.   ■

The reason is clear why this pattern for doing a proof holds. When proving $A \Rightarrow B$, one only needs to consider the case where A is true: if A is false, $A \Rightarrow B$ is trivially true. If A is true, then it may be added to the premises. This shows the soundness of the procedure. Essentially, the argument states that an assumption may be converted into a antecedent of a conditional. This is the content of the deduction theorem, which is now stated as follows:

Theorem 1.3.   Let A and B be two expressions, and let $A_1, A_2, A_3, \ldots$ be premises. If $B, A_1, A_2, A_3, \ldots$ together logically imply C, then $A_1, A_2, A_3, \ldots$ logically imply $B \Rightarrow C$.

Together with this theorem, the rules of inference given in Table 1.25 form a complete system.

Example 1.22   Use the deduction theorem to derive the hypothetical syllogism. Use the modus ponens (MP) as the only rule of inference.

Solution   The problem is to prove that $P \Rightarrow R$, given $P \Rightarrow Q$ and $Q \Rightarrow R$. According to the deduction theorem, P may be assumed, which means that P may be added to the premises. Once this is done, R is proved, which is rather easy. At this point, one is allowed to conclude that $P \Rightarrow R$ and discharge P; that is, P need no longer be true. The complete proof is given in

Figure 1.4. Note that the part where P is used an assumption is indented. Outside this indented part, P can be true or false, and it therefore cannot be used. The point where the implication is asserted and the assumption is discharged is marked by DT, for deduction theorem.   ■

Prove: $P \Rightarrow Q$, $Q \Rightarrow R \vdash P \Rightarrow R$.

Formal Derivation	Rule	Comment
1. $P \Rightarrow Q$	Premise	
2. $Q \Rightarrow R$	Premise	
3. P	Assumption	Assume P.
4. Q	1, 3, MP	
5. R	2, 4, MP	R is now proved.
6. $P \Rightarrow R$	DT	Discharge P, that is, P is no longer to be used, and conclude that $P \Rightarrow R$.

Figure 1.4.   Proof of the hypothetical syllogism

Example 1.23   Prove $\models\ P \Rightarrow P$.

Solution   Instead of deriving $P \Rightarrow P$, one assumes P and derives P. This derivation is trivial. Generally, any premise may be used as a conclusion. Specifically, if P is a premise, we are allowed to conclude P, which means that $P \vdash P$. We now discharge P and obtain $\vdash (P \Rightarrow P)$.   ■

Equivalences are very important within any theory. Once an equivalence has been derived, one can use it for algebraic manipulation within the theory. In mathematics, one uses the following technique for proving that two expressions A and B are equivalent.

1. Assume A.
2. Prove B.
3. Write $A \Rightarrow B$, and discharge A.
4. Assume B.
5. Prove A.
6. Write $B \Rightarrow A$, and discharge B.
7. Conclude $A \Leftrightarrow B$.

The last step follows from lines 3 and 6 because of the equivalence introduction law of Table 1.25. An informal proof following this pattern usually starts with the words "Assume

A," which is followed by some type of proof for B. The fact that this implies $A \Rightarrow B$ is usually omitted. The next step, which involves the introduction of assumption B, often begins with the words "Conversely, assume B," and this is followed by a proof of A. It is now shown how such informal proof can be formalized by logical arguments. This is done by means of an example.

Example 1.24 Prove that $P \wedge (Q \wedge R)$ and $(P \wedge Q) \wedge R$ are logically equivalent. Only the deduction theorem and the laws of combination, simplification, and equivalence introduction may be used. These laws are abbreviated by DT, C, S, and Equ, respectively.

Solution The complete solution is given in Figure 1.5. Note that lines 1 to 7 derive $(P \wedge Q) \wedge R$ from the assumption $P \wedge (Q \wedge R)$. This derivation then allows one to conclude in line 8 that $P \wedge (Q \wedge R)$ implies that $(P \wedge Q) \wedge R$. This constitutes the first half of the proof. Lines 9 to 15 then derive the converse, that the assumption $(P \wedge Q) \wedge R$ can be used to derive $P \wedge (Q \wedge R)$. This allows one to conclude with the implication given in line 16 by making use of the deduction theorem. The last line combines lines 8 and 16 to find the desired equivalence. ∎

A final type of proof is the indirect proof. In the indirect proof, one proves $\neg A$ by demonstrating that the assumption A leads to a contradiction. Here, A can be any expression. Informally, this type of a proof can be worded as follows:

1. Assume A.
2. Prove that this assumption leads to a contradiction.
3. Discharge A, and conclude $\neg A$.

Although this proof is typically used in an informal setting, we will demonstrate it here in a formal way. In formal proof, a contradiction is typically in the form $R \wedge \neg R$, where R is any proposition. In the formal proof, the word Neg is used to indicate that $\neg A$ has been derived and that the assumption A is now dismissed.

Example 1.25 Show that $P \Rightarrow Q$ and $P \Rightarrow \neg Q$ can be used to derive $\neg P$.

Solution The complete solution is given in Figure 1.6, which should be self-explanatory. ∎

Problems 1.6

1. Use the truth table method to show that the following arguments are sound.
    (a) $P \vee Q, \neg P \vee R \models Q \vee R$
    (b) $P \Rightarrow Q, P \Rightarrow R \models P \Rightarrow Q \wedge R$
    (c) $P, P \Rightarrow Q \models P \wedge Q$
    (d) $P \vee Q, P \Rightarrow R, Q \Rightarrow R \models R$
2. Prove the following laws by the truth table method.
    (a) Modus ponens
    (b) Law of cases
    (c) Equivalence elimination

Prove: $\vdash\ P \wedge (Q \wedge R)\ \Leftrightarrow\ (P \wedge Q) \wedge R.$

Formal Derivation	Rule	Comment
1. $P \wedge (Q \wedge R)$	Assumption	Assume $A = P \wedge (Q \wedge R)$.
2. P	1, S	
3. $Q \wedge R$	1, S	
4. Q	3, S	
5. R	3, S	
6. $P \wedge Q$	2, 4, C	
7. $(P \wedge Q) \wedge R$	5, 6, C	This proves that $B = (P \wedge Q) \wedge R.$
8. $P \wedge (Q \wedge R) \Rightarrow (P \wedge Q) \wedge R$	DT	A is discharged, and $A \Rightarrow B$ is asserted.
9. $(P \wedge Q) \wedge R$	Assumption	Assume $B = (P \wedge Q) \wedge R.$
10. R	9, S	
11. $P \wedge Q$	9, S	
12. P	11, S	
13. Q	11, S	
14. $Q \wedge R$	10, 13, C	
15. $P \wedge (Q \wedge R)$	12, 14, C	This proves that $A = P \wedge (Q \wedge R).$
16. $(P \wedge Q) \wedge R \Rightarrow P \wedge (Q \wedge R)$	DT	B is discharged, and $B \Rightarrow A$ is asserted.
17. $P \wedge (Q \wedge R) \Leftrightarrow (P \wedge Q) \wedge R$	8, 16, Equ	The last step consists of combining lines 8 and 16.

Figure 1.5. Proving the associative law for $\wedge$

3. For each of the following arguments, indicate which of the rules of inferences given in Table 1.25 are used.
   (a) If Mr. Smith or Mrs. Smith earns more than \$30,000 per year, the Smith family can afford holidays in Hawaii. Since I know that either Mr. Smith or his wife earns more than \$30,000, I conclude that the family can afford a holiday in Hawaii.
   (b) If Kim discovers that the product you delivered is faulty, she will be furious. Unfortunately, I know for a fact that she did discover that the product is faulty. Thus, Kim is going to be furious.
   (c) If John was at the party yesterday, he will sleep in. John did not sleep in. Consequently, he was not at the party.
   (d) If I do it, I get blamed, and if I don't, I get blamed. Consequently, I get blamed.
   (e) If it is warm and humid, then it is obviously warm as well.
   (f) We know that Frank took either the Ford or the Chevy. Frank did not take the Ford. Consequently, Frank must have taken the Chevy.
   (g) If Bill is at home, then he must certainly be either at home or in his office.
   (h) If $x \geq 0$, then $x^2 \geq 0$, and if it is not true that $x \geq 0$, then $x^2 \geq 0$. Consequently, $x^2 \geq 0$.
4. Give a derivation for the following logical arguments. The rules of inference are restricted to the modus ponens, the modus tollens, and the law of combination. State the laws used at each step.

Prove: $P \Rightarrow Q,\ P \Rightarrow \neg Q\ \vdash\ \neg P.$

Formal Derivation	**Rule**	**Comment**
1. $P \Rightarrow Q$	Premise	
2. $P \Rightarrow \neg Q$	Premise	
3. $\quad P$	Assumption	Assume P in order to derive a contradiction.
4. $\quad Q$	1, 3, MP	
5. $\quad \neg Q$	2, 3, MP	
6. $\quad Q \wedge \neg Q$	4, 5, C	Lines 4 and 5 provide the desired contradiction.
7. $\neg P$	Neg	Since the assumption P leads to a contradiction, one is allowed to conclude that $\neg P$.

Figure 1.6.   An indirect proof

    (a) $P,\ P \Rightarrow (Q \vee R),\ (Q \vee R) \Rightarrow S \vdash S$
    (b) $P \Rightarrow Q,\ Q \Rightarrow R,\ \neg R \vdash \neg P$
    (c) $P,\ P \Rightarrow Q \vdash P \wedge Q$

5. The premises are $P \Rightarrow Q, Q \Rightarrow R$, and $R \Rightarrow P$. Prove that $P \Leftrightarrow Q$. Use only the hypothetical syllogism and the law of equivalence introduction.

6. Prove Theorem 1.2.

7. Use the deduction theorem and the law of addition to prove that $\models\ P \Rightarrow (P \vee \neg P)$ and $\models\ \neg P \Rightarrow (P \vee \neg P)$. What further conclusion can be drawn from $\models\ P \Rightarrow (P \vee \neg P)$ and $\models\ \neg P \Rightarrow (P \vee \neg P)$?

8. Prove that $P \wedge Q \equiv Q \wedge P$. Use the deduction theorem, the law of simplification, the law of combination, and the law of equivalence introduction.

PROBLEMS: CHAPTER 1

1. Translate the following sentences into logic. Abbreviate the propositions and predicates as indicated. Do not introduce new propositional variables or predicates.
   (a) If he is in his office, we will tell him the news; otherwise, we will leave him a message. (P: "he is in his office," Q: "we will tell him the news," and R: "we will leave him a message").
   (b) If the operation succeeds and if she follows the doctor's instructions, she will be fine. (P "the operation succeeds," Q "she follows the doctor's instructions," and R "she will be fine").

   (c) John knows either C, Pascal, or Prolog, and he enjoys working with people. Otherwise, he
        would not be senior programmer. Define P_1 to mean "John knows C," P_2 to mean "John
        knows Pascal," and P_3 as "John knows Prolog." Q stands for "John enjoys working with
        people," and R means "John is a senior programmer."

2. Translate the following regulation into logic using the propositional variables given. An em-
    ployee is eligible for a holiday of a duration of three weeks if (i) he or she is a temporary
    employee who does not receive additional holiday pay and who has been with the company for
    a full calendar year, or (ii) he or she is a permanent employee who has been with the company
    for at least six months.
    - P_1: The employee is eligible for a three-week holiday.
    - P_2: The employee is a temporary employee.
    - P_3: The employee does receive holiday pay.
    - P_4: The employee has been with the company for at least one calendar year.
    - P_5: The employee is a permanent employee.
    - P_6: The employee has been with the company for at least six months.

3. Find the truth tables for the following expressions. State in each case whether the expression is
    contingent, a tautology, or a contradiction.
    (a) $((P \Rightarrow Q) \wedge (Q \Rightarrow R) \wedge (R \Rightarrow P)) \Rightarrow (P \Leftrightarrow Q)$
    (b) $P \vee (\neg(Q \vee R) \wedge \neg P)$
    (c) $(P \wedge Q \wedge R) \vee (\neg P \wedge \neg Q \wedge \neg R)$

4. Simplify the following expressions. State each law that you use.
    (a) $(P \wedge Q \wedge R) \vee (P \wedge \neg Q) \vee (P \wedge \neg R)$
    (b) $(P \wedge Q) \vee (P \wedge R) \vee (P \wedge (Q \vee \neg R))$
    (c) $(\neg P \wedge R \wedge \neg(P \wedge \neg(P \vee Q)))$

5. Consider the following schema:

$$A \Rightarrow (B \Rightarrow C) \equiv B \Rightarrow (A \Rightarrow C)$$

    Prove the correctness of this schema
    (a) By a truth table.
    (b) By statement algebra.

6. Formulate the following laws as tautologies. Eliminate all $\Rightarrow$ and prove them by statement
    algebra.
    (a) Law of simplification
    (b) Modus tollens
    (c) Disjunctive syllogism

7. Prove that $((P \wedge Q) \Rightarrow R) \equiv \neg((P \wedge Q) \wedge \neg R)$. Use both the truth table method and statement
    algebra in your proof.

8. Find the truth table for $((P \vee Q) \Rightarrow R \wedge (R \Rightarrow P))$. Use this truth table to find the full disjunctive
    normal form and the full conjunctive normal form.

9. Consider the following argument. It was either X or Y who committed the crime. X was out
    of town when the crime was committed. If X was out of town, he cannot have been at the
    scene of the crime. If X was not at the scene of the crime, he cannot have committed the crime.
    Consequently, Y must have committed the crime. Formulate this as a formal proof and derive
    the conclusion. Use P_1 for "X committed the crime," P_2 for "Y committed the crime," Q for
    "X was out of town," and R for "X was not at the scene of the crime."

10. The derivation in Figure 1.7 derives R from the premises $P \lor Q$, $P \Rightarrow R$, and $Q \Rightarrow R$. However, all line numbers are missing and so are all rules of inference. Add both pieces of information. The rules used are the hypothetical syllogism (HS), the law of cases (Cs), and $P \lor Q \models (\neg P \Rightarrow Q)$ (DS1), and the deduction theorem.

Prove: $P \lor Q$, $P \Rightarrow R$, $Q \Rightarrow R \models R$

Formal Derivation   **Rule**   **Comment**

1. $P \lor Q$

2. $\neg P \Rightarrow Q$

3. $Q \Rightarrow R$

4. $\neg P \Rightarrow R$

5. $P \Rightarrow R$

6. R

Figure 1.7

11. Given the premise $P \land Q$, prove $P \lor Q$. The laws of inference are restricted to the laws of simplification and addition.

12. Use the deduction theorem to show $Q \models (P \Rightarrow Q)$. Use this result to show $\models Q \Rightarrow (P \Rightarrow Q)$.

13. Show that $P \Rightarrow (Q \Rightarrow R)$ and $P \land Q \Rightarrow R$ are equivalent.

2

Predicate Calculus

There are certain arguments that seem to be perfectly logical, yet they cannot be specified by using propositional calculus. As an example, consider the following statements:

1. All cats have tails.

2. Tom is a cat.

From these two sentences, one should be able to conclude that

3. Tom has a tail.

To show that this argument is sound, we must be able to identify individuals, such as Tom, together with their properties and predicates. This is the objective of predicate calculus, the topic of this chapter.

Generally, predicates are used to describe certain properties or relationships between individuals or objects. For instance, in "Mary and Anne are sisters," the phrase "are sisters" is a predicate. The entities connected in this way, Mary and Anne, are called *terms*. Terms play a similar role in predicate calculus as nouns do in the English language. In addition to terms and predicates, one uses *quantifiers*. Quantifiers indicate how frequently a certain statement is true. Specifically, the *universal* quantifier is used to indicate that a statement is always true, whereas the *existential* quantifier indicates that a statement is sometimes true. For instance, in "All cats have tails," the word "all" indicates that the statement "cats have tails" is universally true. Predicate calculus is a generalization of propositional calculus. Hence, besides terms, predicates, and quantifiers, predicate calculus contains propositions and connectives as part of the language. An important part of predicate calculus is the function, which will be discussed in this chapter. Functions are essential when using equations, and their use is therefore basic to all algebraic manipulations.

For the computer scientist, predicate calculus is important for several reasons. First, it gives the logical underpinnings to the languages of logic programming, such as Prolog. Second, predicate calculus is increasingly used for specifying the requirements of computer applications. This will be shown in more detail in Chapter 8. In the area of proving correctness (see Chapter 9), predicate calculus allows one to precisely state under which conditions a program gives the correct output.

2.1 SYNTACTIC COMPONENTS OF PREDICATE CALCULUS

2.1.1 Introduction

Predicate calculus contains all the components of propositional calculus, including propositional variables and constants. In addition, predicate calculus contains *terms*, *predicates*, and *quantifiers*, which are discussed in this section. Terms are typically used in place of nouns or pronouns. They are combined into sentences by means of *predicates*. In the English sentence "John loves Mary," for instance, the nouns are John and Mary, and the predicate is "loves." The same is true if this sentence is translated into predicate calculus, except that "John" and "Mary" are now called terms. As in English, the predicate is "loves." Predicate calculus uses *quantifiers* to indicate if a statement is always true, if it is sometimes true, or if it is never true. The quantifiers are in this sense used where English would use words such as "all," "some," "never," and related expressions.

2.1.2 The Universe of Discourse

To explain the main concepts of this section, we use the following logical argument:

1. Jane is Paul's mother.

2. Jane is Mary's mother.

3. Any two persons having the same mother are siblings.

4. Paul and Mary are siblings.

The truth of the statement "Jane is Paul's mother" can only be assessed within a certain context. There are many people named Jane and Paul, and without further information the statement in question can refer to many different people, which makes it ambiguous. To remove such ambiguities, we introduce the concept of a *universe of discourse* or a *domain*.

> **Definition 2.1:**   The *universe of discourse* or *domain* is the collection of all persons, ideas, symbols, data structures, and so on, that affect the logical argument under consideration. The elements of the universe of discourse are called *individuals*.

In the argument concerning Mary and Paul, the universe of discourse may, for instance, consist of the people living in a particular house or on a particular block. Many arguments involve numbers and, in this case, one must stipulate whether the domain is the set of natural numbers, the set of integers, the set of real numbers, or even the set of complex numbers. In fact, the truth of a statement may depend on the domain selected. The statement "There is a smallest number" is true in the domain of natural numbers, but false in the domain of integers.

The elements of the domain are called *individuals.* An individual can be a particular person, a number, a data structure, or anything else one wants to reason about. To avoid trivial cases, one stipulates that every universe of discourse must contain at least one individual. Hence, the set of all natural numbers less than 0 does not constitute a universe because there is no negative natural number. Instead of the word individual, one sometimes uses the word *object*, such as in "the domain must contain at least one object." To refer to a particular individual or object, identifiers are used. These identifiers are called *individual constants.* If the universe of discourse consists of persons, the individual constants may be their names. In the case of natural numbers, the individual constants are the digits representing these numbers (i.e., 0, 1, 2, 3, ...). Each individual constant must uniquely identify a particular individual and no other one. For instance, if the universe of discourse consists of persons, there must not be two persons with the same name.

2.1.3 Predicates

Generally, predicates make statements about individuals. To illustrate this notion, consider the following statements:

> Mary and Paul are siblings.
>
> Jane is the mother of Mary.
>
> Tom is a cat.
>
> The sum of 2 and 3 is 5.

In each of these statements, there is a list of individuals, which is given by the *argument list*, together with phrases that describe certain relations among or properties of the individuals mentioned in the argument list. These properties or relations are referred to as *predicates*. In the statement "Mary and Paul are siblings," for instance, the argument list is given by Mary and Paul, in that order, whereas the predicate is described by the phrase "are siblings." Similarly, the statement "Tom is a cat" has an argument list with the single element "Tom" in it, and its predicate is described by "is a cat." The entries of the argument list are called *arguments*. The arguments can be either variables or individual constants, but since we have not discussed variables yet, we restrict our attention to the case when all arguments are individual constants.

In predicate calculus, each predicate is given a name, which is followed by the list of arguments. The list of arguments is enclosed in parentheses. For instance, to express "Jane is the mother of Mary," one would choose an identifier, say "mother," to express the predicate "is mother of," and one would write mother(Jane, Mary). Many logicians use only single letters for predicate names and constants. They would write, for instance $M(j, m)$ instead of mother(Jane, Mary); that is, they would use M as a name for the predicate "is mother of," j for Jane, and m for Mary. To save space, we will often follow this convention.

Note that the order of the arguments is important. Clearly, the statements mother(Mary, Jane) and mother(Jane, Mary) have a completely different meaning. The number of elements in the predicate list is called the *arity* of the predicate. For instance, mother(Jane, Mary) has an arity of 2. The arity of a predicate is fixed. For example, a predicate cannot have two arguments in one case and three in another. Alternatively, one can consider two predicates as different if their arity is different. The following statements illustrate this.

The sum of 2 and 3 is 5.

The sum of 2, 3, and 4 is 9.

To express these statements in predicate calculus, one can either use two predicate names, such as "sum2" and "sum3," and write sum2(2,3,5) and sum3(2,3,4,9), respectively, or one can use the same symbol, say "sum," with the implicit understanding that the name "sum" in sum(2,3,5) refers to a different predicate than in sum(2,3,4,9).

A predicate with arity n is often called an *n-place* predicate. A one-place predicate is called a *property*.

Example 2.1   The predicate "is a cat" is a one-place predicate, or a property. The predicate "is the mother of," as in "Jane is the mother of Mary," is a two-place predicate; that is, its arity is 2. The predicate in the statement "The sum of 2 and 3 is 6" (which is false) contains the three-place predicate "is the sum of."

A predicate name, followed by an argument list in parentheses, is called an *atomic formula*. The atomic formulas are statements, and they can be combined by logical connectives like propositions. For instance, to express the fact that Jane is the mother of Mary, one can use the atomic formula mother(Jane, Mary), and this statement can be part of some compound statement, such as

$$\text{mother(Jane, Mary)} \Rightarrow \neg\text{mother(Mary, Jane)}$$

Similarly, if cat(Tom) and hastail(Tom) are two atomic formulas, expressing that Tom is a cat and Tom has a tail, respectively, one can form

$$\text{cat(Tom)} \Rightarrow \text{hastail(Tom)}$$

If all arguments of a predicate are individual constants, then the resulting atomic formula must either be true or false. This is part of the definition of a predicate. For instance, if the universe of discourse consists of Jane, Doug, Mary, and Paul, we have to know for each ordered pair of individuals whether or not the predicate "is the mother of" (or "mother" for short) is true. This can be done in the form of a table, as in Table 2.1. Any method that assigns truth values to all possible combinations of individuals of a predicate is called an *assignment* of the predicate. For instance, Table 2.1 is an assignment of the predicate "mother." Specifically, the truth value for mother(x, y) is given in row x and column y. For instance, mother(Jane, Paul) is true because in row "Jane" the entry in column "Paul" is T. More generally, if a predicate has two arguments, its assignment can be given by a table in which the rows correspond to the first argument and the columns correspond to the second. The convention that rows correspond to the first argument and columns to the second will be assumed throughout unless explicitly stated otherwise.

TABLE 2.1.
Assignment for the Predicate "mother"

	Doug	Jane	Mary	Paul
Doug	F	F	F	F
Jane	F	F	T	T
Mary	F	F	F	F
Paul	F	F	F	F

Another example of an assignment is as follows. The domain consists of the four numbers 1, 2, 3, and 4. The predicate "greater" is true if the first argument is greater than the second. Hence, greater$(4, 3)$ is true and greater$(3, 4)$ is false. This gives an assignment for all pairs of individuals. For the sake of clarity, we represent this assignment by Table 2.2.

TABLE 2.2.
Assignment for the Predicate "greater"

	1	2	3	4
1	F	F	F	F
2	T	F	F	F
3	T	T	F	F
4	T	T	T	F

In a finite universe of discourse, one can represent the assignments of predicates with arity n by n-dimensional arrays. For instance, properties are assigned by one-dimensional arrays, predicates of arity 2 by two-dimensional arrays, and so on.

Note that the mathematical symbols $\geq$ and $\leq$ are predicates. However, these predicates are normally used in *infix* notations. By this, we mean that they are placed between the arguments. For instance, to express that 2 is greater than 1, we write $2 > 1$, rather than $>(2, 1)$.

2.1.4 Variables and Instantiations

Often, one does not want to associate the arguments of an atomic formula with a particular individual. To avoid this, variables are used. Variables are frequently chosen from the end of the alphabet; that is, x, y, and z, with or without subscripts, suggest variable names. Examples of expressions containing variables include

$\text{cat}(x) \Rightarrow \text{hastail}(x)$

$\text{dog}(y) \wedge \text{brown}(y)$

$\text{grade}(x) \Rightarrow (x \geq 0) \wedge (x \leq 100)$

Clearly, the first and third expressions contain the variable x, and the second expression contains the variable y. As in propositional calculus, expressions can be given names. For instance, one can define A as follows:

$$A = \text{cat}(x) \Rightarrow \text{hastail}(x)$$

which means that when we write A we really mean "$\text{cat}(x) \Rightarrow \text{hastail}(x)$." If the expression represented by A contains the variable x, we say that A contains x. Hence, A, as defined here, contains x, but not y.

Syntactically, one can use variables in any place where one is allowed to use constants. The word *term* is therefore used to refer to either a constant or a variable. More generally, a term is anything that can be used in place of an individual. If A is an expression, one frequently has to replace all occurrences of a particular variable by a term. For example, in the expression $\text{cat}(x) \Rightarrow \text{hastail}(x)$, one may want to replace all instances of x by the term Tom, which yields

$$\text{cat}(\text{Tom}) \Rightarrow \text{hastail}(\text{Tom})$$

Generally, if A is an expression, the expression obtained by replacing all variables x in A by term t is denoted by $S_t^x A$. Specifically, if A is defined as previously, then

$$S_{\text{Tom}}^x A$$

stands for

$$\text{cat}(\text{Tom}) \Rightarrow \text{hastail}(\text{Tom})$$

Definition 2.2: Let A represent an expression, x represent a variable, and t represent a term. Then $S_t^x A$ represents the expression obtained by replacing all occurrences of x in A by t. $S_t^x A$ is called an *instantiation* of A, and t is said to be an *instance* of x.

Example 2.2 Let a, b, and c be individual constants, P and Q be predicate symbols, and x and y be variables. Find

$$S_a^x (P(a) \Rightarrow Q(x))$$
$$S_b^y (P(y) \vee Q(y))$$
$$S_a^y Q(a)$$
$$S_a^y (P(x) \Rightarrow Q(x))$$

Solution $S_a^x(P(a) \Rightarrow Q(x))$ is $P(a) \Rightarrow Q(a)$, and $S_b^y(P(y) \vee Q(y))$ is $P(b) \vee Q(b)$. Since $Q(a)$ does not contain any y, replacing all instances of y by a leaves $Q(a)$ unchanged, which means that

$$S_a^y Q(a) = Q(a)$$

Similarly,

$$S_a^y(P(x) \Rightarrow Q(x)) = P(x) \Rightarrow Q(x)$$ ∎

S_t^x is an operation that can be performed on predicates; therefore, it is not a predicate itself, and this makes $S_t^x A$ a meta-expression.

2.1.5 Quantifiers

Consider the following statements:

1. All cats have tails.

2. Some people like their meat raw.

3. Everyone gets a break once in a while.

All these statements indicate how frequently certain things are true. In predicate calculus, one uses quantifiers in this context. Specifically, we will discuss two quantifiers: the *universal quantifier*, which indicates that something is true for all individuals, and the *existential quantifier*, which indicates that a statement is true for some individuals.

Definition 2.3:   Let A represent an expression, and let x represent a variable. If we want to indicate that A is true for all possible values of x, we write $\forall x A$. Here, $\forall x$ is called the *universal quantifier*, and A is called the *scope* of the quantifier. The variable x is said to be *bound* by the quantifier. The symbol $\forall$ is pronounced "for all."

The quantifier and the bounded variable that follows have to be treated as a unit, and this unit acts somewhat like a unary connective. Statements containing words like "every," "each," and "everyone" usually indicate universal quantification. Such statements must typically be reworded such that they start with "for every x," which is then translated to $\forall x$.

Example 2.3   Express "Everyone gets a break once in a while" in predicate calculus.

Solution   We define B to mean "gets a break once in a while." Hence, $B(x)$ means that x gets a break once in a while. The word "everyone" indicates that this is true for all x. This leads to the following translation

$$\forall x B(x)$$ ∎

Example 2.4   Express "All cats have tails" in predicate calculus.

Solution   We first have to find the scope of the universal quantifier, which is "If x is a cat, then x has a tail." After choosing descriptive predicate symbols, we express this by the following compound formula:

$$\text{cat}(x) \Rightarrow \text{hastail}(x)$$

This expression must be universally quantified to yield the required solution.

$$\forall x(\text{cat}(x) \Rightarrow \text{hastail}(x))$$

Another way to express this statement is

$$\forall y(\text{cat}(y) \Rightarrow \text{hastail}(y))$$

Here, the bound variable is y rather than x. In general, the name of the variable used for quantification is immaterial.   ∎

We now define the existential quantifier as follows.

Definition 2.4:   Let A represent an expression, and let x represent a variable. If we want to indicate that A is true for at least one value of x, we write $\exists x A$. This statement is pronounced "There exists an x such that A." Here, $\exists x$ is called the *existential quantifier*, and A is called the *scope* of the existential quantifier. The variable x is said to be *bound* by the quantifier.

Statements containing such phrases as "some" and "at least one" suggest existential quantification. They should be rephrased as "there is an x such that," which is translated by $\exists x$.

Example 2.5   Let P be the property "like their meat raw." Then $\exists x P(x)$ can be translated as "There exist people who like their meat raw" or "Some people like their meat raw."

Example 2.6   If the universe of discourse is a collection of things, $\exists x \text{blue}(x)$ should be understood as "There exists objects that are blue" or "Some objects are blue."

We mentioned already that $\forall x$ and $\exists x$ have to be treated like unary connectives. The quantifiers are given a higher precedence than all binary connectives. For instance, if $P(x)$ and $Q(x)$ means that x is living and that x is dead, respectively, then one has to write $\forall x(P(x) \vee Q(x))$ to indicate that everything is either living or dead. $\forall x P(x) \vee Q(x)$ means that everything is living, or x is dead.

The variable x in a quantifier is just a placeholder, and it can be replaced by any other variable name not appearing elsewhere in the expression. For instance, $\forall x P(x)$ and $\forall y P(y)$ mean the same thing: they are logically equivalent. The expression $\forall y P(y)$ is called a *variant* of $\forall x P(x)$.

Definition 2.5:   An expression is called a *variant* of $\forall x A$ if it is of the form $\forall y S_y^x A$, where y is any variable name, and $S_y^x A$ is the expression obtained from A by replacing all instances of x by y. Similarly, $\exists x A$ and $\exists y S_y^x A$ are variants of one another.

Quantifiers may be nested, as demonstrated by the following examples.

Example 2.7   Translate the sentence "There is somebody who knows everyone" into the language of predicate calculus. To do this, use $K(x, y)$ to express the fact that x knows y.

Solution   The best way to solve this problem is to go in steps. We write informally

$$\exists x (x \text{ knows everybody})$$

Here "x knows everybody" is still in English and means that for all y it is true that x knows y. Hence,

$$x \text{ knows everybody} = \forall y K(x, y)$$

We now add the existential quantifier and obtain

$$\exists x \forall y K(x, y) \qquad \blacksquare$$

Example 2.8   Translate "Everybody has somebody who is his or her mother" into predicate calculus.

Solution   We define M to be the predicate "mother"; that is, $M(x, y)$ stands for "x is the mother of y." The statement "Someone is the mother of y" becomes

$$\exists x M(x, y)$$

To express that this must be true for all y, we add the universal quantifier, which yields

$$\forall y \exists x M(x, y) \qquad \blacksquare$$

The English statement "Nobody is perfect" also includes a quantifier, "nobody," which is the absence of an individual with a certain property. In predicate calculus, the fact that nobody has property P cannot be expressed directly. To express the fact that there is no x for which an expression A is true, one can either use $\neg \exists x A$ or $\forall x \neg A$. For instance, if P represents the property of perfection, both $\neg \exists x P(x)$ and $\forall x \neg P(x)$ indicate that nobody is perfect. In the first case, we say, in verbal translation, "It is not the case that there is somebody who is perfect," whereas in the second case we say "For everyone, it is not the case that he or she is perfect." The two methods to express that nobody is A must of course be logically equivalent; that is,

$$\neg \exists x A \equiv \forall x \neg A \qquad (2.1)$$

There are many quantifiers in English, such as "a few," "most," and "about a third," that are useful in daily language, but are not precise enough to be used in logic. We do not further consider these approximate quantifiers.

According to Definitions 2.3 and 2.4, the variable appearing in the quantifier is said to be *bound*. For instance, in the expression $\forall x(P(x) \Rightarrow Q(x))$, x appears three times, and each time x is a bound variable. Any variable that is not bound is said to be *free*. Later, we will see that the same variable can occur both bound and free in an expression. For this reason, it is important to also indicate the position of the variable in question.

Example 2.9   Find the bound and free variables in

$$\forall z(P(z) \wedge Q(x)) \vee \exists y Q(y)$$

Solution   Only the variable x is free. All occurrences of z are bound, and so are all occurrences of the variable y.   ∎

Note that the status of a variable changes as expressions are divided into subexpressions. For instance, in $\forall x P(x)$, x occurs twice, and it is bound both times. This statement contains $P(x)$ as a subexpression. Nevertheless, in $P(x)$, the variable x is free.

Instantiations only affect free variables. Specifically, if A is an expression, $S_t^x A$ only affects the free occurrences of the variable x. For instance, $S_y^x \forall x P(x)$ is still $\forall x P(x)$; that is, the variable x is not instantiated. However, $S_y^x(Q(x) \wedge \forall x P(x))$ yields $Q(y) \wedge \forall x Q(x)$. Hence, instantiation treats the variable x differently, depending on whether it is free or bound, even if this variable appears twice in the same expression. Obviously, two things are only identical if they are treated identically. This implies that, if a variable appears both free and bound within the same expression, we have in fact two different variables that happen to have the same name.

We can consider the bound variables to be local to the scope of the quantifier, just as parameters and locally declared variables in Pascal procedures are local to the procedure in which they are declared. The analogy with Pascal can be extended further if we consider the variable name in the quantifier as a declaration. This analogy also suggests that, if several quantifiers use the same bound variable for quantification, then all these variables are local to their scope, and they are therefore distinct. Moreover, when forming variants, one must be careful not to interfere with local definitions. To illustrate this, consider the statement "y has a mother." If M is the predicate name for "is mother of," then this statement translates into $\exists x M(x, y)$. One obviously must not form the variant $\exists y M(y, y)$, which means that y is her own mother. For similar reasons, there are restrictions on instantiations. For example, the instantiation $S_x^y(\exists x M(x, y))$ is illegal, because its result is $\exists x M(x, x)$. In such cases, one tampers with the way in which a variable is defined, and this has undesired side effects. We will refer to instances in which a variable becomes bound, or otherwise changes scope, as *variable clashes*. All variable clashes must be avoided.

If all occurrences of x in an expression A are bound, we say "A does not contain x free." If A does not contain x free, then the truth value of A does not change if x is instantiated to an individual constant. A is *independent* of x in this sense.

2.1.6 Restrictions of Quantifiers to Certain Groups

Sometimes, quantification is over a subset of the universe of discourse. Suppose, for instance, that animals form the universe of discourse. How can one express sentences such as "All dogs are mammals" and "Some dogs are brown"? Consider, first, the statement "All dogs are mammals." Since the quantifier should be restricted to dogs, one rephrases the statement as "If x is a dog, then x is a mammal." This immediately leads to

$$\forall x(\text{dog}(x) \Rightarrow \text{mammal}(x))$$

Generally, the sentence

$$\forall x(P(x) \Rightarrow Q(x))$$

can be translated as "All individuals with property P also have property Q."

Consider now the statement "Some dogs are brown." This statement means that there are some animals that are dogs and that are brown. Of course, the statement "x is a dog and x is brown" can be translated as

$$\text{dog}(x) \wedge \text{brown}(x)$$

"There are some brown dogs" can now be translated as

$$\exists x(\text{dog}(x) \wedge \text{brown}(x))$$

The statement

$$\exists x(P(x) \wedge Q(x))$$

can in general be interpreted as "Some individuals with property P also have property Q." Note that if the universal quantifier is to apply only to individuals with a given property we use the conditional to restrict the universe of discourse. On the other hand, if we similarly restrict application of the existential quantifier, we use the conjunction.

Finally, consider statements containing the word "only," such as "only dogs bark." To convert this into predicate calculus, this must be reworded as "It barks only if it is a dog" or, equivalently, "If it barks, then it is a dog." One has therefore

$$\forall x(\text{barks}(x) \Rightarrow \text{dog}(x))$$

Problems 2.1

1. For each of the following predicates, find a suitable universe of discourse from the following list: real numbers, integers, humans, and animals.
   (a) bird(x)
   (b) ismarried(x)
   (c) even(x)
   (d) negative(x)
   (e) mother(x, y)

2. Express the following statements in predicate calculus. The universe of discourse is the set of all persons.

  (a) If Mary likes Kim, and Kim likes Julie, then Mary likes Julie.
  (b) John is very busy, but Bill is not.
  (c) Ben knows Mr. Smith, but Mr. Smith does not know Ben.

3. Translate the following statements into predicate calculus. The domain is the set of integers.

  (a) If x is between 1 and 2, and if y is between 2 and 3, then the difference between x and y cannot exceed 2. Use the predicate $b(x, y, z)$ if x is between y and z, and use $d(x, y, z)$ if the difference between x and y is greater than z.
  (b) If x is divisible by 3, then x cannot be prime. Use $d(x, y)$ if x is divisible by y, and $p(x)$ if x is prime.
  (c) $x + y = z$ and $x + z = u$. Use $s(x, y, z)$ if $x + y = z$.

4. Suppose that the universe of discourse is a group of people. Translate the statement "Everyone here speaks either English or French" into predicate calculus.

5. Express "No natural number is negative," assuming that the universe of discourse is (a) the set of natural numbers, (b) the set of integers, and (c) the set of real numbers.

6. Instantiate $S_3^x P(x, y)$, $S_y^x P(x, y)$, $S_y^x (P(x) \wedge \forall x Q(x))$, $S_2^y (P(x) \wedge Q(y) \wedge R(x, y))$.

7. Find the free and bound variables in $\forall x \exists y (P(x, y, z) \wedge Q(y, z)) \wedge R(x)$.

8. In the domain of animals, how would you translate the following expressions?

  (a) All lions are predators.
  (b) Some lions live in Africa.
  (c) Only lions roar.
  (d) Some lions eat zebras.
  (e) Some lions only eat zebras.

9. In the domain of natural numbers, how would you translate the following statements? Use $P(x)$ for "x is prime" and $Q(x)$ for "x is even." You may also use $x < y$ for any x and y.

  (a) Some primes are even.
  (b) All even numbers are greater than 1.
  (c) Even numbers are prime only if they are less than 3.
  (d) There is no prime less than 3.

2.2 INTERPRETATIONS AND VALIDITY

2.2.1 Introduction

This section deals with interpretations of logical statements and with the soundness of logical arguments. For instance, we will discuss interpretations of statements such as "all customers have paid" or "someone admires everyone." Interpretations are obviously fundamental to predicate calculus, and they are therefore important in their own right. Moreover, interpretations allow one to distinguish between arguments that are sound and arguments that are not. Soundness is closely related to *validity*. Generally, an expression A is valid if A is true for all interpretations. Valid expressions in predicate calculus play the same role as tautologies in propositional calculus. In particular, logical implications and logical equivalences are defined as valid implications and valid equivalences, respectively. Like tautologies, one can generalize valid expressions by means of schemas. All these issues are now discussed.

2.2.2 Interpretations

An interpretation for a statement must contain sufficient information to determine whether the statement is true or false. To investigate this, consider a concrete statement, such as "all customers have paid." What is needed to decide whether this statement is true? Obviously, one needs to know who the customers are; that is, one needs a universe of discourse. Second, one needs to know who has and who has not paid. This means that one needs some type of assignment of the predicate "has paid." As a second example, let us consider the statement "y has ordered item X." To find whether this statement is true, one has to have again a universe of discourse, such as all customers of a company. In addition to that, one must know who y is, and one must have an assignment of the predicate "has ordered X." These two examples suggest a number of elements that must be present in order to have an interpretation. Formally, an interpretation of a logical expression contains the following components:

1. There must be a universe of discourse.
2. For each individual, there must be an individual constant that exclusively refers to this particular individual, and to no other.
3. Every free variable must be assigned a unique individual constant.
4. There must be an assignment for each predicate used in the expression, including predicates of arity 0, which represent propositions.

We now discuss possible interpretations of $\forall x P(x)$, where P is the predicate "has paid." To find such an interpretation, a list of customers is needed, which provides the required universe of discourse. Let us assume that there are only three customers, Johnson, Miller, and Smith. We abbreviate these names by J, M, and S, respectively. These abbreviations are the individual constants. Next, an assignment for $P(x)$ is needed. If Johnson and Miller have both paid up, but Smith did not, this assignment is given as follows:

$$
\begin{array}{c|ccc}
 & J & M & S \\
\hline
P(x) & \text{T} & \text{T} & \text{F}
\end{array}
\tag{2.2}
$$

The reader should now have no difficulty in determining the truth value of $\forall x P(x)$ at this point. Obviously, the statement is false. In our interpretation, $\forall x P(x)$ is only true if $P(J)$, $P(M)$, and $P(S)$ are all true, and this is not the case. Smith has not paid; that is, $P(S)$ is false.

We now formulate what it means in general for an expression like $\forall x P(x)$ to be true. For the interpretation, we choose a finite domain consisting of the n individuals $a_1, a_2, \ldots, a_n$. $\forall x P(x)$ is only true if $P(a_1)$, $P(a_2)$, $\ldots$, $P(a_n)$ are all true. Hence, $\forall x P(x)$ is given by

$$
\forall x P(x) \equiv P(a_1) \land P(a_2) \land \cdots \land P(a_n)
\tag{2.3}
$$

The left part of this expression would have been exactly the same if we had used a variant of $\forall x P(x)$, such as $\forall y P(y)$ or $\forall z P(z)$. This shows that variants are logically equivalent. Note that $P(a_1)$, $P(a_2)$, $\ldots$, $P(a_n)$ all have only two possible values, T or F, and this makes the expressions $P(a_1)$, $P(a_2)$, ..., $P(a_n)$ propositions. In this sense, one can say that interpretations in finite universes convert the expressions of predicate calculus into propositions.

Given the interpretation expressed by (2.2), one can easily decide if $\exists x P(x)$ is true. Generally, $\exists x P(x)$ is true if there is at least one x for which $P(x)$ holds. In the given interpretation, $P(J)$ is true, and this is sufficient to make $\exists x P(x)$ true. There is a customer, Johnson, who has paid. In a domain with the n individuals $a_1, a_2, \ldots, a_n$, one can find an expression similar to (2.3) for the existential quantifier. $\exists x P(x)$ is true in every interpretation that makes either $P(a_1)$ true, or $P(a_2)$ true, ..., or $P(a_n)$ true. Consequently,

$$\exists x P(x) \equiv P(a_1) \vee P(a_2) \vee \cdots \vee P(a_n) \tag{2.4}$$

The laws given by (2.3) and (2.4) allow one to prove that (2.1) is valid for all finite domains. One has

$$\begin{aligned}
\neg \forall x P(x) &\equiv \neg(P(a_1) \wedge P(a_2) \wedge \cdots \wedge P(a_n)) \\
&\equiv \neg P(a_1) \vee \neg P(a_2) \vee \cdots \vee \neg P(a_n) \\
&\equiv \exists x \neg P(x)
\end{aligned}$$

In other words, (2.1) corresponds to De Morgan's law, a result that is somewhat unexpected.

The assignment of a predicate with two arguments can be expressed using a table in which the rows represent the first argument and the columns represent the second argument. Such assignments must be provided to allow the interpretation of statements containing predicates with two arguments. Consider, for instance, an interpretation of the statement "There is someone who admires everyone." The universe of discourse consists of three people, John, Mary, and Jane. The predicate in question is Q; that is, $Q(x, y)$ means "x admires y." Table 2.3 gives an assignment for this predicate. From this table, one sees that John admires Mary and Jane, that Mary admires only John, and that Jane admires Mary and herself. To find if the statement "There is someone who admires everyone" is true, it is best to convert it into predicate calculus. "There is someone" can be expressed by $\exists x$. Hence, as a first step, we write

$$\exists x \ x \text{ admires everyone}$$

The statement "x admires everyone" can now be translated as $\forall y Q(x, y)$. As a result, "There is someone who admires everyone" can be expressed as

$$\exists x \forall y Q(x, y)$$

To find if this statement is true, we first find the truth value for $\forall y Q(x, y)$. In the given interpretation, this subexpression is false, no matter what value is assigned to x. It is false for John, because John does not admire himself. It is also false for Mary, because Mary does not admire herself. Finally, it is false for Jane, because Jane does not admire John. Hence, $\forall y Q(x, y)$ is false for all x: There is no x for which $\forall y Q(x, y)$ is true, and $\exists x \forall y Q(x, y)$ is consequently false under the given interpretation.

Note that the two propositions $\forall y \exists x Q(x, y)$ and $\exists x \forall y Q(x, y)$ are different. $\forall y \exists x Q(x, y)$ says that for everyone there is somebody he or she admires, which is true

TABLE 2.3.   Example of an Assignment

	John	Mary	Jane	$\forall y Q(x,y)$	$\exists y Q(x,y)$
John	F	T	T	F	T
Mary	T	F	F	F	T
Jane	F	T	T	F	T
$\forall x Q(x,y)$	F	F	F		
$\exists x Q(x,y)$	T	T	T		

under the given interpretation. To see this, consider the expression $\exists x Q(x,y)$. This expression is true for all x: it is true for $x =$ John, $x =$ Mary, and $x =$ Jane. This means that $\forall x \exists y Q(x,y)$ is true for the given interpretation. Under the same interpretation, $\exists x \forall y Q(x,y)$ is false, as shown earlier. Hence, the truth value may change if one interchanges the universal and the existential quantifiers.

Although the universal and the existential quantifiers may not be interchanged, one is allowed to interchange universal quantifiers, and the same is true for existential quantifiers. This is indicated by the following laws:

$$\forall x \forall y Q(x,y) \equiv \forall y \forall x Q(x,y) \tag{2.5}$$

$$\exists x \exists y Q(x,y) \equiv \exists y \exists x Q(x,y) \tag{2.6}$$

Both sides of (2.5) are true if and only if $Q(x,y) =$ T for all possible pairs x and y. The best way to see this is to represent the predicate Q by a table like Table 2.3, with the entry of row x, column y representing the truth value of $Q(x,y)$. $\forall y Q(x,y)$ is true if and only if all entries of row x are true, and $\forall x \forall y Q(x,y)$ is true if this holds for every row. Hence, the left side of (2.5) is true if and only if all table entries are true. Similarly, $\forall x Q(x,y)$ is true if and only if all entries of column y are true, and $\forall y \forall x Q(x,y)$ is true if and only if this holds for each column. The right side of (2.5) therefore is also true if and only if all $Q(x,y)$ are true, which means that the two sides of (2.5) are logically equivalent.

The proof of (2.6) is similar. Both sides of this equivalence are true if there is a single pair of values that makes $Q(x,y)$ true. For instance, in Table 2.3, $Q($Mary, John$)$ is true, which makes both sides of (2.6) true. If there is no T in the table, then, of course, both sides of (2.6) are false.

Consider now the universe of discourse as given in Table 2.3, and suppose that three people are at a garden party and the sun is shining. If the statement "the sun is shining" is denoted by S, one concludes that in this interpretation S is true. Obviously, no matter if x is John, Mary, or Jane, S is true, which makes $\forall x S$ true. The sun is shining, so to say, for everyone. For similar reasons, $\exists x S$ is true. Generally, if R is any proposition, which means that it is either true or false no matter which individual is considered, then $\forall x R$ and $\exists x R$ are both true if R is true, and they are both false if R is false. One concludes that

$$\forall x R \equiv \exists x R \equiv R \tag{2.7}$$

2.2.3 Validity

Logical arguments must hold under all circumstances. If a logical argument is to be sound, it must therefore be true under all interpretations. This leads to the concept of *validity*, which is defined as follows:

> **Definition 2.6:**   An expression is *valid* if it is true under all interpretations. To express that an expression A is valid, we write $\models A$.

All tautologies are valid expressions. In fact, the only difference between the two terms is that tautologies do not involve quantifiers or predicates, whereas valid expressions are not restricted in this way.

By definition, an expression A is not valid if there is a single interpretation that makes A false and $\neg A$ true. To define this more precisely, we introduce the following definitions:

> **Definition 2.7:**   If B is an expression, then any interpretation that makes B yield T is said to *satisfy* B. Any interpretation that satisfies B is called a *model* of B. If B has a model, then B is said to be *satisfiable*.

Hence, an expression A is not valid if $\neg A$ is satisfiable. Equivalently, if $\neg A$ has a model, then A cannot be valid.

> **Definition 2.8:**   An expression B that has no model is said to be *contradictory*.

Consequently, if A is valid, then $\neg A$ is contradictory.

The notion of validity allows one to define logical implication and logical equivalence as follows:

> **Definition 2.9:**   Let A and B represent two expressions. We say that A is *logically equivalent* to B if $A \Leftrightarrow B$ is valid. In this case, we write $A \equiv B$. Moreover, we say A *logically implies* B, or $A \Rightarrow B$, if $A \Rightarrow B$ is valid.

As in propositional calculus, one can use logical equivalences to manipulate expressions, and one can use logical implications as the basis for sound arguments. As in propositional calculus, we write $A_1, A_2, \ldots, A_n \models C$ if $A_1, A_2, \ldots, A_n$ together logically imply C.

The following law will be used for the purpose of demonstration. However, this law is very important in its own right.

$$\forall x(P \Rightarrow Q(x)) \equiv P \Rightarrow \forall x Q(x) \tag{2.8}$$

Here, P is a proposition and Q is a predicate. To prove (2.8), the law of cases is used. Since P is either true or false, this means that, instead of (2.8), one can prove the validity of the following two equivalences:

$$\forall x(\text{T} \Rightarrow Q(x)) \equiv (\text{T} \Rightarrow \forall x Q(x)) \tag{2.9}$$

$$\forall x(\text{F} \Rightarrow Q(x)) \equiv (\text{F} \Rightarrow \forall x Q(x)) \tag{2.10}$$

Both of these equivalences are valid. Since $\text{T} \Rightarrow Q(x)$ is $Q(x)$, the left side of (2.9) reduces to $\forall x Q(x)$. For a similar reason, the right side of (2.9) reduces to $\forall x Q(x)$, and $\forall x Q(x) \Leftrightarrow \forall x Q(x)$ is obviously valid. Moreover, (2.10) is valid because both sides are always true: $\text{F} \Rightarrow Q(x)$ is trivially true, and so is $\text{F} \Rightarrow \forall x Q(x)$. In conclusion, (2.8) is valid no matter whether P is true or false.

As mentioned in Section 1.4.2, tautologies can be converted into schemas in the sense that each logical variable can be made to represent an expression. Of course, the instantiation of the schema must be consistent in the sense that the same symbol must stand for the same expression. For instance, in the schema $A \lor \neg A$, one can replace A by any expression provided that both instances of A are instantiated to the same expression. Hence, $(P \lor Q) \lor \neg(P \lor Q)$ is still a tautology, but $(P \lor Q) \lor \neg R$ is not.

Valid expressions in predicate calculus can also be converted into schemas, except that special attention must be given to bound and free variables. To show this, consider (2.8). In this expression, P can be replaced by any expression as long as the expression does not contain x as a free variable, and this substitution does not affect validity. For instance, let P stand for "the sun is shining," Q for "the weather is nice," and $H(x)$ for "x is happy." Then one has from (2.8)

$$\forall x((P \land Q) \Rightarrow H(x)) \equiv (P \land Q) \Rightarrow \forall x H(x)$$

Hence, if it is true for all people that if the sun is shining and if the weather is nice then they are happy, one may conclude that if the sun is shining and if the weather is nice then everyone is happy. P in (2.8) may even be replaced by a predicate containing a free variable. For instance, let $S(y)$ be the statement "y sings," and let $H(x)$ be "x is happy," as before. Then (2.8) can be instantiated to

$$\forall x(S(y) \Rightarrow H(x)) \equiv S(y) \Rightarrow \forall x H(x)$$

However, the expression replacing P in (2.8) must not contain x as a free variable. To see why not, consider the following equivalence, which is no longer valid.

$$(\forall x(S(x) \Rightarrow H(x))) \Leftrightarrow (S(x) \Rightarrow \forall x H(x)) \tag{2.11}$$

In this case, $\forall x(S(x) \Rightarrow H(x))$ means "everyone who sings is happy," whereas the right side means "if x sings, then everyone is happy," and these two statements are different. Although it may be true that everyone who sings is happy, it may at the same time be true that nobody is happy if x sings. The two statements are different because x at the left of (2.11) is bound, but the first x at the right of (2.11) is free. The two occurrences of x are therefore distinct variables, and (2.11) is no longer an instance of (2.8). In conclusion, if any proposition occurring more than once in a valid expression is replaced by a predicate containing a free variable, then care must be taken that this variable does not become bound in the process. Otherwise, the result may no longer be valid.

Since validity is very important, methods to show that an expression is valid are desirable. Unfortunately, there are no methods that work in all cases. In fact, the problem is what is called *undecidable*. An undecidable problem has no general solution in the sense that there is no method that can reliably provide an answer to the problem. If a problem is undecidable, one either has to look for special cases, or one has to be satisfied with methods that sometimes fail to provide the answer, typically because they run into some type of loop. One method to prove that an expression is valid is to try to show that its negation is contradictory. This approach will be discussed in Section 2.2.5 and in more detail in Section 11.3.2. First, we introduce methods that can be used to show that a given expression is not valid.

2.2.4 Invalid Expressions

An expression A is valid iff there is no interpretation for which A yields F. To find that A is not valid, it is therefore sufficient to give a single counterexample; that is, it is sufficient to give a single interpretation for which A yields F. Equivalently, it is sufficient to find a single model for $\neg A$. In the case when A is of the form $B \Rightarrow C$, $\neg A$ is true if and only if B is true, while C is false. Consequently, to prove that a conditional is not valid, it is sufficient to find a model that makes the antecedent true and the consequent false. To show how to find such a model, consider the following expression, which is obviously not valid.

$$\exists x P(x) \Rightarrow \forall x P(x) \tag{2.12}$$

If $P(x)$ is true for some x, this obviously does not justify the conclusion that $P(x)$ is true for all x. For instance, if a program runs for some input data, this obviously does not allow one to conclude that the program runs for all possible input data.

To prove that (2.12) is not valid, one finds a model that makes the antecedent $\exists x P(x)$ true and the consequent $\forall x P(x)$ false. In other words, one must find a model that satisfies

$$\exists x P(x) \wedge \neg \forall x P(x) \tag{2.13}$$

It is not difficult to find such a model. One chooses an interpretation with two individuals a and b and assigns $P(a)$ the value T and $P(b)$ the value F. Then there is an x, that is, $x = a$, such that $P(x)$ is true, which means that $\exists x P(x)$ yields T. However, $\forall x P(x)$ is obviously false.

Normally, one would argue that if something is true for all x then it is certainly true for $x = y$. Hence, one would normally assume that the following expression is valid.

$$\forall x A \Rightarrow S_y^x A$$

However, if A contains y as a bound variable, this is not necessarily true because of potential variable clashes. In particular, the following expression is not valid.

$$\forall x \exists y P(x, y) \Rightarrow \exists y P(y, y) \tag{2.14}$$

In this case, the variable x is converted to y, where y is the bound variable of the scope of the $\exists y$, and this is illegal. Indeed, (2.14) is not valid. To show this, one must prove that there is a model that makes $\forall x \exists y P(x, y)$ true and $\exists y P(y, y)$ false. In other words, one must find a model for

$$\forall x \exists y P(x, y) \wedge \neg \exists y P(y, y) \tag{2.15}$$

The following interpretation does the job. The universe consists of the two individuals a and b, and the assignment of $P(x, y)$ is given by the following table in which the truth values of $P(x, y)$ are on row x, column y.

	a	b
a	F	T
b	T	F

Here, both $P(a, a)$ and $P(b, b)$ are false, which makes $\exists y P(y, y)$ false. On the other hand, $\exists y P(a, y)$ and $\exists y P(b, y)$ are both true, which means that $\forall x \exists y P(x, y)$ is true. This makes the interpretation a model for (2.15).

To demonstrate that an expression is not valid, one can also use the expression and derive an absurdity from it. In the case of (2.14), for instance, one can define $P(x, y)$ as "the mother of x is y." $\forall x \exists y P(x, y)$ then translates into "everyone has a mother," which is normally true, whereas $\exists y P(y, y)$ means "y is her own mother," which is false.

Finally, we now show that the following expression is not valid.

$$\forall x (P(x) \vee Q(x)) \Rightarrow \forall x P(x) \vee \forall x Q(x) \tag{2.16}$$

We find a model that makes the consequent false and the antecedent true. The interpretation uses a domain in which a and b are the only individuals. The assignment is given by the following table:

x	$P(x)$	$Q(x)$	$P(x) \vee Q(x)$
a	T	F	T
b	F	T	T

The table also gives $P(x) \vee Q(x)$. The reader should be able to see from this table that under this interpretation the left side of (2.16) is true, while the right side of (2.16) is false. To see why (2.16) is not valid, note that the left side indicates that, for all x, either $P(x)$ or

$Q(x)$ is true. The right side, on the other hand, indicates that $P(x)$ is true for all x, or that $Q(x)$ is true for all x. For example, if $P(x)$ and $Q(x)$ mean, respectively, "x has brown shoes" and "x has black shoes," then $\forall x(P(x) \vee Q(x))$ means that everyone has either brown or black shoes, whereas $\forall x P(x) \vee \forall x Q(x)$ means that either everyone has brown shoes or everyone has black shoes. In other words, in addition to the fact that all shoes are either brown or black, all shoes must be of the same color in order to satisfy the right side of (2.16). This is obviously a stronger condition than merely requiring that all shoes be brown or black.

2.2.5 Proving Validity

The problem of proving whether or not an expression is valid is undecidable. It is therefore important to know the cases when such a proof is straightforward. For instance, all tautologies are valid. Consequently, every expression that can be reduced to a tautology is automatically valid. Other expressions are valid by definition. For instance, $\forall x P(x)$ must imply $P(t)$ for every term t. If such simple proofs are available, they should be used. However, since the problem is undecidable, there is no generally applicable method that, given an arbitrary expression, can reliably determine whether or not the expression is valid. However, there are methods that work in a great number of cases. The method most frequently used is based on the fact that if an expression A is valid then $\neg A$ must be contradictory. Techniques to prove that expressions, or sets of expressions, are contradictory are discussed in Section 11.3.2. Here, we give only a few short examples to show how this may be accomplished.

Example 2.10   Show that the following expression is valid.

$$\forall x P(x) \Rightarrow \neg \forall x \neg P(x) \tag{2.17}$$

Solution   To show that (2.17) is valid, one must show that there is no model that makes $\forall x P(x)$ true, while making the negation of the consequent $\forall x \neg P(x)$ false. In other words, one must show that there cannot be a model that makes both $\forall x P(x)$ and $\forall x \neg P(x)$ true. This, however, is easy. It is impossible that $P(x)$ be true for all x and at the same time false for all x. This is a contradiction, which proves that (2.17) is valid.   ∎

Note that even though (2.17) is valid, its converse is not. In other words, the following expression is not valid.

$$\neg \forall x \neg P(x) \Rightarrow \forall x P(x)$$

For instance, if $P(x)$ is "x has a job," then the left side of this expression means "not everyone is unemployed," whereas the right side means "everyone has a job," and the two things are not the same. The statement "not everyone is unemployed" admits the possibility that some people are without jobs, whereas "everyone has a job" does not.

Example 2.11   Show that the following expression is valid.

$$\forall x P(x) \vee \forall x Q(x) \Rightarrow \forall x(P(x) \vee Q(x)) \tag{2.18}$$

Solution   To prove that $\forall x P(x) \vee \forall x Q(x)$ logically implies $\forall x(P(x) \vee Q(x))$, we show that there cannot be an interpretation that makes $\forall x P(x) \vee \forall x Q(x)$ true while making $\forall x(P(x) \vee$

$Q(x))$ false. To do this, assume that $\forall x(P(x) \vee Q(x))$ is false. This means that $\neg\forall x(P(x) \vee Q(x))$ is true or, equivalently, that $\exists x\neg(P(x) \vee Q(x))$ is true. Hence, there must be at least one individual for which $\neg(P(x) \vee Q(x))$ is true. If this individual is called a, then $\neg(P(a) \vee Q(a))$ must be true, and this is only the case if both $P(a)$ and $Q(a)$ are false. If $P(a)$ and $Q(a)$ are both false, then $\forall x P(x)$ and $\forall x Q(x)$ must both be false. This, in turn, implies that $\forall x P(x) \vee \forall x Q(x)$ must be false. A contradiction has been discovered, and (2.18) is therefore valid. ∎

Problems 2.2

1. A universe contains the three individuals a, b, and c. For these individuals, a predicate $Q(x, y)$ is defined, and its truth values are given by the following table:

	a	b	c
a	T	F	T
b	F	T	T
c	F	T	T

   Find $\forall x \exists y Q(x, y)$, $\forall y Q(y, b)$, and $\forall y Q(y, y)$.

2. Given the universe described in Problem 1, determine the truth values for $\exists x \neg Q(a, x)$, $\forall y Q(b, y)$, and $\forall y Q(y, y) \wedge \exists x \forall y Q(x, y)$.

3. A universe of discourse consists of three persons, namely, John, Mary, and Jane. All three persons are students, and none of them is rich. John is male, whereas Mary and Jane are female. S, F, M, and R stand for the properties student, female, male, and rich, respectively.
   (a) Present the assignment of the predicates S, F, M, and R.
   (b) Find $\forall x S(x)$, $\forall x F(x) \vee \forall x M(x)$, $\forall x(F(x) \vee M(x))$, $\exists x R(x)$, and $\exists x(F(x) \Rightarrow R(x))$.
   (c) Let P be T and let Q be F. Find $\forall x P$, $\forall x Q$, $\exists x(P \wedge F(x))$, $\exists x(P \vee F(x))$, and $\exists x(Q \vee F(x))$.

4. Is $P(x) \Rightarrow (P(x) \vee Q(x))$ valid? Give reasons.

5. In a certain interpretation, the domain consists of the individuals a, b, and c, and there is one two-place predicate P. $P(x, x)$ is true for all possible values of x. Moreover, $P(a, c)$ is true. Otherwise, $P(x, y)$ is false. Find the truth values of
   (a) $P(a, b) \wedge P(a, c)$ (c) $P(b, b) \wedge P(c, c)$
   (b) $P(c, b) \vee P(a, c)$ (d) $P(c, a) \Rightarrow P(c, c)$

6. Generate a model for $(P(x) \Rightarrow Q(y)) \wedge \neg Q(y) \wedge P(y)$.

7. Show that $(P(x) \Rightarrow Q(y)) \wedge (Q(y) \Rightarrow R(z)) \Rightarrow (P(z) \Rightarrow Q(z))$ is not valid.

2.3 DERIVATIONS

2.3.1 Introduction

The sections that follow describe how to do derivations in predicate calculus. In particular, they give the necessary rules to insert and remove universal and existential quantifiers. The use of these rules will be demonstrated by a number of examples. We will also introduce a new concept, *unification*. Unification is widely used in logical and functional programming languages.

2.3.2 Universal Instantiation

From $\forall x P(x)$, one should be able to derive $P(t)$ for any term t. For instance, if $P(x)$ stands for "x is sleeping," then $\forall x P(x)$ means "Everyone is sleeping," and from this one should be able to derive, say, that little John is sleeping. More formally, if x represents a variable, t represents a term, and A represents an expression, then the following expression should be valid.

$$\forall x A \Rightarrow S_t^x A \tag{2.19}$$

The validity of this expression derives from the definition of $\forall x$: if A is true for all x, it must be true for $x = t$. There is, however, a slight difficulty that was pointed out earlier. As indicated in Sections 2.1.5 and 2.2.4, the substitution must not lead to a variable clash. In other words, t must not become a bound variable of any quantifier still remaining. The logical implication given by (2.19) can be converted into a rule of inference as follows:

$$\frac{\forall x A}{S_t^x A}$$

This rule of inference is called *universal instantiation*, and it is abbreviated by UI. For example, universal instantiation allows one to conclude that

$$\frac{\forall x (\text{cat}(x) \Rightarrow \text{hastail}(x))}{\text{cat}(\text{Tom}) \Rightarrow \text{hastail}(\text{Tom})}$$

Similarly, $P(y)$ follows from $\forall x P(x)$, because $S_y^x P(x) = P(y)$. We also allow the trivial instantiation $S_x^x A$, which leaves A unchanged. Consequently, from $\forall x P(x)$, we can conclude $P(x)$.

We now give a number of derivations. The first derivation is as follows. The premises are

All human beings are mortal.

Socrates is a human being.

From these premises, we prove that

Socrates is mortal.

To do the derivation, let $H(x)$ indicate that x is human and $M(x)$ that he or she is mortal. Furthermore, let S stand for Socrates. As in Section 1.6, we indicate the line from which each statement is derived, as well as the rule that is applied. In the case of the rule UI, we merely state the instantiation S_t^x. The derivation is shown in Figure 2.1.

As a second example, we derive that Paul is the son of Doug from the following premises:

Prove: $\forall x(H(x) \Rightarrow M(x)), \ H(S) \ \vdash \ M(S)$

Formal Derivation	Rule	Comment
1. $\forall x(H(x) \Rightarrow M(x))$	Premise	All humans are mortal.
2. $H(S)$	Premise	Socrates is human.
3. $H(S) \Rightarrow M(S)$	1, S_S^x	If Socrates is human, he is mortal.
4. $M(S)$	2, 3, MP	Therefore, Socrates is mortal. This follows by applying the modus ponens to lines 2 and 3.

Figure 2.1.   Proof of mortality for Socrates

Doug is the father of Paul.

Paul is not the daughter of Doug.

Any person, whose father is Doug, must either be the son or the daughter of Doug.

The proof is shown in Figure 2.2. To save space in the proof, we use the following abbreviations:

$f(x, y)$: x is the father of y

$s(x, y)$: x is the son of y

$d(x, y)$: x is the daughter of y

Also, we use D for Doug and P for Paul. The derivation uses the disjunctive syllogism (DS), which allows us to write B, once $\neg A$ and $A \vee B$ are established.

2.3.3 Universal Generalization

If A is any expression and if x is a variable that does not appear free in any premise, one has

$$\frac{A}{\forall x A}$$

This rule of inference is called *universal generalization*, or UG for short. Since x becomes bound in the process, we say that the universal generalization is over x, or that one generalizes over x. We will justify universal generalization later. At the moment, it must be pointed out that universal generalization is subject to restrictions. If one generalizes over x, then x must not appear in any premise as a free variable. If x does appear free in any premise, then x always refers to the same individual, and it is fixed in this sense. For instance, if $P(x)$ appears in a premise, then $P(x)$ is only true for x and not necessarily true

Prove: $\forall x(f(D,x) \Rightarrow s(x,D) \vee d(x,D)),$

$\qquad f(D,P), \; \neg d(P,D) \; \vdash s(P,D)$

Formal Derivation	Rule	Comment
1. $\forall x(\; f(D,x)$ $\Rightarrow s(x,D) \vee d(x,D))$	Premise	Any person whose father is Doug must either be the son or the daughter of Doug.
2. $f(D,P)$	Premise	Doug is the father of Paul.
3. $\neg d(P,D)$	Premise	Paul is not the daughter of Doug.
4. $f(D,P)$ $\Rightarrow s(P,D) \vee d(P,D)$	1, S_P^x	This instantiates line 1 to Paul.
5. $s(P,D) \vee d(P,D)$	2, 4, MP	Apply the modus ponens to lines 2 and 4.
6. $s(P,D)$	3, 5, DS	According to the disjunctive syllogism (DS), if Paul is not a daughter (line 3), he must be a son.

Figure 2.2.   Proof that Paul is the son of Doug

for any other individual. If x is fixed, one cannot generalize over x. Generalizations from one particular individual toward the entire population are unsound. If, on the other hand, x does not appear in any premise or if x is bound in all premises, then x is assumed to stand for everyone, and universal generalization may be applied without restriction.

To demonstrate universal generalization, consider the following problem whose domain consists of a group of computer science students. Of course, all computer science students like programming. The derivation must prove that everyone in the domain likes programming. If $P(x)$ and $Q(x)$ stand for "x is a computer science student" and "x likes programming," respectively, the premises become

$$\forall x P(x), \quad \forall x(P(x) \Rightarrow Q(x))$$

The desired conclusion is

$$\forall x Q(x)$$

The proof is given in Figure 2.3. In the proof, $Q(x)$ is derived in line 5, which means that x likes programming. This statement is then generalized to $\forall x Q(x)$. This generalization is only possible because all instances of x in the premises are bound. If the premise $\forall x P(x)$ is replaced by $P(x)$, then universal instantiation over x is no longer sound. This is the case because x is fixed and universal generalization over fixed variables is unsound.

As a second example, we derive $\forall y \forall x P(x,y)$ from $\forall x \forall y P(x,y)$. This derivation is given in Figure 2.4. This confirms (2.5), which states that universal quantifiers may be interchanged. A third example of UG is given in Figure 2.5, which shows that the variable x in a universal quantifier may be changed to the variable y.

Prove: $\forall x P(x),\ \forall x (P(x) \Rightarrow Q(x))\ \vdash\ \forall x Q(x)$

Formal Derivation	Rule	Comment
1. $\forall x P(x)$	Premise	Everyone is a CS major.
2. $\forall x (P(x) \Rightarrow Q(x))$	Premise	CS majors like programming.
3. $P(x)$	1, S_x^x	x is a CS major.
4. $P(x) \Rightarrow Q(x)$	2, S_x^x	If x is a CS major, he or she likes programming.
5. $Q(x)$	3, 4, MP	x likes programming.
6. $\forall x Q(x)$	5, UG	Everyone likes programming.

Figure 2.3. Example of universal generalization

Prove: $\forall x \forall y P(x, y) \vdash \forall y \forall x P(x, y)$

Formal Derivation	Rule	Comment
1. $\forall x \forall y P(x, y)$	Premise	
2. $\forall y P(x, y)$	1, S_x^x	Instantiate line 1 by dropping the first universal quantifier.
3. $P(x, y)$	2, S_y^y	Drop the second quantifier to obtain an expression without quantifiers. Use UG to add the quantifiers back in reverse order.
4. $\forall x P(x, y)$	3, UG	It is sound to generalize, because the premise does not contain x as a free variable. All occurrences of x in the premise are bound.
5. $\forall y \forall x P(x, y)$	4, UG	Generalize again to obtain the desired conclusion.

Figure 2.4. Proof of the validity of interchanging quantifiers

Why is UG sound? Consider the proof before B is obtained, where B is the expression to be generalized. Obviously, if x is not contained in any premise as a free variable, then one can repeat the proof for every individual; that is, B is true for all x. Of course, if any assumption depends on x, B is only valid for this particular value of x, and UG is not appropriate.

Prove: $\forall x P(x) \vdash \forall y P(y)$

Formal Derivation	**Rule**	**Comment**
1. $\forall x P(x)$	Premise	
2. $P(y)$	1, S_y^x	Instantiate the premise for y.
3. $\forall y P(y)$	2, UG	Generalize line 2 to obtain the conclusion.

Figure 2.5.   Proof of variable substitution

A second proof of the soundness of UG is based on (2.8). If B has been derived, then B is logically implied by the conjunction of the premises. Hence, if A is the conjunction of all premises, $A \Rightarrow B$ must be valid. Valid expressions generalize; that is, $\forall x(A \Rightarrow B)$ is valid. Because of (2.8), this means that $A \Rightarrow \forall x B$ is valid. Since A holds, $\forall x B$ follows by the modus ponens. This proves that UG is sound, given that all premises are independent of x.

2.3.4 Deduction Theorem and Universal Generalization

In the deduction theorem, one assumes B, proves C, using the assumption B like a premise, and concludes that $B \Rightarrow C$. Once this is done, B is discharged. The question now is how to treat free variables occurring in B. First, while B is used as an assumption, that is, as long as B is not discharged, B has to be treated like any other premise. In particular, should B contain x as a free variable, then one must not generalize over x. However, as soon as B is discharged, this is no longer true. Once B is discharged, it has no effect whatsoever on the status of any variable. Hence, if x is not free in any other premise, one can universally generalize over x even if x appears free in B.

The deduction theorem is now demonstrated by an example.

Example 2.12   Let $S(x)$ stand for "x studied" and $P(x)$ stand for "x passed." The premise is that everyone who studied passed. Prove that everyone who did not pass did not study.

Solution   The premise "everyone who studied passed" can be translated as $\forall x(S(x) \Rightarrow P(x))$, and the statement "everyone who did not pass did not study" becomes $\forall x(\neg P(x) \Rightarrow \neg S(x))$. Figure 2.6 gives a derivation. To obtain the result, the assumption $\neg P(x)$ is introduced in line 3. As long as this assumption is not discharged, no generalization over x is allowed. To indicate that an assumption is in effect, lines 3 and 4 are indented. However, once the deduction theorem is applied, the indentation is removed and the assumption $\neg P(x)$ is discharged, and one can generalize over x. This is done in line 6. In all other aspects, the proof is self-documenting.

$\forall x(S(x) \Rightarrow P(x))$ is a premise, and $\neg P(x) \Rightarrow \neg S(x)$ follows. To arrive at the desired conclusion, one uses universal generalization. This can be done because x is not free in any premise. ∎

Prove: $\forall x(S(x) \Rightarrow P(x)) \vdash \forall x(\neg P(x) \Rightarrow \neg S(x))$

Formal Derivation	Rule	Comment
1. $\forall x(S(x) \Rightarrow P(x))$	Premise	Everyone who studied passed.
2. $S(x) \Rightarrow P(x)$	S_x^x	If x studied, he passed.
3. $\neg P(x)$	Assumption	Assume that x did not pass.
4. $\neg S(x)$	2, 3, MT	x cannot have studied: this follows from lines 2 and 3 by the modus tollens.
5. $\neg P(x) \Rightarrow \neg S(x)$	DT	Apply the deduction theorem and discharge $\neg P(x)$.
6. $\forall x(\neg P(x) \Rightarrow \neg S(x))$	5, UG	Anyone who did not pass cannot have studied. Note that this generalization is possible because x is not free in any premise.

Figure 2.6.   A proof using the deduction theorem (DT)

2.3.5 Dropping the Universal Quantifiers

In mathematics, universal quantifiers are frequently omitted. For instance, in the statement $x + y = y + x$, both x and y are implicitly universally quantified. This causes problems when such statements are used as premises because, according to our rules, any variable appearing free in a premise is fixed in the sense that throughout the proof it is bound to one and the same individual. To get around this difficulty, we single out certain variables in the premises and explicitly state that these variables are not fixed. All variables that are not fixed will be called *true variables*. A variable may be universally generalized if and only if it is a true variable. If a variable appears in a premise, then it is assumed to be fixed, unless it is explicitly stated that the variable is a true variable.

By using true variables, one can omit many universal quantifiers and this, in turn, simplifies proofs. Moreover, we allow from now on that any true variable can be instantiated to any term. The same effect can, of course, be achieved by using universal generalization first, followed by universal instantiation. However, direct instantiation is shorter and often clearer.

Until now, instantiations were always represented by the symbols S, such as S_y^x. From now on, we will frequently make use of the notation $x := y$ to indicate that x is replaced by y. This notation should be natural to the Pascal programmer, because in Pascal := means "assign to," which is the same as "instantiate to." Hence, instead of writing "instantiate x to a in line n," we will write "line n with $x := a$," or simply "n with $x := a$."

The following example applies the conventions just introduced.

Example 2.13   Let $P(x, y, z)$: $x + y = z$. Given the premises $P(x, 0, x)$ and $P(x, y, z) \Rightarrow P(y, x, z)$, where x, y, and z are true variables, prove that $0 + x = x$; that is, prove $P(0, x, x)$.

Solution   The following derivation is used to prove $P(0, x, x)$. Note that the first two lines are premises and that x, y, and z are explicitly declared as true variables.

   1. $P(x, y, z) \Rightarrow P(y, x, z)$ Premise: $x + y = z \Rightarrow y + x = z$.
                                    x, y, z: true variables.
   2. $P(x, 0, x)$ Premise: $x + 0 = x$. x: true variable.
   3. $P(x, 0, x) \Rightarrow P(0, x, x)$ Line 1 with $x := x$, $y := 0$, $z := x$.
   4. $P(0, x, x)$ Lines 2, 3, modus ponens: $0 + x = x$.   ■

All true variables are strictly local to the line on which they appear. Hence, if the true variable x appears on two different lines, then these two instances of x are really two different variables. For instance, in the proof of Example 2.13, x in line 1 and x in line 2 are two different variables. When doing the proof, one obviously has to establish some type of connection between the variables, and this connection is through instantiation. Of course, instantiations must not be made blindly. Instead, one has to do the instantiations in such a way that progress is made toward the desired conclusion. How this is done in detail depends on the general strategy, and in each proof, some type of strategy should be followed (see Section 11.3.5). However, there are some general principles that are helpful, and one of them is *unification*.

Definition 2.10:   Two expressions are said to *unify* if there are legal instantiations that make the expressions in question identical. The act of unifying is called *unification*. The instantiation that unifies the expressions in question is called a *unifier*.

Example 2.14 $Q(a, y, z)$ and $Q(y, b, c)$ are expressions appearing on different lines. Show that the two expressions unify, and give a unifier. Here, a, b, and c are fixed, and y and z are true variables.

Solution   Since y in $Q(a, y, z)$ is a different variable than y in $Q(y, b, c)$, rename y in the second expression to become y_1. This means that one must unify $Q(a, y, z)$ with $Q(y_1, b, c)$. An instance of $Q(a, y, z)$ is $Q(a, b, c)$, and an instance of $Q(y_1, b, c)$ is $Q(a, b, c)$. Since these two instances are identical, $Q(a, y, z)$ and $Q(y, b, c)$ unify. The unifier is $a = y_1$, $b = y$, $c = z$.   ■

There may be several unifiers. For instance, if a and b are constants, then $R(a, x)$ and $R(y, z)$ have the unifier $y = a$, $z = x$, which yields the common instance $R(a, x)$. However, there is also the unifier $y = a$, $x = b$, $z = b$, which yields the common instance $R(a, b)$. However, $R(a, b)$ is an instance of $R(a, x)$, and the unifier $y = a$, $x = b$, $z = b$ is in this sense less general than the unifier $y = a$, $z = x$. Of course, we always want to find the most general unifier, if one exists. Unification will be discussed further in connection with Prolog in Chapter 4 and in connection with automatic theorem proving in Chapter 11.

The solution of Example 2.13 involved unification. Specifically, to make use of the modus ponens, line 2 was unified with the antecedent of line 1. Generally, unification is performed in such a way that some rule of inference can be applied after unification.

Example 2.15   Clearly, if x is the mother of y and if z is the sister of x, then z is the aunt of y. Suppose now that the mother of Brent is Jane and that Liza is a sister of Jane. Prove that Liza is an aunt of Brent.

Solution   If "mother(x, y)" is the predicate that is true if x is the mother of y, and if "sister(x, y)" and "aunt(x, y)" are defined in a similar fashion, one can state the premises as follows:

   1.   mother$(x, y) \wedge$ sister$(z, x) \Rightarrow$ aunt(z, y)
   2.   mother(Jane, Brent)
   3.   sister(Liza, Jane)

The problem is now to create an expression that unifies with the antecedent of line 1. To do this, one combines lines 2 and 3 to obtain

   4.   mother(Jane, Brent) $\wedge$ sister(Liza, Jane)

This expression can be unified with mother$(x, y) \wedge$ sister(z, x) by setting $x :=$ Jane, $y :=$ Brent, and $z :=$ Liza. This yields

   5.   mother(Jane, Brent) $\wedge$ sister(Liza, Jane) $\Rightarrow$ aunt(Liza, Brent)

The conclusion that Liza is an aunt of Brent now follows from 4 and 5 by the modus ponens.   ∎

The proofs just given were simple in the sense that only a few expressions were available for unification, which made the decision of what to do next relatively easy. In more complex cases, the selection of possible expressions available for unification is wider, and the decision becomes nontrivial. To make a good choice as to which expressions to unify next, one must think about what is to be accomplished. In some cases, one may want to consider all expressions that can be derived and choose the one that is in some sense closest to the conclusion. This is usually a good policy, even though it may fail in some cases. Sometimes, this policy is not applicable because it is not clear how to judge what is closer and what is farther away from the conclusion. In this case, it often helps to set intermediate goals. Moreover, derivations pertaining to similar cases often provide valuable ideas on how to continue.

2.3.6 Existential Generalization

If Aunt Cordelia is over 100 years old, then there is obviously someone, that is, Aunt Cordelia, who is over 100. If there is any term t for which $P(t)$ holds, then one can conclude that some x satisfies $P(x)$. Hence, $P(t)$ logically implies that $\exists x P(x)$. More generally, $\exists x A$ can be derived from $S_t^x A$, where t is any term. This leads to the following rule of inference:

$$\frac{S_t^x A}{\exists x A}$$

This rule of inference is called *existential generalization*, and it is abbreviated EG in formal proofs.

Example 2.16   Let C be Aunt Cordelia, and let $P(x)$ stand for "x is over 100 years old." Then one has

$$\frac{P(C)}{\exists x P(x)}$$

The reason is that if one replaces x by C in $P(x)$ then one finds $P(C)$.

The following example demonstrates how to use existential generalization within a formal proof. The premises of our derivation are

1. Everybody who has won a million is rich.

2. Mary has won a million.

We want to show that these two statements logically imply that

3. There is somebody who is rich.

If somebody were asked to demonstrate that the conclusion follows from the premises, he or she would probably argue as follows. If everybody who wins a million dollars is rich, then Mary is rich if she wins a million. Since we know that Mary has won a million, we apply the modus ponens and conclude that Mary is rich. There is thus somebody, Mary, who is rich. This argument is now formalized. $W(x)$ means that x has won a million, $R(x)$ means that x is rich, and M stands for Mary. The individual steps of the argument are given in Figure 2.7.

It was stated earlier [see (2.1)] that $\neg \exists x P(x)$ is logically equivalent to $\forall x \neg P(x)$. We now prove the first half of this statement by showing that $\neg \exists x P(x) \vdash \forall x \neg P(x)$. This is done in Figure 2.8. We will later prove $\forall x \neg P(x) \vdash \neg \exists x P(x)$. The two proofs together establish (2.1). The proof in Figure 2.8 uses the following rules of inference: modus tollens (MT), universal generalization (UG), existential generalization (EG), and the deduction theorem (DT).

2.3.7 Existential Instantiation

If $\exists x A$ is true, then there must be some term t that satisfies A; that is, $S_t^x A$ must be true for some t. For instance, if $P(x)$ stands for "x does somersaults," then $\exists x P(x)$ means that $S_t^x P(x) = P(t)$ must be true for some t. The problem is that we do not know for which term. If we know that somebody makes somersaults, we still do not know whether it is Aunt Eulalia, Uncle Petronius or even somebody else who makes the somersaults. In a proof, the question must therefore be kept open as to who the individual is who makes somersaults.

Prove: $\forall x(W(x) \Rightarrow R(x)),\ W(M)\ \vdash \exists x R(x)$

Formal Derivation	Rule	Comment
1. $\forall x(W(x) \Rightarrow R(x))$	Premise	Everybody who has won a million is rich.
2. $W(M) \Rightarrow R(M)$	1, S_M^x	Hence, if Mary has won a million, she is rich.
3. $W(M)$	Premise	Mary has won a million.
4. $R(M)$	2, 3, MP	Consequently, Mary is rich.
5. $\exists x R(x)$	4, EG	Somebody (Mary) is rich.

Figure 2.7.   Proof of general existence from specific existence

Prove: $\neg \exists x P(x)\ \vdash\ \forall x \neg P(x)$

Formal Derivation	Rule	Comment
1. $\neg \exists x P(x)$	Premise	There does not exist an x for which $P(x)$ is true.
2. $P(x)$	Assumption	Assume $P(x)$.
3. $\exists x P(x)$	2, EG	Then there must be an x satisfying $P(x)$.
4. $P(x) \Rightarrow \exists x P(x)$	DT	Discharge $P(x)$ and write $P(x) \Rightarrow \exists x P(x)$.
5. $\neg P(x)$	1, 4, MT	Since there is no x such that $P(x)$ is true, and since $P(x)$ implies that there is such an x, $P(x)$ must be false.
6. $\forall x \neg P(x)$	3, UG	Since x does not appear as a free variable in any premise, we can generalize to all x.

Figure 2.8.   Proof of nonexistence expressed using the "for all" notation

To do this, a new variable, say b, is selected to denote this unknown individual. This leads to the following rule of inference:

$$\frac{\exists x A}{S_b^x A}$$

This rule is called *existential instantiation*, and in derivations it is abbreviated EI.

The variable introduced by existential instantiation must not have appeared earlier as a free variable. For instance, when applying EI to the two statements "There exists someone who is over 100 years old" and "There exists someone who makes somersaults," one must not use the same variable b for existential instantiation in both cases. Otherwise, one could conclude that b is both over 100 and makes somersaults, which certainly does not follow logically. Similarly, one cannot use any variable that appears free in any of the premises. Hence, EI must not introduce any variable that has appeared already as a free variable in the derivation. Moreover, the variable introduced is fixed in the sense that one cannot use universal generalization over this variable. For instance, if b makes somersaults, then one cannot use UG to conclude that everyone makes somersaults. Moreover, a variable with an unknown value must not appear in the conclusion, and since any variable introduced by EI is unknown, it must not appear in the conclusion either.

For the purpose of demonstration, suppose that there is someone who won a million dollars, and we want to prove that there is someone who is rich. Hence, the premises are

1. Someone has won a million dollars.

2. Everybody who has won a million is rich.

We want to show that these two statements logically imply

3. There is somebody who is rich.

The detailed proof is shown in Figure 2.9. In this proof, existential instantiation is used on line 2, where the winner is called b. Once this is obtained, the second premise is given in line 3, and this premise is instantiated with $x := b$ in line 4. Note that one must not derive lines 3 and 4 before lines 1 and 2; that is, one must not apply the universal instantiation before the existential instantiation. The reason for this is that, once $W(b) \Rightarrow R(b)$ is obtained, b is no longer a new variable, and it therefore must not be used for existential instantiation. For this reason, it is generally a good idea to apply existential instantiation first.

Prove: $\forall x(W(x) \Rightarrow R(x))$, $\exists x W(x)$ $\vdash \exists x R(x)$

Formal Derivation	Rule	Comment
1. $\exists x W(x)$	Premise	Somebody has won a million.
2. $W(b)$	1, EI	Call the winner b.
3. $\forall x(W(x) \Rightarrow R(x))$	Premise	Anybody who has won a million is rich.
4. $W(b) \Rightarrow R(b)$	3, S_b^x	If $W(x) \Rightarrow R(x)$ holds for everyone, it must hold for the unknown person b.
5. $R(b)$	2, 4, MP	If winning implies being rich and if b won, then b must be rich.
6. $\exists x R(x)$	4, EQ	Somebody is rich.

Figure 2.9.   Proof using EI

As a second example of EI, we prove in Figure 2.10 that $\forall x \neg P(x)$ logically implies $\neg \exists x P(x)$. As shown in Section 1.6.6, proving a negation is typically done by means of an indirect proof: to prove $\neg A$, one assumes A and derives a contradiction. Since we want to prove $\neg \exists x P(x)$, the assumption to be rejected is $\exists x P(x)$. While this assumption holds, the derivation is indented. Existential instantiation allows us to derive $P(b)$ in line 2, which contradicts $\neg P(b)$ derived in line 4 by using universal instantiation. The resulting contradiction is given in line 5. This line allows one to reject the assumption $\exists x P(x)$; that is, $\neg \exists x P(x)$ must be true. Hence, $\forall x \neg P(x)$ implies $\neg \exists x P(x)$, which provides the promised second half of the proof of (2.1).

Prove: $\forall x \neg P(x) \ \vdash \ \neg \exists x P(x)$

Formal Derivation	Rule	Comment
1. $\exists x P(x)$	Assumed	Assumption to be rejected later.
2. $P(b)$	1, EI	Let us say that $P(x)$ is true for b.
3. $\forall x \neg P(x)$	Premise	$P(x)$ is false for all x.
4. $\neg P(b)$	3, S_b^x	Hence, $P(b)$ is false.
5. $P(b) \wedge \neg P(b)$	3, 4, C	We now have the desired contradiction, which implies that the assumption is false.
6. $\neg \exists x P(x)$	5, IP	Since 5 is a contradiction, the assumption is false and can be dismissed.

Figure 2.10.   Proof that existence contradicts nonexistence

Problems 2.3

1. Given P and $\forall x(P \Rightarrow Q(x))$, construct a formal derivation for $\forall x Q(x)$. As rules of inference, use universal instantiation, universal generalization, and modus ponens.

2. Given $\forall x \neg Q(x)$ and $\forall x(P(x) \Rightarrow Q(x))$, give a formal proof for $\forall x \neg P(x)$. Use universal instantiation, universal generalization, and the modus tollens as your rules of inference.

3. Give a formal derivation to show $\exists x \exists y P(x, y)$ logically implies $\exists y \exists x P(x, y)$. Use existential instantiation and existential generalization as your rules of inference.

4. Give a formal derivation to show that $\forall x P(x)$ logically implies $\exists x P(x)$.

5. Let $C(x, y)$ denote the fact that x is a child of y, and let $P(y, x)$ denote the fact that y is a parent of x. Clearly, one has

$$\forall x \forall y (C(x, y) \Rightarrow P(y, x))$$

Give a formal derivation to prove that if Peter is the child of Jane, then Jane must be a parent of Peter. You may use true variables in your premises, which allows you to drop the quantifiers before the derivation starts.

6. Let $L(x, y)$ denote the fact that x and y live in the same city. Clearly,

$$\forall x \forall y \forall z (L(x, y) \land L(y, z) \Rightarrow L(x, z))$$

Using this as one of the premises, give a formal proof that if Peter lives in the same city as Mary, and Mary lives in the same city as Bill, then Peter lives in the same city as Bill. You are allowed to declare true variables in the premises, which allows you to drop the quantifiers before the derivation starts.

7. Find the most general unifier of $Q(a, x, b, x, z)$ and $Q(y, z, u, c, w)$. Here, a, b, and c are constants, and u, w, x, y, and z are variables.

8. Find the most general unifier of $P(a, y) \land R(x, z)$ and $P(x, b) \land R(y, x)$. Here, a and b are constants, and x, y, and z are variables.

2.4 LOGICAL EQUIVALENCES

2.4.1 Introduction

As in the case of propositional calculus, one can use logical equivalences to manipulate logical expressions. Specifically, if A is a logical expression, one can replace any subexpression B of A with a subexpression C, as long as B and C are logically equivalent. To do these types of manipulations, one needs a number of basic logical equivalences. Such a set of basic logical equivalences is given in Section 2.4.2. Section 2.4.3 then shows how to obtain further equivalences from this basic set.

2.4.2 Basic Logical Equivalences

Table 2.4 contains a number of important logical equivalences. Many of these equivalences were derived earlier, but they are given once more for easier reference, often in a generalized form. Note that all the equivalences come in dual pairs. In fact, the existential quantifier turns out to be the dual of the universal quantifier. This is not surprising because in finite domains the universal quantifier is a conjunction, whereas the existential quantifier is a disjunction (see Section 2.2.2). In Table 2.4, all laws are numbered. The duals of each law have the same number, except that a "d" is added for the second law of the pair. The variables that may appear in the expressions in question are restricted for several reasons. Laws 1 and 1d essentially show that if an expression A has a truth value that does not depend on x then one is allowed to universally quantify over x. This generalizes the quantification over propositions stated in (2.7). Laws 2 and 2d say that variants are logically equivalent. However, when forming variants, one must avoid variable clashes, and this is the reason that y must not appear free in A. Laws 3 and 3d can be derived from universal instantiation and existential generalization, respectively. The reader may work out the details. Law 4 can be derived like (2.8), and the fact that x must not appear as a free variable is a genuine restriction. The same is obviously true for the dual law. Laws 5 and 5d are new. Laws 6 and 6d correspond to (2.5) and (2.6) of Section 2.2.2.

TABLE 2.4.   Equivalences Involving Quantifiers

1.	$\forall x A \equiv A$	if x not free in A
1d.	$\exists x A \equiv A$	if x not free in A
2.	$\forall x A \equiv \forall y S_y^x A$	if y not free in A
2d.	$\exists x A \equiv \exists y S_y^x A$	if y not free in A
3.	$\forall x A \equiv S_t^x A \wedge \forall x A$	for any term t
3d.	$\exists x A \equiv S_t^x A \vee \exists x A$	for any term t
4.	$\forall x (A \vee B) \equiv A \vee \forall x B$	if x not free in A
4d.	$\exists x (A \wedge B) \equiv A \wedge \exists x B$	if x not free in A
5.	$\forall x (A \wedge B) \equiv \forall x A \wedge \forall x B$	
5d.	$\exists x (A \vee B) \equiv \exists x A \vee \exists x B$	
6.	$\forall x \forall y A \equiv \forall y \forall x A$	
6d.	$\exists x \exists y A \equiv \exists y \exists x A$	
7.	$\neg \exists x A \equiv \forall x \neg A$	
7d.	$\neg \forall x A \equiv \exists x \neg A$	

Many operations involving quantifiers are easier if distinct variables have distinct names, because this avoids variable clashes. Laws 2 and 2d allow one to rename the variables in order to make them distinct.

Example 2.17   In the expression $\forall x P(x) \vee \forall x Q(x)$, the bound variable x appears under two different scopes. By using law 2, one can change the x of the second universal quantifier to y, which yields

$$\forall x P(x) \vee \forall x Q(x) \equiv \forall x P(x) \vee \forall y Q(y)$$

Example 2.18   In the expression $P(x) \wedge \exists x Q(x)$, the variable x appears first free and then bound. By using law 2d, one can write this as $P(x) \wedge \exists y Q(y)$.

Examples 2.17 and 2.18 show how to standardize the variables apart.

Definition 2.11:   Renaming the variables in an expression such that distinct variables have distinct names is called *standardizing the variables apart*.

Example 2.19   Standardize all variables apart in the following expression:

$$\forall x (P(x) \Rightarrow Q(x)) \wedge \exists x Q(x) \wedge \exists z P(z) \wedge \exists z (Q(z) \Rightarrow R(x))$$

Solution   Use y for x in $\forall x$, u for x in $\exists x Q(x)$, and w for z in $\exists z P(z)$ to obtain

$$\forall y (P(y) \Rightarrow Q(y)) \wedge \exists u Q(u) \wedge \exists w P(w) \wedge \exists z (Q(z) \Rightarrow R(x))$$ ■

Some operations cannot be done for expressions containing negated quantifiers. To remove negated quantifiers, laws 7 and 7d are useful, as the next example demonstrates.

Example 2.20   Apply laws 7 and 7d of Table 2.4 to remove all negations in front of the quantifiers of the following expression:

$$\neg\forall z(\exists x P(x,z) \wedge \neg\forall x Q(x,z))$$

Solution   One has

$$
\begin{aligned}
&\neg\forall z(\exists x P(x,z) \wedge \neg\forall x Q(x,z)) \\
&\equiv \exists z\neg((\exists x P(x,z) \wedge \neg\forall x Q(x,z)) && \text{Law 7d} \\
&\equiv \exists z(\neg\exists x P(x,z) \vee \forall x Q(x,z)) && \text{De Morgan} \\
&\equiv \exists z(\forall x\neg P(x,z) \vee \forall x Q(x,z)) && \text{Law 7}
\end{aligned}
$$
■

2.4.3 Other Important Equivalences

In Section 2.3.5, we showed that it is advantageous to drop universal quantifiers. However, universal quantifiers cannot be dropped unless they are at the beginning of the expression. In this respect, the following equivalence is of interest.

$$\forall x P(x) \vee \forall y Q(y) \equiv \forall x\forall y(P(x) \vee Q(y))$$

To prove this law, rewrite law 4 as

$$(\forall x B) \vee A \equiv \forall x(B \vee A) \tag{2.20}$$

Now, we have

$$
\begin{aligned}
\forall x P(x) \vee \forall y Q(y) &\equiv \forall x(P(x) \vee \forall y Q(y)) && \text{(2.20) with } A := \forall y Q(y) \\
&\equiv \forall x\forall y(P(x) \vee Q(y)) && \text{Law 4 with } A := P(x)
\end{aligned}
$$

Note that the condition that A must not contain the bound variable of the quantifier in question applies. $\forall y Q(y)$ does not contain the bound variable x. Moreover, when $A = P(x)$, the bound variable is y, and $P(x)$ does not contain y.

Consider the statement "If somebody talks, it will be in the news tomorrow." If $C(x)$ stands for "x talks" and if Q stands for "it will be in the news tomorrow," one can translate this sentence as

$$\exists x C(x) \Rightarrow Q \tag{2.21}$$

Another translation, which is not obvious, is

$$\forall x(C(x) \Rightarrow Q) \tag{2.22}$$

Both expressions are logically equivalent, as we will prove next. The first version is the version normally used in English. However, for logical derivations, the second version is preferable. Unfortunately, a verbal translation of the second version is difficult. One would have to say something like this: for each x, it is true that if x talks then it will be in the news.

The following law shows that the expressions given by (2.21) and (2.22) are indeed logically equivalent.

$$\forall x(B \Rightarrow A) \equiv \exists x B \Rightarrow A \tag{2.23}$$

The proof of this relation is as follows:

$$\forall x(\neg B \vee A) \equiv (\forall x \neg B) \vee A \equiv \neg \exists x B \vee A$$

and the last expression is logically equivalent to $\exists x B \Rightarrow A$. Note again that A must not contain x.

Example 2.21   Clearly, if $x < y$ and $y < z$, then $x < z$. If $G(x, y)$ means that $x < y$, then this translates into $G(x, y) \wedge G(y, z) \Rightarrow G(x, z)$. This obviously holds for all x, y, and z; that is

$$\forall x \forall y \forall z(G(x, y) \wedge G(y, z) \Rightarrow G(x, z)) \tag{2.24}$$

On the other hand, one can say that $x < z$ if there is a y such that $x < y$ and $y < z$. This can be expressed as follows:

$$\forall x \forall z(\exists y(G(x, y) \wedge G(y, z)) \Rightarrow G(x, z)) \tag{2.25}$$

Are these two expressions logically equivalent?

Solution   Interchange the second and third quantifier in (2.24) to obtain

$$\forall x \forall z(\forall y(G(x, y) \wedge G(y, z) \Rightarrow G(x, z)))$$

The innermost quantifier is an instance of the left of (2.23), which implies that

$$\forall x \forall z(\exists y(G(x, y) \wedge G(y, z)) \Rightarrow G(x, z))$$

This shows that (2.24) and (2.25) are logically equivalent.

$$\begin{aligned} &\forall x \forall y \forall z(G(x, y) \wedge G(y, z) \Rightarrow G(x, z)) \\ &\equiv \forall x \forall z(\exists y(G(x, y) \wedge G(y, z)) \Rightarrow G(x, z)) \end{aligned} \tag{2.26}$$

∎

Problems 2.4

1. Prove $(\exists x B) \wedge A \equiv \exists x(B \wedge A)$ by using the laws of Table 2.4 and the commutative laws of propositional calculus. Assume that A does not contain x as a free variable.

2. Prove $(\forall x B) \wedge A \equiv \forall x(B \wedge A)$ by using the laws of Table 2.4 and the commutative laws of propositional calculus. Assume that A does not contain x as a free variable.

3. Consider the expression $\forall x P(x) \vee \forall x(Q(x) \Rightarrow P(x))$. Move all universal quantifiers to the front of this expression.

4. Consider the expression $\exists x P(x) \wedge \exists x(Q(x) \wedge P(x))$. Move all universal quantifiers to the front of this expression.

5. In the following expressions, standardize all variables apart.
   (a) $\forall x(\forall y P(x, y) \wedge \forall y Q(x, y)) \Rightarrow R(y)$.
   (b) $\forall z Q(z) \wedge \forall x R(z) \Rightarrow \forall x(Q(x) \wedge R(z))$.
   (c) $\forall u P(u) \Rightarrow (\forall v P(v) \wedge Q(v))$.
   (d) $P(x) \wedge \forall x Q(x) \wedge \forall x R(x)$.

6. Let $F(x)$ stand for "x finds the bug" and let Q stand for "the program error can be corrected." Translate

$$\forall x(F(x) \Rightarrow Q)$$

7. Let P be $\forall y R(y)$. Use the rules of Table 2.4 to show that

$$\forall x(P \vee Q(x)) \equiv \forall x \forall y(Q(x) \vee R(y))$$

State the laws used in your manipulations.

8. Let P be $\exists y R(y)$. Use the rules of Table 2.4 to show that

$$\exists x(P \vee Q(x)) \equiv \exists x \exists y(Q(x) \vee R(y))$$

State the laws used in your manipulations.

2.5 EQUATIONAL LOGIC

2.5.1 Introduction

Obviously, equality is essential for doing algebraic operations. In fact, algebraic operations are used to convert a given expression into some other expression that is either simpler or otherwise more suitable for the purpose at hand. Equality is also important in logic, in particular for indicating that there is only one element satisfying a certain property. This will be shown in some detail. For doing algebraic operations, one needs, of course, an algebra. Generally, an algebra is given by a domain, such as the real numbers, together with some operators, such as + and ×. As it turns out, operators are really functions, and to understand operators, one must know about functions. Generally, functions have values or *images*, and these images depend on their *arguments*. All arguments must be individuals, that is, they must be part of the universe, and so are their images. A function establishes in this sense a relation between individuals. What makes this relation special is that the image of a function is always unique. The fact that the image of a function is always a uniquely identifiable individual makes the image a *term*. In other words, one can use a function wherever one can use a constant or a variable. Moreover, without requiring uniqueness, algebraic manipulations would be rather restricted, as will be shown. Although the standard algebra, that is, the algebra in the domain of the reals using the operators + and ×, is by far the most important algebra, there are other algebras. We mention, in particular, the *statement algebra*, an algebra that has been used extensively in this text. This algebra is discussed as a special case of a *Boolean algebra*.

2.5.2 Equality

Two terms t and r are equal if they refer to the same individual, and to express this, one writes $t = r$. For instance, every child knows that Superman and Clark Kent are one and the same person, which can be expressed as

$$\text{Superman} = \text{Clark Kent}$$

Equality is really a predicate, and $\text{equal}(t, r)$ could be used instead to express that t and r are equal. However, the normal equal sign is more convenient, and it will therefore be used here. Hence, $t = r$ is an atomic formula, which can be combined with other atomic formulas in the usual fashion, as in $(x = y) \wedge (y = z)$ or $\neg(x = y)$. Of course, one typically uses the abbreviation $x \neq y$ for $\neg(x = y)$. It seems to be rather simple to decide whether two objects are equal, but this is not the case. One way to define equality is by an explicit assignment, which means that objects that are equal must be enumerated. This method is obviously restricted to finite domains. Instead of an explicit assignment, one can provide rules to determine whether or not two objects are equal. Here, we use a third approach and postulate a number of axioms that the equality predicate must satisfy.

Obviously, every term t is equal to itself, which leads to the following axiom:

$$\forall x(x = x) \tag{2.27}$$

This axiom is called the *reflexivity axiom*. As will be shown in Section 5.4.3, a predicate $G(x, y)$ is said to be reflexive if $G(x, x)$ is true for all x. Another important property of the equality is the *substitution property*. If t and r, $t = r$, are two terms, the substitution property allows one to substitute t in any expression by r. For instance, if $t = r$, and $P(t)$ is true, then $P(r)$ must also be true. This gives rise to a rule of inference. To formulate this rule, we first introduce a symbol for doing replacements. Specifically, if A is any expression, then $R(n)_r^t A$ is the expression one obtains from A by replacing the nth instance of term t by r. If t occurs fewer than n times in A, then $R(n)_r^t A = A$.

Example 2.22   Determine $R(1)_y^x(x = x)$, $R(2)_y^x(x = x)$, and $R(3)_y^x(x = x)$.

> **Solution**   Since the first instance of x is at the left of the equal sign, $R(1)_y^x(x = x)$ is $y = x$. Similarly, $R(2)_y^x(x = x)$ is $x = y$. Finally, $R(3)_y^x(x = x)$ is $x = x$.   ∎

We now have the following rule:

Substitution Rule: If A and $t = r$ are two expressions that have been derived, one is allowed to conclude that $R(n)_r^t A$ for any $n > 0$. In this case, we say that we *substitute t from $t = r$ into A*.

Example 2.23   Substitute $x + 1$ from $x + 1 = 2y$ into $x < x + 1$.

> **Solution**   According to the substitution rule, $x + 1$ is replaced by $2y$, which yields $x < 2y$. This is the same as $R(1)_{2y}^{x+1}(x < x + 1)$.   ∎

Example 2.24   Show that $x = y$ substituted into $x = x$ yields $y = x$.

Solution   Replace the first x in $x = x$ by y, which yields $R(1)_y^x(x = x) = (y = x)$.   ∎

The equality predicate is symmetric; that is, $x = y$ implies that $y = x$. This is easily proved. We assume $x = y$, and we prove that $y = x$. This is done as follows:

1. $\forall x(x = x)$ Reflexivity axiom
2. $x = y$ Assumption
3. $x = x$ Instantiate line 1 with $x := x$
4. $y = x$ Substitute line 2 into line 3
5. $x = y \Rightarrow y = x$ By the deduction theorem

The proof starts with line 3, which instantiates the reflexivity axiom given in line 1. Specifically, the bound variable x of line 1 is replaced by the free variable x, and this is indicated by writing $x := x$. Line 3 is now obtained by substituting the first x by y. According to the deduction theorem, one concludes that $(x = y) \Rightarrow (y = x)$. This completes the proof. The resulting expression can be universally quantified, which yields

$$\forall x \forall y((x = y) \Rightarrow (y = x))$$

Consequently, given $t = u$, not only can one replace t by u, but one can also replace u by t. (To be completely rigorous, one would have to add the corresponding steps, which are essentially steps 2, 3, and 4 of the preceding derivation. For simplicity, we omit these steps.)

Next we show that $x = y$ and $y = z$ imply that $x = z$. This property is called the *transitive property* of equality. To show that this is the case, $x = y$ and $y = z$ are used as premises. This allows one to derive $x = z$ as follows:

1. $x = y$ Premise
2. $y = z$ Premise
3. $x = z$ Substitute line 2 into line 1

Hence,

$$(x = y) \wedge (y = z) \Rightarrow (x = z)$$

Since this is valid, one can generalize to obtain

$$\forall x \forall y \forall x((x = y) \wedge (y = z) \Rightarrow (x = z))$$

Substitutions of the type used in the proofs for reflexivity and transitivity are frequent. We therefore add the following example:

Example 2.25   After each iteration in a certain loop, the condition $s > 3i$ is true. Moreover, once the loop terminates, i is 10. Prove that when the loop terminates $s > 30$.

Solution   When the loop terminates, both $s > 3i$ and $i = 10$ hold. By using these conditions as premises, one easily finds $s > 30$ by replacing i in $s > 3i$ by 10, which completes the proof. For the sake of completeness, we give the formal proof:

1. $s > 3i$ Premise
2. $i = 10$ Premise
3. $s > 3 \times 10$ Substitute line 2 into line 1   ∎

The substitution rule can be applied to subexpressions embedded in other expressions. No other rule of inference discussed so far can do this. This means that the substitution rule is very efficient. Whereas the application of all other rules requires that the higher layers of a formula must first be peeled away to allow access to the target subexpressions, the substitution rule can be applied directly. This saves many steps in the derivation. The substitution rule is therefore used extensively in mathematical derivations.

2.5.3 Equality and Uniqueness

Consider the statement "The lion is a mammal." Can this statement be expressed as

$$\text{lion} = \text{mammal}$$

The answer to this question is no. To see why, add the statement

$$\text{bear} = \text{mammal}$$

By substituting the first statement into the second, one obtains

$$\text{bear} = \text{lion}$$

and this is obviously false. This shows that the word "is" cannot always be translated to $=$. Generally, the equal sign cannot be used if the left side of the expression can refer to different objects. If $x_1 = y$ and $x_2 = y$, then one can always conclude that $x_1 = x_2$; that is, x_1 and x_2 must also be the same. There can be only one x for each y such that $x = y$. Equality necessitates uniqueness.

Conversely, to express uniqueness, one uses equality. The statement that individual a is the only element with the property P can be reworded as "if x is not a, then $P(x)$ cannot be true." This translates into

$$\forall x(\neg(x = a) \Rightarrow \neg P(x))$$

This is logically equivalent to

$$\forall x(P(x) \Rightarrow (x = a))$$

Example 2.26   Translate "only a could have forgotten the meeting" into logic.

Solution   Let $P(x)$ be "x forgot." Then one has

$$\forall x(P(x) \Rightarrow x = a)$$

In words, "If somebody forgot, then it must have been a."   ∎

The fact that there is one, but only one individual with the property P can now be expressed as

$$\exists x(P(x) \wedge \forall y(P(y) \Rightarrow (y = x)))$$

This expresses the fact that there is an x that makes $P(x)$ true, and $P(y)$ is true only if $y = x$. To avoid having to write such a lengthy expression, one defines $\exists_1 x P(x)$ to indicate that only one element satisfies P. In other words, one has

$$\exists_1 x P(x) \equiv \exists x (P(x) \wedge \forall y (P(y) \Rightarrow y = x)) \tag{2.28}$$

Example 2.27   Translate the statement "The company has exactly one CEO" into logic, both with and without using the quantifier $\exists_1$.

Solution   Let $\text{CEO}(x)$ express the fact that x is CEO. Then one has

$$\exists_1 x \text{CEO}(x) \equiv \exists x (\text{CEO}(x) \wedge \forall y (\text{CEO}(y) \Rightarrow y = x)) \qquad \blacksquare$$

Example 2.28   Let $C(x)$, $S(x)$, and $Q(x)$ indicate that x is a city, x is a capital, and x is a country, respectively. Assume that the universe of discourse is the set of all cities and the set of all countries. Express the statement "All countries have exactly one capital" in terms of logic.

Solution   By using the quantifier $\exists_1$, this statement translates into

$$\forall x (Q(x) \Rightarrow \exists_1 y (S(y) \wedge C(y)))$$

To write this expression without the symbol $\exists_1$, one uses (2.28), except that all variables must be renamed properly. Moreover, care must be taken that the variables do not clash. In the case considered here, one finds that

$$\forall x (Q(x) \wedge \exists y (S(y) \wedge C(y) \wedge \forall z (S(z) \wedge C(z) \Rightarrow z = y))) \qquad \blacksquare$$

There is a second way to express uniqueness. Clearly, if $P(x)$ and $P(y)$ always imply that $x = y$, then there can be at most one x such that $P(x)$ is true. If, in addition to this, there is an element with property P, then this element is unique. Consequently,

$$\exists_1 x P(x) \equiv \exists x P(x) \wedge \forall x \forall y (P(x) \wedge P(y) \Rightarrow x = y) \tag{2.29}$$

This method of expressing uniqueness is logically equivalent to the one given by (2.28). In fact, one can derive (2.28) from (2.29), and one can derive (2.29) from (2.28) (see Chapter Problems 13 and 14).

Example 2.29   Use (2.29) to express "There is exactly one carpenter in the village."

Solution   If $C(x)$ stands for x is carpenter, and if there is at most one carpenter, then $C(x) \wedge C(y)$ implies $x = y$ for all possible x and y. In logic, this can be expressed as $\forall x \forall y (C(x) \wedge C(y) \Rightarrow (x = y))$. If there is exactly one carpenter, one therefore has

$$\exists x C(x) \wedge \forall x \forall y (C(x) \wedge C(y) \Rightarrow (x = y)) \qquad \blacksquare$$

2.5.4 Functions and Equational Logic

Functions are extremely important for doing equational logic. The reader is, of course, familiar with functions as they are used in mathematics. For instance, the absolute value of x is a function of x, and so is the square of x. In both cases, one has a function with one argument, that is x. There are also functions with several arguments. For instance, in $f(x, y) = x^2 + y$, f is a function with the two arguments x and y. For simplicity, we first discuss functions with one argument. To do this, we will use relations in families. In our universe, every individual has exactly one mother and one father.

To define a function, it must first be given a name, such as f. One now has the following:

Definition 2.12:   A *function* f with one argument associates with each individual x a unique individual y, which is referred to as $f(x)$. The value $y = f(x)$ is called the *image* of x.

The fact that each function has a unique image is essential. The reason is that functions are used as terms in logic, and fallacies arise if a term can refer to two different individuals.

Example 2.30   In the domain defined earlier, "mother" is a function, because for each x there is a unique individual, say $y = \text{mother}(x)$, such that y is the mother of x. This mother is the image of x. On the other hand, "child" is not a function, because for each x there may be several individuals, say $y_1, y_2, \ldots$, with $y_1 = \text{child}(x)$, $y_2 = \text{child}(x)$, and so on. The "husband" relation does not establish a function either. There are individuals x for which there is no corresponding y such that $y = \text{husband}(x)$.

Example 2.31   In the domain of real numbers, the square of x is a function of x because for each value of x there is a unique value y such that $y = x^2$. The logarithm of x, on the other hand, is not a function of x, because for negative values there is no logarithm in the domain of real numbers. $x + 1$ is a function of x because for each x there is a unique value $y = x + 1$. The reciprocal of x is not a function of x, because $1/x$ is not defined for $x = 0$. The solution of the equation $x = y^2$ is not a function of x because for each $x \neq 0$ there are two values y such that $f(x) = y$. For instance, for $x = 4$, there are the two values $y_1 = 2$ and $y_2 = -2$, and both values satisfy the equation $y^2 = x = 4$.

Example 2.32   Which of the following expressions represent functions, given that the domain is the set of reals? (a) $f(x) = 3x^2 - 2x + 1.5$, (b) $\sin x$, (c) $f(x) = (2 + x)/(1 - x)$.

Solution $f(x) = 3x^2 - 2x + 1.5$ is a function, because for each value of x there is a unique result $f(x)$. Also, $\sin x$ is a function. However, $f(x) = (2 + x)/(1 - x)$ is not a function on the reals. The reason for this is that one cannot associate any result $f(x)$ with $x = 1$.   ∎

Many functions have several arguments.

Definition 2.13:   A function with n arguments is said to have an *arity* of n. The arity of a function is fixed. A function f with an arity of n associates with each list of n individuals $x_1, x_2, \ldots, x_n$ a unique individual $y = f(x_1, x_2, \ldots, x_n)$. The individual y is said to be the *image* of $x_1, x_2, \ldots, x_n$.

Example 2.33   Consider the domain of integers, and define p, m, and d as follows:

$$p(x,y) = x + y, \qquad m(x,y) = x - y, \qquad d(x,y) = x/y$$

Which of these expressions are functions?

Solution $p(x,y)$ associates with each pair of integers x and y a unique integer $x + y$, which means that $p(x,y)$ is a function. The same is true for $m(x,y)$. However, $d(x,y)$ is not a function. The reason is that for some integers x, y there is no integer x/y. For instance, there is no integer that is equal to $3/7$, and there is certainly no integer that is equal to $3/0$.   ∎

The fact that in function f each x has a unique image y is crucial. Without this condition, one cannot even write $y = f(x)$ without risking fallacies. To demonstrate this, consider the following example, which pretends to show that $1 = -1$.

Example 2.34   Let $y = f(x)$ if $x = y^2$ or, alternatively, $f(x) = \pm\sqrt{y}$. By using $f(x)$ as a function, one can construct a "proof" that $1 = -1$. Find the fallacy in the argument.

$$
\begin{array}{lll}
\textbf{1.}\ 1 = 1 & \text{Reflexivity of equality} & \\
\textbf{2.}\ 1 = f(1) & 1 = \pm\sqrt{1} & \\
\textbf{3.}\ -1 = f(1) & -1 = \pm\sqrt{1} & (2.30) \\
\textbf{4.}\ 1 = -1 & \text{Substitute } f(1) \text{ in line 2 by line 3} &
\end{array}
$$

Solution   Clearly, $f(x)$ is not a function, because for each x there are two values for $f(x)$, that is, $+\sqrt{x}$ and $-\sqrt{x}$. Consequently, one is not allowed to write $1 = f(1)$ or $-1 = f(1)$. When ignoring this fact, fallacies like the preceding can be obtained.   ∎

The definition of a function implies that there is an image for *every* argument or, if the function has n arguments, for every possible n-tuple of arguments. This is essential in logic because, otherwise, predicates that use functions as terms may be undefined. Outside logic, it is sometimes convenient to remove this restriction. For instance, technically, $1/x$ is not a function of x in the domain of the real numbers, yet $1/x$ shares many properties with functions. One therefore defines the following:

Definition 2.14:   A *partial function* f of arity 1 associates with each individual x at most one individual y, which, if it exists, is referred to as $f(x)$. Similarly, a partial function g of arity n associates with each n-tuple of individuals x_1, $x_2, \ldots, x_n$ at most one individual $g(x_1, x_2, \ldots, x_n)$.

Note that every function is a partial function, but not every partial function is a function. For instance, on the domain of integers, $f(x) = x/2$ is a partial function. However, since one cannot associate any individual with $f(x)$ if x is odd, $f(x) = x/2$ is not a function, according to Definition 2.12.

Definition 2.15:   A partial function that is not a function is called a *strict* partial function.

The definitions of the terms *function* and *partial function* are not uniform. Some authors include partial functions among the functions. In this case, they use *total function* for the term function as defined in Definition 2.12. For clarity, we will frequently use the phrase "total function," even though according to our definition, the word "function" alone would suffice.

All floating-point operations on computers that may lead to exponent underflow and overflow are strict partial functions. They are partial functions because for each set of operands, the result, if it exists, will be unique. However, in case of overflow or underflow, there is no result, and the operation is therefore not a total function. Division on the integers is not a total function either. To be total, all divisions have to be defined. However, a division by zero is undefined.

2.5.5 Function Compositions

If f is a function with one variable, then $z = f(y)$ is the image of y under the function f. Since each individual y has an image, one can set $y = g(x)$, where g is some function with one variable. This construct associates with each x a unique value z, where $z = f(g(x))$.

Definition 2.16:   Let f and g be two functions with one argument. The function that associates with each x the value $f(g(x))$ is called the *composition* of f and g, and it is written $f \circ g$. It follows that $f \circ g(x) = f(g(x))$.

We will occasionally use the word composition for the composition of partial functions. Hence, if f and g are two partial functions, $f \circ g$ is the partial function that associates the individual $f(g(x))$ with x, provided such an individual exists.

Example 2.35   Let m be the function that associates with each x his or her mother, and let f be the function that associates with each x his or her father. Then $f(m(x))$ is the father of the mother of x, $f(f(x))$ is the father of the father, $m(f(x))$ is the mother of the father, and $m(m(x))$ is the mother of the mother. Consequently, $f \circ m$ is the function that associates with each individual its maternal grandfather, $f \circ f$ is the function that associates with each individual its paternal grandfather, and so on. Note that $f \circ m$ and $m \circ f$ are distinct. $f \circ m$ is the maternal grandfather function, whereas $m \circ f$ is the paternal grandmother function.

One can also form compositions involving functions with several variables, as shown by the next example.

Example 2.36   Let $s(x, y)$ be the function that associates with each pair (x, y) the sum $x + y$, and let $p(x, y)$ be similarly the function that associates with each pair (x, y) the product $x \times y$. Then $p(s(x, y), z)$, $s(z, p(x, y))$, and $s(s(z, x), y)$ are all function compositions: they all associate with each triple (x, y, z) a unique result, as indicated by the following equations:

$$p(s(x, y), z) \; = \; (x + y) \times z$$
$$s(z, p(x, y)) \; = \; (z + (x \times y))$$
$$s(s(z, x), y) \; = \; (z + x) + y$$

Operators are functions. However, expressions tend to be much clearer when they are written by using operators than when they are written as functions. For instance, $(x + y) \times z$ is much clearer than $p(s(x, y), z)$. Hence, we will now use operator symbols rather than functions. Strictly speaking, only total functions are operators. However, we will occasionally use operators to express partial functions. For instance, the division operator / really corresponds to a partial function. A domain, together with one or more operators constitutes an *algebra*. Since all operators are functions or at least partial functions, algebraic expressions are really compositions, or even compositions of compositions. In this sense, one can say that algebras deal with function compositions.

Algebraic expressions must be unambiguous, because fallacies arise otherwise. For instance, $3 \times 4 + 5$ may be interpreted as either $(3 \times 4) + 5$ or $3 \times (4 + 5)$, and if the wrong interpretation is used, the resulting derivation is erroneous. This is demonstrated by the following derivation, which pretends to show that $12 = 6$.

1. $3 = 2 + 1$ Premise
2. $12 = 3 \times 4$ Premise
3. $12 = 2 + 1 \times 4$ Substitute $2 + 1$ for 3 in line 2

Hence, $12 = 2 + 1 \times 4 = 2 + 4 = 6$, which is obviously false. To avoid such errors, one can use fully parenthesized expressions. If line 1 had been written as $3 = (2 + 1)$, and line 2 as $12 = (3 \times 4)$, the result would have been $12 = ((2 + 1) \times 4)$, which does not admit the erroneous conclusion. Of course, fully parenthesized expressions are cumbersome, and most people do not use them. We note, however, that grouping terms in the wrong way is one of the most frequent errors when doing algebraic manipulations. A reliance on operator precedence will not help, because substitutions often change the groupings implied by the precedence rules. The preceding derivation provides such an example.

Sequences of assignment statements are also function compositions or, if some of the operators used are partial functions, partial function compositions. For instance, consider the following two Pascal statements:

```
y := 2 * x;
z := 3 + y
```

Compare the initial value of x before execution with the final value of z after execution. Clearly, the final value of z is a function of the initial value of x. In fact, after execution, $z = 3 + (2 * x)$. This issue will be discussed further in Section 3.5.3.

2.5.6 Properties of Operators

This section deals with operators and their properties, a topic of great importance in algebra. The properties of the operators in standard algebra are of course well known. For instance, everyone knows that $x + y = y + x$, and that $x \times y = y \times x$. This is the commutative property of addition and multiplication, respectively.

Definition 2.17:   A function f of arity 2 is *commutative* if $f(x, y) = f(y, x)$. Similarly, if $\circ$ is an operator, then $\circ$ is said to be commutative if $x \circ y = y \circ x$.

When speaking about an operator, one always implies a certain domain. Note, however, that the same operator symbol may be used in different domains. This is often called *operator overloading* (see Section 6.1.5). The operator symbol $+$, for instance, may be used for natural numbers, integers, and real numbers. In this case, it is assumed that there are really three different operators, one corresponding to each domain. Often, the domain is given by the context.

Example 2.37   Let $\max$ be the binary operator that selects the maximum value of two integers. For instance, $3 \max 4$ is 4. Is the operator max commutative?

Solution   Since for all x and y, $x \max y = y \max x$, one concludes that the operator max is commutative.   ∎

Example 2.38   Is the operator $-$ on the integers commutative?

Solution   The operator $-$ is not commutative, because $x - y \neq y - x$.   ∎

Operators are not restricted to arithmetic. In fact, all logical connectives can be thought of as operators. Obviously, $\vee$ and $\wedge$ are commutative operators, while $\Rightarrow$ is not commutative. If one deals with files, one can define several operators. A merge operator can be used to indicate that two files are merged. The merge operator is typically commutative.

If $\circ$ is an operator that is defined for a finite universe, one can express the results of the operator by means of a table. Such a table is often called an *operation table*. An example of an operation table is given in Table 2.5. In this case, the domain consists of the four elements a, b, c, and d and the result of $x \circ y$ is given in row x. For instance, $a \circ c = d$, as indicated by the row labeled a and column labeled c. The truth tables of logical connectives are, of course, operation tables as well.

Example 2.39   Is the operation $\circ$ defined in Table 2.5 commutative?

Solution   The operator $\circ$ is *not* commutative. To be commutative, the operator must have the property that $x \circ y = y \circ x$ for all x and y. This is not the case here. For instance, $d \circ a = a$, yet $a \circ d = c$. Generally, an operator is only commutative if its operation table is symmetric, and this is not the case in this example.   ∎

TABLE 2.5.

Operation Table for ∘

∘	a	b	c	d
a	a	b	d	c
b	a	b	c	d
c	d	c	b	a
d	a	b	a	b

Definition 2.18:   If ∘ is an operator, then ∘ is said to be *associative* if, for all x, y, and z, $x \circ (y \circ z) = (x \circ y) \circ z$.

The operator $+$ is associative, because $(x + y) + z = x + (y + z)$ for all x, y, and z. The operator $-$, on the other hand, is not associative, because it is not true that, for all x, y, and z, $x - (y - z) = (x - y) - z$. In fact, the two sides of this equation are almost always different. If an operator ∘ is associative and if an expression contains ∘ as its only operator, one can drop all parentheses. This makes expressions more readable.

To find if an operator ∘ given by its operation table is associative, one must verify that $(x \circ y) \circ z = x \circ (y \circ z)$ holds for each possible combination of x, y, and z. Often the only way to do this is to enumerate all these combinations, and this is a lengthy process. In the case of Table 2.5, for instance, there are $4^3 = 64$ such combinations, and to show that ∘ is associative, one must enumerate all these 64 combinations. This is quite cumbersome. We therefore use a very small universe in our next example to demonstrate how to determine associativity.

Example 2.40   Consider a universe of discourse that contains only the two individuals a and b. In this universe, the operator ∘ is defined by the following operation table:

∘	a	b
a	a	b
b	a	a

Show that ∘ is not associative.

Solution   The values $x \circ (y \circ z)$ and $(x \circ y) \circ z$ are calculated in Table 2.6. Since the values for these two expressions do not always agree, the operator is not associative. For instance, $b \circ (a \circ b)$ evaluates to $b \circ b = a$, but $(b \circ a) \circ b = a \circ b = b$. Hence, $b \circ (a \circ b) = a$, but $(b \circ a) \circ b = b$, and ∘ is not associative.   ∎

Because of rounding errors, floating-point addition is not associative. If $(x + y) + z$ is evaluated, then the operation $x + y$ is done first, and the result is rounded. To this intermediate result, z is added, and rounding takes place once more. In the expression

TABLE 2.6.   Proof of Nonassociativity

$x\,y\,z$	$y \circ z$	$x \circ (y \circ z)$	$x \circ y$	$(x \circ y) \circ z$
$a\,a\,a$	a	a	a	a
$a\,a\,b$	b	b	a	b
$a\,b\,a$	a	a	b	a
$a\,b\,b$	a	a	b	a
$b\,a\,a$	a	a	a	a
$b\,a\,b$	b	a	a	b
$b\,b\,a$	a	a	a	a
$b\,b\,b$	a	a	a	b

$x + (y + z)$, on the other hand, $y + z$ is calculated first, rounded, and then x is added. This final result is then rounded. Typically, the rounding errors in these two evaluations are different, which means that $(x + y) + z$ is not equal to $x + (y + z)$. To minimize rounding errors in floating-point arithmetic, one should make sure that all intermediate results are as small as possible. The reason is that rounding errors tend to be proportional to the numbers that are rounded, and smaller intermediate results therefore typically give rise to smaller absolute errors. Specifically, if x, y, and z are all positive reals and if x and y are both smaller than z, it is best to add x and y first, because the intermediate result $x + y$ is smaller than $y + z$ and the same tends to be true for the rounding error.

If several operators are defined within the same domain, then the relations between the operators become important. In arithmetic, one has addition and multiplication, and in statement algebra, one has $\wedge$ and $\vee$. It is therefore important to consider systems that have two operators, say $\circ$ and $\square$.

Definition 2.19:   Let $\circ$ and $\square$ be two operators. The operator $\circ$ is said to be *left distributive* over $\square$ if, for all x, y, and z, one has

$$x \circ (y \,\square\, z) = (x \circ y) \,\square\, (x \circ z)$$

The operator $\circ$ is said to be *right distributive* over $\square$ if one has, for all x, y, and z,

$$(y \,\square\, z) \circ x = (y \circ x) \,\square\, (z \circ x)$$

An operator which is both right and left distributive is said to be *distributive*.

If $\circ$ is left distributive over $\square$ and if $\circ$ is commutative, then $\circ$ is also right distributive over $\square$. To see this, interchange the order of all terms containing x in the applicable definition.

Example 2.41   Multiplication is distributive over addition, but addition is not distributive over multiplication. This means that $x \times (y + z)$ is equal to $(x \times y) + (x \times z)$, but $x + (y \times z)$ is not equal to $(x + y) \times (x + z)$.

Example 2.42   According to Table 1.16, $\wedge$ is distributive over $\vee$, and $\vee$ is distributive over $\wedge$.

Example 2.43   Let $x \max y$ be the maximum of x and y, and let $x \min y$ be similarly the minimum of x and y. Show that max is distributive over min.

   Solution   To show that max is distributive over min, one has to show that

$$x \max (y \min z) = (x \max y) \min (x \max z) \tag{2.31}$$

Assume first that $y \geq z$. In this case, $y \min z = z$, and equation (2.31) becomes

$$x \max z = (x \max y) \min (x \max z)$$

Since $y \geq z$, $x \max y \geq x \max z$, which proves that both sides of the equation are equal. This settles the case $y \geq z$. The case $y < z$ can be dealt with in the same way. Hence, by the law of cases, (2.31) is always true.   ■

2.5.7 Identity and Zero Elements

In this section we discuss two elements, the *identity element* and the *zero element*, that, if they exist, turn out to be very useful for doing algebraic manipulations. In particular, they allow one to simplify algebraic expressions.

Definition 2.20:   Let $\circ$ be an operator, defined for some universe of discourse. If there is an individual e_r with the property that, for all x, $x \circ e_r = x$, then e_r is called a *right identity*. Similarly, if there is an individual e_l such that, for all x, $e_l \circ x = x$, then e_l is called a *left identity*. An individual e that is both a right and a left identity is called an *identity*.

If an operator possesses both a left and a right identity, then the two identity elements must be equal. In fact, any identity that is both a right and a left identity is unique. This will be proved in the next section. The important point now is that, if we talk about *the* identity, then we implicitly assume that this identity is both a right and a left identity and that it is therefore unique.

Example 2.44   Consider a domain with the three elements a, b, and c, together with the operator $\circ$, as defined by the following operation table:

	a	b	c
a	a	b	c
b	c	b	a
c	a	b	c

Find all left and right identities.

Return Receipt
Liverpool John Moores University
Library Services

Data structures and algorithms in Java.
31117014536682

Total Items: 1
03/11/2017 11:04

Please keep your receipt in case of dispute.

Solution Since $a \circ x = x$, no matter whether x is a, b, or c, a is a left identity. Similarly, c is a left identity. There is no right identity, however. In fact, the existence of two left identities precludes that there be a right identity.   ∎

If an operator is commutative, the right identity must be equal to the left identity. This is shown as follows: If e_l is a left identity of $\circ$, and if $\circ$ is commutative, then $e_l \circ x = x \circ e_l$, which makes e_l a right identity. Hence, for commutative operators there is only one identity, which is both a right and a left identity.

Example 2.45 Find the identities for addition and multiplication.

Solution Since addition and multiplication are both commutative, there is only one identity. Since $\forall x(x + 0 = x)$ is true, 0 is the identity for $+$. Similarly, 1 is the identity for multiplication, because $\forall x(x \times 1 = x)$.   ∎

Suppose that the operator $\circ$ has both a right and a left identity e. Any proper subexpression of the form $x \circ y$ satisfying $x \circ y = e$ can obviously be deleted. To facilitate such deletions, one defines *inverses* as follows:

Definition 2.21: If $\circ$ is an operator and if x is an individual, then y is called a *left inverse* of x if $y \circ x = e$. Similarly, if $x \circ y = e$, then y is called a *right inverse* of x. An individual that is both a left and a right inverse of x is called an *inverse* of x.

Example 2.46 Consider the operator $\circ$ given in Table 2.7. Find the right inverses of all elements. Also, find all left inverses.

TABLE 2.7. Inverses

	a	b	c	d
a	a	b	c	d
b	b	a	a	c
c	c	b	d	c
d	d	a	b	c

Solution In Table 2.7, the identity element is a. To find the inverses, one identifies all combinations $x \circ y$ that yield a. One has

$$a \circ a = a, \quad b \circ b = a, \quad b \circ c = a, \quad d \circ b = a$$

Hence, a is its own inverse, and so is b. In addition, b is the left inverse of c, and c is the right inverse of b. Similarly, d is the left inverse of b, and b is the right inverse of d. There is no right inverse of c; that is, there is no x such that $c \circ x = a$.   ∎

Example 2.47   Let x be a real number. Find the additive inverse, that is, the inverse of $+$ for all numbers x. Also, find the multiplicative inverse of x if it exists.

   Solution   The identity for $+$ is 0. The additive left inverse must therefore satisfy $y + x = 0$. Clearly, $y = -x$, which makes $(-x)$ the left inverse of x. One easily verifies that $-x$ is also the right inverse of x. Hence, there is only one inverse, which is simultaneously the right and the left inverse. The inverse of x in multiplication is the reciprocal of x, which we denote by x^{-1}. The right and the left inverses are again equal, and the inverse, if it exists, is unique. However, the number 0 has no inverse.   ■

The following theorem deals with the uniqueness of an inverse.

   Theorem 2.1.   If the operator $\circ$ has the (right and left) identity e, if $\circ$ is associative, and if the right and left inverses of x are equal, then x has a unique inverse.

Example 2.48   Use Table 2.7 to demonstrate that it is not sufficient to show that the right and left inverses are equal in order to prove that the inverse is unique.

   Solution   In Table 2.7, b is both a right and a left inverse of itself, yet b has other inverses besides b.   ■

In expressions containing inverses, one can rearrange the terms in such a way that the inverses can be combined to yield the identity element, which can then be dropped. For instance, the expression $(x + y) + (-x)$ can be written as $(y + x) + (-x)$ by applying the commutative law, and because of the associative law, this yields in turn $y + (x + (-x))$. Since $x + (-x) = e$, this yields $y + e$, or y. Of course, in an expression like $x + (-x) = e$, nothing is left after e has been dropped, and in this case e must be retained. Except for this case, one can cancel all inverses, provided that the expression contains only one operator and this operator is commutative and associative.

Example 2.49   Simplify $aba^{-1}bcb^{-1}$ in the domain of real numbers, where a^{-1} and b^{-1} denote the inverse of a and b, respectively.

   Solution   Since multiplication is associative and commutative, one can cancel a against a^{-1} and b against b^{-1}, and the result is bc.   ■

Another important element is the *zero element*. To avoid the need to distinguish between a left and a right zero, we assume that the operator in question is commutative.

> **Definition 2.22:**   An element d is called a *zero element* of the commutative operator $\circ$ if, for all x, $x \circ d = d$.

Example 2.50   Find the zero element of multiplication in the domain of integers.

   Solution   Since $\forall x(x \times 0 = 0)$, one concludes that the zero for multiplication is $d = 0$.   ■

Not all operators have zeros. In particular, the operator $+$ for integers has no zero. There is no number d such that $x + d = d$ for all possible values of x.

Suppose that $\circ$ is a commutative and associative operator with the zero element d. Then $x_1 \circ x_2 \circ \cdots \circ x_n = d$ if a single term x_i, $i = 1, 2, \ldots, n$, is equal to the zero element d. The proof of this statement is given in Section 3.1.6.

Example 2.51   In the case of multiplication of integers, $a \times b \times c = 0$ if either a, b, or c is zero.

In expressions with two operators, further possibilities for creating identity elements and zero elements arise. In the domain of the real numbers, the identity of the addition becomes the zero of the multiplications, which allows one to simplify expressions such as $(x + (-x)) \times y$, which simply become 0.

2.5.8 Derivations in Equational Logic

Equational logic will now be used to give formal derivations for results from the previous section. Moreover, some new results will be derived in a formal way. In equational logic, as in other formal derivations, unification plays a major role.

It was stated that if an operator has a right and a left identity then these two identities must be equal, and there is no other identity. Hence, if u and u' are two left identities and if v and v' are two right identities, then $u = u' = v = v'$. This result is now formally derived. We need the following premises, which are direct consequences of the definitions.

1. $u \circ x = x$ u is a left identity, x is a true variable.
2. $x \circ v = x$ v is a right identity, x is a true variable.
3. $u' \circ x = x$ u' is a left identity, x is a true variable.
4. $x \circ v' = x$ v' is a right identity, x is a true variable.

We continue by unifying the left sides of lines 1 and 2 and equate the results.

5. $u \circ v = v$ Instantiate line 1 with $x := v$.
6. $u \circ v = u$ Instantiate line 2 with $x := u$.
7. $u = v$ Substitute $u \circ v$ in line 5 by line 6.

This shows that $u = v$. By unifying lines 3 and 4, one similarly finds $u' = v'$.

8. $u' \circ v' = v'$ Instantiate line 3 with $x := v'$.
9. $u' \circ v' = u'$ Instantiate line 4 with $x := u'$.
10. $u' = v'$ Substitute $u' \circ v'$ in line 8 by line 9.

Finally, we unify lines 1 and 4 to show that $u = v'$.

11. $u \circ v' = v'$ Instantiate line 1 with $x := v'$.
12. $u \circ v' = u$ Instantiate line 4 with $x := u$.
13. $u = v'$ Substitute $u \circ v'$ in line 11 by line 12.

The equalities of lines 7, 10, and 13 imply that u, v, v', and u' are all equal. To be rigorous, one would have to prove equality for each combination of these four variables; that is, one would have to prove that $u = v$, $u = v'$, $u = u'$, $v = v'$, and so on, but since these proofs are easy, they are omitted.

Let d be the inverse of c. Then one has

$$\forall x((x \circ c) \circ d = x) \tag{2.32}$$

For a formal proof, all premises must be stated. They are as follows:

1. $x \circ e = x$ e is an identity, x is a true variable.

2. $c \circ d = e$ d is the inverse of c, where c, d are fixed.

3. $(x \circ y) \circ z = x \circ (y \circ z)$ Operator associative and x, y, z are true variables.

Line 1 indicates that to have an inverse one must have an identity. This identity is denoted by e. Line 2 defines d to be the inverse of c, and line 3 indicates that $\circ$ is associative. To derive $(x \circ c) \circ d = x$, first unify this $(x \circ c) \circ d$ with the left of line 3. This yields line 4. The remaining lines of the derivation are easy to trace.

4. $(x \circ c) \circ d = x \circ (c \circ d)$ Instantiate line 3 with $y := c$ and $z := d$.

5. $(x \circ c) \circ d = x \circ e$ Replace $c \circ d$ in line 4 by line 2.

6. $(x \circ c) \circ d = x$ Substitute line 1 into line 5.

Since x is a true variable, one can generalize the last line to obtain (2.32) as required.

In some cases, there is no premise that can be applied directly. In such cases, one uses the reflexivity axiom to generate an equality that can then be modified by substitutions.

Example 2.52   Show that $a \circ (b \circ e) = a \circ b$. Here, e is the identity.

> **Solution**   The premises are the reflexivity axiom and the definition of the identity. The reflexivity axiom is used first, with x instantiated to the left of the desired conclusion. The following derivation results.
>
> 1. $\forall x(x = x)$ Reflexivity axiom.
>
> 2. $\forall x(x \circ e = x)$ Definition of the identity e.
>
> 3. $a \circ (b \circ e) = a \circ (b \circ e)$ Instantiate line 1 with $x := a \circ (b \circ e)$.
>
> 4. $b \circ e = b$ Instantiate line 2 with $x := b$.
>
> 5. $a \circ (b \circ e) = a \circ b$ Replace second $b \circ e$ in line 3 by line 4.   ∎

It is well known that one can add a constant to both sides of an equation and that one can multiply both sides of an equation by a constant. More generally, from $a = b$, one can conclude that $a \circ c = b \circ c$ for any operator $\circ$. One has

1. $\forall x(x = x)$ Reflexivity axiom.

2. $a = b$ Premise.

3. $a \circ c = a \circ c$ Instantiate line 1 with $x := a \circ c$.

4. $a \circ c = b \circ c$ Substitute from line 2 into line 3.

The step from $a = b$ into $a \circ c = b \circ c$ is called *postmultiplication*. In spite of its name, postmultiplication is not restricted to multiplication; it can involve any operator, including addition, Boolean operators, and others. One can also *premultiply* equations. For instance, if $a = b$, then premultiplication with c leads to $c \circ a = c \circ b$. The proof of this is simple and is left to the reader. Note that a, b, and c are true variables, which means that one can generalize as follows:

$$\forall x \forall y \forall z((x = y) \Rightarrow (z \circ x = z \circ y)) \tag{2.33}$$

$$\forall x \forall y \forall z((x = y) \Rightarrow (x \circ z = y \circ z)) \tag{2.34}$$

2.5.9 Equational Logic in Practice

In practice, algebraic manipulations are often abbreviated. The commutative and associative laws are often used like rules of inference, and to get from one expression to the next, one often applies several rules simultaneously. Moreover, to express the two equations $x = y$ and $y = z$, one frequently writes $x = y = z$. Because of the transitive property of equality, $x = y$ and $y = z$ imply that $x = z$, and for this reason, $x = y = z$ also means that $x = z$. This observation obviously generalizes. For instance, to prove that $x_1 = x_4$, one can show that $x_1 = x_2$, $x_2 = x_3$, and $x_3 = x_4$. One abbreviates this as $x_1 = x_2 = x_3 = x_4$. The following example shows how this idea is applied.

Example 2.53   Simplify $a \circ (b \circ a^{-1})$, where $\circ$ is commutative and associative, and a^{-1} is the inverse of a.

Solution   One has

$$
\begin{aligned}
a \circ (b \circ a^{-1}) &= a \circ (a^{-1} \circ b) && \text{Commutativity} \\
&= (a \circ a^{-1}) \circ b && \text{Associativity} \\
&= e \circ b && \text{Inverse} \\
&= b && \text{Identity}
\end{aligned}
\tag{2.35}
$$

Note the strategy used: in each step, the distance between a and a^{-1} is reduced in some sense, until $a \circ a^{-1}$ is obtained. Generally, one should always try to keep the objective of the derivation in mind and try to narrow the gap between what one has and what one wants to accomplish. It also helps to identify intermediate goals, especially when dealing with derivations. ∎

The process of writing a sequence of expressions, each equal to the previous one, with the objective to reach some type of a goal is used extensively in functional programming languages, where it is known as *rewriting* (see Section 6.5.5). Although rewriting is extremely important, it sometimes leads to inefficiencies because it may require that some subexpression be evaluated repeatedly. To show this, consider the calculation of the

Fibonacci numbers F_n, which are defined as follows: F_0 and F_1 are both 1. For $n > 1$, one finds F_n by adding F_{n-1} to F_{n-2}. For instance, $F_2 = F_1 + F_0$, $F_3 = F_2 + F_1$, and so on. Suppose now that F_4 is to be calculated. This can be done in two ways. First, one can write

$$F_2 = F_1 + F_0 = 1 + 1 = 2 \quad \text{since } F_1 = 1, \ F_0 = 1$$
$$F_3 = F_2 + F_1 = 2 + 1 = 3 \quad \text{since } F_2 = 2, \ F_1 = 1$$
$$F_4 = F_3 + F_2 = 3 + 2 = 5 \quad \text{since } F_3 = 3, \ F_2 = 2$$

Alternatively, one can rewrite F_4 as follows:

$$
\begin{aligned}
F_4 &= F_3 + F_2 \\
&= (F_2 + F_1) + (F_1 + F_0) \quad && \text{Expand } F_3 \text{ and } F_2. \\
&= (F_2 + 1) + (1 + 1) \quad && F_1 = F_0 = 1. \\
&= ((F_1 + F_0) + 1) + 2 \quad && \text{Expand } F_2.\ 1 + 1 = 2. \\
&= ((1 + 1) + 1) + 2 \quad && F_1 = F_0 = 1. \\
&= (2 + 1) + 2 = 5
\end{aligned}
$$

Note that, in the derivation, $F_2 = F_1 + F_0 = 1 + 1 = 2$ has been calculated twice. This can happen unless special precautions are taken. It will be shown in Section 6.3.3 that this can slow down execution times dramatically. Generally, it pays to watch for common subexpressions appearing more than once in order to avoid evaluating them repeatedly.

An important result is the *cancellation rule*, which allows one to infer under certain conditions that $a = b$ from $a \circ c = b \circ c$. To see why restrictions apply, consider the following applications of the cancellation rule. Clearly, if $a + c = b + c$, then $a = b$. In the case $a \times c = b \times c$, one can conclude that $a = b$ only if $c \neq 0$. If $c = 0$, the cancellation law does not apply. This suggests that for the cancellation law to hold the term to be canceled must have an inverse, and this is indeed the case. Hence, the derivation must reflect that c has an inverse, say c^{-1}. The premise is

$$a \circ c = b \circ c$$

Postmultiplying by c^{-1} yields

$$(a \circ c) \circ c^{-1} = (b \circ c) \circ c^{-1}$$

We can now simplify both sides separately. The left side yields

$$
\begin{aligned}
(a \circ c) \circ c^{-1} &= a \circ (c \circ c^{-1}) \quad && \text{Associativity} \\
&= a \circ e \quad && \text{Inverse} \\
&= a \quad && \text{Identity}
\end{aligned}
$$

A similar argument shows that the right side yields b, and we conclude that $a = b$. If c has an inverse and if $\circ$ is associative, then $a \circ c = b \circ c$ implies that $a = b$.

2.5.10 Boolean Algebra

For statement algebra, the notions of commutativity and associativity have already been introduced in Section 1.5.5, Table 1.16. Indeed, $\wedge$ and $\vee$ are both commutative and associative. Moreover, $P \wedge T = P$ for all P, which makes T the identity of $\wedge$. The identity of $\vee$ is similarly F. Also, by the law of domination, $P \wedge F = F$, which makes F the zero of $\wedge$. The zero of $\vee$ is similarly T. Finally, the distributivity laws indicate that $\wedge$ is distributive over $\vee$ and that $\vee$ is distributive over $\wedge$.

There is a difference, however, in the philosophy of statement algebra and other algebras. As the name implies, statement algebra deals with statements, and it distinguishes between atomic and compound statements. Two statements of a different form are not considered equal, even if they are equivalent. For instance, $P \vee Q$ is not considered equal to $Q \vee P$. For this reason, we formally introduce a number of new operators and new symbols. In fact, we introduce a *Boolean algebra*.

A Boolean algebra is an algebra with two operators, which are denoted by $+$ and $\cdot$. Both operators are total in the sense that, for all arguments x and y, $x + y$ and $x \cdot y$ is defined. Moreover, the algebra has the following properties:

1. There is an identity element for $+$, called 0.

2. There is an identity element for $\cdot$, called 1.

3. The operator $+$ is commutative.

4. The operator $\cdot$ is commutative.

5. The operator $+$ is distributive over $\cdot$.

6. The operator $\cdot$ is distributive over $+$.

7. For every individual x, there is an element x', called the *complement* of x, with the property that $x + x' = 1$ and $x \cdot x' = 0$.

8. The domain contains at least two elements.

The operator $\cdot$ is often omitted. In the simplest case, the domain of a Boolean algebra contains two values, 0 and 1. Such a Boolean algebra is called a *two-valued Boolean algebra*. We will show that the following assignment constitutes a two-valued Boolean algebra. In fact, it is the only such algebra.

1. The complements x' are defined as follows:

$$0' = 1, \qquad 1' = 0$$

2. The operations $+$ and $\cdot$ are defined by the following operation tables:

$+$	0	1
0	0	1
1	1	1

$\cdot$	0	1
0	0	0
1	0	1

We now verify that all the conditions of a Boolean algebra are met. Since the operation tables for $+$ and $\cdot$ are both symmetric, the algebra is commutative. Second, 0 is the identity element of $+$ and 1 is the identity element of $\cdot$, as is easily verified from the operation tables. The fact that $+$ is distributive over $\cdot$ is proved by verifying that, for all x and y, $x, y = 0, 1$, one has

$$x + (y \cdot z) = (x + y) \cdot (x + z) \qquad (2.36)$$

This is done in Table 2.8. In this table, $x + (y \cdot z)$ is given in the column labeled 2, and $(x + y) \cdot (x + z)$ in column 5. It is easy to see that the two columns are equal in all cases, which means that (2.36) holds in all cases. The proof that $\cdot$ is distributive over $+$ is done in a similar way. Finally, $0 + 0' = 1$ and $1 + 1' = 1$, which means that, for all x, $x + x' = 1$. The fact that $x \cdot x' = 0$ for all x is shown in a similar way.

TABLE 2.8.   Boolean Algebra Is Distributive

			1	2	3	4	5
x	y	z	$y \cdot z$	$x + y \cdot z$	$x + y$	$x + z$	$(x + y) \cdot (x + z)$
0	0	0	0	0	0	0	0
0	0	1	0	0	0	1	0
0	1	0	0	0	1	0	0
0	1	1	1	1	1	1	1
1	0	0	0	1	1	1	1
1	0	1	0	1	1	1	1
1	1	0	0	1	1	1	1
1	1	1	1	1	1	1	1

We now show that this is the only two-valued Boolean algebra. Since 0 is the identity for $+$, one must have $0 + 0 = 0$ and $0 + 1 = 1 + 0 = 1$, and since 1 is the identity of $\cdot$, one must have $0 \cdot 1 = 1 \cdot 0 = 0$ and $1 \cdot 1 = 1$. Moreover, since a Boolean algebra must satisfy $x + x' = 1$ for all x, $0 + 0'$ must be 1. This rules out that $0' = 0$ because, under this assignment, $0 + 0' = 0 + 0 = 0$. The only alternative assignment for $0'$ is of course 1, and this corresponds to the Boolean algebra given earlier. A similar argument shows that $1' = 0$. The only values still open at this point are $1 + 1$ and $0 \cdot 0$. However, if one does not set $1 + 1 = 1$ and $0 \cdot 0 = 0$, the system no longer has the required distributive properties. The reader may want to verify this. This leaves the Boolean algebra given previously as the only possibility.

If in statement algebra all equivalences are interpreted as equalities, then the statement algebra becomes a Boolean algebra. To see this, redefine 0, 1, $+$, $\cdot$, and complementation as shown in Table 2.9. There is a dual interpretation of two-valued Boolean algebra. In this dual interpretation, one uses 0 for T, 1 for F, $+$ for $\wedge$, and $\cdot$ for $\vee$. This dual interpretation is the explanation for the dual relations discussed earlier.

Example 2.54   Express $p \vee \neg p \Leftrightarrow$ T and $p \wedge \neg p \Leftrightarrow$ F in Boolean algebra.

Solution $p \vee \neg p \Leftrightarrow$ T becomes $p + p' = 1$, and $p \wedge \neg p \Leftrightarrow$ F becomes $p \cdot p' = 0$.   ∎

TABLE 2.9

Boolean algebra	Statement algebra	English equivalent
0	F	false
1	T	true
$+$	$\vee$	or
$\cdot$	$\wedge$	and
x'	$\neg x$	not

In conclusion, equational logic provides a basis for Boolean algebra and indirectly, a basis for propositional calculus. In this sense, it ties together predicate calculus and propositional calculus.

Problems 2.5

1. Use the rules of inference of equality to show that

$$(x = y) \wedge (y \leq z) \Rightarrow (x \leq z)$$

2. Let $E(x)$: x is an electrician, and let j be Jim. Express "Jim is the only electrician" in predicate calculus with and without using $\exists_1$.

3. Express the fact that there are fewer than two elements with a certain property P. Use only the universal and the existential quantifiers.

4. Let $x \circ y$ represent the greatest common divisor of x and y. Show that $\circ$ is commutative and associative. Find the identity element and the zero of $\circ$, if they exist.

5. Prove that $a \circ b = a \circ c$ implies that $b = c$, given that a has a^{-1} as an inverse.

6. Consider the connective $\Rightarrow$ from an algebraic point of view; that is, assume that each expression is either T or F.

    (a) Using Table 2.7 as an example, write the operation table of $\Rightarrow$.

    (b) Formulate the conditions that an element of the set $\{T, F\}$ must satisfy in order to be a left identity of $\Rightarrow$. Is there such a left identity? Is there a right identity?

    (c) Does $\Rightarrow$ have a left zero? Does $\Rightarrow$ have a right zero? Justify your answer.

7. A domain, together with an operation $\circ$ is called a group if and only if $\circ$ is associative, there is an identity, and each element has an inverse. Which of the following are groups?

    (a) Let the set of integers be the domain, and let $\circ$ represent $+$.

    (b) Let the set of integers be the domain, and let $\circ$ represent $\times$.

    (c) Let $0, 1, 2, \ldots$ be the domain, and let $\circ$ represent $+$.

8. Show in a step-by-step fashion that the product $a^{-1}baab^{-1}$ is a.

9. Using the axioms and rules of inference for equality, convert (2.35) into a formal derivation.

10. Find $S(3)$, given $S(0) = 1$ and $S(n + 1) = S(n) + n$. Show all your steps.

11. Given the following two-valued algebra with the operation table for $+$ and $\cdot$, show that the resulting algebra cannot be a Boolean algebra.

$+$	0	1
0	0	1
1	1	0

$\cdot$	0	1
0	0	0
1	0	1

12. A number of people sit around a round table. Define the function $y = f(x)$ if y is immediately to the left of x. What can you say about the number sitting at the table if $f(f(f(x))) = x$?

13. Five books, entitled A, B, C, D, and E are, in that order, on a shelf. If x is a book, then $f(x)$ is the book to the right of x.
   (a) Find $f(f(B))$.
   (b) Is $f(x)$ a total function of x? State why or why not.

PROBLEMS: CHAPTER 2

1. Translate the following sentences into logic. Abbreviate the propositions and predicates as indicated. Do not introduce new propositional variables or predicates.
   (a) All people have ancestors, but not all people have descendants. Use $A(x, y)$ if x is an ancestor of y, and formulate descendant, using the predicate A. The universe of discourse is all people who live or have ever lived.
   (b) Some numbers are greater than others, but all numbers except 0 are greater than 0.

2. Given there are three individuals in the domain, how many possible assignments are there for a predicate of arity 2?

3. Express the following sentences in predicate calculus. Use the universe of discourse as indicated in parentheses.
   (a) There is no lion. (Animals in a Zoo)
   (b) Not every cat has a tail. (Animals)
   (c) Cows give milk only if they have calves, and if they are not fed well, they do not give much milk. Use $\text{cow}(x)$, $\text{calf}(x, y)$, $\text{fedwell}(x)$, $\text{givemilk}(x)$, and $\text{givemuchmilk}(x)$.

4. The translation of a statement into predicate calculus depends on the universe of discourse selected. Consider now the statement "All birds have wings, but some birds cannot fly." Translate this sentence into predicate calculus, given the universe of discourse is as follows:
   (a) The set of all birds
   (b) The set of all animals
   (c) Wings, in addition to the set of all animals

5. Let $P(x, y, z)$ be the predicate that is true if $x + y = z$, and false otherwise. Translate $\exists y \forall x P(x, y, x)$ into English. Find an example which shows that $\exists y \forall x P(x, y, x)$ is true in the domain of real numbers.

6. Which of the following expressions are (i) satisfiable, (ii) valid, (iii) contradictory, and (iv) not valid? Justify your answers.
   (a) $P(x) \lor P(y)$
   (b) $P(x) \Rightarrow (P(x) \lor Q(x))$
   (c) $\forall x (P(x) \lor \neg P(x))$

(**d**) $\exists x (P(x) \land \neg P(x))$

(**e**) $\forall x P(x)$

7. Find an interpretation that satisfies the following three expressions simultaneously:

$$\forall x \neg G(x, x)$$
$$\forall x \forall y \neg (G(x, y) \land G(y, x))$$
$$\exists x \exists y \forall x \exists z (G(x, y) \land G(y, z) \land G(z, x))$$

8. Consider the following premises: $\exists x M(x)$, $\forall x (M(x) \Rightarrow \exists y C(x, y))$, and $\forall x (\exists y C(x, y) \Rightarrow F(x))$. Give a formal derivation to prove that these premises lead one to the conclusion that there must be a y such that $F(y)$ is true. State the rules of inference used in the derivation.

9. Prove the following by derivation.

(**a**) $\forall x P(x) \vdash \forall x (P(x) \lor Q(x))$

(**b**) $\exists x \forall y P(x, y) \vdash \forall y \exists x P(x, y)$

(**c**) $\exists y \forall x P(x, y, z) \vdash \exists z P(z, z, z)$

10. Give a formal derivation for $(P(a) \land \forall x P(x)) \Leftrightarrow \forall x P(x)$. Use only the rules given in Section 1.6.5, the deduction theorem, and UI, UG, EI, and EG.

11. The universe of discourse is birds. The premise is that all birds have wings and feathers, which we translate into

$$\forall x (W(x) \land F(x))$$

Give a formal derivation to prove that all birds have feathers. Use universal instantiation, universal generalization, and the law of simplification.

12. Formulate the following statements in predicate calculus. To do this, you will need the applicable domain (universe of discourse). These domains are given in parentheses at the beginning of each statement. Note that statements consisting of several sentences are the conjunctions of the sentences in question.

(**a**) (Humans) The professor gave an assignment on Monday and the next one on Wednesday. All students complained about this, and some students could not finish their work.

(**b**) (Natural numbers, that is, 0, 1, …) A number is between a and b iff it is greater than a but less than b.

(**c**) (Humans, Days) You can fool all people on some days, and you can fool some people on all days, but you can't fool all people on all days.

(**d**) (Breakfast, Lunch, Supper) There is no such thing as a free lunch.

13. Convert the statement "There is only one goldfish in the pond" into predicate notation.

(**a**) Assume that the universe of discourse consists of the fish in a pond.

(**b**) Assume that the universe of discourse consists of fish in general.

14. Unify the following pairs of expressions. Here, a, b, and c are constants, but x, y, z, u, and v are true variables.

(**a**) $R(x, y) \land R(y, z) \Rightarrow R(x, z)$ with $R(a, b) \land R(b, z) \Rightarrow R(u, a)$

(**b**) $G(x, y) \land G(f(x), f(y))$ and $G(a, b) \land G(z, u)$

15. Which of the following expressions can be unified, and which cannot? Show your work. a, b, and c are constants and x, y, and z are variables.

(**a**) $G(x, a, y, x)$ and $G(b, y, c, y)$

(**b**) $F(c, y, z, y)$ and $F(a, b, x, x)$

(**c**) $G(x, y) \land G(y, z)$ and $G(y, x) \land G(z, z)$

16. Derive $\exists x(P(x) \land \forall y(P(y) \Rightarrow (x = y)))$ from the premise $\exists x P(x) \land \forall x \forall y(P(x) \land P(y) \Rightarrow (x = y))$.

17. Derive $\exists x P(x) \land \forall x \forall y(P(x) \land P(y) \Rightarrow (x = y))$ from the premise $\exists x(P(x) \land \forall y(P(y) \Rightarrow (x = y)))$.

18. Assume that $\exists_1 x P(x)$ and $\forall x(P(x) \Rightarrow Q(x))$ are given. Is the conclusion $\exists_1 x Q(x)$ sound? If not, give a counterexample.

19. A certain job is advertised. Let $P(x)$ stand for the property that x is an applicant. Use universal and existential quantifiers, in conjunction with the predicate for equality, to express the fact that there are at most two applicants. Also, express the fact that there are exactly two applicants.

20. A universe of discourse consists of all people sitting on a committee. Use the universal and existential quantifiers to express that **(a)** there is at most one student on the committee, **(b)** there is at least one student on the committee, and **(c)** there is exactly one student on the committee.

21. Use the laws of Table 2.4 and propositional logic to
   (a) show that $\forall x P(x) \land \forall y(P(y) \lor Q(y)) \equiv \neg \exists x \neg P(x)$
   (b) prove or disprove the validity of $\neg \exists x(\exists y(P(y) \land Q(y)) \Rightarrow R(x)) \Leftrightarrow \forall x \forall y(\neg P(y) \lor \neg Q(y) \lor R(x))$
   Indicate the laws used.

22. Use predicate calculus to formulate the fact that $f(x)$ is a function if for each x, there is exactly one y such that $y = f(x)$.

23. Let $P(x, y)$ stand for "x is the parent of y." Use this predicate to express "x and y are siblings." Note that x is not a sibling of himself or herself.

24. Find a universe in which $f(f(x)) = x$ for all x.

25. Let $x \circ y = x^y$. Determine whether $x \circ y$ is commutative or associative.

26. Let $x \circ y = x + y - xy$. Show that the operation $\circ$ is commutative and associative. Find the identity element and indicate the inverse of each element.

27. An element x is said to be *idempotent* if $x \circ x = x$. Prove that if an operator is commutative, then its identity element is idempotent.

3

Induction and Recursion

There are many domains that contain an unlimited number of elements. For instance, the domain of natural numbers, the domain of logical expressions, and the domain consisting of all programs written in Pascal are all infinite. Whereas in finite domains all elements can be described one by one and their properties can be listed, this is impossible in infinite domains. To generate an infinite number of elements, one must use a finite set of rules, which can be used repeatedly or, to use the technical term, recursively. Another difficulty when dealing with infinite domains arises when one has to prove that all the elements within the domain have a certain property. In this respect, one method has proved to be extremely versatile, and this method is mathematical induction. In this chapter, several aspects of recursion and induction will be discussed, and their application to programming will be described.

The first infinite domain to be described is the domain of natural numbers. In this domain, one starts with the number 0, and then all subsequent numbers are generated through the *successor function*. For each natural number n, there is a *successor* $s(n)$. This successor is merely $n + 1$. To prove that all elements in this domain have a certain property P, one shows that 0 has property P and then demonstrates that if n has property P then $s(n)$ also has property P. Once this is done, the principle of mathematical induction allows one to conclude that P is true for all natural numbers.

Although mathematical induction is frequently connected with natural numbers, it is applicable in many domains that are defined recursively. The way induction works in these domains is as follows: Suppose that it must be proved that all elements within the domain have a certain property P. For some elements of the domain, this follows immediately from their definitions. These elements are the base elements. For other elements, P must be proved. Typically, they will have property P if some simpler elements also have property P. For instance, there may be two elements, x and y, and $P(y)$ is easier to prove than $P(x)$.

In addition to this, one can show that $P(y) \Rightarrow P(x)$. Then a proof of $P(y)$ establishes $P(x)$. For each y, in turn, simpler elements z can again be found, and in this way one can continue until, eventually, the base elements, for which P has already been established, are reached. This form of induction is also called *proof by recursion*. Of great importance is *structural induction*, which is a special form of proof by recursion.

We now give an outline of the main contents of this chapter. First, we introduce natural numbers, and we show how to do mathematical induction in this domain. In the domain of natural numbers, many essential functions, such as addition and multiplication, can be defined, and this will be done by recursive definitions. Induction is then used to prove that these functions have certain properties. In later chapters, we will extensively use sums and similar constructs. In particular, for doing correctness proofs of programs (see Chapter 9), one needs sums. The relevant results in this respect are derived in Section 3.2. To facilitate the derivations, sums are first defined recursively, which then allows one to use mathematical induction to prove some of their properties.

Section 3.3 deals with proofs by recursion. In particular, the interplay between proofs by recursion and recursive definitions is stressed, and important constructs, such as lists and trees, are introduced. Section 3.4 then shows how mathematical induction applies to programming. First, a simplified mathematical model of computation is established. Then, recursive calls in programming are analyzed, and they are proved to be correct, using proofs by recursion. Section 3.5 introduces *recursive functions*, functions that are easy to define, yet powerful enough to express everything that is computable. They allow one to resolve deep theoretical questions, such as whether there exists a solution to the *halting problem*. The halting problem involves finding a procedure that can indicate whether or not a program will ever terminate. As it turns out, this problem is undecidable; that is, no procedure can be guaranteed to work in all cases.

The term *mathematical induction* is somewhat misleading. In philosophy, one distinguishes between *deduction* and *induction*. Deduction involves deriving conclusions based on logical arguments and, in this sense, all mathematical arguments, including mathematical induction, are deductive arguments. Induction, on the other hand, involves the inference of general rules from particular observations. Of course, the number of available observations is always limited, which means that there may always be cases that violate general rules arrived at by inductive arguments. Rules derived from deductive arguments, on the other hand, cannot be false as long as sound arguments are used and the premises are true. Mathematical induction, like induction in a philosophical sense, deals with generalizations. However, whereas normal induction allows exception to the laws derived, mathematical induction does not.

3.1 INDUCTION ON NATURAL NUMBERS

3.1.1 Introduction

In its most common form, mathematical induction or, as it is also called, *complete induction* is used in the domain of natural numbers. Specifically, in induction, it is assumed that a certain property holds for the lowest natural number, which we take here as zero, and that if this property is true for n then it is also true for $n + 1$. If these two conditions are met, then the property in question is true for all natural numbers. To gain a better understanding of induction, it is necessary to learn more about natural numbers. For this reason, we will define addition and multiplication in a recursive way. This allows the derivation of certain important properties of these operations. Later in this section, certain variants of induction will be introduced.

3.1.2 Natural Numbers

The natural numbers, in the sense used here, are the numbers 0, 1, 2, ... This is the predominant definition within computer science. However, outside computer science, 0 is not normally considered as a natural number. In daily life, we use the symbols from 0 to 9 to represent the first 10 natural numbers. All numbers above 9 are constructed according to certain formation rules. In number theory, there is only one special symbol, the symbol 0. All other numbers are written by means of the *successor mechanism*. The *successor* of the number n, written as $s(n)$, is merely the number following n in the sequence of natural numbers. For example, the number 1 is written as $s(0)$, the number 2 as $s(s(0))$, the number 3 as $s(s(s(0)))$, and so on. Of course, $s(n) = n + 1$, but we prefer not to use the addition symbol at this time. This allows us to keep the theory general and to define successor functions outside the area of natural numbers. For instance, the days of the week have successors, as do elements in linked lists.

There are a number of axioms that describe how to work with natural numbers. These axioms were introduced by Giuseppe Peano in 1889, and they are therefore called *Peano axioms*. We will now introduce these axioms one by one and provide some motivation for each of them. The first two axioms provide a definition of a natural number:

1. 0 is a natural number.
2. If n is a natural number, then so is $s(n)$.

Note that $s(n)$ is defined to be a natural number if n is a natural number. This makes the definition of the natural numbers recursive. The third and fourth Peano axioms indicate that all natural numbers are distinct. This means that the $s(n)$ must never be 0 and that $s(n) \neq s(m)$ unless $n = m$. This gives rise to axioms 3 and 4:

3. For all n, $s(n) \neq 0$.
4. If $s(n) = s(m)$, then $n = m$.

Axiom 4 is the counterpositive of $(n \neq m) \Rightarrow (s(n) \neq s(m))$, which says that distinct numbers have distinct successors.

There are four axioms that deal with addition and multiplication. We define 0 to be the right identity of addition; that is

$$\forall m(m + 0 = m) \tag{3.1}$$

The successor function is used for the remaining n. This gives rise to the following recursive definition:

$$\forall m \forall n \, (m + s(n) = s(m + n)) \tag{3.2}$$

To see that this definition is compatible with the standard definition for addition, simply replace $s(n)$ in (3.2) by $n + 1$, which yields

$$\forall m \forall n \, (m + (n + 1) = (m + n) + 1)$$

More importantly, (3.2) can be used to perform the addition of any two natural numbers. Simply apply (3.2) until n is reduced to 0, at which time (3.1) becomes applicable.

Example 3.1 Find $3 + 2 = s(s(s(0))) + s(s(0))$.

  Solution

$$
\begin{aligned}
s(s(s(0))) + s(s(0)) &= s\left(s(s(s(0))) + s(0)\right) & \text{By (3.2)} \\
&= s\left(s\left(s(s(s(0))) + 0\right)\right) & \text{By (3.2)} \\
&= s(s(s(s(s(0))))) & \text{By (3.1)} \qquad \blacksquare
\end{aligned}
$$

Multiplication can be defined in a very similar fashion. One has

$$\forall n(n \times 0 = 0) \tag{3.3}$$

$$\forall m \forall n(m \times s(n) = m \times n + m) \tag{3.4}$$

The reader may verify that multiplication satisfies this definition. Moreover, it may be shown that this defines multiplication between arbitrary natural numbers. Equations (3.1) to (3.4) constitute the Peano axioms 5 to 8. Definitions of this type are used frequently in logical and functional programming languages. They also play an important role in the theory of computability (see Section 3.5).

The last Peano axiom indicates that for each proposition P the following expression is valid.

$$P(0) \wedge \forall n(P(n) \Rightarrow P(s(n))) \Rightarrow \forall n P(s(n)) \tag{3.5}$$

This last axiom formulates, in a rather compact form, the principle of mathematical induction. This principle is very important, and the next section gives many examples for its use.

3.1.3 Mathematical Induction

The previous section showed how to use the successor mechanism to define predicates and functions on natural numbers. If P is any property regarding a natural number, then it is often straightforward to make a statement about $s(n)$, given that a similar statement holds for n. Hence, if $P(n)$ is any predicate regarding the natural number n, it should be relatively easy to prove or disprove the proposition $P(n) \Rightarrow P(s(n))$. Suppose now that $P(n) \Rightarrow P(s(n))$ has been proved for all n. In addition to this, suppose that $P(0)$ is true. Then the following derivation is possible. In this derivation, we use the numbers 1, 2, 3, ... instead of $s(0)$, $s(s(0))$, and so on.

  1. $\forall n(P(n) \Rightarrow P(s(n)))$ Premise

  2. $P(0)$ Premise

  3. $P(0) \Rightarrow P(1)$ Instantiate 1 with $n := 0$

4. $P(1)$ 2, 3, modus ponens

5. $P(1) \Rightarrow P(2)$ Instantiate 1 with $n := 1$

6. $P(2)$ 4, 5, modus ponens

7. $P(2) \Rightarrow P(3)$ Instantiate 1 with $n := 2$

8. $P(3)$ 6, 7, modus ponens

    $\vdots$

It should be noted that this proof can be continued forever. In this way, one first derives $P(1)$ from $P(0)$ and $P(0) \Rightarrow P(1)$, then one derives $P(2)$ from $P(1)$ and $P(1) \Rightarrow P(2)$, and so on. Since this can be continued, the proof essentially shows that, for all n, $P(n)$ is true. This leads to the principle of mathematical induction. We now formulate this rule as a rule of inference. This rule of inference can also be derived from (3.5).

Induction: Assume that the universe of discourse is given by the set of natural numbers, and let $P(n)$ be any property of natural numbers. In this case, one can make use of the following rule of inference:

$P(0)$

$$\frac{\forall n(P(n) \Rightarrow P(s(n)))}{\forall n P(n)}$$

$P(0)$ is called the *inductive base* or the *basis of induction*. The expression $\forall n(P(n) \Rightarrow P(s(n)))$ is called the *inductive step*. Hence, the inductive base and the inductive step together imply that $P(n)$ is true for all n. Note that both the basis of induction and the inductive step are needed to arrive at the conclusion that $P(n)$ is true for all n.

Example 3.2   To illustrate this further, consider the following example. Let's assume that if a tulip is red in year n it will be red in the next year, which is year $s(n)$. This means that the inductive step $\forall n(P(n) \Rightarrow P(s(n)))$ holds. If, in addition to this, $P(0)$ is true, that is, if the tulip is red in year zero, it is possible to conclude that the tulip is red in all years from zero onward. If either $P(0)$ is not true, that is, if the tulip is not red in year zero, or if $\forall n(P(n) \Rightarrow P(s(n)))$ is not true, that is, if the flower can change its color, then one is not allowed to conclude that the tulip will be red in all future years.

Since $s(n) = n + 1$, the inductive step can also be written as

$$\forall n(P(n) \Rightarrow P(n+1))$$

The term *inductive step* is chosen appropriately, because it indicates that, once $P(n)$ is true, one can make the step to the next number $s(n) = n + 1$. It is, in this sense, the step from n to $n + 1$.

In Section 1.6.6, the following approach was used to prove that $A \Rightarrow B$. First A is assumed, and, using A as an assumption, B is derived. If this is possible, we conclude that $A \Rightarrow B$. Note that as soon as $A \Rightarrow B$ has been derived A must be discharged; that is, A is no longer assumed to be true. Moreover, as shown in Section 2.3.4, any free variables in A are fixed while deriving B. On the other hand, once A is discharged, A is no longer a premise, and universal generalization over any true variable in A is possible. This idea is frequently used to perform the inductive step. One assumes $P(n)$ and derives $P(n+1)$ in order to conclude that $P(n) \Rightarrow P(n+1)$. Since $P(n)$ is now discharged, n is no longer fixed (unless, of course, it appears in some other premise), and one is allowed to conclude that $\forall n(P(n) \Rightarrow P(n+1))$. If $P(n)$ is used as an assumption in this fashion, then $P(n)$ is called the *inductive hypothesis*.

In summary, one can prove $\forall n P(n)$ by conducting the following steps:

Inductive base   Prove $P(0)$.

Inductive hypothesis   Assume $P(n)$.

Proof under hypothesis   Fix n, and derive $P(n+1)$.

Discharge hypothesis   Conclude that $P(n) \Rightarrow P(n+1)$, and discharge $P(n)$. Hence, $P(n)$ is no longer necessarily true.

Generalize   By discharging $P(n)$, n becomes a true variable, and one can use universal generalization to conclude from $P(n) \Rightarrow P(n+1)$ that

$$\forall n(P(n) \Rightarrow P(n+1))$$

Conclusion   From the base of induction and the inductive step, one concludes that

$$\forall n P(n)$$

The three steps "proof under hypothesis," "discharge hypothesis," and "generalize" together form the inductive step.

Example 3.3   Let H_n be zero for $n = 0$, and $H_{n+1} = 1 + 2H_n$ otherwise. Prove that $H_n = 2^n - 1$.

Solution   For each n, $H_n = 2^n - 1$ is either true or false. Consequently, $H_n = 2^n - 1$ is a property of n, and we can identify this expression with $P(n)$ in mathematical induction. We now have the following:

Inductive base   First, $P(0)$ must be established; that is, for $n = 0$, $H_n = 2^n - 1$. Since $2^0 - 1$ is 0, this is obviously true.

Inductive hypothesis   Assume $P(n)$; that is, assume that $H_n = 2^n - 1$ is true. This assumption fixes n until $P(n)$ is discharged.

Proof under hypothesis   We must now derive $P(n+1)$, that is, show that $H_{n+1} = 2^{n+1} - 1$. The use of the inductive hypothesis $H_n = 2^n - 1$ is allowed in the construction of this derivation. This yields

$$
\begin{aligned}
H_{n+1} &= 1 + 2H_n &&\text{By definition} \\
&= 1 + 2(2^n - 1) &&\text{Inductive hypothesis: } H_n = 2^n - 1 \\
&= 2^{n+1} - 1 &&\text{Simple algebra}
\end{aligned}
$$

Discharge hypothesis   If $P(n)$ is assumed, it follows that $P(n+1)$ is also true. Hence,

$$(H_n = 2^n - 1) \Rightarrow (H_{n+1} = 2^{n+1} - 1)$$

The inductive hypothesis $H_n = 2^n - 1$ is discharged now; that is, it is no longer assumed to be necessarily true.

Generalization   Since n does not appear in any hypothesis, a generalization over n yields

$$\forall n((H_n = 2^n - 1) \Rightarrow (H_{n+1} = 2^{n+1} - 1))$$

Conclusion   According to the principle of induction, the base and the inductive step together prove that

$$\forall n(H_n = 2^n - 1) \qquad\qquad \blacksquare$$

Since the discharge of the inductive hypothesis and the generalization step are basically always the same, they are often omitted. This means, in effect, that the inductive step has been established as soon as the proof that $P(n+1)$ is true has been completed, using $P(n)$ as an assumption in the proof. It is for this reason that we will omit the steps "discharge hypothesis" and "generalize." However, to indicate that these two steps are tacitly implied, from now on we will call the step "proof under hypothesis" the "inductive step."

Example 3.4   Show that, for all n, $2(n+2) \le (n+2)^2$.

Solution   In this example, $P(n)$ stands for $2(n+2) \le (n+2)^2$. We have the following:

Inductive base   First, we must establish $P(0)$; that is, we must establish that, for $n = 0$, $2(n+2) \le (n+2)^2$. Since for $n = 0$ both sides of the inequality are 4, $P(0)$ is true.

Inductive hypothesis $P(n) : 2(n+2) \le (n+2)^2$ is assumed.

Inductive step   One must now derive $P(n+1)$, that is, show that $2((n+1)+2) \le ((n+1)+2)^2$. This is done as follows:

$$
\begin{aligned}
2((n+1)+2) &= 2((n+2)+1) &&\text{Algebra} \\
&= 2(n+2)+2 &&\text{Algebra} \\
&\le (n+2)^2 + 2 &&\text{Inductive hypothesis} \\
&\le n^2 + 4n + 6 &&\text{Algebra} \\
&\le n^2 + 6n + 9 &&\text{Add the term } 2n+3 \\
&= ((n+1)+2)^2 &&\text{Algebra}
\end{aligned}
$$

Conclusion   Since the inductive base and the inductive step are established, one is allowed to conclude that

$$\forall n(2(n+2) \le (n+2)^2) \qquad\qquad \blacksquare$$

The next proof is somewhat more informal, but closer to a proof by induction as it may appear in a mathematics textbook.

Example 3.5   Show that $n^3 + 2n$ is divisible by 3.

Solution   Let $P(n) : n^3 + 2n$ be divisible by 3. Now $P(0) : 0$ is divisible by 3, so $P(0)$ is true. The inductive hypothesis is that $n^3 + 2n$ is divisible by 3. The inductive step is now as follows:

$$(n+1)^3 + 2(n+1) = n^3 + 3n^2 + 3n + 1 + 2n + 2$$
$$= n^3 + 3n^2 + 5n + 3$$
$$= (n^3 + 2n) + 3(n^2 + n + 1)$$

This means that $(n+1)^3 + 2(n+1)$ can be written as a sum of two terms. The first term is divisible by 3 if the inductive hypothesis holds, and the second term is a multiple of 3 and therefore divisible by 3 as well. Hence, if $P(n)$ is true, so is $P(n+1)$, which completes the inductive step. Since $P(0)$ is true, one concludes that $P(n)$ is true for all n. ∎

3.1.4 Induction for Proving Properties of Addition

From definitions (3.1) and (3.2), all properties of addition can be derived, including the fact that 0 is the left identity of addition and that addition is commutative and associative. These results are, of course, well known to the reader, and he or she may feel that no proofs are needed. However, the proofs are excellent examples for practicing induction. In the proofs, $s(n)$ is used rather than $n + 1$.

We first prove that 0 is the left identity of addition; that is, we prove that

$$\forall n(0 + n = n) \tag{3.6}$$

We identify $P(n)$ with the property of n that $0 + n = n$.

Inductive base $P(0)$, which is $0 + 0 = 0$, is true by (3.1).

Inductive hypothesis   Assume that $P(n)$ is true; that is, assume that $0 + n = n$.

Inductive step   It must be shown that $P(s(n))$ is true; that is, show that, given the inductive hypothesis, $0 + s(n) = s(n)$. This is done as follows:

$$0 + s(n) = s(0 + n) \qquad \text{(3.2) with } m := 0, n := n$$
$$= s(n) \qquad \text{Inductive hypothesis: } 0 + n = n$$

Conclusion   The inductive base and the inductive step together imply (3.6).

Next, we show that addition is commutative. In other words, we prove that, for all m and n, $m + n = n + m$. There are now two variables, m and n, but induction can only be done on one variable. Fortunately, it does not matter in this particular case which variable is picked. We arbitrarily choose to do induction on m. The steps involved are as follows:

Inductive base $P(0)$ is true, because $n + 0 = 0 + n$. This follows from (3.1) and (3.6).

Inductive hypothesis   Assume that $P(m)$ is true; that is, assume that for m fixed $m + n = n + m$.

Inductive step   One must now prove $P(s(m))$, which means that $s(m) + n = n + s(m)$ must be shown. This proof is lengthy, and it will be done later.

Conclusion   The inductive base, together with the inductive step imply that

$$\forall m(m + n = n + m)$$

Since n does not occur in any premise, one can generalize over n, and one concludes that

$$\forall m \forall n(m + n = n + m)$$

We still have omitted the hard part of the proof, the inductive step. In this step, one must show that if $m + n = n + m$ then $s(m) + n = n + s(m)$. Observe that $n + s(m)$ unifies with $n + s(m)$ appearing in (3.2), which allows us to write

$$
\begin{aligned}
n + s(m) &= s(n + m) && \textbf{By (3.2)} \\
&= s(m + n) && \text{By inductive hypothesis}
\end{aligned}
$$

At this stage, we are stuck because no law seems to be applicable. To overcome this difficulty, we make a leap of faith, conjecturing that $s(m + n) = s(m) + n$, which allows us to continue.

$$s(m + n) = s(m) + n$$

This implies that $n + s(m) = s(m) + n$, as required. At this point, however, the conjecture is just wishful thinking, and it needs to be proved before it can be used in the proof. Hence, we must show that, for all m and n,

$$s(m + n) = s(m) + n$$

This equation will also be proved by using mathematical induction. However, this time the induction is on n, because this makes it simpler to prove the inductive base, as shown next.

Inductive base $P(0)$ is $s(m + 0) = s(m) + 0$, which is true, as one can show by applying (3.1) twice.

$$s(m + 0) = s(m) = s(m) + 0$$

Inductive hypothesis   Assume that $s(m + n) = s(m) + n$.

Inductive step   Prove that $s(m + s(n)) = s(m) + s(n)$. This is done as follows:

$$
\begin{aligned}
s(m + s(n)) &= s(s(m + n)) && \text{(3.2) with } m := m, n := n \\
&= s(s(m) + n) && \text{Inductive hypothesis} \\
&= s(m) + s(n) && \text{(3.2) with } m := s(m), n := n
\end{aligned}
$$

Conclusion   The inductive step and the inductive base imply that

$$\forall n(s(m + n) = s(m) + n)$$

To show that this is true for all m, one only needs to generalize over m.

This proves the conjecture used in the proof, completing thus the proof that addition is commutative. Other proofs, including the fact that addition is associative, can be done in a similar fashion.

3.1.5 Changing the Inductive Base

So far, the basis for induction has always been 0, but this is not really a requirement. Indeed, one can start induction with any element n_0 by choosing $P(n_0)$ as the base of induction. This leads to the following rule of inference:

Induction from n_0:

$$P(n_0)$$
$$\frac{\forall n((n \geq n_0) \Rightarrow (P(n) \Rightarrow P(s(n))))}{\forall n((n \geq n_0) \Rightarrow P(n))}$$

The justification of this rule is the same as before. If $P(n_0)$ is true and if, for all $n \geq n_0$, $P(n)$ implies that $P(n + 1)$, then P must also be true for all values following n_0. Specifically, if $P(n_0)$ holds and if $P(n_0) \Rightarrow P(n_0 + 1)$, then $P(n_0 + 1)$ must be true. $P(n_0+1)$, in conjunction with $P(n_0+1) \Rightarrow P(n_0+2)$, implies that $P(n_0+2)$. Continuing in this fashion leads one to conclude that, if the inductive step is established and if $P(n_0)$ is true, then $P(n)$ must be true for all $n \geq n_0$. This rule of inference leads to the following setup:

Inductive base   Establish $P(n_0)$.

Inductive hypothesis   Assume $P(n)$ for $n \geq n_0$.

Inductive step   Prove that $P(n + 1)$ for $n \geq n_0$.

Conclusion   Use the inductive base and the inductive step to conclude that

$$\forall n((n \geq n_0) \Rightarrow P(n))$$

Example 3.6   Show that $2^n < n!$ for $n \geq 4$.

Solution   Let $P(n) : 2^n < n!$. $P(0)$, $P(1)$, $P(2)$, and $P(3)$ are not true, and we do not need them to be true. Now $P(4) : 2^4 = 16 < 4! = 24$, so $P(4)$ holds. We now have the following:

Inductive base $P(4)$ is true; that is, $2^4 < 4!$.

Inductive hypothesis   Assume that $P(n) = 2^n < n!$.

Inductive step   Prove that $P(n + 1) : 2^{n+1} < (n + 1)!$. To do this, multiply both sides of the inductive hypothesis by 2 to get, for $n \geq 4$,

$$2 \times 2^n = 2^{n+1} < 2(n!) < (n + 1) \times (n!) = (n + 1)!$$

Conclusion   The inductive step, together with the fact that $P(4)$ is true, allows one to conclude that, for all $n \geq 4$, $2^n < n!$.   ∎

The starting value n_0 need not even be positive. This allows one to use induction in the domain of integers, which contains negative numbers.

3.1.6 Strong Induction

If one has $P(0)$, and $\forall n(P(n) \Rightarrow P(n+1))$, then one can successively prove $P(1)$, $P(2)$, $P(3)$, and so on. This means that, once $P(n)$ is proved, one can use $P(0)$, $P(1)$, ..., $P(n)$ as the inductive hypotheses to prove that $P(n+1)$. This leads to the following method, which is called *strong induction*:

Inductive base   Establish $P(0)$.

Inductive hypothesis   Assume $P(0)$, $P(1)$, ..., $P(n)$.

Inductive step   Prove $P(n+1)$.

Conclusion   The basis of induction and the inductive step allow one to conclude that

$$\forall n P(n)$$

Note that the method remains applicable if the base of induction is $P(n_0)$ rather than $P(0)$, in the manner indicated in Section 3.1.5. Strong induction can be derived from normal induction.

We now present some examples. The first example deals with the zero of an operator. Generally, d is a zero of the operator ∘ if $\forall x(x \circ d = d)$ and $\forall x(d \circ x = d)$ are both satisfied for all x (see Section 2.5.7). Examples of zeros are the number 0 in multiplication, T in the case of disjunctions, and F in the case of conjunctions. In the case of multiplication, $x \times 0 = 0$ and $0 \times x = 0$ for all x, and in the case of disjunctions, $P \vee \text{T} \equiv \text{T}$ and $\text{T} \vee P \equiv \text{T}$ for all P. The demonstration that F is a zero for the conjunctions is similar.

Example 3.7   Consider expressions using ∘, and no other operator. Define d as the zero of ∘, that is, $x \circ d = d \circ x = d$ for all x. Show that any such expression containing one or more instances of d must be equal to d.

Solution   Let $P(n)$ be the proposition that an expression with n occurrences of ∘ is d, provided that the expression contains a single d.

Inductive base $P(0)$ is true because, among the expressions without operators, only d itself contains d.

Inductive hypothesis   Assume $P(m)$, $m \leq n$; that is, assume that any expression with n or fewer operators that contains a d yields d.

Inductive step   If x is an expression with $n + 1$ operators and if x is of the form $x_1 \circ x_2$, then x contains d only if either x_1 or x_2 contains d. Both x_1 and x_2 have n or fewer operators, and according to the inductive hypothesis, all expressions x_i with n or fewer operators yield d if they contain a single d. It follows that either x_1 or x_2 yields d. If x_1 yields d, so does $x_1 \circ x_2 = d \circ x_2$, and the same is true if $x_2 = d$. Either way, $x = d$, which completes the inductive step.

Conclusion   The basis of induction and the inductive step together imply that $P(n)$ is true for all n. This means that for any n every expression with n operators containing d yields d.
■

Example 3.8   Prove that every nonprime natural number greater than 1 can be written as a product of primes.

Solution   Let $P(n)$ be the statement that n is either prime or can be written as the product of primes. The basis for induction is $P(2)$, which states that 2 is either a prime or can be written as a product of primes. This is correct since 2 is a prime number. The inductive hypothesis is that any number less than or equal to n is either prime or can be written as a product of primes. Based on this inductive hypothesis, we prove that $n + 1$ is either prime or can be written as the product of primes. The only case we have to consider is when $n + 1$ is not prime, because $P(n + 1)$ is certainly true otherwise. If $n + 1$ is not prime, then $n + 1 = n_1 \cdot n_2$ for some numbers $n_1 \leq n$ and $n_2 \leq n$. If n_1 and n_2 are both either primes or products of primes, so is $n_1 \cdot n_2$. This completes the inductive step. Since the basis for induction holds, this proves that $P(n)$ is true for all n.
■

Problems 3.1

1. Let $n = s(s(2))$. Find $s(s(s(n)))$. Express the result as an Arabic number.
2. Define $x \leq y$ for all natural numbers by using only the successor mechanism.
3. Instead of the natural numbers $0, 1, 2, \ldots$, consider the following sequences:
    (a) $0, 1, 2, 0, 1, 2, \ldots$
    (b) $3, 4, 5, \ldots$
    (c) $0, 1, 2, 3, 3, 3, \ldots$
    For each sequence, indicate which of the first 4 Peano axioms holds. Assume $s(n)$ is the next element in the sequence.
4. Construct a model which violates the third Peano axiom, but yet satisfies (3.1) and (3.2).
5. Given the premises $s(n) > n$ and $n > m \Rightarrow s(n) > m$, prove that $\forall n(s(n) > 0)$.
6. Prove that $(n + 2)!$ is even.
7. Prove that addition is associative.
8. Prove that 0 is the left zero for multiplication. In other words, prove that $0 \times n = 0$.
9. Prove that $n^2 \geq 2n + 3$ for $n \geq 3$.
10. Prove that $2^n \geq n^2$ for $n \geq 4$.
11. Show that in propositional calculus every logical expression not containing any negations has an odd number of symbols. Here, every logical variable counts as one symbol, as does every logical constant and every logical connective.

3.2 SUMS AND RELATED CONSTRUCTS

3.2.1 Introduction

There is a special notation for writing sums involving many terms, the $\sum$ or *sigma* notation. This notation will now be discussed at some length. A recursive definition of the sigma notation will be given, and this definition will be applied to prove a number of important results involving sums by mathematical induction. Notations similar to the $\sum$ notation will be introduced to express products, conjunctions, and disjunctions. All these notations make use of *indexes*, which correspond, in some sense, to the bound variables of quantifiers.

3.2.2 Recursive Definitions of Sums and Products

The sigma notation is a compact notation to express sums involving many terms. In particular, if a_m, a_{m+1}, $\ldots, a_n$ are terms, their sum is written as $\sum_{i=m}^{n} a_i$. Hence, one has

$$\sum_{i=m}^{n} a_i = a_m + a_{m+1} + \cdots + a_n \tag{3.7}$$

Here, i is the *index* of the summation, m is its *lower bound*, and n is its *upper bound*. The lower bound together with the upper bound constitutes the *range* of the index. If the upper bound is less than the lower bound, the sum in question contains no terms; it is empty. Empty sums are assigned the value zero. For instance, $\sum_{i=4}^{3} a_i = 0$. The terms a_i constitute a function with the argument i.

Example 3.9   Express the sum of the integers from 3 to 20 by using the $\sum$ notation.

Solution   By replacing the a_i in (3.7) by i, one obtains

$$\sum_{i=3}^{20} i = 3 + 4 + \cdots + 20$$

This solves the problem as required.   ■

Example 3.10   Find the value of $\sum_{4}^{9} 1$.

Solution   Since the sum in question has the 6 terms $a_4, a_5, \ldots, a_9$, all of which are equal to 1, the sum in question is $1 + 1 + 1 + 1 + 1 + 1 = 6$.   ■

The following recursive definition makes it easier to apply induction.

$$\sum_{i=m}^{n} a_i = 0, \qquad\qquad \text{if } m > n$$

$$\sum_{i=m}^{n+1} a_i = \sum_{i=m}^{n} a_i + a_{n+1}, \qquad \text{if } m \leq n+1 \tag{3.8}$$

Example 3.11   Consider the expression $\sum_{i=1}^{3} 2^i$. One has, according to (3.8),

$$\sum_{i=1}^{3} 2^i = \sum_{i=1}^{2} 2^i + 2^3$$

$$= \sum_{i=1}^{1} 2^i + 2^2 + 2^3$$

$$= \sum_{i=1}^{0} 2^i + 2^1 + 2^2 + 2^3$$

$$= 0 + 2^1 + 2^2 + 2^3 = 14$$

Also note that (3.8) implies that

$$\sum_{i=n}^{n} a_i = \sum_{i=n}^{n-1} a_i + a_n = 0 + a_n$$

The definition given in (3.8) can obviously be extended by replacing the $+$ by other operators, such as $\times$, $\wedge$, and $\vee$. In the case when the upper limit is below the lower limit, one replaces the 0 by the identity of the operator. To express a product in this fashion, one uses the symbol $\prod$. The product of the factors $a_m, a_{m+1}, a_{m+2}, \ldots, a_n$ is thus written as

$$\prod_{i=m}^{n} a_i = a_m a_{m+1} \ldots a_n$$

If $m > n$, the product is assigned 1, which is the identity of multiplication. The recursive definition of $\prod$ is therefore

$$\prod_{i=m}^{n} a_i = 1, \qquad\qquad \text{if } m > n$$

$$\prod_{i=m}^{n+1} a_i = \left(\prod_{i=m}^{n} a_i \right) \times a_{n+1}, \qquad \text{if } m \leq n + 1$$

For other operators, one uses a large version of the operator symbol to play the role of $\sum$. For instance, the disjunction of P_i, $i = m, m+1, \ldots, n$, is written as

$$\bigvee_{i=m}^{n} P_i = P_m \vee P_{m+1} \vee \cdots \vee P_n$$

If the upper bound is below the lower bound, the disjunction is assigned the identity of $\vee$, which is F. The conjunction of P_i, $i = m, m+1, \ldots, n$, is similarly expressed as

$$\bigwedge_{i=m}^{n} P_i = P_m \wedge P_{m+1} \wedge \cdots \wedge P_n$$

If the upper bound is below the lower bound, the conjunction is assigned the identity of $\wedge$, which is T. More generally, if $\otimes$ is any operator with identity e, then one has the following recursive definition of $\otimes$:

$$\bigotimes_{i=m}^{n} a_i = e, \qquad\qquad \text{if } m > n$$

$$\bigotimes_{i=m}^{n+1} a_i = \bigotimes_{i=m}^{n} a_i \otimes a_{n+1}, \qquad \text{if } m \leq n + 1$$

The notation introduced is closely related to quantifiers. Specifically, if a universe consists of $a_1, a_2, \ldots, a_n$, then one has

$$\exists x P(x) \equiv \bigvee_{i=1}^{n} P(a_i)$$

$$\forall x P(x) \equiv \bigwedge_{i=1}^{n} P(a_i)$$

If the domain is the set of natural numbers, this yields

$$\exists x P(x) \equiv \bigvee_{x=0}^{\infty} P(x)$$

$$\forall x P(x) \equiv \bigwedge_{x=0}^{\infty} P(x)$$

This shows that the index of a sum is closely related to the bound variable x. Indeed, most of the rules formulated for bound variables also apply to indexes, whether the indexes are in sums, conjunctions, or disjunctions. All the constructs deal with terms a_i, which can be any functions of the index i. These functions form the *scope* of the summation, product, and so on. The index i is strictly local to the scope. In particular, if i appears outside the scope, it must be considered as a distinct variable.

3.2.3 Identities Involving Sums

In this section, we discuss a number of important relations involving sums. These are

$$\sum_{i=m}^{n} (a_i + b_i) = \sum_{i=m}^{n} a_i + \sum_{i=m}^{n} b_i \tag{3.9}$$

$$\sum_{i=m}^{n} (a_i + b) = \left(\sum_{i=m}^{n} a_i \right) + (n - m + 1)b, \quad n \geq m - 1 \tag{3.10}$$

$$\sum_{i=m}^{n} (a_i b) = b \sum_{i=m}^{n} a_i \tag{3.11}$$

$$\sum_{i=1}^{n} a_i = \sum_{i=1+k}^{n+k} a_{i-k} \tag{3.12}$$

Equations (3.9)–(3.12) can easily be proved by induction. We prove (3.9) and (3.12), leaving the proof of (3.10) and (3.11) to the reader. For the proof of (3.9), $P(n)$ is the proposition that (3.9) is true for n. One has the following:

Base for induction   For $n = m - 1$, (3.9) is true. $\sum_{i=m}^{m-1} a_i$ is zero, and the same is true for $\sum_{i=m}^{m-1} b_i$. Hence, for $n = m - 1$, (3.9) instantiates to $0 = 0 + 0$, which is true.

Inductive hypothesis   The inductive hypothesis is given by (3.9), except that n is fixed.

Inductive step   One has

$$\sum_{i=m}^{n+1} (a_i + b_i) = \sum_{i=m}^{n} (a_i + b_i) + (a_{n+1} + b_{n+1}) \qquad \text{By (3.8)}$$

$$= \sum_{i=m}^{n} a_i + \sum_{i=m}^{n} b_i + a_{n+1} + b_{n+1} \qquad \text{Inductive hypothesis}$$

$$= \sum_{i=m}^{n} a_i + a_{n+1} + \sum_{i=m}^{n} b_i + b_{n+1} \qquad \text{Addition commutes}$$

$$= \sum_{i=m}^{n+1} a_i + \sum_{i=m}^{n+1} b_i \qquad \text{By (3.8)}$$

Conclusion   The inductive base and the inductive step imply that (3.9) is valid for all n.

Equation (3.12) can be proved as follows. For $n = 0$, both sides of (3.12) are equal to zero, which settles the base case. We now prove that, for some fixed value of n greater than or equal to 0, (3.12) is true, with n replaced by $n + 1$. One has

$$\sum_{i=1}^{n+1} a_i = \sum_{i=1}^{n} a_i + a_{n+1} \qquad \text{By (3.8)}$$

$$= \sum_{i=1+k}^{n+k} a_{i-k} + a_{n+1} \qquad \text{Inductive hypothesis}$$

$$= \sum_{i=1+k}^{n+k} a_{i-k} + a_{n+k+1-k} \qquad n + 1 = n + k + 1 - k$$

$$= \sum_{i=1+k}^{n+k+1} a_{i-k} \qquad \text{By (3.8)}$$

This completes the inductive step, and since the basis is true, (3.12) must be true for all n. The same result can be obtained by expanding both sides of (3.12). This yields

$$a_1 + a_2 + \cdots + a_n = a_{(1+k)-k} + a_{(2+k)-k} + \cdots + a_{(n+k)-k}$$

It is now immediately evident why both sides of (3.12) are equal.

Indexes are in a sense like bound variables. They, too, are merely placeholders, which means that the name of the index can be changed as long as the new name does not clash with a variable already used. In a sum, for instance, one can use either i, j, or k as the index. Hence,

$$\sum_{i=m}^{n} a_i = \sum_{j=m}^{n} a_j = \sum_{k=m}^{n} a_k$$

As in the case of quantifiers, it is often convenient to standardize variables apart. The following example illustrates this point. At the same time, it demonstrates the use of (3.11).

Example 3.12   Prove that

$$\left(\sum_{i=1}^{n} a_i \right)^2 = \sum_{i=1}^{n} \left(\sum_{j=1}^{n} a_i a_j \right)$$

After the formal proof, verify your result with $n = 3$, $a_1 = 3$, $a_2 = 2$, and $a_3 = 5$.

Solution   One has

$$\left(\sum_{i=1}^{n} a_i \right)^2 = \left(\sum_{i=1}^{n} a_i \right) \left(\sum_{i=1}^{n} a_i \right) \qquad \text{Expand square}$$

$$= \left(\sum_{i=1}^{n} a_i \right) \left(\sum_{j=1}^{n} a_j \right) \qquad \text{Standardize variables apart}$$

$$= \sum_{i=1}^{n} \left(a_i \sum_{j=1}^{n} a_j \right) \qquad \text{(3.11) with } b := \sum a_j$$

$$= \sum_{i=1}^{n} \left(\sum_{j=1}^{n} a_i a_j \right) \qquad \text{(3.11) with } b := a_i$$

To check this result, note that $\sum_{i=1}^{3} a_i = 3 + 2 + 5$, which is 10. $\left(\sum_{i=1}^{3} a_i \right)^2$ is therefore $10^2 = 100$. On the other hand,

$$\sum_{j=1}^{3} a_1 a_j = 3 \times 3 + 3 \times 2 + 3 \times 5 = 30$$

$$\sum_{j=1}^{3} a_2 a_j = 2 \times 3 + 2 \times 2 + 2 \times 5 = 20$$

$$\sum_{j=1}^{3} a_3 a_j = 5 \times 3 + 5 \times 2 + 5 \times 5 = 50$$

Hence,

$$\sum_{i=1}^{3} \sum_{j=1}^{3} a_i a_j = 100 \qquad\qquad \blacksquare$$

The derivation given in Example 3.12 resembles some of the derivations given in Section 2.4, which dealt with equivalences involving quantifiers. This shows again the close relation between indexes and bound variables. A final note concerns instantiations of sums. Instantiations do not affect the index of a sum, because this index is local to the sum. The same rule applies, of course, to quantifiers, as was shown.

We now prove a number of important formulas involving specific functions of i in place of the a_i.

Example 3.13   Use mathematical induction to show that for $n \geq 0$

$$\sum_{i=0}^{n} i = \frac{n(n+1)}{2} \tag{3.13}$$

Solution   Let $P(n)$ be the property of n that (3.13) holds. We now prove by mathematical induction that $P(n)$ is true for all n.

Inductive base $P(0)$ is true because (3.13) holds for $n = 0$. In this case, $\sum_{i=0}^{0} i = 0$, and $0 \cdot \frac{1}{2} = 0$.

Inductive hypothesis   The inductive hypothesis is given by (3.13) with n fixed.

Inductive step   One has

$$\sum_{i=0}^{n+1} i = \sum_{i=0}^{n} i + n + 1 \qquad \text{By (3.8)}$$

$$= \frac{n(n+1)}{2} + n + 1 \qquad \text{Inductive hypothesis}$$

$$= \frac{n(n+1) + 2(n+1)}{2} \qquad \text{Add fractions}$$

$$= \frac{(n+1)(n+2)}{2} \qquad \text{Right side of (3.13) with } n := n+1$$

Conclusion   The inductive base and the inductive step imply that (3.13) is valid for all n.

■

Example 3.14   Prove that

$$\sum_{i=0}^{n} a^i = \frac{a^{n+1} - 1}{a - 1}, \qquad a \neq 1, \quad n \geq 0 \tag{3.14}$$

Solution $P(n)$ is given by (3.14). One has the following:

Inductive base   For $n = 0$, both sides of (3.14) evaluate to 1, which proves the basis of induction.

Inductive hypothesis   Assume that (3.14) is true for n fixed.

Inductive step   One has

$$\sum_{i=0}^{n+1} a^i = \sum_{i=0}^{n} a^i + a^{n+1} \qquad\qquad \text{By (3.8)}$$

$$= \frac{a^{n+1} - 1}{a - 1} + a^{n+1} \qquad\qquad \text{Inductive hypothesis}$$

$$= \frac{a^{n+1} - 1 + a^{n+2} - a^{n+1}}{a - 1} \qquad\qquad \text{Add fractions}$$

$$= \frac{a^{n+2} - 1}{a - 1} \qquad\qquad \text{Right side of (3.14) with } n := n + 1$$

Conclusion   The inductive base and the inductive step together imply that (3.14) is valid for all n.   ■

3.2.4 Double Sums and Matrices

In Example 3.12, an expression with nested sums appeared. Such expressions will now be discussed. To deal with double sums, one uses terms of the form a_{ij}. If i is in the range $1, 2, \ldots, m$, and j is in the range $j = 1, 2, \ldots, n$, then one can arrange the a_{ij} in rows and columns as follows: In the first row, one writes all $a_{1j}, j = 1, \ldots, n$, in the second row, one writes all $a_{2j}, j = 1, \ldots, n$, and so on, and in the mth row, one writes all a_{mj}. This yields

$$\begin{bmatrix} a_{11} & a_{12} & a_{13} & \cdots & a_{1n} \\ a_{21} & a_{22} & a_{23} & \cdots & a_{2n} \\ \vdots & \vdots & \vdots & & \vdots \\ a_{m1} & a_{m2} & a_{m3} & \cdots & a_{mn} \end{bmatrix} \qquad (3.15)$$

A table such as this is called a *matrix*. If the matrix has the elements a_{ij}, one denotes this matrix by $[a_{ij}]$. In other words, (3.15) represents the matrix $[a_{ij}]$. Matrices can be given names, such as A, B, and so on. We will normally use capital letters to denote matrices. If $A = [a_{ij}]$, then a_{uv} is the element in row u, column v of A. In other words, the first subscript always denotes the row, and the second always denotes the column.

Example 3.15   What are the values of a_{21} and a_{13} in the following matrix, and what is the value of $\sum_{k=1}^{3} a_{2k}$?

$$A = \begin{bmatrix} 3 & 5 & 2 \\ 4 & 3 & 1 \end{bmatrix} \qquad (3.16)$$

Solution a_{21} is the element in row 2, column 1, which is 4. In a similar fashion, the element a_{13} is found to be 2. $\sum_{k=1}^{3} a_{2k}$ refers to row 2. In fact, one has

$$\sum_{k=1}^{3} a_{2k} = a_{21} + a_{22} + a_{23} = 4 + 3 + 1 = 8 \qquad\qquad ■$$

Although we frequently denote the row by i and the column by j, this is incidental. For instance, a_{ji} denotes the element one finds in row j, column i.

Given the terms a_{ij}, $i = 1, \ldots, m$ and $j = 1, 2, \ldots, n$, the form $\sum_{i=1}^{m} \sum_{j=1}^{n} a_{ij}$ is merely the sum of the following sums:

$$\sum_{i=1}^{m} \sum_{j=1}^{n} a_{ij} = \sum_{j=1}^{n} a_{1j} + \sum_{j=1}^{n} a_{2j} + \cdots + \sum_{j=1}^{n} a_{mj}$$

Since the sum $\sum_{j=1}^{n} a_{ij}$ is the total of row i, the double sum $\sum_{i=1}^{m} \sum_{j=1}^{n} a_{ij}$ is the sum of the row totals. It is intuitively clear that the sum of the row totals is the overall total and that this overall total can also be obtained by taking the sums of the column totals, which is $\sum_{j=1}^{n} \sum_{i=1}^{m} a_{ij}$. Hence, it is intuitively clear that

$$\sum_{i=1}^{m} \sum_{j=1}^{n} a_{ij} = \sum_{j=1}^{n} \sum_{i=1}^{m} a_{ij} \tag{3.17}$$

We now prove by induction that this is indeed the case. Induction is done on n. If $n = 0$, both sides of (3.17) are zero. We now assume that (3.17) is true for n fixed and prove that this equation also holds for $n + 1$.

$$\sum_{i=1}^{m} \sum_{j=1}^{n+1} a_{ij} = \sum_{i=1}^{m} \left(\sum_{j=1}^{n} a_{ij} + a_{i\,n+1} \right) \qquad \text{By (3.8)}$$

$$= \sum_{i=1}^{m} \sum_{j=1}^{n} a_{ij} + \sum_{i=1}^{m} a_{i\,n+1} \qquad \text{By (3.9)}$$

$$= \sum_{j=1}^{n} \sum_{i=1}^{m} a_{ij} + \sum_{i=1}^{m} a_{i\,n+1} \qquad \text{Inductive hypothesis}$$

$$= \sum_{j=1}^{n+1} \sum_{i=1}^{m} a_{ij} \qquad \text{By (3.8)}$$

Two matrices A and B can be *multiplied*, and this operation is denoted as $A \cdot B$. If $C = A \cdot B$, with $C = [c_{ij}]$, then c_{ij} is given as

$$c_{ij} = \sum_{k=1}^{r} a_{ik} b_{kj}, \qquad i = 1, 2, \ldots, m, \quad j = 1, 2, \ldots, n \tag{3.18}$$

Here, m is the number of rows of A, and r is the number of columns of A. The number of rows of B must match the number of columns of A; that is, B must have r rows. If this is not the case, then matrix multiplication is not defined. The number of columns of B is denoted by n. Note that, to find c_{ij} for given i and j, only row i of matrix A and column j of matrix B are involved. Specifically, one multiplies the elements of row i of matrix A with the corresponding elements of column j on matrix B and adds the products.

Example 3.16   Multiply A with B, where A is as given in (3.16) and B is

$$B = \begin{bmatrix} 1 & 2 \\ -1 & 1 \\ 2 & 3 \end{bmatrix}$$

Solution   The result of the matrix multiplication is

$$A \cdot B = \begin{bmatrix} 3 & 5 & 2 \\ 4 & 3 & 1 \end{bmatrix} \begin{bmatrix} 1 & 2 \\ -1 & 1 \\ 2 & 3 \end{bmatrix} = \begin{bmatrix} 2 & 17 \\ 3 & 14 \end{bmatrix} = C$$

In the resulting matrix, the entries c_{ij} are found by using (3.18). To find c_{12}, for example, one has to use row 1 of matrix A and column 2 of matrix B. The result is

$$c_{12} = 3 \cdot 2 + 5 \cdot 1 + 2 \cdot 3 = 17$$

The other c_{ij} can be found in a similar fashion.   ■

Problems 3.2

1. Prove (3.10) by mathematical induction.
2. Prove (3.11) by mathematical induction.
3. Use mathematical induction to prove that $\neg \bigwedge_{i=1}^{n} P_i = \bigvee_{i=1}^{n} \neg P_i$.
4. Use mathematical induction to prove that $\left(\prod_{i=1}^{n} a_i\right)^2 = \prod_{i=1}^{n} a_i^2$.
5. Prove (3.17) by induction on m.
6. Use mathematical induction to prove that $\sum_{i=0}^{n} i(i+1) = n(n+1)(n+2)/3$, if $n \geq 0$.
7. Enumerate all terms of $\sum_{i=0}^{3} \left(\sum_{j=0}^{2} a_{ij}\right)$.
8. Simplify $\sum_{i=1}^{n} a_i - \sum_{i=1}^{n} a_{i+1}$.
9. Multiply the following two matrices:

$$A = \begin{bmatrix} 1 & -1 & 1 \\ 2 & -1 & 3 \end{bmatrix}, \quad B = \begin{bmatrix} -2 & 3 \\ 1 & -1 \\ 3 & -2 \end{bmatrix}$$

3.3 PROOF BY RECURSION

3.3.1 Introduction

Proofs by recursion are inductive proofs that are not necessarily tied to the natural numbers. In fact, proofs by recursion are more general than mathematical induction on natural numbers, and it is possible to prove mathematical induction in the domain of natural numbers by using a proof by recursion. On the other hand, one can always introduce natural numbers when conducting arguments, which means that all proofs by recursion can also be reduced to a proof by induction on the natural numbers. This, however, often lengthens the argument.

To see how induction can be generalized, consider a logical expression from propositional calculus. All logical expressions share certain properties. For instance, given an assignment, their truth values can be evaluated. Proof by recursion could be used to show that all logical expressions have this property. The argument would be as follows: to prove that every logical expression yields a truth value, pick any expression x and show that it has a truth value. Two cases must be distinguished. In the case when x is atomic, the truth value of x is given by assignment. This is an example of a base case. Otherwise, x can be divided into components, and x has a truth value if all its components y have truth values. The proof that x has a truth value is thus reduced to a similar proof for all its components y. Hence, two cases must be distinguished, (1) that y is atomic, in which case y has a truth value by assignment, or (2) that y is compound. If y is compound, then it must be shown that each component of y has a truth value. In this way, the problem can be recursively reduced to smaller and smaller subproblems until a base case is finally reached. If this happens in all possible cases, then it logically follows that every x has a truth value.

Of course, what has been said for the property that x has a truth value can be applied for any other property P of x. Hence, to prove $P(x)$, pick an arbitrary x. Two cases must be distinguished. If $P(x)$ can be shown directly, then x is basic. For logical expressions, this typically means that x is atomic. Otherwise, x is compound, and it must be shown that $P(x)$ is true, provided that $P(y)$ is true for every component y of x. If this can be established, then the principle of induction allows one to conclude that $P(x)$ is true for all x. This essentially outlines the nature of proofs by recursion.

Of course, proofs by recursion are not limited to expressions from propositional calculus. In fact, induction on the natural numbers can easily be derived by the same principles. Pick an arbitrary n and show that $P(n)$ is true. Again, two cases arise. Either $n = 0$, in which case $P(n)$ must be proved directly, or one shows that $P(n - 1)$ implies that $P(n)$. If this can be done, then $P(n - 1)$ must be proved. Again, one must distinguish two cases: either $n - 1 = 0$, in which case the proof is complete, or one has to prove that $P(n - 2)$. In this case, one continues until finally the problem is reduced to finding the truth value of $P(0)$.

Both in the case of proving a property of a formula and in mathematical induction on the natural numbers, the problem is reduced to one that is somehow simpler or smaller, and this continues until the base case is reached. The question that arises is: under which conditions will this method work? First, note that this procedure does not work for the integers. Of course, in the integers one can recurse to smaller and smaller numbers, but this recursion will never end. The fact that the recursion will come to an end is crucial, and proofs by recursion are unsound unless this fact can be established. Hence, one has to introduce the notion of a *descending sequence*, which is, generally, a sequence of smaller and smaller elements, say shorter formulas or smaller numbers. As will be shown in Section 3.3.4, proofs by recursion are not applicable unless all descending sequences end and unless one can prove the property of interest for all minimal elements of the sequence.

A domain in which all descending sequences are finite is said to be *well founded*. Hence, proofs by recursion are only sound if the domain is well founded and, for this reason, they are also called proofs by *well-founded induction*. In the examples given previously, it was very easy to show that the domain in question was well founded. In other cases, this may not be easy to establish. For instance, when using recursion in programming, one must, in some sense, reduce the problem with each recursive call; otherwise, the program may never end. The decision as to whether a program will ever end is an undecidable problem. This suggests that the problem of finding whether a domain is well founded is also undecidable.

3.3.2 Recursive Definitions

In all cases, proofs by recursion have to rely on the definition of the domain, because these definitions provide the necessary premises upon which proofs can be built. For this reason, we discuss recursive definitions next, which we do by means of an example. If x is a person, then, by definition, all the following persons are descendants of x.

1. All children of x are descendants of x.
2. If y is a child of x, then all descendants of y are descendants of x.
3. Nobody else is a descendant of x.

To find if a particular individual is a descendant of x, it may be necessary to apply this definition repeatedly or, to use the technical term, recursively. It is assumed that this recursion always comes to an end. The answer can be either positive, in which case the individual in question is a descendant, or negative, in which case there are no persons left for whom the definition applies.

In a recursive definition, one distinguishes between a *base rule* and a *recursive rule*. The base rule of a recursive definition describes the term to be defined directly. In the case of the descendants, the base rule states that all children of x are descendants of x. The recursive rule contains the term to be defined in its own definition. In the example, every child of a descendant is defined to be a descendant. This feature, which introduces a kind of circularity, makes it possible to use the recursive part of the definition repeatedly, or recursively.

We now give a number of recursive definitions from the area of mathematical reasoning and computer applications. First, we define something to be called an SL-expression, for *simplified logical* expression. One has

1. All propositional variables are SL-expressions.
2. If A and B are two SL-expressions, so are $(A \wedge B)$, $(A \vee B)$, and $\neg A$.
3. Nothing else is an SL-expression.

Note that this definition first asserts explicitly that certain items, that is, all propositional variables, are SL-expressions. These are the base cases of the definition. All other items are defined recursively in terms of expressions that are presumably SL-expressions. By applying the recursive part repeatedly, expressions of arbitrary complexity can be created.

Mathematical formulas can be defined in the same fashion. Hence, a *simplified mathematical* expression, or an SM-expression might be defined as follows:

1. All integers and all variable names are SM-expressions.

2. If A and B are two SM-expressions, then $-A$, $(A + B)$, $(A - B)$, $(A \times B)$, and (A/B) are also SM-expressions.

3. Nothing else is an SM-expression.

A structure that is often defined recursively is a *binary tree*. Before giving a definition for a binary tree, we first give a short discussion about trees. A more extensive discussion on trees can be found in Chapter 7. The general concept of a tree should be familiar to the reader because of parse trees, family trees, organizational diagrams of companies, and similar constructs. Generally, trees consist of *nodes*, which are connected by *arcs*. All arcs have a *direction* associated with them and are directed arcs. In an organizational diagram, for instance, the nodes represent the different individuals, together with their position. Moreover, there is an *arc*, or a straight line, from each superior to his or her immediate subordinate. In a parse tree of a logical expression, the nodes are themselves expressions, and there is an arc from each expression to its immediate subexpression. In a graphical representation of a tree, the direction of an arc is, by convention, from the top to the bottom. In an organization chart, there is usually a top node, that is, a node with no incoming arc. This top node corresponds to the chief executive officer. Similarly, the parse tree of a logical expression contains a top node, which represents the original expression to be parsed. Trees that contain such a top node are called *rooted trees*. Here, we only discuss rooted trees; however, as will be shown in Chapter 7, not all trees are rooted trees.

A number of terms relating to rooted trees are now introduced. To have a concrete example, we refer to Figure 3.1. In this figure, each node is represented by a circle, and it contains a label from a to f.

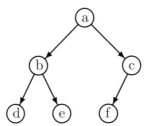

Figure 3.1.   Graphical representation of a tree

If there is an arc from node i to j, then i is said to be a *parent* of j, and j is said to be a *child* of i. In Figure 3.1, for instance, b is a parent of d and e, and a is a parent of b and c. The node c has only one child, f. Nodes without children are called *leaves*, whereas all other nodes are called *branch nodes*. In Figure 3.1, a, b, and c are branch nodes, and d, e, and f are leaves. The root is the only node without a parent. Here, we restrict our attention to *binary trees*. By definition, a binary tree is a rooted tree in which each node can have at most two children. More generally, an *m-ary* tree is a tree in which each node has at most m children. The order in which the children are listed is relevant. Specifically, in a binary tree, one must distinguish between a *left child* and a *right child*.

Example 3.17   Draw a graph of the binary tree that has d as its root, the leaf c as its left child, and f as its right child. f has e as its left child, but it has no right child. Nodes c and e are leaves.

Solution   The solution is given in Figure 3.2.   ∎

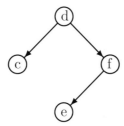

Figure 3.2.   Graphical representation of a tree

We now define a binary tree recursively, as follows:

1. The empty tree is a binary tree. This tree is denoted by $()$.

2. If A and B are two binary trees, and if c is a node, then (A, c, B) is a binary tree. This binary tree is denoted by (A, c, B). Here, A is the *left subtree* and B is the *right subtree*.

3. Nothing else is a binary tree.

According to this definition, the tree consisting of the node c only would be represented as $((), c, ())$. This notation is clumsy, and we abbreviate $((), c, ())$ by (c). Hence, the tree consisting only of c is denoted by (c).

Example 3.18   Use the recursive definition of a binary tree to describe the binary tree given in Figure 3.2.

Solution   The tree has the form $((c), d, B)$ since the left subtree is (c). The subtree B has the left subtree (e). There is no right subtree, which means that the right subtree is $()$. B, the right subtree of d is therefore $((e), f, ())$. Consequently, the tree rooted at d is $((c), d, ((e), f, ()))$.   ∎

Finally, we define the term *list*. Like a tree, each list consists of *nodes*. The nature of the nodes is of no interest here. One has the following:

1. The empty list is a list. This list is denoted by $[\,]$.

2. If a is a node and B is a list, then $a.B$ is a list. Here, the node a is called the *head* of the list $a.B$, and B is the *tail* of the list.

3. Nothing else is a list.

Lists are very useful for storing information in sequence, and they are used extensively in logic programming and functional programming. To store the three numbers 15, 3, and 22, for instance, one can write $15.(3.(22))$. Since this is hard to read, one normally writes all the elements of a list side by side and encloses them in square brackets. For instance, instead of $15.(3.(22))$, one writes $[15, 3, 22]$. Moreover, instead of using the notation $a.B$, where

a is the head and B is the tail, one typically writes $[a|B]$. These issues will be discussed in further detail in Sections 4.4.5 and 6.5.4. The lists, as they are defined here, must be distinguished from the linked lists used in algorithmic languages. Lists, in the sense defined here, may by implemented as linked lists, but other implementations are conceivable.

Recursive rules can be classified as either *shortening rules* or *lengthening rules*. In a lengthening rule, the new object to be defined is longer than the components used for its definition. All rules discussed up to now are lengthening rules. For instance, the rule "If A and B are two SL-expressions, so is $(A \wedge B)$" is a lengthening rule: the new expression, $(A \wedge B)$, is longer than its components A and B. However, not all rules are lengthening rules. For instance, the rule "If (A) is an SL-expression, A is also an SL-expression" is a shortening rule. Another case when shortening rules are used is in the definition of what constitutes a theorem of a theory. A theory is the set of all conclusions derivable from a certain set of premises. Given the set of premises, and a set of rules of inference, a theorem can be defined recursively as follows:

1. Every premise is a theorem.
2. Every expression that can be derived from a theorem by means of a valid rule of inference is a theorem.
3. Nothing else is a theorem.

Hence, the rules of inference become the recursive rules, and, typically, there are shortening and lengthening rules among the rules of inference. For instance, the rule of combination states that if A and B are theorems then $A \wedge B$ is a theorem, and this is a lengthening rule. The law of simplification, on the other hand, which allows one to derive the theorem A from $A \wedge B$, is a shortening rule.

Other areas where recursive definitions are used extensively are grammars. Grammars will be discussed in detail in Chapter 10. Here, we only note that grammars can contain both shortening and lengthening rules.

3.3.3 Descending Sequences

Proofs by recursion require that the domain in question be well founded; that is, all descending sequences in the domain must be finite. The question now is what is meant by a descending sequence. In the domain of natural numbers, $15, 5, 0$ is clearly a descending sequence, and so is $22, 21, 20$. Moreover, the sequence $-2, -4, -6$ is a descending sequence in the domain of integers, and $1, \frac{1}{2}, \frac{1}{4}$ is a descending sequence in the domain of real numbers. Similarly, the following is a descending sequence of SL-expressions.

$$(P \wedge Q) \wedge \neg R, \ P \wedge Q, \ P \tag{3.19}$$

Before dealing with descending sequences, we introduce what we call *G-sequences*. A G-sequence requires a two-place predicate $G(x, y)$, with the stipulation that, for any two successive terms x and y in the sequence, $G(x, y)$ must hold. In the case of numbers, a sequence of descending numbers is a G-sequence provided that $G(x, y)$ is true for $x > y$. For instance, $15, 5, 0$ is a G-sequence because $15 > 5$ and $5 > 0$. Similarly, the sequence given in (3.19) is a G-sequence provided that $G(x, y)$ means "y is a proper subexpression of x." Every descending sequence is a G-sequence for some predicate G, but not every G-sequence is a descending sequence. In fact, a sequence in which all elements are identical is a G-sequence, provided that G is the predicate for equality, but not a descending sequence. By using the notion of a G-sequence, one can define well-founded domains as follows:

Definition 3.1:   A domain is well founded with respect to the predicate G if all G-sequences terminate.

We should mention here that most authors define well-founded domains as domains in which all descending sequences end. Our definition seems to be more general, because the G-sequences are generalizations of descending sequences. This issue will be addressed in Section 5.4.8, where descending sequences are defined in a formal way. At this point, it will also be shown that if all G-sequences end then the G-sequences can be interpreted as descending sequences.

In well-founded domains, there are elements x such that $G(x, y)$ is false for all y. These elements are called *minimal elements*. In the case of natural numbers, 0 is a minimal element because there is no y such that $0 > y$. In the domain of SL-expressions, all atomic expressions are minimal elements, because atomic expressions do not have subexpressions. Once a G-sequence reaches a minimal element, it must terminate.

In recursive definitions, there is a natural way to define $G(x, y)$: $G(x, y)$ merely means that y is needed to generate x from its definition. If all recursive rules are lengthening rules, then all G-sequences will end, and the minimal elements are the basic elements. For example, to define the SL-expression $(P \wedge Q)$, P is needed, which means that $G((P \wedge Q), P)$ is true. Since $(P \wedge Q)$ is longer than both P and Q, the formation rule in question is a lengthening rule. In general, all G-sequences defined by using lengthening rules in the way indicated will eventually end, typically with atomic expressions, but possibly with empty expressions. No expression can be shorter than the empty expression. If the recursive definition contains both lengthening and shortening rules, this is no longer true. In this case, the G-sequence may contain expressions that are longer or shorter than the previous expression. There is no compelling reason anymore why such a sequence should end. In conclusion, if the recursive definition only contains lengthening rules, then any G-sequence defined by this definition will end, and the domain is well founded. If, on the other hand, there are shortening rules, the domain may not be well founded.

3.3.4 The Principle of Proofs by Recursion

To prove that $\forall x P(x)$ by recursion, the following steps must be taken:

Domain well founded   A predicate G must be chosen, and it must be shown that all G-sequences are finite.

Inductive base   If x is minimal, it must be proved that $P(x)$ is true.

Inductive hypothesis   Pick an arbitrary x, and assume that $P(x)$ holds for all y satisfying $G(x, y)$. For instance, if $G(x, y)$ is $x > y$, assume that, for all $y < x$, $P(y)$ holds.

Inductive step   Prove $P(x)$.

Conclusion   Conclude that $\forall x P(x)$.

Note that if the domain of the natural numbers is selected, and if $G(x, y)$ is true for $x > y$, then proofs by recursion instantiate to strong induction. Since 0 is the only minimal element, the basis for induction in the new scheme requires one to prove $P(0)$, just as before. Moreover, to prove $P(x)$, one may assume $P(y)$ for all $y < x$.

Formally, proofs by recursion can be cast as follows. Given that the domain is well founded, one can conclude that $\forall x P(x)$, provided the following holds:

$$\forall x(\forall y(G(x, y) \Rightarrow P(y)) \Rightarrow P(x)) \tag{3.20}$$

This equation includes both the base case and the inductive step. In the base case, there are no elements y for which $G(x, y)$ holds, which makes the expression $\forall y(G(x, y) \Rightarrow P(y))$ trivially true. Hence, (3.20) is true as long as $P(x)$ is true, which settles the base case. If x is not basic, then (3.20) instantiates to

$$\forall y(G(x, y) \Rightarrow P(y)) \Rightarrow P(x)$$

To prove this equation, one assumes that $P(y)$ for all y satisfying $G(x, y)$, and one derives $P(x)$.

We now show that in well-founded domains (3.20) implies that $\forall x P(x)$. The idea of the proof is as follows: We assume that, for some x, $\neg P(x)$ is true and derive a contradiction. In fact, we show that if $\neg P(x)$ is true an infinite G-sequence can be constructed, and since this is impossible in well-founded domains, $\neg P(x)$ must be false for all x. In other words, $P(x)$ must be true for all x. To construct an infinite G-sequence, one starts with the individual x for which $\neg P(x)$ holds. This value of x, we call it x_0, is the first element of the infinite G-sequence. To find the next element in the sequence, instantiate (3.20) to

$$\forall y(G(x_0, y_0) \Rightarrow P(y)) \Rightarrow P(x_0)$$

Since $\neg P(x_0)$ is true by assumption, the modus tollens yields

$$\neg \forall y(G(x_0, y) \Rightarrow P(y))$$

This is logically equivalent to

$$\exists y(G(x_0, y) \wedge \neg P(y))$$

Hence, there must be a y, say $y = x_1$, for which $G(x_0, y) = G(x_0, x_1)$ and $\neg P(y) = \neg P(x_1)$ are both true. Choose x_1 as the second element of the G-sequence. Now, $\neg P(x_1)$ implies, using the same argument as before, that there is an element x_2, such that $G(x_1, x_2) \wedge \neg P(x_2)$ is true. This element is the third element in the sequence. This process can obviously be continued forever, which means that one finds an infinite descending sequence. However, a well-founded domain does not allow infinite G-sequences. Consequently, if the domain contains no infinite descending sequences, then it cannot contain any x for which $\neg P(x)$ holds. In other words, $P(x)$ must be true for all x, as stated.

If the domain is not well founded, proofs by recursion are unsound. For instance, the domain of integers, with $G(x, y) = (x > y)$, does not allow for proofs by recursion. The following example shows this.

Example 3.19   Will a call to the procedure fact(n) always terminate?

```
function fact(n : integer) : integer;
  begin
    if n = 0 then fact := 1
      else fact := n * fact(n−1);
  end;
```

Solution   The call to fact(n) obviously terminates if $n \geq 0$. However, it will not terminate if $n < 0$. To see how proofs by recursion work, we now show this in detail. Hence, let $P(n)$ be true if fact(n) terminates.

Domain well founded   If n is restricted to be a natural number, then the domain is well founded under $>$. In this case, the minimal element is 0.

Inductive base   The program terminates for $n = 0$, which settles the base case.

Inductive hypothesis   To prove that the program ends for any $n \geq 0$, assume that the program terminates for all $m < n$.

Inductive step   To show that fact(n) will eventually return a value, we use the inductive hypothesis, which indicates that fact(m) terminates for all $m < n$. In particular, the call fact(n − 1) will eventually return a value, and at this point, the else clause will terminate. We conclude that under the inductive hypothesis the call terminates.

Conclusion   We conclude that the call terminates for all $n \geq 0$.

This settles the case where n is a natural number. If n can be any integer, the proof is no longer valid, because there are infinite descending sequences in the domain of integers. Specifically, if $n = -1$, the call fact(n − 1) instantiates to fact(−2). To find fact(−2), fact(−3) has to be calculated, and for doing this fact(−4) is required. Hence, an infinite loop is started from which the program will never return.   ■

In domains generated through the use of recursive rules, one has to distinguish between two cases. If the recursive rules do not contain any shortening rules, then the domain is well founded and the basic elements are identical to the minimal elements. In this case, one speaks about *structural induction*. If, on the other hand, there are shortening rules, then it is not obvious how to find a predicate G for which all G-sequences terminate. We will therefore no longer deal with domains generated by recursion containing shortening rules.

3.3.5 Structural Induction

Structural induction is applicable in domains generated by recursive definitions without shortening rules. In this case, one does not need to show that the domain is well founded. Moreover, the predicate $G(x, y)$ is true whenever y is needed to generate x. In this sense, one can say that y is a component of x. To prove that $P(x)$ holds for all x, one therefore must first prove $P(x)$ for all base cases. Then one picks an arbitrary compound x and makes the assumption that $P(y)$ is true for all y that are proper components of x. Once this is done, the conclusion $\forall x P(x)$ follows. This procedure can be expressed as follows:

Inductive base   If x is a basic object, it must be proved that $P(x)$ is true.

Inductive hypothesis   To prove $P(x)$, assume that $P(y)$ is true for all proper components of x.

Inductive step   Prove $P(x)$.

Conclusion   Conclude that $\forall x P(x)$.

Example 3.20   Show that in all SL-expressions the number of opening parentheses is equal to the number of closing parentheses.

Solution   The recursive rules for SL-expressions were stated earlier as follows:

1. All propositional variables are SL-expressions.
2. If A and B are SL-expressions, then $(A \wedge B)$, $(A \vee B)$, and $\neg A$ are also SL-expressions.
3. No other expressions are SL-expressions.

Structural induction is now used to prove $P(x)$, where $P(x)$ means that the SL-expression x has as many opening as closing parentheses.

Inductive base   By definition, the base objects are exactly the expressions consisting of a single variable, and these expressions have no parentheses. Hence, it is true that there are as many opening as closing parentheses, that is, none.

Inductive hypothesis   Let C be any compound expression, and assume that all components of C have as many opening as closing parentheses.

Inductive step   If $C = (A \wedge B)$ and if A and B have as many opening as closing parentheses, then C also has as many opening parentheses as closing parentheses. In fact, to form C, one opening and one closing parenthesis are added. The same argument applies if $C = (A \vee B)$. The last case involves $C = \neg A$. $\neg A$ and A have exactly the same open and closing parentheses. By the inductive hypothesis, the number of opening parentheses in A matches the number of closing ones. The same must therefore hold for $C = \neg A$.

Conclusion   One concludes that all SL-expressions have as many opening as closing parentheses.   ■

Example 3.21   Define the *dual* of an SL-expression C as follows:

$$\operatorname{dual}(A \wedge B) = \operatorname{dual}(A) \vee \operatorname{dual}(B), \quad \text{if } C = A \wedge B$$
$$\operatorname{dual}(A \vee B) = \operatorname{dual}(A) \wedge \operatorname{dual}(B), \quad \text{if } C = A \vee B$$
$$\operatorname{dual}(\neg A) = \neg \operatorname{dual}(A), \quad\quad\quad\quad \text{if } C = \neg A$$
$$\operatorname{dual}(C) = C, \quad\quad\quad\quad\quad\quad\quad \text{otherwise}$$

Show that in $\operatorname{dual}(C)$ all $\wedge$ are replaced by $\vee$, and all $\vee$ are replaced by $\wedge$.

Solution   Let $P(C)$ be true if in $\operatorname{dual}(C)$ all $\wedge$ are replaced by $\vee$, and all $\vee$ by $\wedge$. Now use structural induction as follows:

Inductive base   According to the definition of an SL-expression, the only expressions for which the default clause dual$(C) = C$ applies are the basic expressions, and since these expressions do not contain any connectives, $P(C)$ is true for all basic expressions. In other words, if C is basic, then all $\wedge$ in C, of which there are none, are replaced by $\vee$ and, similarly, all $\vee$ are replaced by $\wedge$.

Inductive hypothesis   Assume that $P(X)$ is true for all X that are proper subexpressions of C; that is, all $\wedge$ in X are replaced by $\vee$, and all $\vee$ are replaced by $\wedge$.

Inductive step   Three cases must be distinguished for the inductive step. If C is $A \wedge B$, then dual(C) is dual$(A) \vee$ dual(B) by definition. According to the inductive hypothesis, all $\wedge$ in both A and B are replaced by $\vee$ and all $\vee$ by $\wedge$, and since a similar statement can be made regarding B and dual(B), one only needs to replace the main connective in C by $\vee$, and the statement becomes true for C. A similar argument holds if $C = A \vee B$. In the case $C = \neg A$, the inductive hypothesis ensures that A, and with it $\neg A$, has all its $\wedge$ and $\vee$ interchanged.

Conclusion   One concludes that, if dual is defined as here, all $\wedge$ are replaced by $\vee$ and all $\vee$ by $\wedge$.   ■

Example 3.22   Define the complement of C, denoted by comp(C) as follows:

$$\text{comp}(A \vee B) = (\text{comp}(A) \wedge \text{comp}(B)), \quad \text{if } C = (A \vee B)$$

$$\text{comp}(A \wedge B) = (\text{comp}(A) \vee \text{comp}(B)), \quad \text{if } C = (A \wedge B)$$

$$\text{comp}(\neg A) = \neg\text{comp}(A), \qquad\qquad\quad \text{if } C = \neg A, \; A \; compound$$

$$\text{comp}(\neg P) = P, \qquad\qquad\qquad\qquad \text{if } C = \neg P, P \; atomic$$

$$\text{comp}(P) = \neg P, \qquad\qquad\qquad\qquad \text{if } C = P, P \; atomic$$

Show that comp$(C) \equiv \neg C$.

Solution

Inductive base   The only cases for which the definition of the complement can be applied directly are the last two definitions, and in both definitions, comp$(C) \equiv \neg C$. These are the base cases.

Inductive hypothesis   Assume that, for any subexpression A of C, $\neg A = \text{comp}(A)$.

Inductive step   If $C = (A \wedge B)$, one has

$$
\begin{aligned}
\text{comp}(A \wedge B) &\equiv (\text{comp}(A) \vee \text{comp}(B)) && \text{By definition} \\
&\equiv (\neg A \vee \neg B) && \text{Inductive hypothesis} \\
&\equiv \neg(A \wedge B) && \text{De Morgan} \\
&\equiv \neg C && \text{Since } C = (A \wedge B)
\end{aligned}
$$

The proof for the case $C = (A \vee B)$ is similar. If $C = \neg A$, then the result comp$(C) \equiv \neg C$ is as follows:

$$
\begin{aligned}
\text{comp}(\neg A) &\equiv \neg\text{comp}A && \text{By definition} \\
&\equiv \neg\neg A && \text{Inductive hypothesis} \\
&\equiv \neg C && \text{Since } C = \neg A
\end{aligned}
$$

This completes the inductive step.

Conclusion   One concludes that comp$(C) = \neg C$ in all cases.   ■

We now prove a number of important theorems regarding binary trees. To do this, we first define a number of terms. These terms apply to every tree.

Definition 3.2: Each node in a tree has a *level*. The root node of a tree has level 1. The children of a node of level n have a level of $n + 1$.

Hence, the children of the root are in level 2, the grandchildren of the root are in level 3, and so on. Generally, each generation corresponds to a level.

Definition 3.3: The *height* of a tree is the maximum level found in the tree. The empty tree has a height of zero.

This definition of the word height is convenient for mathematical proofs, and it is used in texts with a mathematical flavor. However, in texts on computer science, the height of a tree is typically defined as the number of steps one has to make, starting from the root, to reach the node with the highest level. This measure is one less than the height as defined here. The only exception to this is the empty tree, for which the level is equal to zero under either definition.

Theorem 3.1. The maximum number of nodes in a binary tree of height n is $2^n - 1$, and there is a binary tree of height n with $2^n - 1$ nodes.

The proof is by structural induction. The word "tree" in this proof always refers to a binary tree.

Inductive base The only tree of height 0 is the empty tree, and the theorem is true for the empty tree: it has $2^0 - 1$ nodes, and this is 0.

Inductive hypothesis Assume that the theorem is true for all proper subtrees of any tree.

Inductive step A proper subtree of a tree of height n can be at most of height $n - 1$ and, by the inductive hypothesis, it can have at most $2^{n-1} - 1$ nodes. Since there are two such subtrees, this yields twice $2^{n-1} - 1$ nodes or, together with the root, $2(2^{n-1} - 1) + 1$ nodes. Hence, under the inductive hypothesis, there are up to $2^n - 1$ nodes in total. This value can be obtained if both subtrees are of the same height.

Conclusion All trees of height n have at most $2^n - 1$ nodes, and this maximum is reached by some binary trees with height n.

The theorem is important because it allows one to find the minimum height required to include a given number of nodes. A tree of height 10 can contain up to $2^{10} - 1 = 1027$ nodes, and a tree of height 20 can contain up to $1028^2 - 1$ nodes, which makes about a million nodes. Generally, one can increase the maximum number of nodes by a factor of approximately 1000 by adding 10 levels to the tree.

Theorem 3.2.   A nonempty binary tree of n nodes contains exactly $n + 1$ empty subtrees.

The proof is again by structural induction.

Inductive base   The tree with $n = 1$ node contains $(n + 1) = 2$ empty subtrees.

Inductive hypothesis   It is assumed that the theorem holds for all proper subtrees of any tree. Hence, if the tree in question has n nodes, then it is assumed that any proper subtree with $m < n$ nodes contains $m + 1$ empty subtrees.

Inductive step   We now prove that a tree with $n > 1$ nodes has $n+1$ empty subtrees. Such a tree has a root, which leaves $(n - 1)$ nodes to the immediate subtrees. At this point, two cases must be considered.

(**a**) If one immediate subtree is empty, the other immediate subtree has $n - 1$ nodes, where $n > 0$. By inductive hypothesis, a tree with $n - 1$ nodes has n empty subtrees. The entire tree therefore has $n + 1$ empty subtrees, as required.

(**b**) If no immediate subtree is empty, then there are m_1 nodes in the left subtree and m_2 in the right subtree, where $m_1 + m_2 = n - 1$. According to the inductive hypothesis, the left subtree has $m_1 + 1$ empty subtrees, and the right subtree has $m_2 + 1$ empty subtrees. Hence, both subtrees together have $(m_1 + m_2 + 2) = (n - 1) + 2$ empty subtrees. The original tree therefore has $n + 1$ empty subtrees, as required.

Conclusion   All trees with n nodes have $n + 1$ empty subtrees.

Problems 3.3

1. Show that if $G(x, x)$ is true there are infinite G-sequences.
2. Show that if there is a value y such that $G(x, y)$ and $G(y, x)$ are both true then there are infinite G-sequences.
3. Define E-expressions as follows:
   (**a**) $2, 3, 4, \ldots$ is an E-expression. These E-expressions are called atomic.
   (**b**) If x and y are E-expressions, so is $(x + y)$ and $(x \times y)$.
   E-expressions can be evaluated in the normal way.   Show that the value of every E-expression is at least $2n$, where n is the number of atomic expressions.
4. Define strings of the form $a^m b a^m$ recursively. Use the formation rules to prove that all strings generated in this way have an odd number of characters.
5. Show that the minimum number of nodes in a tree of height h is h.

6. An m-ary tree is a tree where every node can have up to m children. Show that the maximum number of leaves in an m-ary tree of height n is $m^{(n-1)}$.

7. Use the laws of commutativity and associativity of addition to show that

$$a_1 + (a_2 + (a_3 + \cdots + (a_{n-1} + a_n)\ldots)) = a_n + (a_{n-1} + (\cdots + (a_2 + a_1)\ldots))$$

8. Consider the following Pascal function:

```
function power(n : integer) : integer;
  begin
    if n = 0 then power := 1
    else power := 2 * power(n − 1)
  end
```

Prove that this function returns 2^n for $n \geq 0$. Can the proof be generalized to hold for all integer values of n?

3.4 APPLICATIONS OF RECURSION TO PROGRAMMING

3.4.1 Introduction

The following subsections deal with the application of recursion to programming. To do this, it is best to interpret executable statements and groups of executable statements as functions. These functions affect the *state* of the system, where the state is essentially defined by the declarations. The method of well-founded induction will be used to give informal proofs for the correctness of certain recursive procedures. Note that in this section the term "well-founded induction" rather than the term "proof by recursion" is used. This is done to minimize confusion with the term "recursive procedure." Moreover, the term "well-founded induction" stresses the need for actually proving that the domain is well founded, and in programming this is an important issue. If the domain is not well founded, the procedure may never terminate. A preliminary example of how this can happen was given in Section 3.3.4, where a program was discussed that ran into an infinite loop for some inputs. These issues will be discussed further in this section. Recursive procedures are particularly important for structures that are defined recursively. For this reason, we discuss a number of procedures involving trees.

3.4.2 Programming as Function Composition

Programs, and even parts of programs, are functions that affect the state of the system. To indicate what this means, we first consider a simple program, such as the one given in Figure 3.3. Like every other program, this program contains *declarations* and *executable statements*. It also contains other information, such as a program name, comments, and so on, which are of no interest here. In our context, the declarations are important because they identify the variables to be used, and the values of these variables determine the *state* of the system. In Figure 3.3, for instance, the state of the system is given by the values of the variables n, sum, and i. The executable statements work with the variables and, in this way, they change the state. Generally, an executable statement converts the initial state of the statement (the state before its execution) into some final state. The final state may be identical to the initial state, in which case the statement has no effect. Groups of executable statements also convert an initial state into a final state. Generally, we will call every statement or every group of statements that affects the state a *piece of code*. Not all initial states can be handled by all pieces of code. For instance, the statement $1/x$ cannot be handled if $x = 0$. However, if the initial state can be handled by the piece of code and if the piece of code does not run into an infinite loop, then the final state is deterministic. Hence, a piece of code expresses a *partial function* in the sense that for each initial state there is at most one final state. The piece of code does not represent a total function, however, because there are no final states corresponding to initial states that cause the code to bomb or that cause it to run into an infinite loop.

```
program sum(input, output);
    {This program calculates the sum of the integers from 1 to 5.}
    var
        n, sum, i : integer;
    begin
        n := 5;
        sum := 0;
        i := 0;
        while i <> n do
            begin
                i := i + 1;
                sum := sum + i
            end
    end.
```

Figure 3.3.   Calculating the sum of integers from 1 to 5

Example 3.23   Given the initial state n = 5, i = 2, and sum = 1, what is the final state of the piece of code sum := sum + i? What is the final state of the piece of code i := i + 1; sum := sum + i?

Solution   The code sum := sum + i associates with the given initial state the final state n = 5, i = 2, and sum = 3. The final state of the code i := i + 1; sum := sum + i is, similarly, n = 5, i = 3, and sum = 4.   ∎

Since each piece of code represents a function, the final state of a code A corresponding to the initial state x is the image of A at x, which will be denoted by $A(x)$. Note that x is a state, which is defined by the values of all variables occurring in the program, and the same is true for $A(x)$.

If there are two pieces of code A and B that are executed in sequence, one typically writes $A; B$, and the result is called the *concatenation* of A and B. Actually, concatenation is essentially function composition. To see this, let x be the initial state of $A; B$, that is, the state before A is executed. By definition, the state after A is executed is $A(x)$, and this is the state before B is executed. Since $B(y)$ is the final state of B, given an initial state y, and since $y = A(x)$, this means that the final state of $A; B$ is $B(A(x))$. Consequently,

$$A; B = B \circ A$$

The final state of $A; B$ can also be written as $(A; B)(x)$.

Example 3.24   Consider the following code, in which the state consists of a single variable, a.

$$a := a + 3; \; a := a * 2$$

Find $(a := a + 3)(x)$, $(a := a * 2)(y)$, and $(a := a + 3; \; a := a * 2)(x)$.

Solution

$$(a := a + 3)(x) = x + 3$$
$$(a := a * 2)(y) = 2 * y$$
$$(a := a + 3; \; a := a * 2)(x) = 2 * (x + 3)$$ ∎

If A, B, and C are three pieces of code, one can concatenate A and B, forming $A; B$. Since $A; B$ is also a piece of code, one can form $(A; B); C$. Similarly, $B; C$ is a piece of code, which means that one can also form $A; (B; C)$. As it turns out, both $(A; (B; C))(x)$ and $((A; B); C)(x)$ are $C(B(A(x)))$, which means that $A; (B; C)$ and $(A; B); C$ are equal. One can therefore drop parentheses and write $A; B; C$. Hence, the semicolon works like an associative operator. Of course, in Pascal one does not use parentheses. Instead, one uses the keywords **begin** and **end** to accomplish the grouping of statements. Concatenation is not commutative; that is, $A; B$ and $B; A$ are normally different. However, concatenation has an identity element. In fact, any statement that does not change the state can be considered as an identity element. We use I for any statement that leaves the state unchanged. Hence, $I(x) = x$.

Every code is at least a partial function, but in many cases there are different functions that solve a given problem. To see this, consider again the program given in Figure 3.3. In this code, the only thing that matters is that in the end the variable sum has the correct value, and there are many functions that satisfy this condition. Moreover, the program could in principle accept any initial state, because the values n, i, and sum are assigned values by the program that override the original values. As another example, consider the code given by Figure 3.3 without the statement n := 5. In this case, one considers as correct any function that converts any initial state satisfying $n \geq 0$ into a final state satisfying $sum = \sum_{i=1}^{n} i$. To handle these issues, one has to consider the *properties* of the possible states. For instance, the state n := 3, i := 2, sum := 1 has the property $n > 0$.

Definition 3.4:   If x is a state and if $g(x)$ is either true or false, then g is called a *condition* or *assertion*. An assertion concerning the initial state of a piece of code is called a *precondition*, and an assertion regarding the final state of a piece of code is called a *postcondition*.

Preconditions and postconditions will be discussed extensively in Chapter 9.

Example 3.25   The code given by Figure 3.3 without the statement n := 5 must satisfy the precondition $n \geq 0$ because, otherwise, it will run into an infinite loop, as the reader may verify. The postcondition of the same code is given by the condition $sum = \sum_{i=1}^{m} i$, where m is the initial value of n. The entire program, including the statement n := 5, has no preconditions, and its postcondition is $sum = \sum_{i=1}^{5} i$.

By concentrating on preconditions and postconditions, one removes information irrelevant to a program's correctness and concentrates on the essential facts. Generally, one can describe what is to be done by stating the preconditions and postconditions. A piece of code can then be considered as correct if any state satisfying the preconditions also satisfies the postconditions. In this sense, one can say that the preconditions and postconditions jointly describe what the code has to accomplish.

The pieces of code are functions, but these functions must be distinguished from the built-in functions and operators. Pieces of code work on states, and they convert an initial state to a final state. Built-in functions and operators only affect individual variables, and not necessarily the same variable in each invocation. To mark the difference between these two types of functions, capital letters will be used for pieces of code and lowercase letters for all other functions. We also use lowercase letters for conditions.

The if-statement can also be expressed as a function. To see this, let H be an if-statement with condition $g(x)$, which executes A if $g(x)$ is true and B if $g(x)$ is false. H can be expressed in a Pascal-like way as follows:

$$H = (\textbf{if } g(x) \textbf{ then } A \textbf{ else } B) \tag{3.21}$$

In a similar fashion, one can express the one-armed if-statement. In fact, in a one-armed if-statement, the else clause is missing or, in other words, the else clause is replaced by the identity I. One therefore has, if H_1 is a one-armed if-statement,

$$H_1 = (\textbf{if } g(x) \textbf{ then } A) = (\textbf{if } g(x) \textbf{ then } A \textbf{ else } I)$$

For many mathematical arguments, it is convenient to replace logical variables by integers. This can be done by defining 0 to be false and 1 to be true. If H is a one-armed if-statement, then the image of H for the argument x can be expressed as

$$H(x) = (A(x) * g(x) + B(x) * (1 - g(x)))$$

Here, $A(x) * g(x)$ is the result one obtains by multiplying all variables of the state x by $g(x)$. For the one-armed if-statement, one similarly finds

$$(\textbf{if } g(x) \textbf{ then } A(x)) = (A(x) * g(x) + (1 - g(x)) * x)$$

The while-loop statement is also a function, but this function is recursive. To see this, let W be a while-loop statement, which is written as follows, following closely the syntax of Pascal.

$$W = (\textbf{while } g(x) \textbf{ do } A)$$

Here, A is the code corresponding to the body of the loop. Note that if $g(x)$ is false then A is not done at all, and $W(x)$ is the identity $I(x)$. Otherwise, A is done once and, after that, W is done again. Hence, if $g(x)$ is true, then W is $A; W$. Hence,

$$W(x) = (\textbf{if } g(x) \textbf{ then } (A; W)(x))$$

or

$$W(x) = g(x) * (W(A(x))) + (1 - g(x)) * x$$

At this point, we can conclude that the assignment statements, if-statements, and while-loop statements all can be expressed as functions. Can every statement be converted into a function statement? In a sense, this is true. In fact, as will be shown later, everything that is in some sense computable can be computed using function composition applied to a small number of given primitive functions. This fact is used by the functional programming languages (see Section 6.5 for details). On the other hand, the conversion is not always simple. In particular, the meaning of the goto statement may vary, depending on where it occurs. Goto statements can be used to form loops, in which case they should be replaced by while-loop statements. In other cases, they can be removed by merely rearranging the code, in which case they serve a syntactic, as opposed to semantic, function. In still other cases, they may replace an else clause. Hence, transfer statements, like the goto, make reasoning about programs more difficult, which is a further reason why they should be avoided.

Another logical difficulty is caused by parameter passing to functions and procedures. In fact, by passing a different set of variables to procedures, one changes the way the procedure affects the state. This means that, as far as the state is concerned, one really creates a set of different functions. Further complications arise due to the fact that a call may affect the state within a procedure. This happens if, in the call, two variables passed by reference are identical. These are advanced topics, and they are as such beyond the scope of this text.

3.4.3 Recursion in Programs

Recursive procedures are procedures that call themselves, either directly or indirectly. As such, they are similar to recursive definitions, which also make reference to the term to be defined. In fact, if a recursive procedure is to work properly, then its structure must be similar to a recursive definition. There must be one or more base cases, which are cases that can be resolved without using a recursive call. In all other cases, one makes use of recursive calls. This parallelism can be exploited when writing procedures manipulating structures such as trees, which are defined recursively.

We will use informal methods to prove that the recursive procedures discussed here work as specified. To do this, preconditions and postconditions of the procedures are formulated. The main proof is then done through well-founded induction. It is demonstrated that the program works correctly for all base cases. The inductive hypothesis is then that all recursive calls of the procedure work correctly, provided the initial states meet the preconditions of the procedure. In other words, one assumes that all calls made with states satisfying the preconditions will be handled appropriately. Given this hypothesis, the inductive step requires one to show that, under the inductive hypothesis, the program works correctly. Once this is done, one can conclude that the program runs correctly, given that the domain is well founded. We already discussed the problem of well-foundedness in connection with a recursive procedure in Section 3.3.4. It was found that if the domain is not well founded the program runs into an infinite loop. This is true in general. Hence, the issue of well-foundedness becomes essential. If the domain is not well founded, one can still guarantee that the procedure works correctly, provided it does not run into an infinite loop. This issue leads to the notion of *partial correctness*, as opposed to *total correctness*.

Definition 3.5:   A piece of code is said to be *partially correct* with respect to the precondition P and the postcondition Q if the final state of the code always satisfies Q, provided that the code starts in P and provided that it terminates. The code is said to be *totally correct* if for each state satisfying P the code terminates with a final state satisfying Q.

Hence, a code that, given termination, will yield the correct result is at least partially correct. To be totally correct, it must also terminate for all initial states satisfying the precondition.

Recursive procedures are partially correct if the base cases are correct and if the inductive step can be established. To prove total correctness, one must also show that the domain is well founded. As it turns out, it is convenient to prove well-foundedness last. Our setup is therefore changed accordingly. To prove well-foundedness, one can sometimes make use of the fact that the structures being considered are finite. In a finite domain, any G-sequence will terminate, given that no sequence can contain the same element more than once. In fact, if the domain contains m elements, where m is finite, and if no G-sequence contains the same element twice, then the longest possible G-sequence contains m elements.

All proofs given in this section are informal. There are also formal correctness proofs, and these will be discussed in Chapter 9. However, due to the difficulties of formulating procedure calls as functions, no formal correctness proofs involving procedure calls are given in this text. All proofs involving recursive procedures must therefore remain informal.

After these remarks, we present two examples. The first example deals with the binary search, a procedure readers are likely to know. This makes it easier for them to cast the problem in terms of well-founded induction. The problem input consists of a vector of natural numbers, called Vector, a single natural number, called Key, and a range over which to search, specified by the parameters First and Last. Two parameters, Found and Position, are used to return the result of the search. The vector is sorted in ascending order; that is, Vector[i+1] > Vector[i], except, of course, if i is the last element of the vector. This is a precondition for the procedure to work. If there is a value i such that Vector[i] = Key, where First $\leq$ i $\leq$ Last, then this value must be returned by the variable named Position. At this point, the variable Found must be set to true. If it can be proved that there is no i in the range such that Vector[i] = Key, then Found must be set to false. The procedure is given the name search(First, Last, Key, Vector, Found, Position). Since we are frequently dealing with ranges, let us agree that a range from a to b always includes both a and b as part of the range. Hence, the procedure search(First, Last, Key, Vector, Found, Position) must determine if Key is in a position with the range from First to Last and, if so, return its Position. Found indicates whether the search was successful.

We now have to identify the possible base cases, that is, the cases for which the solution is immediate. The solution is immediate if First > Last. In this case, the interval is empty, and Key cannot be found in an empty interval. Therefore, Found must be set to false. Cases when the interval consists of exactly one entry could also be used as base cases, but this is not necessary. Next, we come to the inductive, or recursive, cases. The problem is to construct a procedure for finding Key between First and Last under the assumption that such a procedure already exists for all proper subintervals of this interval. The interval is divided into half by the variable Middle defined by

$$\text{Middle} := (\text{First} + \text{Last}) \ \textbf{div} \ 2$$

If Key < Vector[Middle], then Key can only be in the interval between First and Middle−1, and this is a proper subinterval of the interval from First to Last. Hence, search(First, Middle−1, Key, Vector, Found, Position) works correctly according to our inductive hypothesis. Similarly, if Key > Vector[Middle], then Key is within the interval from Middle+1 to Last, and in this case search(Middle+1, Last, Key, Vector, Found, Position) works correctly. Finally, if Key=Vector[Middle], then Found:= true, and the variable Position must be set to Middle. This completes the inductive case. Hence, the procedure is at least partially correct; that is, if it ever returns a result, this result must be correct. The procedure is given in Figure 3.4.

```
procedure search(First, Last, Key : integer;
                 Vector : Vec_Type;
                 var Found : boolean;
                 var Position : integer);
begin
  if First > Last
    then  {Base Case}
      Found := false
    else
      begin  {Recurrent Cases}
        Middle := (First + Last) div 2;
        if Key < Vector[ Middle]
          then search(First, Middle−1, Key, Vector, Found, Position);
          else
            if Key > Vector[ Middle]
              then
                search(Middle+1, Last, Key, Vector, Found, Position);
              else
                begin
                  Position := Middle;
                  Found := true
                end
      end
end;
```

Figure 3.4.   Binary search

To show that the procedure is totally correct, it must be shown that it will always terminate or, equivalently, that the domain is well founded. The domain consists of all intervals from a to b, where a and b can assume any values inside the range of the array Vector. The predicate $G(x, y)$ to be selected is "y is a proper subinterval of x," and the sequence obtained by this predicate is a sequence of proper subintervals. Since any such sequence must be finite, the domain is well founded.

At this point, we can express the fact that the procedure works as specified in a more standardized format.

Inductive base   The base case is the case in which there is no element between First and Last, and in this case Found := **false** as in Figure 3.4.

Inductive hypothesis   Choose any values for First and Last. These two values provide an interval, and the inductive hypothesis is that the procedure works for all proper subintervals of this interval.

Inductive step   Since the array Vector is sorted, Key must be in the interval from First to Middle−1 if Key < Vector[Middle], and it must be in the interval from Middle+1 to Last if Key > Vector[Middle]. Both intervals are proper subintervals of the interval with the end points First and Last and, according to the inductive hypothesis, the procedure is correct for these intervals. Under this hypothesis, the procedure works correctly.

Domain well founded   The domain is well founded because all sequences in which the subsequent element is a proper subinterval of the preceding element must be finite.

Conclusion   The procedure is correct in the base cases. If the inductive hypothesis holds, it also works correctly in the recursive cases. Consequently, the procedure works correctly for all initial states satisfying the precondition that the vector is sorted.

The proof can be used to design a test procedure. If the procedure gives wrong answers in the base case, then there must be an error in that part of the procedure. If the procedure runs into an infinite loop, then the domain is not well founded. Any other errors must be in the recursive case, and the inductive hypothesis, together with the inductive step, has to be checked once more.

The next example deals with the merge sort. The problem can be described as follows: One has again a vector Vector of natural numbers, and the problem is to bring all elements of the vector from positions First to Last into an ascending order. The procedure is to have the name sort(First, Last, Vector). To write this procedure, use should be made of an already existing procedure, merge(First, Middle, Last, Vector). This procedure merges the values in the two intervals from First to Middle and from Middle+1 to Last, provided that the values in these two intervals are already sorted. Merging is then used to assure that, once merge is completed, all numbers in the combined interval from First to Last are sorted. Hence, the procedure merge has the precondition that the values in the two subintervals must be sorted, and it has the postcondition that the values in the combined interval are sorted. A program that accomplishes this is easily written, but is not provided here.

A subinterval consisting of a single entry is sorted. This is the base case, and no action is required for this base case. The inductive hypothesis is that sort(Start, Finish,

Vector) works for all cases in which the subinterval from Start to Finish is strictly smaller than the subinterval from First to Last. If this is the case, then one can divide the interval into nonempty subintervals, say from First to Middle and from Middle+1 to Last, and sort the values in both intervals independently by using sort(First, Middle, Vector) and sort(Middle+1, Last, Vector). According to the inductive hypothesis, they both give the correct result. It is best to split the interval in the middle, that is, at Middle := (First+Last) **div** 2. Once the values in both intervals are sorted, they can be merged by merge(First, Middle, Last, Vector). This setup leads to the program in Figure 3.5.

```
procedure sort(First, Last : integer;
                    var Vector : Vec_Type);
    var
        Middle : integer;
    begin
        {No code for base case}
        if First < Last then
          begin  {Recurrent cases}
              Middle := (First + Last) div 2;
              sort(First, Middle, Vector);
              sort(Middle+1, Last, Vector);
              merge(First, Middle, Last, Vector)
          end {Recurrent cases}
end; {sort}
```

Figure 3.5. A recursive merge–sort routine

We now justify the program by well-founded induction.

Inductive base   If First = Last, nothing needs to be done. Hence, no code is needed for the base case.

Inductive hypothesis   Assume that an interval extending from First to Last is given and that, for all subintervals of this interval, sort works correctly. In other words, one must require that the initial value a be greater than or equal to First and that the final value b be less than or equal to Last, and that either a be different from First or that b be different from Last.

Inductive step   The two subintervals given from First to Middle and from Middle + 1 to Last are both proper subintervals of the original interval. This is true because in the recursive case Middle is always less than Last, and Middle + 1 is always greater than First. According to the inductive hypothesis, sort(First, Middle, Vector) and sort(Middle+1, Last, Vector) sort the elements of the vector within each of the two subintervals. This satisfies the precondition of the procedure merge, and merge therefore satisfies the postcondition that the array from First to Last be sorted. This completes the inductive step.

Well-founded domain   Sequences of proper subintervals always end, which means that the domain is well founded.

Conclusion   By proving that the procedure handles the base cases correctly and by proving that, under the inductive hypothesis, the recursive case is handled correctly, we have shown that the procedure works correctly for all possible initial states.

3.4.4 Programs Involving Trees

Since trees can be defined recursively, recursive procedures are ideally suited to deal with them. Before we begin our discussion on trees, some general remarks on trees and their internal representation are in order, and an appropriate storage structure for dealing with trees must be created. The corresponding Pascal type will be called TreeType. Essentially, a variable of type TreeType is a pointer to the root of the tree. If the tree is empty, this pointer is **nil**. Trees are mainly used to store and access a certain type of information. For instance, the set of all customers of a company could be organized in the form of a tree. In this case, the information to be stored by each node would be the name of the customer, the address, and similar information. In an organization diagram, the information to be stored is the job designation, together with the name of the official. In the parse tree of a logical expression, the root contains the main connective, possibly together with some description, and the children contain immediate subconnectives, again possibly followed by some description, and so on. To deal with nodes of trees in general, whether they are customers, positions, connectives, or other persons or objects, we create a Pascal type InfoType. The actual information stored by the node will be called Info.

The arcs of a tree are implemented by pointers. The pointers representing the arcs are by convention considered as part of the nodes from which they emanate. Consequently, each node in a binary tree has two pointers, which may be called Ltree and Rtree and which point to the left and right subtree, respectively. (Strictly speaking, the correct terms are immediate left and right subtrees, but we will omit the word immediate as this should cause no confusion.) This leads to the following Pascal type for representing a tree.

```
type
    TreeType = ↑ NodeType;
    NodeType = record
        Info : InfoType;
        Ltree, Rtree : TreeType
    end;
```

One frequently has to do some operation on a tree that involves every node. In some applications, one may want to print every node, in others, copy the entire tree node by node, and in others still, merely count the number of nodes in a tree. To do these operations, a *tree traversal* is required. Tree traversals can be done in different orders. We discuss a *preorder traversal* first. In a preorder traversal, one processes the root first, then the left subtree, and finally the right subtree. This leads to the following general algorithm:

1. If the tree is empty, return. This is the base case.
2. Process the root node.
3. Traverse the left subtree in preorder.
4. Traverse the right subtree in preorder.

Depending on the application, the word process can mean different things. For instance, if the tree is to be printed, then "process node" merely means "print contents of node." If the nodes are to be counted, then process means "add one to total," and so on. The procedure DisplayPreorder given in Figure 3.6 prints all nodes of a tree in preorder. The procedure uses a subprocedure PrintNode(Root), which prints the information contained in the node pointed at by Root. We now prove that this program in fact prints the contents of all nodes of the tree.

```
procedure DisplayPreorder(Root : TreeType);
  begin  {Step 1: Base case: Tree empty. Simply return}
    if Root <> nil then
      begin
          {Step 2: Display information of root node}
          PrintNode(Root);

          {Step 3: Traverse left subtree}
          DisplayPreorder(Root↑.Ltree);

          {Step 4: Traverse right subtree}
          DisplayPreorder(Root↑.Rtree);
      end
  end;
```

Figure 3.6. Procedure for printing a tree in preorder

Inductive base   If the tree is empty, there are no nodes, and the fact that all these nodes are printed is trivially true.

Inductive hypothesis   Pick any tree, and assume that the procedure DisplayPreorder prints all nodes of any subtree of this tree.

Inductive step   The procedure prints the root first. All other nodes must either be in the left subtree, in which case they are printed in step 3 of the program, or they must be in the right subtree, in which case they are printed by step 4 of the program.

Domain well founded   Let $G(x, y)$ be true if node x contains a pointer to y, either Ltree or Rtree. Since computer memory is finite, all G-sequences must eventually end, unless some nodes are repeated. Hence, if Root points to a genuine tree, then all G-sequences end. If, however, some pointers point to previously used nodes, this is no longer true.

Conclusion   The procedure is correct for all genuine binary trees.

Traversing trees in inorder is similar. The general algorithms for traversing a tree in inorder is as follows:

1. If the tree is empty, return. This is the base case.
2. Traverse the left subtree in inorder.
3. Process the root node.
4. Traverse the right subtree in inorder.

Hence, compared to the preorder traversal, steps 2 and 3 are interchanged. To print a tree in inorder, one can similarly change steps 3 and 4 in the procedure DisplayPreorder. Finally, trees can be traversed in postorder. In postorder, the tree is processed in the order "left subtree, right subtree, root," as indicated by the following general algorithm:

1. If the tree is empty, return. This is the base case.
2. Traverse the left subtree in postorder.
3. Traverse the right subtree in postorder.
4. Process the root node.

The next example involves copying a tree. The copy must have nodes that are at different storage locations from the ones in the original; otherwise, the new tree cannot be updated without affecting the original tree. To do the copying, the root is copied first, then the left subtree, and then the right subtree. This leads to the following general algorithm:

1. If the tree is empty, return an empty pointer. This is the base case.
2. Copy the root: Create a new tree, and copy the information of the root into the new root node.
3. Copy the left subtree, which becomes the left pointer of the new root.
4. Copy the right subtree, which becomes the right pointer of the new root.

The Pascal implementation of this algorithm is given in Figure 3.7. We now prove that this procedure is correct.

```
function Copy(Tree : TreeType) : TreeType;
    { This function returns a copy of the tree Tree}
  var
    NewTree : TreeType;       {Pointer to new tree}
  begin
    if Tree = nil then
      Copy := nil    { Tree empty}
    else
      begin
        { Create and Copy Root}
        new(NewTree);
        NewTree↑.Info := Tree↑.Info;
        Copy := NewTree;

        { Copy left and right subtrees}
        NewTree↑.Ltree := Copy(Tree↑.Ltree);
        NewTree↑.Rtree := Copy(Tree↑.Rtree)
      end
  end;
```

Figure 3.7.   Function to copy a tree

Inductive base   The function Copy works for the empty tree, which is the base case.

Inductive hypothesis   Given the tree Tree, assume that the function Copy works correctly for the subtrees of Tree. In particular, Copy(Tree↑ .Ltree) returns a copy of the left subtree of Tree, and Copy(Tree↑.Rtree) returns a copy of the right subtree of Tree.

Inductive step   Note that the root node of Tree is correctly copied by the assignment statement. By the inductive hypothesis, Copy(Tree↑.Ltree) and Copy(Tree↑.Rtree) return pointers to copies of the left and right subtrees of Tree. The values returned by Copy must still be stored by the appropriate pointers of the new tree. This is done by the corresponding assignment statements. This completes the inductive step.

Domain   The domain is well founded if the pointers do, in fact, represent a binary tree. In particular, no pointer to a child must point toward itself or to an ancestor. If this is the case and if the descending sequences in question are proper subtrees, then all descending sequences must end with the empty tree.

Conclusion   From the basis and the inductive step, one concludes that Copy works for any tree. To be a tree, of course, no pointer must point to an ancestor.

Problems 3.4

1. Can you define preorder and postorder for lists? If so, what is the effect of printing [a, b, c] in (a) preorder? (b) postorder?

2. A 3-tree is a tree in which every node can have up to three children. Write a program that copies a 3-tree.

3. Write a recursive procedure to implement the function $C(n, k)$, which is given as follows:

$$C(n, 0) = 1$$
$$C(n, k + 1) = C(n, k) * (n - k)/(k + 1)$$

Indicate for which values of k your procedure will terminate. Do this by means of a proof by recursion.

4. Write a procedure that compares two binary trees. The procedure should return true if both trees are equal in structure and in the information stored in corresponding nodes. Otherwise, it should return false.

5. Write a recursive procedure to print all elements of a list.

6. Write a recursive procedure to make a copy of a list.

3.5 RECURSIVE FUNCTIONS

3.5.1 Introduction

A basic question in computer science, as well as in logic and mathematics, is what can be computed and what cannot be computed. More specifically, if a particular computational problem is given, the question is whether there is an effective method to solve this computational problem. By this, we mean that there is some computer that can solve the problem within a finite time period. How much time is needed to find such a solution, whether it is minutes, days, years or centuries, is irrelevant for the present discussion. We are only concerned with the possibility of a solution in principle. Whether this procedure is feasible in practice will be addressed in Chapter 6 (see Section 6.3).

It is important to distinguish between a problem in general and a particular instance of a problem. For instance, determining the validity of an arbitrary logical expression is a problem in general. A particular instance of this problem is the determination of the validity of a specific logical expression. A computer procedure corresponding to this problem would then have a particular instance of the problem as input, and it would indicate in its output whether the particular instance is valid. The question is now if there are procedures that can solve all instances of a certain problem within a finite time span. For our example, this question becomes "Is there a procedure which can decide for all logical expressions whether they are valid?" For logical expressions from propositional calculus, there is such a procedure. The problem of validity in propositional calculus is in this sense *decidable*. However, there is no procedure that can determine for arbitrary expressions from predicate calculus whether they are valid. The problem of proving the validity of a general expression from predicate calculus is *undecidable*. It is important to see what this means. It is very well possible that a procedure can be designed that will indicate for most logical expressions containing predicates whether they are valid. However, it is a fact that there are some expressions for which validity cannot be proved by any procedure. Specifically, the procedure will never terminate. Note that we have used the word procedure rather than algorithm in our discussion. The difference between a procedure and an algorithm is that an algorithm always terminates, whereas a procedure may not. Hence, a problem is undecidable if and only if there is no algorithm for it.

To put the notion of undecidability on a firmer footing, one has to indicate what operations are allowed. There are several ways to state in a mathematical way what can be computed and what cannot be computed. In 1936, Alan M. Turing defined a machine, which is now called a *Turing machine*, and he showed that everything that can be computed in an intuitive sense can also be computed on a Turing machine. In 1931, Gödel introduced a set of functions that he called *recursive*. These functions were generalized by Kleene in 1934 and this gave rise to what Kleene called *generalized recursive functions*. It turns out that everything that can be calculated on a Turing machine can be expressed in terms of generalized recursive functions. This later led Church to formulate his famous thesis that indicates that everything one can calculate in an intuitive sense can be expressed in terms of generalized recursive functions. For details and further references, the reader may consult G. Hunter [26].

Today, the term recursive function is used to describe what Kleene called generalized recursive functions, and for the functions discovered by Gödel, the term primitive recursive function is used. This section deals with recursive functions. It is shown that everything that can be expressed in Pascal can also be expressed in terms of recursive functions. Recursive functions are useful in other ways as well. In particular, to show that a programming language has the same power as Pascal, one merely has to prove that in this language, it is possible to program all recursive functions. Recursive functions provide, in this fashion, a standard against which every programming language can be compared.

The recursive functions contain an important subclass of functions, the *primitive recursive functions*. The primitive recursive functions will be discussed first. It can be proved that all primitive recursive functions are decidable; that is, there is an algorithm that can evaluate any primitive recursive function, no matter what its arguments. There are recursive functions that are not primitive recursive. These functions are formed by adding one additional operation, *minimalization*, to the primitive recursive functions. Unfortunately, procedures containing minimalization may never terminate. In fact, some problems that can only be solved by making use of minimalization turn out to be undecidable. One problem belonging to the class of undecidable problems is the problem of finding if an arbitrary logical expression is valid. In Section 3.5.4, the *halting problem* will be explored. The halting problem involves finding whether an arbitrary given computer program will ever terminate. As it turns out, the halting problem is undecidable.

3.5.2 Primitive Recursive Functions

Essentially, primitive recursive functions are functions that can be generated by what is called *primitive recursion* from a set of initial functions, employing methods similar to the ones used in the definition of addition and multiplication [see equations (3.1)–(3.4)]. For the sake of simplicity, we define primitive recursive functions somewhat informally by using what we call *primitive recursive expressions*. Instead of expressions, most authors use *function compositions*, but this really amounts to the same thing. The basic building blocks of primitive recursive expressions are the constant 0, a set of variables, say x_1, x_2, . . . , and a set of function symbols. Moreover, there is an arity connected with each function symbol. The formation rules for the primitive recursive expressions can now be formulated as follows:

1. 0 is an expression.
2. Every variable is an expression.
3. If f is a function symbol of arity d and if t_1, t_2, . . . , t_d are expressions, then $f(t_1, t_2, \ldots, t_d)$ is an expression.

One function is given initially, and this is the successor function $s(n)$. Every other function must formally be defined before it can be used in primitive recursive expressions. Functions may be defined either by *explicit definitions* or by *primitive recursion*. An explicit definition of function f with the arguments $x_1, x_2, \ldots, x_d$ is given as

$$f(x_1, x_2, \ldots, x_d) = e$$

Here, e is any expression not containing f and not containing functions not yet defined.

Example 3.26   Give an explicit definition of the function $f(x_1, x_2, \ldots, x_d)$, which yields x_1 for all values of $x_1, x_2, \ldots, x_d$.

   Solution

$$f(x_1, x_2, \ldots, x_d) = x_1$$ ∎

Any function that yields just one of its n arguments is called a *projection function*. A function of d arguments has d projection functions. A function of one argument therefore has one projection function, $f(x) = x$. This function is called the *identity function*.

If a function f is to be defined by primitive recursion, one must first select one variable of the function, say x_d, as the *recursion variable*. The definition of f consists of a *base part* and a *recursive part*. These two parts can be written as

$$f(x_1, x_2, \ldots, x_{d-1}, 0) = e_0 \tag{3.22}$$

$$f(x_1, x_2, \ldots, x_{d-1}, s(x_d)) = e_r \tag{3.23}$$

The base part of the definition is given by (3.22). The function to be defined is on the left, and it is to be taken at $x_d = 0$. The right side of (3.22), which is e_0, must be an expression not containing the function f. The recursive part is given by (3.23), and the function f on the left contains $s(x_d)$ in place of x_d. The function is defined in terms of an expression e_r. This expression may contain $f(x_1, x_2, \ldots, x_d)$ as a term, but it cannot contain function f with any other arguments. Moreover, neither e_0 nor e_r may contain any functions not yet defined.

Example 3.27   Find e_0 and e_r in the case of addition. To do this, use the function symbol *add* for addition.

Solution   One can write (3.1) and (3.2) to yield

$$add(x, 0) = x$$
$$add(x, s(y)) = s(add(x, y))$$

By comparing the first of these equations with (3.22), one finds that $e_0 = x$, and according to the formation rules given, this is well formed. By comparing the second of these equations with (3.23), one finds that $e_r = s(add(x, y))$, which is again a well-formed expression.   ∎

Example 3.28   Find e_0 and e_r in the case of multiplication. To do this, use the function symbol *mult* for multiplication. Assume that the function *add* is already defined.

Solution   One has

$$mult(x, 0) = 0$$
$$mult(x, s(y)) = add(mult(x, y), x)$$

By comparing the first of these equations with (3.22), one finds that $e_0 = 0$ and $e_r = add(mult(x, y), x)$. Both of these expressions are well formed according to our formation rules.   ∎

The set of primitive recursive functions is the set that can be formulated either explicitly or by primitive recursion. Most of the functions used in practical computations are primitive recursive functions. A number of important primitive recursive functions are now formally defined. To make this easier, we use the normal operator symbols; that is, we again use $+$ for addition and $\times$ for multiplication.

The first function to be defined is the predecessor function $p(x)$. The predecessor of x is the number immediately preceding x, or $x - 1$. Of course, $s(x)$ has x as predecessor; that is, $p(s(x)) = x$. 0 has no predecessor, but to convert $p(x)$ into a total function, one arbitrarily sets $p(0) = 0$. The result is the following definition of $p(x)$ by primitive recursion.

$$p(0) = 0$$
$$p(s(x)) = x$$

As the next function, we define subtraction within the domain of natural numbers. Since there are no negative natural numbers, subtraction is undefined if the minuend is smaller than the subtrahend. We therefore introduce a different form of subtraction, called *proper subtraction*, which is denoted by $\dot{-}$. Specifically, $x \dot{-} y$ is $x - y$ if $x \geq y$, and $x \dot{-} y = 0$ if $x \leq y$. Proper subtraction is primitive recursive, as one can see from the following two equations:

$$x \dot{-} 0 = x$$
$$x \dot{-} s(y) = p(x \dot{-} y)$$

The use of the infix notation should not detract from the fact that the operator $\dot{-}$ is in fact defined by primitive recursion. To see this, the reader may want to rewrite the definition, using $sub(x, y)$ instead of $x \dot{-} y$. As in the earlier recursive definitions, one has a base case, which deals with $y = 0$, and a recursive case, which contains $s(y)$ on the left of the definition and which contains $x \dot{-} y$ as part of the expression on the right. The recursive part of the definition is correct because $x - (y + 1) = (x - y) - 1$.

If 0 is used to denote false and if 1 is used to denote true, then all logical connectives turn out to be primitive recursive. To ensure that the numbers in question are always 0 and 1, one must make use of the signum function $sg(x)$ and the inverse signum function $isg(x)$. These two functions are defined as follows:

$$sg(0) = 0$$
$$sg(s(x)) = 1$$
$$isg(0) = 1$$
$$isg(s(x)) = 0$$

Note that $sg(x)$ is really the truth value of $x > 0$, and $isg(x)$ is the truth value of $x = 0$. Suppose now that p and q are Boolean variables; that is, they can only be 0 or 1. We now define the following Boolean functions:

$$and(p, q) = sg(p \times q)$$
$$or(p, q) = sg(p + q)$$
$$not(p) = isg(p)$$

Hence, all Boolean functions are primitive recursive as stated.

Frequently, one needs functions that are defined by different expressions in different ranges. To allow this, a function *if* is defined as follows:

$$if(x, y, z) = (y \times sg(x)) + (z \times isg(x))$$

Hence, $if(x, y, z) = y$ if $x > 0$ and $if(x, y, z) = z$ if $x = 0$. If x is Boolean, then $if(x, y, z)$ is y if x is true and z otherwise. For instance, the absolute difference is given as

$$absdiff(x, y) = if(x \dot{-} y, x \dot{-} y, y \dot{-} x)$$

Of course, two numbers x and y are equal if their absolute difference is zero.

$$eq(x, y) = isg(absdiff)(x, y)$$

Computers use integers rather than natural numbers. Integers can be represented by the pair (s, x), where s is the sign and x is the absolute amount. All operations on integers must be implemented by two functions, one function for the sign s and the other for the absolute amount x. For instance, if the two integers (s_1, x_1) and (s_2, x_2) must be added, one must have one function to express the sign of the result, while another function is needed to find its absolute amount. If $sadd$ and $aadd$ denote these two functions, one has

$$sadd(s_1, x_1, s_2, x_2) = if(x_1 \dot{-} x_2, s_1, s_2)$$
$$aadd(s_1, x_1, s_2, x_2) = if(eq(s_1, s_2), absdiff(x_1, x_2), x_1 + x_2)$$

The first of these definitions expresses the fact that the sign of the result must be equal to the sign of the larger number. The second definition calculates the absolute amount of the sum as follows: If the signs of the two integers to be added are the same, then the absolute amount of the result is obtained by adding the absolute amount of the summands. Otherwise, one must take the absolute difference of the two amounts.

In *parallel recursion*, one defines several functions simultaneously by recursion. For instance, suppose that one wants to define the following function g recursively:

$$g(n) = \sum_{i=0}^{n-1} i!$$

The most natural way to define g is as follows: Let $f(n) = n!$, and define both g and f simultaneously as

$$g(0) = 0, \qquad\qquad\qquad f(0) = s(0)$$
$$g(s(n)) = g(n) + f(n), \qquad f(s(n)) = s(n) \times f(n)$$

In this way, one defines the two functions g and f concurrently, that is, by parallel recursion. Formally, k functions in d variables can be defined by parallel recursion with the following scheme:

$$f_i(x_1, x_2, \ldots, x_{d-1}, 0) = e_0^i, \qquad i = 1, 2, \ldots, k$$
$$f_i(x_1, x_2, \ldots, x_{d-1}, s(x_d)) = e_r^i, \qquad i = 1, 2, \ldots, k$$

Although parallel recursion looks very much like primitive recursion, it is a different technique. All functions that can be defined by parallel recursion can also be defined by primitive recursion, however. The proof that this is the case is difficult and will not be given here.

3.5.3 Programming and Primitive Recursion

Given enough computer time, it is possible to compute the value of any primitive recursive function for any configuration of its arguments. The easiest way to do this is by using recursion. If e_0 and e_r are the expressions defining f according to (3.22) and (3.23), respectively, one has

```
function f(x₁, x₂, ..., x_d)
   begin
      if x_d = 0 then
      f := e₀
      else
      f := S^{x_d}_{p(x_d)} e_r
   end
```

The code must be provided for all functions used in e_0 and e_r. The resulting recursion is well founded because, for calculating e_0, one only needs functions that have been defined previously. Also, for calculating e_r one only needs $f(x_1, x_2, \ldots, p(x_d))$, in addition to the previously defined functions. Hence, the calculations always come to a conclusion. The implementation is not particularly efficient because the function f is evaluated several times for the same arguments. To avoid this at least partially, one can first calculate all previously defined functions needed for e_0 and e_r and proceed as follows:

```
evaluate all functions appearing in e₀
f(x₁, x₂, ..., x_{d-1}, 0) := e₀
for i := 1 to x_d do
    evaluate all functions appearing in S^{x_d}_{i-1} e_r
    f(x₁, x₂, ..., x_{d-1}, i) := S^{x_d}_{i-1} e_r
```

Here, the functions on the left sides of the assignment statements must still be replaced by some variable name, say f_X. In conclusion, all primitive recursive functions can be effectively evaluated in Pascal in the sense that, for a given combination of arguments, one can guarantee that the program will eventually halt and return the required answer.

The next question that arises is whether every Pascal procedure can be converted into a set of primitive recursive functions. The answer to this question is negative, and the reason is simple: All primitive recursive functions can be implemented by Pascal procedures that terminate. Hence, if it were possible to convert every Pascal procedure into a set of primitive recursive functions, then one could convert every Pascal program into a program that terminates. Since there are Pascal programs that do not terminate, this is false. Consequently, nonterminating Pascal programs cannot be converted into primitive recursive functions. Since both recursive procedure calls and uncounted-loop statements, such as the while-loop statements and the repeat-loop statements, can result in nonterminating programs, one concludes that none of these two constructs is expressible in terms of primitive recursion.

If recursive procedures and uncounted-loop statements are eliminated, then one can express every Pascal program in terms of primitive recursive functions. We will not prove this here, but we will give some indication as to how such a program can be converted into a set of primitive recursive functions. First, any Pascal program containing nonrecursive procedures can be converted into a program containing no procedures. In other words, procedures are not needed, even though their use makes programming much more modular. Arrays are not really needed either: Instead of an array of integers, one can use a single variable that is of such a length that it can accommodate all the digits of all the integers in the array. Of course, using such a device is extremely inconvenient, but it can be done. For a similar reason, one can eliminate pointers. All variables can be represented as integers, which means that one only needs the standard operations on integers, that is, addition, multiplication, subtraction, and division. All these operations are primitive recursive. Since sequences of assignments statements can be expressed as function compositions, they, too, are primitive recursive. As was shown, the if-statement is also primitive recursive. The case statement can be converted into a nested if-statement. At this point, the only statement left is the for-statement. To complete our discussion, we now show how the for-statement can be converted into a primitive recursive function. To do this, suppose that the counted loop goes from 1 to m and uses n as its index. Moreover, assume that all variables besides m and n are combined into a single variable y. The body of the for-statement can now be expressed by a single function, say $y = g(y, n)$, where n is the index of the loop. The function corresponding to this loop is f. In other words, f can be expressed as

$$f = (\textbf{for } n := 1 \textbf{ to } m \textbf{ do } y := g(y, n))$$

It is possible to express f in terms of primitive recursion. In fact, one has

$$f(y, 0) = y$$
$$f(y, s(m)) = g(f(y, m), s(m))$$

The reader may prove that this construct indeed implements f. This shows that counted-loops statements can also be converted to primitive recursive functions, as claimed. In conclusion, primitive recursive functions are extremely powerful, and almost every Pascal program can be converted into a set of primitive recursive functions.

3.5.4 Minimalization

Primitive recursive functions are sufficient to express any algorithm that contains neither uncounted-loop statements nor recursive procedure calls. Since all recursive procedure calls can be converted into while statements, as the reader may verify, all that is needed to make the set of primitive recursive functions complete is an equivalent to the Pascal while statement. This facility is provided through *minimalization*. Minimalization involves finding the smallest number satisfying a certain condition. For instance, finding the smallest prime greater than 1000 is a problem of minimalization, and so is finding the smallest number $n > 2$ satisfying $x^n + y^n = z^n$, where x, y, and z are integers. Unfortunately, there is no n with this property, as was recently proved in the literature. Hence, when searching for such an n, an infinite loop results. In some cases, nobody knows if there is a smallest n satisfying a certain condition, and in this case the search for such an n may or may not come to a conclusion. One such problem is the question of whether there exists an odd perfect number. A number n is perfect if the sum of its factors, including 1 but not n, is equal to n. For instance, 28 is a perfect number, because 28 has the factors 1, 2, 4, 7, and 14, and $1 + 2 + 4 + 7 + 14 = 28$, which proves that 28 is a perfect number. To find the smallest odd perfect number, if one exists, would involve the use of minimalization. Unfortunately, if no odd perfect number existed, then this search would never end.

The class of *general recursive functions* comprises all functions that are either primitive recursive or that can be obtained using minimalization involving only primitive recursive functions. All general recursive functions can be evaluated in Pascal. For minimalization, one merely needs a while statement. In fact, the smallest n satisfying a certain condition $P(n)$ can be found by the following program:

```
n := 0;
while not P(n) do   n := n + 1
```

Conversely, all Pascal programs can be reduced to general recursive functions. In other words, the general recursive functions comprise exactly the functions for which it is possible to design procedures in Pascal.

Pascal programs containing while statements may never end, and the same is true for general recursive functions. General recursive functions may, in this sense, turn out to be undecidable. This brings us back to the problem of undecidability. A famous undecidable problem, the *halting problem*, is now discussed.

It would be desirable if one could determine for each program whether it terminates for all possible inputs. If such a procedure existed, one could tell in advance whether there is a risk that the program will never end. Unfortunately, there is no such procedure. This is the famous halting problem. Specifically, the halting problem involves the design of a procedure that can predict for any arbitrary program if it will ever come to a conclusion. Unfortunately, we now show that the halting problem is undecidable.

Consider a function, say stops(x), that has a program as its argument and yields 0 if the program does not stop and 1 otherwise. Of course, programs can be considered as strings, and each character in the string has a numerical representation. In this sense, one can say that x is an integer given by the internal computer representation of the program x. To prove that stops(x) is undecidable, we assume that stops always terminates, and we derive a contradiction. To this end, consider the following program skeleton, which we denote as B.

```
read(x);
while stops(x) = 1 do   print("still running")
```

We are free to set the input x to any value, including x = B. Hence, we execute B, with B itself as input. Assume now that stops is a decidable function. This means that B, which has itself as input, will either terminate or not terminate. If B terminates, then stops(B) must return 1, which means that the while statement runs forever. Hence, the very termination of B implies that B will not terminate. This is a contradiction, and we conclude that the premise that B terminates is false. Unfortunately, if B does not terminate, then stops(B) is 0, and if stops(B) = 0, the while statement in B terminates. Again, a contradiction is derived. Hence, whether B terminates or not, a contradiction can be derived. This means that the assumption that stops(B) terminates must be false and therefore the halting problem is undecidable.

The fact that there are undecidable problems has the consequence that one is never sure how long to wait for a procedure to return with an answer. For instance, if one wants to find the smallest odd perfect number, one could merely design a procedure to determine whether each individual odd number is perfect. However, nobody can tell whether this search will ever end. No matter how long the search lasts, the very next number may be the perfect number. Moreover, because of the halting problem, the problem of determining if the procedure for finding the smallest odd perfect number will end is also undecidable. Of course, one may say that finding a perfect number may not be all that relevant, but this misses the point. To see this, just change the problem. Suppose, for instance, that the search is for the smallest encryption method having certain desirable properties. With the growth of computer crimes, such schemes may be of great economic importance. One could search for such a scheme, but this problem may turn out to be undecidable. Worse still, it cannot be decided whether this problem *is* decidable. The fact that there are undecidable problems thus imposes a real limit as to what can ever be computed.

Problems 3.5

1. Let $f(n) = 0 + 1 + \cdots + n$. Express $f(n)$ as a primitive recursive function. Assume that $s(n)$ and $plus(n, m)$ are the only given functions.

2. Express x^y (x to the power of y) as a primitive recursive function. You may assume $+$ and $\times$ have already been defined.

3. Express integer multiplication in terms of primitive recursive functions. To do this, represent each factor by its sign and its amount.

4. Let $H(0) = 0$ and $H(n + 1) = 2 + H(n)$. Is $H(n)$ a primitive recursive function?

5. The following program calculates c, using a as input. Express c as a function of a.

$$b := 3 + a;$$
$$c := 2 + a;$$
$$c := c * b;$$

6. Express the final value of the variable sum as a primitive recursive function.

$$\text{for } i := 1 \text{ to } n \text{ do} \quad \text{sum} := \text{sum} + i * i$$

7. Use the if-function to express $f(n) = n \bmod 3$ as a primitive recursive function.

PROBLEMS: CHAPTER 3

1. Prove that for all $n \geq 0$, $7^n - 2^n$ is divisible by 5.

2. Try all integers from 0 onward until you find an integer $n = n_0$ satisfying $3n + 2 \leq n^2$. Prove that for all $n > n_0$ this inequality holds.

3. Suppose that all atomic formulas are T and all compound expressions are of the form $(A \wedge B)$ or $(A \vee B)$. Show that all expressions obtained in this fashion yield T. What can you say if instead of T, the only atomic formulas are F? Use this observation to show that if the only atomic formulas are propositional variables the resulting expression cannot be a tautology.

4. Use a proof by recursion to show that all logical expressions can be written using $\Rightarrow$ and $\neg$ as their only connectives. To do this, use the fact that, for any propositions P and Q, $P \vee Q$ can be expressed as $\neg P \Rightarrow Q$, and $P \wedge Q$ can be expressed as $\neg(P \Rightarrow \neg Q)$.

5. Convert the following Pascal fragment into a primitive recursive function.

```
j := 1;
sum := 0;
for i := 1 to n do
    begin
        sum := sum + j;
        j := j * 2
    end
```

6. Write a program containing a recursive procedure to count all empty subtrees within a binary tree.

7. Suppose that each node of a tree contains a number. Write a recursive program that finds the sum of all these numbers.

8. Describe the data structure needed to represent well-formed formulas of propositional calculus as binary trees. Suppose that a truth value is associated with each leaf node. Write an algorithm to find the truth value of the expression.

9. Express $f(m) = \sum_{n=0}^{m} n(n+1)$ in terms of primitive recursion.

10. Prove that if $a_i < b_i, i = 1, 2, \ldots, n$, then $\sum_{i=1}^{n} a_i < \sum_{i=1}^{n} b_i$.

11. Let $\circ$ be an associative binary operator, and let a be a variable. Consider the U-expressions formed as follows:
   (1) a is a U-expression.
   (2) If A and B are two U-expressions, so is $(A \circ B)$.
   (3) Nothing else is a U-expression.

   Prove that all U-expressions can be written as a^n, where $a^1 = a$ and where $a^{n+1} = a \circ a^n$.

12. Prove that all logical expressions without constants, and containing no variable more than once, must be contingent.

13. Use induction, together with $(A \Rightarrow B) \Rightarrow C \equiv (A \wedge B) \Rightarrow C$ to prove for $n \geq 1$ that $P_1 \Rightarrow P_2 \Rightarrow \ldots \Rightarrow P_n \equiv (P_1 \wedge P_2 \wedge \ldots \wedge P_{n-1}) \Rightarrow P_n$.

14. A company has constructed a tree to keep track of the amounts each customer owes. Specifically, one has

```
type
    InfoType = record
        Name : string;
        AmountOwing : real;
    end;
```

Section 3.4.4 contains a Pascal declaration for the nodes of the tree.
   (a) Write a program that prints all customers owing over $1000.00.
   (b) Write a program that constructs a tree containing only the customers owing over $1000.00.

15. Student records are entered in a tree as follows. There is a data structure InfoType, defined as

type
    InfoType = **record**
        Name : **string**;
        AverageForYear : **real**;
        NumberOfClasses : **integer**;
    end;

where the node type is given in Section 3.4.4. The records are in lexicographical order in the sense that all names in the left subtree of a node precede the name in the node, which in turn precedes all names in the right subtree.

(a) Write a program that prints the student record, given his/her name.

(b) Write a program that finds the average number of classes taken by the students.

4

Prolog

Prolog is a language based on the rules of logic. A Prolog program is expressed in a *database* that consists of *facts* and *rules*. Facts state basic data, such as properties of certain entities. Rules, on the other hand, allow one to draw inferences from these facts. Information can be extracted from the database by means of *queries*. Queries correspond, in this sense, to the conclusion of a derivation, whereas the database provides the premises. The salient point is that all premises can be considered as executable, which makes Prolog a programming language and the database a program. Indeed, every problem solvable in a traditional programming language can also be solved in Prolog. There is, however, a big difference between Prolog and the more traditional programming languages, such as Pascal. Whereas Pascal programmers must indicate all the steps needed to convert the input into the output, Prolog merely states the conditions that must be met by the program in order to be correct. Prolog is in this sense a *declarative language*, as opposed to the *procedural languages*, such as Pascal. Declarative languages do not solve the problem by a sequence of steps as procedural languages do. Instead, the computer generates the steps needed to convert the input into the output from a number of declarative statements, such as rules and facts. This obviously saves time for the programmer, and it makes the programs much more compact. On the other hand, the computer may not identify the most efficient solution.

Since facts and rules are logical statements, their order should be irrelevant. In practice, this ideal is not quite met. The order of the statements in a Prolog program does influence efficiency, and some facilities even require that statements be placed in a certain order.

4.1 BASIC PROLOG

4.1.1 Introduction

Prolog stands for Programming in Logic, and this concisely expresses the nature of Prolog. On one hand, Prolog is based on logic and, indeed, it can be viewed as a system for doing proofs. On the other hand, it is a computer language with the same computational power as, say, Pascal. In other words, everything that can be calculated in Pascal can also be calculated in Prolog. These two aspects of Prolog are closely connected. Indeed, in order to do a proof, certain calculations have to be done, and these calculations can be used for solving whatever problem needs to be resolved.

A Prolog program consists of a *database*, and from this database, one can extract information by means of *queries*. The database contains, in this sense, the premises of a proof, whereas the query is the conclusion. The database has two types of statements, *facts* and *rules*. The query, the facts and rules are *clauses*. The clauses form, in this sense, the statements of a Prolog program. A clause in Prolog always ends with a period. However, when clauses are discussed in this text, these periods will be omitted in order to preserve the natural flow of the text. The clauses, in turn, consist of *atomic formulas*, that is, of predicates, followed by argument lists. A fact contains a single atomic formula, whereas rules contain more than one atomic formula. Section 4.1.2 discusses the structure of facts, rules, and queries, and Sections 4.1.3 and 4.1.4 show how these facts and rules are used to derive answers for queries. This process may require *backtracking*, as will be shown in Sections 4.1.4 and 4.1.6. The main engine of Prolog, however, is unification, which will be discussed in Section 4.1.5.

4.1.2 Facts, Rules, and Queries

Derivations consist of a number of premises from which certain conclusions can be derived. The premises of a derivation correspond to the facts and rules in the Prolog database, and the conclusion can be thought of as an inquiry or *query* regarding the database. For instance, the premises that Aunt Rosalia likes chocolate, that lions are mammals, and that John is married to Linda can all be expressed as *facts*. *Rules* correspond to statements such as all humans are mortal, and aunts are sisters of parents. The database is read or *consulted* by Prolog at the beginning of a session. Once Prolog has consulted the database, queries regarding this database can be entered. A typical query, when expressed in English, may be "Who is the aunt of Jimmy?" In answer to this query, the system retrieves the appropriate information from the database, replying, say, that Minnie is the aunt of Jimmy. Facts, rules, and queries are jointly known as *clauses*.

Facts contain one atomic formula only, and they are used to convey information such as "lions are mammals," "Ottawa is the capital of Canada," or "Aunt Rosalia likes chocolate." In Prolog, these facts are expressed as follows:

```
mammal(lion).
capital(canada, ottawa).
loves(rosalia, chocolate).
```

As one can see, these facts are expressed very much like atomic formulas in predicate calculus. They start with a predicate name, such as mammal or capital. After the name, all arguments are listed. The list of arguments is enclosed in parentheses. As in logic, a term is anything that can be an argument of a predicate. In Prolog, all terms must be either *atoms*, *numbers*, *variables*, or *structures*. Atoms are individuals.

In the preceding database, lion, canada, ottawa, rosalia, and chocolate are all atoms. Atoms always start with a lowercase letter. Any identifier starting with an uppercase letter or with an underscore is considered to be a *variable*. Variables are used to indicate that a clause holds for every possible individual. For instance, the statement that everybody likes chocolate can be expressed as

loves(X, chocolate).

Here, X is a variable because it starts with a capital letter. Other examples of variables include Being, Individual, Person, but not person or x. Every clause is universally quantified over all variables appearing in the clause. This means that a variable can be instantiated to any term. In fact, Prolog variables are identical to *true variables* as they were introduced in Section 2.3.5.

Atoms are never capitalized, since this would make the atom a variable. However, everything enclosed in quotes is also considered an atom. Consequently, if one really wants to capitalize the words Canada and Ottawa, one can write "Canada" and "Ottawa". Hence, whereas capital(Canada, Ottawa) is not normally appropriate, capital("Canada", "Ottawa") is acceptable.

Prolog also knows about numbers, both floating point and integer. The numbers behave somewhat like atoms. However, there are several predicates especially designed to deal with numbers. These predicates are discussed in Section 4.3.5. Numbers can be used in predicates like any other term. The statement $X + 0 = X$, for instance, can be expressed by the fact plus(X, 0, X). This completes our discussion of terms. However, one additional term, structures, has not been discussed yet. Structures will be described in Section 4.3.3.

In addition to facts, Prolog uses *rules*. Rules are really conditionals. For instance, the statement "if Aunt Rosalia eats chocolate, then she gains weight" can be expressed as

gainsweight(rosalia) :− eatschocolate(rosalia).

Rules consist of a *head*, which expresses the consequent of the conditional, and a *body*, which forms the antecedent. In this example, the head is gainsweight(rosalia), and the body is eatschocolate(rosalia). The head of the rule is always a single atomic formula. The body of a rule is frequently a conjunction. However, instead of ∧, Prolog uses the comma, as shown in the following clause:

aunt(X,Y) :− sister(X,Z), parent(Z,Y).

Translated into English, this means that X is an aunt of Y if X is a sister of Z, and Z is a parent of Y. The atomic formulas appearing in the body of a rule are often called *goals*. In the rule describing aunt, the goals are therefore sister(X,Z) and parent(Z,Y). In rules, as in any other clause, the variables used in the rule can be instantiated to any term, but all instances of a variable appearing in a rule must be instantiated to the same term.

Example 4.1 Consider the rule mortal(X) :− human(X). Which of the following instantiations can be derived from this rule?

mortal(socrates) :− human(socrates).
mortal(fido) :− human(fido).
mortal(rosalia) :− human(socrates).

Solution   The first two instantiations are legal because both occurrences of X are replaced by the same term. The third instantiation cannot be derived from the rule in question because X is replaced first by rosalia and then by socrates.   ∎

Rules can also contain disjunctions. In this case, one uses a semicolon instead of the ∨. For instance, one can define parent(X,Y) by the rule

parent(X,Y) :− mother(X,Y); father(X,Y).

However, the use of disjunctions in not encouraged. Instead of the semicolon, different clauses should be used. For instance, the predicate parent should be defined as follows:

parent(X,Y) :− mother(X,Y).
parent(X,Y) :− father(X,Y).

A list of clauses that together describes a predicate is also called a *procedure*.

One can derive conclusions from a database through *queries*. Essentially, a query is a rule without a head. Like a rule, it can contain one or several goals. The goals must be separated by either a comma or a semicolon, depending on whether the query is a conjunction or disjunction. The following are some sample queries.

| ?- loves(X, chocolate).
| ?- capital(canada, Capital).
| ?- aunt(X,Y), loves(X, chocolate).

All these queries are in interactive mode, or in *query mode*. In query mode, the prompt is | ?-.

As indicated, there are three types of clauses in Prolog: facts, rules, and queries. These three types of clauses are closely connected. Whereas a fact can be thought of as a rule without a body, a query can be thought of as a rule without a head. Indeed, one often treats the body of a rule in the same way as a query. Moreover, the syntactical rules for queries and bodies of rules are identical. All clauses, including the queries, must terminate with a full stop.

For increasing clarity, one can use *comments*. In Prolog, comments start with the symbol % and terminate with the end of the line, as illustrated by the following example:

mammal(lion). % is a fact
human(X) :− mortal(X). % is a rule
% and this is a comment on a line by itself

4.1.3 Derivations Involving Facts

Queries are derived from the database in essentially the same way as one would derive a conclusion from its premises. Such derivations may be trivial in the sense that the conclusion is one of the premises. In the terminology of Prolog, this means that the query is identical to one of the clauses of the database. In nontrivial cases, Prolog uses rules of inference to derive the query from the database. Essentially, Prolog only uses *resolution*, a generalization of the modus ponens, together with *unification* for its derivations. For our initial discussion, suppose that the present database is

```
country(usa).
country(canada).
country(mexico).
capital(usa, washington).
capital(canada, ottawa).
capital(mexico, mexicocity).
```

This database is made known to the system by means to be discussed later. Once this is done, the system turns to query mode, as indicated by the prompt | ?-. A number of queries are now given, and each reply of the system is recorded.

```
| ?- country(canada).
yes
| ?- capital(mexico, mexicocity).
yes
| ?- capital(usa, ottawa).
no
| ?- country(japan).
no
```

The first three queries can be matched immediately with facts in the database, and their derivations are trivial. To indicate that a derivation can be done, Prolog merely prints a yes as a reply. There is no possibility of deriving the fact country(japan) from the database. For this reason, Prolog replies with no. Note that no really means "this query cannot be derived from the database." This has nothing to do with the truth and falsity of the query. Of course, Japan is a country, but this cannot be derived from the database. A query that can be derived from the present database is said to *succeed*. A query for which Prolog can establish that it cannot be derived is said to *fail*. Unless Prolog runs into an infinite loop, all queries will either succeed or fail.

A query containing variables succeeds if there is any instance of the query that succeeds. If such an instance is found, the values to which the variables instantiate are printed. A query fails if all instantiations of the query fail and, in this case, a no is printed. To see how this works with the previous database, consider the following queries, together with the replies of the system.

```
| ?- capital(canada, City).
City = ottawa ?
| ?- capital(ottawa, X).
no
| ?- county(X).
X = usa ?
```

The first query unifies with capital(canada, ottawa), because the variable City can be instantiated to ottawa. Prolog therefore prints City = ottawa. The variable in the query can actually be thought of as existentially quantified. The question to the system is, in this sense, "is there a city that is the capital of Canada," to which Prolog replies, "the city is Ottawa." After such an answer, one can get into the query mode again by simply pressing the return key. The system responds by a prompt, and a new query can be entered. In the example, the new query is capital(ottawa, X). Since there is no clause that matches with this query, a no is printed in return. Ottawa is a city, and cities have no capitals. The next query is country(X). There are three ways to satisfy this query. It can be unified with the fact country(usa), with the fact country(canada), or with the fact country(mexico). Prolog always picks the first possible match, which is country(usa). Consequently, X instantiates to usa, as indicated. If this instantiation is satisfactory, then the user merely presses return to get back into query mode. To obtain other solutions, a semicolon is typed, followed by a return. This causes the system to undo the instantiation and to search the database for additional clauses with which the query can be unified. This leads to the following dialog:

```
| ?– country(X).
X = usa ?;
X = canada ?;
X = mexico ?;
no
```

After the last answer, no more answers can be found, which is indicated by no. Whenever the user types a semicolon, the query is *redone*. Redoing a query with a single goal in a database involving facts proceeds as follows. A marker is placed in the line containing the fact that caused the query to succeed. When the query is redone, all clauses up to and including this marker must no longer be used. If there still is a clause that can be matched with the query, then redoing the query succeeds. Otherwise, redoing fails.

A query may contain several goals. Typically, these goals are separated by commas, which means that all goals must succeed for the query to succeed.

Example 4.2   Find the result of the query article(X), noun(Y) in connection with the following database:

```
article(the).
article(a).
noun(tree).
noun(chair).
```

In this case, the first goal is article(X), which succeeds with X = the. The next goal is noun(Y), which succeeds with Y = tree. Since both goals succeed, so does the query, and Prolog prints X = the, Y = tree.

4.1.4 Derivations Involving Rules

As indicated in the previous section, a goal succeeds if it unifies with a fact. A goal also succeeds if it unifies with the head of a rule, and the body of the rules succeeds. This is explained now. Consider the following database:

```
human(socrates).
human(plato).
mortal(X) :– human(X).
```

Our first query is mortal(plato). The answer to this query is yes, as expected. Formally, the proof is conducted by instantiating mortal(X) :– human(X) to

```
mortal(plato) :– human(plato).
```

From the point of view of logic, mortal(plato) is true if human(plato) is true, and since human(plato) is asserted as a fact, the query follows because of the modus ponens. Prolog arrives at this conclusion in the following way. The goal to be satisfied is mortal(plato). The only clause that allows one to derive this goal is the third clause, which has mortal(X) as its head. Prolog therefore unifies the head with the query, which yields mortal(plato) :— human(plato), as above. The problem thus reduces to satisfying the goal human(plato). This goal succeeds because it matches with a fact, and the query succeeds.

Basic to derivations in Prolog are the *goals*. The goals can come from the query, but they can also be generated by clauses. Prolog tries to satisfy any given goal in the following way: it finds the first clause in the database that may be applicable or that *matches* the goal in question. A fact matches a goal if it can be unified with the goal and, in this case, the goal succeeds. A rule matches with a goal if the head of the rule can be unified with the goal, and the rule succeeds. This will now be elaborated further.

Consider the query mortal(Y). The matching rule is mortal(X) :— human(X). To unify the head of this rule with the query, X and Y must be unified. This yields the rule mortal(Y) :— human(Y). Note that both instances of X in the rule have been changed to Y. The new goal is now human(Y), and this goal unifies with human(socrates). Hence, Y instantiates to socrates, and the query succeeds. Prolog indicates this by typing Y = socrates. As in the earlier examples, it is possible to obtain further solutions for the query mortal(Y), as indicated by the following dialog:

```
| ?— mortal(Y).
Y = socrates ?;
Y = plato ?;
no
```

Here, a first solution to the query is Y = socrates. After this, a semicolon is typed, which means that the query is redone. A new solution is obtained, Y = plato. A quest for additional solutions ends in failure, and this is indicated by the reply no.

If the query mortal(X) is used instead of the query mortal(Y), the result is the same, except that the system prints X = socrates rather than Y = socrates. Note, however, that X in the query is distinct from X in mortal(X) :— human(X). In general, variables are local to the clauses in which they appear, and two variables appearing in different clauses are in fact distinct. To make this clear, it is a good idea to rename the variables in such a way that the two clauses to be unified do not contain variables with the same name. For instance, before the goal mortal(X) is unified with the head of the rule mortal(X) :— human(X), the variable name X appearing in the query should be changed, say to Y or Z.

Consider now the query aunt(U,V) in connection with the database given in Figure ??. This query contains one goal, aunt(U,V). The query matches with the rule on the last line of the database; that is, the head of the rule can be unified with the query. The result is

$$\text{aunt(U, V) :— sister(U,Z), parent(Z,V).}$$

To satisfy aunt(U,V), the two goals sister(U,Z) and parent(Z,V) must be satisfied. These new goals replace, so to say, the query. The first goal matches with the fact sister(andrea, paul) and it succeeds. In the process, U is instantiated to andrea and Z is instantiated to paul. The rule with the head aunt(U,V) therefore becomes

aunt(andrea, V) :− sister(andrea, paul), parent(paul, V).

The only goal left to satisfy is parent(paul, V), and this goal unifies with the fact parent(paul, jimmy) and succeeds. The unification is accomplished by the instantiation of V to jimmy. The system therefore replies with U = andrea, V = jimmy. It can be easily checked that this is indeed correct. One merely has to instantiate the aunt rule with X = andrea, U = andrea, V = jimmy, Y = jimmy, and Z = paul, and one obtains

aunt(andrea, jimmy) :− sister(andrea, paul), parent(paul, jimmy).

The goals of this rule are all facts, and therefore they imply the head aunt(andrea, jimmy). In fact, the rule in question is really an implication, and the head of the rule is the consequent of the implication. The consequent is therefore true by the modus ponens, together with the rule of combination.

```
sister(andrea, paul).
sister(minnie, jane).
parent(paul, jimmy).
parent(jane, kathy).
female(kathy).
male(jimmy).
aunt(X, Y) :− sister(X,Z), parent(Z,Y).
```

Figure 4.1. Example of a database

In queries, as well as in bodies of rules, the goals are typically separated by commas, in which case all goals must be satisfied in conjunction. Goals separated by semicolons, on the other hand, can be satisfied as a disjunction. Here, we restrict ourselves to the conjunctive case. In the conjunctive case, a query succeeds if all its goals succeed. Similarly, a rule succeeds if all the goals contained in the body of the rule succeed. To satisfy a query or a rule, Prolog attempts to satisfy one goal after the other, going from left to right. The attempt to satisfy a goal is a *call*. Sometimes Prolog finds a solution that satisfies some initial goals, but that turns out to be unsuitable to satisfy the remaining goals. In this case, the partial solution arrived at is abandoned in favor of a new solution, which may be better suited to satisfy all goals of the query. To this end, the goal preceding the first goal that failed is *redone*; that is, a new solution for this goal is derived. This is accomplished in the same fashion as in the case when the user prompts the system for additional solutions by typing a semicolon. The process of abandoning partial solutions for a potentially better partial solution is called *backtracking*. Backtracking is used extensively in Prolog.

Backtracking is only needed if at some point a rule is selected that later turned out to be the wrong choice. From the point of view of clarity, it is better to omit all these false starts and to only present the steps that eventually lead to the solution. This idea leads to the concept of a *nondeterministic machine*. A nondeterministic machine is a machine that always guesses the right thing to do in order to avoid backtracking. It has, in this sense, an infinite foresight. Algorithms done on such nondeterministic machines are, of course, called *nondeterministic algorithms*. Because all dead-end choices are omitted, nondeterministic algorithms concentrate on the steps that lead to success, and they omit all the steps that lead nowhere and only cloud the issue. This makes nondeterministic algorithms much easier to understand. This is the reason why we use them here. In particular, we assume that if there is a choice as to which of the possible matching clauses is selected, Prolog will select the correct one. The fact that there is no nondeterministic machine need not detract us at this point.

For illustration, consider the query aunt(U,V), female(V). This query succeeds if both goals within the query succeed. As shown, the first goal, which is aunt(U,V), succeeds with U = andrea and V = jimmy. This solution does not satisfy the second goal female(X), because jimmy is not a female. At this point, a standard algorithm would have to backtrack, finding a new solution to the problem. In other words, the goal aunt(U,V) is redone, and a new solution is derived. In fact, there is a second solution to aunt(U,V), and this is U = minnie, V = kathy. This solution satisfies the second goal, which is female(V), which becomes female(kathy). If there had not been such a second solution to the goal aunt(U,V), then the query would have failed.

A nondeterministic algorithm solves the query aunt(X,Y), female(Y) in the following way: As in the preceding derivation, the algorithm searches for a clause that matches with the first goal of the query, and this clause is the aunt clause. Hence, it concludes

aunt(U, V) :− sister(U,Z), parent(Z,V).

In its next step, the algorithm tries to find a clause that matches the first goal of the rule, which is sister(U,Z). Normally, the first such clause, which is sister(andrea, paul), would be selected. A nondeterministic algorithm, in its infinite foresight, will not make this choice, because it knows that this choice will not lead to a solution. It therefore tries the second available clause, which is sister(minnie, jane). The aunt clause becomes under this choice

aunt(minnie, V) :− sister(minnie, jane), parent(jane,V).

The second goal of the aunt clause is parent(jane, V), which unifies with parent(jane, kathy) and succeeds, instantiating V to kathy in the process. The call aunt(U,V) therefore succeeds with U = minnie, V = kathy. Consequently, the query instantiates to

aunt(minnie, kathy), female(kathy).

This leaves the goal female(kathy), which unifies with a fact and succeeds. Since all goals of the query succeed, the query also succeeds.

4.1.5 Instantiations and Unification

All variables in Prolog are implicitly universally quantified, and they can be replaced by any term, provided that this is done consistently. For instance, in the clause foo(X,Y) :− roo(X, Z), soo(Z,Y), one is allowed to replace all occurrences of X by, say, minnie. This does not affect any X occurring in any other clause. The variables are, in this sense, local to the clause in which they occur. In fact, they only ensure that, *within the clause*, all instantiations are consistent. Consistency is, of course, only important if the variable in question appears more than once. Any variable appearing only once does not need a name. A variable that does not need a name may be replaced by an underscore (_). The underscore is also called an *anonymous variable*. This is partially a misnomer because the underscore can represent different variables in the same clause, all of which can assume different instantiations.

Example 4.3   There is a predicate father(X,Y) in the database, which is true if X is the father of Y. Write a predicate isfather(Z) that succeeds if Z is a father.

Solution   The desired predicate is

$$isfather(Z) :- father(Z, \_).$$

Instead of the anonymous variable _, any other variable name could have been used. However, since this variable appears in only one place, it can be left anonymous.   ■

Variables are instantiated in order to unify the goal with a fact or a clause head. It is important for the reader to see exactly how this is done. To demonstrate this, consider the predicate plus(X, Y, Z), which is to succeed if $X + Y = Z$. The statement $X + 0 = X$ can then be expressed by the fact plus(X, 0, X). Consider now the query plus(4, Z, Y), which must be unified with this fact. To accomplish this, the expressions to be unified are written one below the other.

$$plus(X, 0, X).$$
$$plus(4, Z, Y).$$

Whenever a variable in one of the two expressions matches with an atom or a number, then the variable is set to the atom or the number, and all occurrences of the variable within the expression it appears in must be replaced by this atom or this number. This is best done starting from the left and going to the right. First, X is instantiated to 4, and next Z is instantiated to 0. The result of these two instantiations is

$$plus(4, 0, 4).$$
$$plus(4, 0, Y).$$

Now Y is instantiated to 4. Hence, plus(X, 0, X) and plus(4, Z, Y) can be unified, and the unifier is $X = 4$, $Z = 0$, and $Y = 4$.

The query plus(3, Z, 4) cannot be unified with plus(X, 0, X). One has

$$plus(X, 0, X).$$
$$plus(3, Z, 4).$$

Unifying X with 3 and Z with 0 leads to

$$plus(3, 0, 3).$$
$$plus(3, 0, 4).$$

At this point, 3 would have to be unified with 4, and this is impossible.

Next consider the goal blah(X, X, c, u). Can this goal be unified with the fact blah(u, Y, X, Y)? To decide this, note that the two variables X in these two atomic formulas are distinct variables. Hence, instantiating the X in the goal to, say, u has no effect on the instantiation of the variable X in the clause head. This was pointed out earlier, and it was suggested to use a different name for the X in the clause head. With this in mind, we write the two expressions, one on top of the other.

$$\text{blah}(X, X, \ c \ , u \ ).$$
$$\text{blah}(u, Y, X1, Y \ ).$$

An attempt is now made to unify the variables. Starting at the left, one first instantiates X to u, which yields

$$\text{blah}(u, u, \ c \ , u \ ).$$
$$\text{blah}(u, Y, X1, Y \ ).$$

Now Y can be instantiated to u and c to X1. This yields

$$\text{blah}(u, u, c, u \ ).$$
$$\text{blah}(u, u, c, u \ ).$$

Since the two expressions in question are identical, unification succeeds.

Of course, if the clause head is unified, all goals in the clause may be affected. Specifically, consider the goal blah(X, X, c, u) once more and suppose that the rule it is matched with is

$$\text{blah}(u, Y, X, Y,) :- \text{new}(Y, Z), \text{old}(Z, X), \text{top}(X, Y).$$

As indicated, the goal can be unified with the head of the clause by setting Y = u and X = c. The result is

$$\text{blah}(u, u, c, u) :- \text{new}(u, Z), \text{old}(Z, c), \text{top}(c, u).$$

Hence, instantiations must be consistent. The only exception to this consistency is the underscore, which can assume different values.

The fact that a variable within an expression can only bind to one atom means that there cannot be a proper assignment statement in Prolog. For instance, a statement such as $X := X + 1$ requires that X have two different values, and this is inconsistent with Prolog. The only way variables can assume values is through instantiations.

4.1.6 Backtracking

The purpose of this section is to explain how Prolog finds its solutions or, if there is none, how it proves that no solution exists. Before the problem is solved in its full complexity, it is analyzed under simplifying assumptions. It is assumed that the query contains a single goal only and that this goal matches with a rule. Furthermore, it is assumed that no goal of this rule matches another rule, that is, if the goals of the rule match any clause at all, this clause must be a fact. The general case will be discussed later.

To explain the theory, consider the following database:

```
% The next clause will be called the buy-rule
buys(X, Y) :- forsale(Y), likes(X, Y), good(Y).
```

```
forsale(dress).
forsale(hat).
forsale(shoes).
likes(jim, shoes).
likes(mary, dress).
likes(mary, hat).
good(hat).
```

For convenience, we will refer to the rule at the start of this database as the *buy-rule*. The buy-rule indicates under which conditions X will buy product Y. Assume that the query is buys(mary, hat). This query matches with the buy-rule, and after performing the proper instantiations, the buy-rule becomes

```
buys(mary, hat) :− forsale(hat), likes(mary, hat), good(hat).
```

This rule succeeds if all its goals succeed, which they do: they can all be matched with facts. Next consider the query buys(Z,U). Of course, there is a solution to this query: to see this, instantiate Z to mary and U to hat, in which case the query buys(Z,U) reduces to buys(mary, hat), and it was shown that this query succeeds. Hence, if the system had enough foresight to guess the right clauses, that is, the clauses that made the query buys(mary, hat) succeed, then the query could be resolved directly and without backtracking. However, such nondeterministic machines do not exist, which necessitates a systematic search. Prolog tries to satisfy the goals of a query from left to right, and for each goal it tries the matching clauses from top to bottom. In the case considered, this works as follows: First, the query buys(Z, U) is unified with the buy-rule, which yields

```
buys(Z, U) :− forsale(U), likes(Z, U), good(U).
```

The first goal is forsale(U), and the first matching clause is forsale(dress). To use this clause, U in the buy-rule must be instantiated to dress. We refer to this instantiation as the *dress instantiation* of the buy-rule. The dress instantiation is

```
buys(Z, dress) :− forsale(dress), likes(Z, dress), good(dress).
```

The next goal to be satisfied is likes(Z, dress). This goal matches with likes(mary, dress), and Z instantiates to mary. The result of this instantiation is

```
buys(mary, dress) :− forsale(dress), likes(mary, dress), good(dress).
```

The final goal is good(dress). This goal fails because there is no rule that matches this goal. This does not mean that buys(Z, U) fails. It only means that clauses were selected that do not lead to a solution. Before the search can continue, backtracking is required. In many cases, part of the solution that was derived is still useful. For this reason, only the goal preceding the goal that was causing failure is redone. Hence, Prolog tries to redo the goal likes(Z, dress). If redoing this goal fails, and this is the case here, then further backtracking is required. Here the goal forsale(dress) must also be redone. With this, all instantiations caused by calling this goal have to be undone. Hence instead of dress the system reverts to U, which essentially means that one reverts to the buy-rule. To find a new solution of the buy-rule, forsale(U) is instantiated to forsale(hat). This leads to the following instantiation of the buy-rule:

$$\text{buys(Z, hat) :} - \text{forsale(hat), likes(Z, hat), good(hat).}$$

Next the goal likes(Z, hat) is instantiated to likes(mary, hat) and succeeds. This leads to

$$\text{buys(mary, hat) :} - \text{forsale(hat), likes(mary, hat), good(hat).}$$

Finally, good(hat) is tried, and it succeeds.

Essentially the same method can be used in general. To see this, we summarize the essential steps. To make a rule succeed, one has to make all goals of the rule succeed. Hence, one goal after the other is tried or, to use the technical term, *called*. If all calls succeed, then the rule succeeds; otherwise, backtracking is used, which means that some goals are redone. If a call can be matched with a fact, then it can be decided immediately if the call succeeds or fails: if the call unifies with the fact, it succeeds; otherwise, it fails. This idea can be extended to the case when some goals match with rules. The only assumption needed is that one can decide whether or not the matching clause succeeds or fails. As long as this can be decided, the method developed before remains viable. Whether the matching clause succeeds or fails can be decided recursively, as the reader may verify. For this recursion to work, one must assume that the predicate appearing in the head is not called by any of its goals, neither directly nor indirectly. Otherwise, an infinite loop may result. We do not further explore these complications.

Many queries contain several goals, and the query succeeds if all its goals succeed. The methods used to make a query succeed are the same as to make the body of a rule succeed. Prolog calls one goal after another and backtracks if a goal fails. This continues until either all goals succeed, in which case the query succeeds, or until no further backtracking is possible, in which case the query fails.

4.1.7 Resolution

We briefly discuss the term *resolution* here. Generally, two clauses a and b can be *resolved*, provided a contains a goal that matches b. In this case, one can, after proper instantiations, replace the goal of a being matched by the body of b. The resulting clause is called the *resolvent* of a with b.

Example 4.4   Consider the following database:

```
buys(X, Y) :- opportunity(Y), likes(X, Y).
opportunity(Z) :- lowprice(Z), highquality(Z).
opportunity(Z) :- recommended(Z).
```

Resolve the first clause with the second clause. Also, resolve the first clause with the third clause.

Solution   The goal of the first clause that matches with the second clause is opportunity(Y). To complete the match, this goal is unified with the head opportunity(Z). Z is therefore instantiated to Y, which yields

$$opportunity(Y) :- lowprice(Y), highquality(Y).$$

The body of this rule now replaces the goal opportunity(Y) in the first rule, which leads to

$$buys(X, Y) :- lowprice(Y), highquality(Y), likes(X, Y).$$

Similarly, the first clause can be unified with the third clause, which yields

$$buys(X, Y) :- recommended(Y), likes(X, Y). \qquad ∎$$

Resolution is particularly simple if a rule is resolved with a fact. Since facts have no bodies, the goal to be matched by the rule essentially disappears. To see how this works, consider the following example:

Example 4.5   Find the resolvent of the following two clauses:

$$buys(Z, U) :- forsale(U), likes(Z, U), good(U).$$
$$forsale(hat).$$

Solution   To find the resolvent, one unifies the goal forsale(U) in the buy-rule with for-sale(hat). This is accomplished by the instantiation U = hat. All instances of the variable U are replaced by hat, and the goal forsale(hat) is dropped. The result of this is the following rule:

$$buys(Z, hat) :- likes(Z, hat), good(hat). \qquad ∎$$

The engine used by Prolog is resolution. To show this, the derivation involving the query buys(U,V), which was discussed earlier, is repeated in terms of resolution. To derive the query, one has to resolve all goals of the buy-rule. To do this, it is assumed that the right clauses for doing resolution can always be determined in advance; that is, a nondeterministic algorithm is used. First, by resolving the buy-rule with the fact forsale(hat), one finds, as shown before, that

$$buys(Z, hat) :- likes(Z, hat), good(hat).$$

Next one resolves likes(Z, hat) with likes(mary, hat) to obtain

$$buys(mary, hat) :- good(hat).$$

This clause can now be resolved against good(hat), which removes the last remaining goal from the clause in question. The body of the goal therefore becomes empty, and a clause with an empty body merely becomes a fact. This means that buys(mary, hat) :- good(hat) is converted into the fact buys(mary, hat). This fact agrees with the query, and the query succeeds.

It is natural to treat the query like a rule, except that this rule has no head. Nevertheless, one can resolve all goals of the query. In fact, this is the method actually applied by Prolog. The query succeeds if all goals can be resolved. Once all goals have been resolved, nothing remains: The empty clause has been derived. Hence, a query succeeds if it leads to the empty clause. Otherwise, the query fails. This is shown by the following example:

Example 4.6   Consider the database given in Figure 4.1, together with the query aunt(U, V), female(V). By repeatedly applying resolution, derive the empty clause.

Solution   The rule for aunt is given as

$$\text{aunt}(X, Y) :- \text{sister}(X, Z), \text{parent}(Z, Y).$$

Resolving this rule with the query leads to

$$:- \text{sister}(U, Z), \text{parent}(Z, V), \text{female}(V).$$

Here, we start the clause with :− in order to indicate that, like the query, this is a clause without a head. The first goal of this clause can be resolved with sister(minnie, jane), which yields

$$:- \text{parent}(\text{jane, kathy}), \text{female}(\text{kathy}).$$

Since parent(jane, kathy) and female(kathy) are both facts, they can easily be resolved, which means that all goals disappear. Therefore, the query succeeds.   ∎

The derivation of the *empty clause* has a logical interpretation. Generally, a rule with head A and body B can be interpreted as $A \Leftarrow B$ or, equivalently, as $A \vee \neg B$. The query is a rule with no head, and it can in this sense be interpreted as $\neg B$. In other words, when resolving the query, one really adds the negation of the query to the database. Moreover, the empty clause is a disjunction in which both terms A and B are missing, and this empty disjunction is false. Under this interpretation, deriving the empty clause really means deriving a contradiction from the database and from the negation of the query. As indicated in Chapter 1, this is equivalent to deriving the query from the database. These ideas will be explained later in greater detail.

Problems 4.1

1. Convert the following statements into clauses.
   (a) Everyone who is young and enterprising likes to travel. Use the one-place predicates young, enterprising, and travel for the properties in question.
   (b) Everybody likes books that are well written and have lots of pictures, and everybody likes movies with a happy ending. Use the two-place predicates likes(X, Y) for X likes Y, and the one-place predicates book(X), movie(X), haspictures(X), wellwritten(X), and endshappy(X).

2. Let times(X, Y, Z) represent the statement that $X \times Y = Z$. How can you state $X \times 0 = 0$, $X \times 1 = X$, and $X \times Y = Y \times X$ in Prolog?

3. Convert the following information into facts and rules.
   (a) Suppose that we have the foods carrots, beef, ham, lettuce, and spinach. Use facts to express which of these foods are vegetables and which are meat.
   (b) We are having Jim over for supper. Jim is a vegetarian who likes every vegetable. Write a rule that defines what Jim will eat.

4. Jane likes men who are tall and dark. Joe likes women who are tall and blonde. Suzy likes men who are tall and blonde. Express this information as rules in a database.

5. Consider a database that contains a list of persons, together with their phone numbers. For instance, phone_number(miller, 4746) indicates that miller has the phone number 4746. What query is needed to extract the phone number of miller from the database? Also, give the query necessary to find if the number 4883 is in use.

6. The following is a list of fact–query pairs. For each pair, indicate the output given by Prolog when the query is entered, assuming that the fact is part of the database.
    (a) Fact: pred1(abc, cde), query: pred1(X, Cde).
    (b) Fact: whoever(jim, beth, Y, john), query: whoever(X, Y, X, Z).
    (c) Fact: last(mary, beth, mary, Y), query: last(X, Y, Z, Z).

7. A database describes who is a child of whom for a number of people, and it also shows the sex of each person in the database. This information is recorded in the form parent(a, b), where a is the parent of b, and in the form male(a) and female(b) when a is male and b is female. Formulate rules describing grandparents, grandchildren, grandmothers, and greatgrandfathers.

8. For a number of people, the database contains facts about pairs of people, indicating whether they are brothers, sisters, or married. Design a predicate that can extract an in-law relation. In particular, you should design a predicate for brother-in-law and sister-in-law.

9. Consider the following Prolog database:

    ```
    abc(X, Y) :- cde(X, U), efg(V, U), hij(V, Y).
    cde(a, b).
    cde(a, c).
    efg(d, b).
    efg(h, c).
    hij(h, b).
    ```

    Suppose the query is abc(a, b). Trace the execution of the query abc(a, b). The trace should indicate in which order the different goals are attempted, together with an indication whether or not they succeed.

10. In a Prolog database, there is a fact for each English word indicating whether it is a noun, a verb, an article, and so on. For instance, there is a fact noun(dog) to indicate that "dog" is a noun, there a fact verb(run) to indicate that "run" is a verb, and there is a fact article(the) to indicate that "the" is an article. Design a rule sentence(X, Y, Z) which must succeed if X is an article, Y is a noun, and Z is a verb.

4.2 RUNNING AND TESTING PROGRAMS

4.2.1 Introduction

Databases essentially correspond to programs, and the question now is how to run and test these programs. The databases are recorded as textfiles, which are read or, to use the technical term, *consulted* by Prolog. Once the database is consulted, queries about the database can be entered. This is discussed in the sections that follow. We first mention some of the Prolog compilers and interpreters, in particular the ones that we have used. We then show how a database is consulted. Well-constructed Prolog programs typically run without much testing. Nevertheless, debugging facilities are provided, and they will be discussed as well.

4.2.2 Prolog Compilers and Interpreters

A number of good Prolog translators are available, many of them free of charge. For up-to-date information, the reader may want to consult the Internet newsgroup comp.lang.prolog. The earlier Prolog systems were interpreters. Nowadays, however, many systems compile the clauses of the database for better efficiency. Even in systems that compile the programs, it is possible to enter *dynamic* clauses, which are interpreted. In fact, many modern Prolog systems combine interpreted and compiled clauses seamlessly.

Prolog is available for almost every computer. For preparing this book, a C-based Prolog was used, SICStus Prolog. This product is available from the Swedish Institute of Computer Science. We also used Quintus Prolog, a commercial product from the Quintus Corporation in Mountain View, California. Both SICStus and Quintus Prolog compile code. They run on many different Unix-based systems. In both systems, it is possible to add interpreted rules during a session. We have also used AAIS Prolog, produced by Advanced A. I. Systems, Inc. in Mountain View, California. AAIS Prolog runs on Macintoshes. Some of the programs were also run on personal computers, using the ESL Prolog. Both AAIS and ESL Prolog are interpreters.

Although the different versions of Prolog all share the main features of the language, they are not fully compatible. The main differences are that in some systems certain predicates are built in, whereas in others they are not. Also, AAIS Prolog has more elaborate methods to handle strings. Finally, ESL Prolog has different conventions for writing comments. Whereas the standard way of starting a comment is the percentage sign (%), ESL Prolog uses the comment conventions of C. To find all these differences, the respective manuals should be consulted.

4.2.3 Consulting a Database

Prolog programs are implemented as databases. The databases are textfiles, which are *consulted* by Prolog. This means that the information is read and, depending on the system used, either compiled at the beginning of the session or interpreted as needed. The file containing the database can be prepared by a normal text editor. Once the database is prepared, the available Prolog system can be activated. At the beginning, Prolog is in query mode, and the first query to be entered is typically a query consulting the database. For instance, consult(aunt) will cause the database contained in file aunt to be read, possibly compiled, and stored for later use. Like any query, the line containing consult(aunt) must be terminated by a full stop. At the time of compilation, any errors and warnings will be indicated. If there are no errors that prevent the query consult(aunt) from succeeding, a yes will be printed. Instead of consult(aunt), one can also type [aunt]. The effect is exactly the same. To exit Prolog, the predicate halt is used.

Any names starting with capital letters are interpreted as variables. This causes certain problems when consulting files that start with a capital letter. For instance, if the file Minnie contains the database, then consult(Minnie) will fail. In a case like this, the file name must be enclosed in quotes; that is, one has to use consult("Minnie"). Names containing special characters must also be quoted. For instance, consult("/prolog/aunt") reads the file /prolog/aunt.

After a file has been successfully consulted, queries involving the database in question can be entered. For instance, suppose that the file aunt contains the database given in Figure 4.1, and this file has been successfully consulted. At this point, the query aunt(X, Y) will lead to the following interactive session:

```
| ?- aunt(X, Y).
X = minnie, Y = jimmy
```

It is possible to add clauses to the database during query mode. To do this, one types consult(user) or simply [user]. The system is then ready to accept facts and rules and, to indicate this, the prompt is | instead of | ?-. To terminate input from the terminal, an appropriate end-of-file marker is entered. On Unix systems, this is $^\frown D$. One can also add further predicates by consulting other files. When this is done, care must be taken that no predicate already in the database is defined in the file to be consulted. In Quintus and SICStus Prolog, a warning message appears if a predicate defined in the file to be consulted is already in the database, and a number of options are given on how to proceed. The choice is typically among accepting the old definition, replacing the old definition by the new one, or aborting the consultation. There is one exception to this rule: If one consults a file with the same name twice, no warning appears, and the new predicate overrides the old one. This is useful for developing programs, because at the development stage the programmer should be able to correct or otherwise change predicates in the file containing the original database. After the corrections are done, the file can then be reconsulted, which means that the corrections override the original definitions. Interpretive systems tend to distinguish between consult, in which case the database is merely incremented, and reconsult, in which case the new definition completely overrides the old one. In such systems, the predicate reconsult must be used when developing programs. Otherwise, both the original and the corrected version become part of the database.

During a session, the user sometimes wants to know all clauses that jointly define a specific procedure. To find these clauses, the predicate listing is used. This predicate has one argument, which is the predicate name whose clauses are to be listed. For instance, the query listing(aunt) results in the following output:

```
| ?- listing(aunt).

aunt(A, B) :-
        sister(A, C),
        parent(C, B).
yes
| ?-
```

When listing a procedure, the system uses its own variable names instead of the variable names provided by the programmer. For instance, X and Y are replaced by A and B, respectively. The variable names in Prolog are merely placeholders, which are only used to enforce consistency, and their names are irrelevant otherwise.

It is also possible to list all procedures in the database. To do this, one types listing. The procedures are then all listed, sorted according to their arity, and, within each arity, in alphabetical order. Some systems do not allow one to list any compiled procedures.

Sometimes certain actions are to be taken at the moment that a database is consulted. For instance, one may list certain procedures, print instructions for the user, or consult another file before starting execution. To accomplish this, one uses rules in the database that do not have any head. In other words, the rule starts with :−, and it lists all the goals in sequence. For instance, a file file1, containing a first database, may contain the rule

$$:- \text{consult(file2)}.$$

In this case, the command consult(file2) is executed immediately when file 1 is consulted, which has the effect that file file2 is consulted right after file file1.

4.2.4 Debugging and Tracing

A well-designed Prolog program is much more likely to run the first time than a program written in an algorithmic language, such as Pascal. In a Prolog program, the proof of all clauses should be self-evident. For instance, the clause to express who is an aunt is almost a verbal translation of the definition as expressed in English, and any result derived from this definition cannot possibly be faulty. Nevertheless, a number of features can trip the inexperienced. One of the most frequent errors is missing periods or, worse, periods that have been mistyped as commas. Another frequent cause for errors is capitalizing identifiers that are meant to denote atoms. If a query does not run as expected, it pays to check for these types of errors. Once this is done, one should check if the rules really reflect the definitions. As a last resort, one can trace a program. In query mode, Prolog can be put into trace mode by typing trace followed by a period. The predicate notrace turns the trace off. The trace prints the present goal, together with the actions taken. The goal is either attempted, which is indicated by the word Call; done successfully, as indicated by the word Exit; redone, as indicated by the word Redo; or it fails, as indicated by the word Fail. To demonstrate how a trace looks, consider the database of Figure 4.1, together with the query aunt(minnie, jimmy). The trace looks as follows:

```
| ?− aunt(minnie, jimmy).
 + 1 1 Call: aunt(minnie, jimmy) ?
 + 2 2 Call: sister(minnie, _333) ?
 + 2 2 Exit: sister(minnie, jane) ?
 + 3 2 Call: parent(jane, jimmy) ?
 + 3 2 Fail: parent(jane, jimmy) ?
 + 2 2 Redo: sister(minnie, jane) ?
 + 2 2 Fail: sister(minnie, _333) ?
 + 1 1 Fail: aunt(minnie, jimmy) ?

        no
```

In the trace, each goal is identified by a number, which is printed in the first column. For instance, goal number 1 is aunt(minnie, jimmy), goal number 2 is sister(minnie, _333), and so on. Beside the goal number is the *depth* of the present goal. The query corresponds to a depth of 1, any goal called by the query has a depth of 2, any goal called by a goal of depth 2 has a depth of 3, and so on. As stated earlier, the words Call, Exit, Fail, and Redo signify, respectively, "attempt to satisfy goal," "goal succeeded," "goal failed," and "goal is redone." Note that the system uses an internal notation to represent variables. The variable name starts with an underscore, followed by some number selected by the computer. In the preceding trace, _333 is a variable, which eventually unifies with jimmy. Of course, Prolog cannot use the original variable names because variables can, by definition, instantiate to any term, and different calls to the same clause typically instantiate the variables to different terms.

Whenever a line of the trace is printed, the user has a number of options. The user can merely press return to obtain the next line of the trace. This was the option used here. Other possible commands are a for abort and / for leap. The option a aborts the present query, whereas the option / completes the query without tracing. To get a complete listing of all options, one can type h for help. Often, one only wants to see the goals belonging to a particular procedure, such as parent. To do this, one types spy(parent) instead of trace while in query mode. In this case, the trace is off initially, turning on at the moment the procedure parent is encountered. This is the first spypoint. By typing / for leap, the system jumps to the next spypoint, which is the next goal involving the procedure parent.

Problems 4.2

1. Suppose that you have a database in a file called Assign1. Give the query that consults this file.
2. The following fact is erroneous. State why, and correct it.

    son(Charles, Elizabeth).

3. Consider the database given in Figure 4.1, and do a manual trace of the query aunt(X, Y), female(Y). Give the number of each goal, and indicate its level. Run the query to check your results.

4.3 ADDITIONAL FEATURES OF PROLOG

4.3.1 Introduction

Prolog contains a variety of built-in predicates that allow one to input and output data, to do arithmetic, and to compare terms. These predicates are described in the following sections. There is also a section on *structures*, which are terms useful for grouping data. The reason why structures are discussed here is that arithmetic expressions are really structures, and to understand how Prolog handles arithmetic expressions, a knowledge of structures is useful.

4.3.2 Input and Output

To read and write terms, the predicates read and write are available. For instance, the query write(hello) will cause the word hello to appear on the screen, followed by yes to indicate that the query has succeeded. Note that hello is interpreted as a term or, more specifically, as an atom. This is important, because one is tempted to use the goal write(Hello). In this case, Hello starts with a capital letter, which makes it a variable. The system outputs the internal representation of this variable name, say _41, which is probably not what was intended. To indicate that Hello is not a variable, one can use single quotes. In particular, the goal write('Hello') works as expected. To make the output look nice, one can use the predicates nl and tab. The predicate nl stands for newline, and it causes a linefeed. The predicate tab is used for tabbing. For instance, tab(4) tabs 4 spaces to the right from its present position. To see how this all works together, consider the following rule:

```
greeting :– nl, tab(4), write('This is Prolog'), nl, tab(6),
           write(greetings).
```

Once this rule is in the database, it can be called by typing greeting, followed by a period.

Statements without a head are executed immediately. This allows one to write instructions at the beginning of the program. One can also write any other message in this way. The following rule, for instance, causes the program to type This is Prolog when the file is consulted.

```
:– nl, tab(4), write('This is Prolog').
```

In writing games, one would like the name of the player. The predicate name asks for the name, and it echoes it.

```
name :– write('Your name '), read(Name), nl,
        write('nice to meet you '), write(Name).
```

Note that the input must be an atom; that is, it must start with a lowercase letter, and there must be a period after the name.

4.3.3 Structures

Many objects are best described by several data items. A date, for instance, consists of a year, a month, and a day. A book in a library is identified by call number, author, and title. To deal with objects consisting of several data items, Prolog uses structures. For instance, date(Day, Year, Month) is a structure. To see how to work with structures, consider the following database:

```
birthday(peter, date(23, aug, 1970)).
birthday(john, date(15, nov, 1973)).
birthday(anne, date(20, jan, 1971)).
```

The query birthday(peter, X) will result in the reply

$$X = date(23, aug, 1970)$$

Moreover, this database allows one to extract all people with birthdays in January. The appropriate query is birthday(X, date(Y, jan, Z)).

Structures, like variables and atoms, are terms. In contrast to atoms, structures can have arguments. The arguments of structures must be terms. These terms can even be structures themselves. For instance, a structure point(X, Y) can be used to represent a point in a Euclidean plane. The structure point can then be used to formulate a structure circle(R, C), where R is the radius and C is the center, which must be a structure of the type point(X, Y). Structures can also be used to express operators. For instance, the logical operator $\wedge$ can be represented by the structure and(α, β) and the operator $\vee$ can similarly be represented by the structure or(α, β), where α and β are some arbitrary terms. These structures allow one to express and manipulate logical expressions. For example, the expression $P \wedge (Q \vee R)$ can be expressed as and(p, or(q, r)). The structure and can even have another instance of the structure and as one of its arguments. In this way, a recursive structure is created. Once the structures are formulated, clauses to manipulate the structures can be designed. For instance, one can express the idempotent laws of logic as follows:

idempotent(and(A, A), A).
idempotent(or(A, A), A).

For illustration, consider the query idempotent(and(p, p), X). The system will respond to this query with X = p as required. The query idempotent(Y, r) has the result Y = and(r, r). This solution is obtained by unifying Y with r. When redoing the query, the result becomes Y = or(r, r).

4.3.4 Infix Notation

Suppose that John loves Mary. So far, this fact had to be expressed in prefix notation; that is, one had to write loves(john, mary). However, Prolog can be instructed to use infix notation, which means that one can also write john loves mary. Infix notation can also be used for structures. Assume, in particular, that the structure and is used to define conjunctions and the structure or to define disjunctions. These structures allow one, in principle, to write all logical expressions containing only the connectives $\wedge$ and $\vee$. When written in prefix notation, these structures are hard to read. Infix notation makes them much more readable. For instance, $P \wedge (Q \vee R)$ is p and (q or r) in infix notation, and this is much more readable than and(p, or(q, r)).

To represent operators by structures, all the information needed to manipulate the operator must be provided. In particular, the system must know the precedence of the operator and whether it is left or right associative. For the definition of these terms, the reader may consult Section 1.3.4. This information must also be provided for any other structure or predicate used in infix notation. The built-in predicate op has been created for this purpose, and this predicate must be executed before any structure or predicate in infix notation can be used. The predicate op has three arguments. The first argument is the precedence, the second argument describes the arity and associativity, and the third gives the name of the structure or predicate. The precedence is given as an integer, and the lower this number, the higher the precedence. It is suggested that the programmer use only integers between 200 and 700. To express that a binary operator is left associative, the symbol yfx is used. To make an operator right associative, one uses the symbol xfy. Unary operators, like $\neg$, can also be defined. To do this, one uses fy instead of yfx.

The predicate op must be done when the database is read in, which means that the clauses containing this predicate must start with :−. We now demonstrate op by defining loves, neg, and, and or, as infix operators. Here, neg stand for $\neg$, and for $\wedge$, and or for $\vee$.

```
:− op(250, yfx, loves).
:− op(200, fy, neg).
:− op(300, yfx, and).
:− op(400, yfx, or).
```

Once these definitions are executed, the operators in question can be used as infix operators.

4.3.5 Arithmetic

To do arithmetic, Prolog contains a number of built-in predicates that can handle arithmetic expressions. The most important of these predicates is the is predicate, which is used in infix notation and which evaluates arithmetic expressions. The following query uses an is predicate to evaluate $3 + 4$.

$$| \ ?- \ Y \ is \ 3 + 4.$$

The result of this query is

$$Y = 7$$

The is predicate has two arguments: the result of the calculation at the left and the expression to be evaluated at the right. The is predicate works on operators, which are really structures in infix notation. All the standard arithmetic operators are available. In the preceding query, the operator + was used, but −, *, and / are also implemented. Real and integer arithmetic can be mixed. The result of the expression in question is in integer mode, provided that all its subexpressions are in integer mode. Otherwise, the result is real.

As an illustration of the is predicate, suppose that the database contains the rule

$$sum(A, B, C) :- C \ is \ A + B.$$

Here are example queries, together with their results.

```
| ?− sum(3, 4, 7).
yes
| ?− sum(3, 4, X).
X = 7 ?
yes
| ?− sum(X, 4, 7).
{INSTANTIATION ERROR: in expression}
```

The first two of these queries should be obvious. The third query would fit very well into the framework of Prolog, and one might expect that X is unified with 3. In some implementations, this is exactly what happens. Most implementations of Prolog cannot handle this case, and some error message results. Moreover, even the most elaborate implementation cannot handle all possible unifications arising with the is predicate, because this would involve the ability to solve equations of arbitrary complexity. Consider, for instance, the rule

$$expr(X, Y) :- Y \ is \ X*X*X + 2*X*X + 3*X.$$

The query expr(X, 2) would then have to find a value for X such that Y = 2, and this involves solving an equation of third degree. Even more difficult equations can be found.

Prolog also provides relational operators for numerical expressions. For instance, 4 < 5 succeeds, while 4 > 5 fails. Other relational operators are =:=, =\=, =<, and >=, which represent equality, inequality, less than or equal, and greater than or equal, respectively. All these operators are restricted to numbers.

Arithmetic operators are structures, and they can be manipulated in the same way as other structures. For example, consider the fact

$$\text{combine}(A - A + B, B).$$

With this fact in the database, the query combine(2 − 2 + c, X) yields X = c.

4.3.6 Equality Predicates

Equality means different things in different applications, and to account for this, there are different predicates in Prolog to express equality. The most important equality predicate is the =. If A and B are two terms, then A = B means that A and B can be unified. For instance, if A is an uninstantiated variable, then A = minnie is true. Moreover, as a side effect of the =, A is unified with minnie. The predicate = has to be distinguished from the predicate ==. Whereas the predicate A = B succeeds if A and B can be unified, the predicate A == B only succeeds if A and B are the same without making use of unification. For instance, if A is an uninstantiated variable, then A == minnie fails. For the predicate ==, uninstantiated variables are always different from constants. Also, if A and B are two uninstantiated variables, then A == B fails, unless A and B are identical, as is the case if they have been unified earlier. In simple cases, the == is seldom used, and beginners are well advised not to use == at all unless they know exactly what they are doing.

Note that = and is are two very different predicates, which must not be confused. The predicate is means "evaluate," whereas = has no such meaning. To see the difference, compare the following two queries:

$$\text{A is } 3 + 4.$$
$$\text{A} = 3 + 4.$$

The first query causes 3 + 4 to be evaluated before unification. It then unifies A with the result and succeeds. The output is therefore A = 7. In the second case, 3 + 4 is *not* evaluated, but it is taken to be a structure. Hence, the program succeeds by unifying A with the expression 3 + 4, which leads to the following output:

$$\text{A} = 3 + 4$$

This shows that Prolog distinguishes between expressions and results. To stress this point, consider the query 3 + 4 = 7. This query fails because 3 + 4 is an expression and, as such, it cannot be unified with 7. However, if A has been unified with an expression, it can be manipulated like an expression. This is shown by the following query:

| ?– 3 + 4 = A, B is A.
A = 3 + 4,
B = 7 ?

Here, A first unifies with 3 + 4, as indicated. Since A is now the expression 3 + 4, it can be manipulated like an expression. Hence, B is A really means that A is evaluated, and the result is unified with B.

The equality predicate is extensively used in arithmetic. Unfortunately, this use of equality is neither implied by =, ==, nor is. Instead, Prolog has still another predicate for comparing arithmetic expressions, and this is the =:=. The operator =:= indicates that both sides must first be evaluated and then compared. For instance, 3 + 4 =:= 7 succeeds. On the other hand, it was indicated earlier that 3 + 4 = 7 fails.

There is no assignment in Prolog. Consequently, statements like A is A + 1 will always fail. There is no way to unify A with A + 1. Similarly, the following query will always fail.

| ?– 3 + 4 = A, A is A.

The reason for the failure is that A unifies with 3 + 4, and once this unification is in force, A cannot be unified with 7.

There are a number of cases in which a test for inequality is needed. In this case, the predicate \== is used. As an application of the predicate \==, consider the definition of "sibling." The definition that two people are siblings if they have the same mother or the same father is incomplete because this definition implies that everybody is his or her own sibling. In the definition, this case must be explicitly excluded. This leads to the following procedure:

sibling(X, Y) :– father(Z, X), father(Z, Y), X \== Y.
sibling(X, Y) :– mother(Z, X), mother(Z, Y), X \== Y.

As another example where \== is needed, let bilingual be the predicate expressing that someone speaks two languages. If speaks(X, Y) is the predicate that X speaks language Y, then the predicate bilingual can be implemented as

bilingual(X) :– speaks(X, Y), speaks(X, Z), Z \== Y.

In this rule, the \== is essential. Otherwise, Y and Z could refer to the same languages, which means that everyone speaking a single language is bilingual, and this is definitely not what was intended.

Problems 4.3

1. Write a program to compute the sum and product of two numbers and print these results. A program description should be printed when the database is consulted, and the user should be prompted for the input. The results, together with any suitable explanations, should appear as output.

2. A company has a list of all its employees consisting of a name and an employee number. The name is a structure, consisting of a first and a last name. For example, the record of john smith might look as follows:

employee(name(john, smith), 13345)

Give a rule called id that succeeds for valid surname and employee number pairs. For example, id(smith, 13345) must succeed.

3. If A is a logical expression, $A \wedge \neg A$ always yields false. Write a rule that converts any expression of the form $A \wedge \neg A$ to the logical constant f.

4. Write a predicate convert(H, M, Result) that converts a time interval of H hours, M minutes into Result minutes.

5. Write a program that finds, for each date given by a month and a day, how many days are left in the year. The data should be given as a structure.

6. Consider the following rule:

calc :– read(X), Y is X, write(X), tab(1), write('is'), tab(1), write(Y).

If the input to this program is $3 + 5$, what is the output?

7. Suppose a Prolog implementation does not have the built-in predicate =:=. Write a predicate equ(X, Y) that accepts two arithmetic expressions and that succeeds if these expressions yield the same result when evaluated.

8. A warehouse has an item name and a price for each item that it carries. The name and price for each item are given by a number of facts, called costs. For instance, costs(vcr, 400) indicates that a vcr costs $400. Write a Prolog rule named howmuch that has an item name as its only argument and that prints The α costs β, where α is the name for which a price is requested by the query and β is the price.

4.4 RECURSION

4.4.1 Introduction

In Section 3.3.2 the term "descendant" was defined recursively. To convert this definition into Prolog, one needs recursive rules, that is, rules in which the predicate to be defined appears on both sides of the :–. Such rules will be discussed later. Generally, any procedure that contains recursive rules is called a recursive procedure. Recursive procedures contain both recursive clauses and nonrecursive clauses. The nonrecursive clauses form the basis of the recursion, and they will be referred to as *base clauses*.

In Chapter 3, it was pointed out that loops can be replaced by recursion. Loops are in this sense redundant. In fact, Prolog uses recursion where procedural languages would use loops, which means that recursion is fundamental for writing programs in Prolog. To do recursion effectively, Prolog allows one to define recursive structures. Some recursive structures were already introduced in Section 4.3.4 for representing logical connectives. Here, we concentrate on a very basic new recursive structure, the *list*. In Prolog, lists take the place of arrays as used in procedural languages, which makes them extremely important. Almost every nontrivial program contains lists in some form or other.

4.4.2 Recursive Predicates

As an example of a recursive definition, consider the definition of a descendant. We can define a descendant of Y as someone who is either a child of Y or who is a child of a descendant of Y. This yields

```
% base case
descendant(X, Y) :− child(X, Y).

% recursive case
descendant(X, Y) :− child(X, Z), descendant(Z, Y).
```

Note that any query that unifies with either the base clause or the recursive clause of the procedure descendant must, by definition, be correct. X is a descendant of Y either if X is a child of Y or if there is a child of Z, where Z is a descendant of Y. All descendants satisfy this definition, and everybody satisfying the definition is a descendant. There is nobody satisfying the definition who is not a descendant, and there is no descendant who does not satisfy this definition. Queries that succeed must satisfy these definitions, and they must be correct in this sense.

To illustrate the procedure, the abstract family trees given in Figure 4.2 are used. This figure results in the following database:

```
child(b, a).
child(c, a).
child(d, b).
child(e, b).
child(f, e).
child(q, p).
child(r, q).
```

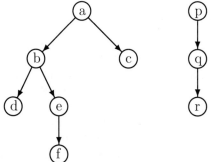

Figure 4.2.   Family trees

Consider now the query descendant(f, a), a query that succeeds. In fact, the following is the derivation as arrived at by Prolog.

```
descendant(f, a) :− child(f, e), descendant(e, a).
descendant(e, a) :− child(e, b), descendant(b, a).
descendant(b, a) :− child(b, a).
```

The first line of this derivation instantiates the recursive rule with X = f, Y = a, and Z = e. Similarly, the second line instantiates the recursive rule with X = e, Y = a, and Z = b. Note that in these two lines the same variables are instantiated by different atoms. This is allowed since the variables are strictly local to the rule in which they occur. For instance, X can be replaced by any term, provided, of course, that all occurrences of X in the rule are replaced by the same term. The lines of the derivation are written in the order in which they are derived by Prolog. To verify that the derivation is valid, it is better to start with the last line, however. The third line of the derivation indicates that descendant(b, a) is true if child(b, a) is true. Since child(b, a) is a fact, one concludes descendant(b, a). The second line of the derivation indicates that descendant(e, a) is true of child(e, b) and descendant(b, a) are both true. Since child(e, b) is a fact and since descendant(b, a) has just been derived, descendant(e, a) follows. This result, together with the fact child(f, e) finally implies the query descendant(f, a) by the first line. This completes the proof.

The derivation is obtained as follows by Prolog. The query descendant(f, a) is unified first with the base rule descendant(X, Y) :− child(X, Y), but this rule fails. The recursive rule is then tried, which yields

descendant(f, a) :− child(f, Z), descendant(Z, a).

The goal child(f, Z) is attempted, and it succeeds with Z = e. If the corresponding instantiation is carried through, the rule in the first line of the derivation is obtained. This rule has two goals. The first goal is child(f, e), which is a fact, and it succeeds. The second goal is descendant(e, a), and this goal is resolved in a similar way as for the original query. First, the base rule for descendant is tried, and it fails. Consequently, the head of the recursive rule is unified with descendant(e, a), and it instantiates to

descendant(e, a) :− child(e, Z), descendant(Z, a).

The goal child(e, Z) unifies with the fact child(e, b) and succeeds. This unifies Z with b, and the second line of the derivation in question is obtained. It remains to prove descendant(b, a). This goal unifies with the base rule and yields

descendant(b, a) :− child(b, a).

Since child(b, a) is a fact, the query succeeds and a yes is printed.

4.4.3 Termination

When defining procedures in Prolog, care must be taken that the resolution of queries does not cause infinite loops. This can be approached by defining the appropriate well-founded sets, as indicated in Section 3.3. Here we will use similar arguments to show that certain definitions for procedures will not lead to infinite loops. However, since the halting problem is undecidable, there are no generally applicable methods to prove termination. Still, for many important cases, termination proofs exist.

First, a termination proof is given for the procedure descendant as defined previously. There is no query that will cause this procedure to run into an infinite loop. To see this, consider the recursive rule of the procedure, which is

$$\text{descendant}(X, Y) :- \text{child}(X, Z), \text{descendant}(Z, Y).$$

If the database is really a database of descendants, then no atom can appear twice in the family tree. Hence, if a and b are two atoms, then once the fact child(a, b) is used by the recursive rule, it cannot be so used again. Hence, whenever the recursive rule succeeds, there is one fewer individual and, eventually, none remains. At this time, the procedure stops. Hence, no matter what query is used, as long as the database is a correct family tree, there cannot be an infinite loop.

This proof breaks down if the recursive rule is written as

$$\text{descendant}(X, Y) :- \text{descendant}(Z, Y), \text{child}(X, Z).$$

In this case, the goal descendant(Z, Y) is always done first, and the goal child(X, Z) is never used. In this case, the procedure may run into an infinite loop. For instance, the query descendant(q, a) will never be resolved. This can be easily verified by a trace. This example shows that with recursive clauses the goal order is important. One has the following definition:

> **Definition:** A rule is called *left-recursive* if the recursive predicate occurs first. If the recursive predicate occurs last, the rule is called *right-recursive*.

Due to Prolog's problem-solving strategy, left-recursive rules may lead to infinite loops, and they should be avoided.

The original procedure descendant terminates because the same child cannot occur twice in the family tree. This assumption is valid for all trees. However, many problems involve situations for which the same elements can occur more than once. This is demonstrated by the following example: An airline serves three cities, a, b, and c. In some cases, there are direct connections between the cities. In particular, as can be seen from Figure ??, there are direct connections from a to b, from b to a, from b to c, and from c to b. It is obvious that there is also a connection between a and c, but not a direct one. We should be able to write a Prolog program that can determine which cities are connected, either directly or indirectly. This, however, is more difficult than it seems. First, consider the following attempt:

```
directconnection(a, b).
directconnection(b, a).
directconnection(b, c).
directconnection(c, b).
connection(X, Y) :- directconnection(X, Y).
connection(X, Y) :- directconnection(X, Z), connection(Z, Y).
```

The clauses defining connection indicate that there is a connection between X and Y if and only if one of the following two alternatives applies:

1. There is a direct connection between X and Y.

2. There is a direct connection between X and some place Z and a connection between Z and Y.

Hence, if either the base case or the recursive case applies, we do in fact have a connection. Thus, if Prolog provides us with an answer, the answer will be correct. The question now is whether Prolog will provide an answer or go into an infinite loop. To see what happens, consider the query connection(a, d). This query should fail since there is no atom d. This is not what happens, as is revealed by the following trace:

```
| ?− connection(a, d).
+ 1 1 Call: connection(a, d) ?
+ 2 2 Call: directconnection(a, d) ?
+ 2 2 Fail: directconnection(a, d) ?
+ 2 2 Call: directconnection(a, _316) ?
+ 2 2 Exit: directconnection(a, b) ?
+ 3 2 Call: connection(b, d) ?
+ 4 3 Call: directconnection(b, d) ?
+ 4 3 Fail: directconnection(b, d) ?
+ 4 3 Call: directconnection(b, _864) ?
+ 4 3 Exit: directconnection(b, a) ?
+ 5 3 Call: connection(a, d) ? a
Execution aborted
| ?−
```

One sees that goals 1 and 5 are identical, and this means that an infinite loop results. The trace was therefore aborted by entering an a after the question mark. If the trace had not been aborted, the trace output following goal 5 would have been almost identical to the trace from 1 to 5, and it, too, would eventually have resulted in the call connection(a, d). Hence, connection(a, d) is called over and over again, and the program will never arrive at an answer. This problem is not easily fixed. To keep track of the cities already visited, lists must be used. We will not do this here, however.

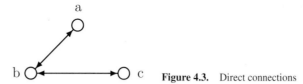

Figure 4.3. Direct connections

4.4.4 Loops and Prolog

All algorithmic languages have loops, such as for loops and while loops. Loops are seldom used in Prolog, even though there are some built-in predicates that have the effect of a loop. Most Prolog programmers use recursion in place of loops. This matches with the theory developed in Section 3.5.3, where it was shown that all loops can be converted to recursive function calls. It is now demonstrated how to program a loop in Prolog. Suppose that one would like to define a predicate listintegers that has two arguments, say M and N, and that prints all integers from M to N. The base case is the case when M > N, in which case nothing needs to be done, except that one may type a new line. This yields

$$\text{listintegers(M, N) :– M > N, nl.}$$

In all other cases, one divides the problem into two subproblems: first, M is printed, and then the numbers from M + 1 to N are printed. To do this, one must calculate M + 1 and assign the result to, say, NewM. This decomposition approach yields to the following clause:

$$\text{listintegers(M, N) :– M <= N, nl, write(M), NewM is M + 1,}$$
$$\text{listintegers(NewM, N).}$$

For instance, to print the numbers from 4 to 6, the query listintegers(4, 6) can be used. This query first prints the number 4, and it calls listintegers(5, 6), which prints the remaining two numbers.

Even though every loop can thus be converted into a recursive Prolog procedure, it is normally not wise to do this. It is much safer to use definitions as guides, which can be directly converted to Prolog. What is important, however, is that, as far as loops are concerned, procedural languages do not have any greater computational power than Prolog.

4.4.5 Lists

Every programming language needs constructs to deal with collections of data. There are no arrays in Prolog. Prolog programmers typically use lists instead. As the name implies, a list is merely a list of terms, such as constants, variables, and structures. Lists can even contain other lists. The following are examples of lists.

1. [mary, jack, john]
2. [3, 5.5, newyork]
3. [3, [jack, john], 6.2]
4. [ ]

Note that lists are enclosed in square brackets. The first of these four lists contains the atoms mary, jack, and john, in that order, as its members. The second list similarly contains the numbers 3 and 5.5, and the atom newyork. The third list has three elements: the number 3, the list [jack, john], and the number 6.2. List number 4 has no elements in it; it is called the *empty list*.

It was mentioned in Section 3.3.2 that lists can be constructed recursively. In Prolog, lists are constructed in the same way. To this end, we define the *head* of a list to be the first element of a nonempty list. The remaining elements form a list of their own, and this list is called the *tail* of the list. For instance, in the list [a, b, c, d] the head is a, and the tail is the list [b, c, d]. A list with head A and tail B can be written as [A | B]. For instance, instead of [a, b, c, d], one can write [a | [b, c, d]]. These two expressions are completely equivalent, and one may be exchanged freely with the other. When using this convention, one can define a list as follows:

1. The empty list [ ] is a list.

2. If A is any term, and B is a list, then [A | B] is a list. This list contains A as its first element, while the list B forms the tail of the new list.

3. Nothing else is a list.

Note that the empty list has neither a head nor a tail. The list [a] has a as its head, but the tail is the empty list [ ].

Actually, the notation involving the tail is more general than indicated here. All entries before the | are considered as individual members, whereas all elements following the | collectively form a list, called the tail. For instance, the list [a, b, c, d] can also be written as [a, b | [c, d]]. Here a and b are the first two elements, and the list of the remaining elements is [c, d].

Some simple examples now demonstrate the head–tail mechanism. The first predicate is called "firstel(X, Y)." Here Y is a list, and the predicate must succeed if X is the first element of Y. This predicate is implemented by the following fact:

$$\text{firstel}(X, [X \mid \_]).$$

For instance, the goal firstel(c, [c, a, d, e]) succeeds. To see this, consider the following two lines:

$$\text{firstel}(X, [X \mid \quad \_ \quad]).$$
$$\text{firstel}(X, [c \mid [a, d, e]]).$$

Here c unifies with X, and since the anonymous variable can be anything, the query succeeds. The goal firstel(Head, [2, 5, 1]) also succeeds.

$$\text{firstel}(X \quad , [X \mid \quad \_ \quad]).$$
$$\text{firstel}(\text{Head}, [2 \mid [5,1]]).$$

Here X unifies with Head, and since X later unifies with 2, Head must unify with 2. The query therefore succeeds with Head = 2.

The second predicate we define is called strip. This predicate has two arguments, both of them lists, and it succeeds if the second list is equal to the first list, except that its first element is deleted. One finds

$$\text{strip}([\_ \mid \text{Tail}], \text{Tail}).$$

Here Tail is merely a variable name. The query strip([3, 2, 1], X) yields X = [2, 1]:

$$\text{strip}([\_ \mid \quad \text{Tail}], \text{Tail}).$$
$$\text{strip}([3 \mid [2,1]], \quad X).$$

The next predicate, we call it firstsequal, is to succeed for all lists in which the first two elements are equal. One has

$$\text{firstsequal}([X \mid [X \mid \_]]).$$

This predicate splits the list into a head and a tail and, in order to succeed, the head must match the first element of the tail. If this is true, then the first two elements of the list must obviously match. For example, the goal firstsequal([X, a, b, c]) succeeds with X being instantiated to a. There is a simpler version of the same predicate, which is

$$\text{firstsequal}([X, X \mid \_]).$$

4.4.6 Recursive Predicates Involving Lists

By definition, the empty list is a list, and all other lists are constructed by combining a term A with a list B to form the new list [A | B]. In this new list, A becomes the head and B becomes the tail. If a procedure is to be defined involving lists, it is best to adopt the following procedure, which merely reflects the definition of a list.

1. Deal with the empty list.
2. Deal with nonempty lists. This, in turn, involves two steps.
   (a) Deal with the head.
   (b) Deal with the tail. This is usually done recursively.

It is good to follow these steps, even though not all steps necessarily result in clauses. Moreover, the clause dealing with the empty list often follows the clause or the clauses dealing with the nonempty list. The application of this scheme is now demonstrated by a number of examples.

     A very important procedure is the predicate member. This predicate is used to find if a certain element X is within a list A. Specifically, the predicate member(X, A) succeeds if X is an element of the list A. The procedure suggested can now be applied as follows:

1. In regard to the empty list, note that member(X, [ ]) always fails. To make a predicate fail, one merely has to omit any clause with which it can unify.
2. If the list is nonempty, then one has to deal with the head and the tail, as follows:
   (a) The predicate member(X, A) must succeed if X is the first element of A. This can be accomplished through the following clause:

$$\text{member}(X, [X \mid \_]).$$

   (b) The predicate member(X, A) must succeed if X is in the tail of A. To find whether this is true, the predicate member is used recursively. One finds

$$\text{member}(X, [\_ \mid Z]) :- \text{member}(X, Z).$$

In summary, the predicate member is defined by the following two clauses:

$$\text{member}(X, [X \mid \_]).$$
$$\text{member}(X, [\_ \mid Z]) :- \text{member}(X, Z).$$

We now prove that if X is a member of A then member(X, A) will succeed, and that it will fail otherwise.

Inductive base If A is [ ], then member(X, A) cannot be matched with any rule, which means that it always fails. This is correct because X can never be an element of an empty list.

Inductive hypothesis Choose a list A = [B | C], and assume that member(X, C) works correctly.

Inductive step To be in A, X must either be the head of A or it must be part of B, where B is the tail of A. If X is equal to the head of A, then the query unifies with member(X, [X | _]) and succeeds. If X is not equal to the head of A, the recursive clause applies, and this clause succeeds if X is in B. According to the inductive hypothesis, member(X, B) succeeds if and only if X is in B. This establishes the inductive step for both the case when X is the head and when X is part of the tail.

Domain well founded The list in the body of the recursive clause is always smaller than the list in the head of the recursive clause. Hence, with each call the goal deals with a smaller list, and this cannot continue forever. Sooner or later the query must either succeed or the empty clause must be reached, in which case the query fails.

Conclusion In all cases, the query succeeds if X is in A, and it fails otherwise.

Let us now try the member procedure, using the following two queries:

> member(mary, [mary, john, jack]).
> member(jack, [mary, john, jack]).

The first query unifies with the first clause as follows:

> member(X , [X | _   ]).
> member(mary, [mary|[john, jack]]).

Hence, X = mary and the anonymous variable is [john, jack]. The second query also succeeds. To see this, consider the following two instantiations of the recursive rule:

> member(jack, [mary | [john, jack]]) :− member(jack, [john, jack]).
> member(jack, [john | [jack]]) :− member(jack, [jack]).

Both clauses are instantiations

> member(X, [_ | Z]) :− member(X, Z).

In the first clause, X = jack and Z = [john, jack], and in the second clause, X = jack as before, and Z = [jack]. Note that the lists [mary, john, jack] and [mary | [john, jack]] are equal, and the same is true for [john, jack] and [john | [jack]]. It is easily verified that the goal member(jack, [jack]) unifies with member(X, [X | _]) and succeeds. In the preceding derivation the second line therefore succeeds, and this makes the first line succeed. Since the head of the first line is equal to the query, one concludes that the query succeeds. Prolog finds the derivation almost as it is given here, except that in its incomplete foresight it always tries the first clause first. This clause fails until the element in question is first in the list.

An important property of a list is its length. Hence, let length(List, N) be the predicate that succeeds if List contains exactly N elements. Although this predicate is usually predefined, it is instructive to formulate it here.

1. The empty list has a length of zero. Consequently,

$$\text{length}([\,], 0).$$

2. If the list is not empty, one has to deal with the head and the tail of the list.

   (a) The head of the list has exactly one element.
   (b) The length of the tail can be found by using the predicate length. This leads to

$$\text{length}([\_ \mid \text{Tail}], \text{Length}) :- \text{length}(\text{Tail}, \text{TailN}),$$
$$\text{Length is } 1 + \text{TailN}.$$

This procedure works correctly by construction. It is correct for the empty list. Moreover, the inductive hypothesis is that, given any list A, it works correctly for the tail of A. This inductive hypothesis ensures that length(Tail, TailN) succeeds with the correct value TailN, and the inductive step is derived by noting that N is one more than TailN. Moreover, since Tail is smaller than the original list, one concludes that the domain is well founded.

Another important predicate is append. This predicate has three lists as arguments, which we call As, Bs and Cs, and it succeeds if the Cs combine the As and Bs into one list containing first the As and then the Bs, both in their original order. There are three lists in this predicate now, and the question arises as to which list to use for recursion. This can essentially be only found by trial and error. In the case considered here, recursion on the As works. This was found by the authors by relying on their brains, which were working in nondeterministic modes. Other people may have to use backtracking. After these remarks, one proceeds according to the standard method.

1. Consider the case when As is empty. In this case, the Bs and Cs must be identical. Hence,

$$\text{append}([\,], \text{Bs}, \text{Bs}).$$

   (a) If the list As is not empty, then the head of the As must be equal to the head of the Cs.
   (b) The tail of the Cs must be equal to the tail of the As, followed by the Bs.

These considerations suggest the following rule:

$$\text{append}([\text{A} \mid \text{TailAs}], \text{Bs}, [\text{A} \mid \text{TailsCs}]) :-$$
$$\text{append}(\text{TailAs}, \text{Bs}, \text{TailCs}).$$

This, together with the base clause append([ ], Bs, Bs), provides the required procedure. To prove that the procedure puts all elements of the As in front of the Bs, one uses induction.

Inductive base The procedure is correct if the list As is empty.

Inductive hypothesis Choose two arbitrary lists As and Cs, and assume that append works for the tail of As and the tail of Cs.

Inductive step Since the first element of the Cs must be equal to the first element of the As, and since the tail of the Cs is the combination of the tail of the As and Bs, the inductive step follows.

Domain well founded The recursion ends because the lists on the right of the recursive rule are smaller than the ones on the left.

Conclusion Since the As, Bs, and Cs can all be chosen arbitrarily, the proof works for all As. The same holds true for the Bs and Cs.

As the last example of this section, consider the problem of printing all elements in a list. After the list is printed, a carriage return should be executed. Let printlist(A) be the procedure that prints all elements of A. Again, one must consider the empty list and the head and tail of the nonempty lists. If the list is empty, one only needs to print the carriage return. This is accomplished by the following clause:

$$\text{printlist}([\ ]) :- \text{nl}.$$

If the list is not empty, one must first write the head of the list and then its tail. The following rule accomplishes this task:

$$\text{printlist}([A \mid B]) :- \text{write}(A), \text{printlist}(B).$$

4.4.7 Successive Refinement

In programming, one usually starts with a general problem description, which is then refined in a stepwise fashion. Specifically, one has a general objective, which is then divided into subobjectives. These objectives can be mapped directly onto Prolog goals, as will be shown by the following example.

The teacher of the class Prolog 101 has just marked the midterm for her class, and she now would like to calculate the grade average. All the grades are given in a list. To find the average, she needs the number of students who wrote the midterm and the sum of all the grades. This leads to the following rule:

$$\text{average}(\text{List}, \text{Av}) :- \text{sum}(\text{List}, T), \text{length}(\text{List}, N), \text{Av is } T/N.$$

Two procedures still need to be written: sum(List, T) and length(List, N). The second procedure has already been programmed, and it is frequently built in. We therefore do not consider it further. However, the procedure sum(List, T) must still be defined. The question is thus what to do if List is empty and how to process the head and tail of List if the list is not empty. One has

```
sum([ ], 0).
sum([Grade | Grades], T) :- sum(Grades, NewT),
        T is NewT + Grade.
```

This completes the task as stated.

To make the problem more challenging, suppose that the list contains both students and their grades, and suppose that the procedure must list the student names and their grades, followed by the class average. To accomplish this, we assume that the list of grades has pairs as members. Each pair consists of a name and a grade. For instance, to express that Jim got an 80, the pair jim, 80 is used. These pairs are then combined to form a list. If the class has only two students, Jim and Mary, who score 80% and 90%, respectively, then this list looks as follows:

$$[[\,jim,\ 80],\ [mary,\ 90]]$$

Now for the program. The problem can be divided into two steps, the calculation of the average and the printing of the class list. The procedure to accomplish this is called marks(X), where X is the class list. This procedure has only one clause. This clause has two goals, mean(X, Av) for calculating the class average and list(X, Av) for listing the students, together with their grades and the class average. This yields

$$marks(X) :- mean(X, Av), list(X, Av).$$

To calculate a mean, one needs to have the total of all marks and the number of students.

$$mean(A, Av) :- total(X, T), length(X, N), Av\ is\ T/N.$$

This clause requires total(X, T), which is to calculate the total of all marks and return it as T. This procedure is recursive. If X is empty, then T = 0. Otherwise, T is the total of the tail, plus the grade of the name–grade pair forming the head.

```
total([ ],0).
total([[ _, Grade] | B], T) :- total(B, TTail),
                               T is Grade + TTail.
```

Consider now the procedure list(X, Av). If the list is empty, a carriage return should be printed. Otherwise, the record given by the head of the list must be printed, and then all other records must be printed.

```
list([ ], _) :- nl.
list([A | B], Av) :-
        nl, printRec(A, Av), list(B, Av).
```

The procedure printRec(A, Av) is as follows:

```
printRec([Name, Grade], Av) :- write(Name), tab(5),
                               write(Grade), tab(5),
                               write(Av).
```

Problems 4.4

1. The predicate is_a(X, Y) describes the relation that is true if X is a Y. For instance, is_a(lion, bigcat) is true because the lion is a big cat. Moreover, is_a(lion, predator), is_a(lion, mammal), and is_a(lion, animal) are all true. Of course, is_a(lion, predator) follows from is_a(bigcat,

predator), and is_a(lion, animal) similarly follows from is_a(predator, animal). Use this idea to implement a database describing the predicate is_a with the least number of facts.

2. Write the procedure power(N, X, Y), which is supposed to succeed if $Y = X^N$. The procedure must work for $N \geq 0$, and it must find the power through multiplication.

3. Write a procedure last(X, Y) that succeeds if X is the last element of the list Y.

4. Trace member(3,[1,2,3,4]) manually. Check your result by using the trace facility of Prolog.

5. Write a procedure that removes the last element from a list.

6. Write a procedure that multiplies all elements of a list of numbers by a given number Y.

7. Write a Prolog program that accepts a list of items and calculates the total bill for all items. The list of items consists of triplets: the name of the item, its price, and the quantity delivered. This information must be printed line by line, followed by the product of the price with the quantity. These products must be added to find the total bill before taxes. This amount is increased by a tax of 14%.

8. Write a procedure that counts the number of times a given word X is appearing in a sentence. The sentence is given in the form of a list.

4.5 NEGATION IN PROLOG

4.5.1 Introduction

Prolog contains a number of logical connectives, including the comma to express $\wedge$, the semicolon to express $\vee$, and the :— to express $\Leftarrow$. However, Prolog does not contain a proper equivalent for negation. The reason for this is that, at the time Prolog was designed, all known methods for implementing negation were unacceptably inefficient. Although this has improved somewhat, the standard implementations of Prolog still do not contain a proper negation. In its place are a number of substitutes. These may not be completely satisfactory from an esthetic point of view, but they are unavoidable. These substitutes will now be discussed. Before this is done, it is shown how Prolog relates to logic and why negation is needed to make Prolog complete.

4.5.2 Prolog as a Logic Language

In Prolog there are three types of statements: facts, rules, and queries. All these statements are clauses or, more specifically, *Horn clauses*. Horn clauses will be defined later.

For the discussion, consider the following clause:

uncle(X, Y) :— brother(X, Z), child(Y, Z).

This statement means that X is an uncle of Y if X is a brother of Z and Y is a child of Z. In predicate calculus, this statement becomes

$$(brother(X, Z) \wedge child(Y, Z)) \Rightarrow uncle(X, Y)$$

or

$$\neg(brother(X, Z) \wedge child(Y, Z)) \vee uncle(X, Y)$$

We apply De Morgan's law and obtain

$$\neg brother(X, Z) \vee \neg child(Y, Z) \vee uncle(X, Y)$$

This expression is a disjunction of *literals*. In the context of predicate calculus, a literal is an atomic formula or the negation of an atomic formula. Nonnegated literals are said to be *positive*, and negated literals are said to be *negative*. Any disjunction of literals is called a *clause*. So far, we have used the word clause in a different sense. For us, any Prolog statement was a clause. The two meanings are related because every conjunctive Prolog clause can be converted into a clause in the more technical sense. The preceding example shows how to convert rules into clauses. Generally, rules are implications; that is, they can be written as $B \Rightarrow A$, where A is the head of the rule and B the body. If the clause is conjunctive, then B is a conjunction. Because of De Morgan's law, the negation of B can be written as a disjunction of literals. Hence, if the goals of B are $G_1, G_2, \ldots, G_n$, then the rule can be written as

$$A \vee \neg B \equiv A \vee \neg G_1 \vee \neg G_2 \vee \cdots \vee \neg G_n$$

This shows how to convert rules into clauses. However, not every clause can be written as a Prolog rule. From the preceding analysis, it follows that all goals of a Prolog rule result in negative literals, and the only positive literal is given by the head of the rule. In fact, the only clauses expressible in Prolog are *Horn clauses*.

Definition 4.1: A clause is called a *Horn* clause if it contains at most one nonnegated literal.

Consequently, rules are Horn clauses. Facts are also Horn clauses. They only consist of a single positive literal. Facts are in this sense rules that have a body that is empty. Moreover, the next paragraph shows that queries are also Horn clauses.

In Chapter 1, it was explained that if a set of premises logically implies the conclusion then a contradiction results if the negation of the conclusion is added to the premises. This philosophy is implemented in Prolog as indicated in Section 4.1.7. Every query is negated, and the resulting clause is added to the database with the objective to derive a contradiction. Suppose, for example, that the database contains the fact son(a, b). The query son(X, Y) generates the clause ¬son(X, Y), which is added to the database. By instantiating X to a and Y to b, this new clause leads to ¬son(a, b), which contradicts the fact son(a, b). In fact, as indicated in Section 4.1.7, Prolog uses resolution, which means that any goal in the query that matches with the head of a clause is, after proper instantiations, replaced by the body of the clause. Since facts do not have a body, any goal that unifies with a fact is removed after the appropriate instantiations are completed. If resolution is applied to the query ¬son(a, b) and if son(a, b) is a fact, resolution removes all goals from the query, and the empty clause results. The empty clause is therefore really the disjunction without any term, and this disjunction is false. Consequently, deriving the empty clause means deriving a contradiction. As mentioned in Section 4.1.7, a query succeeds if the empty clause can be derived. In terms of logic, this means that the query follows from the database if the negation of the query, when added to the database, leads to a contradiction. If it is not possible to derive a contradiction, then the query does not follow from the database. In this case, one cannot derive the empty clause, and the query fails.

The negation of the query and the empty clause are both Horn clauses. A clause is a Horn clause if it contains at most one positive literal, which is represented by the head of the clause. Since the query contains no head, it is a Horn clause. The empty clause, which contains no literal at all, is also a Horn clause. In conclusion, Prolog deals with a set of Horn clauses, and it tries to derive a contradiction from this set.

It can be shown that all logical expressions can be written in clause form. However, many expressions cannot be written in the form of Horn clauses. For instance, suppose that somebody is unemployed if he or she holds no job. This can be expressed as

$$\neg hasjob(X) \Rightarrow unemployed(X)$$

which leads to the clause

$$hasjob(X) \vee unemployed(X)$$

Since this clause contains two nonnegative literals, it is not a Horn clause. Consequently, it cannot be expressed in Prolog. More generally, the logical equivalent of the "if-then-else" construct cannot be expressed in terms of Horn clauses. The statement "If P then Q, else R" can be translated into logic as

$$(P \Rightarrow Q) \wedge (\neg P \Rightarrow R)$$

This leads to the two clauses $\neg P \vee Q$ and $P \vee R$, but $P \vee R$ is not a Horn clause. This clause therefore cannot be handled in a declarative way within Prolog. It can, however, be handled in a procedural way. Specifically, one can make use of the order in which Prolog resolves the clauses to implement else constructs. This solution destroys the logical basis of Prolog, which means that seemingly correct definitions may become erroneous when used in unexpected ways. One should therefore avoid having to make use of the clause order whenever possible.

Although it is not related to negation, the following discussion is of interest here. In Prolog, all facts and rules are universally quantified, which allows unification to work. For example, the rule

uncle(X, Y) :− brother(X, Z), child(Y, Z).

holds for all values X, Y, and Z. This yields, if the scope of the quantifiers is expressed in clause form,

$$\forall\, X\, \forall\, Y\, \forall\, Z(uncle(X,Y) \vee \neg brother(X,Z) \vee \neg child(Z,Y))$$

By using the relations given in Table 2.4, it is easy to prove that this is logically equivalent to

$$\forall X \forall Y \, (uncle(X,Y)) \Leftarrow \exists Z (brother(X,Z) \wedge child(Y,Z))$$

This is how one normally thinks about Prolog rules. X is an uncle of Y if there is a Z such that X is a brother of Z, and Y is a child of Z. The negation of the query is also universally quantified. In other words, Prolog attempts to show that for all possible elements the query is not true. This amounts to proving that there exists no term satisfying the query.

4.5.3 Negation as Failure

In a finite universe, one can say that a goal is not true if there is no individual in the universe that makes the goal true. This leads to the idea to implement the negation as failure. Indeed, Prolog has a predicate, \+, that has a predicate as its argument and that succeeds only if the predicate in question fails. To demonstrate this predicate, consider the following database:

```
red(rose).
green(grass).
white(daisy).
```

Consider now the query red(poppy). The answer to this query is "no." Note that the meaning of no is not that poppies are not red. The reply no only indicates that there is nothing in the database to indicate that poppies are red. According to the definition of \+, the Prolog statement \+(red(poppy)) will thus succeed. However, \+(white(daisy)) will fail, because white(daisy) succeeds. Instead of \+(X), one may also write \+ X.

The predicate \+ is not a true negation. True negation instantiates variables under the scope of the negation, and the predicate \+ does not do this. A query \+(red(X)) never results in an answer such as X = daisy. Indeed, once a goal fails, all instantiations are undone. To see what this implies, consider the following database:

```
jobhunter(X) :− \+ hasjob(X), needsmoney(X).
needsmoney(bill).
hasjob(joe).
```

In this case, bill has no job, and he needs money. One would therefore expect jobhunter(X) to succeed with X = bill. However, this is not the case. The goal hasjob(X) succeeds with X = joe, and this means that \+ hasjob(X) fails. Consequently, the query fails, and the system replies with no instead of with X = bill. To make the rule work properly, one has to place the negated clause at the end like this:

```
jobhunter(X) :− needsmoney(X), \+ hasjob(X).
```

Now, the query jobhunter(X) succeeds with X = bill. Generally, the \+ works as expected if the variables are instantiated. This means that by ordering the clauses properly, one can often use \+ for expressing negation. This, however, inserts a procedural element into Prolog.

4.5.4 Use of the Clause Order

In order to implement an if-then-else construct, one can, in principle, make use of the order in which Prolog scans the clauses in the database. For instance, consider the predicate youpay(Income, Tax). This predicate is to work as follows: If Income is below 10000, then Tax is 10% of Income. If Income is 10000 or more, but less than 50000, then Tax is 1000 plus 15% of the amount that Income exceeds 10000. If Income is above 50000, then Tax is 7000 plus 20% of the income over 50000. We tentatively implement this problem by the following clauses:

```
youpay(Income, Tax) :− Income < 10000,
                       Tax is 0.1∗Income.
youpay(Income, Tax) :− Income < 50000,
                       Tax is 1000 + 0.15∗(Income − 10000).
youpay(Income, Tax) :− Tax is 7000 + 0.2∗(Income − 50000).
```

This works fine unless the goal youpay is redone. For instance, the query youpay(20000, Tax) matches with the second rule head and yields Tax = 2500, which is correct. Similarly, the query youpay(1000, Tax) matches with the first clause and yields Tax = 100. However, if any goal is redone, then problems arise. To see this, consider again the goal youpay(1000, Tax). After the first answer, which is Tax = 100, a semicolon is printed, which means that the system searches for other answers. Since clause 1 has already been tried, the system now tries clause 2, and the tax is calculated accordingly. This yields Tax = −350. If the query is redone once more, the answer is calculated according to the third clause, and the result is Tax = −2800. Hence, the clause order can only be used if goals are not redone. This is hard to enforce, and programs using the clause order are therefore unreliable.

The clause order can be used to express a default option. This is frequently necessary in recursive procedures, as the following example demonstrates:

Example 4.7   A library has a list of all its books, and it wants to condense this list by removing all duplicate entries. Write a procedure to accomplish this task.

Solution   Let condense(As, Bs) be the predicate in question. Specifically, condense has two lists as arguments, and it succeeds if the Bs contain the same books as the As, except that the Bs contain no duplicates. Following the normal procedure, consider first the case, for which As is the empty list. In this case, one finds

```
condense([ ], [ ]).
```

Nonempty lists have to be divided into a head and a tail. When doing this, two cases must be distinguished. If the head of the first list is a member of the tail, then the head is a duplicate, and it must not appear in the second list. Otherwise, the head of the first list is equal to the head of the second list. These two alternatives lead to two different clauses. The second clause is, so to say, the "otherwise" clause; that is, the second clause is to be used only if the first clause fails. One has

```
condense([A|As], Bs) :− member(A, Bs), condense(As, Bs).
condense([A|As], [A|Bs]) :− condense(As, Bs). % Otherwise
```
∎

4.5.5 Cuts

The cut, which is abbreviated by !, is a predicate that, once called, fixes all choices made up to this point by preventing backtracking. The cut always succeeds. Once the cut is executed, all past choices regarding the call in question are fixed. No goal preceding the cut can be redone. Backtracking is allowed only among the goals following the cut. Moreover, if a cut is executed, the choice of the clause is fixed, and if the clause fails, no other clause of the same predicate is tried. To illustrate this, consider the following clause:

$$\text{melody}(X) :- \text{la}(X), !, \text{re}(X), \text{mi}(X).$$

Once la(X) succeeds, the cut is done, which fixes all choices. Specifically, the goal la(X) cannot be redone. Moreover, any other rule with the head melody(X) is excluded from further consideration.

It was shown earlier that if the clause order is used then errors result from redoing clauses. Cuts are useful because they prevent redoing. To see how this works, consider again the procedure youpay(Income, Tax). In this procedure, only one tax bracket must be selected, and to fix the choice of this bracket, a cut should be used. This leads to the following procedure:

```
youpay(Income, Tax) :- Income < 10000, !,
                       Tax is 0.1*Income.
youpay(Income, Tax) :- Income < 50000, !,
                       Tax is 1000 + 0.15*(Income - 10000).
youpay(Income, Tax) :- Tax is 7000 + 0.2*(Income - 50000).
```

Generally, cuts should be used whenever a selection must be made among alternatives that are mutually exclusive.

Consider now the predicate condense(As, Bs) discussed earlier. In this predicate, the head of the As is either a member of the Bs or it is not a member of the Bs. One cannot have both. Hence, it is a good idea to freeze the choice of the clause as soon as one finds that the head of the As is a member of the Bs. This leads to the following improved version of the predicate.

```
condense([ ], [ ]).
condense([A|As], Bs) :- member(A, Bs), !, condense(As, Bs).
condense([A|As], [A|Bs]) :- condense(As, Bs). % Otherwise
```

Cuts can also be used to prune irrelevant choices, which reduces execution time. Consider, for instance, the predicate member(X, A). One can say that, once it is found that X is a member of A, there is no need to redo this predicate. Hence, as soon as a member is found, this choice is frozen. This leads to the following predicate:

```
member(X, [X|_ ]) :- !.
member(X, [_|A]) :- member(X, A).
```

This new version of member(X, A) is more efficient than the original version because, once an X is found in A, the search stops. On the other hand, the old version can find all occurrences of X in the list should there be several of them. The new version, which stops as soon as the first X is found, obviously cannot do this.

Cuts cannot prevent all errors that are caused by using the clause order. To show this, consider the following problem: There exists a database that matches individuals to their occupations (jobs). For any individual who does not have a job, we want to have a default clause that will return none when queried for a job. This problem is implemented by the following database:

```
job(a, carpenter) :− !.
job(b, lawyer) :− !.
job(X, none) :− !.
```

In this example, job(X, none) is the default clause, which should only apply if X does not have a job. This database works correctly for the query job(X, Y). Unfortunately, the query job(a, none) yields yes, despite the rule job(a, carpenter) :− !.

Problems 4.5

1. Consider the following database

```
dog(X) :− barks(X).
barks(fido).
```

The query in dog(fido). As mentioned in Section 4.5.2, the query is negated and added to the database. Formulate the resulting database in logic, and derive a contradiction. What logical law is used by Prolog to accomplish this?

2. Write a procedure to find if two lists do not share common elements. For this end, define a predicate exclusive(As, Bs) that succeeds if the list As has no common elements with the list Bs. Make use of the predicate \+ when writing the procedure.

3. Design a predicate max(X, Y, Z) that succeeds if Z is the maximum of X and Y. Use the cut to avoid useless attempts to resatisfy clauses.

4. Given the predicate carpenter(X, Y) is already part of the database, and it succeeds if X is a carpenter in city Y, write a predicate find(X, Y) that returns a carpenter living in Y if there is one and that prints otherwise "no carpenter in city."

5. Below there are two predicates, namely member(X, Y) and memberc(X, Y). Test these two predicates with the query member(1, [1,3,1]) and memberc(1, [1,3,1]). When testing, try for multiple solutions. In which way are the two programs different?

```
member (X, [X|_]) .
member (X, [_|Z]) :− member(X, Z).
memberc(X, [X|_]) :− !.
memberc(X, [_|Z]) :− memberc(X, Z).
```

6. Write a procedure yielding a list of the common elements for two different lists As and Bs.

7. Write a procedure yielding a list containing all elements that are either in As or Bs or both.

4.6 APPLICATION OF PROLOG TO LOGIC

4.6.1 Introduction

In the following sections, we show how some of the definitions given in Sections 1.3.3 and 1.3.5 can be converted into Prolog and made executable. For doing this, two methods are used. The first method relies on lists, whereas the second one uses structures. Both methods are similar. We also show how Prolog can be used to move negations in, a topic discussed in Section 1.5.7. The reader should have no difficulty finding other applications for the methods explored in the following sections. In particular, she or he may want to investigate the possibility of manipulating algebraic expressions.

4.6.2 Lists as Logical Expressions

In Prolog, one can represent logical expressions as lists. In fact, the negation can be represented as a list of two elements, the first of which is the word neg, an abbreviation for negation, and the second is any logical expression. Similarly, conjunctions, disjunctions, implications, and equivalences can be represented as lists of three elements. The first element of the list is the left scope of the expression, the second element of the list is the connective, and the third element of the list is the right scope of the expression. We now can give definitions as to what constitutes a logical expression in Prolog. The form of the formation rules differs slightly from the ones presented in Section 1.3.3.

1. The binary connectives are and, or, then, and iff.
2. The logical constants t and f are expressions.
3. All logical variables are expressions.
4. If A is an expression, so is [neg, A].
5. If A and B are expressions and if Con is a binary connective, then [A, Con, B] is an expression.

These definitions can be converted into Prolog and made executable. To this end, we abbreviate *logical constant* by logc, *logical variable* by logv, and *binary connective* by bicon. We also have to list all logical variables that can be used. For simplicity, we assume that the only symbols that can represent logical variables are p, q, r, and s. The predicate expr represents, of course, an expression. The resulting Prolog program looks as follows:

```
bicon(and).
bicon(or).
bicon(then).
bicon(iff).
logc(t).
logc(f).
logv(p).
```

```
logv(q).
logv(r).
logv(s).
expr(X) :— logc(X).
expr(X) :— logv(X).
expr([neg, A]) :— expr(A).
expr([A, Con, B]) :— expr(A), bicon(Con), expr(B).
```

We can now try different expressions, and see whether they are well formed. We can, for instance, enter the query

$$| ?— expr([p, and, [q, then, [neg, r]]]).$$

The result of this query is yes, and the method used to obtain this result is very similar to the one discussed in Section 1.3.3. First, it is established that p is an expression, which is true because of the fact logv(p). Next, and is recognized as a binary connective because of the fact bicon(and). B unifies with [q, then, [neg,r]], and the next goal is expr([q, then, [neg,r]]). This goal is satisfied in a similar fashion as the original goal; that is, q is recognized as an expression, then as a binary connective, and [neg, r] is also eventually recognized as an expression. Hence, the list given initially is an expression, and the system answers yes.

On the other hand, if we use the query

$$| ?— expr([and, q, [p, or, r]]).$$

the system will answer with no. No rule is applicable in this case, as the reader may verify.

In Section 1.3.5, it was shown how to find the truth value of an expression. The method described there can be made executable by Prolog. Here, we use slightly different methods to define truth. In particular, we assume that what cannot be shown to be true is by definition false.

1. The logical constant t is true.
2. There is a list of logical variables, which are all defined to be true.
3. [neg, A] is true if A is not true.
4. [A, and, B] is true if A is true and B is true.
5. [A, or, B] is true if A is true.
6. [A, or, B] is true if B is true.
7. [A, then, B] is true if A is not true.
8. [A, then, B] is true if B is true.
9. [A, iff, B] is true if [A, then, B] and [B, then, A] are true.

We assume that the propositional variables q and r are true, whereas all other variables are false. These definitions yield the following Prolog program:

```
true(t).
true(q).
true(r).
true([neg, A]):− \+(true(A)).
true([A, and, B]) :− true(A), true(B).
true([A, or, B ] ) :− true(A).
true([A, or, B]):− true(B).
true([A, then, B]) :− true(B).
true([A, then, B]) :− \+(true(A)).
true([A, iff, B]):− true([A, then, B]), true([B, then, A]).
```

A typical query might look like this:

```
| ?− true([[ p, or, q], and, [q, then, r]]).
```

The answer to this query is yes. Note that p is not specified as true, which makes p false.

4.6.3 Representing Logical Expressions as Structures

As indicated in Section 4.3.4, one can use structures in infix notation to express logical connectives. To do this, one has to use the op predicate first to declare the structure and define its precedence. For example, the connectives and and or can be declared as infix operators by using the following clauses:

```
:− op(300, yfx, and).
:− op(400, yfx, or).
```

As indicated in Section 4.3.4, these clauses have no head, and they are executed immediately.

The predicates used when expressing logical formulas as structures are very similar to the ones given earlier. For instance, the database that finds whether a logical expression is true can be written as

```
:− op(200, fy, neg).
:− op(300, yfx, and).
:− op(400, yfx, or).
:− op(500, yfx, then).
:− op(600, yfx, iff).
true(neg A) :− \+ true(A).
true(A and B) :− true(A), true(B).
true(A or B) :− true(A).
true(A or B) :− true(B).
true(A then B) :− true(B).
true(A then B) :− \+ true(A).
true(A iff B) :− true( A then B), true(B then A).
```

Moreover, one would have to add a number of facts, declaring which propositional variables are true.

As our next application, we define the predicate movenegin(A, B). This predicate accepts a logical expression A that does not contain any then and iff, and it moves all negations in as far as possible. The resulting expression is given by B. In the implementation that follows, the predicate uses the built-in predicate atom(X), which succeeds if X is an atom.

```
:— op(200, fy, neg).
:— op(300, yfx, and).
:— op(400, yfx, or).
% Base case
movenegin(A, A):— literal(A).
% Remove double negation
movenegin(neg neg A, D) :— movenegin(A, D).
% De Morgan's laws
movenegin(neg (A and B), D or C) :—
        movenegin(neg A, D),
        movenegin(neg B, C).
movenegin(neg (A or B), D and C) :—
        movenegin(neg A, D),
        movenegin(neg B, C).
% All other cases
movenegin(A or B, D or C) :—
        movenegin(A, D),
        movenegin(B, C).
movenegin(A and B, D and C) :—
        movenegin(A, D),
        movenegin(B, C).
literal(X) :— atom(X).
literal(neg X) :— atom(X).
```

To illustrate the predicate movenegin, consider the following query:

$$| ?— movenegin(neg (p and neg (p or q)), D).$$

The result of this query is

$$D = neg\ p\ or\ (p\ or\ q)$$

The reader should verify that this is indeed the correct result.

The program works as follows. If we are given an expression A, and A is a literal, then we leave it as is. This yields

$$movenegin(A, A):— literal(A).$$

If the expression begins with a double negation, we must remove the double negation and apply the predicate movenegin to the doubly negated expression. Specifically, if we have the expression neg neg A, we have to remove the two negs, and we also have to move all the negations of A inward. This is implemented by the rule

$$movenegin(neg\ neg\ A\ , D) :— movenegin(A, D).$$

In case neither of these two rules applies, we use the theorems of De Morgan. This is accomplished by the following rules:

```
movenegin(neg (A and B), D or C) :-
        movenegin(neg A, D),
        movenegin(neg B, C).
movenegin(neg (A or B), D and C) :-
        movenegin(neg A, D),
        movenegin(neg B, C).
```

When the main connective is the and or the or, it is still possible that one of the sub-connectives is a negation. For this reason, one must apply movenegin to the subexpressions A and B if the expression is of the form A or B or A and B. The program shows how to do this. The predicates require the predicate literal. The predicate literal succeeds if the expression is either a propositional variable or the negation of a propositional variable.

In the same way as we implemented movenegin, we can also implement a predicate moveandin, which moves all the conjunctions in. Applying first movenegin and then moveandin converts any expression into its disjunctive normal form.

Problems 4.6

1. Write a Prolog program that uses the idempotent law, the identity laws, and the laws of domination to simplify logical expressions.

2. Write a Prolog program that first moves all negations in and then uses the distributive laws to move all conjunctions in.

3. Write a Prolog program that removes any term of 0 and any factor of 1 from a given arithmetic expression. Assume that $+$, $-$, $*$, and $/$ are the only operators.

PROBLEMS: CHAPTER 4

1. Enter the following facts in a Prolog database:

```
likes(mary, john).
likes(john, tony).
likes(mary, tony).
```

Once this is done, do the following queries.
   (a) likes(mary, tony).
   (b) likes(mary, Who).
   (c) likes(Who, mary).
   (d) likes(Who, john).
   (e) likes(Anybody, Somebody).

2. Frequently, young readers only remember the title of a book they liked and they want to read another book by the same author. The library has already established a database for its books, containing instances of the form book(Author, Title), where Author is the author of the book, and Title is the title. Write a predicate morebooks(X, Y), where X is a title, and $Y \neq X$ is another title by the same author.

3. The pay rate for a marker depends on the year he or she is in, and these rates are determined by the predicate payrate(Year, Wage), where Year is the year and Wage the hourly wage. There is also a predicate student(Name, Year, Hours), where Name is the student's name, Year the year he or she is in, and Hours are the hours worked. The predicates payrate and student are expressed as facts within the database. Write a Prolog predicate earned which has three arguments: the first for the name, the second for the rate, and the third one for the amount earned. The amount earned is the appropriate rate multiplied by the hours worked. For example, the query earned(smith, X, Y) should unify X with the wage rate of smith and Y with the amount he earned. A sample database follows.

> payrate(2, 6.00).
> payrate(3, 6.50).
> payrate(4, 7.00).
> student(john, 2, 15).
> student(jack, 3, 10).
> student(mary, 2, 15).

4. A Prolog database gives the year of birth for a number of people in the form birth(Person, Year). For example, the fact birth(jim, 1950) indicates that jim was born in 1950. Write a predicate age(X, Y, Z) that succeeds if person X has the age Y at the end of year Z. For instance, age(jim, Y, 1960) should succeed with $Y = 10$.

5. The following database is given

> supplier(abm, cars).
> supplier(cdf, trucks).
> supplier(ggm, cars).

Define a predicate twosuppliers(X) that succeeds if two suppliers carry item X. Hence, in the database, twosuppliers(cars) should succeed, and twosuppliers(trucks) should fail.

6. A database contains a predicate of the form supplier(Snum, Name), where Snum represents a supplier number and Name a supplier name. Similarly, there is a predicate item(Name, Snum), where Name represents the name of the item and Snum the supplier number. Write a predicate supphas that has two arguments: an item name and a supplier name. This predicate should succeed if the supplier in question has the item.

7. Write a predicate in_sequence(X, Y, Z) that succeeds if Z is a list that contains X and also contains Y immediately following X. For instance, in_sequence(a, b, [c, a, b]) should succeed, whereas in_sequence(a, b, [c, a, d, b]) should fail.

8. Design a predicate that prints $x + x^2$ for x from 1 to 20.

9. A family wants to move to another city. They want to write a database that lists their expenses. Among them are a king-sized bed, $1000, a bunk bed, $200, a washer and dryer, each $600, and a piano for $2000. This information is given in a database as follows:

> costs(king, 1000).
> costs(bunk, 200).
> costs(washer, 600).
> costs(dryer, 600).
> costs(piano, 2000).

Write a predicate called so_much that has two parameters, one being a list and the other one being the total cost of all items in the list.

10. A database contains the predicate name(First, Last), where First is the first name and Last the last name. No two persons have the same last name. Write a predicate fullname(X, Y), where X and Y are both lists pertaining to the same individuals. However, whereas X contains only the last name, Y contains two elements for each individual, the first name followed by the last name. You may assume that the database contains all names in question.

11. Consider the Prolog statement

$$abc(X, a, C, B) :- de(X, X, B, C).$$

(a) Express this statement in the form of predicate calculus.
(b) Does the body of this statement unify with the fact de(X, C, a, Y)? Justify your answer and give the correspondence between variables and atoms.

12. Given X is a list of people, write a procedure numbered (X) that lists all people in X, each on a separate line, with the lines numbered consecutively, starting with line number 1.

13. Suppose that husband(X, Y) succeeds if X is the husband of Y. This predicate is defined by a number of facts in the database, such as husband(john, mary), husband(mark, anne), and so on. Write a predicate convert that has two arguments, both of them lists, and that succeeds if the first list contains the husbands of the second list.

14. A list contains a name, a price, and a quantity for each item, which are, in turn, represented by structures called item. For instance, the list

$$[item(pencil, 2, 0.50), item(pens, 4, 0.75)]$$

means that there are two pencils costing 50 cents each and 4 pens costing 75 cents each. Write a predicate called bill with two arguments, the first argument being a list of the format described and the second being the total value. For instance, bill([item(pencil, 2, 0.50), item(pens, 4, 0.75)], Bill) should succeed with Bill = 4.00.

15. A person has friends she knows from sports and friends she knows from business. She wants to invite both groups to a party. The two groups overlap, but each friend should get only one invitation. Write a predicate all that has three arguments, all of them lists. The first argument is the list of sports friends, the second the list of business friends, and the third argument is a joint list of all friends, with no one appearing twice.

16. A database contains a French translation for each English word. For instance, french(tree, arbre) indicates that the word "tree" is "arbre" in French. A sentence is given in the form of a list, and this sentence must be translated word by word. Write the appropriate Prolog procedure.

17. For the following programs, indicate the types of arguments that they require (lists, atoms, numbers), and try to find out what they do. It may help if you test these programs with a number of sample queries.
(a) abc([ ], [ ]).
abc([X | Xs], [Y | Ys]) :- Y is 2 * X.
(b) cde(1, [Z | _], Z).
cde(N, [_ | Zs], Z) :- N1 is N - 1, cde(N1, Zs, Z).

18. Air Blue-Sky, a local carrier, maintains a Prolog database containing all pairs of cities that are connected by a direct Blue-Sky flight. Write a procedure connection(X, Y), where X and Y are two cities, which should succeed if Y can be reached from X, flying only Blue-Sky. Since there may be several legs making up this connection, your procedure should be recursive. However, to avoid going in circles, you need a list of the cities already included in the tour.

19. Let Expression be an arithmetic expression that contains, besides numbers and operators, the atom x. Let Value be a numerical value, and let Result be the result obtained by evaluating Expression with x replaced by Value. Write the appropriate predicate.

20. Let Expression be an arithmetic expression possibly containing the atom x. Write a procedure derivative (Expression, Derivative), where Derivative is the derivative of Expression with respect to x. For simplicity, you may restrict yourself to expressions containing only the operators $+, *, -, /$, and not containing any functions.

21. A sequence of points is given, each expressed by the two coordinates of the point. The sequence can be interpreted as an open polygon. Write a Prolog procedure to find the length of the open polygon by using the formula

$$\text{length of polygon} = \sum_{i=2}^{n} \sqrt{(x_i - x_{i-1})^2 + (y_i - y_{i-1})^2}$$

Here, n is the number of points describing the polygon.

5

Sets and Relations

Any collection of objects or individuals is referred to as a set. Therefore, one can speak of the set of all Canadians, the set of all nations, or the set of all integers. In the domain of languages, there are many sets, including the set of all characters, the set of all vowels, the set of all conjunctions, and the set of all words. Here we are mainly interested in mathematical objects, such as the set of all natural numbers, the set of all premises of a derivation, or the set of all free variables in an expression. There is an important connection between sets and predicates: all individuals having a certain property collectively form a set. Many programming languages make use of types, such as integers, reals, and characters. As it turns out, types are sets.

As the name indicates, relations describe relationships between objects. For instance, there is a relation between people and their belongings, and there are relations between producers and the goods that they produce. Mathematically, relations are defined as sets of n-tuples. For instance, the relations between producers and goods produced can be represented by a set of pairs, with the first element of the pair indicating the producer and the second element indicating the product. Likewise, there is a close link between relations and predicates. For instance, the set of all pairs (x, y) satisfying a certain predicate $G(x, y)$ forms a relation.

Both sets and relations are important for specifying computer applications. In fact, the specification language Z (see Chapter 8) makes extensive use of these constructs.

5.1 SETS AND SET OPERATIONS

5.1.1 Introduction

In the following sections, sets are introduced, and the operations available to combine sets are discussed. There are several ways to combine two sets. For instance, if A and B are two sets, then the set of all individuals belonging to either A or B forms a new set, called the *union* of A and B. There is a close relation between the connectives of propositional calculus and the way different sets can be combined to form new sets. All these issues will be explored.

5.1.2 Sets and Their Members

Any collection of individuals forms a set. If a certain individual x is part of set A, then x is said to be an *element* of A. On the other hand, a set is fully determined by its members, that is, by the individuals that jointly form the set. To describe a particular set, it is therefore sufficient to enumerate all the elements of the set in some order. To indicate that this enumeration is a set, it is customary to put curly braces around the list of elements. For instance, the set consisting of the numbers 3, 6, and 7 can be described by $\{3, 6, 7\}$. This representation of a set is called *roster* notation. Obviously, only finite sets can be explicitly enumerated. To provide a roster notation for infinite sets, one extends the notation by using an ellipsis. For instance, the set of all even natural numbers can be represented as $\{0, 2, 4, 6, \ldots\}$. Here the elements given so far suggest some formation rule, and the ellipsis "$\ldots$" indicates that all other individuals obtainable by applying this formation rule should be added to the set. Instead of an ellipsis, which implicitly suggests a formation rule, one can also state the formation rules explicitly. Typically, the formation rules are recursive and, in this case, the set is said to be a *recursively defined set*. An example of a recursively defined set is the set of all well-formed logical expressions.

Not all sets can be represented in roster notation. To be representable in roster notation, a set must be *enumerable*. Details on this topic are given in Sections 6.2.2 and 6.2.3. As will be shown there, the set consisting of all real numbers between 0 and 1 is an example of a set that cannot be enumerated.

Some sets are so important that they are represented by special symbols. In particular, the set of natural numbers is represented by $\mathbb{N}$, the set of integers by $\mathbb{Z}$, and the set of reals by $\mathbb{R}$. Both $\mathbb{N}$ and $\mathbb{Z}$ are enumerable, but $\mathbb{R}$ is not. The set containing all integers from n to m is denoted by $n..m$. For instance, $3..5 = \{3, 4, 5\}$.

Often, a universe of discourse is given, and all sets to be considered must be formed from the individuals of this universe. In this case, one can consider the universe of discourse as a set, the set that includes all individuals of the universe. This set is called the *universal set*, and it is denoted by E. The universal set is, in this sense, the largest set. At the other extreme is the *empty set*, the set that contains no element at all. The empty set is denoted by $\{\}$ or, alternatively, by $\emptyset$.

If a certain individual x is an element of a set A, one writes $x \in A$. Here, $x \in A$ is stated as "x is an element of A" or simply "x is in A." For instance, if A is the set $\{3, 5, 8\}$, then $3 \in A$ is true, while $6 \in A$ is false. If a certain element y is not in A, one writes $y \notin A$, which says "y is not an element of A," or "y is not in A." For instance, $3 \notin \{2, 4, 6\}$ is true. Of course, $x \in \emptyset$ is always false, and $x \in E$ is always true.

A set is completely described by its elements. In fact, two sets are equal if they have the same elements, and they are different if a single element of one set is not also an element of the other set. This is expressed by the following axiom:

Axiom of Extensionality: Let A and B be two sets. Then A and B are equal if and only if they have the same members. If A and B are equal, we write $A = B$.

Example 5.1   Consider the sets $A = \{a, b, c\}$, $B = \{1, 2, 3\}$, $C = \{c, b, a\}$, $D = \{3, 1, 2, 2\}$, and $F = \{1, 2, 2\}$. Which of these sets are equal?

Solution   The sets A and C are equal because they contain exactly the same members, the characters a, b, and c. The fact that the order of the characters in A and C differs is irrelevant as long as each member of A is a member of C and each member of C is a member of A. For similar reasons, $B = D$. The fact that the elements in D are listed in a different order than in B and that the number 2 appears twice in D is irrelevant. The sets B and F, on the other hand, are different, because B contains the number 3 as one of its members, whereas F does not. For similar reasons, $D \neq F$. Of course, $A \neq B$, $A \neq D$ and $A \neq F$, $B \neq C$, $C \neq D$, and $C \neq F$.   ∎

If P is some property, then all elements of the universe satisfying P form a set.

Definition 5.1:   Let P be a property. The *extension* of P, written $\{x \,|\, P(x)\}$ is the set of all objects x for which $P(x)$ holds. The notation $\{x \,|\, P(x)\}$ is called *setbuilder* notation.

For instance, if the set of natural numbers is the universe of discourse and if P stands for "x is even," then the extension of P is $\{0, 2, 4, \ldots\}$, which can also be written in setbuilder notation as $\{x \,|\, P(x)\}$. Note that the extension of a property critically depends on the universe of discourse. For instance, if $P(x)$ is defined as before to stand for "x is even," then $\{x \,|\, P(x)\}$ is $\{0, 2, 4, \ldots\}$ if the universe of discourse is given by the natural numbers; but in the universe consisting of the integers, this set has the extension $\{0, \pm 2, \pm 4, \ldots\}$. Formally, sets in setbuilder notation can be defined as follows:

$$\forall x (x \in \{x \,|\, P(x)\} \Leftrightarrow P(x)) \tag{5.1}$$

On the other hand, if one is given a set, say in roster notation, then this set immediately describes a property, that is, the property of belonging to the set. This property is called the *characteristic condition* of the set. If A is a set, the characteristic condition of A is denoted by $\chi_A(x)$. Here χ is the Greek letter chi.

Example 5.2   The universe of discourse is given by the letters of the alphabet, and the set $A = \{a, u, v\}$. Find the assignment for χ_A.

Solution $\chi_A(a)$, $\chi_A(u)$, and $\chi_A(v)$ are all true, and $\chi_A(x)$ is false in all other cases.   ∎

In the set $\{x \mid P(x)\}$, $P(x)$ is, of course, the characteristic condition, which is also referred to as the *intension* of the set.

To each logical connective, there corresponds an operation on sets. Specifically, the equality between two sets A and B corresponds to equivalence of the predicates $x \in A$ and $x \in B$. Hence,

$$(A = B) \equiv (x \in A \Leftrightarrow x \in B) \tag{5.2}$$

Here the symbol $\equiv$ indicates that this relation is always true by definition. In this sense, the $\equiv$ represents a logical equivalence. Of course, (5.2) is merely a mathematical restatement of the axiom of extensionality.

Sets can be represented graphically by *Venn diagrams*, named after John Venn, an English mathematician. In a Venn diagram, the universal set is represented by a rectangle (or any other figure), and the set of interest, say A, is represented by the interior of a circle or some other simple closed curve inside the rectangle. Venn diagrams give a pictorial representation of how sets can interact with each other and can make the visualization of sets much easier. Figure 5.1 shows two sets, A and B, represented by two circles. The set B is completely inside A, which means that all members of B are also members of A. Venn diagrams are also useful for visualizing logical connectives. This is the case because to every logical connective there is a corresponding operation on sets.

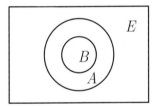

Figure 5.1. B is a subset of A

5.1.3 Subsets

Clearly, the European countries are a subset of all countries, just as the integers are a subset of the real numbers. Generally, *subsets* are defined as follows:

> **Definition 5.2:**   Let A and B be two sets. Set A is said to be a *subset* of B if every element of A is also an element of B. However, not every element of B needs to be an element of A. We express the proposition that A is a subset of B by $A \subseteq B$.

For instance, $\{a, b\}$ is a subset of $\{a, b, c\}$, and so is $\{a, b, c\}$. However, $\{a, d\}$ is not a subset of $\{a, b, c\}$. In a Venn diagram, a set B is a subset of A if it is completely enclosed by A (see Figure 5.1).

Let $A \subseteq B$. If x is in A, then x must also be in B. This fact can be used to formulate Definition 5.2 as follows:

$$(A \subseteq B) \equiv (x \in A \Rightarrow x \in B) \tag{5.3}$$

This definition links the notion of a subset with logical implication. It was shown in Section 1.5.2 that logical equivalence can be expressed in terms of logical implication. In a similar fashion, one can express the equality between sets by using the subset relation, which leads to

$$(A = B) \equiv (A \subseteq B \wedge B \subseteq A)$$

This is proved as follows:

$$
\begin{aligned}
A = B \ &\equiv x \in A \Leftrightarrow x \in B & &\text{By (5.2)} \\
&\equiv (x \in A \Rightarrow x \in B) \wedge (x \in B \Rightarrow x \in A) & &\text{Eliminate } \Leftrightarrow \\
&\equiv (A \subseteq B) \wedge (B \subseteq A) & &\text{By (5.3)}
\end{aligned}
$$

Consequently, two sets A and B are equal if and only if A is a subset of B and B is a subset of A.

The empty set is a subset of all sets. This can be shown by replacing A in (5.3) by $\emptyset$: the expression $x \in \emptyset$ is false for every x, which implies that $x \in \emptyset \Rightarrow x \in B$ is true for every set B.

Some subsets are *proper subsets*.

Definition 5.3: A is a *proper subset* of B if A is a subset of B, but A is not equal to B. If A is a proper subset of B, we write $A \subset B$.

A proper subset is essentially only a part of a set. For instance, most subcommittees are proper subsets of committees, because not all members of the committee form part of the subcommittee. The logical expressions that are tautologically true form a proper subset of the set of all logical expressions: All tautologies are logical expressions, but there are logical expressions that are not tautologies. The set of natural numbers is a proper subset of the set of integers because all natural numbers are integers, but not all integers are natural numbers.

The opposite of the subset is the *superset*.

Definition 5.4: A is a *superset* of B if B is a subset of A. The proposition that A is a superset of B is expressed as $A \supseteq B$. Moreover, A is a *proper superset* of B if A is a superset of B that is different from B. In this case, we write $A \supset B$.

This definition implies that the set of all animals is a superset of the set of all vertebrates, which is in turn a superset of the set of all mammals.

To express that A is not a subset of B, one writes $A \nsubseteq B$. Actually, $A \nsubseteq B$ is just another notation for $\neg(A \subseteq B)$. The symbols $\not\subset$, $\nsupseteq$, and $\not\supset$ can be similarly interpreted as the negations of $\subset$, $\supseteq$, and $\supset$, respectively.

If A is a proper subset of B, then A has fewer elements than B. To make this precise, we define *cardinality* as follows:

Definition 5.5:   Let A be a set with a finite number of elements. The *cardinality* of A, denoted by $|A|$ or $\#A$, is equal to the number of elements in A.

This definition only applies to sets with a finite number of elements, or finite sets. However, it can be extended to hold for infinite sets (see Section 6.2.3). It is clear that, if A and B are both finite sets with $A \subset B$, then A must have fewer members than B; that is

$$(A \subset B) \Rightarrow (\#A < \#B)$$

5.1.4 Intersections

All writers taken together constitute a set, as do all Americans. The intersection of these two sets is the set of all American writers. Generally, an intersection is defined as follows:

Definition 5.6:   Let A and B be two sets. The set $A \cap B$, called the *intersection* of A and B, is the set containing all elements common to both A and B.

Example 5.3   Let $A = \{a, b, 1\}$, $B = \{a, 1, 2\}$, and $C = \{2, 3, 4\}$. Find the intersections $A \cap B$, $A \cap C$, and $B \cap C$.

Solution

$$A \cap B = \{a, b, 1\} \cap \{a, 1, 2\} = \{a, 1\}$$
$$A \cap C = \{a, b, 1\} \cap \{2, 3, 4\} = \{\} = \emptyset$$
$$B \cap C = \{a, 1, 2\} \cap \{2, 3, 4\} = \{2\}$$

∎

Example 5.4   If A is the set of all programmers of a company, and B is the set of all employees in the accounting department, then $A \cap B$ is the set of all programmers in the accounting department.

The intersection of two sets A and B in a Venn diagram is given by the area where the two sets overlap (see Figure 5.2). Note in particular that the intersection between two

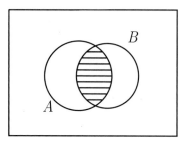

Figure 5.2.   The intersection of two sets A and B

sets A and B is $\emptyset$ if and only if A and B have no elements in common. If the sets are represented by a Venn diagram, this means that the corresponding sets do not overlap.

Intersections correspond to conjunctions: object x belongs to $A \cap B$ if x belongs to A and x belongs to B, which yields

$$x \in (A \cap B) \equiv (x \in A) \wedge (x \in B) \tag{5.4}$$

Since the set $A \cap B$ is the set of all objects that belong to A and B, one can use setbuilder notation to express the intersection as follows:

$$A \cap B = \{x \mid x \in A \wedge x \in B\} \tag{5.5}$$

Both (5.4) and (5.5) express Definition 5.6 in a formal fashion.

By definition, $A \cap B$ contains no element that is not also an element in A. Consequently,

$$(A \cap B) \subseteq A \tag{5.6}$$

This relation holds as an equality if all elements of A are also elements of B, that is, if $A \subseteq B$.

5.1.5 Unions

Committees are sets of people, and if there are two committees A and B, the set of all people belonging to either committee form the *union* of A and B. Generally, the union of two sets is defined as follows:

Definition 5.7:   Let A and B be two sets. The set $A \cup B$, called the *union* of A and B, is the set containing all elements belonging either to A or to B.

Example 5.5   Let $A = \{a, b, 1\}$, $B = \{a, 1, 2\}$, and $C = \{2, 3, 4\}$. Find the unions $A \cup B$, $A \cup C$, and $B \cup C$.

Solution

$$A \cup B = \{a, b, 1\} \cup \{a, 1, 2\} = \{a, b, 1, 2\}$$
$$A \cup C = \{a, b, 1\} \cup \{2, 3, 4\} = \{a, b, 1, 2, 3, 4\}$$
$$B \cup C = \{a, 1, 2\} \cup \{2, 3, 4\} = \{a, 1, 2, 3, 4\}$$

∎

Example 5.6   Let A be the set of all physicians, and B be the set of all lawyers. In this case, $A \cup B$ includes everyone who is either a physician or a lawyer or both.

The Venn diagram of a union of two sets A and B is given by the area covered by A or B or both (see Figure 5.3).

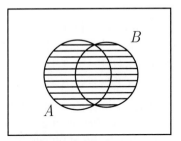

Figure 5.3.   The union of two sets A and B

Unions can also be used to add a single individual to a set. For instance, if an employee, say x, joins a company, and if the set of employees before x joins is A, then the set of employees after x joins is $A \cup \{x\}$. If x is already in A, however, then $A \cup \{x\} = A$.

The formal definition of a union is similar to the formal definition of an intersection, except that one has to take disjunctions rather than conjunctions: x is in the union of A and B if x is in A or B, which means that

$$x \in (A \cup B) \equiv (x \in A) \vee (x \in B)$$

Alternatively, the set $A \cup B$ is the set of all points that are either in A or B; that is,

$$A \cup B = \{x \,|\, x \in A \vee x \in B\}$$

This shows that unions are closely related to the logical connective $\vee$.

By definition, $A \cup B$ contains all elements in A; that is,

$$A \subseteq A \cup B$$

This relation holds as an equality if all elements of A are also elements of B.

5.1.6 Differences and Complements

If A is the set of all workers that are employees of a company and if B is the set of all workers that have been assigned to a project, then all workers of the company that have not been assigned to any project form a set, that is, the *difference* of the two sets A and B.

Definition 5.8:   Let A and B be two sets. The set $A - B$, called the *difference* of A and B, is the set consisting of all elements of A that do not belong to B.

Instead of $A - B$, some authors write $A \backslash B$, that is, they use $\backslash$ as the operator for expressing a set difference.

Example 5.7   Let $A = \{a, b, 1\}$, $B = \{a, 1, 2\}$, and $C = \{2, 3, 4\}$. Find the differences $A - B$, $B - A$, $A - C$, and $B - C$.

Solution

$$A - B = \{a, b, 1\} - \{a, 1, 2\} = \{b\}$$
$$B - A = \{a, 1, 2\} - \{a, b, 1\} = \{2\}$$
$$A - C = \{a, b, 1\} - \{2, 3, 4\} = \{a, b, 1\}$$
$$B - C = \{a, 1, 2\} - \{2, 3, 4\} = \{a, 1\}$$

∎

A Venn diagram for the difference of two sets A and B is given in Figure 5.4. Here are some examples involving differences. The set of composite numbers is the difference between the set of natural numbers and the set of primes. If one works with a windowing system on a computer, the pixels of a window jointly form a set. If window A is behind window B, then the difference $A - B$ is exactly the set of the pixels of A that are visible. Differences can also be used to remove a single element x from a set. For example, if A is the set of all employees of a company on December 31, 1994, and if x quits at this date, then after December 31 the set of employees is $A - \{x\}$. Generally, $A - \{x\} = A$ if and only if x is not an element of A.

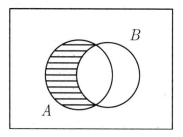

Figure 5.4.   Venn diagram for $A - B$

Formally, x is in $A - B$ if x is in A and not in B.

$$x \in (A - B) \equiv (x \in A) \wedge (x \notin B)$$

Alternatively, the set $A - B$ is the set of all objects that are in A, but not in B.

$$A - B = \{x \mid x \in A \wedge x \notin B\}$$

If E is the universal set, then $E - A$ is the set of all elements that are not in A. The set $E - A$ is called the *complement* of A.

Definition 5.9:   Let A be a set. The complement of A, written $\sim A$, is the set of all objects not belonging to A.

A Venn diagram of $\sim A$ is given in Figure 5.5. There the complement of set A is given by the points lying outside the enclosure representing A. The complement corresponds to negation. Specifically, x is in the complement of A iff x is not in A.

$$x \in \sim A \equiv \neg(x \in A)$$

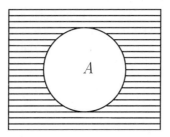

Figure 5.5.   Venn diagram of $\sim A$

Consequently, the complement of A is the set of all points not in A; that is,

$$\sim A = \{x \mid x \notin A\}$$

If sets are given in roster notation or some other explicit form, then the results of forming unions, intersections, and set differences do not depend on the universal set. In other words, the results of these operations do not change when the universal set E is changed. For instance, the union of $\{3, 6, 9, \ldots\}$ and $\{4, 7, 10, 11, \ldots\}$ is the same, whether the universal set is the set of all natural numbers, the set of all integers, or the set of all real numbers. This is no longer true when forming the complement. For instance, the complement of $\{0, 2, 4, \ldots\}$ is $\{1, 3, 5, \ldots\}$ if E is the set of natural numbers, but it is $\{1, 3, 5, \ldots, -1, -2, \ldots\}$ if E is the set of integers. When forming complements, the universal set must therefore be known. Frequently, this universal set is implied, and it is not explicitly mentioned.

The following are some examples of complements. If the universal set is the set of all natural numbers and if A is the set of even numbers, $\sim A$ is the set of odd numbers. If B is the set of primes, then $\sim A$ is the set of composite numbers. If the universe of discourse is the set of people, then the set of nonsmokers is the complement of the set of smokers.

5.1.7 Expressions Involving Sets

Unions and intersections are themselves sets, and they can be used to build further unions and intersections. In this way, complex expressions involving sets can be obtained. For example, let B be the set of all British subjects, let C be the set of all French citizens, and let A be the set of all professors. Then $A \cap B$ is the set of all British professors, and $A \cap C$ is the set of all French professors. The union of $A \cap B$ with $A \cap C$ is the set of all professors that are either French or English. This set can be written as

$$(A \cap B) \cup (A \cap C)$$

The parentheses indicate the order in which the operations are to be carried out. Alternatively, one can use precedence rules. We will give $\sim$ the highest precedence, followed by

∩ and ∪. The lowest precedence is given to −. If an expression involving sets contains the symbol ∈, then we give this symbol a higher priority than ∼.

Table 5.1 gives some basic identities that are useful to manipulate expressions involving sets. As mentioned earlier, all relations involving complementation require a specific universal set E. We therefore assume that E is defined. Since complements, intersections, and unions correspond to negations, conjunctions, and disjunctions, respectively, it follows that the laws involving negations, conjunctions, and disjunctions have their analog in set theory. Specifically, every law given in Table 5.1 corresponds to a law of Table 1.16. In both cases, the laws come in pairs of duals. The only exception to this is the double complementation law, which is its own dual.

TABLE 5.1.   Basic Set Identities

Set Algebra	Name
$A \cup {\sim} A = E$ $A \cap {\sim} A = \emptyset$	Complementation law Exclusion law
$A \cap E = A$ $A \cup \emptyset = A$	Identity laws
$A \cup E = E$ $A \cap \emptyset = \emptyset$	Domination laws
$A \cup A = A$ $A \cap A = A$	Idempotent laws
${\sim}({\sim} A) = A$	Double complementation law
$A \cup B = B \cup A$ $A \cap B = B \cap A$	Commutative laws
$(A \cup B) \cup C = A \cup (B \cup C)$ $(A \cap B) \cap C = A \cap (B \cap C)$	Associative laws
$A \cup (B \cap C) = (A \cup B) \cap (A \cup C)$ $A \cap (B \cup C) = (A \cap B) \cup (A \cap C)$	Distributive laws
${\sim}(A \cap B) = {\sim} A \cup {\sim} B$ ${\sim}(A \cup B) = {\sim} A \cap {\sim} B$	De Morgan's laws

The laws of Table 5.1 can be proved by using the formal definitions of unions and intersections and the laws of Table 1.16. We illustrate this for the case of the commutative law involving an intersection.

$$
\begin{aligned}
x \in (A \cap B) &\equiv x \in A \wedge x \in B &&\text{By (5.4)}\\
&\equiv x \in B \wedge x \in A &&\wedge \text{ is commutative}\\
&\equiv x \in (B \cap A) &&\text{By (5.4)}
\end{aligned}
$$

It is instructive to verify the laws given in Table 5.1 by Venn diagrams. For instance, Figure 5.6 shows how to verify the second distributive law. In this figure, the area of

$A \cap (B \cup C)$ is shaded. It is easily verified that this area can also be obtained by taking the union of $A \cap B$ and $A \cap C$. De Morgan's laws can similarly be obtained from Venn diagrams.

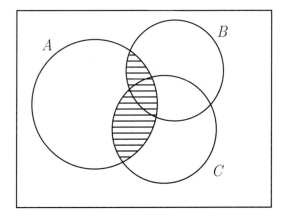

Figure 5.6.   Venn diagram of
$A \cap (B \cup C)$

As in the case of propositional calculus, one can derive further equations of importance from the ones given in Table 5.1. Of particular importance are the absorption laws, which are given as

$$A \cup (A \cap B) = A \tag{5.7}$$
$$A \cap (A \cup B) = A \tag{5.8}$$

The truth of the absorption laws is almost immediately obvious when considering the Venn diagrams of the intersection and the union (see Figures 5.2 and 5.3, respectively). In particular, to see that (5.7) is true, note in Figure 5.2 that $A \cap B$ is given by the shaded area, and the union of the shaded area with A is A. Equation (5.8) can be illustrated in a similar fashion by using Figure 5.3.

Venn diagrams do not prove any relations; they merely illustrate them. We therefore add a formal proof of (5.7).

$$
\begin{aligned}
A \cup (A \cap B) &= (A \cap E) \cup (A \cap B) && \text{Identity} \\
&= A \cap (E \cup B) && \text{Distributive} \\
&= A \cap E && \text{Domination} \\
&= A && \text{Identity}
\end{aligned}
$$

This completes the proof of (5.7). To prove (5.8), one proceeds the same way, except that the dual relations are used.

Two further equations that are useful are

$$(A \cap B) \cup (A \cap {\sim} B) = A \tag{5.9}$$
$$(A \cup B) \cap (A \cup {\sim} B) = A \tag{5.10}$$

The reader may want to derive these two equations from the equations given in Table 5.1.

To simplify expressions involving sets, one can use similar strategies as were described in Section 1.5. In particular, it normally pays to move all complementations in. At each stage, one should check if any of the first nine laws of Table 5.1 can be used to simplify the expressions. The laws given by (5.7) to (5.10) can also be used to simplify expressions.

Example 5.8   Simplify $\sim((\sim A \cup \sim C) \cap B) \cup \sim (A \cup \sim (C \cap \sim B) \cup C)$.

   Solution   By repeated applications of De Morgan's and the double negation laws, one finds

$$((A \cap C) \cup \sim B) \cup (\sim A \cap (C \cap \sim B) \cap \sim C)$$

The term $(\sim A \cap (C \cap \sim B) \cap \sim C)$ is an intersection that contains both C and $\sim C$, and such intersections always yield the empty set. Because of the identity law, the empty set may be dropped. What remains is the expression

$$(A \cap C) \cup \sim B$$

This cannot be simplified further, and the problem is solved.   ∎

Problems 5.1

1. Which of the following propositions are true? Here $S = \{2, a, 3, 4\}$, $R = \{a, 3, 4, 1\}$, and E is the universal set.
   (a) $a \in S$ (e) $\{a\} \in S$
   (b) $a \in R$ (f) $\emptyset \subset R$
   (c) $R = S$ (g) $\emptyset \subseteq R$
   (d) $\{a\} \subseteq S$ (h) $\emptyset \in R$

2. Which of the following propositions are true? Here $R = \{1, 2, 3\}$, $S = \{3, 4, 5\}$, and E is the universal set.
   (a) $R \subseteq S$ (c) $\emptyset \subset R$
   (b) $R \not\subseteq S$ (d) $E \not\subset R$

3. Let $A = \{1, 1, 2, 2, 2\}$, $B = \{1, 2, 3\}$, and $C = \{1, 2, 3, 4\}$. Moreover, E is the universal set. Arrange A, B, C, $\emptyset$, and E such that the set to the right is a subset of the set to the left.

4. Find the cardinality of the sets $\emptyset$, $\{a, b\}$, $\{999\}$, and $\{1, 2, 3\}$.

5. Find the union, the intersection, and the set difference of A and B, where $A = \{1, 3, 4, 5\}$ and $B = \{3, 5, 7, 8\}$.

6. Let $E = \{0, 1, 2, \ldots, 9\}$, $X = \{2, 3, 4\}$, $Y = \{1, 2, 5\}$, and $Z = \{2, 5, 7\}$. Find $\sim(X \cup Y)$, $\sim X \cup \sim Y$, $X \cup (Y \cap Z)$, and $(X \cup Y) \cap Z$.

7. Let the universal set be all people. Within this universal set, A is the set of all systems analysts, B is the set of all accountants, C is the set of all women, and D is the set of all people of age 40 and older. Find the following:
   (a) The set of all female systems analysts who are simultaneously accountants.
   (b) The set of all male accountants of age 40 and older.
   (c) The set including all female systems analysts under 40 as well as all accountants under 40.

8. Suppose that the universe of discourse consists of the days of the week. Let A represent the workdays, which are Monday through Friday. Find $\sim A$.

9. Show that
$$(R \subseteq S) \wedge (S \subset Q) \Rightarrow R \subset Q$$
   Is it correct to replace $R \subset Q$ by $R \subseteq Q$? Explain your answer.

10. Draw a Venn diagram for two sets A and B. Find $A \cap B$ and $A \cap \sim B$. What is the union of $A \cap B$ and $A \cap \sim B$?

11. Prove (5.9) and (5.10) using the relations given in Table 5.1.

12. Simplify $(\sim(A \cup \sim(B \cup C) \cap \sim A \cap \sim(B \cap C) \cup A))$.

13. Give examples of sets A, B, and C such that $A \cup B = A \cup C$, but $B \neq C$.

14. Prove the identities
$$A \cap A = A, \quad A \cap \emptyset = \emptyset, \quad A \cap E = A, \quad \text{and} \quad A \cup E = E$$

15. Use the relations of Table 5.1 to prove that
$$A \cap (\sim A \cup B) = A \cap B$$

16. Show that $(A \cap B) \cup C = A \cap (B \cup C)$ iff $C \subseteq A$.

17. Draw Venn diagrams showing that
$$
\begin{array}{lll}
A \cup B \subset A \cup C & \text{but} & B \not\subseteq C \\
A \cap B \subset A \cap C & \text{but} & B \not\subseteq C \\
A \cup B = A \cup C & \text{but} & B \neq C \\
A \cap B = A \cap C & \text{but} & B \neq C
\end{array}
$$

18. Draw Venn diagrams and show the sets
$$\sim B, \quad \sim(A \cup B), \quad B - (\sim A), \quad \sim A \cup B, \quad \sim A \cap B$$
   where $A \cap B \neq \emptyset$.

19. Show that $(A \cup B) \cup C = A \cup (B \cup C)$.

20. Prove that $A \cap \sim A = \emptyset$ and $A \cup \emptyset = A$.

21. Show that $(A - B) - C = (A - C) - (B - C)$.

22. Prove that $(A \cap B) \cap (C \cap D) = (A \cap C) \cap (B \cap D)$. Use the relations of Table 5.1.

5.2 TUPLES, SEQUENCES, AND POWERSETS

5.2.1 Introduction

In the following sections, a number of structures somewhat related to sets is discussed. Section 5.2.2 introduces n-tuples, or tuples for short. Such n-tuples consist of n objects, which are arranged in a certain order. They are useful for organizing data, and many files contain records that are, mathematically speaking, n-tuples. The most important n-tuple is the pair, which is really a 2-tuple. If (x, y) is such a pair, then one often restricts x to be from a set A and y to be from a set B. The set of all possible pairs obtainable in this way is called the *Cartesian product* of A and B. Tuples are related to sequences and strings, which will be discussed in Section 5.2.3. Of course, strings and string manipulations are very important in computer science. Section 5.2.4 discusses powersets. The powerset of set D is the set of all sets that one can form by using only elements from D. Both Cartesian products and powersets are important for defining types, a topic addressed in Section 5.2.5.

5.2.2 Tuples and Cartesian Products

Two objects, given in a certain order, form a pair. A married couple is a pair, consisting of husband and wife. In the phone directories of some companies, each entry consists of a pair, a name and a phone number. A point on the Northern hemisphere can be expressed by a pair of numbers, its longitude and its latitude. Generally, if x and y are two objects, one can form the pair consisting of x and y, and this pair is denoted by (x, y). Note that the order of the pair is important. For instance, a point on the Northern hemisphere can be represented by (x, y), where x is the latitude and y is the longitude. Consequently, $(41, 74)$ is a point with a latitude of 41 and a longitude of 74, and this point happens to be close to New York City. The point $(74, 41)$, on the other hand, is a point with latitude 74 and longitude 41, and this point is in the middle of Greenland. Generally, two pairs (x_1, y_1) and (x_2, y_2) are equal if and only if $x_1 = x_2$ and $y_1 = y_2$. The two elements forming the pair need not belong to the same set. For instance, the pairs found in the telephone directory consist of a name, which is an element from the set of names, together with a number, which is an element from the set of numbers. More generally, in the pair (x, y), x may belong to set A and y to set B. In applications, both A and B are often types.

Pairs consist of two objects. More generally, one can create n-tuples, which consist of n objects, arranged in a certain order. Instead of n-tuples, the term tuples is also used. The following is an application of a tuple for describing an inventory system. In this system, there are many different parts, and each part is associated with a 4-tuple, consisting of a part number, a part name, a supplier number, and the quantity on hand. Again, the order in which these items are listed is important. Tuples are also used in geometry. There, a point in the n-dimensional space is given by its n coordinates, which together form an n-tuple. For instance, in three-dimensional space, a point is given by its x, y, and z coordinate, and this 3-tuple can be written as (x, y, z). More generally, in the n-dimensional space a point is described by the n-tuple $(x_1, x_2, \ldots, x_n)$, in that order.

When dealing with n-tuples, one is naturally interested in the set of all n-tuples. This leads to the concept of a *Cartesian product*. First, the Cartesian product is defined for pairs as follows:

Definition 5.10:   Let A and B be any two sets. The set of all ordered pairs, such that the first member of the ordered pair is an element of A and the second member is an element of B, is called the *Cartesian product* of A and B and is written as $A \times B$. Accordingly,

$$A \times B = \{(x, y) \mid (x \in A) \wedge (y \in B)\}$$

An important example of a Cartesian product is given by the set of all Cartesian coordinates in a plane. This product is given as $\mathbb{R} \times \mathbb{R}$, or $\mathbb{R}^2$ for short. The Cartesian coordinates were introduced by Descartes in the early seventeenth century. Following the customs of his time, he latinized his name to Cartesius. The Cartesian coordinates were named in his honor. The Cartesian products derive their name from the Cartesian coordinates. Other examples of Cartesian products follow.

Example 5.9   Let $A = \{$Marie, Rose$\}$ and $B = \{$Jim, Bill, Brent$\}$. Find the set $A \times B$.

Solution   The set $A \times B$ is the set of all couples in which the first element is either Marie or Rose and in which the second is either Jim, Bill, or Brent. We can write this either in the form of a set, as is done in the following equation, or in tabular form, as in Table 5.2.

$$
\begin{aligned}
A \times B = \ & \{(\text{Marie, Jim}), (\text{Marie, Bill}), (\text{Marie, Brent}), \\
& (\text{Rose, Jim}), (\text{Rose, Bill}), (\text{Rose, Brent})\}
\end{aligned}
$$

■

TABLE 5.2.   Table Representation of a Cartesian Product

	Jim	Bill	Brent
Marie	(Marie, Jim)	(Marie, Bill)	(Marie, Brent)
Rose	(Rose, Jim)	(Rose, Bill)	(Rose, Brent)

Example 5.10   If $A = \{\alpha, \beta\}$ and $B = \{1, 2, 3\}$, what are $A \times B$, $B \times A$, $A \times A$, and $B \times B$?

Solution

$$
\begin{aligned}
A \times B = \ & \{(\alpha, 1), (\alpha, 2), (\alpha, 3), (\beta, 1), (\beta, 2), (\beta, 3)\} \\
B \times A = \ & \{(1, \alpha), (2, \alpha), (3, \alpha), (1, \beta), (2, \beta), (3, \beta)\} \\
A \times A = \ & \{(\alpha, \alpha), (\alpha, \beta), (\beta, \alpha), (\beta, \beta)\} \\
B \times B = \ & \{(1, 1), (1, 2), (1, 3), (2, 1), (2, 2), (2, 3), \\
& (3, 1), (3, 2), (3, 3)\}
\end{aligned}
$$

■

The set of all n-tuples $(x_1, x_2, \ldots, x_n)$, with $x_1 \in A_1$, $x_2 \in A_2, \ldots, x_n \in A_n$, is denoted by $A_1 \times A_2 \times \cdots \times A_n$. Mathematically, this n-fold Cartesian product is defined as follows:

$$A_1 \times A_2 \times \cdots \times A_n = \{(x_1, x_2, \ldots, x_n) \mid x_1 \in A_1, x_2 \in A_2, \ldots, x_n \in A_n\}$$

Example 5.11   Let A_1 be the set of all consonants and A_2 the set of all vowels. In this case, the set of all words with three letters, which start and end with a vowel, is given by the set $A_2 \times A_1 \times A_2$.

Frequently, the sets in a Cartesian product are all identical. For instance, the points in three-dimensional space are given by the Cartesian product $\mathbb{R} \times \mathbb{R} \times \mathbb{R}$, and this product is usually written as $\mathbb{R}^3$. Generally, if A is a set, A^n is the set of all n-tuples in which each element of the n-tuple is chosen from A.

Example 5.12   A byte is an 8-tuple of bits. Express the set of all bytes as a Cartesian product.

Solution   A bit is of the type $\{0, 1\}$. The set of all bytes is therefore $\{0, 1\}^8$.   ∎

The cardinality of the Cartesian product $A \times B$ is $\#A$ multiplied by $\#B$. This can be shown by writing the product in form of a table, such as Table 5.2. Each element of A corresponds to a row in this table, and since A has $\#A$ elements, this yields $\#A$ rows. The number of columns is similarly $\#B$. Since there is an entry for each row and column, this yields $\#A \cdot \#B$ elements in total. More generally, an n-fold Cartesian product $A_1 \times A_2 \times \cdots \times A_n$ has $m_1 m_2 \ldots m_n$ elements, where m_i is the cardinality of A_i, $i = 1, 2, \ldots, n$. The number of different n-tuples therefore increases exponentially with n.

5.2.3 Sequences and Strings

To form sequences, one needs an initial set A, where A may be the set of natural numbers, the set of integers, the set $\{0, 1\}$, or the set of ASCII characters. Generally, if A is any set, one can form a sequence in which all elements are members of A. In this case, the sequence is said to be *over* A. Typically, one delimits sequences by angle brackets. For instance, $\langle 3, 5, 2 \rangle$ is a sequence over $\mathbb{N}$. Other sequences over $\mathbb{N}$ include

$$\langle 2, 4, 8, 16, \ldots \rangle$$
$$\langle 1, 2, 1, 2, 1, \ldots \rangle$$
$$\langle 3, 5, 3, 7 \rangle$$
$$\langle \rangle$$

The first two of these sequences are *infinite sequences*; that is, they do not end. The third sequence, on the other hand, is finite. In fact, it has a *length* of 4. Generally, the length of a sequence is the number of objects in the sequence, counting each object with the multiplicity with which it appears. The last of these sequences is the *empty* sequence, which does not contain any element. The empty sequence is also denoted by ϵ.

In sequences, the order in which the elements appear is important. For instance, $\langle 2, 3, 4 \rangle$ and $\langle 3, 2, 4 \rangle$ are two different sequences. Repetitions of elements are significant. For instance, $\langle 1, 1, 2 \rangle$ and $\langle 1, 2 \rangle$ are two different sequences. Generally, a sequence of length n is an n-tuple. As such, one can say that each sequence of length n over A must be a member of the Cartesian product A^n. The set of all nonempty sequences of length n or less is given by

$$A^1 \cup A^2 \cup \cdots \cup A^n$$

The set of all finite nonempty sequences, usually denoted by A^+, is therefore

$$A^+ = A^1 \cup A^2 \cup A^3 \ldots$$

If A^0 is the set that contains the empty sequence $\langle \rangle$ as its only member, then the set of all sequences, including the empty sequence, becomes

$$A^* = A^0 \cup A^+ = A^0 \cup A^1 \cup A^2 \cup A^3 \ldots$$

Example 5.13   Let $B = \{0, 1\}$ be the set of bits. A byte, which is 8 bits, is then a sequence of length 8 over B; that is, it is a member of B^8.

Sequences of characters are obviously of great importance in computer science, and they are called *strings*. The set of characters over which the string is formed is called, naturally, the *alphabet*. Typically, strings are not delimited by angle brackets like other sequences, but by single quotes, as is the case in the string '$abdx$.' This string consists of the four characters a, b, d, and x, in that order. The empty string is denoted by ϵ, as in the case of any other sequence, or by ' '.

We now consider finite sequences, including finite strings. An important operation on such sequences is the *concatenation*. To concatenate two sequences x and y, one first writes all elements of x, followed by all elements of y. The symbol for concatenation is $\frown$. For instance, the concatenation of $\langle 3, 5, 4 \rangle$ and $\langle 4, 2, 7 \rangle$ is given by

$$\langle 3, 5, 4 \rangle \frown \langle 4, 2, 7 \rangle = \langle 3, 5, 4, 4, 2, 7 \rangle$$

Similarly, one has

$$'abs' \frown 'def' = 'absdef'$$

Formally, the concatenation is an operation over sequences, including strings. This operation is not commutative; that is, $x \frown y = y \frown x$ is not necessarily true. For instance, 'ab' $\frown$ 'cd' is '$abcd$', yet 'cd' $\frown$ 'ab' is '$cdab$'. Concatenation is associative, however; that is, it does not matter in which order one performs concatenation. For instance, both $('ab' \frown 'cd') \frown 'ef'$ and $'ab' \frown ('cd' \frown 'ef')$ yield '$abcdef$.' Consequently, one is allowed to omit parentheses. One can also formulate equations involving sequences. For instance, the solution of $X \frown 'ful' = 'beautiful'$ for X yields $X = 'beauti'$. More generally, if there are values X and Y such that $X \frown Z \frown Y = U$, then Z is said to be a *substring* of U. For instance, 'tif' is a substring of '$beautiful$' because there are values of X and Y, that is, $X = 'beau'$ and $Y = 'ul'$, such that $X \frown 'tif' \frown Y = 'beautiful'$.

5.2.4 Powersets

If one is given a set E, then one can consider all sets that can be formed by using only elements of E. All these sets are by definition subsets of E. This leads to the following definition:

Definition 5.11:   Let A be some set. The powerset of A, written as $\mathbb{P}\,A$, is the set of all subsets of A.

Some authors use $\rho(A)$ or 2^A instead of $\mathbb{P}\,A$ to denote the powerset.

Example 5.14   Construct the powerset of $A = \{a, b, c\}$.

> **Solution**   The powerset of A is the set of all sets that one can construct by using only the objects a, b, and c; that is,
>
> $$\mathbb{P}A = \{\emptyset, \{a\}, \{b\}, \{c\}, \{a, b\}, \{b, c\}, \{a, c\}, \{a, b, c\}\}$$ ∎

Example 5.15   If $\mathbb{N} = \{0, 1, 2, \ldots\}$ is the set of natural numbers, then the powerset $\mathbb{P}\,\mathbb{N}$ contains all sets of natural numbers.

If A and B are elements from some powerset D, then $A \cap B$, $A \cup B$, and $D - A$ are also elements of the powerset of D. Given any set D, one can form a closed system for set operations, provided that for each set $A \in \mathbb{P}\,D$ the complement $\sim A$ is defined as $D - A$. In this way, D takes the place of the universal set.

If D is a finite set of cardinality n, then one can express all subsets of D by bit strings of length n. A bit string is, of course, a string over the alphabet $\{0, 1\}$. Specifically, let $D = \{d_1, d_2, \ldots, d_n\}$, and suppose that the set A is to be represented as a bit string. In this case, one sets the ith bit of the string representing A to 1 if A contains d_i. Otherwise, the ith bit is set to 0. For instance, if $D = \{d_1, d_2, d_3, d_4, d_5\}$, then the set $A = \{d_3, d_5\}$ can be represented by 00101. Here, the 1 in the third position indicates that $d_3 \in A$, and the 1 in the last position similarly indicates that d_5 is in A. Once two sets are given in bit notation, one can easily form unions, intersections, set differences, and complements with respect to D. For instance, if D is a set of five elements and if A is given by 01100 and B by 10100, then $A \cup B$ is 11100 and $A \cap B$ is 00100. These results can easily be obtained by using a bitwise "or" or a bitwise "and."

As shown, each subset of D can be expressed as a bit string of length n. Conversely, each bit string of length n represents a subset of D. Consequently, there are as many elements in the powerset of D as there are bit strings of length n. Since there are 2^n such strings, there are 2^n sets in the power set of D. This is the reason some authors use 2^D instead of $\mathbb{P}\,D$.

5.2.5 Types and Signatures

In Chapters 1 to 3, the universe of discourse was assumed to be uniform in the sense that all individuals could be treated in the same way. In reality, this is not the case. There are many different types of individuals, such as persons, things, and ideas, and operations that are meaningful for some of these types are not meaningful for others. The same is true in mathematics and computer science. In these areas, one deals with natural numbers, integers, real numbers, strings, expressions, and so on. Typically, each predicate requires that its arguments be of a certain type, and the same is true for functions. A type is also associated with the result of a function. To show that some predicates require that their arguments be of a certain type, consider the predicate $\mathrm{prime}(x)$, which stands for "x is prime." This predicate is clearly not applicable when x is an arbitrary real number, such as 2.5 or 3.14159. In fact, to be a prime, the argument x must be an integer. Similarly, for $x < y$ to be meaningful, both x and y must be real numbers. They cannot be imaginary numbers or vectors. To show that the arguments of functions must be of a certain type, consider the function $\sin x$. In this case, x must be a real number, and the result is also a real number. Operators are functions, and their operands are also restricted to be of a certain type. For instance, in $x + y$, both x and y must be real numbers. They cannot be characters. The result of the operation is also real.

Generally, a type is a set, and for any new predicate and function to be defined, it must be stated from what set the arguments must be selected. Any set used in this way can be considered as a type. In many cases of practical importance, there are a number of initial types, which can in turn be used to create additional types. The following types are particularly important.

1. The set of natural numbers, which is denoted by $\mathbb{N}$

2. The set of integers, which is denoted by $\mathbb{Z}$

3. The set of real numbers, which is denoted by $\mathbb{R}$

4. The set consisting of T and F, which constitutes the type Boolean

5. The set of bits, that is, the set $\{0, 1\}$

6. The set of characters

Additional types can be introduced as the application warrants. For instance, one can create a type for all cities, which could be denoted by C. Objects of these initial types can be combined into sets and sequences, which allows one to form new types. An example of such a type is a pair of natural numbers, and this type is given by the set $\mathbb{N} \times \mathbb{N}$. If an address consists of a number, a street name, and a city name, then the type "address" is given by $\mathbb{N} \times S \times C$, where S is the set of streets and C is the set of cities. If E is some universal set, then all subsets of E are of type $\mathbb{P}\,E$.

Two elements of a different type are always considered to be different. For instance, one has to distinguish between the integer 4 and the set $\{4\}$, which contains 4 as its only element. For instance, the statement $3 \in \{4\}$ is meaningful, though false, whereas $3 \in 4$ is meaningless. On the other hand, the number 4 has a successor, but no successor is defined for sets. Consequently, if $s(x)$ is the successor of x, then $s(4)$ is meaningful, but $s(\{4\})$ is not.

In many computer languages, all variables must be *declared*; that is, their type must be indicated before the variable is used. This is done through a *declaration*. Another word for declaration is *signature*. Signatures can also be used when dealing with a universe of discourse in which some predicates are restricted to variables of certain types. Such logic systems are called *many-sorted logics*. Moreover, types are used in many mathematical systems. They are also very important in specification languages such as Z (see Chapter 8).

Generally, a signature restricts a variable to belong to a certain type. The format of the declarations used here is as follows: The variable name is given first; then, separated by a colon, the type is given. For instance, the signature $x : \mathbb{N}$ indicates that the variable x is a natural number.

In programming languages, the types are usually given at the beginning of the procedure with the understanding that, throughout the procedure, each variable always has its associated declared type. In many-sorted logic and in declarative languages, signatures frequently have a very narrow scope. In fact, the signature may form part of the description of a set, or it may be used for declaring the variable of a quantifier.

If a variable appearing in a quantifier contains a signature, it can only range over the values of the type specified by the signature. For instance, if the variable x in a universal quantifier is declared to be an integer, then universal quantification over x means that the expression that is quantified must be true for all integers. The notation used for declaring bound variables in quantifiers varies. Here the method used by Z is adopted. In Z, the bound variable following the quantification is replaced by a signature, and the signature terminates with a bullet, as indicated by the following example.

Example 5.16   Consider the statement "if x is a natural number, then $x \geq 0$." This statement can be expressed as

$$\forall x : \mathbb{N} \bullet (x \geq 0)$$

Logicians sometimes consider everything that is meaningless as false. This convention allows them to avoid declarations. To indicate that a certain property is true for all $x \in X$, one can write

$$\forall x (x \in X \Rightarrow P(x)) \tag{5.11}$$

The convention that meaningless means false is required. To see this, write (5.11) as

$$\forall x (P(x) \lor \neg(x \in X))$$

If $P(x)$ is meaningless for x not in X, then this becomes meaningless, even though it is logically equivalent to (5.11).

Setbuilder notation uses expressions such as $A = \{x \mid P(x)\}$ to indicate that all x satisfying $P(x)$ belong to set A. Again, x is a variable, and it is appropriate to declare its type. To do this, one replaces the first x by its signature. If x is of type A, this leads to expressions of the form $\{x : A \mid P(x)\}$.

Example 5.17   To describe the set consisting of all integers below 3, one uses

$$\{x : \mathbb{Z} \mid x < 3\}$$

This set has the extension $\{2, 1, 0, -1, \ldots\}$.

Problems 5.2

1. Let $A = \{1, 2, 3\}$. Find $A \times (A - \{2\})$ and A^2.
2. Given $\#A = 4$. Find $\#(A^2)$, $\#(A^3)$, and $\#(2^A)$.

3. Let $A = \{0, 1, 2\}$. Find $(A^2 - \{(0, 0)\}) \times A$.

4. Let $A = \{3, 5, 7\}$ and $B = \{a, b\}$. Find A^2, B^2, $A \times B$, and $B \times A$.

5. Let A represent the English alphabet, and let $0..9$ be the set of digits from 0 to 9.
   (a) Find the number of strings of length 3 over the alphabet $A \cup 0..9$.
   (b) How many of these strings start with a letter?
   (c) How many strings of length 4 or less start with a letter?

6. Use setbuilder notation to prove that $A \times (B \cap C) = (A \times B) \cap (A \times C)$.

7. Let A be of type customer and let B be of type product. There is a list of orders, containing the customer name, together with the product ordered. Given that this list is to be treated as a set, give the type of this list.

8. Let $A = 1..3$ and let $B = 2..4$. Find $\mathbb{P}(A \cup B)$.

9. Let $A = \{a, 2, 3\}$ and $D = \{(x, x) \mid x \in A\}$. Find $A^2 - D$.

10. Suppose that the universe of discourse is $\mathbb{R}$. Is the following statement true?

$$\forall x (\text{even}(x) \vee \text{odd}(x))$$

If not, correct this statement.

11. The universe of discourse consists of people and things. Consider the statements "some people own nothing" and "nobody owns everything." Express these statements in predicate calculus, declaring the bound variables as either type O for object or P for person.

5.3 RELATIONS

5.3.1 Introduction

An n-ary relation is a set of n-tuples. Relations are useful for many reasons. All *relational databases* use n-ary relations for the storage and access of data. Databases are the subject of Chapter 12, and they will not be discussed in this chapter. Here we will concentrate on *binary* relations. Hence, the word relation, when used in the following sections, will by default refer to binary relations. Binary relations are sets of pairs, and they occur in many contexts. There are also many methods to represent binary relations. For visualizing relations, graphical methods are particularly useful. To do mathematical operations involving relations, on the other hand, it is often more convenient to represent them as matrices. Of course, relations are sets, and methods to represent sets are also available to represent relations. There are a number of operations dealing with relations. Of particular importance is the *composition* of two relations. Moreover, all operations available for sets are also available for relations.

Previously, the connection between sets and properties was discussed. It was mentioned that all sets correspond to properties in the sense that each property defines a set, and each set defines a property. Predicates and n-ary relations are connected in a similar way. All sets of n-tuples satisfying a certain n-ary predicate define a relation, and every n-ary relation R defines the predicate "belongs to R." From this, it follows that many results of predicate calculus have an equivalent in the theory of relations.

5.3.2 Relations and Their Representation

Formally, binary relations can be defined as follows:

Definition 5.12: Let A and B be two sets. A *relation* from A to B is any set of pairs (x, y), $x \in A$ and $y \in B$. If $(x, y) \in R$, we say that x is R-related to y. To express that R is a relation from A to B, we write $R : A \leftrightarrow B$.

The Cartesian product $A \times B$ was defined earlier as the set of all pairs (x, y), $x \in A$, $y \in B$. Consequently, a relation $R : A \leftrightarrow B$ is always a subset of $A \times B$. In fact, $A \times B$ is itself a relation, the *universal relation*. The universal relation contains all possible pairs. The opposite of the universal relation is the *empty relation*, which contains no pair at all. All other relations must be somewhere between these two extreme cases.

There are many ways to express relations. Since relations are sets, one can, of course, use a roster notation, which lists all elements in the relation. For instance, suppose that A is a set of suppliers and B is a set of products. Specifically, suppose that the suppliers in question are S_1 and S_2, and the products are P_1, P_2, and P_3. Hence, $A = \{S_1, S_2\}$ and $B = \{P_1, P_2, P_3\}$. A relation C can now be defined as the list of all pairs (x, y), where x is a supplier, y is a product, and x carries y. For instance, if S_1 carries P_1 and P_3, and S_2 carries P_2 and P_3, C becomes

$$C = \{(S_1, P_1), (S_1, P_3), (S_2, P_2), (S_2, P_3)\} \tag{5.12}$$

Every property defines a set. Similarly, every n-ary predicate defines an n-ary relation. In particular, predicates of arity 2 define binary relations. For instance, the predicate "married(x, y)" is true when x and y are married. Consequently, one may define a set, say M, such that

$$M = \{(x, y) \mid \text{married}(x, y)\}$$

Since M is a set of pairs, M is a relation.

The previous paragraph showed that every predicate defines a relation. Conversely, every relation R defines a predicate. Specifically, if (x, y) is a pair, one can define a predicate P_R for each relation R that is true if $(x, y) \in R$, and false otherwise. This predicate is frequently expressed as xRy, that is, one defines

$$xRy \equiv (x, y) \in R$$

For instance, in the universe of discourse S_1, S_2, P_1, P_2, and P_3, relation C defined by (5.12) defines the predicate xCy. Clearly, $S_1 C P_1$, $S_1 C P_3$, $S_2 C P_2$, and $S_2 C P_3$ are all true, whereas xCy is false in all other cases. Sometimes one uses the notation $x\cancel{R}y$ for $\neg(xRy)$. For instance, $S_1 \cancel{C} P_2$ is true, but $S_1 \cancel{C} P_1$ is false.

Consider now the expression $x < y$. In this expression, $<$ can be interpreted as a relation. In this sense, $<$ is a set, which contains all pairs (x, y) with the property that

$x < y$. Hence, if $R = (<)$, then xRy instantiates to $x < y$. The expressions $x = y, x > y$, and so on, can be interpreted in a similar way. Another example of the notation xRy is the subset notion, as in $A \subseteq B$. Here $\subseteq$ is a relation among pairs of sets.

In Chapter 2, predicates were defined by tables. For instance, Table 2.1 defined the predicate "mother." Relations can be defined similarly. For instance, the relation in (5.12) is given by Table 5.3. In this table, there is a row for each supplier and a column for each product. The numeral 1 in a particular cell of the table indicates that the supplier carries the product, whereas a zero indicates that he does not.

TABLE 5.3.
Relation in Tabular Form

	P_1	P_2	P_3
S_1	1	0	1
S_2	0	1	1

Tables are closely related to matrices. To convert a table into a matrix, one merely drops the headings. In the case of Table 5.3, this yields

$$M_C = \begin{bmatrix} 1 & 0 & 1 \\ 0 & 1 & 1 \end{bmatrix}$$

If R is a relation, we will use the symbol M_R to denote the matrix of this relation. The entry of row i, column j is denoted by m_{ij}^R. The superscript R is omitted if this cannot cause any ambiguity.

Relations can also be expressed graphically. To represent a relation from A to B, draw a circle for each element of A at the left, and draw a circle for each element of B to the right. If the pair $x \in A, y \in B$ is in the relation, the corresponding circles, or *nodes* as they are called, are connected by straight lines called *arcs*. The arcs start at the first element of the pair, and they go to the second element of the pair. This direction is indicated by an arrow. All arcs with an arrow are called *directed arcs*. The resulting figure is called a *directed graph* or *digraph*. For instance, Figure 5.7 graphically represents the relation given in (5.12).

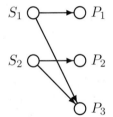

Figure 5.7.   Digraph of a relation

5.3.3 Domains and Ranges

When $R : A \leftrightarrow B$ is a relation, then A is called the domain space, and B is called the range space. The domain itself is a subset of the domain space, and the range is similarly a subset of the range space. This is a consequence of the following definitions:

> **Definition 5.13:**   Let R be a relation from X to Y. The *domain* of R, abbreviated by $\operatorname{dom} R$, is the set of all elements $x \in X$ that occurs in at least one pair $(x, y) \in R$. This can be expressed as
>
> $$\operatorname{dom} R = \{x \mid \exists y((x, y) \in R)\}$$

> **Definition 5.14:**   The *range* of R, abbreviated by $\operatorname{ran} R$, is the set of all $y \in Y$ that occurs in at least one pair $(x, y) \in R$. This can be expressed as
>
> $$\operatorname{ran} R = \{y \mid \exists x((x, y) \in R)\}$$

For instance, in the supplier–product relation, the domain consists of all suppliers that carry at least one product, and the range consists of all products offered by at least one supplier in the relation. In the pair (x, y), x can be thought of as the origin, and y as the destination. Hence, the domain consists of all origins and the range of all destinations.

Example 5.18   Find the domain and the range of relation R from the set $\{1, 2, 3, 4\}$ to the set $\{a, b, c, d\}$ given by

$$R = \{(2, c), (1, d), (3, d), (2, a)\}$$

Solution   The domain of this relation is the set of values that appears first in any pair of the relation. The set of the second elements of the pairs similarly provides the range. This yields

$$\operatorname{dom} R = \{1, 2, 3\}, \qquad \operatorname{ran} R = \{a, c, d\} \qquad\qquad \blacksquare$$

Domain and range can be found easily if the relation is represented by a graph. The graph of R as defined previously is given in Figure 5.8. Every node to the left with an outgoing arc belongs to the domain, and every node to the right with an incoming arc belongs to the range. One sees at a glance that the domain is $\{1, 2, 3\}$ and that the range is $\{a, c, d\}$.

If a relation is given by a matrix, as is the case for C defined by (5.12), then the domain is given by the rows that contain at least one nonzero entry, and the range is similarly given by the columns with at least one nonzero entry. In the case of Table 5.3, the domain is $\{S_1, S_2\}$, whereas the range is $\{P_1, P_2, P_3\}$.

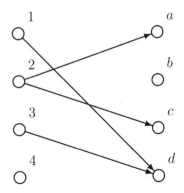

Figure 5.8.   Digraph for finding domain and range

Sometimes the domain space and the range space are identical, in which case one speaks about a relation *on* some set.

Definition 5.15:   Let A be a set, and let R be a relation from A to A. Then R is said to be a relation *on* A.

For instance, the relation $<$, which includes all pairs of reals (x, y) satisfying $x < y$, is a relation on the reals.

5.3.4 Some Operations on Relations

With every relation R from X to Y, one can associate an *inverse relation* $R^\sim$ from Y to X. Essentially, the inverse relation has the pair (y, x), where the original relation has (x, y), as is indicated by the following definition:

Definition 5.16:   If $R : X \leftrightarrow Y$ is a relation, then the *inverse* relation $R^\sim : Y \leftrightarrow X$ is defined as $\{(y, x) \mid (x, y) \in R\}$. Consequently, $xRy \equiv yR^\sim x$.

Some authors use the terms *converse* instead of inverse. For instance, the inverse (or converse) of the relation $<$ is $>$, because $x < y$ iff $y > x$. As a second example, consider the parent relation P, which includes the set of all pairs (x, y) such that x is a parent of y. The inverse relation consists of the same set of pairs, except that the order is now (y, x); that is, the child comes first. In this sense, one can say that the inverse of the parent relation is the child relation.

All set operations can be done on relations. The resulting sets contain ordered pairs and are therefore relations. If R and S denote two relations, then $R \cap S$ defines a relation such that

$$x(R \cap S)y \equiv xRy \wedge xSy$$

Similarly, $R \cup S$ is a relation such that

$$x(R \cup S)y \equiv xRy \vee xSy$$

Also,

$$x(R - S)y \equiv xRy \wedge x\cancel{S}y$$

and

$$x(\sim R)y \equiv x\cancel{R}y$$

To be completely formal, one would have to indicate how these operations affect the domain space and the range space of the result. Most authors, including the authors of this text, omit this step. The interested reader is referred to Problem 5 at the end of this section in this regard.

Example 5.19   Let $R : X \leftrightarrow Y$ and $S : U \leftrightarrow V$ be two relations. The domain spaces are $X = \{a, b, c\}$ and $U = \{a, b\}$, and the range spaces are $Y = \{A, B, C\}$ and $V = \{B, C\}$. Moreover, $R = \{(a, A), (a, B), (b, C)\}$ and $S = \{(a, B), (b, C)\}$. Find $\sim S$, $R \cup S$, $R \cap S$, and $R - S$.

Solution   The complement of S consists of all pairs of the Cartesian product $U \times V$ that are not in S. This yields

$$\sim S = \{(a, C), (b, B)\}$$

For $R \cup S$, $R \cap S$, and $R - S$, one finds

$$
\begin{aligned}
R \cup S &= \{(a, A), (a, B), (b, C)\} \\
R \cap S &= \{(a, B), (b, C)\} \\
R - S &= \{(a, A)\}
\end{aligned}
$$

∎

Relations can be *extended* and *restricted*. Any increase in the domain space or the range space is called an *extension*. The extension can be trivial in the sense that the domain and the range of the relation remain unaffected. For instance, if $R : X \leftrightarrow Y$, where $X = \{a, b, c\}$ and $Y = \{A, B, C\}$ is given as $R = \{(a, A), (a, B), (b, C)\}$, then one can find a trivial extension of R, replacing X with the set of lowercase letters and Y with the set of uppercase letters. Trivial extensions leave the roster notation of the relation intact. This means that there is really no difference between the relation and its trivial extension, except when taking complements.

Normally, however, when extending the domain space and the range space, one also extends the domain and the range. For instance, let L be the relation $<$ on the natural numbers. When extending this relation to become a relation L' on the integers, one not only adds the negative numbers to the domain space and the range space, but one also adds new pairs to the relation L by indicating how the relation affects negative numbers. The same is true for other extensions. When extending $R : A \leftrightarrow B$ to $R' : A' \leftrightarrow B'$, new pairs with $x \in (A' - A)$ and/or $y \in (B' - B)$ are frequently added. In this sense, the extended relation can be interpreted as the union of the original relation with some other relation. A trivial extension may be necessary before taking this union.

Any decrease of the domain space and the range space is called a *restriction*. A restriction is trivial if it leaves the domain and the range of the relation unaffected. Many restrictions remove some pairs (x, y) from the relation because either x is no longer part of the domain space or y is no longer part of the range space. For instance, if L' is the relation for $<$ on the integers, then restricting this relation to the natural numbers affects the domain and range of L'. Specifically, all pairs (x, y) in which either x or y are negative are removed because negative numbers do not form part of the domain space, and neither do they form part of the range space. Formally, a restriction of a relation $R : X \leftrightarrow Y$ to $R' : X' \leftrightarrow Y'$ can be thought of as the intersection of R with the relation $X' \times Y'$.

$$R' = R \cap (X' \times Y')$$

We now provide a number of examples for demonstrating in more detail the operations discussed so far.

Example 5.20   Let S be the sister relation on the human beings, and let B similarly be the brother relation. $S \cup B$ contains all pairs (x, y) such that x is either a sister or a brother of y. In other words, the relation $S \cup B$ is the sibling relation.

Example 5.21   Let I be the relation on $\mathbb{R}$ which contains all pairs of reals that are equal, and let G be the relation that includes all pairs (x, y) with $x > y$. In this case, $I \cup G$ includes all pairs (x, y) with $x \geq y$. If we write $=$ for I and $>$ for G, then one can express the relation $\geq$ as

$$(\geq) = (> \cup =)$$

Example 5.22   As an example of a complement, consider the $<$ relation involving reals. The pair (x, y) belongs to the complement of $<$ iff $x \not< y$ or, equivalently, if $x \geq y$. This means that $\sim(<)$ is given by the relation $\geq$.

Note the difference between $\sim R$ and $R^\sim$. The relation $R^\sim$ contains all elements not in R, and $R^\sim$ contains all elements of R, except that their order is reversed. For instance, if R is the relation $<$, then $R^\sim$ is the relation $>$, because $x < y$ if and only if $y > x$. On the other hand, $\sim R = \geq$ because if $x < y$ is false then $x \geq y$. Or, to take another example, if R is the empty relation, then $\sim R$ is the universal relation, whereas $R^\sim$ is the empty relation.

5.3.5 Composition of Relations

Compositions were defined in Section 2.5.5 for functions. However, the definition of a composition for functions is not quite the same as for relations, so it is best to start from scratch. An example of a composition of two relations is the aunt relation. An aunt is a sister of a parent, and it involves in this fashion two relations, the sister relation and the parent relation. In fact, an aunt is a sister of a parent. Mathematically, a composition of two relations is given as follows:

Definition 5.17:   Let $R : X \leftrightarrow Y$ and $S : Y \leftrightarrow Z$ be two relations. The *composition* of R and S, denoted by $R \circ S$, contains the pairs (x, z) if and only if there is an intermediate object y such that (x, y) is in R and (y, z) is in S. Consequently,

$$x(R \circ S)z = \exists y(xRy \wedge yRz)$$

The definition implies that (x, z) is in the composition of the sister and the parent relation if there is an individual y such that x is a sister of y and y is a parent of z. This is exactly the aunt relation. It follows that the aunt relation is the composition of the sister and the parent relation as claimed. Generally, to find if (x, z) is in the relation $R \circ S$, one always needs an intermediary y, the sister in the case of the aunt relation, such that xRy and yRz both hold.

Example 5.23   There are five people, A, B, C, D, and E. C owns the truck called Rusty and E owns the truck called Roaring. A has the friends B and D, B has the friend C, and C has the friend E. Let R be the relation "x has friend y" and let S be the relation "y owns truck z." Find the relation $R \circ S$.

Solution   If R is the relation "x has y as a friend" and if S is the relation "y owns truck z," then the pair (x, z) is in $R \circ S$ if there is an intermediary y such that x has y as a friend and y owns truck z. The relations R and S are given as

$$R = \{(A, B), (A, D), (B, C), (C, E)\}$$
$$S = \{(C, \text{ Rusty}), (E, \text{ Roaring})\}$$

Now

$$R \circ S = \{(B, \text{ Rusty}), (C, \text{ Roaring})\}$$

B has access to a truck through the intermediary C, who owns Rusty, and C has access to a truck through the intermediary E, who owns Roaring.   ■

The composition of two relations can be illustrated by means of a graph. Let $R : X \leftrightarrow Y$ and $S : Y \leftrightarrow Z$. One now draws all nodes of X at the left, all nodes of Z to the right, and all nodes of the intermediate set Y in the middle. This is done in Figure 5.9. It is assumed that the members of X are x_1 through x_4, the members of Y are y_1 through y_4, and the members of Z are z_1 through z_5. According to what has been said before, the pair (x_i, z_k) is in $R \circ S$ iff there is an intermediary y_j such that an arc goes from x_i to y_j and from there to z_k. For instance, (x_1, z_4) is in $R \circ S$ because there is an arc from x_1 to y_2, and from there there is an arc to z_4. On the other hand, (x_1, z_3) is not in $R \circ S$ because there is no intermediary y_j through which x_1 could gain access to z_3. All pairs of $R \circ S$ are now systematically enumerated as follows:

$$(x_1, z_2) \in S \circ R \text{ through intermediary } y_1$$
$$(x_1, z_1) \in S \circ R \text{ through intermediary } y_2$$

$$(x_1, z_4) \in S \circ R \text{ through intermediary } y_2$$
$$(x_4, z_3) \in S \circ R \text{ through intermediary } y_3$$

The resulting relation is thus

$$R \circ S = \{(x_1, z_1), (x_1, z_2), (x_1, z_4), (x_4, z_3)\}$$

This result is given at the right of Figure 5.9.

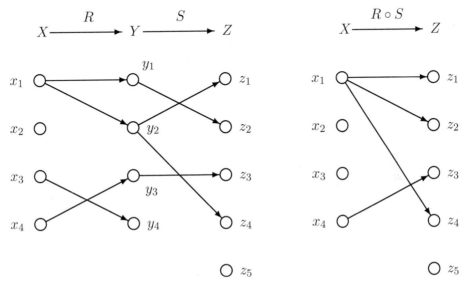

Figure 5.9.   Relations R, S, and $R \circ S$

Composition is an associative operation; that is, if R, S, and P are three relations, then the following holds:

$$(R \circ S) \circ P = R \circ (S \circ P) \tag{5.13}$$

Parentheses can therefore be dropped in cases involving the composition of several relations.

Proof. As the first step, we show that $x(R \circ (S \circ P))w$ is true if and only if there are two intermediaries y and z such that $xRy \wedge ySz \wedge zPw$. Afterward, we show that if there are two such intermediaries then $x((R \circ S) \circ P)w$ holds.

$$
\begin{aligned}
x(R \circ (S \circ P))w &\Leftrightarrow \exists y(xRy \wedge y(S \circ P)w) \\
&\Leftrightarrow \exists y(xRy \wedge \exists z(ySz \wedge zPw)) \\
&\Leftrightarrow \exists y \exists z(xRy \wedge ySz \wedge zPw)
\end{aligned}
$$

This completes the first step of the proof. We now interchange the existential quantifiers to obtain

$$\exists z \exists y(xRy \wedge ySz \wedge zPw) \equiv \exists z(\exists y(xRy \wedge ySz) \wedge zPw)$$

It is easy to see that the last expression is logically equivalent to $x((R \circ S) \circ P)w$. Consequently,

$$x(R \circ (S \circ P))w \Leftrightarrow x((R \circ S) \circ P)w$$

Equation (5.13) now follows easily.   ∎

If S_1, S_2, and S_3 are three relations, then $xS_1 \circ S_2 \circ S_3 y$ is true iff there are exactly two intermediaries through which x has access to y. This was essentially the idea of the proof of (5.13). Hence, the composition of three relations gives the set of all pairs (x, y) such that x can reach, so to say, the object y in exactly three steps. More generally, the composition of the n relations S_1, S_2, ..., S_n contains the set of all pairs (x, y) such that x can reach y in exactly n steps. If $R : A \leftrightarrow A$ is a relation on A, then $R \circ R$ is the set of pairs (x, y) such that x can reach y in exactly two steps. Normally, one abbreviates $R \circ R$ by R^2, $R \circ R \circ R$ by R^3, and so on. Clearly, R^n is the set of pairs (x, y) such that x can reach y in exactly n steps.

Example 5.24   Define

$$R = \{(1, 2), (3, 4), (2, 2)\}$$
$$S = \{(4, 2), (2, 5), (3, 1), (1, 3)\}$$

Find $R \circ S$, $S \circ R$, $R \circ (S \circ R)$, $(R \circ S) \circ R$, $R \circ R$, $S \circ S$, and $R \circ R \circ R$.

Solution

$$R \circ S = \{(1, 5), (3, 2), (2, 5)\}$$
$$S \circ R = \{(4, 2), (3, 2), (1, 4)\} \neq R \circ S$$
$$(R \circ S) \circ R = \{(3, 2)\}$$
$$R \circ (S \circ R) = \{(3, 2)\} = (R \circ S) \circ R$$
$$R \circ R = \{(1, 2), (2, 2)\}$$
$$S \circ S = \{(4, 5), (3, 3), (1, 1)\}$$
$$R \circ R \circ R = \{(1, 2), (2, 2)\}$$
                                                                                            ∎

If two relations are given in matrix form, then their composition can be found through a variant of matrix multiplication. The relation matrices only contain the numbers 0 and 1. To deal with these matrices in terms of logic, 0 is interpreted as false and 1 is interpreted as true. If u and v are two variables that are either 0 or 1, $u \vee v$ is 1 unless u and v are both 0, and $u \wedge v$ is 0 unless u and v are both 1. In fact, $u \wedge v$ is exactly the product of u and v; that is, $u \wedge v = vu$. Moreover, as mentioned in Section 2.2.2, in finite domains, existentially quantified expressions can be written as disjunctions. In particular, if the universe consists of the individuals $y_1, y_2, \ldots, y_m$, then one has

$$\exists x P(x) \equiv P(y_1) \vee P(y_2) \vee \cdots \vee P(y_m) \equiv \bigvee_{k=1}^{m} P(y_k) \qquad (5.14)$$

This theory is now applied to find the compositions of two relations R and S. R is assumed to be a relation from X to Y, with $X = \{x_1, x_2, \ldots, x_n\}$ and $Y = \{y_1, y_2, \ldots, y_m\}$. S is a relation from Y to $Z = \{z_1, z_2, \ldots, z_n\}$. The relation matrices of S, R, and $S \circ R$ are denoted by M^R, M^S, and $M^{R \circ S}$, respectively. The entries of M^R, M^S, and $M^{R \circ S}$ are, respectively, m_{ij}^R, m_{ij}^S, and $m_{ij}^{S \circ R}$. From (5.14), it follows that

$$m_{ij}^{R \circ S} \equiv \exists y(x_i R y \wedge y S z_j) \equiv \bigvee_{k=1}^{m} (x_i R y_k \wedge y_k S z_j)$$

According to the preceding convention, $x_i R y_k$ may be equated with m_{ik}^R and $y_k S z_j$ with m_{kj}^S. This yields, if the equal sign is used as in algebra,

$$m_{ij}^{R \circ S} = \bigvee_{k=1}^{m} (m_{ik}^R m_{kj}^S) \tag{5.15}$$

To clarify this, note that $m_{ij}^{R \circ S} = 1$ if there is a path from x_i to z_j. This is true if there is a point y_k such that $m_{ik}^R m_{kj}^S = 1$ or, equivalently, if $\bigvee_{k=1}^{m} (m_{ik}^R m_{kj}^S) = 1$. Equation (5.15) is identical with matrix multiplication as described in Section 3.2.4, except that the summation sign is replaced by $\bigvee$. Hence, (5.15) can be interpreted as some type of matrix multiplication, and this matrix multiplication is denoted by $\odot$. Consequently,

$$M_{R \circ S} = M_R \odot M_S$$

Example 5.25   For the relations R and S given in Example 5.24 over the set $\{1, 2, \ldots, 5\}$, obtain the relations matrices for $R \circ S$ and $S \circ R$.

Solution

$$\begin{bmatrix} 0 & 1 & 0 & 0 & 0 \\ 0 & 1 & 0 & 0 & 0 \\ 0 & 0 & 0 & 1 & 0 \\ 0 & 0 & 0 & 0 & 0 \\ 0 & 0 & 0 & 0 & 0 \end{bmatrix} \odot \begin{bmatrix} 0 & 0 & 1 & 0 & 0 \\ 0 & 0 & 0 & 0 & 1 \\ 1 & 0 & 0 & 0 & 0 \\ 0 & 1 & 0 & 0 & 0 \\ 0 & 0 & 0 & 0 & 0 \end{bmatrix} = \begin{bmatrix} 0 & 0 & 0 & 0 & 1 \\ 0 & 0 & 0 & 0 & 1 \\ 0 & 1 & 0 & 0 & 0 \\ 0 & 0 & 0 & 0 & 0 \\ 0 & 0 & 0 & 0 & 0 \end{bmatrix}$$

$$\qquad M_R \qquad\qquad\qquad M_S \qquad\qquad\qquad M_{R \circ S}$$

$$\begin{bmatrix} 0 & 0 & 1 & 0 & 0 \\ 0 & 0 & 0 & 0 & 1 \\ 1 & 0 & 0 & 0 & 0 \\ 0 & 1 & 0 & 0 & 0 \\ 0 & 0 & 0 & 0 & 0 \end{bmatrix} \odot \begin{bmatrix} 0 & 1 & 0 & 0 & 0 \\ 0 & 1 & 0 & 0 & 0 \\ 0 & 0 & 0 & 1 & 0 \\ 0 & 0 & 0 & 0 & 0 \\ 0 & 0 & 0 & 0 & 0 \end{bmatrix} = \begin{bmatrix} 0 & 0 & 0 & 1 & 0 \\ 0 & 0 & 0 & 0 & 0 \\ 0 & 1 & 0 & 0 & 0 \\ 0 & 1 & 0 & 0 & 0 \\ 0 & 0 & 0 & 0 & 0 \end{bmatrix}$$

$$\qquad M_S \qquad\qquad\qquad M_R \qquad\qquad\qquad M_{S \circ R} \qquad \blacksquare$$

5.3.6 Examples

If R and S are two relations, one can form $R^{\sim}$, $R \cup S$, $R \cap S$, $R - S$, $\sim R$, and $R \circ S$. These operations can be combined in several ways to express a number of new relations that are of interest. This will be illustrated through a number of examples. To solve these examples, the following relation $I_A : A \leftrightarrow A$ is needed.

$$I_A = \{(x, x) \mid x \in A\}$$

I_A is called the *identity relation* on A. All members of I_A are pairs of identical objects.

Example 5.26   Let S be a set of suppliers, and let P be a set of products. xCy indicates that supplier x carries product y, which makes C a relation from S to P. Define a new relation $B : S \leftrightarrow S$, where xBy means that suppliers x and y carry at least one common product.

Solution   Two suppliers x and y both carry z if x carries z and y carries z. This translates into $xCz \land yCz$ or $xCz \land zC^\sim y$. According to the question, there must be at least one such z, that is, $\exists z(xCz \land zC^\sim y)$. The relation formed in this way is by definition $C \circ C^\sim$. One concludes that

$$B = (C \circ C^\sim)$$

Note that B also includes all pairs (x, x). To remove these elements from the relation, the set difference can be used. Specifically, if the pairs (x, x) are to be excluded, one must write

$$B = (C \circ C^\sim - I_S)$$ ∎

Example 5.27   Let H be the set of all humans, and let $P : H \leftrightarrow H$ be the parent relation. In words, xPy is true if x is a parent of y. Define the cousin relation. Note that two people are cousins if their parents are siblings.

Solution   To solve the problem, we proceed in steps and derive S, the sibling relation, first. Two siblings have the same parents, but are not in the identity relation. Now zPx means that z is the parent of x, and $xP^\sim z$ means that x has z as parent. Consequently, x and y have the same parents if $xP^\sim z$ and $yP^\sim z$ hold or, equivalently, if $xP^\sim z$ and zPy hold. Since x and y must not be in the identity relation, one therefore has

$$S = P^\sim \circ P - I_H$$

The desired relation can now be formulated by noting that two persons are cousins if their parents are siblings, which yields, if C is the cousin relation,

$$C = P^\sim \circ S \circ P = P^\sim \circ (P^\sim \circ P - I_H) \circ P$$ ∎

Problems 5.3

1. Let
$$P = \{(1, 2), (2, 4), (3, 3)\} \quad \text{and} \quad Q = \{(1, 3), (2, 4), (4, 2)\}$$
Find $P \cup Q, P \cap Q$, $\text{dom}(P)$, $\text{dom}(Q)$, $\text{dom}(P \cup Q)$, $\text{ran}(P)$, $\text{ran}(Q)$, and $\text{ran}(P \cap Q)$. Show that

$$\text{dom}(P \cup Q) = \text{dom}(P) \cup \text{dom}(Q)$$

and

$$\text{ran}(P \cap Q) \subseteq \text{ran}(P) \cap \text{ran}(Q)$$

2. What are the ranges of the relations

$$S = \{(x, x^2) \mid x \in \mathbb{N}\} \quad \text{and} \quad T = \{(x, 2x) \mid x \in \mathbb{N}\}$$

where $\mathbb{N} = \{0, 1, 2, \ldots\}$?

3. Let L denote the relation "less than or equal to" and D denote the relation "divides," where xDy means "x divides y." Both L and D are defined on the set $\{1, 2, 3, 6\}$. Write L and D as sets, and find $L \cap D$.

4. Some people are sitting around a table. You are given the relation xRy, which is true if x is sitting to the right of y. Express the relation B, which is true if x is either to the right or to the left of y.

5. Let $R : X \leftrightarrow X$ and $S : U \leftrightarrow V$. Find the domain spaces of $R \cap S$, $R \cup S$, and $\sim R$.

6. Let xPy be true if x is the parent of y, and let xSy be true if x is a sister of y. Express the aunt relation as the compositions of these two relations.

7. Let S stand for the relation $\subset$ as defined for sets. Find $\sim S$ and $S^\sim$.

8. Let $R = \{(1, a), (2, b), (1, c)\}$ and $S = \{(a, A), (a, B), (c, D)\}$. Find $R \circ S$.

9. Let $R = \{(1, 2), (1, 3), (3, 4)\}$. Find R^2 and R^3.

10. Let xRy be true iff $x = y - 1$. Find xR^2y and $xRy \vee xR^2y$.

11. Let $R = \{(1, 2), (2, 3), (3, 4), (4, 1)\}$. Find M_R and M_{R^2}.

5.4 PROPERTIES OF RELATIONS

5.4.1 Introduction

A number of important properties associated with relations are discussed in the subsequent sections. To illustrate these, consider the set of integers. Three important relations on this set are the relations $=$, $<$, and $\leq$. One important property of the equality relation is its *reflexivity*. By this, we mean that for every integer x one has $x = x$. This property is not shared with the less-than relation; that is, $x < x$ is false. On the other hand, $x \leq x$ holds; that is, the less-than or equal relation is also reflexive. Another property of equality is its symmetry. This means that from $x = y$ one is allowed to conclude $y = x$. This property is neither shared by the relation $<$ nor by the relation $\leq$, that is, $x < y$ does not imply $y < x$, and $x \leq y$ does not imply $y \leq x$. A third property of equality is *transitivity*. This means that from $x = y$ and $y = z$ one is allowed to conclude that $x = z$. This property also holds for $<$ and $\leq$. In other words, from $x < y$ and $y < z$ one is allowed to conclude that $x < z$, and a similar result holds for $\leq$. The following sections explore the reflexivity, symmetry, and transitivity for a number of relations.

All relations discussed in the following sections are relations on some sets, which means that the domain space and the range space are identical. In the next sections, these relations are discussed, and a new type of graph to represent them is introduced.

5.4.2 Relations on a Set

The domain and the range spaces in $R : A \leftrightarrow A$ coincide, and this makes R a relation on A. The fact that R is on A makes it possible to draw the graph of R as follows: Each element of A is represented by a node, and two nodes are connected by a directed arc if the pair (x, y) is in R. Consider, for example, the relation $\leq$ restricted to the set $A = \{1, 2, 3, 4\}$, which is represented graphically in Figure 5.10. There is a node for each element of A in the figure, and a directed arc connects node i to node j if $i \leq j$. For instance, there is an arc from node 1 to node 2, indicating that $1 \leq 2$. The arcs from 1 to 3 and 1 to 4 similarly represent the fact that $1 \leq 3$ and $1 \leq 4$. Of course, $1 \leq 1$, which means that there is a directed arc starting and ending in 1. Generally, arcs starting and ending with the same node are called *loops* or *slings*.

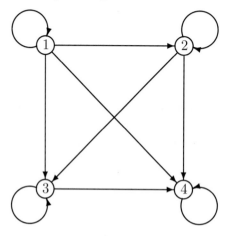

Figure 5.10.   Graph of the less-than or equal, $\leq$, relation

As another example of a relation defined on a set, consider a number of computers, say $P_1, P_2, \ldots, P_n$, that are connected by a local area network. The relation to be defined can be called the *connectivity relation*. Two computers are in the connectivity relation if there is a direct connection from computer x to computer y. Figure 5.11 gives a graphical representation of several possible configurations. Note that connections need not be symmetric; that is, if x has a direct connection to y, it does not follow that y has a direct connection to x. Consequently, if there is a connection from x to y and a connection from y to x, two arcs are needed, one for each direction. In this case, one often draws only one arc and indicates that the flow can go in both directions by means of two arrows, one for each direction.

If X is a set with n elements, then any relation R on X can be expressed by a matrix with n rows and n columns. Consequently, M_R is a square matrix.

5.4.3 Reflexive Relations

Equality is a reflexive relation; that is, $x = x$ is true for all x. The following definition extends this notion to arbitrary relations.

Definition 5.18:   A binary relation R on X is *reflexive* if, for every $x \in X$, the pair (x, x) is in the relation. Consequently,

$$R \text{ is reflexive } \equiv \forall x (x R x)$$

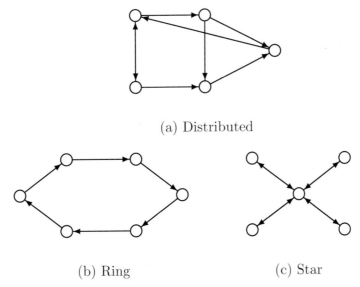

(a) Distributed

(b) Ring   (c) Star

Figure 5.11.   Configurations for a computer network

The simplest example of a reflexive relation is the identity relation I_X, whose only elements are (x, x), $x \in X$. The relation $\leq$ is reflexive on the set of real numbers since, for any x, we have $x \leq x$. Similarly, the relation of inclusion, $\subseteq$, is reflexive on the family of all subsets of a universal set. The relation $<$ is not reflexive on the set of real numbers because $x < x$ is false.

In some cases, R contains *no* element of the type (x, x). In this case, R is called irreflexive, as follows:

Definition 5.19:   A relation R on X is *irreflexive* if, for every $x \in X$, $(x, x) \notin R$. In other words, there is no $x \in X$ such that xRx.

Note that the terms reflexive and irreflexive are extreme cases. Reflexive means that xRx is true for all x, and irreflexive means that xRx is true for no x. If xRx is true for some x and false for others, then the relation is neither reflexive nor irreflexive.

The relation $<$ on the set of real numbers is irreflexive because $x < x$ is false for every x. Similarly, the relation of proper inclusion on the set of all nonempty subsets of a universal set is irreflexive. The sibling relation is also irreflexive, because nobody can be his or her own sibling. The relation $\{(1, 2), (2, 1), (1, 1)\}$ on the set $\{1, 2\}$ is neither reflexive—it does not contain the element $(2, 2)$—nor irreflexive, because it contains the element $(1, 1)$.

If a relation contains (y, y) for a certain value y, then the graph of the relation has a loop for node y. A loop is an arc that starts and ends at the same node. If all nodes have a loop, then the relation is reflexive, and if no node has a loop, then the relation is irreflexive. The relation is neither reflexive nor irreflexive if some nodes have a loop while others do not. For instance, the relation given by Figure 5.10 is reflexive, while all relations given in Figure 5.11 are irreflexive.

Reflexivity is also visible when a relation is given in matrix form. If $M_R = [m_{ij}]$ is the matrix of relation R, then the *main diagonal* of the matrix M_R is the set of all m_{ii}. If all $m_{ii} = 1$, then the relation is reflexive, and if all $m_{ii} = 0$, then the relation is irreflexive. Hence, reflexivity and irreflexivity depend only on the diagonal. Of course, if some elements of the diagonal are 1 while others are 0, then the relation is neither reflexive nor irreflexive. The following three matrices illustrate these concepts.

$$
\begin{matrix}
\text{Reflexive} & \text{Irreflexive} & \text{Neither} \\
\begin{bmatrix} 1 & 0 & 1 \\ 0 & 1 & 0 \\ 1 & 0 & 1 \end{bmatrix} &
\begin{bmatrix} 0 & 0 & 1 \\ 0 & 0 & 0 \\ 1 & 1 & 0 \end{bmatrix} &
\begin{bmatrix} 1 & 0 & 1 \\ 0 & 0 & 0 \\ 0 & 1 & 1 \end{bmatrix}
\end{matrix}
$$

5.4.4 Symmetric Relations

The relation $=$ is symmetric, while $<$ is not. Generally, one has the following:

> **Definition 5.20:** A relation R on a set X is *symmetric* if, for every x and y in X, xRy implies yRx. Consequently,
>
> $$R \text{ is symmetric} \equiv \forall x \forall y (xRy \Rightarrow yRx)$$

The sibling relation is symmetric because, if x is a sibling of y, then y is a sibling of x. In the set of human beings, the brother relation is an example of a nonsymmetric relation. If x is male while y is female, then x can be a brother of y, yet y is certainly not a brother of x. However, on the set of males, the brother relation is symmetric. The relation of similarity on the set of triangles in a plane is both reflexive and symmetric.

The relation $<$ is definitely not symmetric. In fact, $<$ is an example of an *antisymmetric* relation, as indicated by the following definition:

> **Definition 5.21:** A relation R on X is *antisymmetric* if, for every $y \neq x$, xRy excludes yRx. In other words, if xRy and yRx both hold, then $x = y$. Consequently,
>
> $$R \text{ antisymmetric} \equiv \forall x \forall y (xRy \land yRx \Rightarrow x = y)$$

An example of an antisymmetric relation is the subset relation. In fact, if A and B are two sets, and if A is a subset of B and if B is a subset of A, then $A = B$. In the case where a relation is irreflexive, Definition 5.21 means that xRy excludes yRx. For instance, the relation "mother of" is antisymmetric because "x is mother of y" precludes "y is mother of x." The "loves" relation, such as "x loves y," is neither symmetric nor antisymmetric. There is unrequited love, as indicated by many novels and poems.

In the digraph of a symmetric relation, all arcs are bidirectional; that is, if there is a directed arc from x to y, there is also an arc in the opposite direction. An example of such a digraph is given by the star network given in Figure 5.11. When representing a symmetric relation, one often draws only a single arc between any two nodes, and this arc represents both directions. The arrows are then usually omitted. If this is done, the resulting figure is no longer a directed graph, since there are no directions any more, but an undirected graph, or simply a graph.

If a relation is antisymmetric, then no arc has a mate that goes in the opposite direction. The ring network of Figure 5.11b is an example of an antisymmetric relation. Some relations are neither symmetric nor antisymmetric. An example is given by Figure 5.11a. In this case, there is one arc that is bidirectional, whereas all other arcs are unidirectional.

If $M_R = [m_{ij}]$ is the matrix of a relation, then $m_{ij} = m_{ij}$ iff the matrix is symmetric. This means that the matrix M_R must be symmetric with respect to the diagonal. A matrix is antisymmetric iff $m_{ij} = 1$ necessitates that $m_{ji} = 0$. This is slightly more difficult to spot. The following matrices illustrate the notions of symmetry and antisymmetry. Note that in antisymmetric relations both m_{ij} and m_{ji} can be zero.

$$
\begin{array}{ccc}
\text{Symmetric} & \text{Antisymmetric} & \text{Neither} \\
\begin{bmatrix} 1 & 0 & 1 \\ 0 & 1 & 0 \\ 1 & 0 & 0 \end{bmatrix} &
\begin{bmatrix} 0 & 0 & 1 \\ 0 & 1 & 0 \\ 0 & 1 & 0 \end{bmatrix} &
\begin{bmatrix} 1 & 0 & 1 \\ 0 & 0 & 0 \\ 1 & 1 & 1 \end{bmatrix}
\end{array}
$$

5.4.5 Transitivity

If x is taller than y and y is taller than z, we immediately conclude that x is taller than z. Similarly, if x is an ancestor of y and y is an ancestor of z, then x must be an ancestor of z. Finally, if $x = y$ and $y = z$, then $x = z$. All these arguments depend on *transitivity*, a property that is defined as follows:

Definition 5.22:   A relation R on X is *transitive* if, for every x, y, and z on X, whenever xRy and yRz, then xRz. That is,

$$R \text{ is transitive} \equiv \forall x \forall y \forall z (xRy \land yRz \Rightarrow xRz)$$

Examples of transitive relations include the taller than relation, the ancestor relation, and the equality relation. In addition, the $<$ relation is transitive because $x < y$ and $y < z$

imply $x < z$. Finally, the subset relation on sets is transitive because if $A \subseteq B$ and if $B \subseteq C$ then $A \subseteq C$.

In many cases, $x = y$ and $y = z$ are abbreviated by $x = y = z$. A similar notation can be used for other transitive relations. For instance, $x < y < z$ really means that $x < y$ and $y < z$. In the case of the subset relation, one writes $A \subseteq B \subseteq C$ to express $A \subseteq B$ and $B \subseteq C$.

Transitivity is related to composition. According to (2.26), Section 2.4.3, one can express the equation given in Definition 5.22 as follows:

$$\forall x \forall z (\exists y (xRy \land yRz) \Rightarrow xRz) \tag{5.16}$$

Now $xR^2z \equiv \exists x(xRy \land yRz)$, and one has the following definition for transitivity:

$$\forall x \forall z (xR^2z \Rightarrow xRz)$$

Consequently, a relation is transitive iff all pairs of objects that can be reached through an intermediary can also be reached directly. The relation R^2 must therefore be a subset of R. Hence, a relation R is transitive if and only if $R^2 \subseteq R$. This can be exploited to prove transitivity.

Example 5.28   Is the relation R with the following matrix M_R transitive?

$$M_R = \begin{bmatrix} 0 & 1 & 1 & 0 & 0 \\ 0 & 0 & 0 & 0 & 0 \\ 0 & 0 & 0 & 1 & 1 \\ 0 & 0 & 0 & 0 & 0 \\ 0 & 0 & 0 & 0 & 0 \end{bmatrix}$$

Solution R is transitive if xR^2y implies xRy. To find if this is the case, R^2 is calculated by evaluating $M_R \odot M_R$, where M_R is the matrix of R. One has

$$M_R \odot M_R = \begin{bmatrix} 0 & 1 & 1 & 0 & 0 \\ 0 & 0 & 0 & 0 & 0 \\ 0 & 0 & 0 & 1 & 1 \\ 0 & 0 & 0 & 0 & 0 \\ 0 & 0 & 0 & 0 & 0 \end{bmatrix} \odot \begin{bmatrix} 0 & 1 & 1 & 0 & 0 \\ 0 & 0 & 0 & 0 & 0 \\ 0 & 0 & 0 & 1 & 1 \\ 0 & 0 & 0 & 0 & 0 \\ 0 & 0 & 0 & 0 & 0 \end{bmatrix} = \begin{bmatrix} 0 & 0 & 0 & 1 & 1 \\ 0 & 0 & 0 & 0 & 0 \\ 0 & 0 & 0 & 0 & 0 \\ 0 & 0 & 0 & 0 & 0 \\ 0 & 0 & 0 & 0 & 0 \end{bmatrix}$$

From this matrix, one finds that the first object cannot reach the fourth object directly, but it can reach the fourth object through an intermediary. This follows from the fact that $m_{14}^R = 0$ while $m_{14}^{R \circ R} = 1$. The relation is therefore not transitive.   ∎

Transitivity is often invoked repeatedly. For instance, if $x = y_1$, $y_1 = y_2$, $y_2 = y_3, \dots, y_n = z$, one concludes that $x = z$. Similarly, if $x < y_1$, $y_1 < y_2$, $y_2 < y_3, \dots, y_n < z$, one concludes that $x < z$. Generally, if R is transitive, and if xRy_1, $y_1Ry_2, \dots, y_nRz$ all hold, one concludes xRz. What this really means is that xR^nz implies xRz or, equivalently, $R^n \subseteq R$. This can be shown by complete induction. The proof is left to the reader.

The examples used to illustrate transitivity can be divided into two groups depending on whether the relations involved are symmetric or antisymmetric. Transitive relations that are neither symmetric nor antisymmetric are relatively rare, and they will not be considered here. The equality relation is the most important representative of the symmetric transitive relations. There are several antisymmetric transitive relations, including the ancestor relation, the subset relation, and the $<$ relation. The presence of symmetry or antisymmetry can therefore be used to classify transitive relations. First, however, we consider closures, which are the topic of the next section.

5.4.6 Closures

The reflexive closure $R^{(r)}$ of a relation R is the smallest reflexive relation that contains R as a subset. The symmetric closure $R^{(s)}$ is similarly the smallest symmetric relation that contains R as a subset. R^+ is the smallest transitive relation that includes R as a subset, and R^* is the smallest reflexive transitive relation that includes R as a subset. In other words, to obtain the reflexive (symmetric, transitive) closure, one adds as few elements as possible to make the relation in question reflexive (symmetric, transitive).

To find the reflexive closure, one therefore has to know what pairs have to be added to the relation to make it reflexive. Clearly, since for reflexive closures xRx must be true for all x, one must add all pairs (x, x) that do not yet form part of R. For a relation R on A, this can be accomplished by taking the union of R with the identity relation I_A. The identity relation is, of course, the relation that contains all (x, x), $x \in A$. Hence,

$$R^{(r)} = R \cup I_A$$

Example 5.29   The reflexive closure of the $<$ relation on reals is obtained by adding the identity relation for reals to $<$. The identity relation on reals is the equality relation, and the union of $<$ and $=$ is $\leq$. It follows that the $\leq$ relation is the reflexive closure of the $<$ relation.

A symmetric relation contains (x, y) if it contains (y, x). Since the inverse relation $R^\sim$ contains (y, x) if (x, y) is in R, one finds the symmetric closure as the union of R and $R^\sim$.

$$R^{(s)} = R \cup R^\sim$$

Example 5.30   On the reals, the symmetric closure of $<$ is the $\neq$ relation. This is true because the inverse of $<$ is $>$, and the union of $<$ and $>$ is $\neq$.

The transitive relation R^+ contains all pairs (x, y) such that one can go from x to y directly or through one or more intermediaries. If R is on A and if A contains m elements, one never needs more than m steps. Consequently, to make a relation R transitive, one has to add all pairs of R^2, all pairs of R^3, ..., all pairs of R^m, unless these pairs are already in R. Consequently,

$$R^+ = R \cup R^2 \cup R^3 \cup \cdots \cup R^m$$

This equation could, in principle be used to find the transitive closure of a relation. However, there are more efficient methods to do this. These methods are given in Chapter 7. Once the transitive closure is obtained, one finds the reflexive transitive closure by adding the identity relation, a relation that is sometimes denoted by R^0. This yields

$$R^* = R^+ \cup R^0 = R^0 \cup R^1 \cup R^2 \cup \cdots \cup R^m$$

Example 5.31   The transitive closure of parent is ancestor.

Example 5.32   The reflexive transitive closure of $\{(n, n+1) \mid n \text{ integer}\}$ is the $\leq$ relation. To see this, note that $R^2 = \{(n, n+2)\}$, $R^3 = \{(n, n+3)\}$, and so on.

5.4.7 Equivalence Relations

Most transitive relations are either symmetric or antisymmetric. As it turns out, symmetric relations tend to be reflexive. In fact, if the domain and the domain space of a relation are identical, symmetry implies reflexivity. This will be proved in Chapter 11.

Definition 5.23:   A relation R is an equivalence relation iff it is reflexive, symmetric, and transitive.

The most important equivalence relation is the equality relation. This relation is reflexive because $x = x$ for all x. It is symmetric because $x = y$ implies that $y = x$. Finally, it is transitive because $x = y$ and $y = z$ imply that $x = z$. Generally, two objects x and y are in an equivalence relation if they are equal in a certain aspect. In geometry, for instance, all figures with the same area form an equivalence relation and so do all figures with the same circumference. All people with the same height are in an equivalence relation and so are all people of the same weight.

To illustrate equivalence relations, let $C : \mathbb{N} \leftrightarrow \mathbb{N}$ be the relation that associates x with y if x and y have the same remainder if divided by 3. In other words, if $x \bmod 3 = y \bmod 3$, then xCy. Hence, $1C4$ is true, because both have a remainder of 1 if divided by 3. $4C5$, on the other hand, is false, because the remainder of 4 divided by 3 is 1, and the remainder of 5 div 3 is 2. One can now form sets that contain all elements which are related. We define $[x]$ to be the set of all elements that are related to x:

$$y \in [x] \Leftrightarrow xRy \qquad (5.17)$$

For instance, the set of all elements that have the same remainder as 4 div 3, which is 1, is given by $[4]$. Of course, 1 div 3 also has the remainder 1, which means $[1] = [4]$. This yields

$$[0] = [0, 3, 6, \ldots]$$
$$[1] = [1, 4, 7, \ldots]$$
$$[2] = [2, 5, 8, \ldots]$$

Moreover, $[0] = [3] = [6] = \cdots$, $[1] = [4] = [7] = \cdots$, and $[2] = [5] = [8] = \cdots$. Note that the three sets $[0]$, $[1]$, and $[2]$ include all elements of the relation and that each natural number is in exactly one of these three sets. In fact, the sets in question form a *partition*.

Definition 5.24:   A *partition* of a set S is the division of the set into subsets A_i, $i = 1, 2, \ldots, m$, called *blocks*, such that each element of S is in exactly one of the blocks.

For instance, a partition for set $\{a, b, c, d\}$ is given by

$$\{\{a, b\}, \{c\}, \{d\}\}$$

However, $\{\{a, b\}, \{b, c\}, \{d\}\}$ is not a partition because b appears in two different blocks. The set $\{\{a, b\}, \{d\}\}$ is not a partition either because c is in no block. The sets $[0]$, $[1]$, and $[2]$ discussed earlier partition the set $\mathbb{N}$, because every natural number is in one of the three sets, and no natural number is in two distinct sets. Mathematically, the blocks A_i, $i = 1, 2, \ldots, m$, together *cover* S in the sense that their union is equal to S.

$$A_1 \cup A_2 \cup \cdots \cup A_m = S$$

On the other hand, no element is in more than one block. Consequently, different blocks do not intersect; that is,

$$(A_i \neq A_j) \Rightarrow (A_i \cap A_j = \emptyset)$$

Usually, one uses the contrapositive of this implication, which is

$$(A_i \cap A_j \neq \emptyset) \Rightarrow (A_i = A_j) \tag{5.18}$$

The main result of this section is that every equivalence relation induces a partition and every partition induces an equivalence relation. This is expressed by the following theorem:

Theorem 5.1.   Let $R : S \leftrightarrow S$ be an equivalence relation, and let $[x]$ be the set containing all elements y that are related to x.

$$[x] = \{y \mid xRy\}$$

Then $\{[x] \mid x \in S\}$ is a partition. On the other hand, if $\{A_i, i = 1, 2, \ldots, m\}$ is a partition, then one can define the relation P that contains (x, y) iff x and y are in the same block. In this case, P is an equivalence relation.

The blocks generated by an equivalence relation on S are called *equivalence classes*. Hence, every equivalence relation partitions S into equivalence classes. Two elements x and y are related iff they belong to the same equivalence class.

Proof. The fact that elements belonging to the same block in a partition establish an equivalence relation P can be verified as follows. Every element x is in the same block as x, which makes P reflexive. If x and y are in the same block, then y and x are in the same block, which makes P symmetric. Finally, P is transitive because, if x is in the same block as y and y is in the same block as z, then x is in the same block as z. This implies that P is an equivalence relation.

We now prove that every equivalence relation on S partitions S into equivalence classes. By definition, $x \in [x]$, which establishes that every element is in at least one block. To show that no element is in two different blocks, we need

$$xRy \Rightarrow [x] = [y] \tag{5.19}$$

Hence, xRy implies that the sets $[x]$ and $[y]$ are identical. To prove this, assume that xRy holds and prove that $[x] = [y]$. Because R is transitive, xRy and yRz imply that xRz holds. Since xRy is assumed, this means that $yRz \Rightarrow xRz$. Moreover, xRy implies yRx, and from yRx and xRz, one concludes yRz. Hence, $xRz \Rightarrow yRz$. From $yRz \Rightarrow xRz$ and $xRz \Rightarrow yRz$, one concludes that $xRz \Leftrightarrow yRz$. Now

$$
\begin{aligned}
xRz \Leftrightarrow yRz \quad &\equiv \quad z \in [x] \Leftrightarrow z \in [y] \qquad \text{By (5.17)} \\
&\equiv \quad [x] = [y] \qquad\qquad\quad \text{Axiom of extensionality}
\end{aligned}
$$

The assumption xRy is now discharged, which leads to (5.19). We now prove that if $u \in [x]$ and $u \in [y]$ then $[x] = [y]$. In fact, if xRu, then $[x] = [u]$, and if yRu, then $[y] = [u]$. Now $[x] = [u]$ and $[y] = [u]$ obviously imply that $[x] = [y]$. This shows that if $[x]$ and $[y]$ have any element u in common they must be identical. Any two blocks sharing a common element must therefore be identical.   ■

5.4.8 Partial Orders

Earlier, we mentioned that transitive relations are typically either symmetric or antisymmetric. Symmetric transitive relations lead to equivalence relations, as discussed in the previous section. Antisymmetric transitive relations lead to *partial orders*. In fact, there are two types of partial orders, as indicated by the following definition:

Definition 5.25:   A relation $R : S \leftrightarrow S$ is called a *weak partial order* if it is reflexive, antisymmetric, and transitive. R is called a *strict partial order* if it is irreflexive, antisymmetric, and transitive.

Strict and weak partial orders are closely related. In fact, if R is a strict partial order, its reflexive closure is a weak partial order. On the other hand, if S is a weak partial order on some set A and if I_A is the identity relation on A, then $S - I_A$ is a strict partial order.

Example 5.33   The relation $<$ is a strict partial order: it is irreflexive, antisymmetric, and transitive. The reflexive closure of $<$ is $\leq$, and this relation is a weak partial order: it is reflexive, antisymmetric, and transitive. On the other hand, one obtains the strict partial order $<$ from the weak partial order $\leq$ by removing all pairs of the form (x, x).

Example 5.34   If E is some universal set, then the relation $\subset$ on $\mathbb{P}E$ is a strict partial order. The corresponding weak partial order is, of course, the $\subseteq$ relation.

All strict partial orders are noncyclic; that is, it is impossible to find a sequence y_1, $y_2, \ldots, y_n$ such that $xRy_1, y_1Ry_2, y_2Ry_3, \ldots, y_nRx$; consequently, it is impossible to find a path that starts and ends in x. The reason is that a path of length n from x back to x would establish xR^nx, and for transitive relations, $xR^nx \Rightarrow xRx$. Hence, in a transitive relation, every path starting and ending in the same point x implies that xRx is true, and since a strict partial order is irreflexive, this is impossible. This has an important implication for function calls. If xRy means that function x can call function y and if it can be established that these function calls form a partial order, then it is impossible that the calls cause a loop.

The notion of a *partially ordered set* is very important.

Definition 5.26:   A set A together with a partial order R is called a *partially ordered set* or a *poset*. The poset (A, R) is the set A together with the partial order R.

Note that R in Definition 5.26 is a weak partial order. The strong partial order associated with (A, R) will be denoted by R_1, where $R_1 = R - I_A$. A number of important partial orders are now given.

- The set of reals together with $\geq$ form the partially ordered set $(\mathbb{R}, \geq)$.
- The set of all subsets of A together with $\supseteq$ form the poset $(\mathbb{P}A, \supseteq)$.
- If x and y are strings and R is true if y is a substring of x, and if A is the set of all substrings of some string a, then (A, R) is a poset.
- If x is an expression, if A is the set of all subexpressions of x, and if uRv is true if u has v as a subexpression, then (A, R) is a poset.

Additional posets can be derived by using the following two rules:

- If (A, R) is a poset, so is $(A, R^\smile)$. For instance, the inverse of $\geq$ is $\leq$. Consequently, $(\mathbb{R}, \leq)$ is a partial order. The poset $(A, R^\smile)$ is called the *dual poset* of (A, R).
- If (A, R) is a partial order and if $B \subseteq A$, and R' is the restriction of R to B, then (B, R') is a poset. For instance, since $(\mathbb{R}, \geq)$ is a poset, so is $(\mathbb{Z}, \geq)$ and $(\mathbb{N}, \geq)$.

Any two real numbers x and y are *comparable* in the sense that either $x \geq y$ or $y \geq x$ or both. In the poset $(\mathbb{P}A, \supseteq)$, on the other hand, not all elements are comparable. For instance, if $A = \{a, b, c\}$, then it is neither true that $\{c\} \supseteq \{a, b\}$ nor $\{a, b\} \supseteq \{c\}$. One defines the following:

Definition 5.27:   A partial order R is a *total order* or *linear order* iff, for all x and y, $xRy \lor yRx$ is always true. In this case, one calls the poset (A, R) a *totally ordered set* or a *chain*.

The natural numbers, the integers, and the real numbers, together with the $\geq$ relation, are all chains. The set of English words, together with the lexicographical order, is also a chain.

Partial orders describe some order, even when not all elements are comparable. To convey the idea of an ordering, one uses the symbol $\succ$ for strict partial orders, and the symbol $\succeq$ for weak partial orders. The inverse of $\succ$ is denoted by $\prec$, and the inverse of $\succeq$ by $\preceq$. Hence, if $(A, \succeq)$ is a poset, its dual is $(A, \preceq)$.

If xRy and yRz, then y is said to be *between* x and z. For instance, in $\{\mathbb{N}, \geq\}$, 3 is between 5 and 2, because $5 \geq 3$ and $3 \geq 2$. Similarly, in the poset $\{\mathbb{P}\{a, b, c\}, \supseteq\}$, the set $\{b\}$ is between the $\{\}$ and $\{b, c\}$. If xRz is true, but there is no y between x and z, then x *directly covers* z. For instance, in $(\mathbb{N}, \geq)$, the number 4 directly covers the number 3. There is nothing between 4 and 3. Similarly, the set $\{b, c\}$ directly covers the set $\{b\}$. There is no set between $\{b, c\}$ and $\{b\}$. The *transitive reduction* of R_1 is the relation one obtains from R_1 by only retaining the pairs (x, y), where y is directly covered by x. The transitive reduction of R_1 is equal to $R_1 - R_1^2$; that is, one finds the transitive reduction by removing from R_1 all pairs (x, y) for which there is some intermediary z such that xR_1z and zR_1y, and this is exactly R_1^2.

Example 5.35   Describe the transitive reduction on the set of natural numbers under the partial order $\geq$.

Solution   To find the transitive reduction of $(\mathbb{N}, \geq)$, one removes all pairs (x, y) for which there is a z such that $x > z$ and $z > y$. This leaves only pairs of the form $(x + 1, x)$, because in all other cases there is a number between x and y. The transitive reduction of $\geq$ is therefore given by the relation $\{(x, y) \mid x = y + 1\}$. ∎

Example 5.36   Find the transitive reduction of the descendant relation.

Solution   The transitive reduction of the descendant relation is the child relation, because only children are immediate descendants. ∎

When drawing a graph of a poset, one can omit all arcs from x to y in case x does not directly cover y, because all these arcs can be inferred. When doing this, one obtains *Hasse diagrams*.

Definition 5.28:   Let (A, R) be a poset. The *Hasse diagram* of the poset (A, R) is the graph of the transitive reduction of R_1. In this graph, there is an arc from x to y iff x directly covers y.

The Hasse diagrams of linear orders are extremely simple: they merely constitute a sequence of nodes, one below the other, and there is an arc from each node to the one immediately below. If the order is not linear, more interesting pictures result. Figure 5.12a gives a Hasse diagram of the subset relation on the powerset of $A = \{a, b, c\}$. Right below each set, one finds the sets that are directly covered, that is, the subsets that differ by a single element only. The empty set, which has no proper subset, is at the bottom of the Hasse diagram. Figure 5.12b is the Hasse diagram of a different poset. It is the poset of (A, R), where

$$A = \{2, 4, 5, 7, 10, 14, 28, 35, 70\}$$

and R is the relation "is a multiple of." Specifically, (x, y) is in the relation if x is a multiple of y. Of course, Hasse diagrams are only possible if the number of elements in the poset is finite. Hence, one cannot draw a Hasse diagram of $(\mathbb{N}, \geq)$ or $(\mathbb{R}, \geq)$.

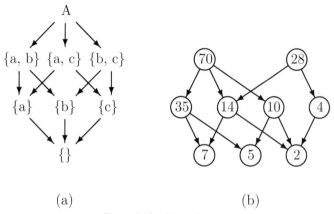

Figure 5.12.   Hasse diagrams

Strict partial orders cannot have loops. Since R_1 is a strict partial order and since the Hasse diagram is based on R_1, it cannot have a loop either. This means that there is a direction in the Hasse diagram: there are some elements that are higher in the diagram, whereas others are lower down. Hasse diagrams represent, in this fashion, a clearly defined order. If the partial order is not a chain, then the order illustrated by the Hasse diagram is partial in the sense that not all elements of the set can be compared. In Hasse diagrams, it makes sense to speak about *descending sequences*.

Definition 5.29:   Let (A, R) be a poset, and let $R_1 = R - I_A$ be the strict partial order of the poset. A *descending sequence* in this poset is any sequence $\langle x_1, x_2, \ldots, x_n \rangle$ satisfying $x_i R_1 x_{i+1}$ for $i < n$.

In the poset $(\mathbb{Z}, \geq)$, the sequence $\langle 2, 1, -4, -16 \rangle$ constitutes a descending sequence. In the case of $(\mathbb{P}\{a, b, c\}, \supseteq)$, which is represented in Figure 5.12a, an example of a descending sequence is $\langle\{a, b, c\}, \{b\}, \{\}\rangle$. Note that the elements of a descending sequence need not necessarily cover the subsequent element in the sequence directly; that is, one is allowed to skip elements in between. If the poset has a finite number of elements only, all descending sequences must be finite. All elements in a descending sequence must be in R_1, and since R_1 is acyclic, no element can be repeated. However, if the number of elements in the poset is infinite, then infinite descending sequences are possible. For instance, in the poset $(\mathbb{Z}, \geq)$, the sequence given by the numbers $x_0 = 0$, $x_{i+1} = x_i - 2$ is an infinite descending sequence. Similarly, the sequence $x_i = 1/i, i \geq 0$ is an infinite descending sequence on the set $(\mathbb{R}, \geq)$.

Definition 5.30: A poset (A, R) is said to be *well founded* if (R, A) has no infinite descending sequences.

The notions of descending sequences and well-foundedness were already introduced in Section 3.3 in connection with proofs by recursion. We now show that the definition given there is closely related to the definition here. In Section 3.3, we discussed sequences $y_1, y_2, y_3, \ldots$ such that, for each pair (y_i, y_{i+1}) in the sequence, $y_i G y_{i+1}$ holds. These sequences were called G-sequences. A set was said to be well founded with respect to G if there were no infinite G-sequences. To make this true, xGy must be irreflexive because, otherwise, the sequence $x, x, x, \ldots$ would be an infinite G-sequence. G must also be antisymmetric, because if xGy for any value of $x \neq y$ then $x, y, x, y, \ldots$ would be an infinite G-sequence. If A is the domain and if G^+ is the transitive closure of G, then (A, G^+) is a partial order, as one can verify. All descending sequences of this partial order are also descending G-sequences. Hence, if (A, G^+) is well founded, then all descending G-sequences are finite.

There are some other terms that are sometimes used in the context of posets.

Definition 5.31: Let (A, R) be a poset. Any $x \in A$ satisfying xRy is said to be an *upper bound* of y. Similarly, any $y \in A$ satisfying xRy is said to be a *lower bound* of x.

For instance, in the poset $(\mathbb{N}, \geq)$, the number 6 is an upper bound for the number 3, and so is any other number $x \geq 3$. Even 3 is an upper bound of 3. In general, R is reflexive and yRy is true, which makes y an upper bound of y. In a Hasse diagram, all upper bounds of y except y itself are above y.

Maximal and *minimal* elements can now be defined as follows:

Definition 5.32:   Let (A, R) be a poset. Then, y is a *maximal* element if there is no x such that $x R_1 y$. Moreover, x is a *minimal* element if there is no y such that $x R_1 y$.

The maximal and minimal elements in a poset can easily be identified from the Hasse diagram. In a Hasse diagram, every element that has no incoming arc is a maximal element, whereas every element with no outgoing arc is a minimal element. For instance, the nodes 28 and 70 in Figure 5.12b are both maximal elements. The minimal elements are similarly the nodes representing 7, 5, and 2. The Hasse diagram of the power set $\{a, b, c\}$ given in Figure 5.12a has only one maximal element, $\{a, b, c\}$, and only one minimal element, $\{\ \}$. There are posets that have no maximal element, but such posets must be infinite. For instance, in the poset $(\mathbb{N}, \geq)$, no n can be a maximal element because $n+1$ is always greater than n. However, $(\mathbb{N}, \geq)$ has a minimal element, 0. The poset $(\mathbb{Z}, \geq)$, on the other hand, has neither a maximal nor a minimal element.

A descending sequence of a poset can also be considered as a poset. Hence, such a sequence may have a maximal and a minimal element. If a descending sequence has a minimal element, then this element terminates the sequence, making the sequence finite. Conversely, every finite descending sequence must have a minimal element. Consequently, a poset is well founded if and only if all its descending sequences have a minimal element.

It frequently happens that one forms the Cartesian product of two sets for which partial orders are defined. For instance, one can form $\mathbb{R}^2$, the set of all points in the Cartesian coordinate system, together with the relation $\geq$. In this case, one can create a new partial order on the product set as follows: One sets

$$(x_1, x_2) \succ (y_1, y_2) \qquad \text{if } x_1 \geq y_1 \quad \text{and} \quad x_2 \geq y_2$$

In a similar fashion, one can define partial orders on any Cartesian products of sets for which partial orders are defined. Note that, even if one has total orders on all factors of the Cartesian product, the induced partial order is no longer total. For instance, the order on points in the Cartesian product induced by $\geq$ is no longer total, even though $(\mathbb{R}, \geq)$ is a totally-ordered set. Specifically, there is no relation between (x_1, x_2) and (y_1, y_2) if $x_1 > y_2$ and $x_2 < y_2$. Concretely, one clearly has $(4, 6) \succ (2, 5)$, but no ranking is specified between $(4, 6)$ and $(5, 5)$. To get around this difficulty, one often uses a *lexicographical order*. If $\succ_l$ is a lexicographical order, then the first element of the pair provides the primary order criterion, in the sense that the second element of the pair is used only in case of a tie. Hence, if $\succ_l$ is the lexicographical ordering of $\mathbb{R} \times \mathbb{R}$, then $(5, 4) \succ_l (4, 99)$, and $(5, 5) \succ_l (5, 4)$. The lexicographical ordering of n-tuples, $n > 2$, can be established in a similar manner.

Example 5.37   Let $A_1 = \{1, 2, 3\}$, $A_2 = \{a, b\}$, and $A_3 = \{0, 1\}$. Find the lexicographical order on $A_1 \times A_2 \times A_3$. Use the natural order for numbers and letters; that is, $3 > 2 > 1 > 0$ and $b > a$.

Solution   The following is the lexicographical order of the elements of the Cartesian product $A_1 \times A_2 \times A_3$.

$$(1, a, 0), (1, a, 1), (1, b, 0), (1, b, 1),$$
$$(2, a, 0), (2, a, 1), (2, b, 0), (2, b, 1),$$
$$(3, a, 0), (3, a, 1), (3, b, 0), (3, b, 1)$$ ■

Problems 5.4

1. List all the properties of the following relations. Which of the relations are equivalence relations, which are weak partial orders, and which are strict partial orders?
    (a) $R_1 = \{(1, 2), (2, 2)\}$ (c) $R_3 = \{(1, 1), (2, 2), (2, 3), (3, 2), (3, 3)\}$
    (b) $R_2 = \{(1, 2), (2, 3), (1, 3), (2, 1)\}$ (d) $R_4 = \{(1, 2), (2, 3)\}$

2. List all properties of the empty relation and the universal relation. Are these relations equivalence relations? Are they weak partial orders or strict partial orders? Note that these categories may overlap.

3. List all properties of the identity relation. Is the identity relation an equivalence relation, a weak partial order, or a strict partial order?

4. Give an example of a relation that is both symmetric and antisymmetric.

5. If relations R and S are both reflexive, show that $R \cup S$ and $R \cap S$ are also reflexive.

6. If relations R and S defined on the same set are reflexive, symmetric, and transitive, show that $R \cap S$ is also reflexive, symmetric, and transitive.

7. Let $R = \{(1, 2), (4, 3), (2, 2), (2, 1), (3, 1)\}$ be a relation on $S = \{1, 2, 3, 4\}$. Find the symmetric closure, the reflexive closure, and the transitive closure of R.

8. Given $S = \{1, 2, \ldots, 10\}$ and the relation $R = \{(x, y) \mid x + y = 10\}$ on S, what are the properties of R?

9. What nonzero entries are there in the relation matrix of $R \cap R^{\sim}$ if R is an antisymmetric relation?

10. Given the relation matrix M_R of a relation R on the set $\{a, b, c\}$, find the relation matrices of $R^{\sim}, R^2 = R \circ R, R^3 = R \circ R \circ R$, and $R \circ R^{\sim}$.

$$M_R = \begin{bmatrix} 1 & 0 & 1 \\ 1 & 1 & 0 \\ 1 & 1 & 1 \end{bmatrix}$$

11. Two equivalence relations R and S are given by their relation matrices M_R and M_S. Show that $R \circ S$ is not an equivalence relation.

$$M_R = \begin{bmatrix} 1 & 1 & 0 \\ 1 & 1 & 0 \\ 0 & 0 & 1 \end{bmatrix}, \quad M_S = \begin{bmatrix} 1 & 0 & 0 \\ 0 & 1 & 1 \\ 0 & 1 & 1 \end{bmatrix}$$

12. Prove that $S^{\sim}$ is an equivalence relation iff S is an equivalence relation.

13. Prove that the composition of two equivalence relations R and S is an equivalence relation if R and S have the same equivalence classes.

14. For each of the following relations, indicate its properties. State whether the relation is reflexive, irreflexive, symmetric, antisymmetric, or transitive. If the relation is an equivalence relation, a partial order, or a strict partial order, state this also. All relations are on the set of humans.
   (a) xRy stands for x is a child of y.
   (b) xRy stands for x is a descendant of y.
   (c) xRy stands for x is a spouse of y.
   (d) xRy stands for x is the wife of y.
   (e) xRy stands for x is the immediate superior of y.
   (f) xRy stands for x is a superior (not necessarily the immediate superior) of y.
   (g) xRy stands for x and y have the same parents.
   (h) xRy stands for x is of the same size or smaller than y.

15. A matrix M is called upper triangular if its elements $m_{ij} = 0$ for $i > j$. What can you say about relation R if you find that its relation matrix is upper triangular?

16. Is the set $\mathbb{N}$ well founded under the relation $\leq$? Is it well founded under the relation $\geq$?

17. Let $\mathbb{N}$ together with $\leq$ be a poset. Is this set well founded? Is the set $\mathbb{N}$ with the relation $\geq$ well founded?

18. For each of the following relations, indicate whether or not the relation is reflexive (r), irreflexive (i), symmetric (s), or transitive (t). For instance, for the relation $<$ on the set of integers, one has i, t because $<$ is both irreflexive and transitive.
   (a) Let x and y be integers, and let xRy be true if x divides y without a remainder.
   (b) Let x and y be human beings, and let xRy be true if x and y belong to the same family.
   (c) Let x and y be boys, and let xRy be true if x is a brother of y or if $x = y$.
   (d) Let x and y be human beings, and let xRy be true if x is related to y. When answering this question, assume that everyone is related to himself or herself.

19. Are any of the relations described in Problem 18 equivalence relations? If so, give reasons.

20. Use complete induction to prove $xR^ny \Rightarrow xRy$ provided R is transitive.

21. Let $R = \{(a, b), (b, d), (c, b), (d, a)\}$. Find the relation matrix. Moreover, find (i) the matrix of the reflexive closure, (ii) the matrix of the symmetric closure, and (iii) the matrix of the transitive closure.

PROBLEMS: CHAPTER 5

1. Given that $A = \{\, x \,|\, x$ is an integer, $1 \leq x \leq 5\}$, $B = \{3, 4, 5, 17\}$, and $C = \{1, 2, 3, \ldots\}$, find $A \cap B$, $A \cap C$, $A \cup B$, and $A \cup C$.

2. Simplify
$$((A \cup B) \cap \sim(C \cup A)) \cup ((C \cap B) \cup A)$$

3. Let A, B, and C be three sets, and let E be the universal set. Simplify
$$\sim(\sim(A \cup \sim C) \cap B) \cup \sim(A \cap \sim(C \cap \sim B)) \cup C$$

4. Let $A = \{x \,|\, (x \geq 0) \wedge (x^2 \leq 10)\}$, and let the universal set be the set of natural numbers. Give A explicitly, and find $\#A$, the cardinality of A.

5. Let A, B, and C be three sets. Simplify
$$\sim((\sim A \cup \sim C) \cap B) \cup \sim(A \cup (C \cup \sim B) \cap C)$$

6. Prove the following:
   - **(a)** $A \cap B \subseteq A \cup B$
   - **(b)** $A - B \subseteq A$
   - **(c)** If $A \subseteq C$ and if $B \subseteq C$, then $A \cup B \subseteq C$.

7. Let $A = \{a, \{a\}\}$. Give the power set of A.

8. Given the two relations $A = \{(a, b), (b, a), (b, c), (c, b), (c, d)\}$ and $B = \{(a, 1), (a, 3), (b, 2), (c, 2)\}$. Find
   - **(a)** $A \circ A^\sim$
   - **(b)** $(A \circ A) - A \circ A^\sim$
   - **(c)** $(A \circ A) \cup A$
   - **(d)** $A \circ B$

9. Use quantifiers and logical connectives to eliminate $\circ$, $\cup$, $\cap$, and $-$ from the following expressions. Here P and R are arbitrary relations, and I is the identity relation.
   - **(a)** $(P \circ P) \cup (S - (S \circ S))$
   - **(b)** $P \cup ((P \cap R) \circ R)$
   - **(c)** $P^3 - P^2$
   - **(d)** $(P \cup R) \circ (P \cap R)$

10. A relation R is a function if for every x there is only one y such that xRy. Can R be reflexive? Can R be symmetric? Can R be transitive? Give examples.

11. Let xPy be the relation that x is the parent of y, let $P^\sim$ be the inverse of this relation, and let I be the identity relation.
   - **(a)** The relations $P^\sim$, $P^\sim \circ P^\sim$, and $P \circ P$ all have names in a family context. Give these names.
   - **(b)** If you get older, you may get a $(P^\sim \circ P - I) \circ P$. What is that?
   - **(c)** Let xRy be the relation that x and y have at least one ancestor in common. Define R in terms of P, $P^\sim$, and their transitive closures.

12. For each of the following relations, state if they are equivalence relations, strict partial orders, or weak partial orders. Some relations may be classified under several categories, others under no category. In all cases, the relations are on the set $A = \{1, 2, 3, 4\}$.
   - **(a)** $\{(1, 1), (2, 2), (3, 3), (4, 4)\}$
   - **(b)** $\{(1, 1)\}$
   - **(c)** The Cartesian product $A \times A$
   - **(d)** $\{(1, 1), (1, 2), (2, 1), (4, 4), (2, 2)\}$
   - **(e)** $\{(1, 4), (1, 2), (4, 3)\}$

13. Find the matrices of all equivalence relations with three equivalence classes that can possibly be defined on the set $1..4$.

14. Prove the following statement: If R is reflexive and if S is the universal relation, then $R \circ S = S$.

15. Draw a graph of the relation A given in Problem 8, and give the relation matrix.

16. Define the sibling relation on the set of all persons. Use S for the sister relation and B for the brother relation. Only symbols involving sets are allowed.

17. Show that if a relation R is reflexive then $R^\sim$ is also reflexive. Show also that similar remarks hold if R is transitive, irreflexive, symmetric, or antisymmetric.

18. Find the smallest nonempty set that has a relation defined on the set that is neither reflexive nor irreflexive.

19. Find the smallest nonempty set that has a relation defined on the set that is neither symmetric nor antisymmetric.

20. Prove that $A \circ A^\sim$ is reflexive if and only if $\text{dom } A = A$.

— 6 —

Functions

Functions were introduced in Section 2.5 for doing equational logic, and their importance in this regard was mentioned. In this chapter, functions and their representations will be discussed in further detail, and additional application areas for functions will be introduced. There is a close connection between functions and relations. In fact, functions can be interpreted as relations, which makes it possible to apply many results derived for relations to functions. Functions can also be used for counting and for establishing the cardinality of sets. Functions are related to isomorphisms, which are, roughly speaking, correspondences between functions. The growth of functions is very important in many applications. In particular, it has ramifications on computational complexity, an issue that will be addressed in some detail. As was pointed out in Section 3.5, it is possible to express everything that is computable in terms of functions. This result is exploited by *functional programming languages*. In this chapter, one of these languages, Miranda, will be discussed in some detail.

6.1 REPRESENTATIONS AND MANIPULATIONS INVOLVING FUNCTIONS

6.1.1 Introduction

This section gives some basic properties of functions, and it introduces some new notations. The approach is to interpret functions as relations and to apply the results derived for relations to functions. For instance, we will talk about compositions of functions and inverses of functions. However, these new constructs are no longer necessarily functions. In fact, an inverse of a function is deemed not to exist unless this inverse is a function. Like all other relations, one can represent functions by graphs, which helps to visualize functions and the operations involving functions.

There are several methods to deal with functions having more than one argument. One method is to combine all arguments into a single argument. In this way, one reduces a function with several variables into a function with one variable, except that the variable is now an element of a Cartesian product. Alternatively, one can make use of λ-calculus, which maps variables into functions. Here both approaches will be treated.

6.1.2 Definitions and Notation

Functions are relations. In particular, a function with one variable is a binary relation because it associates with each argument a unique result, called the *image of the function*. A function with n variables can similarly be interpreted as an $(n + 1)$-ary relation that associates a unique image with all possible argument values. The following definition expresses this in a more precise form.

Definition 6.1:   A binary relation $f \subseteq X \times Y$ is called a *function* if for each $x \in X$ there is a unique $y \in Y$ such that (x, y) is an element of f. An $(n + 1)$-ary relation $g \subseteq (X_1 \times X_2 \times \cdots \times X_n \times Y)$ is called a function if for each $x_1 \in X_1, x_2 \in X_2, \ldots, x_n \in X_n$, there is a unique $y \in Y$.

This definition captures the uniqueness of the image, which is essential for doing equational logic as discussed in Section 2.5.

There are different notations to represent functions. Of course, one can write functions as relations, as indicated in the following example:

Example 6.1   Which of the following relations from $A = \{a, b, c\}$ to $B = \{1, 2, 3, 4\}$ are functions?

> **1.** $f_1 = \{(a, 3), (b, 1), (c, 2)\}$
> **2.** $f_2 = \{(a, 3), (b, 3), (c, 2)\}$
> **3.** $f_3 = \{(a, 3), (a, 4), (b, 1), (c, 2)\}$
> **4.** $f_4 = \{(a, 3), (b, 1)\}$

Solution   Relations f_1 and f_2 are both functions because each x is related to exactly one y. In f_2, the image 3 is associated with both a and b, but this does not violate the requirements for being a function. On the other hand, f_3 is not a function because a relates to both 3 and 4, and this contravenes the requirement that each x must have a unique y. The relation f_4 is not a function either because there is no element that is related to c.   ∎

The most common notation to indicate that the pair (x, y) belongs to some function f is $y = f(x)$. Hence, $(x, y) \in f$ and $y = f(x)$ are synonymous. Instead of $f(x)$, one occasionally uses fx. This notation is common when dealing with trigonometric functions, as indicated by expressions such as $\sin x$. An alternative way to express a function is $f : x \to y$. The notations $f(x)$ and $f : x \to y$ can be extended to cover functions with several variables. For instance, one writes $y = f(x_1, x_2, \ldots, x_n)$ to indicate that with each n-tuple $(x_1, x_2, \ldots, x_n)$ there is a unique y. This can also be expressed as

$$f : (x_1, x_2, \ldots, x_n) \to y$$

This notation suggests a different way to look at functions with several variables. In a function with several variables, one associates with each n-tuple $(x_1, x_2, \ldots, x_n)$ a unique y. The n-tuple $(x_1, x_2, \ldots, x_n)$ can be given a name, say x, and treated as an individual. This allows one to consider the function $y = f(x_1, x_2, \ldots, x_n)$ as a function with one argument x, except that x is the n-tuple $(x_1, x_2, \ldots, x_n)$. All functions with several arguments can be converted in this way into functions with one argument. The n-tuple representing the arguments is called the *preimage*. Every function associates in this sense a unique image with each preimage.

If a relation $f : X \leftrightarrow Y$ is a function, one writes $f : X \to Y$. As in relations, X is the domain space and Y is the range space. To be a function, there must be a $y \in Y$ for each $x \in X$ such that (x, y) is in f. Consequently, the domain and the domain space are identical, and one frequently does not distinguish between the two. If $f : X \to Y$, then the domain $\text{dom } f$ is equal to X. The *range* of a function is the set of all points $y \in Y$ for which there is a preimage satisfying $y = f(x)$. If $\text{ran } f$ denotes the range of f, one therefore has

$$\text{ran } f = \{y \mid \exists x (y = f(x))\}$$

The range must be a subset of the range space, as the reader may verify. A function $f : X \to Y$ is said to *map* X to Y or to *transform* X to Y. Many authors also use map in a different sense. They say that f maps the preimage x into the image y. This also explains why functions are called *transformations* or *mappings*.

Example 6.2   Give the domains and ranges for f_1 and f_2 in Example 6.1.

> **Solution**   The domains of f_1 and f_2 are equal to A, the domain space. The range of f_1 is equal to $\{1, 2, 3\}$, which is a proper subset of the range space B. The range of f_2 is $\{2, 3\}$. ∎

If a function has several arguments, then the domain space is given as a Cartesian product. For instance, the arithmetic operators $+$ and $\times$ really represent functions. In particular, $x + y$ associates with each pair (x, y), $x, y \in \mathbb{R}$, a unique image $(x + y) \in \mathbb{R}$. This makes $+$ a function from $\mathbb{R} \times \mathbb{R}$ to $\mathbb{R}$:

$$+ : \mathbb{R} \times \mathbb{R} \to \mathbb{R}$$

The multiplication operator can be defined in a similar fashion. It, too, maps $\mathbb{R} \times \mathbb{R}$ into $\mathbb{R}$. We also have discussed several operations on sets, and these operations are also functions, except that the operands are subsets of some universal set E. For instance, when taking the union of sets A and B, one really obtains a function in which the arguments are sets and in which the result is also a set. Given a universal set E, the union is therefore a function that maps $\mathbb{P}E \times \mathbb{P}E$ into $\mathbb{P}E$.

$$\cup : (\mathbb{P}E \times \mathbb{P}E) \to \mathbb{P}E$$

The intersection can be characterized in a similar fashion.

There are many functions of the type $A^2 \rightarrow A$ or, more generally, $A^n \rightarrow A$. For instance, arithmetic operators $+$, $-$, and $\times$ are of type $\mathbb{R}^2 \rightarrow \mathbb{R}$, and operators $\cup, \cap$, and $-$ on sets are of type $(\mathbb{P}E)^2 \rightarrow \mathbb{P}E$. Generally, in the function $A^n \rightarrow A$, the domain is A^n. However, frequently, one considers A as the domain of a function $f : A^n \rightarrow A$. In this sense, one can say that $+$ is an operator in the domain $\mathbb{R}$, and the $\cup$ is an operator in the domain $\mathbb{P}E$.

The domain space is often distinct from the range space. The cardinality of a set, for instance, has the $\mathbb{P}E$ as its domain space, yet the range space is $\mathbb{N}$. Hence, the symbol $\#$, as in $\#A$, stands for a function that maps sets into natural numbers. Hence, $\# : \mathbb{P}E \rightarrow \mathbb{N}$. The length of a string is also a function. If X is a string over the alphabet A and if length is the function that associates each string with the number of characters that it contains, then length is a function from A^* to $\mathbb{N}$. Every predicate, too, can be thought of as a function. Specifically, if E is the universe of discourse and if G is a predicate with n arguments, then G maps every n-tuple of individuals into the set $\{T, F\}$.

$$G : E^n \rightarrow \{T, F\}$$

Hence, a predicate is a function, which has E^n as its domain and $\{T, F\}$ as its range. For predicates, the universe of discourse is therefore equal to the domain.

The real numbers, natural numbers, sets, and strings all describe certain types. Obviously, algebraic systems are easier to handle if they have only a few types and if all functions have one of these types as their domain and range spaces. Using the same type for domain and range spaces is convenient, but it cannot be implemented in all cases. For instance, in the set of reals, one cannot define the reciprocal as a total function because the number 0 has no reciprocal. To deal with cases like this, one can either change the domain space, which means introducing additional sets, or one can relax the condition that there must be an image for every element from the domain. This leads to the notion of a *partial function*, a term already introduced in Section 2.5.4. Here we define the following:

> **Definition 6.2:**   A relation f is a partial function from X to Y if for every $x \in X$ there is at most one $y \in Y$. In this case, one writes $f : X \nrightarrow Y$.

In other words, some elements of the domain space X may have no corresponding $y = f(x)$. In this case, the domain can be a proper subset of the domain space.

Two very simple functions are the *identity function* and the *constant function*. The identity function is really the same as the identity relation; that is, if $I_A : A \rightarrow A$ denotes the identity function, one has, for all x,

$$I_A = \{(x, x) \mid x \in A\}$$

The identity function therefore maps each value into itself; that is, $x = I_A(x)$. The constant function has only a single image, say c, and all arguments relate to the same image. If C denotes the constant function, one has therefore

$$C = \{(x,c) \mid x \in A\}$$

Hence, the image of each x is c; that is, $c = C(x)$ for all x.

6.1.3 Representations of Functions

There are different methods to represent functions. One can, of course, use the roster notation to represent a function, but it is usually more convenient to expand the roster notation into a table. For instance, Table 6.1 gives the capitals of the nations on the North American continent. This function is from the set of North American countries into the North American cities. This function has a country as its argument and a city as its image. Tables can also be used for representing functions with two variables. In fact, in Section 2.5.6, the functions corresponding to operators were given by tables, the *operator tables*. Similar tables are certainly familiar to every reader. As an example, Table 6.2 gives a remaining lifetime for every age group and sex for people in the United States. Hence, if $S = \{male, \ female\}$ is the sex, and if $A = \{0, \ 10, \ 20, \ 40, \ 60\}$ is the set of age groups, then the remaining lifetime given in Table 6.2 maps $S \times A$ into R, where R is the remaining lifetime.

TABLE 6.1.   North American Capitals

Country	Capital
Canada	Ottawa
United States	Washington
Mexico	Mexico City

TABLE 6.2.   Life Expectancy in the United States

	Age Group				
Sex	0	10	20	40	60
Male	71.5	62.5	53.0	34.8	18.2
Female	78.3	69.2	59.4	40.2	22.5

Functions on reals can be represented by curves. In these curves, there must be only one value $f(x)$ for each x. Otherwise, the curve does not represent a function. For instance, Figure 6.1 is the graphical representation of the function with the images $f(x) = x^2 + 2$. To represent functions with two variables graphically, one needs three dimensions, which is more difficult to accomplish. One can use *contour maps* in this case, but these will not be discussed here.

Like relations, functions have graphs, and these graphs are helpful in understanding the properties of functions. Furthermore, graphs aid in visualizing the operations that can

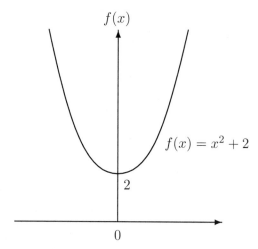

Figure 6.1.   The function $f(x) = x^2 + 2$

be performed on functions. The graphs of functions are constructed exactly like the graphs of relations.

6.1.4 The Lambda Notation

In mathematics, $f(x)$ can mean two things. So far in this text, it was used only to describe the image corresponding to the argument x, which is a single value. However, in the phrase "$f(x)$ is a function of x," one obviously does not refer to a single image, but to the entire function or to a set of pairs. Hence, $f(x)$ is also used instead of f, except that $f(x)$ contains additional information. If $f(x)$ is used to denote a function, as opposed to an image, then it is obviously a function with one argument, x. For certain applications, this ambiguity is unacceptable and an alternative notation is needed. This leads to the introduction of λ-calculus, or *lambda calculus*. In λ-calculus, one uses $f(x)$ strictly for a single value, the image of x. To denote the function f, one writes $\lambda x.f(x)$.

Generally, if e is an expression, then $\lambda x.e$ indicates that e is now to be considered as a function of x. For instance, $\lambda x.(x + y)$ indicates that $x + y$ is to be considered as a function of x and that y is only a parameter. The expression $\lambda x.(x + y)$ is an example of a λ-expression. More generally, if e is any expression, one can form the λ-expression $\lambda x.e$. To express that the function represented by the λ-expression $\lambda x.e$ is to be taken at a certain value t, one writes $(\lambda x.e)\, t$. For instance, $(\lambda x.(x+3))\, 1$ is the image of the function $\lambda x.(x + 3)$ for $x = 1$. In other words, $(\lambda x.(x + 3))\, 1$ is $1 + 3$, or 4. Similarly, $(\lambda x.x^2)\, 3$ evaluates to 9. Generally, the conversion of a function into an image is called a λ-*reduction*. Hence, 3^2 is a λ-reduction of $(\lambda x.x^2)\, 3$.

Example 6.3   Find $(\lambda x.(x + y))(2y)$, $(\lambda x.(x + y))2$, and $(\lambda x.(\lambda y.(x + y))3)2$.

Solution

$$\begin{aligned}
(\lambda x.(x + y))\, (2y) &= 2y + y = 3y \\
(\lambda x.(x + y))\, 2 &= 2 + y \\
(\lambda x.(\lambda y.(x + y))\, 3)\, 2 &= (\lambda y.(2 + y))\, 3 = 2 + 3 = 5
\end{aligned}$$

∎

In Figure 6.1, the function $\lambda x.(x^2 + 2)$ was given in the form of a curve. One can draw on the same figure a family of functions of the form $\lambda x.(x^2 + y)$. All these functions have the same general shape, but when compared to $\lambda x.(x^2 + 2)$, they are shifted by $y - 2$. The function $\lambda x.(x^2 + 3)$, for instance, is the curve $\lambda x.(x^2 + 2)$ given in Figure 6.1 shifted upward by an amount of 1. In λ-calculus, one can express the identity function as $\lambda x.x$. The function that yields the value c for all x is similarly $\lambda x.c$.

Technically, functions expressed in λ-calculus can only have one argument. To convert a function with several variables into λ-calculus, one creates functions in which the range consists of functions. For example, consider the expression $x^2 + y$. For each given y, one can use this expression to find a function for x, that is, $\lambda x.(x^2 + y)$. One can therefore consider the new function $f = \{(y, \lambda x.(x^2 + y))\}$. The function f has the single argument y, and the image for a given y is $\lambda x.(x^2 + y)$. To indicate this, one writes $\lambda y.(\lambda x.(x^2 + y))$. If $x : \mathbb{R}$, $y : \mathbb{R}$, then $\lambda x.(x^2 + y)$ is of type $\mathbb{R} \to \mathbb{R}$, and $\lambda y.(\lambda x.(x^2 + y))$ is therefore of the type $\mathbb{R} \to (\mathbb{R} \to \mathbb{R})$.

To reinforce the notion of a function that maps into a function, let us return to Table 6.2. In this table, one can associate a remaining lifetime with each age–sex combination. Alternatively, one can consider the table as consisting of two functions, one for each sex. The first row gives the function for males and the second for females. Since there is a unique life-expectancy function for each sex, this constitutes a function in which the argument is the sex and the image is a function.

Earlier, functions with two arguments were given a type of the form

$$f : (X_1 \times X_2) \to Y$$

In λ-calculus, a function with two variables is modeled as a function from X_1 into the range $X_2 \to Y$; that is, such functions have the type

$$f : X_1 \to (X_2 \to Y)$$

Parentheses are frequently omitted; that is, $X_1 \to (X_2 \to Y)$ is often written as $X_1 \to X_2 \to Y$. There are thus two different methods to express functions with two arguments $x_1 \in X_1, x_2 \in X_2$. On one hand, one can consider the function as a mapping from $X_1 \times X_2$ into Y, and, on the other, one can consider the same function as a mapping from X_1 into $X_2 \to Y$. These two representations are essentially equivalent. Both methods can easily be extended to hold for functions with n arguments. As was shown earlier, a function with the n variables $x_1 \in X_1, x_2 \in X_2, \ldots, x_n \in X_n$ has the type

$$f : (X_1 \times X_2 \times \cdots \times X_n) \to Y$$

In λ-calculus, one maps functions into functions, which means that the function f must be declared as

$$f : X_1 \to X_2 \to \cdots \to X_n \to Y$$

6.1.5 Restrictions and Overloading

Some languages use overloading as a programming technique. In overloading, several different functions are given the same name and which function is chosen depends on the type of the arguments. Overloading can be modeled as the union of two functions.

As was mentioned in Section 5.3.4, one can restrict a relation to any domain space U or range space V by intersecting it with the universal relation $U \times V$. This type of restriction is also applicable when dealing with functions. For instance, if $f(x) = x^2$ is a function from the reals into the reals, then one can restrict this function to, say, the integers by forming $f \cap (\mathbb{N} \times \mathbb{N})$. In a similar fashion, any partial function can be restricted in such a way that its domain is equal to its domain space, which makes the partial function a total function. For instance, the partial function $f : \mathbb{R} \twoheadrightarrow \mathbb{R}$ given by $f(x) = 1/x$ can be restricted to $f_1 : (\mathbb{R} - \{0\}) \rightarrow \mathbb{R}$ by forming the intersection $f \cap ((\mathbb{R} - \{0\}) \times \mathbb{R})$; however, this restriction makes the domain space irregular, and this may cause difficulties if f_1 is to be used in the context of other calculations. One can restrict the range space to the range, but, again, the range space may become irregular as a result.

Function extensions are much more frequent than function restrictions. For instance, one can introduce addition for the natural numbers and, from there, extend the operation to the integers, to the rational numbers, and finally to the real numbers. Actually, on computers the operations on integers are handled by different routines than the operations involving floating-point numbers, and the two functions are in this sense different. This means that the addition symbol $+$ is *overloaded* in the sense that two different operations have the same name. The attractiveness of overloading is easy to see. If one has objects of different types, one typically needs different routines to handle them. However, some of the operations are very similar, even though the types involved may be quite different. This is the case in integer and floating-point addition. However, similar operations arise in other contexts. In graphics, in particular, there are often many types of different figures, such as triangles, rectangles, and circles. For all these figures, one has some basic operations, such as translating, rotating, flipping over, and displaying. It is reasonable to use the same name for these operations, regardless of the figure to which the operation is applied. Consider, for instance, the operations rotate(fig,angle), which rotates a figure called fig by an angle angle. For each figure, a separate procedure has to be written to accomplish this operation, yet it is obviously advantageous to call all these procedures by the same name.

Overloading is not restricted to computer languages. All natural languages use over-loading extensively. In fact, many abstract terms have their origin in the physical world: They took on new meanings, and they were in this sense overloaded. Examples of this pro-cess are the words "bright" and "colorful" as in "a bright person" and "a colorful person." Mathematicians also overload operators. For instance, instead of writing $A \cup B$, some mathematicians write $A + B$, which overloads the operator $+$. Overloaded operators were used in this text. The set difference between the sets A and B is written as $A - B$, even though nothing is subtracted, at least not in an arithmetic sense.

When functions or partial functions are overloaded, what is formed is the union of the two functions, one of which is renamed to represent this union. For example, let $f : X \twoheadrightarrow Y$ and $g : U \twoheadrightarrow V$ be two partial functions, and let $h = f \cup g$ be their union. If, instead of h, one uses the name f for the union of f and g, then f is said to be overloaded.

If there is any case where $(x, y) \in f$ and $(x, z) \in g$, then y must be z because, otherwise, the $f \cup g$ is no longer a function. If $y \neq z$, then the result of the overloaded construct is ambiguous. Practically, this means that in different implementations different results will occur, and this is something to be avoided. Hence, when overloading, one must check that the result of forming the union of two partial functions is again a partial function.

6.1.6 Composition of Functions

Functions are relations and, as with any other relation, one can form a composition of functions. For function composition, the symbol $\circ$ is used again, but its meaning is different. In fact, it is customary to invert the order of the operands when dealing with functions. Specifically, what used to be $f \circ g$ is now $g \circ f$. The inversion of the order of operands in function composition is confusing, but, unfortunately, it represents the present standard. There have been some attempts to rectify the situation, but none of these attempts has met with general acceptance. For instance, Z uses the symbol $\overset{\circ}{9}$ for the type of composition discussed in connection with relations and $\circ$ for function composition. In this case,

$$g \circ f = f \overset{\circ}{9} g$$

We will not use the symbol $\overset{\circ}{9}$ here, however.

To clarify the issue, pretend that f and g are relations, which allows us to use the notation $f \circ g$ instead of $g \circ f$. The relation f contains all pairs (x, y) with $y = f(x)$, and the relation g contains all pairs (y, z) with $z = g(y)$. A pair (z, x) is therefore in $f \circ g$ if there is a y such that $y = f(x)$ and $z = g(y)$. Since there is only one value $y = f(x)$, one can replace y in $g(y)$ by $f(x)$, which yields $z = g(f(x))$. Hence, if the two functions are treated like relations, then their composition is given by the pairs $(x, g(f(x)))$. We now stop pretending that f and g are relations, which means that $f \circ g$ becomes $g \circ f$, or

$$g \circ f = \{(x, g(f(x)))\}$$

Note that this is fully compatible with function composition as defined in Section 2.5.5, where it was indicated that the composition of the two functions f and g associates with each value x the value $g(f(x))$. In this sense, we will write $g \circ f(x)$ for the value of $g \circ f$ at x; that is,

$$g \circ f(x) = g(f(x))$$

The ordering of g and f on the left side of the equation is consistent with their ordering on the right side of the equation. This is the reason why function composition is expressed in a manner inconsistent with relation composition.

To fully specify a function, one must define its domain space and its range space. Hence, let $f : X \to Y$ and let $g : Y \to Z$. In this case, one has $g \circ f : X \to Z$. Hence, the composition $g \circ f$ first applies f to map X into Y, and it then employs g to map Y into Z. In other words, the range space of f becomes the domain space of g. Figure 6.2 illustrates the composition of the two functions f and g. In this figure, x_2 can reach z_1 by passing through y_1, which implies that the image of $g \circ f$ for $x = x_1$ is z_1. Algebraically, one has

$$y_1 = f(x_2)$$
$$z_1 = g(y_1)$$
$$\overline{z_1 = g(f(x_2))}$$

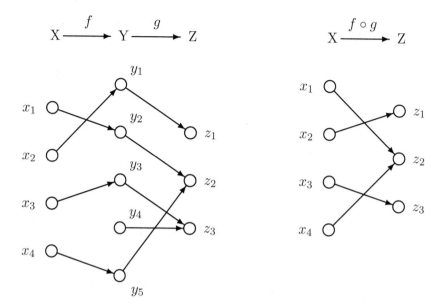

Figure 6.2.   Composition of functions

Similarly, (x_4, z_2) is in $g \circ f$ because there is a connection from x_4 to z_2 via y_5. This leads to

$$y_5 = f(x_4)$$
$$z_2 = g(y_5)$$
$$\overline{z_2 = g(f(x_4))}$$

The remaining pairs (x, z) in the function $g \circ f$ can be found in a similar fashion.

The following are additional examples of function compositions. All the functions have the reals as their domain.

Example 6.4   Find $g \circ f(x)$ if $f(x) = x + 4$ and $g(y) = y^2$.

Solution   Since $g(y) = y^2$, $g(f(x))$ is $f(x)^2$, and since $f(x) = x + 4$, $f(x)^2$ is $(x + 4)^2$. Consequently,

$$(g \circ f)(x) = g(f(x)) = (x + 4)^2$$
∎

Example 6.5   Let f and g be $x + 3$ and $2x$, respectively. Find $g \circ f$ and $f \circ g$.

Solution $g(y)$ is $2y$ and $g(f(x))$ is therefore $2f(x)$. With $f(x) = x + 3$, this yields $2(x + 3)$. Consequently,

$$(g \circ f)(x) = g(f(x)) = 2(x + 3)$$

The function $f \circ g$, on the other hand, yields

$$(f \circ g)(y) = f(2y) = 2y + 3$$
∎

Example 6.5 shows that $g \circ f$ and $f \circ g$ are not necessarily equal.

Function compositions are important in the context of data-processing applications as the following example indicates.

Example 6.6   In a company, each purchase is identified by a unique purchase number. Each purchase number has one supplier, and each supplier has one account at which he or she is credited. The purchase numbers are 204, 217, 451, and 233. The suppliers are ABM Manufacturing, CC Producers, MDM Metalworks, and RMB Suppliers. The function f, which matches each purchase with a supplier, is given as

$$f = \{(204, ABM), (217, RMB), (451, CC), (233, RMB)\}$$

The suppliers and accounts are as follows:

$$g = \{(ABM, 124), (CC, 321), (MDM, 214), (RMB, 113)\}$$

Find $g \circ f$, which matches each purchase number with an account.

Solution   The solution is readily established as follows:

$$\{(204, 124), (217, 113), (451, 321), (233, 113)\} \qquad \blacksquare$$

As in the case of relations, one may compose more than two functions. Hence, if $f : X \to Y, g : Y \to Z$, and $h : Z \to W$ are three functions, one can form $(h \circ g) \circ f$ and $h \circ (g \circ f)$. Since functions are relations, equation (5.13) applies. It yields

$$(h \circ g) \circ f = h \circ (g \circ f)$$

The parentheses may therefore be dropped when forming the composition of three functions, and the same is true when dealing with more than three functions.

Example 6.7   Let $f(x) = x + 2, g(x) = x - 2$, and $h(x) = 3x, x \in \mathbb{R}$. Find $f \circ h, (f \circ h) \circ g$, $h \circ g$, and $f \circ (h \circ g)$.

Solution

$$f \circ h(x) = f(h(x)) = h(x) + 2 = 3x + 2$$
$$(f \circ h) \circ g(x) = f \circ h(g(x)) = 3g(x) + 2 = 3(x - 2) + 2 = 3x - 4$$
$$h \circ g(x) = h(g(x)) = 3(g(x)) = 3(x - 2)$$
$$f \circ (h \circ g)(x) = f(h \circ g(x)) = h \circ g(x) + 2 = 3(x - 2) + 2 = 3x - 4 \qquad \blacksquare$$

Compositions are closely related to computing. To see this, consider the following two statements:

```
y := x + 3;
z := 2 * y
```

The final value of z depends, of course, on the initial value of x, which makes z a function of x. This function can be interpreted as a composition of the two functions $f(x) = x + 3$ and $g(y) = 2 * y$. To see this, note that z can be obtained by replacing y in the second statement by $f(x) = x + 3$. The overall result is the function $z = 2 * (x + 3)$. More generally, sequences of assignment statements can be interpreted as function compositions. This has been discussed in Section 3.5.3.

Up to now, it was assumed that the composition $g \circ f$ can only be formed if the range space of f matches the domain space of g. This condition can be relaxed, but one has to require at least that the range space of f be a subset of the domain space of g. In computer science and in mathematics, this is ensured by setting the domain and range spaces to certain types, such as $\mathbb{N}$, $\mathbb{Z}$, or $\mathbb{R}$. In this way, one partially avoids having to check for each composition if the ranges and the domains of the functions being composed match.

It is also possible to compose partial functions. For instance, on the reals, the composition of $f(x) = x + 1$ and $g(y) = 1/y$ involves the partial function $1/y$. In this case, the composition is obviously a partial function as well. In fact, $g(f(x)) = 1/(x + 1)$ is undefined for $x + 1 = 0$.

6.1.7 Injections, Surjections, and Inverses

Given the equation

$$(x + 2)^3 = (y + 1)^3$$

one is allowed to conclude $x + 2 = y + 1$. However, from the equation

$$(x + 2)^2 = (y + 1)^2$$

no such conclusion can be derived. In fact, if $x = 0$ and $y = -3$, then $x + 2 = 2$ and $y + 1 = -2$, which is not the same, yet $(x + 2)^2 = (y + 1)^2$. There are thus some functions for which $f(x_1) = f(x_2)$ implies that $x_1 = x_2$, whereas for other functions it does not. The following definition is crucial in this respect.

Definition 6.3:   A function $f : X \to Y$ is called *one-to-one* (*injective*, or *1–1*) if, for all $x_1, x_2 \in X$, one has

$$f(x_1) = f(x_2) \Rightarrow x_1 = x_2 \qquad\qquad (6.1)$$

The contrapositive of (6.1) is

$$x_1 \neq x_2 \Rightarrow f(x_1) \neq f(x_2)$$

This, of course, indicates that a function is an injection if different arguments have different images. This is also suggested by the name one-to-one. If a function is many-to-one, then many arguments share the same image and the function is no injection. In this case, one cannot conclude that $x_1 = x_2$ if $f(x_1) = f(x_2)$.

To demonstrate the distinction between one-to-one functions and many-to-one functions, consider Figure 6.3. The functions f and h are both injections, but g is not an injection, it is not one-to-one, but many-to-one. Another example of a one-to-one function

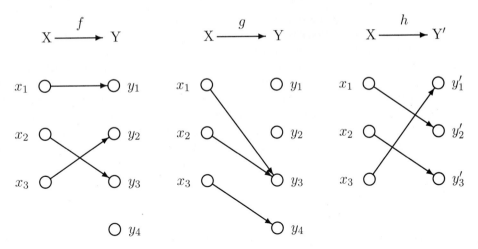

Figure 6.3.   Graphs of functions

or injection is given by Table 6.1. This function is an injection because no city can be the capital of two different countries. In data processing, one often identifies each record on a file with a key. The key is a function of the record and, typically, this function is injective; otherwise, one could not recover the record in an unambiguous way once the key is given. Each record must have a unique key, which makes the key a function of the record, and different records must have different keys, which makes the key an injection. The concept of using key values to uniquely identify a record is explored in Section 12.2.5 and is used throughout Chapter 8 when declaring uniquely identifiable items from a set.

In the domain of reals, all functions that are *strictly increasing* are injections. A function is strictly increasing if

$$(x_1 > x_2) \Rightarrow (f(x_1) > f(x_2))$$

Strictly decreasing functions are also injections. A function is strictly decreasing if

$$(x_1 > x_2) \Rightarrow (f(x_1) < f(x_2))$$

A function that is strictly increasing always increases, whereas a function that is strictly decreasing always decreases.

Example 6.8   Show that $f(n) = n^2$ and $f(n) = n^3$, where n is a natural number, are injections.

Solution   The two functions are injections because they are both strictly increasing.   ∎

The concepts of strictly increasing and strictly decreasing functions can easily be generalized to any totally ordered sets.

Functions are relations and, as such, they have inverses. This inverse is certainly a relation, but not necessarily a function. For instance, the relation $\{(1, 2), (2, 3), (3, 3)\}$ is a function because there is exactly one image for each argument. The inverse of this relation, which is $\{(2, 1), (3, 2), (3, 3)\}$ is not a function, however: the number 3 relates to both 2 [through $(3, 2)$] and 3 [through $(3, 3)$]. In a case such as this, the function is said to have no inverse. A function has an inverse only if its relational inverse is a function. This means that for every y there must be a unique x that satisfies the equation $y = f(x)$.

If a function is not injective, then the same image is shared by different arguments, and no inverse exists. For instance, the function $f(x) = x^2$, $x : \mathbb{R}$ has no inverse because there are two different values of x, that is, $x = 2$ and $x = -2$, that have the same image $y = 4$. It follows that only injections can have inverses. However, not all injections have inverses. For an inverse of a function $f : X \rightarrow Y$ to exist, it is also necessary that every element of Y be the image of some $x \in X$. This is addressed by the following definition:

> **Definition 6.4:**   A mapping of $f : X \rightarrow Y$ is called *onto* (*surjective*, a *surjection*) if ran $f = Y$. Hence, for each $y \in Y$, there must be an $x \in X$ such that $y = f(x)$.

A function is not onto if there are some elements of the range space left over once all arguments are matched with their images. The graph of the function in question helps to discover this situation. Figure 6.3 gives the graphs of the three functions $f : X \rightarrow Y$, $g : X \rightarrow Y$, and $h : X \rightarrow Y'$. Both f and g are from X into, but not onto Y. Function h is onto Y'.

Generally, when deciding whether a function $f : X \rightarrow Y$ is onto or not, one has to know Y. If Y is restricted to ran f, then the function is automatically onto.

When speaking of subsets of integers, it is intuitively clear what is meant by the word gap, and the same is true for subsets of the real numbers. Generally, gaps can be defined whenever a function has a range space that is a partially ordered set. Hence, consider a function $f : X \rightarrow Y$, where Y has the partial order R. Let y_1 and y_2 be two elements from Y. Then the range of f has a gap between y_1 and y_2 if there is a $z \in Y$ that does not belong to the range of f such that $y_1 \prec z \prec y_2$. Obviously, any function f in which ran f contains gaps is not onto.

Example 6.9   Let $f(n) = 3n$ be a function from $\mathbb{N}$ to $\mathbb{N}$. Show that $f(n)$ is not onto.

Solution $f(n) = 3n$ is not onto because the function f contains gaps. For instance, there is a gap between 3 and 6; that is, there is a number z, say $z = 4$, such that $3 < z < 6$.   ∎

If $f : X \rightarrow Y$, where Y is a partially ordered set and ran f does not contain gaps, it is still possible that f is not onto. This happens if there is any maximal or minimal element of Y that is not in ran f.

Example 6.10   Let $f : \mathbb{N} \rightarrow \mathbb{N}$ be given as $f(n) = n + 3$. Since 0 is a minimal element of $\mathbb{N}$, but not of ran f, the function $n + 3$ is not onto.

If A is some set, and $f : A \rightarrow \mathbb{R}$, then f maps A into $\mathbb{R}$. Such functions can have *upper bounds* and *lower bounds*. Function f has an upper bound z if, for all $x \in A$, $f(x) \leq z$. Similarly, f has a lower bound t if, for all $x \in A$, $f(x) \geq t$. Functions with

upper and lower bounds obviously cannot be onto. Upper bounds and lower bounds can
also be defined for functions into $\mathbb{Z}$. Again, any function with an upper or lower bound
cannot be onto. Functions into $\mathbb{N}$ cannot be onto if they have an upper bound or if they have
a lower bound greater than zero.

Example 6.11   If $f : \mathbb{R} \to \mathbb{R}$ is given as $2x - x^2$, then f is not onto, because f has no value of x
that exceeds 1. Consequently, $f(x) = 2x - x^2$ is not onto.

To have an inverse, a function must both be one-to-one and onto. This observation
leads to the following definition:

Definition 6.5:   A mapping $f : X \to Y$ is called *one-to-one onto* or *bijective*
if it is both one-to-one and onto. Such a mapping is called a *bijection* between
X and Y.

According to this definition, a function f has an inverse if and only if it is a bijection.
Hence, in Figure 6.4, only h has an inverse. The function g has no inverse because y_2 can
be associated with both x_2 and x_3, and f has no inverse because there is no x such that
$f(x) = y_3$.

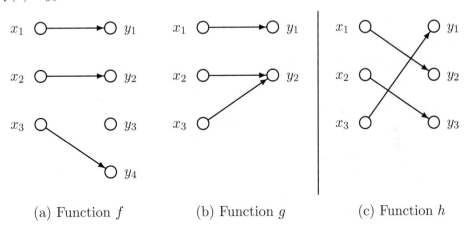

(a) Function f (b) Function g (c) Function h

Figure 6.4.   Conditions for the existence of an inverse

We now describe how the inverse of a function is found. If the function is given in
roster notation, merely change the order of each pair. If f is given by its graph, invert the
direction of the arrows and read the graph from right to left. For instance, in Figure 6.4,

$$h^{-1} = \{(y_1, x_3), (y_2, x_1), (y_3, x_2)\}$$

If a function is in the form of a table, the inverse function can be found by a table
look-up. For instance, an employee number is a function of the employee as indicated by

particulars such as his or her name, address, age, and so on. The inverse of this function has the employee number as its argument and the employee as its range. To find this particular inverse, it helps, of course, if the list is sorted according to the employee numbers rather than the employee names.

If a function is given as a mathematical expression, the inverse can be found by solving an equation. For instance, let $f(x)$, $x \in \mathbb{R}$ be the function that associates $y = 4x$ with each value of x. The value of x that corresponds to a given y is then $x = y/4$. The inverse function of $f(x) = 4x$ is therefore $f^{-1}(y) = y/4$. More generally, if $y = f(x)$ is a function, then $x = f^{-1}(y)$ is the solution of the equation $y = f(x)$ for y, provided that this solution is unique.

In the case where $h = g \circ f$, one finds the inverse by going backward in the composition graph. This means that

$$h^{-1} = (g \circ f)^{-1} = f^{-1} \circ g^{-1}$$

Moreover, if $f : X \to Y$ and if I_X and I_Y are the identity functions of the sets X and Y, respectively, one has

$$f^{-1} \circ f = I_X, \qquad f \circ f^{-1} = I_Y$$

To show this, one uses the graph of f. In the composition $f^{-1} \circ f$, one first uses f to go from X to Y on arc (x, y) and then returns on the same arc (x, y), but in the opposite direction. This obviously brings one to the starting point, and every point $x \in X$ is matched by itself. This explains that $f^{-1} \circ f = I_X$. Similarly, if f^{-1} is done first, as is the case in the composition $f \circ f^{-1}$, then one starts and returns to the same point in Y, which yields the function I_Y.

6.1.8 Creating Inverses by Creating Types

In this section, we consider injections that are not onto and that (strictly speaking) have no inverses. In this case, one can associate at most one x with each y such that $y = f(x)$. The pairs (y, x) thus form a partial function, and this partial function may be called a partial inverse. It is interesting to note that such partial inverses have repeatedly led to the creation of new types. First, consider functions of the form $f(x) = a + x$, where x and a are natural numbers. For $a > 0$, this function has a lower bound greater than 0, and it is therefore not onto. Consequently, it only has a partial inverse. To convert this partial inverse into a total inverse, the negative numbers were invented. Next, if x and a are natural numbers with $a \geq 1$, then the function $f(x) = ax$ has no inverse because the range has gaps. To create a total inverse, the rational numbers were created. In the domain of the rational numbers, x^2 has a lower bound of zero, and therefore it has no inverse. This led to the creation of the complex numbers.

Another method to create functions that can be used as substitutes for inverses is to associate with each value in the range space a value in the range. To demonstrate this approach, consider functions $f : \mathbb{N} \to \mathbb{N}$ given by $y = f(x) = ax$, where a is a natural number. The range of this function is the multiples of a, that is, $\{0, a, 2a, \ldots\}$. For $a > 1$, this is a proper subset of the range space $\mathbb{N}$, and there is therefore no inverse. To get around this, one associates any y between ax and $a(x + 1) - 1$ with x. This yields the integer division, which is denoted by y div a. Since y div a associates several values of y with a given value x, the function div is no longer one-to-one. To make it one-to-one, use is made of the *remainder* $y - ax$. The remainder r must always satisfy $0 \leq r < a$. Note that even though y div a is not one-to-one the pair $(r, y$ div $a)$ is.

The notion of a remainder can be generalized. Let $x \in \mathbb{N}$, and let $f(x)$ be an increasing function from $\mathbb{N}$ to $\mathbb{N}$. For each y, one can find an x such that $f(x) \leq y < f(x+1)$. The remainder is then $r = y - f(x)$. Clearly, $0 \leq r < f(x+1) - f(x)$. For instance, consider the square function $f(x) = x^2$, $x : \mathbb{N}$. Every y can be associated with some x by means of the inequality $x^2 \leq y < (x+1)^2$. The remainder is then $y - x^2$. For instance, $y = 13$ is associated with $x = 3$, and the remainder is 4, because $3^2 \leq 13 < 4^2$ and $13 - 9 = 4$.

In many cases, one does the calculation in real numbers and then uses the floor function to convert the result into integers. The floor function is defined as follows:

Definition 6.6:   The *floor* of x is the highest integer that is below x. The floor of x is written as $\lfloor x \rfloor$.

In other words, $\lfloor x \rfloor$ is x if x is an integer, and it is the integer just below x otherwise. For instance, $\lfloor 3.5 \rfloor$ is 3, whereas $\lfloor -3.5 \rfloor$ is -4. The floor must be distinguished from truncation. The truncation function merely truncates any fraction. For positive x, this is the same as the floor of x, but for negative x the results are different. For instance, $\lfloor -2.3 \rfloor$ is -3 and not -2. One can now define the integer division in terms of the real division as

$$x \text{ div } y = \lfloor x/y \rfloor$$

This is, at least, how mathematicians define integer division. Unfortunately, most computer languages use truncation instead of the floor, which means that there is a discrepancy for negative quotients. The remainder of x div y is written as x mod y. One has

$$x \text{ mod } y = x - \lfloor x/y \rfloor y$$

Again, programming languages tend to define the mod operator differently. This is unfortunate because there is an elegant theory of modular arithmetic (to be discussed later) that could be very useful in some applications.

Problems 6.1

1. Which of the following relations are functions and which are partial functions from $X = \{1, 2, 3\}$ to $Y = \{1, 2, 3\}$? In each case, indicate the domain and the range of the relation.

$$f_1 = \{(1,3),(2,3),(3,3)\}$$
$$f_2 = \{(1,3),(1,1),(2,3),(2,4)\}$$
$$f_3 = \{(1,1),(2,2),(3,3)\}$$
$$f_4 = \{(1,2)\}$$

2. Do the following sets define functions? If so, give their domain and their range.

$$\{(1,(2,3)),(2,(3,4)),(3,(1,4)),(4,(2,4))\}$$
$$\{((1,2),3),((2,3),4),((3,3),2)\}$$
$$\{(1,(2,3)),(2,(3,4)),(1,(2,4))\}$$
$$\{(1,(2,3)),(2,(2,3)),(3,(2,3))\}$$

3. Show that if A is in $\mathbb{P}D$, where D is some set, then $\sim A$ is a function, and provide this function's declaration. Is the subset relation between two sets also a function? If so, how could it be declared?

4. Reduce the following expressions:

$$(\lambda a.(\lambda b.(x + 2b))3)4$$
$$(\lambda x.(\lambda y.((x + y)(x - y)))2)2$$
$$(\lambda x.(\lambda y.(x + y))a)b$$
$$(\lambda x.4)2$$

5. Let $f = \{(1,c),(2,a),(3,b),(4,c)\}$, $g = \{(a,2),(b,1),(c,4)\}$, and $h = \{(1,A),(2,B),(3,D),(4,D)\}$. Find $g \circ f$, $h \circ (g \circ f)$, $h \circ g$, and $(h \circ g) \circ f$. Also find $f \circ g$.

6. Let $f(x) = 3x^2 + y$, $g(x) = x/y + 2$, and $h(x) = (x + y)^2$. Find $f \circ g$, $g \circ f$, $f \circ h$, and $f \circ g \circ h$.

7. Determine which of the following functions are one-to-one, which are onto, and which are one-to-one onto.
   (a) $f : 1..3 \rightarrow \{a,b,c\}$ $f = \{(1,a),(2,b),(3,c)\}$
   (b) $f : 1..3 \rightarrow \{a,b,c,d\}$ $f = \{(1,a),(2,b),(3,c)\}$
   (c) $f : 1..3 \rightarrow \{1..2\}$ $f = \{(1,2),(2,1),(3,2)\}$

8. Determine which of the following functions are one-to-one, which are onto, and which are one-to-one onto.
   (a) $f : \mathbb{N} \rightarrow \mathbb{N}$ $f(j) = j^2 + 2$
   (b) $f : \mathbb{N} \rightarrow \mathbb{N}$ $f(j) = j(\bmod\ 3)$
   (c) $f : \mathbb{N} \rightarrow \mathbb{N}$ $f(j) = \begin{cases} 1, & j \text{ is odd} \\ 0, & j \text{ is even} \end{cases}$
   (d) $f : \mathbb{N} \rightarrow \{0,1\}$ $f(j) = \begin{cases} 0, & j \text{ is odd} \\ 1, & j \text{ is even} \end{cases}$

9. Let $\mathbb{N}_1$ be the set of positive integers, and let $(0..p) = \{0, 1, 2, .., p\}$. Determine which of the following functions are one-to-one, which are onto, and which are one-to-one onto.

   (a) $f : \mathbb{Z} \to \mathbb{Z}$ $\qquad\qquad$ $f(j) = \begin{cases} j/2 & j \text{ is even} \\ (j-1)/2 & j \text{ is odd} \end{cases}$

   (b) $f : \mathbb{N}_1 \to \mathbb{N}_1$ $\qquad\qquad$ $f(x) = \text{greatest integer} \leq \sqrt{x}$

   (c) $f : (0..6) \to (0..6)$ $\qquad\quad$ $f(x) = 3x(\text{mod } 7)$

   (d) $f : (0..3) \to (0..3)$ $\qquad\quad$ $f(x) = 3x(\text{mod } 4)$

10. If X and Y are finite sets, find a necessary condition for the existence of one-to-one mappings from X to Y.

11. List all possible functions from $X = \{a, b, c\}$ to $Y = \{0, 1, 2\}$ and indicate in each case whether the function is one-to-one, onto, or one-to-one onto.

12. If $A = \{1, 2, \ldots, n\}$, show that any function from A to A that is one-to-one must also be onto, and conversely.

13. Show that the functions f and g that are both from $\mathbb{N} \times \mathbb{N}$ to $\mathbb{N}$ given by $f(x, y) = x + y$ and $g(x, y) = xy$ are onto but not one-to-one.

14. Let $f : \mathbb{R} \to \mathbb{R}$ be given by $f(x) = x^3 - 2$. Find f^{-1}.

15. Show that there exists a one-to-one mapping from $A \times B$ to $B \times A$. Is it also onto?

16. Let $X = \{1, 2, 3, 4\}$. Define a function $f : X \to X$ such that $f \neq I_X$ and is one-to-one. Find $f \circ f = f^2, f^3 = f \circ f^2, f^{-1}$, and $f \circ f^{-1}$.

17. Find a one-to-one function $g : X \to X$ such that $g \neq I_X$ but $g \circ g = I_X$.

18. Find $\lfloor 5/2 \rfloor$, $\lfloor (5/2)^2 \rfloor$, and $(\lfloor 5/2 \rfloor)^2$.

19. Find $4 - \lfloor 5/2 \rfloor$ and $4 + \lfloor -5/2 \rfloor$.

20. For $i = 1..10$, find $i \bmod 3$ and $(-i) \bmod 3$.

21. Let $y = x(x + 1)$, $x \in \mathbb{N}$. Define $h(y)$ to be z if $z(z + 1) \leq y < (z + 1)(z + 2)$. Express $h(y)$ by using the floor function. Also, define the remainder associated with this $h(y)$.

22. Prove that $x^2 - \lfloor x \rfloor^2 < 2\lfloor x \rfloor + 1$.

6.2 ENUMERATIONS, ISOMORPHISMS, AND HOMOMORPHISMS

6.2.1 Introduction

When a shepherd counts her flock, she associates a number with each animal. If she has n animals, then counting establishes a one-to-one correspondence between numbers from 1 to n and the animals of her flock. In this sense, one can say that every enumeration is a bijection, an idea that is elaborated on in Section 6.2.2. It is clear that all finite sets can be counted. Again, a set is of size n, or of *cardinality* n, if there is a bijection between each element of the set and $1..n$. Bijections can also be used to find the cardinality of infinite sets. If there is a bijection between a set and some subset of the integers, the set is called *countable*. As it turns out, not all sets are countable. This result will be proved in Section 6.2.3. Section 6.2.4 gives an introduction to permutations and combinations and shows how to count the number of permutations and combinations. In Section 6.2.5, isomorphisms and homomorphisms are discussed. Two functions are said to be isomorphic if they can be converted into one another by means of a bijection. For instance, addition in Roman and Arabic numerals is isomorphic. It is also true that addition and multiplication are isomorphic. Homomorphisms are in some sense similar to isomorphisms, except that the transformation does not need to be a bijection.

6.2.2 Enumerations

Sets can obviously be enumerated. For instance, the set $A = \{a, b, c\}$ can be enumerated as follows:

1. a
2. b
3. c

Of course, this enumeration can also be expressed as $\{(1, a), (2, b), (3, c)\}$. It is easy to see that this is a bijection. Generally, any enumeration of a set is a bijection, as indicated by the following definition:

Definition 6.7:   Let A be a set with the cardinality $n = \#A$. An enumeration of A is a bijection from $1..n$ to A.

The enumeration $\{(1, a), (2, b), (3, c)\}$ can also be written as $\langle a, b, c \rangle$, with the understanding that first comes a, then b, and then c. More generally, when arranging a set into a sequence, one implicitly obtains an enumeration, which is a bijection. Functions are sets and they can therefore be enumerated. Enumerations of functions have, in fact, already been used. For instance, Table 6.1 enumerates all North American countries together with their capitals.

In a sequence, the number 1 can be associated with the first element of the sequence, the number 2 with the second, and so on. Given that the sequence is of length n, one can therefore associate an element of the sequence with each $i \in 1..n$, namely, the element in the nth position. In this sense, one can say that a sequence of length n is a function from $1..n$ to the elements of the sequence. However, unless no element in the sequence is repeated, this function is not injective. For instance, $\langle 3, 2, 6, 3 \rangle$ is a sequence in which the number 3 appears both in the first place and in the fourth place. Consequently, both 1 and 4 have 3 as their image. Arrays can be interpreted as sequences, except that the first element of the array need not correspond to the number 1. Generally, if an array contains the elements from m to n, then the array is a function that maps every $i \in m..n$ into an array element.

A two-dimensional array or matrix can also be thought of as a function. Specifically, if the matrix has n_1 rows and n_2 columns and if a_{ij} is of type X, then $a_{ij}, i : 1..n_1, j : 1..n_2$ represent a function from $(1..n_1) \times (1..n_2)$ into X. This function associates with each pair (i, j) a value a_{ij}. To store the elements of the array, they have to be brought into a linear sequence; that is, they have to be enumerated: all pairs (i, j), $i = 1, 2, \ldots, n_1$, $j = 1, 2, \ldots, n_2$ have to be given a number, which means that the Cartesian product $(1..n_1) \times (1..n_2)$ must be enumerated. To accomplish this, all pairs (i, j) are listed in lexicographical order. Listing in lexicographical order is accomplished by first listing all pairs in which the first element of the pair is 1, then all pairs in which it is 2, and so on. This yields the following list:

$$\langle (1, 1), (1, 2), \ldots, (1, n_2), (2, 1), (2, 2), \ldots, (2, n_2)$$
$$\ldots (n_1, 1), (n_1, 2), \ldots, (n_1, n_2) \rangle$$

To make use of this enumeration, a position k must be associated with each pair (i, j). This position can be calculated as follows:

$$k = g(i, j) = (i - 1)n_2 + j \tag{6.2}$$

To see that the function $g(i, j)$ gives the correct position, note that the first n_2 positions correspond to $(1, 1)$ to $(1, n_2)$, which makes the formula correct for $i = 1$. This leaves positions $n_2 + 1$ to $n_2 + n_2$ for the pairs $(2, 1)$, $(2, 2)$, and $(2, n_2)$, and the formula is again correct. Continuing this way, one finds that the formula is correct for all i. The highest value for $k = g(i, j)$ is reached when both i and j are at their upper limit, in which case $g(n_1, n_2)$ is $(n_1 - 1)n_2 + n_2 = n_1 n_2$.

The function $g(i, j)$ allows one to find the position for any given i and j. Often, one needs the inverse of $g(i, j)$; that is, one is given a position k, and the problem is to find the corresponding pair (i, j). To accomplish this, an integer division with a remainder is used. Essentially, the result of the integer division corresponds to the i and the remainder to the j. However, adjustments have to be made. If m div n yields q with a remainder r, then $m = qn + r$, with $0 \le r \le n - 1$. Consequently, if $i - 1 = q$, $j - 1 = r$, $n = n_2$, and $m = k - 1$, then $k - 1 = (i - 1)n_2 + j - 1$, with $0 \le j - 1 \le n_2 - 1$, in agreement with (6.2). Consequently, to find $i - 1$, one divides $k - 1$ by n_2. The remainder of this division is $j - 1$. Alternatively, one can write

$$i = \lfloor (k - 1)/n_2 \rfloor + 1 \tag{6.3}$$
$$j = k - (i - 1)n_2 \tag{6.4}$$

Example 6.12   A matrix with 8 rows and 10 columns is stored by rows in consecutive storage locations. Find the row and column of the 27th matrix entry.

Solution   According to (6.3), the 27th entry is on row $\lfloor 26/10 \rfloor + 1$, that is, on the third row. The third row extends from $k = 21$ to $k = 30$. The 27th entry is therefore in column 7, in accordance with (6.4).   ■

The way the Cartesian product is enumerated is unfair in the sense that i is not increased before all values of j are enumerated. Fairness may be an issue for searches. For instance, if f is a function with two integer arguments i and j with a certain range and if a function value $f(i, j)$ must be identified that satisfies certain conditions, then it is likely that such a value is found sooner if i and j are treated equitably. Moreover, if j is not bounded, then the enumeration suggested by (6.2) would never change i at all. For cases like this, a different enumeration is required, and this is *Cantorian diagonalization*. We only discuss the case where both n_1 and n_2 are infinite, even though Cantorian enumeration also works for finite n_1 and n_2.

Cantorian diagonalization is based on the pairs (s, j) instead of (i, j), where s is $i + j$. The pairs (s, j) are then enumerated in lexicographical order. This means that one first enumerates all pairs $i + j = 2$, then all pairs with $i + j = 3$, and so on. To do the enumeration, one needs to know the number of pairs in which $s = i + j$. Clearly, $i + j = s$ and $i \geq 1$ imply that $j \leq s - 1$. Consequently, $1 \leq j \leq s - 1$; that is, given s, there are $s - 1$ different values for j or $s - 1$ different pairs (s, j). Hence, if $s = 2$, there is one pair (s, j), that is, $(2, 1)$, if $s = 3$, there are two such pairs, $(3, 1)$ and $(3, 2)$, and so on. In total, there are $1 + 2 + \cdots + (s - 1)$ pairs $(m, j), m \leq s$, which makes $(s - 1)(s - 2)/2$ such pairs. To make sure that there are no gaps, and that the enumeration is therefore bijective, one must list all the pairs (s, j) after the $(s - 1)(s - 2)/2$ pairs with $m < s$, and this leads to the following formula:

$$k = (s - 1)(s - 2)/2 + j \tag{6.5}$$

This equation provides a position k for each s and j. If, on the other hand, one is given a position k, one has to find the pair (s, j) by a remainder mechanism. One first finds s and, once s is found, j is merely the remainder obtained by subtracting $(s - 1)(s - 2)/2$ from k. Without going into details, one has

$$s = \lfloor \tfrac{1}{2}(3 + \sqrt{8k - 7}) \rfloor$$
$$j = k - (s - 1)(s - 2)/2$$

6.2.3 Countable and Uncountable Sets

Bijections can be used to extend the notion of cardinality to infinite sets. If A is a finite set, then the cardinality of A is n if and only if there is a bijection between A and $1..n$. It is not correct to merely let n go to infinity, because, as it turns out, there are several clearly distinguishable sets that have an infinite number of elements. Instead, one uses bijections to classify the infinite sets into several equivalence classes. In particular, it turns out that there are *countable infinite* sets and *noncountable infinite sets*.

Two sets are *equipotent* if there is a bijection between them. Equipotence is an equivalence relation; that is, equipotence is reflexive, symmetric, and transitive. The relation is reflexive because each set is equipotent with itself, and it is symmetric because, if there is a bijection g between two sets A and B, then g^{-1} is a bijection between B and A. Finally, equipotence is transitive because, if A is equipotent to B and B is equipotent to C, then A is equipotent to C. It follows that two equipotent sets belong to the same equivalence class. All sets of this equivalence class are said to have the same cardinality. If A and B are equipotent, one writes $\#A = \#B$. This notion generalizes the concept of cardinality introduced earlier for finite sets. There is, however, a difference between finite and infinite sets with regard to the cardinality of subsets. If set A is a finite set, then $A \subset B$ implies that the cardinality of A is strictly less than the cardinality of B. If A is infinite, it is possible that $A \subset B$, yet A and B still may have the same cardinality.

Example 6.13   Show that the sets $X = \{1, 2, 3, \ldots\}$ and $Y = \{0, 4, 8, \ldots\}$ are equipotent, even though $X \subset Y$.

Solution   The sets X and Y are equipotent iff there is a bijection between them. The function $f(x) = 4(x - 1)$, $x \in X$ is a bijection. $f(x)$ is increasing and therefore injective. Moreover, $f(x)$ does not omit any elements from Y, which makes f onto. Any function that is injective and onto is a bijection.   ∎

Example 6.14   Show that $\mathbb{Z}$, the set of integers, has the same cardinality as $\mathbb{N}_1$, the set of natural numbers greater than 0.

Solution $\mathbb{Z}$ and $\mathbb{N}_1$ are equipotent iff there is a bijection between them. The following function is such a bijection.

$$f(n) = 2n + 1, \qquad n \geq 0$$
$$f(n) = -2n, \qquad n < 0$$

The function f is a bijection because it leaves no gaps and because there is no natural number that is not matched with an integer. Consequently, $\mathbb{Z}$ and $\mathbb{N}_1$ have the same cardinality.   ∎

It turns out that all subsets of $\mathbb{Z}$ that are infinite have the same cardinality as $\mathbb{Z}$. We do not prove this here, however. All these sets belong to the same equivalence class. They are said to be *countably infinite*.

Definition 6.8:   Any set that is equipotent with $\mathbb{N}_1$ is said to be *countably infinite*.

The fact that we chose $\mathbb{N}_1$ as the representative of its equivalence class is arbitrary, and any other set from this class would do as well. For instance, we could have chosen $\mathbb{N}$ or even $\mathbb{Z}$. However, it turns out that it is easier to show that a given set is equipotent to $\mathbb{N}_1$ than to $\mathbb{N}$ or $\mathbb{Z}$. One surprising fact is that the set $\mathbb{N}_1 \times \mathbb{N}_1$ is equipotent to $\mathbb{N}_1$. In other words, all pairs of nonnegative integers are countable. This is true because $c(i, j)$ given in (6.5) is a bijection between $\mathbb{N}_1$ and $\mathbb{N}_1^2$.

The fact that $\#\mathbb{N}_1 = \#\mathbb{N}_1^2$ has a number of implications. First, the set of rational numbers can be represented by pairs of natural numbers and is therefore countable. The set of all n-tuples consisting of rational numbers is also countable. To see this, note that if A and B have the same cardinality, then $A \times C$ has the same cardinality as $B \times C$. This fact can be used to prove inductively that $\mathbb{N}_1^n$ has the same cardinality as $\mathbb{N}_1$. Clearly, $\mathbb{N}_1^1 = \mathbb{N}_1$ has the same cardinality as $\mathbb{N}_1$, which establishes the basis for induction. If $\mathbb{N}_1^n$ has the same cardinality as $\mathbb{N}_1$, then $\mathbb{N}_1^{n+1} = \mathbb{N}_1^n \times \mathbb{N}_1$ has the same cardinality as $\mathbb{N}_1 \times \mathbb{N}_1$ and consequently the same cardinality as $\mathbb{N}_1$. One concludes that $\mathbb{N}_1^n$ has the same cardinality as $\mathbb{N}_1$, which implies that all n-tuples of natural numbers are countable. This implies that all Cartesian products of countable sets are countable.

Since the natural numbers are countable and since all n-tuples consisting of natural numbers are countable, the question arises as to whether there are any sets that are not countable. In answer to this question, we now prove

$$\#\mathbb{P}A > \#A \tag{6.6}$$

This indicates that the cardinality of the powerset of a set A always exceeds the cardinality of A. By setting A in (6.6) to $\mathbb{N}_1$, one concludes that there is a set that is greater than $\mathbb{N}_1$, namely $\mathbb{P}\mathbb{N}_1$. By definition, this set is no longer countable.

To prove (6.6), the following technique is used. A set X is of lower cardinality than Y if no function from X to Y is onto. To prove this, one has to find at least one element of Y that is not in the range of f. To show that an element $u \in Y$ is not the range of f and consequently not the image of any $x \in X$, one uses an indirect proof; that is, one shows that if there is an x such that $f(x) = u$ then a contradiction results.

This proof idea is now fleshed out to show that $\#A < \#\mathbb{P}A$. One takes any function $G : A \to \mathbb{P}A$ and constructs an element $U \in \mathbb{P}A$ such that there cannot be an $x \in A$ with $G(x) = U$. Since $G(x) \in \mathbb{P}A$, $G(x)$ is a subset of A. Consequently, every $z \in A$ is or is not in $G(x)$. To form U, one takes every x in A that are not elements of $G(x)$.

$$U = \{x \mid x \in A \text{ and } x \notin G(x)\}$$

Equivalently,

$$(x \in U) \Leftrightarrow (x \notin G(x)) \tag{6.7}$$

This equation implies that there cannot be an $x \in A$ such that $U = G(x)$, for if there were one a contradiction could be derived as follows: Assume that, for $x = a$, $G(x) = G(a) = U$. Since (6.7) holds for all $x \in A$, it must hold for $x = a$, which yields

$$(a \in U) \Leftrightarrow (a \notin G(a))$$

The desired contradiction is now at hand because $G(a) = U$, which yields

$$(a \in U) \Leftrightarrow (a \notin U)$$

One concludes that there is no $a \in A$ such that $G(a) = U$; that is, U is not in the range of G. Consequently, $\#\mathbb{P}A > \#A$ as claimed.

Another example of an uncountable set is the set of all infinite sequences over an alphabet A. Note, however, that the set of all finite sequences is countable. To show that the infinite sequences are not countable, let A^∞ be the set of all infinite sequences over A. The idea is to show that if g is a function from $\mathbb{N}_1$ to A^∞ then g cannot be onto. In this respect, the proof is similar to the one just given. Again, it is shown that the assumption that there is an enumeration of A^∞ leads to a contradiction. To derive the contradiction, suppose that x_i, $i = 1, 2, 3, \ldots$, is an enumeration of the set A^∞. The jth element of string x_i is denoted by x_{ij}; that is,

$$x_i = \langle x_{i1}, x_{i2}, x_{i3}, \ldots \rangle$$

We now construct a string $u = \langle u_1, u_2, u_3, \ldots \rangle$ such that there is no i with $x_i = u$. Hence, $u \neq x_1$, $u \neq x_2$, $u \neq x_3$, $\ldots$. To find such a u, one only needs to make u_i different from x_{ii}, as follows:

$$
\begin{array}{lll}
u_1 \neq x_{11} & \text{hence} & u \neq x_1 = \langle x_{11}, x_{12}, x_{13}, \ldots \rangle \\
u_2 \neq x_{22} & \text{hence} & u \neq x_2 = \langle x_{21}, x_{22}, x_{23}, \ldots \rangle \\
u_3 \neq x_{33} & \text{hence} & u \neq x_3 = \langle x_{31}, x_{32}, x_{33}, \ldots \rangle \\
\vdots
\end{array}
$$

Since u is different from all the x_i in the set of sequences, it cannot be an element of the set. Consequently, the set of infinite sequences is not countable.

As an application of this result, we show that the number of real numbers between 0 and 1 is not countable. First, there is a bijection between the number of reals from 0 to 1 and the infinite decimal fractions expressing these reals. The string consisting entirely of 9s is meant to represent the real number 1. All other strings are read as decimal fractions. Consequently, the number of reals from 0 to 1 must have the same cardinality as the number of infinite strings. They, too, are not countable.

6.2.4 Permutations and Combinations

Permutations are arrangements of objects, and combinations are selections of objects. Specifically, one has the following definition:

Definition 6.9:   Given n different objects, then any ordered arrangement of these objects is called a *permutation*. An ordered arrangement of r of the n objects is called an *r-permutation*.

Example 6.15   List all permutations of the letters a, b, and c. List all 2-permutations of these letters.

Solution   There are six permutations of the three letters a, b, and c: abc, bac, bca, acb, cab, and cba. There are also six 2-permutations: ab, ac, bc, ba, ca, and cb.   ∎

For many purposes, it is important to count the number of permutations. To do this, we need the *sum rule* and the *product rule*. The sum rule indicates that if there are two sets, containing n_1 and n_2 elements each, and if there are no elements common to both sets, then there are $n_1 + n_2$ elements in both sets together. The *product rule* indicates that the number of ordered pairs that one can form from the two sets A and B, such that the first element comes from A and the second element comes from B, is $n_1 \times n_2$, where n_1 and n_2 are the number of elements in sets A and B, respectively. The sum law should be obvious, and the product law was proved in Section 5.2.5.

Example 6.16   There are 3 tasks and 4 computers. Each task has to be allocated to a single computer, and no computer should be allocated more than one task. In how many ways can this be done?

Solution   There are 3×4 ways to allocate 3 tasks to 4 computers. To see this, imagine a table with 3 rows, one for each task, and 4 columns, one for each computer. This table has 12 entries, one for each task–computer pairing.   ∎

The product rule can be used to prove the following theorem:

Theorem 6.1.   The number of r-permutations of n distinct objects is given as
$$P(n, r) = n(n-1)(n-2)\dots(n-r+1), \quad r \leq n$$

Proof. In an r-permutation, there are r positions to be filled. Since there are n elements in total, one has n ways to fill the first position. Once the first element is chosen, there are $n-1$ elements left to fill the second position. Similarly, there are $n-2$ elements to fill the third position, and so on, until there are $n-r+1$ elements left to fill the last position. Hence, by the product rule, there are n ways to select the first element, $n(n-1)$ ways to select the first two elements, $n(n-1)(n-2)$ ways to select the first three elements, and so on. Continuing in this fashion, one finds the number of ways to fill r positions as given in Theorem 6.1.   ∎

Example 6.17   In how many ways can one arrange 3 letters from the English alphabet.

Solution   The English alphabet has 26 letters. Selecting 3 letters can therefore be done in $P(26, 3)$ ways, or $26 \cdot 25 \cdot 24 = 15600$ ways.   ∎

The total number of permutations is $P(n, n)$, the number of ways to arrange all n of the n elements. Theorem 6.1 yields

$$P(n, n) = n \cdot (n-1)\cdots(n-1) \cdot 2 \cdot 1 = n!$$

There are thus $n!$ ways to arrange n elements. Also note the following result, which is easily verified.

$$P(n, r) = n \cdot (n-1)\cdots(n-r+1) = n!/(n-r)! \tag{6.8}$$

Definition 6.10:   An r-*combination* from a set of n elements is an unordered selection of r elements from the set. In other words, an r-combination is the number of subsets of size r of a set of size n.

Example 6.18   A department consists of 4 people, A, B, C, and D. List all committees of size 2 that can be formed within this department.

Solution   There are six ways to form a subcommittee consisting of two department members. They are $\{A, B\}, \{A, C\}, \{A, D\}, \{B, C\}, \{B, D\}$, and $\{C, D\}$.   ∎

Theorem 6.2.   The number of r-combinations from a set with n elements is given as

$$C(n, r) = \frac{n!}{r!(n - r)!}$$

Proof. To obtain an r-permutation, one can first select r elements, which can be done in $C(n, r)$ ways, and then arrange them, which can be done in $r!$ ways. Consequently, $P(n, r) = C(n, r)r!$ or $C(n, r) = P(n, r)/r!$. Since $P(n, r)$ is given by (6.8), one finds

$$C(n, r) = \frac{P(n, r)}{r!} = \frac{n!}{(n - r)!} \frac{1}{r!} = \frac{n!}{r!(n - r)!} \qquad \blacksquare$$

For any $r \geq 0$, one has, as one can easily verify,

$$C(n, r) = C(n, n - r)$$

There is another common notation for $C(n, r)$, that is, $\binom{n}{r}$, read as "n choose r."

Example 6.19   There are 10 decimal digits. How many sets can be formed containing exactly 3 such digits?

Solution   The number of sets of size 3 that one can form out of a set of size 10 is a 3-combination. The number of such combinations is $C(10, 3)$. This yields $(10 \cdot 9 \cdot 8)/(1 \cdot 2 \cdot 3)$, or 120 such sets. $\qquad \blacksquare$

6.2.5 Isomorphisms and Homomorphisms

Roman and Arabic numerals are different, yet any normal arithmetic operation on $\mathbb{N}_1$ that can be performed with Arabic numbers has its analog in Roman numerals. Specifically, any addition carried out with Arabic numerals corresponds to the same addition done in Roman numerals. Consider now some function, say $f(x) = x + 2$. For each argument x, we can find the image $f(x)$ in our system. For instance, if $x = 1$, then $f(x) = 3$. In Roman numerals, one would write $g(z) = z + \text{ii}$. Note that the function name was changed to g to indicate that g accepts Roman rather than Arabic numerals. If z is equal to the Roman number i, then $g(\text{i}) = \text{iii}$, and this is the same result as was obtained before by using Arabic numerals. There are in total three functions, $f(x)$, which is expressed in Arabic numerals, $g(z)$, which is essentially the same as $f(x)$, except that it is expressed in Roman numerals, and a function t, which maps every Arabic numeral into a Roman numeral. Technically, the functions f and g are *isomorphic*, and the function t is an *isomorphism* with respect to the functions f and g. This follows from the following definition:

Definition 6.11: Let $f : A \to A$ and $g : B \to B$ be two functions on A and B, respectively. If there is a mapping t that converts all x in A into some $u = t(x)$ in B such that $g(t(x)) = t(f(x))$, then the two functions f and g are *homomorphic*. The function t is called a *homomorphism*. If, in addition to this, t is bijective, then the two functions f and g are *isomorphic* and the function t is called an *isomorphism*.

To see that our example leads indeed to an isomorphism, consider the meaning of $g(t(x))$ and $t(f(x))$. By definition, $t(x)$ is the Roman numeral corresponding to an Arabic x, and $g(t(x))$ is adding ii to $t(x)$. The expression $t(f(x))$, on the other hand, means that the image $f(x)$ is converted into Roman numerals. Obviously, both methods must lead to the same result; that is, $g(t(x)) = t(f(x))$.

Every isomorphism is reversible. For instance, one can switch from Arabic numerals to Roman numerals and back again without losing any information. Homomorphisms, on the other hand, are not reversible. For instance, let $f(x) = x + 2$, $x : \mathbb{N}_1$ as before, but let $t(x) = \lfloor x/2 \rfloor$ and $g(x) = x + 1$. Here $t(x)$ is a homomorphism, but not an isomorphism. The transformation t is not an isomorphism because it is not bijective. However, one can verify that $g(t(x)) = t(f(x))$, which makes t a homomorphism. Thus, one has

$$g(t(x)) = g(\lfloor x/2 \rfloor) = \lfloor x/2 \rfloor + 1$$
$$t(f(x)) = t(x + 2) \ = \lfloor (x + 2)/2 \rfloor = \lfloor x/2 \rfloor + 1$$

Notice that there is no longer a one-to-one correspondence between x and $t(x)$. The transformation t reduces the amount of information. This makes the inverse transformation t^{-1} impossible, and the transformation $t(x)$ is therefore irreversible.

Note that $g(t(x))$ is the composition of the functions g and t and that $t(f(x))$ is the composition of t and f. Consequently, if $u = t(x)$, $y = f(x)$, and $z = g(u)$, there are two ways to find z. One can apply t to x, which yields $t(x) = u$, and then g to $t(x)$, which yields $g(t(x))$. Alternatively, one can obtain $f(x)$ first and then translate $f(x)$ into $z = t(f(x))$. The following figure depicts this graphically.

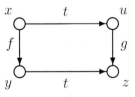

Example 6.20   Show that the two functions $f(x) = x + x$, $x \in \mathbb{R}$ and $g(u) = u \times u$, $u \in \mathbb{R}^+$ are isomorphic. Here $\mathbb{R}^+$ denotes all positive numbers, excluding 0.

Solution   The function $t(x) = 2^x$ is a bijection from $\mathbb{R}$ onto $\mathbb{R}^+$. Moreover, this bijection is an isomorphism because $t(f(x)) = g(t(x))$. This is shown by the following derivation:

$$t(f(x)) = 2^{f(x)} \qquad t(x) = 2^x$$
$$= 2^{x+x} \qquad f(x) = x + x$$
$$= 2^x \times 2^x \qquad \text{Arithmetic}$$
$$= g(2^x) \qquad u \times u = g(u)$$
$$= g(t(x)) \qquad 2^x = t(x) \qquad \blacksquare$$

Isomorphisms and homomorphisms are especially important when dealing with functions with two arguments. If f and g are two functions in two arguments, they are homomorphic iff there is a function t satisfying

$$t(f(x, y)) = g(t(x), t(y)) \tag{6.9}$$

In words, the transformation of the image of f is equal to the image g of the transformations. If t is bijective, then f and g are isomorphic.

Example 6.21   Show that the functions $f(x, y) = x + y$ and $g(u, v) = u \times v$ are isomorphic if $x, y \in \mathbb{R}$ and $u, v \in \mathbb{R}^+$.

Solution   As in Example 6.20, set $t = 2^x$. It is easy to verify that this is a bijection. Moreover,

$$t(f(x, y)) = 2^{x+y} = g(t(x), t(y)) = 2^x \times 2^y$$

Consequently, the addition of real numbers is isomorphic to the multiplication of positive real numbers. $\blacksquare$

Example 6.22   Let x and y be two strings, and let $x \frown y$ be their concatenation. Furthermore, let m and n be two natural numbers. Show that $x \frown y$ and $m + n$ are homomorphic.

Solution   Let $length(x)$ be a function from the set of strings to the set of natural numbers. Obviously,

$$length(x \frown y) = length(x) + length(y)$$

This shows that $length$ is a homomorphism, and we are done. The function $length$ is not a bijection because many strings have the same length. Consequently, the homomorphism in question is not an isomorphism. $\blacksquare$

The domain and range of an isomorphism may be identical, in which case one speaks of an *automorphism*. One defines the following:

Definition 6.12:   An isomorphism $t : A \to A$ is called an *automorphism*. A homomorphism $t : A \to A$ is called an *endomorphism*. Two functions f and g are said to be *automorphic* if there is an automorphism between them, and they are *endomorphic* if there is an endomorphism between them.

Example 6.23   Show that Boolean $+$ and Boolean $\cdot$ as defined in Section 2.5.10 are automorphic operators.

Solution   Let $t(x)$ be x', which is the negation of x. It is easily verified that t is a bijection from $\{T, F\}$ to $\{T, F\}$. Moreover, (6.9) instantiates to

$$(x + y)' = x' \cdot y'$$

This holds because of De Morgan's theorem, and the proof is complete.   ∎

Example 6.24   Let $x + y$ be the normal addition involving natural numbers. Furthermore, let u and v represent two digits between 0 and 9, and let $u +_{10} v$ be equal to $u + v$ if $u + v < 10$ and $u + v - 10$ if $u + v > 10$. In other words, after addition one only retains the least significant digit. Show that $+$ and $+_{10}$ are endomorphic.

Solution   Let x be written in decimal notation, and let $t(x)$ be the least significant digit of x. Then
$$t(x + y) = t(x) +_{10} t(y)$$

Consequently, (6.9) applies, which makes $t(x)$ a homomorphism. Moreover, t is a function from $\mathbb{N}$ to $\mathbb{N}$, which makes t an endomorphism.   ∎

This last example can be generalized. As in Section 6.1.8, let $x \bmod y$ be the remainder of the integer division $x \operatorname{div} y$, and let $+_a$ be equal to $(x + y) \bmod a$. Then $u +_a v$ is homomorphic to $x + y$, $x, y \in \mathbb{Z}$, and so is $x \times y$ and $u \times_a v$. The homomorphism is the $\bmod_a$ function. Unfortunately, on most computers the homomorphism will not work for negative y because the implementation of the mod function does not conform with the mathematical definition.

Problems 6.2

1. Consider the set D^3, where $D = \{1, 2, 3\}$.
    (a) Enumerate this set in lexicographical order.
    (b) Enumerate this set by Cantorian diagonalization, which means that all elements of D^3 with a sum of 3 must be enumerated first, then all elements with a sum of 4, and so on.

2. Let $A = 1..n_1$, $B = 1..n_2$, and $C = 1..n_3$. Find a formula that maps all $(i, j, k) \in (A \times B \times C)$ into $1..n_1 n_2 n_3$. Find the reverse transformation; that is, find a formula that finds the triple for each integer in $1..n_1 n_2 n_3$.

3. Find the formulas for doing a Cantorian enumeration of $A \times B$, where A and B are finite sets.

4. Let A be the set of all real numbers between 0 and 1, and let B be the set of all real numbers between 0 and 10. Show that A and B have the same cardinality.

5. Show that the set of rational numbers is enumerable.

6. In how many ways can you place six people around a table, given that one of the six has a fixed seat?

7. In how many ways can you form strings of length 4 over the English alphabet, given that no character in the string can be repeated?

8. In how many ways can you divide a group of 10 people into two groups, provided that each group must contain at least one person?

9. Let $X = \{a, b, c, d\}$ and $V = \{1, 2, 3, 4\}$. Find an isomorphism between $f : X \rightarrow X$ and $g : V \rightarrow V$, with

$$f = \{(a, b), (b, a), (c, c), (d, b)\}, \quad g = \{(1, 2), (2, 1), (3, 3), (4, 2)\}$$

10. Let $X = \{a, b, c, d\}$ and $V = \{1, 2, 3\}$. Find a homomorphism between f and g, $f : X \rightarrow X$ and $g : V \rightarrow V$, with

$$f = \{(a, b), (b, a), (c, c), (d, b)\}, \quad g = \{(1, 1), (2, 2), (3, 1)\}$$

11. Let $A = 1..3$, and let $f, g : A \rightarrow A$ be as follows

$$f = \{(1, 2), (2, 1), (3, 3)\}, \quad g = \{(1, 3), (2, 2), (3, 1)\}$$

Find an isomorphism between the functions f and g.

12. Let $t(x)$ be the function that associates with each $x \in \mathbb{N}$ the sum of its digits, provided that x is written in the decimal system. Is t an isomorphism with respect to addition? If so, show that $t(x + y) = t(x) + t(y)$. Otherwise, give a counterexample.

6.3 COMPUTATIONAL COMPLEXITY

6.3.1 Introduction

Computers are not free resources, and estimates of how much time it takes to solve a problem of a particular size is therefore important. Besides the time, the amount of memory used to execute a program is often relevant. The question of how much time a particular algorithm needs involves the *time complexity* of the algorithm. The *space complexity*, on the other hand, deals with the amount of memory required to solve a particular task. Here we only deal with the time complexity.

The size of a problem can often be measured by the size of the input. For instance, the time required to sort a file or update all records of a file certainly increases with the number of the records in the file and consequently with the size of the input. Moreover, the size of the input can easily be established. However, in some cases other measures are more appropriate. When solving systems of linear equations, for instance, one typically uses the number of equations or, equivalently, the number of variables as a measure of the size of the problem.

Since computers are no longer very expensive, and since they work very fast, one may believe that the time needed to run a particular algorithm on a computer is of no great relevance. Even if the problem takes too long on today's equipment, in a few years, new and better equipment will overcome these problems. Unfortunately, for some algorithms, no matter how fast the computer runs, the execution times are intolerably long for problems of nontrivial size. One may think that in this case better algorithms can be developed, but this may not be possible. In fact, for some problems no efficient algorithms are known, and there are problems for which no efficient algorithms can ever be invented. As will be shown, technological advances will never overcome these obstacles either.

In this section, we first present a number of algorithms and inquire about their computational complexity. A typical computational complexity analysis is independent of the computer used. Instead, it relies on an operation count of some type. In principle, one could count all operations, broken down according to their type, and associate execution times with each. Typically, such a detailed analysis is not needed and approximate methods are sufficient. In numerical analysis, for instance, one only counts the number of floating-point operations, ignoring calculations needed for initializations of for and while loops, increments of indexes, and so on. The number of floating-point operations is frequently referred to as FLOPs. In computer science literature, FLOPs are less popular, even though they are also used to some extent. Authors in computer science typically select certain commands or operations localized at a specific location in the algorithm and count how often these operations are executed. An operation that is selected for the purpose of performing time complexity analysis is called an *active operation*. The active operations must be chosen such that they occur at least as often as any other operation. Otherwise, they would not give a true picture of the complexity. The number of active operations must then be expressed as a function of the variable that describes the problem size. This variable is typically given by the size of the input.

The function relating problem size to time complexity can take different forms. In Section 6.3.2 we consider the case when the complexity can be expressed as a polynomial function of the problem size, and in Section 6.3.3 we consider the case when this function is exponential. Section 6.3.4 compares a number of functions with regard to complexity, a discussion that is continued in Section 6.3.5. It turns out that if the time complexity of an algorithm increases exponentially with the problem size then only small problems can be solved, no matter what technical progress is being made in increasing the speed of computers. Section 6.3.6 describes a useful approach to reduce the complexity, the divide-and-conquer approach. Finally, Section 6.3.7 describes NP-complete problems, a class of problems for which no efficient algorithm is known and for which it is unlikely that an efficient algorithm exists.

6.3.2 Polynomials and Polynomial-time Algorithms

In complexity analysis, polynomials play a major role. They are defined as follows:

Definition 6.13:   Any function of the following form is called a *polynomial*.

$$f(x) = a_0 + a_1 x + a_2 x^2 + \cdots + a_n x^n$$

If a_n is nonzero, then the polynomial is said to be of *degree n*.

For computational complexity issues, one finds that polynomials of degree 3 or less are of particular importance and they are given special names. One has

Constant functions	$f_0(x) = a$
Linear functions	$f_1(x) = a + bx$
Quadratic functions	$f_2(x) = a + bx + cx^2$
Cubic functions	$f_2(x) = a + bx + cx^2 + dx^3$

Note that the constant function is a polynomial of degree 0. Algorithms that increase linearly with problem size are called *linear time algorithms* or *linear algorithms* for short. The terms *quadratic algorithm* and *cubic algorithm* must be understood in a similar way. Of course, linear algorithms, quadratic algorithms, and cubic algorithms are all *polynomial algorithms*. Linear algorithms are extremely frequent. For instance, a program that calculates the sales tax for n items multiplies each of the n numbers by some percentage, and the number of multiplications needed to do this must obviously be n as well. More generally, the complexity of an algorithm is linear if some operation has to be done individually for each element appearing in the input. In these cases, one cannot really reduce the complexity of the algorithm.

As another example of a linear algorithm, consider the addition of n numbers. If the numbers to be added are given as a[1] to a[n], one can use the following algorithm expressed as a Pascal fragment:

```
sum := 0;
for i := 1 to n do   sum := sum + a[i]
```

This algorithm does not have the minimal possible time complexity. To see this, note that the following Pascal algorithm accomplishes this job with n $-$ 1 additions.

```
sum := a[1];
for i := 2 to n do   sum := sum + a[i]
```

The question is, of course, if anyone would care about the saving of one operation, especially since the new algorithm has the disadvantage that it will fail when n $=$ 0. The question as to how to weigh savings in computer time against other considerations, such as clarity and reliability, is an issue to be addressed later. Generally, if an algorithm has a complexity that can be expressed as a polynomial, and if by changing the algorithm, one can lower the degree of the polynomial, then it normally pays to do so. This is not the case here, and the saving does not seem to be worthwhile. In other words, the first, less efficient algorithm for evaluating a sum is almost always preferable.

The Fibonacci numbers $F_0, F_1, F_2, \ldots$ crop up in many computer applications, and they are defined as follows: $F_0 = 0$ and $F_1 = 1$, while for $n > 1$ one has

$$F_n = F_{n-1} + F_{n-2} \qquad (6.10)$$

Sometimes F_m must be found for a given m, such as $m = 10$. To find F_m, (6.10) can be used to find F_2, F_3, and so on, until finally F_m is found. If $m = 2$, one addition is needed, and if $m = 3$, two additions are needed, one to calculate F_2 and one to add F_1 and F_2.

Generally, to find F_m, $m - 1$ additions are needed, and this is a linear function of m. An alternative algorithm is available in this case, however. As will be shown in Section 6.4.2, F_n is equal to

$$F_n = \frac{1}{\sqrt{5}} \left(\frac{1 + \sqrt{5}}{2} \right)^n - \frac{1}{\sqrt{5}} \left(\frac{1 - \sqrt{5}}{2} \right)^n \tag{6.11}$$

Since exponentiation can be done in constant time, this formula can easily be converted into a constant-time algorithm. Note that for small n the resulting algorithm is less efficient than the one based on (6.10). However, as n increases, the constant time algorithm becomes more and more attractive until, at some point, it outperforms the algorithm based on (6.10).

Actually, the amount of calculation required to obtain the Fibonacci numbers can be reduced even further. As a matter of fact, the second term is negligible. The basis of the second term is $(1 - \sqrt{5})/2$, and this evaluates to 0.618. Now, 0.618^2 is 0.382, which is less than 0.5, and 0.618^n is smaller than that, as $n > 2$. This makes the contribution of the second term less than a possible rounding error for $n \geq 2$. Hence, for $n \geq 2$, one has

$$F_n = \frac{1}{\sqrt{5}} \left(\frac{1 + \sqrt{5}}{2} \right)^n$$

rounded to the closest integer.

As an example of a quadratic algorithm, consider the case when the input consists of $2n$ numbers, a_i, $i = 1..n$ and b_i, $i = 1..n$, and the problem is to evaluate the products $a_i b_j$ for all i and j. All these calculations can obviously be done through a double loop, one for i and the other one for j, and the total operations needed are n^2. Hence, if n is taken as a measure of the problem difficulty, then the time complexity $f(n) = n^2$.

The multiplication of two matrices of size n by n has a time complexity of n^3. To see this, let the given matrices be $A = [a_{ij}]$ and $B = [b_{ij}]$. Furthermore, let $C = AB$. According to Section 3.2.4, the elements of C, we call them c_{ij}, are given by

$$c_{ij} = \sum_{k=1}^{n} a_{ik} b_{kj}$$

This sum has n terms, which means that one needs at least $n - 1$ additions. For simplicity, we assume that n additions are needed. Since all terms are products, one also needs n multiplications, which together yield $2n$ FLOPs. Since C has n^2 entries and since this calculation must be done for each entry, this yields $2n^3$ FLOPs. Again, the question arises if a better algorithm exists. As it turns out, if n is high enough, one can multiply two matrices with less than $2n^3$ operations. However, this advanced topic will not be discussed here.

The next example involves the solution of n linear equations in n variables. For doing such a job, one frequently makes use of canned programs. This obviously makes it unnecessary to know the details of these programs. Nevertheless, their complexity is of interest. To discuss this issue, consider Gauss's method, which can be considered as the classical method for solving linear equations. This method consists of two phases, the elimination phase and the backsubstitution phase. If n is the number of variables, then the elimination phase requires $\frac{1}{2}n(n-1)$ divisions, $\frac{1}{3}(n+1)n(n-1)$ multiplications, and the same number of additions. The backsubstitution phase requires $\frac{1}{2}n(n-1)$ additions and the same number of multiplications. If n is small, the computer time to do this many operations is insignificant, and the emphasis should therefore be placed on the case when n is large. For the sake of concreteness, let $n = 100$. Consider now the number of divisions in the elimination phase, which is $\frac{1}{2}n(n-1) = \frac{1}{2}n^2 - \frac{1}{2}n$. For $n = 100$, $\frac{1}{2}n$ is only 1% of $\frac{1}{2}n^2$, and this is negligible. Hence, for values of n of 100 and over, the number of divisions is for all practical purposes $\frac{1}{2}n^2$. The number of multiplications is $\frac{1}{3}(n+1)n(n-1)$, or $\frac{1}{3}n^3 - \frac{1}{3}n$. The term $\frac{1}{3}n$ is again insignificant and can be dropped. One concludes that the number of multiplications in the elimination phase is approximately $\frac{1}{3}n^3$. Since the number of additions is equal to the number of multiplications, there are therefore approximately $\frac{2}{3}n^3$ FLOPs. For $n = 100$, this exceeds the number of divisions by more than a factor of 100; that is, compared to $\frac{2}{3}n^3$, $\frac{1}{2}n^2$ is negligible. The same is true for the operations needed to perform the backsubstitution phase, and one concludes that for high values of n one needs for all practical purposes $\frac{2}{3}n^3$ FLOPs.

More generally, in polynomial algorithms, one only needs to concentrate on the term with the highest power. Unless the terms with a lower power have extremely high coefficients, they can safely be ignored. We will call the term with the highest power the *most significant* or *dominant term* of the polynomial. This should be kept in mind when designing algorithms. A reduction of the degree of a polynomial should be the first priority in designing nontrivial procedures. If this is not possible, one should try to increase the efficiency of all parts of the algorithm that directly contribute to the coefficient of the most significant term. Typically, these are the statements in the innermost loop. The efficiency of all other statements is of minor concern, and other considerations, such as clarity and simplicity, become more relevant.

Sometimes, the number of operations cannot be obtained before running an implementation of the algorithm. One such problem is a sequential search for a particular element in an unordered list. The active operation in such a search is a comparison to see if the desired element has been found. If the list contains n elements and if the item searched for is the head of the list, then one comparison is needed to find the search item, and this is considered to be the *best-case* scenario for the procedure. If, on the other hand, the item is at the end of the list or not in the list at all, then the number of comparisons needed to find the item is to n, and this is considered to be the *worst-case* scenario. One therefore has different complexities, depending on the input data. In this context, one can either analyze the best case or the worst case or one can calculate some type of an average. The average time complexity is in a sense the most meaningful measure, but it has major disadvantages. First, the frequency at which the different cases arise is often difficult to estimate ahead of time. In the example about the search, for instance, one may not know how often one searches for an item that is not in the list. In this case, one would somehow have to rely on estimates, but such estimates may not be obtainable. Sometimes simplifying assumptions can help. For instance, if the item searched for is with certainty in the list, then one can assume that it is in any of the n possible positions with equal likelihood. In addition to the difficulty of obtaining meaningful estimates, one needs stochastic models in order to find the expected number of operations needed, and probabilistic models are often very difficult to analyze and, in some cases, completely intractable by analytical means. It is therefore often impossible to derive appropriate formulas to express the average time complexity.

In cases when it is impractical to find the average time complexity, one still has the choice between the worst-case and the best-case analysis. Of the two, one usually chooses the worst-case analysis, because it is usually more prudent to overestimate rather than to underestimate the computational complexity.

6.3.3 Functions and Algorithms Related to Exponentials

A function is said to be exponential if it is of the form $f(x) = ca^x$. Here a is called the *basis* and x is called the *exponent*. If n is a natural number, one can define the exponential function $f(n) = a^n$ recursively as follows:

$$a^0 = 1$$
$$a^{n+1} = a^n \times a$$

This expression epitomizes the essential feature of the exponential function: its value for $n+1$ is a times its value for n. For instance, population growth tends to be exponential, because the number of people in year $n+1$ tends to be a multiple of the present population size. The increase of computer power seems to be proportional to the present computer power, which makes the increase of computer power an exponential function over time. Another variable that increases exponentially is an investment in which all interest is reinvested. In this case, the investment grows at a rate of $a = 1 + i/100$, where i is the interest rate. In computer science, the exponential function $f(n) = 2^n$ is of particular importance. For instance, in a full binary tree, every branch node has two subtrees, and the number of nodes doubles in each generation. In the last generation, there are 2^{n-1} leaf nodes, where n is the height of the tree.

It is said the inventor of chess, when asked by a king what he would like as a reward, replied that he had only one small wish. He would like to have one grain of wheat for the first square of the board, two for the second, four for the third, eight for the fourth, and so on, until the 64th square. The king granted the wish without much further thought. The king certainly did not keep his word. 2^{64} makes 1.8446×10^{19}, and at 0.0648 grams a grain, this makes 1.1953×10^{12} tons. This exceeds the current world production of wheat by far. The example demonstrates dramatically how fast an exponential function grows. Of course, this growth is not restricted to the example. The truth table of an expression with 64 variables has more rows than there are grains of wheat in the world. Moreover, a binary tree with 64 levels is sufficiently large to have a node for every single grain of wheat harvested in the world.

The exponential function can be extended to hold for negative, rational, and even real exponents. Since these results are well known, they are not proved here. The following relations hold for all x and y.

$$a^x a^y = a^{x+y}$$
$$a^{-x} = 1/a^x \tag{6.12}$$
$$(a^x)^y = a^{xy}$$

It is mathematically convenient to use the number e as the basis for the exponential function, where e is defined as

$$e = \sum_{n=0}^{\infty} 1/n!$$

When this basis is used, one has, as shown in books on calculus,

$$f(x) = e^x = \sum_{n=0}^{\infty} x^n/n! \tag{6.13}$$

Often, e^x is simply referred to as *the* exponential function.

The inverse of the exponential a^x is the base a logarithm, as indicated by the following definition:

Definition 6.14:   Let x be the solution of the equation $a^x = y$. Then x is called the base a logarithm of y, and it is written as $\log_a y$.

For instance, the base 2 logarithm of 8 is $\log_2 8 = 3$ because $2^3 = 8$. Definition 6.14 implies that

$$a^{\log_a y} = y \tag{6.14}$$

Moreover, the base e logarithm of a number y is normally denoted by $\ln y$. Consequently, $\ln y = \log_e y$. The following equation relating logarithms of different bases is well known, and it will not be proved here.

$$\log_b y = \log_a y/(\log_a b) \tag{6.15}$$

Logarithms allow one to convert between bases. One has

$$b^y = a^{y \log_a b}$$

To see why this is the case, note that by (6.14) $b = a^{\log_a b}$. Hence, $b^y = (a^{\log_a b})^y$, and this is $a^{y \log_a b}$ according to (6.12).

The factorial is closely connected to the exponential by the following formula, known as *Stirling's formula*:

$$n! \approx \sqrt{2\pi n}(n/e)^n$$

More precisely, one has

$$\sqrt{2\pi n}(n/e)^n \le n! \le \sqrt{2\pi n}(n/e)^{n+1/(12n)}$$

Some algorithms having exponential or factorial complexity are now discussed. The first algorithm, which uses recursion to calculate the Fibonacci numbers, shows that it is important to avoid exponential algorithms. In fact, the algorithm that follows shows what *not* to do, and it should never be used.

```
function fib(n: integer): integer;
  begin
    if n < 2 then fib := n  {fib(0) = 0, fib(1) = 1}
      else fib := fib(n − 1) + fib(n − 2)
      end
  end
```

We now calculate the number of calls generated by the else clause; that is, these function calls are considered to be the active operations. However, any other choice of active operation will lead to similar results. The number of active operations needed to calculate $F_n = \text{fib}(n)$ is denoted by $T(n)$. Clearly, $T(0) = T(1) = 0$. Moreover, to obtain fib(n), n $\geq$ 2, two calls are executed immediately, one to fib(n − 1), the other one to fib(n − 2). By the definition of $T(n)$, these calls require $T(n − 1)$ and $T(n − 2)$ active operations, respectively. To find F_n, one therefore needs $2 + T(n − 1) + T(n − 2)$ calls; that is,

$$T(n) = 2 + T(n − 1) + T(n − 2) \tag{6.16}$$

One can calculate $T(n)$ as follows:

$$T(2) = 2 + T(1) + T(0) = 2$$
$$T(3) = 2 + T(2) + T(1) = 2 + 2 = 4$$
$$T(4) = 2 + T(3) + T(2) = 2 + 4 + 2 = 8$$
$$\vdots \qquad \vdots$$

It turns out that for $n = 100$ there are roughly 10^{21} recursive function calls. At a rate of a million function calls per second, this yields roughly 10^{15} seconds, or 10 million years. In other words, the calculation of F_n by the algorithm just given is infeasible for high values of n. Note that (6.16) represents a nonhomogeneous recurrence relation, and its solution will be derived in Section 6.4.3. One has

$$T(n) = -2 + \frac{2}{\sqrt{5}} \left(\frac{1 + \sqrt{5}}{2} \right)^{n+1} - \frac{2}{\sqrt{5}} \left(\frac{1 - \sqrt{5}}{2} \right)^{n+1} = 2F_{n+1} - 2$$

Consequently, $T(n)$ increases exponentially like F_n.

In Chapter 1, the truth table method was used extensively to determine whether an expression is a tautology. In a truth table, one evaluates the truth value of a logical expression for each assignment. Since the number of assignments doubles with each variable added, this means that the computational effort also doubles, and it reaches 2^n for expressions containing n variables. Hence, even if the number of operations for evaluating the truth value under a particular assignment can be as low as $n - 1$, the number of operations to find the truth value of all 2^n assignments is $(n-1)2^n$, and (except for the factor $n-1$) this is an exponential function. The truth table method can be used as the basis for an algorithm to find whether an expression is a tautology. For the sake of completeness, we provide part of a Pascal program to find all assignments in Figure 6.5. The possible assignments are stored in a Boolean vector Assignment, and the entries of this vector are determined recursively by the procedure Assign. As soon as a new assignment is found, a separate function Evaluate is called, which then obtains the truth value. This function must be tailored to the expression in question, and it is therefore not given here.

```
program TruthTable(input,output);
    var
        Assignment : array [1..n] of boolean;
        n : integer;
    procedure Assign(first : integer);
        begin
            if first > n then Evaluate
            else
                begin
                    Assignment[first] := true;
                    Assign(first + 1);
                    Assignment[first] := false;
                    Assign(first + 1)
                end
        end
    begin
        Assign(1)
    end.
```

Figure 6.5.   Truth table program

This algorithm has acceptable execution times for n up to 10, at which time the truth table contains 1028 rows. However, according to (6.12), 2^{10n} is equal to $(2^{10})^n$. Hence, for $n = 20$ one has truth tables of 1028^2, for $n = 30$ truth tables of size 1028^3, and so on. At this rate, it does not take long until even the fastest computer needs years to come up with an answer. In all these calculations, the time to do the procedure Evaluate has been ignored. If the complexity of Evaluate is taken into account, the complexity increases by a factor of at least $n - 1$, but compared to the effect of exponential growth, this factor is insignificant.

There are obvious ways to speed up the algorithm. Since one only needs a single assignment that makes the expression false, one could obviously stop as soon as such an assignment is encountered. Although this does not affect the worst case, it certainly improves the average performance of the algorithm. However, even the average complexity may grow exponentially. In fact, if the probability is high that the expression in question is indeed a tautology, then the worst case is the most frequent case. Other methods can also be used to speed up the algorithm, but all algorithms known so far for finding if an expression is a tautology have a worst-case complexity that increases exponentially with the number of variables. In fact, there is strong evidence that no good algorithm to solve this problem can ever be found. This will be discussed later.

The last algorithm of this section deals with the *traveling salesperson problem*, a famous problem in the area of complexity. This problem can be described as follows: A salesperson has to visit n cities by car. The person starts in one of the cities, and he or she has to return to the starting point at the end of the tour after having visited all the cities. He or she wants to do the visits in such an order that the total distance of the tour is minimized. One way to approach this problem is to enumerate all possible orders to visit the cities and calculate the length of each possible tour. This length can be found by adding the n distances between the n cities, which requires $n - 1$ additions. There are $(n - 1)!$ possible tours, corresponding to the $(n - 1)$ ways the order of the cities can be permuted, given that the starting point and the end point are fixed. Hence, this algorithm requires $(n - 1)(n - 1)!$ additions, or approximately $n!$ additions. Unfortunately, as will be shown in the next section, the computational cost for doing $n!$ operations is prohibitive unless n is very small. There are better methods to solve this problem. However, all algorithms known so far for solving the traveling salesperson problem have an exponential time complexity and, for reasons to be discussed later, it is unlikely that better algorithms exist.

6.3.4 The Limits of Computability

If there is an algorithm with a complexity of $f(n)$, where $f(n)$ is some function of the size n of the problem, the question arises as to how large n can be such that the problem remains solvable within a given time frame. Of course, technology improves, which leads to the second question: To what extent does n increase as computers become faster? We now explore different functions $f(n)$ with a view as to what is the highest value of n for which solutions can reasonably be found. To make this question concrete, we assume that $f(n)$ measures the time complexity in FLOPs and that the algorithm must complete within m FLOPs, where m varies from 10^3 to 10^{15}. The results are given in Table 6.3. The values for m were chosen to reflect the entire spectrum of the possible computational requirements. Whereas it is possible to do 10^3 operations manually, it is not possible to do 10^6 operations manually. However, 10^6 operations are no problem for a computer. In fact, we assume that a computer can do 10^6 operations in a second. If this is true, then 10^9 operations take 17 minutes, which is still acceptable for an occasional calculation. 10^{12} operations take about twelve 24-hour days, and 10^{15} operations take 32 years. Of course, 32 years is a long time, but if there is a real need for it, this time can be compressed by using parallelism and/or supercomputers. This means that all the times are still within the grasp of humankind, although not necessarily in the grasp of an individual researcher. We now discuss some results of Table 6.3. For $f(n) = n^4$, large problems become quite costly to solve. If $f(n) = 2^n$, there is no way that large problems can be solved in a practical way. The maximum problem size in this case that is solvable within 10^{15} FLOPs is $n = 49$. If $f(n) = n!$, the situation is still worse.

The problems that are hard to solve at present will remain hard to solve in the future. In fact, the benefits from increasing the computing power by a factor of 1000 decrease as problem difficulty increases. If $f(n) = n^2$, a thousand-fold increase in computer power increases the maximum problem that one can solve by a factor of roughly 32. If $f(n) = n^4$, this factor shrinks to 5.6. If the function is exponential, as is the case with $f(n) = 2^n$

TABLE 6.3.   Problem Size Solvable with m Operations

	Maximum Size of Problem				
$f(n)$	10^3	10^6	10^9	10^{12}	10^{15}
n	10^3	10^6	10^9	10^{12}	10^{15}
n^2	31	1000	31,622	1,000,000	31,622,777
n^3	10	100	1000	10,000	100,000
n^4	5	31	177	1,000	5,623
2^n	9	19	29	39	49
10^n	3	6	9	12	15
$n!$	6	9	12	14	17

or 10^n, an increase of computer power by a factor of 1000 increases the largest solvable problem only by 10 and 3, respectively. Hence, even though the size of the problems with linear and even quadratic complexity that become solvable in the future increases greatly, no such progress is possible for problems that have an exponential complexity. This tendency becomes more pronounced as computer speed increases. The increase of computer speed mainly benefits the simple problems, such as problems with polynomial complexity, whereas hard problems, that is, problems with exponential complexity, barely benefit from technological progress. This means that hard problems will comparatively become even harder.

The next question that must be addressed is what happens if, instead of $f(n)$, the complexity is $cf(n)$, where c is some constant. To explore this question, consider the function $f(n) = 4n^2$. A problem with this complexity requires four times the execution time of a problem with $f(n) = n^2$, which seems to be much larger. Yet, in terms of the size of the maximum problem solvable, the difference is much smaller. If the maximum number of FLOPs allowed is b, then the largest problem that can be solved if $f(n) = n^2$ is equal to $\sqrt{b}$, whereas in the case where $f(n) = 4n^2$, the largest problem that can be solved is $\sqrt{b}/2$. In other words, the largest problem that can be solved decreases only by a factor of 2, rather than a factor of 4. As the degree of $f(n)$ increases, this tendency becomes even more pronounced. When comparing $f(n) = 4n^4$ with $f(n) = n^4$, one finds that the largest problem solvable in b operations increases only by a factor of approximately 1.41. Hence, if the complexity of an algorithm is polynomial, the degree of the polynomial alone already gives important indications about the maximum size of the problems that can be solved by the algorithm.

6.3.5 Asymptotic Analysis

In this section, we consider the behavior of a function $f(n)$ as n becomes large. This type of analysis is done frequently within mathematics. Moreover, it has become increasingly important in computational complexity. Indeed, for almost every algorithm published, an asymptotic analysis is provided. Of course, such an analysis has its limitations. By definition, it is only valid if n is large.

In a polynomial, the term with the highest power dominates. This will now be explored in a rigorous way. Let $f(n)$ be a polynomial of degree d; that is, $f(n)$ can be written as

$$f(n) = \sum_{i=0}^{d} a_i n^i$$

The term with the highest power, or the dominant term of $f(n)$, is $a_d n^d$, and for high enough n, all other terms become negligible. To show this, we demonstrate how to find a value n_c for each $c > 0$ such that, for all $n > n_c$, the dominant term is c times as large as all other terms combined. Specifically, it will be shown how to find a threshold value n_c such that, for $n > n_c$, the following inequality is satisfied.

$$c \left| \sum_{i=0}^{d-1} a_i n^i \right| \le |a_d| n^d \tag{6.17}$$

For $n > 1$, this implies that

$$\frac{c}{|a_d|} \left| \sum_{i=0}^{d-1} a_i n^{i-d+1} \right| < n$$

Since the absolute value of a sum is less than or equal to the sum of the absolute values, one has

$$\left| \sum_{i=0}^{d-1} a_i n^{i-d+1} \right| \le \sum_{i=0}^{d-1} |a_i| n^{i-d+1}$$

Furthermore, for $n > 1$,

$$\sum_{i=0}^{d-1} |a_i| n^{i-d+1} \le \sum_{i=0}^{d-1} |a_i|$$

It follows that, by choosing $n > (c/|a_d|) \sum_{i=0}^{d-1} |a_i|$, (6.17) is always satisfied. Hence,

$$n_c = \frac{c}{a_d} \sum_{i=0}^{d-1} |a_i| \tag{6.18}$$

Example 6.25   Find the threshold value n_c such that, for $n > n_c$, the dominant term of $10 - 25x + 2x^2 + 2x^3$ exceeds all other terms by a factor of 10. Hence, since $c = 10$, $n_c = n_{10}$.

Solution   According to (6.18), one must add the absolute values of all nondominant coefficients and multiply the result by $c/|a_d| = \frac{10}{2}$. This yields $5(10 + 25 + 2) = 185$. Hence, for $n > 185$, the term $2x^3$ exceeds all other terms by a factor of 10. In other words, for $n > 185$, $2x^3$ does not deviate from $f(n)$ by more than 10%. ∎

If the dominant term exceeds all other terms, it will determine the sign of the polynomial. Hence, if $f(n) = \sum_{i=0}^{d} a_i n^i$, then for large n the sign of $f(n)$ is given by the sign of a_d. If $a_d > 0$, then $f(n) > 0$ for large enough n, and if $a_d < 0$, then $f(n) < 0$ for large enough n.

If there are two polynomials $f(n)$ and $g(n)$ and if the degree of $f(n)$ is lower than the degree of $g(n)$, then one can always find a threshold n_c such that, for $n > n_c$, $c|f(n)| < |g(n)|$. To see this, assume that the dominant terms of $f(n)$ and $g(n)$ are both positive. Clearly, $cf(n) < g(n)$ iff $g(n) - cf(n) > 0$, and since $g(n)$ contains the dominant term of the difference, one can always find a value n_c such that, for $n > n_c$, $g(n) - f(n) > 0$. If, on the other hand, the dominant term of $f(n)$ is negative, then $f(n) < 0$ for large enough n, and $|f(n)| = -f(n)$. The dominant term of $-f(n)$ is now positive and the previous analysis applies. Similarly, if $g(n)$ has a negative dominant term, then $|g(n)| = -g(n)$ for large enough n. In either case, one concludes that there is an n_c such that, for $n > n_c$, $c|f(n)| < |g(n)|$.

Example 6.26   Let $f(n) = 25 + 10x + 4x^2$ and let $g(n) = 10 - 10x - 20x^2 + x^3$. Find a value $n_c = n_1$ such that, for $n > n_1$, $f(n) < g(n)$.

Solution   One has

$$g(n) - f(n) = -15 - 20x - 16x^2 + x^3$$

By (6.18), one can find an n_c such that, for $n > n_c$, $x^3 > |-15 - 20x - 16x^2|$. In fact, (6.18) yields $n_c = 51$. For $n > 51$, both $f(n)$ and $g(n)$ are positive, and 51 is the desired threshold.   ∎

We now extend the concepts developed for polynomials to other functions. To this end, let $f(n)$ and $g(n)$ be two functions. If, for each $c > 0$, one can find an n_c such that, for $n > n_c$, $c|f(n)| < |g(n)|$, we write $f(n) \prec g(n)$. Consider the functions of the form $f(n) = n^a$, where a is some real number. In this case

$$n^a \prec n^b \quad \text{if} \quad a < b$$

This follows because $cn^a < n^b$ means $c < n^{b-a}$, and since $b > a$, one can always find an n_c such that, for all $n > n_c$, n^{b-a} exceeds c. Similarly, one has for exponentials

$$a^n \prec b^n \quad \text{if} \quad a < b$$

To see this, note that $ca^n < b^n$ iff $c < (b/a)^n$, and it is now immediately obvious that there must be an n_c such that, for $n > n_c$, this inequality is always satisfied. Next, if n^d and a^n are functions of n, then

$$n^d \prec a^n \quad \text{for all } d \text{ and } a \tag{6.19}$$

The proof of (6.19) is left to the reader. A summary of the main results, together with some new ones, follows:

1. $k \prec \log_a n$ for all k, and all $a > 0$

2. $\log_a n \prec n^b$ for all $a > 0, b > 0$

3. $n^b \prec n^c$ for all $b < c$

4. $n^c \prec d^n$ for all $c, d > 0$

5. $d^n \prec h^n$ for all $0 < d < h$

6. $h^n \prec n!$ for all h

Consider now two programs for doing a particular task, F and G. Suppose further that F has the complexity $f(n)$, and G has the complexity $g(n)$, where n is the problem size. If $f(n) \prec g(n)$, one would tend to prefer F, because for large problems $f(n) < g(n)$. For instance, if F solves a problem of size n with $10n^2$ FLOPs, whereas G needs n^3 FLOPs, then F seems to be preferable to G. In fact, for $n > 10$, $f(n) < g(n)$. This is indeed a good guideline, even though there are exceptions. For instance, if $f(n) = n^{10}$ and if $g(n) = 1.1^n$, then $f(n) \prec g(n)$, yet there is no practical case when F would run faster than G. In fact, the threshold n_c is far beyond what can be calculated by computers.

We now consider sums of functions. If $f(n)$ and $g(n)$ are two functions and if both $f(n)$ and $g(n)$ are eventually positive, then the sum $f(n) + g(n)$ is dominated by $f(n)$ if $f(n) \succ g(n)$ and by $g(n)$ if $g(n) \succ f(n)$. Hence, if $f(n) \succ g(n)$, one can say that $f(n)$ *dominates* $g(n)$. For instance, a polynomial can be thought of as a sum of monomials, that is, as a sum of expressions of the form $a_i n^i$. In this case, the monomial with the highest power dominates, and all other terms become comparatively insignificant. The same is true for any other sum of functions. If $f(n) \prec g(n)$, then one can always choose a value n_c such that $cf(n) < g(n)$, which implies that $f(n) + g(n) \le (1/c)g(n) + g(n)$, and by making c large, one can make the contribution of $f(n)$ negligible.

Example 6.27   Show that $4 \cdot 2^n + 2 \cdot 3^n$ approaches $2 \cdot 3^n$.

Solution   Since $d^n \prec h^n$ for $d < h$, $2^n \prec 3^n$, and one only needs to retain the terms containing 3^n. Hence, as n increases, $4 \cdot 2^n + 2 \cdot 3^n$ approaches $2 \cdot 3^n$.   ∎

Example 6.28   Show that $2n \log_2 n + 4n$ approaches $2n \log_2 n$.

Solution   Since $1 \prec \log_2 n$, $n \log_2 n \prec n$, and the result follows.   ∎

We now introduce the *big O* or order notation. Technically, if $g(n)$ is some function, then $O(g(n))$ is the set of all functions $f(n)$ such that for $n > n^*$ there exists a $c > 0$ satisfying $f(n) \le cg(n)$, assuming that $f(n)$ and $g(n)$ are nonnegative. In mathematical symbols, this translates to the following:

$$O(g(n)) = \{f(n) \mid \exists c \exists n^* \forall n((n > n^*) \Rightarrow f(n) \le cg(n))\}$$

Compared to the definition of $\prec$, there are two differences. First, the factor c is no longer with $f(n)$, but with $g(n)$. This, however, is easily corrected. One merely replaces c by $1/c$. The big change is that $f(n) \prec g(n)$ if $cf(n) < g(n)$ for all c, whereas for $f(n) \in O(g(n))$, it is sufficient that there be one c. Hence, whereas for defining $\prec$ the c was universally quantified, it is now existentially quantified. For this reason, the symbol n^*, rather than n_c, is used to denote the threshold.

Clearly, if $f(n)$ and $g(n)$ are positive, $f(n) \le cg(n)$ if and only if $f(n)/g(n) \le c$. Hence, to find if $f(n) \in O(g(n))$, it must be established that for $n > n^*$ the ratio $f(n)/g(n)$ is bounded in the sense that it never exceeds the value c. This discussion can be generalized to the case where either $f(n)$ or $g(n)$ is negative. In this case, one merely uses absolute values, and one concludes that $|f(n)| \le c|g(n)|$ if and only if $|f(n)/g(n)| \le c$.

Example 6.29   Which of the following functions are of order $O(n^3)$? (a) n^2, (b) $3n^2$, (c) $n^2 + 2n^3$, (d) n^4.

Solution   To find if $f(n) \in O(g(n))$, check if $|f(n)/g(n)|$ is bounded when n is above some threshold value n^*. In detail, one has the following:

(a) $n^2 \in O(n^3)$ because n^2/n^3 is bounded by 1 for $n \geq 1$.

(b) $3n^2 \in O(n^3)$ because $3n^2/n^3$ is bounded by 3 for $n \geq 1$.

(c) $n^2 + 2n^3 \in O(n^3)$ because $(n^2 + 2n^3)/n^3$ is $(1/n) + 2$, and this is less than or equal to 3 for $n \geq 1$.

(d) n^4 is not in $O(n^3)$ because $n^4/n^3 = n$ is not bounded.   ∎

Generally, all polynomials of degree d or less are in $O(x^d)$.

Example 6.30   Is $2^n \in O(3^n)$? Is $3^n \in O(2^n)$?

Solution 2^n is of order $O(3^n)$ because $2^n/3^n$ is $(2/3)^n$, and this is bounded. On the other hand, 3^n is not of order $O(2^n)$ because $(3/2)^n$ is 1.5^n, and this diverges.   ∎

Example 6.31   Is $\log_2 n \in O(\log_3 n)$? Is $\log_3 n \in O(\log_2 n)$?

Solution   According to (6.15), $\log_2 n = \log_3 n/(\log_3 2)$. Consequently, $(\log_2 n)/(\log_3 n)$ is bounded, which means that $\log_2 n \in O(\log_3 n)$. Similarly, $(\log_3 n)/(\log_2 n)$ is bounded, which means that $\log_3 n \in O(\log_2 n)$. Hence, $\log_2 n \in O(\log_3 n)$ and $\log_3 n \in O(\log_2 n)$ both hold.   ∎

If $f(n) \prec g(n)$, then $f(n) \in O(g(n))$, as one can easily verify. However, the set $O(g(n))$ contains functions $h(n)$ for which $h(n) \prec g(n)$ does not hold. For instance, $2x^3$ is in $O(x^3)$, but $2x^3 \prec x^3$ is not true. This leads to the notion of functions that are asymptotically similar.

Definition 6.15:   Two functions $f(n)$ and $g(n)$ are said to be *asymptotically similar* if $|f(n)/g(n)|$ and $|g(n)/f(n)|$ are both bounded. In this case, one writes $f(n) \asymp g(n)$.

In other words, two functions are asymptotically similar if for large n they only deviate by a constant. By definition, the set $O(g(n))$ contains all functions $f(n)$ that are either dominated by $g(n)$ or that are asymptotically similar to $f(n)$.

Frequently, one uses the set $\Theta(g(n))$, which is given by

$$\Theta(g(n)) = \{f(n) \mid f(n) \asymp g(n)\}$$

Hence, $\Theta(g(n))$ is the set of all functions asymptotically similar to $g(n)$. It follows that $f(n) \in \Theta(g(n))$ iff $f(n) \asymp g(n)$. Alternatively, $f(n) \in \Theta(g(n))$ iff $f(n) \in O(g(n))$ and $g(n) \in O(f(n))$. For instance, if $f(n)$ has the same degree as $g(n)$, then $f(n) \in \Theta(g(n))$.

Example 6.32   The function $f(n) = 2n^3 + 10$ is in $\Theta(n^3)$, because n^3 and $f(n)$ both have a degree of 3. However, $g(n) = 2n^2 + n$ is not in $\Theta(n^3)$, even though it is in $O(n^3)$.

If $f(n) = \log_a n$, and if $g(n) = \log_b n$, then $f(n) \asymp g(n)$. This follows directly from (6.15).

6.3.6 Divide and Conquer

In the divide and conquer approach, one divides a given problem into two subproblems, solves the subproblems separately, and combines the results to find the solution of the entire problem. This approach is demonstrated by the following two examples. The divide and conquer approach is usually used recursively, but to keep the explanation simple, this will be shown later.

Example 6.33   Two programs are given, and they are called sort and merge. The program sort can sort a file with n records in time an^2. The program merge can combine two sorted files into a single sorted file. The time needed by merge is bn, where n is the total number of records of both files combined. Outline how to use the divide and conquer approach to sort a file, and show how the complexity of the sort may be reduced in this fashion.

Solution   If the file is divided into two subfiles, sorting the subfiles needs $a(n/2)^2$ time units each, or $2a(n/2)^2$ time units for both subfiles. The merge of these two subfiles has a complexity of bn. Consequently, one must compare the original algorithm, which requires an^2 operations, with $2a(n/2)^2 + bn$, the time needed to sort and subsequently merge two files of size $n/2$. Hence, divide and conquer is advantageous as long as

$$2a(n/2)^2 + bn < an^2$$

or

$$n > 2b/a$$

Hence, as long as $n > 2b/a$, the divide and conquer approach is always advantageous.   ∎

Example 6.34   Let A be a matrix, and assume that A^n must be evaluated, where n is even. Use the divide and conquer approach to solve this problem, and show under which conditions it reduces the complexity. You may assume that a procedure is available to calculate the product of two matrices.

Solution   The first method to find A^n is to determine first A^2, then A^3, and so on. To find A^n, one multiplies A^{n-1} by A. This method will be referred to as the direct method. The direct method requires $n-1$ matrix multiplications. In the divide and conquer approach, one calculates $D = A^{n/2}$ first and then multiplies D by itself. If D is found by the direct method, this requires $n/2 - 1$ matrix multiplications. One additional matrix multiplication is used to find D^2, which means that the divide and conquer approach requires $n/2$ matrix multiplications. This is almost half of the direct method. The divide and conquer approach is only effective if n is at least 2. In fact, for $n = 2$ the divide and conquer approach and the direct approach become identical. The divide and conquer approach can be adapted to the case when n is odd. In this case, one evaluates A^{n-1} as before and multiplies the result by A, which leads to $(n-1)/2 + 1$ matrix multiplications.   ∎

Clearly, when the subproblem is large enough to be divided again, then it should be so divided. Hence, problems can be divided recursively into smaller and smaller subproblems. Moreover, for low values of n the solution of the original problem often becomes so trivial that no specific routine is needed any more to solve it. For instance, for $n = 1$, $A^n = A$. Also, files containing one record only are always sorted. This fact was exploited in the merge sort given in Section 3.4.3.

We now analyze the computational complexity of the merge sort. Before doing this, we summarize the merge sort as follows. In this description, n is the size of the file.

1. Split the file into two subfiles: the first subfile contains the records from 1 to $\lfloor n/2 \rfloor$, and the second subfile the records from $\lfloor n/2 \rfloor + 1$ to n.

2. Recursively sort both subfiles.

3. Merge the two subfiles.

It is assumed that a program is available to merge two subfiles with a combined length of n by using n active operations. The time needed to split files and the time needed for procedure calls are ignored. The problem is to find $T(n)$, the number of active operations for sorting the entire file. Clearly, if $T(n)$ is the complexity for sorting a file of length n, the complexity of sorting the files containing the records from 1 to $\lfloor n/2 \rfloor$ and from $\lfloor n/2 \rfloor + 1$ to n is $T(\lfloor n/2 \rfloor)$ and $T(n - \lfloor n/2 \rfloor)$, respectively. Recursively sorting these two subfiles therefore requires $T(\lfloor n/2 \rfloor) + T(n - \lfloor n/2 \rfloor)$ active operations. The third step of the procedure, the merge, requires by assumption n active operations. $T(n)$, the complexity of sorting a file with n records, is therefore, for $n \geq 2$,

$$T(n) = T(\lfloor n/2 \rfloor) + T(n - \lfloor n/2 \rfloor) + n \tag{6.20}$$

Since files of size $n = 1$ are always sorted, $T(1) = 0$.

The relation (6.20), together with $T(1) = 0$, allows one to find $T(n)$ for all n. For instance, $T(10)$ equals

$$T(10) = T(5) + T(5) + 10$$

Here

$$T(5) = T(2) + T(3) + 5$$

Since $T(2) = T(1) + T(1) + 2 = 2$ and since $T(3) = T(2) + T(1) + 3 = 5$, $T(5) = 12$ and $T(10) = 34$.

Even though (6.20) allows one to find $T(n)$ for all n, a mathematical formula for finding $T(n)$ is preferred. To derive such a formula, (6.20) must be simplified. To make this possible, it is assumed that n is divisible by 2. Equation (6.20) can then be written as

$$T(n) = 2T(n/2) + n \tag{6.21}$$

For this to hold until $n = 1$, n must be a power of 2. Specifically, let $n = 2^i$. By definition, $i = \log_2 n$. After defining

$$G(i) = T(2^i) \tag{6.22}$$

(6.21) becomes

$$G(i) = 2G(i - 1) + 2^i \tag{6.23}$$

This holds for all $i > 0$. If $i \geq 2$, one can replace i by $i - 1$, which yields

$$G(i - 1) = G(i - 2) + 2^{i-1}$$

The right side of this expression can be used to replace $G(i - 1)$ in (6.23), which yields

$$G(i) = 2(2G(i - 2) + 2^{i-1}) + 2^i = 2^2 G(i - 2) + 2^i + 2^i$$

From this equation, one can eliminate $G(i - 2)$ by using (6.23) with i replaced by $i - 2$, which yields

$$G(i) = 2^3 G(i - 3) + 2^i + 2^i + 2^i = 2^3 G(i - 3) + 3 \cdot 2^i$$

This suggests that

$$G(i) = 2^j G(i - j) + j \cdot 2^i$$

If $i = j$, this guess leads to the following formula:

$$G(i) = 2^i G(0) + i \cdot 2^i \tag{6.24}$$

This tentative formula will now be proved by induction. To establish the base of induction, set $i = 0$, in which case (6.24) becomes $G(0) = 2^0 G(0) + 02^i$, and this is $G(0)$. For the inductive step, assume (6.24) for i fixed and prove that

$$G(i + 1) = 2^{i+1} G(0) + (i + 1)2^{i+1}$$

This is done as follows:

$$
\begin{aligned}
G(i + 1) &= 2G(i) + 2^{i+1} & &\text{(6.23) with } i := i + 1 \\
&= 2(2^i G(0) + i \cdot 2^i) + 2^{i+1} & &\text{By inductive hypothesis} \\
&= 2^{i+1} G(0) + (i + 1)2^{i+1} & &\text{Algebra}
\end{aligned}
$$

This completes the proof of (6.24). To use (6.24), one still needs $G(0)$. By definition [see (6.22)], $G(0) = T(2^0) = T(1)$ and $T(1) = 0$. Consequently, $G(0) = 0$ and

$$G(i) = i \cdot 2^i$$

or

$$T(2^i) = i \cdot 2^i$$

Since $2^i = n$ and $i = \log_2 n$, this yields

$$T(n) = n \log_2 n$$

One concludes that the complexity is $\Theta(n \log_2 n)$.

The type of analysis used here is typical for a complexity analysis in the divide and conquer approach. If the problem is divided into two parts in each step, one finds an expression for $n = 2^i$, which can then be analyzed. Since $i = \log_2 n$, this typically leads to an expression involving logarithms.

6.3.7 Nondeterministic Polynomial

As an example of an exponential algorithm, we discussed earlier a method to find if a logical expression is a tautology. This problem is closely related to the *CNF satisfiability problem*. In the CNF satisfiability problem, one is given a logical expression in conjunctive normal form (CNF form), and the problem is to find whether there is an assignment that makes the expression true. Of course, if there is no such assignment satisfying a given conjunctive normal form, then the expression in question is a contradiction, and its negation is a tautology. Hence, the CNF satisfiability problem can be solved by the truth table method. However, as was shown earlier, the truth table method results in an exponential algorithm. More specifically, the algorithm is exponential in the number of variables. Here, it turns out to be more convenient to measure the problem size in terms of the input size. However, if we assume that the number of variables increases approximately linearly with the input, and this is a reasonable assumption, then either measure leads to a complexity that is exponential.

The CNF satisfiability problem is a *decision problem*. This means that for any given expression there either is or is not an assignment satisfying the expression. Moreover, if the decision is positive, there is an assignment, and it can be verified in polynomial time if the assignment in fact satisfies the expression. Many problems have a similar structure as the CNF satisfiability problem, and all these problems are called *nondeterministic polynomial* or NP. All NP problems are decision problems. Moreover, in NP problems the decision as to whether or not a tentative solution satisfies the problem can be made in polynomial time.

Another NP problem involves scheduling final exams. In this problem, a number of students write final exams, and the schedule must be designed in such a way that no student has two final exams at the same time. A schedule that meets this constraint is called *feasible*. The decision problem now is whether there exists a feasible schedule in a given situation. The problem is again NP, because if a timetable is given, its feasibility can be verified in polynomial time. This problem will be referred to as the *scheduling problem*.

There is an additional important relationship between the CNF satisfiability problem and the scheduling problem: there is a polynomial time algorithm C_1 that can convert any scheduling problem into a CNF satisfiability problem, and there is a polynomial time algorithm C_2 that can convert any CNF satisfiability problem into a scheduling problem. We will not give these algorithms here, but they do exist. This fact has a major implication that is now explored. Since C_1 is polynomial and since the output produced by any algorithm must be less than the number of operations it performs, the output of C_1 is polynomial. Hence, if t is the size of the output produced by C_1, then t is given as

$$ t = \sum_{i=0}^{d} a_i n^i $$

Suppose now that an algorithm P_1 is discovered that can solve the CNF satisfiability problem in polynomial time. Specifically, if m is the size of the input, then the complexity of P_1 is given as

$$\sum_{j=0}^{g} b_j m^j$$

We can now use C_1 and P_1 together in order to construct a polynomial algorithm to solve the scheduling problem in polynomial time. One uses C_1 first to convert the scheduling problem into the CNF satisfiability problem and feed the output into P_1. This means that $t = m$, and the complexity of P_1 (expressed in terms of n) becomes

$$\sum_{j=0}^{g} b_j \left(\sum_{i=0}^{d} a_i n^i \right)^j$$

It is not difficult to see that this is again a polynomial. In fact, its degree is dg. Consequently, by stringing C_1 and P_1 together, a polynomial algorithm results. This means that if there is a polynomial algorithm to solve the CNF satisfiability problem then there is also a polynomial algorithm to solve the scheduling problem. In a similar fashion, one can show that if there is an algorithm P_2 that solves the scheduling problem in polynomial time then one can combine C_2 with P_2 to construct a polynomial time algorithm for solving the CNF satisfiability problem in polynomial time.

A great number of different problems are in the same relation to the CNF satisfiability problem as is the scheduling problem. All these problems together form an equivalence class, and every problem in this equivalence class is called NP-complete. Specifically, problem X is NP-complete if X is NP, and if there exist two polynomial time algorithms, say D_1 and D_2, such that D_1 converts X into the CNF satisfiability problem and such that D_2 converts the CNF satisfiability problem into X. There are many different NP-complete problems, some of them quite important, and hundreds of researchers have analyzed NP-complete problems in detail; yet nobody has found a polynomial time algorithm for any of them. Hence, more and more people suspect that no such algorithm exists. As a consequence, if a problem is found to be NP-complete, most people will assume that no efficient algorithm for its solution exists, and they look for approximate methods, special cases, or whatever other means exist to make the problem tractable.

Not all NP problems are NP-complete. For instance, the fact as to whether there is a path from some node a in a graph to some other node b is an NP problem. It is a decision problem; that is, either there is a path or there is none. Moreover, given an instance of a sequence of nodes, it can be verified in polynomial time if this sequence is or is not a path from a to b. However, this problem can be solved in polynomial time, which is an indication that it is not NP-complete. There is no method that would allow one to convert the CNF satisfiability problem, or any other NP-complete problem for that matter, into the problem of finding if there is a path between two nodes of a graph.

Many decision problems are closely related to minimization problems. For instance, in the traveling salesperson problem, one wants to find the shortest tour visiting a number of cities, with the condition that the tour must start and end in the same city. The decision problem associated with the traveling salesperson problem asks if there is a tour that is below some given value. As it turns out, this decision problem is NP-complete. Another example is the integer programming problem. The solution to this problem involves the determination of n variables $x_1, x_2, \ldots, x_n$, all of which must be integer. The solution must minimize a certain linear function, called the objective, and the solution must also satisfy a set of linear inequalities. One can now formulate a decision problem, in which case one has the same restrictions; but instead of minimizing the objective, one requires that the objective be at most equal to a given value. This decision problem is also NP-complete. Of course, if there is a polynomial algorithm to solve the minimization problem, there is also a polynomial algorithm to solve the corresponding decision problem. If the minimum is below the given bound, then this minimum provides the desired solution. If the minimum is above the given bound, then the decision problem has no solution. Hence, the optimization problem is at least as difficult as the decision problem. Problems that are at least as difficult as an NP-complete problem are called NP-hard. For instance, the traveling salesperson problem and minimization problems like it are NP-hard.

As it turns out, the optimization problem is not significantly more difficult than the decision problem. In fact, the optimization problem can be reduced to the decision problem as follows: one finds a solution satisfying all restrictions of the minimization problem. This provides an upper bound for the desired minimum. A lower bound is also found readily in most cases. For instance, in the traveling salesperson problem, the lower bound for the best solution is zero. One can now take the average between the upper and the lower bound and ask if there is a solution that is less than or equal to this average. This is, of course, a decision problem. If there is one, this solution provides a new upper bound, and the process can continue. If there is no solution, then the average between the upper and lower bound provides a new lower bound, and one can proceed with this lower bound. Hence, no matter what the outcome of the decision problem, the difference between the upper and the lower bound decreases. By repeating this procedure, one can make the difference between the upper and the lower bound as small as desired. This essentially means that, if there were a polynomial algorithm to solve the decision problems, there also would be a polynomial algorithm to solve the corresponding optimization problem.

Problems 6.3

1. Write the following functions as sums of powers of n and identify the dominant terms:
   (a) $f_1(n) = (n+2)^2 + 1/n^3$ (c) $f_3(n) = 1 + 1/n$
   (b) $f_2(n) = (n+1)n(n-1)$ (d) $f_4(n) = 3000 + 15n + 0.0001n^2$

2. Find a value n_c such that $n^4 - n^2 + 1 > 10n^3 + n + 15$ for $n > n_c$.

3. Which of the following functions are in $O(n^2)$ and which are not?
   (a) $3n^2 + 2n + 2$ (c) $n^2 \log_2 n$
   (b) $n \log_2 n$ (d) $20 \frac{n^2}{\log_2(4n)}$

4. Prove that $n! \prec (n+1)!$.

5. Suppose that you are given an algorithm that needs n^5 operations, where n is the size of the input. If the time complexity must not exceed 10^{12} operations, what is the largest problem that can be solved?

6. Use the divide and conquer approach to write a procedure for calculating A^n. You may assume the availability of a procedure for multiplying two matrices. Find the minimal number of matrix multiplications needed for n ranging from 3 to 16.

7. Is the problem of finding if a system of linear equations has a solution an NP problem? Is this problem an NP-complete problem?

8. A company advertises a computer program for finding the best schedule for final exams. The advertisement claims that the time needed to find this schedule is equal to n^2, where n is the number of exams. Is this claim believable? State why or why not.

6.4 RECURRENCE RELATIONS

6.4.1 Introduction

Sequences such as H_n, $n = 0, 1, 2, \ldots$, are often given recursively as follows: The initial values $H_0, H_1, \ldots, H_{c-1}$ are stated explicitly, whereas for all $n \geq c$, H_n is obtained from the relation

$$H_n = f(H_{n-1}, H_{n-2}, \ldots, H_{n-c}, n)$$

Relations such as this are called *recurrence equations*. A number of recurrence equations have been used earlier in this text. For instance, the Fibonacci sequence F_n, $n \geq 0$, was defined in Section 6.3.2 as follows. The initial values are $F_0 = 0$ and $F_1 = 1$, and for $n > 1$, F_n is given by

$$F_n = F_{n-1} + F_{n-2} \tag{6.25}$$

In Section 6.3.3, we discussed an algorithm to calculate the Fibonacci numbers. In this algorithm, the number of active operations was given by $T(0) = 0$, $T(1) = 0$, and

$$T(n) = 2 + T(n-1) + T(n-2) \tag{6.26}$$

This is a sequence that is different from, but related to, the Fibonacci sequence.

Recurrence relations not only occur in complexity analysis, but in many other areas, both inside and outside computer science. For instance, the growth of a population is often given in terms of a recurrence relation. The simplest growth model is based on the assumption that the population increases at a constant factor a from year to year. Hence, if H_n is the number in the population at time n, then one has

$$H_n = aH_{n-1}$$

We know that such a population increases exponentially with n or, if $n < 1$, it decreases exponentially with n (see Section 6.3.3). More complex recurrence equations result if the reproduction rate in the populations varies with age. This is shown by the following model: Let H_n be the average number of females born in year n in some population. (For population studies, males are usually ignored because population growth is essentially independent of the number of males.) If a female of the population can live up to c years and gives birth to a_1 female offspring on average in year 1, a_2 female offspring on average in year 2, and so on, then H_n has the form

$$H_n = a_1 H_{n-1} + a_2 H_{n-2} + \cdots + a_c H_{n-c}$$

Given the initial conditions $H_0, H_1, \ldots, H_{c-1}$, this allows one to calculate all H_n, $n \geq c$. This model is more complex than the one in which the population increases by a constant factor each year. However, in one important aspect it resembles the earlier model: The population H_n still increases or decreases exponentially.

Populations in the wild are subject to predation, which reduces the growth. This leads to a revised model, as follows: If c females die of predation, then H_n becomes

$$H_n = -c + a_1 H_{n-1} + a_2 H_{n-2} + \cdots + a_c H_{n-c}$$

Unless the predator wipes out the population, predation does not prevent the population from growing exponentially. In fact, if H_n is large enough, the value of c becomes insignificant and can be ignored. The same is even true if predation increases polynomially with time.

We now can classify the recurrence relations. All the recurrence relations given so far are *linear*; that is, H_n depended on $H_{n-1}, H_{n-2}, \ldots, H_{n-c}$ in a linear way. This leads to the following general form of a linear recurrence relation.

$$H_n = f(n) + a_1 H_{n-1} + a_2 H_{n-2} + \cdots + a_c H_{n-c}$$

Potentially, each a_i, $1 \leq i \leq c$, could depend on n. If they do not, then one speaks about a *recursion with constant coefficients*. For simplicity, we restrict our attention to recursions with constant coefficients. Any recurrence relation that is not linear is called *nonlinear*. For instance, $H_n = H_{n-1} \times H_{n-2}$ is a nonlinear recurrence relation. We will not deal with nonlinear recurrence relations further.

A linear recurrence relation is called *homogeneous* if it is of the form

$$H_n = a_1 H_{n-1} + a_2 H_{n-2} + \cdots + a_c H_{n-c}$$

Homogeneous recurrence relations are dealt with in Section 6.4.2. For instance, population growth without predation leads to a homogeneous recurrence relation. The same is true for the Fibonacci numbers. With each homogeneous recurrence relation, one can associate a *nonhomogeneous* recurrence relation. Generally, if $f(n)$ is a function of n that is not identically 0, then a nonhomogeneous recurrence relation has the form

$$H_n = f(n) + a_1 H_{n-1} + a_2 H_{n-2} + \cdots + a_c H_{n-c}$$

For instance, a population subject to constant predation c is a nonhomogeneous recurrence relation with $f(n) = -c$. Another example of a nonhomogeneous recurrence relation is given by (6.26), in which case $f(n) = 2$. To solve nonhomogeneous recurrence relations, one first must solve the associated homogeneous recurrence relation. This will be shown in Section 6.4.3. We note that the following two sections are mathematically challenging, and they should be omitted at the first reading.

6.4.2 Homogeneous Recurrence Relations

If a recurrence relation for H_n is given that involves $H_{n-1}, H_{n-2}, \ldots, H_{n-c}$, then one can always find H_n directly, given that one has the initial conditions $H_0, H_1, \ldots, H_{c-1}$. This direct method is demonstrated next for the case of the Fibonacci numbers.

Example 6.35   Find the Fibonacci numbers F_2 to F_6. Use the fact that $F_0 = 0$ and $F_1 = 1$.

Solution
$$F_2 = F_1 + F_0 = 1 + 0 = 1$$
$$F_3 = F_2 + F_1 = 1 + 1 = 2$$
$$F_4 = F_3 + F_2 = 2 + 1 = 3$$
$$F_5 = F_4 + F_3 = 3 + 2 = 5$$
$$F_6 = F_5 + F_4 = 5 + 3 = 8$$
■

In the theory of recurrence relations, one tries to find nonrecursive or *explicit* solutions, that is, solutions that do not require the calculation of all H_m, $m < n$, before it is possible to find H_n. The problem is now to find an explicit formula for H_n, that is, a formula that does not involve H_{n-1}, H_{n-2}, and so on. Such an explicit formula is now derived for homogeneous recurrence relations, that is, recurrence relations that are given as

$$H_n = a_1 H_{n-1} + a_2 H_{n-2} + \cdots + a_c H_{n-c} \tag{6.27}$$

Since H_n can be used to model population growth and since populations tend to increase exponentially, one would suspect that H_n increases exponentially as well. To illustrate this point, consider the Fibonacci numbers again. We conjecture that $F_n = bx^n$ for some x and b, and the problem is to find the right value of both x and b. To do this, we obviously need some equations. First, we try the recurrence equation, and we merely replace F_n by bx^n. This yields

$$bx^n = bx^{n-1} + bx^{n-2} \tag{6.28}$$

After dividing both sides of (6.28) by bx^{n-2}, we obtain

$$x^2 = x + 1$$

This is a quadratic equation for x, and its solutions can be obtained without difficulty. One has

$$x_1 = \tfrac{1}{2}(1 + \sqrt{5}), \qquad x_2 = \tfrac{1}{2}(1 - \sqrt{5})$$

This gives not only one, but two potential solutions, $F_n^1 = b_1 x_1^n$ or $F_n^2 = b_2 x_2^n$. The two solutions can be combined. Generally, if H_n^1 and H_n^2 are two different solutions of a homogeneous recurrence relation, so is $H_n^1 + H_n^2$. This follows directly from (6.27), as the reader may want to verify. In our case, this leads to

$$F_n = b_1 x_1^n + b_2 x_2^n$$

At this point, b_1 and b_2 still remain to be determined. This can be done by observing that the explicit solution must yield $F_0 = 0$ and $F_1 = 1$. Consequently, the following two equations must be satisfied.

$$F_0 = b_1 x_1^0 + b_2 x_2^0 = b_1 + b_2 \qquad = 0$$
$$F_1 = b_1 x_1^1 + b_2 x_2^1 = b_1 x_1 + b_2 x_2 = 1$$

To solve these equations, we replace x_1 and x_2 by $\frac{1}{2}(1 + \sqrt{5})$ and $x_2 = \frac{1}{2}(1 - \sqrt{5})$, respectively, which yields

$$
\begin{aligned}
b_1 \qquad\qquad + b_2 \qquad\qquad &= 0 \\
b_1 \tfrac{1}{2}(1 + \sqrt{5}) + b_2 \tfrac{1}{2}(1 - \sqrt{5}) &= 1
\end{aligned}
$$

The solution of these two equations is found to be

$$
b_1 = 1/\sqrt{5} \qquad b_2 = -1/\sqrt{5}
$$

Hence, F_n is given as

$$
F_n = (1/\sqrt{5}) \left(\tfrac{1}{2}(1 + \sqrt{5})\right)^n - (1/\sqrt{5}) \left(\tfrac{1}{2}(1 - \sqrt{5})\right)^n
$$

The function F_n is thus expressed as the sum of exponential functions. As indicated in Section 6.3.5, the term with the highest exponent dominates. Table 6.4 shows this for the case of Fibonacci numbers. There, x_1 is larger than x_2, and $F_n^1 = b_1 x_1^n$ alone, when rounded to the closest integer, yields the correct result for all values observed.

TABLE 6.4.   Fibonacci Numbers

n	F_n	$b_1 x_1^n$	$b_2 x_2^n$
0	0	0.4472	−0.4472
1	1	0.7236	0.2764
2	1	1.1708	−0.1708
3	2	1.8944	0.1055
4	3	3.06525	−0.0652
5	5	4.9597	0.0403
6	8	8.0249	−0.0249
7	13	12.9846	−0.0154
8	21	21.0095	−0.0095
9	34	33.9941	0.0059
10	55	55.0036	−0.0036
11	89	88.9978	−0.0022

The method is now generalized to solve any recurrence relation of the form (6.27). The following steps lead to the desired explicit solution for H_n.

1. Set H_n to x^n.

2. Replace H_n in (6.27) by x^n and divide by x^{n-c}. The result is the *characteristic equation*, which can be solved for x.

3. Solve the characteristic equation for x. Since the characteristic equation is of degree c, this yields c roots. Some roots may be repeated roots. If this happens, special methods are needed, but since this case is rare, it will not be discussed here.

4. Set $H_n = b_1 x_1^n + b_2 x_2^n + \cdots + b_c x_c^n$, where the b_i, $i = 1, 2, \ldots, c$, are determined by using the following equations. In these equations, $H_0, H_1, \ldots, H_{c-1}$ are the initial conditions that are assumed to be known.

$$H_0 = b_1 + b_2 + \cdots + b_c$$
$$H_1 = b_1 x_1 + b_2 x_2 + \cdots + b_c x_c$$
$$H_2 = b_1 x_1^2 + b_2 x_2^2 + \cdots + b_c x_c^2$$
$$\vdots = \vdots$$
$$H_{c-1} = b_1 x_1^{c-1} + b_2 x_2^{c-1} + \cdots + b_c x_c^{c-1}$$

To see how this works, we solve a second example.

Example 6.36   Let $H_0 = 0$, $H_1 = 1$, and $H_n = H_{n-1} + 2H_{n-2}$. Give an explicit solution for H_n.

Solution   We set $H_n = x^n$ to obtain

$$x^n = x^{n-1} + 2x^{n-2}$$

If both sides are divided by x^{n-2}, this yields the following characteristic equation:

$$x^2 = x + 2$$

Both $x_1 = 2$ and $x_2 = -1$ satisfy the characteristic equation. Hence, the solution is $H_n = b_1 2^n + b_2 (-1)^n$, with b_1 and b_2 to be determined by using the initial conditions, which results in the following equations:

$$0 = b_1 + b_2$$
$$1 = b_1(2) + b_2(-1)$$

The solutions of these two equations are $b_1 = \frac{1}{3}$ and $b_2 = -\frac{1}{3}$. Hence, the explicit solution of H_n is given by

$$H_n = (\tfrac{1}{3})2^n - (\tfrac{1}{3})(-1)^n$$ ∎

The fact that the solution of a homogeneous recurrence equation is a sum of exponentials is important for asymptotic analysis. Specifically, if x_1 is the largest root according to its absolute value, then H_n approaches $b_1 x_1^n$ as n increases, provided, of course, that b_1 is not zero. This remains true if some roots are negative or complex.

6.4.3 Nonhomogeneous Recurrence Relations

A recurrence relation is nonhomogeneous if it is of the form

$$H_n = f(n) + a_1 H_{n-1} + a_2 H_{n-2} + \cdots + a_c H_{n-c} \tag{6.29}$$

Here $f(n)$ must not be equal to zero for all n. For our discussion, we assume that $f(n)$ is a polynomial. To explore the possible solutions, it is assumed that H_n grows exponentially with a growth exceeding unity. In this case, H_n soon dominates the polynomial $f(n)$, which can be ignored at this point. This leads to the following homogeneous recurrence relation, which is called the *complementary* recurrence relation.

$$H_n = a_1 H_{n-1} + a_2 H_{n-2} + \cdots + a_c H_{n-c}$$

This relation is certainly relevant when H_n reaches high values, but possibly otherwise as well. This suggests that the solution of the complementary relation forms part of the final solution. Besides this complementary solution, it is well known that nonhomogeneous recurrence equations have *particular solutions*. These remarks should motivate the form we now provide for H_n. It should be noted, however, that the discovery of this formula was a long and tortuous process.

$$H_n = H_n^p + b_1 x_1^n + b_2 x_2^n + \cdots + b_c x_c^n \tag{6.30}$$

Here the x_i, $1 \leq i \leq n$, are the roots of the characteristic equation of the complementary problem, and H_n^p is the particular solution. This means that H_n^p must satisfy the recurrence relation.

$$H_n^p = f(n) + a_1 H_{n-1}^p + a_2 H_{n-2}^p + \cdots + a_c H_{n-c}^p$$

As it turns out, if $f(n)$ is a polynomial in n, so is H_n^p. In fact, H_n^p typically has the same degree as $f(n)$. Consequently, if $f(n)$ is a constant, so is H_n^p. If $f(n)$ is linear, so is H_n^p, and so on. (There are some exceptions to this rule, in which case the degree of H_n^p must be increased.) At this point, the only problem that needs to be addressed is how to find the coefficients of the polynomial H_n^p and, of course, how to find the b_i, $i = 1, 2, \ldots, c$.

We now try this method to solve the following simple example:

$$H_n = c + a H_{n-1} \tag{6.31}$$

where H_0 is given. We first find the particular solution H_n^p, which, according to what has been said, should be constant; that is, $H_n^p = d$. The constant d must now be found. By replacing H_n by d in (6.31), one obtains

$$d = c + ad$$

This can be solved for d, which yields

$$H_n^p = d = c/(1-a)$$

Next, one has to determine the x by solving the complementary system $H_n = a H_{n-1}$. Replacing H_n in this expression by x^n leads to $x^n = ax^{n-1}$, which leads to the characteristic equation $x = a$. This characteristic equation directly provides the solution for x. Equation (6.30) can now be used with $H_n^p = c/(1-a)$ and $x = x_1 = a$:

$$H_n = c/(1-a) + ba^n$$

Here $b = b_1$ must be determined from the initial condition H_0. One has

$$H_0 = c/(1-a) + b$$

Hence,

$$b = H_0 - c/(1-a)$$

Consequently,

$$H_n = c/(1-a) + (H_0 - c/(1-a))a^n$$

This is the desired explicit solution. For instance, if $H_0 = 1$ and $H_n = 1 + 2H_{n-1}$, then $c = 1$ and $a = 2$. Hence, $b = H_0 - c/(1-a) = 1 - 1/(1-2)$, and this is 2. H_n is therefore $c/(1-a) + 2 \times 2^n$, or $2^{n+1} - 1$.

In general, the nonhomogeneous equation as given in (6.29) can be solved by completing the following steps:

1. Set $H_n^p = d(n)$, where $d(n)$ is a polynomial of the same degree as $f(n)$.

2. Use the recurrence equations to find all coefficients of $d(n)$. [If this is impossible, increase the degree of $d(n)$ by 1. We will not further discuss this eventuality here.]

3. Drop the term $f(n)$ to obtain the complementary recurrence relation.

4. Find the characteristic equation of the complementary recurrence equation.

5. Find the roots of the characteristic equation. It is assumed there are c distinct roots, $x_1, x_2, \ldots, x_c$.

6. The solution is now $H_n = d(n) + b_1 x_1^n + b_2 x_2^n + \cdots + b_c x_c^n$, where the b_i, $i = 1, 2, \ldots, c$ must be determined by using the initial conditions. This is accomplished by solving the following equations for $b_1, b_2, \ldots, b_n$.

$$H_0 = d(0) + b_1 + b_2 + \cdots + b_c$$
$$H_1 = d(1) + b_1 x_1 + b_2 x_2 + \cdots + b_c x_c$$
$$H_2 = d(2) + b_1 x_1^2 + b_2 x_2^2 + \cdots + b_c x_c^2$$
$$\vdots = \vdots$$
$$H_{c-1} = d(c-1) + b_1 x_1^{c-1} + b_2 x_2^{c-1} + \cdots + b_c x_c^{c-1}$$

Example 6.37   Solve the recurrence relation $H_n = H_{n-1} + 2H_{n-2} + n$, given that $H_0 = 0$ and $H_1 = 1$.

Solution   We first find the particular solution. Since $f(n) = n$ is linear, the following linear equation for H_n^p is used.

$$H_n^p = d_0 + d_1 n$$

To find d_0 and d_1, this solution is substituted into the recurrence relation, which yields

$$d_0 + d_1 n = (d_0 + d_1(n-1)) + 2(d_0 + d_1(n-2)) + n$$

We simplify this equation to obtain

$$0 = 2d_0 - 5d_1 + 2d_1 n + n$$

This equation must hold for all n, and this is only possible if the constant terms and the linear terms are both zero. This yields the following equations:

$$2d_0 - 5d_1 = 0 \qquad \text{Constant term is } 0$$
$$2d_1 + 1 = 0 \qquad \text{Linear term is } 0$$

From these two equations, one finds that

$$d_0 = -5/4, \quad d_1 = -1/2$$

Hence $H_n^p = -5/4 - n/2$. We now solve the complementary equation

$$H_n = H_{n-1} + 2H_{n-2}$$

Set $H_n = x^n$ to obtain

$$x^n = x^{n-1} + 2x^{n-2}$$

Dividing both sides by x^{n-2} yields the characteristic equation

$$x^2 = x + 2$$

The roots of the characteristic equation are $x_1 = 2$ and $x_2 = -1$. H_n is therefore of the form

$$\begin{aligned} H_n &= H_n^p + b_1 2^n + b_2(-1)^n \\ &= -5/4 - n/2 + b_1 2^n + b_2(-1)^n \end{aligned}$$

For $n = 0$ and $n = 1$, this yields

$$\begin{aligned} H_0 &= -5/4 + b_1 + b_2 \\ H_1 &= -5/4 - 1/2 + 2b_1 - b_2 \end{aligned}$$

Since $H_0 = 0$, $H_1 = 1$, this gives rise to the following equations for b_1 and b_2.

$$\begin{aligned} 0 &= -5/4 + b_1 + b_2 \\ 1 &= -7/4 + 2b_1 - b_2 \end{aligned}$$

The solution of these two equations is $b_1 = \frac{4}{3}$ and $b_2 = -\frac{1}{12}$. Consequently,

$$H_n = -\frac{5}{4} - \frac{n}{2} + \frac{4}{3}2^n - \frac{1}{12}(-1)^n \qquad\qquad \blacksquare$$

We are now ready to derive an explicit formula for (6.26) with $T(0) = T(1) = 0$. Since (6.26) is given as $T(n) = 2 + T(n-1) + T(n-2)$, $f(n) = 2$, and this is a constant. The particular solution should therefore also be a constant, say d.

$$T(n)^p = d$$

Inserting this solution into (6.26) yields

$$d = 2 + d + d$$

Hence, $T(n)^p = d = -2$. The complementary equation of (6.26) is $T(n) = T(n-1) + T(n-2)$, and this is the recurrence defining the Fibonacci series. Consequently,

$$x_1 = \frac{1 + \sqrt{5}}{2}, \qquad x_2 = \frac{1 - \sqrt{5}}{2}$$

$T(n)$ is therefore of the form

$$T(n) = -2 + b_1 \left(\frac{1 + \sqrt{5}}{2} \right)^n + b_2 \left(\frac{1 - \sqrt{5}}{2} \right)^n$$

Using this equation, together with the initial conditions $T(0) = 0$, $T(1) = 0$, one finds the following equations for b_1 and b_2.

$$0 = -2 + b_1 + b_2$$

$$0 = -2 + b_1 \left(\frac{1 + \sqrt{5}}{2} \right) + b_2 \left(\frac{1 - \sqrt{5}}{2} \right)$$

We solve these equations to obtain

$$b_1 = \frac{\sqrt{5} + 1}{\sqrt{5}}, \qquad b_2 = \frac{\sqrt{5} - 1}{\sqrt{5}}$$

Consequently,

$$T(n) = -2 + \frac{\sqrt{5} + 1}{\sqrt{5}} \left(\frac{1 + \sqrt{5}}{2} \right)^n + \frac{\sqrt{5} - 1}{\sqrt{5}} \left(\frac{1 - \sqrt{5}}{2} \right)^n$$

or

$$T(n) = -2 + \frac{2}{\sqrt{5}} \left(\frac{1 + \sqrt{5}}{2} \right)^{n+1} - \frac{2}{\sqrt{5}} \left(\frac{1 - \sqrt{5}}{2} \right)^{n+1}$$

This last relation equals $2F_{n+1} - 2$.

The fact that the particular solution is a polynomial as long as $f(n)$ is polynomial has an important consequence. As mentioned earlier, the complementary solution approaches $b_1 x_1^n$, provided that $b_1 \neq 0$. Hence, if $x_1 > 1$, the complementary solution dominates. This means that for the asymptotic analysis no distinction need be made between homogeneous and nonhomogeneous recurrence relations, given that the conditions just mentioned apply: $f(n)$ must be polynomial, $|x_1| > 1$ and $b_1 \neq 0$.

Problems 6.4

1. Consider the recurrence relation $H_n = 5H_{n-1} - 6H_{n-2}$ with the initial conditions $H_0 = 1$, $H_1 = 0$.

(a) Evaluate H_n directly for $n = 2, 3, 4$.

(b) Find the explicit solution for H_n.

(c) Evaluate H_n by using the explicit solution for n from 0 to 4. Compare the results.

2. Given that $H_n = 5 \times 1.5^n + 2^n$. How large must n be such that the term 2^n is greater than $0.9H_n$?

3. Given that $H_n = 1 + n + H_{n-1}$, find an explicit solution of H_n in terms of H_0.

4. Consider the recurrence relation $H_n = 2n + 5H_{n-1} - 6H_{n-2}$, with the initial conditions $H_0 = H_1 = 0$.

(a) Evaluate H_n directly for $n = 2, 3, 4$.

(b) Find the explicit solution for H_n.

(c) Evaluate H_n by using the explicit solution for n from 0 to 4. Compare the results.

5. The solutions of the characteristic equation may be complex. The method of this section remains applicable in this case, however. With this in mind, find an explicit solution of $H_n = 2(H_{n-1} - H_{n-2})$ with $H_0 = 0$, $H_1 = 1$.

6.5 MIRANDA

6.5.1 Introduction

As was shown in Section 3.5, function evaluation is sufficient to evaluate everything that can reasonably be evaluated. This suggests the use of function evaluations as a programming paradigm, an idea that is implemented by the *functional programming languages* or *functional languages*. The main functional languages are Miranda, ML, Haskel, and its more popular subset Gofer, but there are many others. In this section, we concentrate on Miranda. The programs in Miranda are called *scripts*. Scripts consist of function definitions and possibly declarations. Before a script can be used by the system, it must be compiled and loaded. Once this is done, the user merely has to provide the actual arguments of the function, and everything else is done automatically. Miranda is strongly typed, but the types are inferred, and it is not necessary to write them down explicitly. However, if the programmer declares the types, the compiler can provide better diagnostics. Section 6.5.2 describes commands that are needed to edit and run scripts. Section 6.5.3 indicates how to write scripts. Section 6.5.4 describes the data types available and how to declare them. Finally, Section 6.5.5 presents the computational engine underlying Miranda.

6.5.2 Command Level

At command level, instructions can be given for loading and editing scripts, soliciting on-line help, terminating a session, and similar activities. At command level, one can also type any expression not involving variables. For instance, typing $3 + 4$ will return the answer 7. The expressions may contain functions and, indeed, the way to run a script is by typing expressions that contain functions defined in the script.

Three basic types of constants can be used: num, bool, and char. Constants of type num must be either real or integer. Reals and integers are stored using different internal representations, but they can be freely mixed in calculations. Examples of constants of type num include 45, 0.5, and 3.45e-5. There are only two constants of type bool, True and

False. Constants of type char include all ASCII characters. To be used as constants, they must be enclosed in single quotes. For instance, 'a' and 'X' are both constants of type char.

Generally, Miranda expressions are not much different from expressions of other programming languages. One has the normal numerical operators, including $+$, $-$, $*$, $/$, and $\hat{\ }$, and the normal relational operators, including $<, =, >, <=$, and so on. The Boolean operator $\backslash$ / stands for logical or, and the Boolean operator & for logical and. For instance, the expression $(3 < 4)$ & $(3 > 4)$ yields False. The usual built-in functions are available. However, the arguments of a function need not be enclosed in parentheses. For instance, to calculate the square root of, say, 4, one can write sqrt 4. The result of this expression is, as one would expect, equal to 2. If the user needs functions that are not available, he or she writes a *script* containing the definition of the new functions. In fact, all programs must be formulated as function definitions.

To write a script, one invokes the editor, which is done by typing /editor. By setting the appropriate parameters, the user can choose different editors and write scripts for the functions that he or she intends to use. Once the editing session is terminated, the system returns immediately to the command level and compiles the definitions of all functions that have been defined during the editing session. In other words, all definitions are now available to be used in expressions.

6.5.3 Function Definitions

If a function is to be used by other functions or by expressions in command level, it must be defined by a script. For example, the function definition for the function square would look as follows:

$$\text{square } x = x * x$$

Function definitions start with a function name, such as square in the example, followed by zero or more formal parameters. After the last parameter, there is an equal sign, followed by an expression.

Example 6.38   Write a function with the two arguments x and y that calculates the square root of $x^2 + y^2$.

   Solution   The function in question must be given a name, say distance. The name indicates that the function returns the distance of the point (x,y) from the origin. The function distance is defined as follows:

$$\text{distance } x\ y = \text{sqrt}(x * x + y * y)$$

   Alternatively, one can make use of the function square given earlier and write

$$\text{distance } x\ y = \text{sqrt}(\text{square } x + \text{square } y)$$ ∎

Function names and variable names are examples of *identifiers*. Identifiers consist of digits, letters, and the underscore, and they must start with a letter. Upper- and lowercase letters are treated as distinct. For our purposes, the first letter must always be lowercase. Identifiers starting with uppercase letters are reserved for special purposes, and their use is beyond the scope of this text. The order in which the function definitions are given is irrelevant. In particular, there is no need to define a function before its first use.

Miranda contains an if-clause to distinguish between different cases. The definition of the absolute difference between two values x and y, for instance, can be expressed as follows:

$$\text{absdiff } x\ y = x - y, \textbf{ if } x > y$$
$$= y - x, \textbf{ if } x <= y$$

The expressions following the word **if** are called *guards*. Guards must be Boolean expressions. Miranda chooses the first alternative for which the guard yields True. The last alternative may contain the word **otherwise** in place of the **if**, and it gives the definition to be applied if all other guards fail. Both **if** and **otherwise** are preceded by a comma.

Example 6.39   Taxes are 0 for incomes under \$5000, 0.2 * (income − 5000) for incomes between \$5000 and \$20,000, and 0.3 * (income − 20000) + 3000 for incomes above \$20,000. Write a script that does the tax calculation.

   Solution

$$\text{tax income} = 0, \qquad\qquad\qquad\qquad\qquad \textbf{if } \text{income} < 5000$$
$$= 0.2 * (\text{income} - 5000), \qquad\qquad \textbf{if } \text{income} < 20000$$
$$= 0.3 * (\text{income} - 20000) + 3000, \ \texttt{otherwise} \qquad \blacksquare$$

It is normally considered good style to formulate the guards in such a way that their order is irrelevant. To accomplish this, all cases must be mutually exclusive. The preceding script does not adhere to this rule.

In Miranda, the indentation is syntactically significant. Without going into details, we must mention one rule of particular importance. Any part of a definition must be underneath or to the right of the first character of the right side of the function definition. Otherwise, a new definition is assumed to start.

Expressions in function definitions may in turn contain definitions. To show why this is useful, consider the following definition:

$$\text{myfunction } x\ y = (3 * x + 2 * y)\ /\ (3 * x + 2 * y + 1)$$

The expression $3 * x + 2 * y$ appears twice, and it must be evaluated twice. This may increase the execution time and, more importantly, it may reduce the clarity of the definition. To avoid this, one uses a where-clause and writes

$$\text{myfunction } x\ y = z\ /\ (z + 1)$$
$$\textbf{where } z = 3 * x + 2 * y$$

Note that the where-clause must be underneath the first z because of the indentation rules. Otherwise, a new definition is deemed to have started.

Example 6.40   A supplier gives a rebate of \$5 if the quantity is 100 or more and a rebate of \$50 if the quantity is 500 or more. Given the price and the quantity, give the function definition of the amount to pay.

Solution

to_Pay price quant
= price ∗ quant − rebate
  where
  rebate = 0,   **if** quant < 100
         = 5,   **if** (quant >= 100) & (quant < 500)
         = 50,   **if** quant>=500   ∎

6.5.4 Types, Functions, and Declarations

Miranda is a strongly typed language in the sense that each expression and each variable have a type that can be deduced from the program at compile time. As stated earlier, there are three atomic built-in types, num, bool, and char. From these atomic types, programmers can create new types: *functions*, *lists*, and *n-tuples*. An n-tuple consists of n elements enclosed by parentheses. For instance, (3, 4), (5, 'a', 3), and ('b', True) are all n-tuples. The types of the elements in an n-tuple can be different. Tuples with the same elements arranged in different order are considered to be different. For instance, the expression (3, 4) = (4, 3) yields False. In general, the n-tuples in Miranda reflect the n-tuples as they were defined in Section 5.2.5.

Lists are sequences of elements, all of which must be of the same type. The elements of the list can be stated explicitly, as in [2, 4, 4]. Note that the elements of a list are enclosed in brackets. To describe lists, one can also use two points. For instance, the list [0..100] is the list of all integers between 0 and 100. Infinite lists can also be used. For instance, [0..] is the list of all natural numbers. Strings are lists of characters. However, for a string, one has the option of just writing all characters of the string in sequence. For instance, "abc" is a string, and this string is equal to ['a', 'b', 'c']. In fact, "abc" = ['a', 'b', 'c'] yields True.

Both n-tuples and lists are polymorphic, and the same is true for functions, which are discussed later. This means that the constituent elements of n-tuples, lists, and functions may be of any type. Expressions such as [(1, 'a'), (2, 'b')] and ["hello", "stranger"] are acceptable. Note, however, that the list [(1,'a'), (2, 3)] is not acceptable, because in a list all elements must be of the same type, and (1, 'a') and (2, 3) violate this rule. Two n-tuples are only of the same type if all the types of corresponding elements match.

In Miranda, it is possible to use functions as arguments of functions. Consider, for instance, the function twice, which is defined as follows:

twice f x = f (f x)

Here, f is a function and it can be instantiated to any function name. For instance, if square x is the function evaluating the square of x, then one can write the following expression in command level.

twice square 3

In this case, the function twice instantiates to

twice square 3 = square (square 3)

Since square 3 is 9, square square 3 is equal to square 9, which is 81. Built-in functions can be passed just like any other function, as indicated by the following definition:

$$\text{twice sqrt 16}$$

In this case, the computer gives the answer 2, because the square root of the square root of 16 is 2.

Consider now the following definition:

$$\text{add x y} = \text{x} + \text{y}$$

Note that add 3 instantiates x to 3, while leaving y uninstantiated. The result is the function 3 + y, and this function can be used like any other function. For instance, one can issue the command

$$\text{twice (add 3) 2}$$

Since add 3 is the function that adds 3, twice (add 3) adds 6. Adding 6 to 2 yields 8. Generally, if f x y is a function with the formal parameters x and y, then f x is a function with one parameter, and it can be used as such. In fact, f x y is really a shorthand for (f x) y; that is, for each y, there is a function f x. Each y is mapped into a function with one variable. This idea will now be elaborated by using λ-notation.

If e is some expression, then f x y = e defines a function. By convention, f x y must be understood as (f x) y, and f x in λ-notation is λ x.e. Hence, (f x) y becomes

$$f = \lambda \ y.\lambda \ x.e$$

To see how this works, consider again the function add x y = x + y. In λ-notation, this function becomes

$$\text{add} = \lambda \ y.(\lambda \ x.(x + y))$$

Wherever the name add appears, one can therefore write the λ-expression on the right. For instance, add 3 4 expands to

$$\lambda \ y.\lambda \ x.(x + y) \ 3 \ 4$$

This yields

$$\lambda \ y.(3 + y) \ 4 = 3 + 4$$

Note that in Miranda all operators are functions, and they too can be passed as arguments to a function. In this case, the name of the operator must be enclosed in parentheses. It is therefore unnecessary to define a function add, as we did previously. We could as well have written (+). For instance, twice ((+) 2) 3 adds twice 2 to 3.

Functions can be declared explicitly. Although declarations are not mandatory, it is a good idea to declare all functions used in a program. Unfortunately, functions appearing in a where or if-clause cannot be declared. As an example of a declaration, consider the function square defined earlier, which calculates the square of its argument. This function maps a number into a number, and it is therefore declared as

$$\text{square :: num } -> \text{ num}$$

Here the two colons are to be read as "is of type," which means that the function square is of type num $->$ num. This declaration may be placed anywhere in the program, but for improved readability it is suggested to declare a function immediately before its definition.

Consider now functions with two arguments. For example, absdiff is the function that calculates the absolute difference between two numbers, and it should be declared as

$$\text{absdiff :: num } ->(\text{ num } -> \text{ num})$$

This definition indicates that the function absdiff maps a number into a function of the type num $->$ num. Since $->$ associates to the right, one is allowed to omit the parentheses and write

$$\text{absdiff :: num } -> \text{ num } -> \text{ num}$$

In function declarations, any type is admissible. A binary logical connective, for instance, is of type bool $->$ bool $->$ bool. Consequently, if xor implements the exclusive or, it is declared as

$$\text{xor :: bool } -> \text{ bool } -> \text{ bool}$$

Lists of numbers are of type [num], and lists of characters are of type [char]. In an n-tuple, all elements must be declared. For instance, (num, char) is a pair in which the first element is a number and the second element is a character. Pairs of numbers have the declaration (num, num). Sometimes, functions are polymorphic; that is, they work for all types. As an example of a polymorphic function, consider the function that calculates the length of a list. This function should work for numbers, characters, strings, and any other type. To declare such a function, one uses an asterisk in place of a type. The length function for lists, which is provided by the system, has the declaration

$$\text{length [*] } -> \text{ num}$$

Since built-in functions are not declared, no such declaration will appear in the script. The declaration is given here strictly for illustration.

6.5.5 Pattern Matching and Rewriting

In Section 3.5.2, multiplication was defined as

$$\text{mult}(m, 0) = 0$$
$$\text{mult}(m, n+1) = \text{mult}(m, n) + m, \ n > 0$$

Miranda allows such definitions; that is, one can write

$$\text{mult m } 0 = 0$$
$$\text{mult m n} + 1 = (\text{mult m n}) + m$$

To interpret this definition, a pattern-matching algorithm is used. The subexpressions 0 and n + 1 on the left sides of the definition are examples of *patterns*, and the respective right sides are executed only if the patterns match. For instance, mult 3 0 will match the first clause, because the second argument is zero in both cases. The expression mult 3 2, on the other hand, only matches with the second clause. Essentially, a pattern is an expression that can be used on the left side of a definition. However, not all expressions can be used as a pattern. For instance, one cannot use expressions involving real numbers as patterns. In fact, the only numerical expressions allowed as patterns are of the form n + k, where n and k must be natural numbers and k must be a constant. For instance, myvar + 2 is a legal pattern, but myvar + 2.5 is not. Generally, pattern matching is really a version of unification.

Patterns are extremely useful in connection with compound types. In particular, the empty list [] can be used as a pattern. There is also a pattern to divide a list into a head and a tail. In fact, if as is a list of some type x and if a is an object of type x, then a:as is the list which "hitches" a to as in the sense that a is now the first element of the list, followed by the elements of as. The hitch operator is a normal operator that can be used anywhere in a script. It also can be used as a pattern, and it allows one to separate a list into a head and a tail. The head, as indicated in Section 3.3.2, is the first element of a list, whereas the tail is a list that includes all the remaining elements. Hence, the pattern [a:as] only matches with a nonempty list, and it unifies a with the first element of the list, whereas the variable as unifies with the list of all the remaining elements. To illustrate this, consider the sum of a list of numbers. Even though this function is built in, it is instructive to see how it is defined. One has

$$\text{sum } [] = 0$$
$$\text{sum a:as} = a + \text{sum(as)}$$

We will trace this program shortly. Before doing this, however, we need to discuss rewrite rules.

When an expression is given at command level, the expression is rewritten in the sense that all instantiations that are possible at this moment are performed, if necessary, recursively. The end result is either a constant, which can be printed, or it is an expression, in which case Miranda merely gives an appropriate diagnostic.

Example 6.41   A script contains the following two definitions:

$$\text{square } x = x \char`\^ 2$$
$$\text{twice f } x = f \ (f \ x)$$

How is the command twice square 3 rewritten, and what is the final result?

Solution   In the first stage, the function twice is instantiated with f = square and x = 3. This leads to

$$\text{twice square 3} = \text{square (square 3)}$$

Since square 3 is 9, square (square 3) is $9^2 = 81$. In detail, one has the following derivation:

$$\text{twice square } 3 = \text{square (square 3)}$$
$$= \text{square } (3\,\widehat{\ }\,2)$$
$$= \text{square } 9$$
$$= 9\,\widehat{\ }\,2 = 81 \qquad\blacksquare$$

One issue was glossed over in the preceding example, even though it is of some theoretical importance. In the solution, the inner argument was instantiated first. However, one could also have instantiated the outer argument first. In the example, this does not make any difference, but in general it can make quite a difference. There are some deep theorems about this issue, which are beyond the scope of this book.

If the left side of a definition contains patterns, it may be necessary to change the expression such that the patterns can be accommodated. The next two examples show how this works. Both examples are recursive.

Example 6.42   Given that mult is defined as previously, indicate how Miranda finds mult 3 2.

Solution

$$\text{mult } 3\ 2 = \text{mult } 3\ (1+1)$$
$$= (\text{mult } 3\ 1) + 3$$
$$= (\text{mult } 3\ (0+1)) + 3$$
$$= ((\text{mult } 3\ 0) + 3) + 3$$
$$= (0 + 3) + 3$$
$$= 6 \qquad\blacksquare$$

Example 6.43   Indicate how Miranda finds sum [3, 2, 4], given that the function sum is as defined before.

Solution

$$\text{sum } [3,\ 2,\ 4] = \text{sum } (3\ :\ [2,4])$$
$$= 3 + \text{sum } [2,4]$$
$$= 3 + \text{sum } (2\ :\ [4])$$
$$= 3 + (2 + \text{sum } [4])$$
$$= 3 + (2 + \text{sum } (\ 4\ :\ []))$$
$$= 3 + (2 + (4 + \text{sum } []))$$
$$= 3 + (2 + (4 + 0))$$
$$= 9 \qquad\blacksquare$$

6.5.6 A Programming Problem

A somewhat larger problem is now described. According to Newton's method, one can find the root of the equation $f(x) = 0$ as follows: One starts with an initial guess x_0, which is then used to obtain successive improved approximations $x_1, x_2, \ldots$ Specifically, if $f'(x)$ is the derivative of $f(x)$, then the new approximation x_{i+1} is given by

$$x_{i+1} = x_i - f(x_i)/f'(x_i) \qquad (6.32)$$

A Miranda script to solve this problem will now be presented for the case when $f(x)$ is a polynomial of the form $a_0 + a_1 x + a_2 x^2 + \cdots + a_n x^n$. The coefficients a_i are stored in a list of the form $[a_1, a_2, \ldots, a_n]$.

First, the value of the function is evaluated for any given value x. The following script accomplishes this task.

```
value [] m = 0
value (first : rest) m = first * x ^ m + value rest (m + 1)
```

The function value has two arguments, the list of coefficients that has not been dealt with and the power of x corresponding to the first coefficient of the list of coefficients. Initially, value list 0 is used to find $a_0 + a_1 x + a_2 x^2 + \cdots + a_n x^n$, then value list 1 is used to find $a_1 x + a_2 x^2 + \cdots + a_n x^n$, and so on. The derivative can be evaluated in a similar way.

```
derivative [] m = 0
derivative (first : rest) m = first * m * x ^ m +
                              derivative rest m − 1
```

The functions value and derivative allow one to program the main procedure, newton. This procedure has three arguments: a precision eps, the polynomial, appropriately called polynomial, and an initial approximation, called x. The polynomial is represented by a list. If the value of the polynomial is less than eps, the value of newton is merely the next value of x_{i+1} as calculated according to (6.32). Otherwise, Newton's method is applied with the new approximation.

```
newton eps polynomial x
= newx,                          if abs(fx) < eps
= newton eps polynomial newx, otherwise
  where
  fx = value polynomial x 0
  fpx = derivative polynomial x 0
  newx = x − fx / fpx
```

This completes our discussion of functional languages. In general, functional languages require only very few concepts, yet they are very concise.

Problems 6.5

1. Define the function proper_root x, which returns the square root of x for all nonnegative values of x and 0 otherwise.

2. An instructor gives three exams in his class. Normally, all exams have an equal weight, and the final mark is merely the average of these three exams; however, if one exam is more than 20% below the average of the other two exams, then the worst exam is only given half its normal weight. Write a script to calculate the final grade, given that the grades in the different exams are e1, e2, and e3.

3. Consider the function sumrec, which has n as its only argument and which is given as $\sum_{i=1}^{n} 1/i$. Write a function definition for sumrec in Miranda.

4. Let this x y be defined as x + 2 * y and let twice be defined as in the text. How will the function call twice (this 2) 4 be evaluated?

5. Program Newton's method to solve $f(x) = 0$, where $f(x) = exp(a + bx)$. The derivative of this function is $f'(x) = b \times exp(a + bx)$.

6. Write a script in Miranda that accepts 3 numbers a, b, and c, and that constructs a list containing the elements $a + bi + ci^2$, where i ranges from 1 to 10.

PROBLEMS: CHAPTER 6

1. Let f be a mapping from A to B, where $A = \{a, 2, 4\}$ and $B = \{2, 5, a, b\}$. Moreover, $f(a) = 2$, $f(2) = 2$, and $f(4) = 5$.

   (**a**) Find the domain, the range-space, and the range of this function.

   (**b**) Is the function onto, into, or bijective?

2. Consider the following graphs of relations. Which of the graphs are (a) partial functions, (b) total functions, (c) onto, (d) one-to-one, and (e) one-to-one onto?

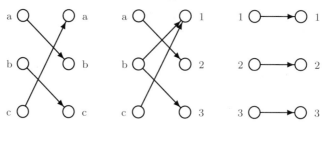

Relation 1   Relation 2   Relation 3

3. The social security number is a partial function, but not a total function from the set of U.S. residents into the set of integers. Explain what this statement implies. How does the mathematical meaning change if the word into is replaced by onto? Similarly, how does the mathematical meaning change if we replace function by relation?

4. Let $f : \mathbb{R} \to \mathbb{R}$ and $g : \mathbb{R} \to \mathbb{R}$, where $\mathbb{R}$ is the set of real numbers. Find $f \circ g$ and $g \circ f$, where $f(x) = x^2 - 2$ and $g(x) = x + 4$. State whether these functions are injective, surjective, or bijective.

5. If $f : X \to Y$ and $g : Y \to Z$ and both f and g are onto, show that $g \circ f$ is also onto. Is $g \circ f$ one-to-one if both g and f are one-to-one?

6. Prove that the composition of two total functions is a total function.

7. Let $f(x)$ and $g(x)$ be strictly two partial functions. Is it possible that $g \circ f$ is a total function? Justify your answer.

8. Let X and Y be two sets, and consider all possible functions from X to Y. Furthermore, let $m = \#X$ and $n = \#Y$.

   (**a**) For any given $x \in X$, in how many ways can you choose an image $y \in Y$?

   (**b**) Find the number of functions from X to Y.

9. For each of the following relations, indicate whether it is a partial function, a total function, or no function at all. If it is a partial or total function, indicate if the function is into or onto. If it is a total function, indicate if it is bijective or not. If the relation is neither a partial nor a total function, give the two pairs from the relation that you use to justify your answer. All relations are from A to A, where $A = \{a, b, c, d\}$.

(a) $\{\ \}$ (this is the void relation, that is, the relation that contains no element).

(b) $\{(a, b), (b, a), (c, d), (d, c)\}$

(c) $\{(a, b), (a, a), (c, d), (d, a)\}$

(d) $\{(a, b), (b, b), (c, b), (d, b)\}$

(e) $\{(b, c), (c, b), (d, b)\}$

10. Some of the following relations are functions, but others are not. If the relation is not a function, remove the minimum number of elements such that the relation becomes a function. If there is a choice as to which element to remove, always remove the last one or the last ones of the roster. State for each function obtained or given in the question whether it is into or onto and whether it is one-to-one or not. All relations are from $A = \{a, b, c\}$ to $B = \{1, 2, 3\}$.

(a) $\{(a, 3), (c, 2), (a, 1), (b, 1)\}$

(b) $\{(a, 3), (c, 2), (b, 1)\}$

(c) $\{(c, 1), (c, 2), (b, 3), (b, 1), (a, 3)\}$

11. Find the following:

(a) $\lambda x.(\lambda y.(x + y)x)z$

(b) $\lambda x.(\lambda y.(x + y)(x + z))(y + z)$

(c) $\lambda u.(\lambda x.(\lambda y.f(x, y)u)v)z$

12. Variables in λ-calculus can be instantiated to functions. For instance, $(\lambda f.(f\ x))(\lambda y.(y + 2))$ yields $(\lambda y.(y + 2))x$, which in turn yields $x + 2$. Find the following:

(a) $\lambda g.(g(g\ 2))(\lambda x.(x + 3))$

(b) $\lambda h.(h\ 2)(\lambda x.(x + 3))$

13. Let $f : X_1 \to Y_1$ and $g : X_2 \to Y_2$ be two functions. How can you modify the definitions of an isomorphism and a homomorphism such that the statements "f is isomorphic to g" and "f is homomorphic to g" are meaningful?

14. Generalize the definition of a homomorphism such that it applies to functions with n variables.

15. Prove (6.19).

16. Given the following functions:

(a) $f(n) = (n + 3)(4n^2 + 5)$

(b) $f(n) = n^3 + 100/n$

(c) $f(n) = 105 + (n^3/2 + n)/(1 + 1/n)$

For each $f(n)$, find $\Theta(g(n))$, where $g(n)$ is of the form n^m.

17. Prove that if $g(x) \prec f(x)$ and if $h(x) \prec f(x)$, then $ah(x) + bg(x) \prec f(x)$ for all values a and b.

18. Let $H_n = 2H_{n-1} - 0.99H_{n-2}$.

(a) Find an explicit formula for H_n, given that $H_0 = H_1 = 1$.

(b) Assume that $H_1 = 1$ is not known, but you know that $H_3 = 1.5$. Find an explicit solution for H_n.

19. If $H_n > 0$ for all n and all positive initial conditions, why would you expect that the largest root of the characteristic equation be positive?

20. The BCD wholesalers have a number of customers, and for each customer there is an amount payable. This information is contained in a list of pairs, where the first element of the pair is a name and the second is the amount outstanding. A similar list contains payments. Specifically, the list consists of pairs, with the first element of the pair being a name and the second element of the pair being the amount paid. Define a Miranda function having the lists just described as their arguments and that returns a third list, which is an updated list of the outstanding amounts. In other words, the amounts of the list that is returned must be the ones of the first list, reduced by the payments. You may assume that the two lists used as arguments are in the same order. However, the payment list may contain gaps in the sense that there may not be a payment for every customer.

21. Write a Miranda script that can count the number of occurrences of a specified word in a text. To this end, define a function count, which has the word whose number of occurrences must be counted as its first argument and which has the text, given as a list of strings, as its second argument.

22. Write a Miranda script to solve Problem 21 of Chapter 4.

23. Let X and Y be two sets, with $\#X = m$ and $\#Y = n$.
   (a) Given $m = n$, how many bijections $f : X \to Y$ are there?
   (b) Given $m < n$, how many injections $f : X \to Y$ are there?

24. Let $x \circ y = y \circ x$ be an operation between x and y and let e be its identity element. Consider the homomorphism f which maps $x \circ y$ into $u \square v$. Show that the operator $\square$ is commutative. Also show that the identity element is preserved.

7

Graphs and Trees

Recall that in Section 5.3.2 a binary relation on a set V was defined as a subset of $V \times V$. It was shown that such a relation could be represented at least in some cases by a diagram that was called the graph of the relation. Of course, there a graph was used as a pictorial device to represent a relation and, as such, it served only a limited purpose. An alternative method given of representing a relation was by means of a relation matrix or incidence matrix. In this chapter, we shall use these ideas, along with some new ones, to develop a more general and extended graph theory.

While a graph will be introduced as an abstract mathematical system, a significant part of its attractiveness derives from the intuitions that can be obtained from pictorial representations. Nevertheless, what seems obvious from the pictures of graphs must be mathematically verified or erroneous conclusions will result.

Graph theory is applied in such diverse areas as social sciences, linguistics, physical sciences, communication engineering and others. Because of this diversity of application, it is useful to develop and study the subject in abstract terms and to interpret its results in terms of the objects of any particular system in which one may be interested. Graph theory also plays an important role in several areas of computer science, such as switching theory and logical design, artificial intelligence, formal languages, computer graphics, operating systems, compiler writing, and information organization and retrieval.

In this chapter, first we introduce and give some examples of where graphs are used. Section 7.2 focuses on basic terminology of graph theory. Section 7.3 introduces further graph terminology to deal with the notions of paths, reachability, and connectedness. The representation of graphs, as a basis of computing paths, is the primary topic of the next two sections. Sections 7.4 and 7.5 deal with the representations of a graph by an adjacency matrix and adjacency lists, respectively. The emphasis is on computing paths, minimum paths, and minimum weighted paths. Section 7.6 deals with the notion of a free or unrooted tree. This section also examines how to generate a free tree that spans a given graph, as well as a minimum spanning tree for weighted graphs. The chapter concludes with a discussion of the application of graphs to project planning and management.

7.1 INTRODUCTION AND EXAMPLES OF GRAPH MODELING

Graphs are used widely to model problems in many different application areas. Constructing a model is essentially a process of deciding what features or aspects of a real-world problem or application are to be represented for analysis or study. A model of an application will vary widely depending on the view or purpose that a modeler has of the application. Good models capture the essence of the real world that is of interest (that is, a modeler's view) and ignore irrelevant details to this view. Also, good models are robust; that is, they have the ability to remain relevant as applications evolve.

This section focuses on several problems to which graph theory has been applied successfully. Although many problems could have been chosen, the ones presented in this section illustrate the diversity of application areas that can be modeled with graph structures.

Some graphs may be considered as graphs of certain relations, but others cannot be interpreted in this manner. Several diagrams are shown in this section. For our purpose here, these diagrams represent graphs. Notice that every diagram consists of a set of points (nodes, vertices), which are shown by circles or small dots or other icons that are sometimes labeled $v_1, v_2, \ldots$ or $1, 2, \ldots$. In graph terminology, these are called points, nodes, or vertices. Also, in every diagram certain pairs of such points are connected by lines, edges, or arcs. The other details, such as geometry of the arcs, their lengths, and the position of the points, are of no importance at present. Notice that every line starts at one point and ends at another point. If this line has a specific direction as indicated by an arrowhead, then it is called *directed*. Otherwise, it is called *undirected*. A definition of the graph, which is essentially an abstract mathematical system, will be given in the next section.

The definition of a graph contains no reference to the length or the shape and positioning of the edge or arc joining any pair of nodes, nor does it prescribe any ordering of positions of the nodes. Therefore, for a given graph, there is no unique diagram that represents the graph, and it can happen that two diagrams that look entirely different from one another may represent the same graph.

Example 7.1   One of the most frequent uses of a graph occurs when you plan a vacation. If travel is by car, road maps representing the roadway systems available for the trip are used. A road map is a graph in which the vertices are the towns and cities in some geographical area, and the edges represent the roads connecting these cities or towns.

Figure 7.1 shows a map of the main highway arteries connecting the major cities in western Canada. Each edge or road in the graph allows travel in either direction. The number (or weight) associated with each edge denotes the distance in kilometers between the two cities. When planning a trip, a traveler may be interested in the distance between two cities (e.g., between Winnipeg and Victoria via Edmonton). A traveler, because of time constraints, may also be interested in the minimum distance between two given cities.

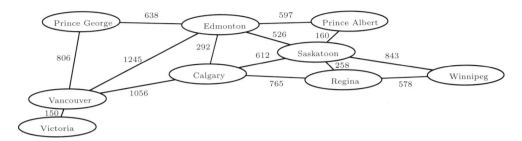

Figure 7.1.   Graph of major highway arteries in western Canada

A tool frequently used by transportation and city planners is the computer simulation of traffic systems. The systems modeled range from the traffic network of a nation, a city, or area of a city right down to the traffic flow in one bridge or intersection. The models are used to pinpoint present or future bottlenecks and to suggest and test proposed changes or new systems.

In a city the street system can be modeled as a graph in which the intersections are represented as nodes and the street segments between intersections are the edges. Two-way streets are represented as undirected edges (i.e., edges without an arrowhead), while one-way streets are directed edges.

Figure 7.2a exhibits a part of such a city street system in which the arrows denote one-way streets. Police and fire departments in cities are interested in the shortest paths from a police or fire station to the location of a 911 call for help. Figure 7.2b is an abstraction of part of a city street system in which the edges would be labeled with street names, traffic densities, or the like.

Example 7.2   A more recent use of graphs has been in the modeling of computer networks. A computer network typically consists of a variety of elements, such as computers and communication lines. In a graph representation of a computer network, each vertex is a device, such as a computer or terminal, and each edge or link denotes a communication medium, such as a telephone line or communication cable. Many industrial firms and universities have one or more local-area networks, which typically cover an area less than 1 square kilometer. However, there are several long-haul or wide-area networks whose vertices, from a geographical viewpoint, can cover one or more countries. Graphs are important in modeling these networks with respect to reliability and efficiency.

A graph representation of a computer network appears in Figure 7.3. The subnet part of the network represents the communication part of the network. The other devices around the communications subnet can be viewed as external devices. Although we have taken the liberty of representing terminals and personal computers by icons, it should be emphasized that in an abstract graph these components would be nodes or vertices. The arrangement of subnet vertices (nodes) and links in the subnet is not arbitrary. In wide-area networks, subnet nodes can be far apart and arranged in arbitrary ways. In a local-area network, however, the subnet nodes are often arranged as a ring or star structure (see Figure 7.4).

Example 7.3   Many software products consist of modules that invoke each other. A *call graph* represents modules by nodes. A directed line from a node x to a node y indicates that x invokes y. In the call graph of Figure 7.5, for example, module A invokes modules B, C, and

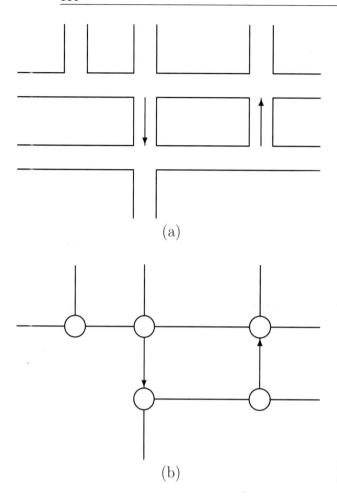

(a)

(b)

Figure 7.2.   Graphical representation of a city street system

D. Modules B and C invoke module E. When one module invokes another, there must be communication between two modules through an interface. An interface usually consists of a list of parameters. Some researchers have attempted to use graphs to evaluate the overall quality of a software system by modeling the system with an extended call graph that shows the module interfaces.

Example 7.4   In several applications in computer science, it is convenient to model (represent) an algorithm or a computer program by a graph. An example of such an application arises in the context of generating test cases for a program module. These tests are derived by analyzing the structure of a program module with respect to control flow. The control flow of a module is modeled by a *flow graph*. Each vertex in a flow graph represents one or more procedural statements. A sequence of procedural statements followed by a conditional statement, such as a while statement or case statement, is mapped into a single vertex. The arcs (directed edges) in the flow graph represent flow of control.

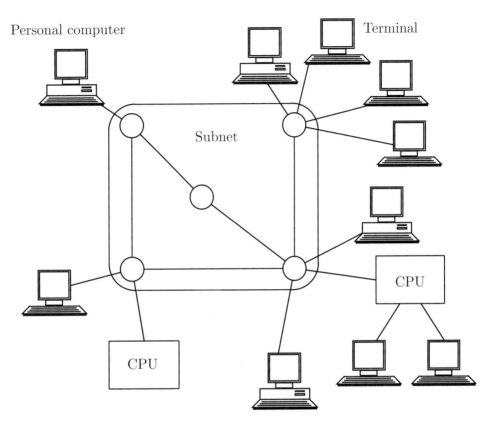

Figure 7.3.   Graph representation of a computer network

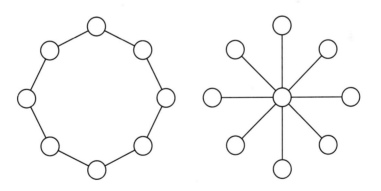

(a) Ring arrangement   (b) Star arrangement

Figure 7.4.   Specialized node arrangements in local-area networks

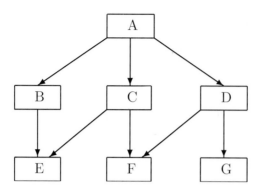

Figure 7.5.   Graph representation of a software system

Any module that is specified in some procedural language can be translated into a flow graph. For example, Figure 7.6 shows the flow graph representations of certain familiar constructs that are available in most procedural languages. The label T on an edge denotes a true branch, while F denotes a false branch.

To avoid unnecessary complexity, we assume that each condition is a simple condition, that is, a condition that does not contain logical operators such as "and" and "or."

A program module contains a sequence of language constructs from a base set such as that given in Figure 7.6. An example flow graph for a module skeleton in Figure 7.7 appears in Figure 7.8, where vertices 1 and 9 denote the start vertex and finish vertex of the module, respectively. The statements s_1 through s_7 in Figure 7.7 are assumed to be noncontrol statements such as assignment statements.

One control testing approach is to use the flow graph of a module to generate a set of independent paths that must execute to assure that all statements (and branches) have been executed at least once.

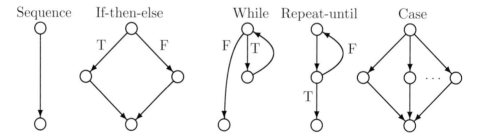

Figure 7.6.   Flow graph notation for various constructs

Example 7.5   One of the first computerized applications of graphs was concerned with project scheduling. A graph with directed edges is a natural way of describing, representing, and analyzing complex projects that consist of many interrelated activities. Such a project might be, for example, the design and construction of a power dam or the design and erection of a house.

To illustrate project management concepts, let us consider, as an example, a house construction firm that manufactures prefabricated houses that are moved and placed on concrete basements on purchased building lots. To put a prefabricated house on a concrete basement on

```
procedure Whatever (...);
  begin
    while (...) do
      begin
        s1;
        if Flag1 = 0
          then begin
                 s2;
                 s3;
                 s4
               end
          else begin
                 if Flag2 = 0
                   then s5
                   else s6;
                 s7
               end;
      end;
  end;
```

Figure 7.7. Module skeleton

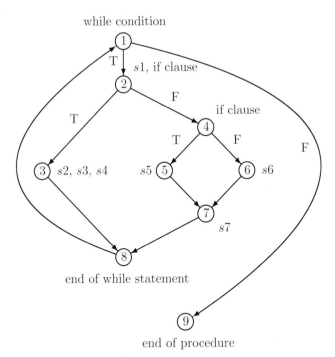

Figure 7.8. Modeling a module with a flow graph

some designated building lot involves breaking the project into a set of interrelated activities or tasks as shown in Table 7.1. The project begins with the selection and purchase of a building lot and the selection of a desired house design (and the preparation of blueprints). The third activity in the table consists of the acquisition of a building permit. In the fourth activity, bids are accepted from potential subcontractors for installing plumbing and heating equipment, electrical wiring, and the like. The next activity involves the selection of subcontractors. The remaining activities in the table should be self-explanatory.

TABLE 7.1.   House Contractor Activities

Edge	Activity	Preceding Activities	Estimated Duration (days)
a_1	Select and purchase lot	—	2
a_2	Select house design	—	5
a_3	Obtain a building permit	(a_1, a_2)	1
a_4	Accept bids from subcontractors	(a_1, a_2)	14
a_5	Select subcontractors	(a_4)	2
a_6	Excavate the basement	(a_3, a_5)	1
a_7	Construct the basement	(a_6)	7
a_8	Order windows and doors	(a_1, a_2)	3
a_9	Ship and receive windows and doors	(a_8)	10
a_{10}	Frame walls and roof	(a_3, a_5)	12
a_{11}	Perform roughing-in of plumbing	(a_{10})	5
a_{12}	Perform roughing-in of electrical	(a_{10})	3
a_{13}	Install roofing material	(a_{10})	4
a_{14}	Install doors and windows	(a_9, a_{11}, a_{12})	7
a_{15}	Design advertising promotion for sale of house	(a_2)	3
a_{16}	Advertise	(a_1, a_2, a_{15})	14
a_{17}	Install vapor barrier and insulation	(a_{13}, a_{14})	2
a_{18}	Put on drywall	(a_{17})	5
a_{19}	Paint interior walls, etc.	(a_{18})	3
a_{20}	Install cupboards	(a_{19})	3
a_{21}	Put down carpets and flooring	(a_{19})	5
a_{22}	Install light fixtures	(a_{19})	2
a_{23}	Install plumbing fixtures and heating equipment	(a_{19})	6
a_{24}	Install exterior siding	(a_{13}, a_{14})	2
a_{25}	Perform exterior painting	(a_{24})	3
a_{26}	Move house to building site	$(a_{20}, a_{21}, a_{22}, a_{23}, a_{25})$	1
a_{27}	Perform on-site connections/touch-ups	(a_7, a_{26})	2
a_{28}	Landscape lot	(a_7, a_{26})	4

The second column of the table contains *precedence requirements.* For example, it is necessary to complete activities a_1 (select and purchase lot) and a_2 (select house design) before starting activity a_3 (obtain a building permit). Observe that activities a_3, a_4, and a_8 have the same precedence requirements; this means that these three activities can proceed simultaneously. Similarly, activities a_{20}, a_{21}, a_{22}, and a_{23} can be done concurrently providing activity a_{19} is done.

The third column of the table contains an estimate for the time required to complete each activity, in days.

The creation of a table for a project such as that given in Table 7.1 can be a nontrivial task for medium-sized projects. For large projects, however, the task can be formidable. The level of detail (granularity) in identifying the activities and the associated precedence requirements is based on experience, "art," and current practice in the industry. The modeling of a project by identifying meaningful precedence requirements can affect the usefulness of the approach.

The project description of Table 7.1 can be represented by a simple directed graph called a scheduling network such as that given in Figure 7.9. The nodes or vertices represent events and the directed edges the activities. For simplicity, the events are numbered sequentially.

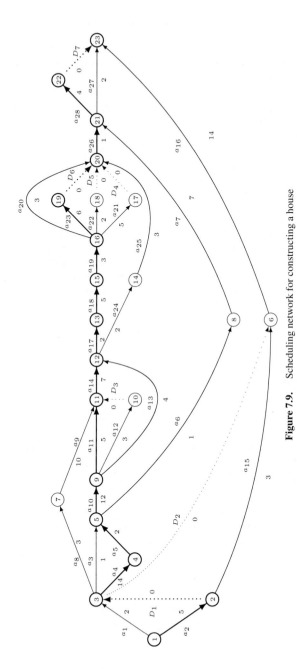

Figure 7.9. Scheduling network for constructing a house

Furthermore, each directed edge in the graph is assigned a weight (time) value. As just stated, the directed edges are meant to represent activities, or tasks, with the directed edge connecting vertices that represent the start time and the finish time of the activity. The weight value of each edge is taken to be the time it takes to complete the activity. A *source* is a vertex with no incoming edge, and a *sink* is a vertex with no outgoing edge. In a scheduling graph for a project, there is exactly one source and one sink.

Some special activities (dotted edges) in the figure represent dummy activities. The nature and purpose of these dummy activities will be discussed in Section 7.7.

An important use of the graph of a project is to find a critical path, that is, to find a longest path between the source to the sink. A critical path is a path from the source to the sink such that, if any activity on the path is delayed by an amount t, then the entire project is delayed by t. The critical path for the graph in Figure 7.9 (shown with heavier lines) consists of the activity sequence

$$a_2 \ a_4 \ a_5 \ a_{10} \ a_{11} \ a_{14} \ a_{17} \ a_{18} \ a_{19} \ a_{23} \ a_{26} \ a_{28}$$

We have ignored the dummy activities in this sequence. The earliest completion time is 66 days. A delay in the completion of any activity on the critical path will result in a delay in completing the project. To reduce the earliest completion time for the project, we find that only those activities on the critical path must be speeded up. Since in practice the number of activities that lies on the critical path in large graphs is a small percentage of the total number of activities, say 10%, only those 10% need be speeded up.

7.2 BASIC DEFINITIONS OF GRAPH THEORY

Graph theory has been applied in many fields, which has resulted in a great diversity in terminology. Different authors often use different terms for the same concept and, what is worse, the same term for different concepts. We will use computer science notation, and wherever possible we shall indicate the common alternatives that are also used in the literature of computer science.

In this section we shall define a graph as an abstract mathematical system. However, to provide some motivation for the terminology used and also to develop some intuitive feelings, we shall represent graphs diagrammatically. Any such diagram will also be called a graph. Our definitions and terms are not restricted to those graphs that can be represented by means of diagrams, even though this may appear to be the case because these terms have strong associations with such a representation. We shall see later that a diagrammatic representation is only suitable in some very simple cases. Alternative methods of representing graphs will be discussed in Sections 7.4 and 7.5. After introducing the terminology, we shall also discuss some of the basic results and theorems of graph theory.

Graphs consist of nodes, which are connected by edges. A mathematical definition of a graph must therefore rely on V, the set of vertices, and E, the set of edges. Each edge is associated with two nodes, that is, there is a mapping from the edges to the ordered or unordered pair of nodes. This is summarized in the following definition:

Definition 7.1:   A *graph* $G = (V, E, f)$ consists of a nonempty set V called the set of *nodes* (*points, vertices*) of the graph, E is said to be the set of *edges* of the graph and f is a mapping from the set of edges E to a set of ordered or unordered pairs of elements of V. If an edge is mapped to an ordered pair, it is called a *directed edge*; otherwise, it is called an *undirected edge*.

Notice that the definition of a graph implies that to every edge of the graph G we can associate an ordered or unordered pair of nodes of the graph. If an edge $e \in E$ is thus associated with an ordered pair (u, v) or an unordered pair $\{u, v\}$, where $u, v \in V$, then we say that the edge e connects or joins the nodes u and v. Any pair of nodes that is connected by an edge in a graph is called *adjacent* nodes. We shall assume throughout that both the sets V and E of a graph are finite. Often it will be convenient to write a graph G as (V, E), or simply as G. In the former case, each edge is directly represented as the pair that it is mapped to, which eliminates the need to specify f if f is a one-to-one mapping.

A graph in which every edge is directed is called a *digraph*, or a *directed graph*. A graph in which every edge is undirected is called an *undirected graph*. If some edges are directed and some are undirected in a graph, the graph is called *mixed*.

In the diagrams the directed edges (arcs) are shown by means of arrows, which also show the directions. The graphs given in Figure 7.10a and c are directed graphs. The one given in Figure 7.10b is undirected. Notice that the edges e_1, e_2, and e_3 in Figure 7.10a are associated with the ordered pairs $(1, 2)$, $(2, 3)$, and $(3, 1)$, respectively. The set representation of the graph is $(\{1, 2, 3\}, \{(1, 2), (2, 3), (3, 1)\})$. The edges e_1, e_2, and e_3 in Figure 7.10b are associated with the unordered pairs $\{1, 2\}$, $\{2, 3\}$, and $\{3, 1\}$, respectively. In Figure 7.10b, the node 1 is adjacent to nodes 2 and 3.

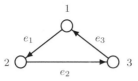

(a) Directed graph with edge labels

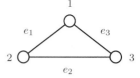

(b) Undirected graph with edge labels

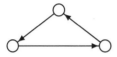

(c) Directed graph without edge labels

Figure 7.10.   Pictorial representations of graphs

Let $G = (V, E)$ be a graph and let $e \in E$ be a directed edge associated with the ordered pair of nodes (u, v). Then the edge e is said to be *initiating* or *originating* in the node u and *terminating* or *ending* in the node v. The nodes u and v are also called the *initial* and *terminal* nodes of the edge e. An edge $e \in E$ that joins the nodes u and v, whether it be directed or undirected, is said to be *incident* to the nodes u and v.

An edge of a graph that joins a node to itself is called a *loop* or *sling* (not to be confused with a loop in a program). The direction of a loop is of no significance; hence it can be considered either a directed or undirected edge. Some authors do not allow any loops in the definition of a graph.

The graphs given in Figure 7.10 have no more than one edge between any pair of nodes. In the case of directed edges, the two possible edges between a pair of nodes that are opposite in direction are considered distinct. In some directed as well as undirected graphs, we may have certain pairs of nodes joined by more than one edge, as shown in Figure 7.11a and b. Such edges are called *parallel*. Note that there are no parallel edges in the graph of Figure 7.11c. In Figure 7.11a there are two parallel edges joining the nodes 1 and 2, two parallel edges joining the nodes 2 and 3, while there are two parallel loops at 2. In Figure 7.11b there are two parallel edges associated with the ordered pair (v_1, v_2).

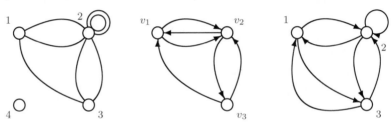

(a) Undirected multigraph   (b) Directed multigraph   (c) Directed graph with a loop
    with loops

Figure 7.11.   Multigraphs and loops

Any graph that contains some parallel edges is called a *multigraph*. In this case, the mapping between edge pairs and nodes is not one-to-one. The shorthand notation $G = (V, E)$ is not sufficient for representing a multigraph and the full notation $G = (V, E, f)$ is needed. On the other hand, if there is no more than one edge between a pair of nodes (no more than one directed edge in the case of a directed graph), then such a graph is called a *simple graph*. In this chapter, we deal primarily with simple graphs. The example graphs given in Section 7.1 are all simple graphs.

We may have graphs in which the numbers on the edges show the weights of the edges. For example, a graph representing a system of pipelines in which the weights assigned indicate the amount of some commodity transferred through the pipe is an example of a weighted graph. Similarly, a graph of city streets may be assigned weights according to the traffic density on each street. A graph in which weights are assigned to every edge is called a *weighted graph*.

In a graph a node that is not adjacent to any other node is called an *isolated node*. A graph containing only isolated nodes is called a *null graph*. In other words, the set of edges in a null graph is empty.

Definition 7.2:   In a directed graph, for any node v the number of edges that have v as their initial node is called the *outdegree* of node v. The number of edges that have v as their terminal node is called the *indegree* of v, and the sum of the outdegree and the indegree of a node v is called its *total degree*. In the case of an undirected graph, the *total degree* or the degree of node v is equal to the number of edges incident to v.

The total degree of an isolated node is 0, and that of a node with a loop and no other edges incident to it is 2.

A simple result involving the notion of the degree of nodes of a graph is that the sum of the degrees (or total degrees in the case of a directed graph) of all the nodes of a graph must be an even number that is equal to twice the number of edges in the graph.

Some graph applications are concerned with only parts of a graph. The notion of a subset in sets is useful in formalizing what we mean by a part of a graph.

Definition 7.3:   Let $V(H)$ be the set of nodes of a graph H and $V(G)$ be the set of nodes of a graph G such that $V(H) \subseteq V(G)$. If, in addition, every edge of H is also an edge of G, then the graph H is called a *subgraph* of the graph G, which is expressed by writing $H \subseteq G$.

Naturally, the graph G itself and the null graph obtained from G by deleting all the edges of G are also subgraphs of G. Other subgraphs of G can be obtained by deleting certain nodes and edges of G.

An example of a subgraph is given in Figure 7.12. The graph in part (b) is a subgraph of the graph in part (a).

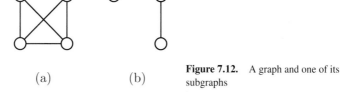

(a)   (b)   **Figure 7.12.**   A graph and one of its subgraphs

There are several special classes of simple graphs that frequently arise. We now briefly describe a few of these.

A graph (V, E) is said to be *complete* if every node is adjacent to all nodes in the graph. A complete graph of n nodes is denoted by K_n. Figure 7.13 shows the first five complete graphs.

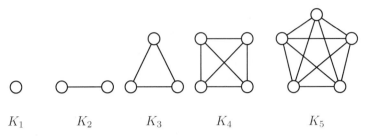

K_1 K_2 K_3 K_4 K_5

Figure 7.13.   Examples of complete graphs

Another type of simple graph is a bipartite graph. A simple graph $G = (V, E)$ is called a *bipartite graph* if V can be partitioned into subsets V_1 and V_2 such that no two nodes of V_1 are adjacent, and if the same is true for V_2. Consequently, an edge cannot connect two nodes in V_1 or two nodes in V_2. The graph in Figure 7.14a is a bipartite graph since the two disjoint subsets of nodes are $V_1 = \{v_1, v_2\}$ and $V_2 = \{v_3, v_4, v_5\}$, and all the edges only connect nodes in V_1 to nodes in V_2. Figure 7.14b, however, is not a bipartite graph since the nodes cannot be partitioned into two nonempty disjoint subsets where edges only connect a node from one subset to another.

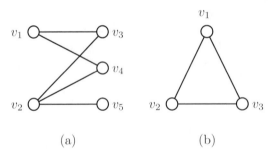

(a)   (b)

Figure 7.14.   Bipartite and nonbipartite graphs

We shall end this section by showing how the theory of binary relations given in Chapter 5 is closely linked to the theory of simple digraphs.

Let $G = (V, E)$ be a simple digraph. Every edge of E can be expressed by means of an ordered pair of elements of V; that is, $E \subseteq V \times V$. On the other hand, any subset of $V \times V$ defines a relation in V. Accordingly, E is a binary relation in V whose graph is the same as the simple digraph G. This observation permits us to carry over the terminology and the results developed in Chapter 5 to binary relations.

A simple digraph $G = (V, E)$ is called *reflexive, transitive, symmetric, antisymmetric,* and so on, if the relation E is reflexive, transitive, symmetric, antisymmetric, and so on. A simple undirected graph is irreflexive and symmetric.

If a simple graph $G = (V, E)$ is reflexive, symmetric, and transitive, then the relation E must be an equivalence relation on V, and hence V can be partitioned into equivalence classes. If we consider any such equivalence class of nodes along with the edges that join these nodes, we have subgraphs of G. These subgraphs are such that every node in the subgraph is adjacent to every other node of that subgraph. However, no node of any subgraph is adjacent to any node of another subgraph. In this sense, the graph G is partitioned into subgraphs that are disjoint. Such subgraphs are called *components* of the graph.

Problems 7.2

1. Show that the sum of indegrees of all the nodes of a simple digraph is equal to the sum of the outdegrees of all its nodes and that this sum is equal to the number of edges of the graph.

2. Because graphs can be drawn in an arbitrary manner, it can happen that two diagrams that look entirely different from one another may represent the same graph, as in Figure 7.15a and a'. Two graphs are *isomorphic* if there exists a one-to-one correspondence between the nodes of the two graphs that preserves adjacency of the nodes as well as the directions of the edges, if any.

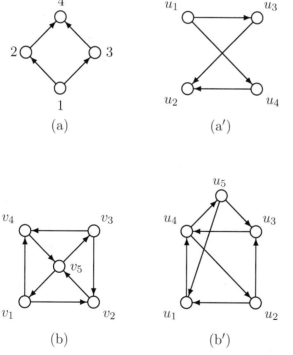

(a)   (a')

(b)   (b')   **Figure 7.15**

According to the definition of isomorphism, we note that any two nodes in one graph that are joined by an edge must have the corresponding nodes in the other graph also joined by an edge; hence, a one-to-one correspondence exists between the edges as well. The graphs given in

Figure 7.15a and a′ are isomorphic because of the existence of a mapping

$$1 \rightarrow u_1, \ 2 \rightarrow u_3, \ 3 \rightarrow u_4, \ \text{and } 4 \rightarrow u_2$$

Under this mapping, the edges $(1,3)$, $(1,2)$, $(2,4)$, and $(3,4)$ are mapped into (u_1, u_4), (u_1, u_3), (u_3, u_2), and (u_4, u_2), which are the only edges of the graph in part (a). Show that the digraphs given in Figure 7.15b and b′ are isomorphic.

3. Draw all possible simple digraphs having three nodes. Show that there is only one digraph with no edges, one with one edge, four with two edges, four with three edges, four with four edges, one with five edges, and one with six edges. Assume that there are no loops and that isomorphic graphs are not distinguishable. State the properties of these digraphs, such as symmetric and transitive.

4. Show that the graphs given in Figures 7.16a and b are isomorphic.

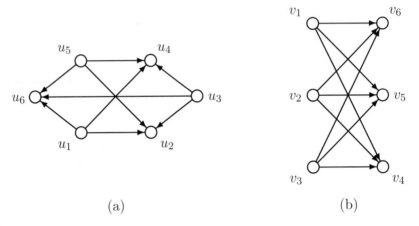

(a)   (b)

Figure 7.16

5. Show that the digraphs in Figure 7.17 are not isomorphic.

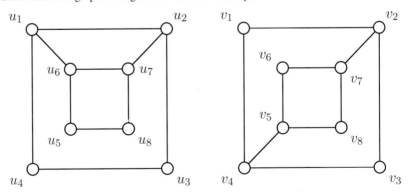

Figure 7.17

6. Show that a complete digraph with n nodes has the maximum number of edges, $n(n-1)$, assuming that there are no loops.

7.3 PATHS, REACHABILITY, AND CONNECTEDNESS

In this section we introduce some additional terminology associated with a simple digraph. During the course of our discussion we shall also indicate how the same terminology and concepts can be extended to simple undirected graphs. In graphs, one has *paths*, which are defined as follows:

Definition 7.4:   Let $G = (V, E)$ be a simple digraph. A sequence of edges is called a *path* of G iff the terminal node of each edge in the path is the initial node of the next edge, if any, in the path.

An example of such a path is

$$\langle (v_{i1}, v_{i2}), (v_{i2}, v_{i3}), \ldots, (v_{ik-2}, v_{ik-1}), (v_{ik-1}, v_{ik}) \rangle$$

where it is assumed that all the nodes and edges appearing in the path are in V and E, respectively. It is customary to write such a path as

$$\langle v_{i1}, v_{i2}, \ldots, v_{ik-1}, v_{ik} \rangle$$

Note that not all edges and nodes appearing in a path need be distinct. Also, for a given graph any arbitrary set of nodes written in any order does not necessarily give a path. In fact each node appearing in the path must be adjacent to the nodes appearing just before and after it, except in the case of the first and last nodes. We now elaborate on this notion.

A path is said to *traverse* through the nodes appearing in the sequence, *originating* in the initial node of the first edge and *ending* in the terminal node of the last edge in the sequence.

Definition 7.5:   The number of edges appearing in the sequence of a path is called the *length* of the path.

Consider the digraph of a module given in Figure 7.8 on page 359. Some of the paths originating in node 1 and ending in node 9 are

$$P_1 = \langle 1, 9 \rangle$$
$$P_2 = \langle 1, 2, 3, 8, 1, 9 \rangle$$
$$P_3 = \langle 1, 2, 4, 5, 7, 8, 1, 9 \rangle$$

From a testing viewpoint, in the flow graph of a module we are interested in the set of independent paths for the graph. An independent path must contain at least one edge that is not contained in a previous path. For example, the following are the independent paths for the flow graph in Figure 7.8:

$$path_1 : \langle 1, 9 \rangle$$
$$path_2 : \langle 1, 2, 3, 8, 1, 9 \rangle$$
$$path_3 : \langle 1, 2, 4, 5, 7, 8, 1, 9 \rangle$$
$$path_4 : \langle 1, 2, 4, 6, 7, 8, 1, 9 \rangle$$

From these paths, test cases can be designed to force execution of these paths.

Definition 7.6:   A path in a digraph in which the edges are all distinct is called a *simple path*. A path in which all the nodes are distinct is called an *elementary path*.

According to the definition, a path is called simple if no edge is repeated (edge simple), and a path is called elementary if no node is repeated (node simple). Naturally, every elementary path of a digraph is also simple. The paths P_1, P_2, and P_3 of the digraph in Figure 7.8 are all simple, but the paths P_2 and P_3 are not elementary. We shall show here that if there exists a path from a node, say u, to another node, say v, then there must also be an elementary path from u to v.

Definition 7.7:   A path that originates and ends in the same node is called a cycle (circuit). A cycle is called *simple* if no edge in the cycle appears more than once in the path. A cycle is called *elementary* if it does not traverse through any node more than once.

Note that in a cycle the initial node appears at least twice even if it is an elementary cycle. The following are some of the cycles in the graph of Figure 7.8:

$$C_1 = \langle 1, 2, 3, 8, 1 \rangle$$
$$C_2 = \langle 1, 2, 4, 5, 7, 8, 1 \rangle$$
$$C_3 = \langle 1, 2, 3, 8, 1, 2, 3, 8, 1 \rangle$$

Observe that any path that is not elementary contains cycles traversing through those nodes that appear more than once in the path. By deleting such cycles, one can obtain elementary paths. For example, in the path C_3, if we delete the cycle $\langle (1, 2), (2, 3), (3, 8), (8, 1) \rangle$, we obtain the path C_1, which also originates at 1 and ends in 1 and is an elementary path. Some authors use the term "path" to mean only the elementary paths, and they likewise apply the notion of the length of a path to only elementary paths. The digraphs generated by many applications never contain cycles. For instance, scheduling graphs never contain cycles. This type of graph has led to the following definition:

Definition 7.8:   A simple digraph that does not have any cycles is called *acyclic*.

Naturally, acyclic graphs cannot have any loops.

Examples of acyclic graphs are given in Figure 7.18. We consider a class of acyclic digraphs called free trees in Section 7.6.

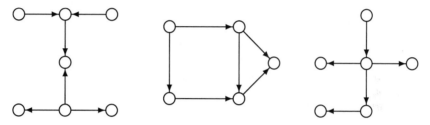

Figure 7.18.   Examples of acyclic graphs

In the previous section we viewed a graph as an ordered pair (V, E), where E is a binary relation in V; that is, $E \subseteq V \times V$. This view can be used to formalize the notion of what vertices can be "visited" from some starting vertex.

Definition 7.9:   Let $G = (V, E)$ be a simple digraph. The *path relation*, P, of G is defined as $P = \{(u, v) \mid \text{there exists a path from node } u \text{ to node } v\}$.

Note that the concept of the path relation is independent of the number of alternative paths from u to v and also of their lengths. For the graph given in Figure 7.8, we have given paths P_1 to P_3 from node 1 to node 9. Any one of these paths is sufficient to establish the reachability of node 9 from node 1. For the sake of completeness, we shall assume that every node is reachable from itself.

If a node v is reachable from the node u, then a path of minimum length from u to v is called a *minimum path length*. The length of a minimum path length from the node u to the node v is called the distance and is denoted by $d(u, v)$. It is assumed that $d(u, u) = 0$ for any node u.

It is clear from the definition that reachability is a binary relation on the set of nodes of a simple digraph. By our definition, reachability is reflexive. If there exists a path from node u to node v and a path from node v to a node w, then clearly there is a path from u to

w. In other words, reachability is also a transitive relation. In general, it is not true that if there is a path from u to v then there exists a path from v to u; therefore, reachability is not necessarily symmetric, nor is it necessarily antisymmetric.

The distance $d(u, v)$ from a node u to a node v satisfies the following properties:

$$d(u, v) \geq 0$$
$$d(u, u) = 0$$
$$d(u, v) + d(v, w) \geq d(u, w)$$

The last inequality is called the *triangle inequality*. If v is not reachable from u, then it is customary to write $d(u, v) = \infty$. Note also that, if v is reachable from u, and u is reachable from v, then $d(u, v)$ is not necessarily equal to $d(v, u)$.

The following theorem about the length of an elementary path between two nodes of a simple digraph will be used in Section 7.4.

Theorem 7.1. In a simple digraph, the length of any elementary path is less than or equal to $n - 1$, where n is the number of nodes in the graph. Similarly, the length of any elementary cycle does not exceed n.

Proof. The proof is based on the fact that in any elementary path the nodes appearing in the sequence are unique. The number of distinct nodes in any elementary path of length k is $k + 1$. Since there are only n distinct nodes in the graph, we cannot have an elementary path of length greater than $n - 1$. For an elementary cycle of length k, the sequence contains k distinct nodes; hence the result. ∎

Let us now briefly consider how the concepts of path and cycle can be extended to undirected graphs.

Definition 7.10: In a simple undirected graph, a sequence $\langle v_1, v_2, \ldots, v_d \rangle$ forms a *path* if for $i = 2, 3, \ldots, d$ there is an undirected edge $\{v_{i-1}, v_i\}$. The edge $\{v_{i-1}, v_i\}$ is said to be on the path. The length of the path is given by the number of edges on the path, which is $d - 1$. If $v_1 = v_d$, then the path forms a *cycle*.

Cycles in undirected graphs are different from those found in digraphs. For example, in an undirected graph we do not consider the path $\langle v_1, v_2, v_1 \rangle$ a cycle if $\{v_1, v_2\}$ is an edge. More generally, we do not consider the traversal of a sequence of edges in a forward and then in a reverse direction a cycle. A simple cycle in an undirected graph is a simple path that must have at least three distinct edges where only the initial and last nodes in the sequence are repeated.

Notice that the definition of a path requires that the edges appearing in the sequence must have a definite initial and terminal node. In the case of a simple undirected graph, an edge is given by an unordered pair, and any one of the nodes in the ordered pair can be considered as the initial or terminal node of the edge. To apply the same definition of a path to an undirected graph, we can consider every edge in an undirected graph to be replaced by two directed edges in opposite directions. Once this is done, we have a directed graph, and the definitions of path, cycle, elementary path, path relation, and so on, are carried over to undirected graphs. In the case of an undirected graph, the path relation is symmetric and so also is the distance. Theorem 7.1 holds for undirected graphs.

We shall now introduce an important concept, the connectedness of nodes in a graph.

Definition 7.11:   An undirected graph is said to be *connected* if for any pair of nodes of the graph the two nodes are reachable from one another.

The notion of connectedness induces a partition on an undirected graph by partitioning a given graph into disjoint subgraphs. Each subgraph is called a connected component of the graph. For example, the graph in Figure 7.19 contains the following three components:

$$G_1 = (\{v_1, v_2, v_3, v_4, v_5\},$$
$$\{\{v_1, v_2\}, \{v_1, v_4\}, \{v_2, v_3\}, \{v_3, v_4\}, \{v_2, v_5\}, \{v_3, v_5\}\})$$
$$G_2 = (\{v_6, v_7, v_8\}, \{\{v_6, v_7\}, \{v_7, v_8\}\})$$
$$G_3 = (\{v_9, v_{10}\}, \{\{v_9, v_{10}\}\})$$

This definition cannot be applied to directed graphs without some further modifications, because in a directed graph, if a node u is reachable from another node v, the node v may not be reachable from u. To overcome this difficulty, we call a digraph *connected* or *weakly connected* if it is connected as an undirected graph in which the direction of the edges is neglected, that is, if each directed edge is converted to an undirected edge.

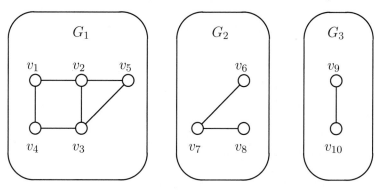

Figure 7.19

Definition 7.12:   A simple digraph is said to be *unilaterally connected* if for any pair of nodes of the graph at least one of the nodes of the pair is reachable from the other node. If for any pair of nodes of the graph both nodes of the pair are reachable from one another, then the graph is called *strongly connected*.

Observe that a unilaterally connected digraph is weakly connected, but a weakly connected digraph is not necessarily unilaterally connected. In fact, in a weakly connected digraph we may find that for any pair of nodes, say u and v, neither u is reachable from v nor v is reachable from u. A strongly connected digraph is both unilaterally and weakly connected.

For the digraphs given in Figure 7.20, the digraph in part (a) is strongly connected, in part (b) it is weakly connected but not unilaterally connected, while in part (c) it is unilaterally connected but not strongly connected.

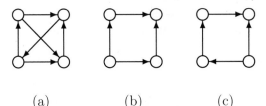

(a)   (b)   (c)   **Figure 7.20.**   Connectivity in digraphs

Let $G = (V, E)$ be a simple digraph and $X \subseteq V$. A subgraph whose nodes are given by the set X and whose edges consist of all those edges of G that have their initial and terminal nodes in X is called the *subgraph induced by X*.

Definition 7.13:   A subgraph G_1 is said to be *maximal* with respect to some property if no other subgraph has the property and also includes G_1. For a simple digraph, a maximal strongly connected subgraph is called a *strong component*. Similarly, a maximal unilaterally connected or maximal weakly connected subgraph is called a *unilateral* or *weak component*, respectively.

For the digraph given in Figure 7.21, $\{1, 2, 3\}$, $\{4\}$, $\{5\}$, $\{6\}$ are the strong components. $\{1, 2, 3, 4, 5\}$ and $\{5, 6\}$ are the unilateral components, and $\{1, 2, 3, 4, 5, 6\}$ is the weak component.

Theorem 7.2.   In a simple digraph $G = (V, E)$, every node of the digraph lies in exactly one strong component.

Proof. Let $v \in V$ and S be the set of all those nodes of G that are mutually reachable with v. The set S naturally contains v and is a strong component of G. This shows that every node of G is contained in a strong component.

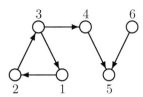

Figure 7.21

Assume now that a node v is in two strong components. It would imply that any node in the strong component that contains v is reachable from any node in the other strong component that also contains v, because every such path is easily established through v. This, however, is impossible. Hence, every node is contained in exactly one strong component. Thus, the strong components partition V.   ∎

Note that any edge $e \in E$ of a simple digraph may or may not be contained in a strong component. If $e = (u, v)$ and both u and v are in a strong component S, then e is in a strong component.

A similar result can be proved for weak components.

Theorem 7.3.   Every node and edge of a simple digraph is contained in exactly one weak component.

The proof is similar to the one given for Theorem 7.2, and it is omitted.

We shall now show how a simple digraph can be used to represent the resource allocation status of an operating system. This example was chosen because of the fundamental importance of the operating system in almost every conceivable computer system.

In a multiprogrammed computer system, it appears that several programs are executed at one time. In reality, the programs are sharing the resources of the computer system, such as tape units, disk devices, the central processing unit, main memory, and compilers. A special set of programs called the operating system controls the allocation of these resources to the programs. When a program requires the use of a certain resource, it issues a request for that resource, and the operating system must ensure that the request is satisfied.

It is assumed that all resource requests of a program must be satisfied before that program can complete execution. If any requested resources are unavailable at the time of the request, the program will assume control of the resources that are available, but must wait for the unavailable resources.

Let $P_t = \{p_1, p_2, \ldots, p_m\}$ represent the set of programs in the computer system at time t. Let $A_t \subseteq P_t$ be the set of active programs, or programs that have been allocated at least a portion of their resource requests at time t. Finally, let $R_t = \{r_1, r_2, \ldots, r_n\}$ represent the set of resources in the system at time t. An allocation graph G_t is a directed graph representing the resource allocation status of the system at time t and consisting of a set of nodes $V = R_t$ and a set of edges E. There is a directed edge from node r_i to r_j if and only if there is a program p_k in A_t that has been allocated resource r_i but is waiting for r_j.

It may happen that requests for resources are in conflict. For example, program A may have control of resource r_1 and require resource r_2, but program B has control of resource r_2 and requires resource r_1. In such a case, the computer system is said to be in a state known as *deadlock*. The nature of deadlock is best explained through graph theory.

First, a program x must wait for program y to complete if y holds a resource that x needs. Program y can also block x indirectly because y blocks a program z, either directly or indirectly, that requires a resource that x needs. In a graph, this means that x must wait for y to complete if there is a path from y to x. If x waits for y and if y waits for x, that is, if there is a cycle involving x and y, then there is a deadlock. Graph theory can thus help to detect deadlock, and it makes the removal of deadlock easier.

For example, let $R_t = \{r_1, r_2, r_3, r_4\}$, $A_t = \{p_1, p_2, p_3, p_4\}$, and the resource allocation status be

p_1 has resource r_4 and requires r_1.

p_2 has resource r_1 and requires r_2 and r_3.

p_3 has resource r_2 and requires r_3.

p_4 has resource r_3 and requires r_1 and r_4.

Then the allocation graph at time t is given in Figure 7.22. Observe that this graph is like a bipartite graph. Furthermore, if there are no cycles in the graph, there is no deadlock. In Figure 7.22, there is a deadlock, as the reader may verify.

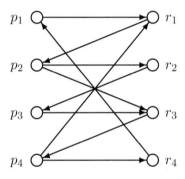

Figure 7.22.   Allocation graph for detecting deadlocks

Problems 7.3

1. Give three different elementary paths from v_1 to v_3 for the digraph given in Figure 7.23. What is the shortest distance between v_1 and v_3? Is there any cycle in the graph? Is the digraph transitive?

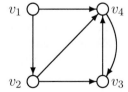

Figure 7.23

2. Find all the indegrees and outdegrees of the nodes of the graph given in Figure 7.24. Give all the elementary cycles of this graph. Obtain an acyclic digraph by deleting one edge of the given digraph. List all the nodes that can reach every other node in the digraph.

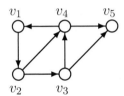

Figure 7.24

3. Given a simple digraph $G = (V, E)$, under what conditions is the equation

$$d(v_1, v_2) + d(v_2, v_3) = d(v_1, v_3)$$

satisfied for v_1, v_2, and $v_3 \in V$?

4. For the digraphs given in Figures 7.23 and 7.24, determine whether they are strongly, weakly, or unilaterally connected.

5. Show that a simple digraph is strongly connected iff there is a cycle in G that includes each node at least once.

6. The diameter of a simple digraph $G = (V, E)$ is given by δ, where

$$\delta = \max d(u, v), \qquad \text{where } u, v \in V$$

Find the diameter of the digraphs given in Figures 7.23 and 7.24.

7. Find the strong components of the digraph given in Figure 7.24. Also find its unilateral and weak components.

8. Show that every node and edge of a graph are contained in exactly one weak component.

7.4 COMPUTING PATHS FROM A MATRIX REPRESENTATION OF GRAPHS

A diagrammatic representation is only possible when the number of nodes and edges is reasonably small. In this subsection we shall present an alternative method of representing graphs using matrices. Such a method of representation has several advantages. It is easy to store and manipulate matrices and hence the graphs represented by them in a computer. Also, well-known operations of matrix algebra can be used to calculate paths, cycles, and other characteristics of a graph.

Given a simple digraph $G = (V, E)$, it is necessary to assume some kind of ordering of the nodes of the graph in the sense that a particular node is called a first node, another the second node, and so on. Our matrix representation of G depends on the ordering of the nodes.

Definition 7.14: Let $G = (V, E)$ be a simple digraph in which $V = \{v_1, v_2, \ldots, v_n\}$ and the nodes are assumed to be ordered from v_1 to v_n. The $n \times n$ matrix A whose elements a_{ij} are given by

$$a_{ij} = \begin{cases} 1, & \text{if } \langle v_i, v_j \rangle \in E \\ 0, & \text{otherwise} \end{cases}$$

is called the *adjacency matrix* of the graph G.

The adjacency matrix is the same as the relation matrix of the relation E in V. Any element of the adjacency matrix is either 0 or 1. Any matrix whose elements are either 0 or 1 (false or true) is called a *bit matrix* or a *Boolean matrix*. Note that the ith row in the adjacency matrix is determined by the edges that originate in the node v_i. The number of elements in the ith row whose value is 1 is equal to the outdegree of the node v_i. Similarly, the number of elements whose value is 1 in a column, say the jth column, is equal to the indegree of the node v_j. An adjacency matrix completely defines a simple digraph.

For a given digraph $G = (V, E)$, the adjacency matrix depends on the ordering of the elements of V. For different orderings of the elements of V, we get different adjacency matrices of the same graph G. However, any one of the adjacency matrices of G can be obtained from another adjacency matrix of the same graph by interchanging some of the rows and the corresponding columns of the matrix. We shall neglect the arbitrariness introduced in an adjacency matrix due to the ordering of the elements of V and take any adjacency matrix of the graph to be the adjacency matrix of the graph. A Pascal data type for a graph using the matrix representation is given in Figure 7.25. Note that in the remainder of the chapter we represent **true** and **false** in Pascal by 1 and 0, respectively.

```
const
   {The maximum number of vertices or nodes}
   MaxVert = 20;
type
   {The type definition of the graph}
   BoolMatrix = array [1..MaxVert, 1..MaxVert] of boolean;
var
   A: BoolMatrix;
```

Figure 7.25.   Graph data structure in Pascal

As an example, consider the digraph given in Figure 7.26 in which first we order the nodes as v_1, v_2, v_3, v_4, and v_5. Its adjacency matrix is

$$A = \begin{pmatrix} 0 & 0 & 0 & 1 & 0 \\ 0 & 0 & 1 & 0 & 0 \\ 0 & 1 & 0 & 0 & 0 \\ 0 & 1 & 1 & 0 & 1 \\ 1 & 1 & 0 & 0 & 0 \end{pmatrix}$$

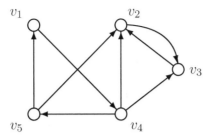

Figure 7.26

Some of the properties of a simple digraph are immediately seen from its adjacency matrix. If a digraph is reflexive, then the diagonal elements of the adjacency matrix are 1's. For a symmetric digraph and for an undirected graph, the adjacency matrix is also symmetric; that is, $a_{ij} = a_{ji}$ for all i and j.

Consider now the powers of an adjacency matrix. Naturally, an entry of 1 in the ith row and jth column of A shows the existence of an edge (v_i, v_j), that is, a path of length 1 from v_i to v_j. Let us denote the elements of A^2 by a_{ij}^2. Then

$$a_{ij}^2 = \sum_{k=1}^{n} a_{ik} a_{kj}$$

For any fixed k, $a_{ik} a_{kj} = 1$ iff both a_{ik} and a_{kj} equal 1; that is, iff (v_i, v_k) and (v_k, v_j) are the edges of the graph. For each such k we get a contribution of 1 in the sum. Now (v_i, v_k) and (v_k, v_j) imply that there is a path from v_i to v_j of length 2. Therefore, a_{ij}^2 is equal to the number of different paths of exactly length 2 from v_i to v_j. This argument is similar to the argument presented in Section 5.4.5, except that in that section the sum was interpreted in a Boolean sense.

The following matrices are obtained from A.

$$A^2 = \begin{pmatrix} 0 & 1 & 1 & 0 & 1 \\ 0 & 1 & 0 & 0 & 0 \\ 0 & 0 & 1 & 0 & 0 \\ 1 & 2 & 1 & 0 & 0 \\ 0 & 0 & 1 & 1 & 0 \end{pmatrix}, \quad A^3 = \begin{pmatrix} 1 & 2 & 1 & 0 & 0 \\ 0 & 0 & 1 & 0 & 0 \\ 0 & 1 & 0 & 0 & 0 \\ 0 & 1 & 2 & 1 & 0 \\ 0 & 2 & 1 & 0 & 1 \end{pmatrix}$$

$$A^4 = \begin{pmatrix} 0 & 1 & 2 & 1 & 0 \\ 0 & 1 & 0 & 0 & 0 \\ 0 & 0 & 1 & 0 & 0 \\ 0 & 3 & 2 & 0 & 1 \\ 1 & 2 & 2 & 0 & 0 \end{pmatrix}, \quad A^5 = \begin{pmatrix} 0 & 3 & 2 & 0 & 1 \\ 0 & 0 & 1 & 0 & 0 \\ 0 & 1 & 0 & 0 & 0 \\ 1 & 3 & 3 & 0 & 0 \\ 0 & 2 & 2 & 1 & 0 \end{pmatrix}$$

For the graph given in Figure 7.26, we see that there is one path of length 2 from v_1 to v_2, hence the entry 1 in the first row and second column of A^2. Similarly, there are three paths of length 4 from v_4 to v_2, $\langle 4, 5, 2, 3, 2 \rangle$, $\langle 4, 5, 1, 4, 2 \rangle$, and $\langle 4, 3, 2, 3, 2 \rangle$, hence the corresponding entry in A^4. This can be generalized as follows:

Theorem 7.4. Let A be the adjacency matrix of a digraph G. The element in the ith row and the jth column of A^n, $n > 0$, is equal to the number of paths of length n from the ith node to the jth node.

Theorem 7.4 can be proved for any positive integer n by using mathematical induction and an argument similar to the one given previously.

Given a simple digraph $G = (V, E)$, let v_i and v_j be any two nodes of G. From the adjacency matrix of A, we can immediately determine whether there exists an edge from v_i to v_j in G. Also, from the matrix A^r, where r is some positive integer, we can establish the number of paths of length r from v_i to v_j. If we add the matrices A, A^2, A^3, ..., A^r to get B_r, then

$$B_r = A + A^2 + \cdots + A^r$$

From the matrix B_r, we can determine the number of paths of length less than or equal to r from v_i to v_j. If we wish to determine whether v_j is reachable from v_i, it would be necessary to investigate whether there exists a path of any length from v_i to v_j. To decide this, with the help of the adjacency matrix, we would have to consider all possible A^r for $r = 1, 2, 3, \ldots$. This method is neither practical nor necessary, as we shall show.

Recall that in a simple digraph with n nodes the length of an elementary path or cycle does not exceed n (see Theorem 7.1). Any path can be converted to an elementary path by eliminating all cycles of the path, and any cycle can similarly be converted into an elementary cycle. To determine whether there exists a path from v_i to v_j, we therefore only need to examine the elementary paths of length less than or equal to $n - 1$. If $v_i = v_j$, the path is a cycle, and only elementary cycles need to be examined, and elementary cycles are at most of length n. All paths and cycles are counted by B_n, where

$$B_n = A + A^2 + A^3 + \cdots + A^n$$

The element in the ith row and jth column of B_n shows the number of paths of length n or less that exists from v_i to v_j. If this element is nonzero, then it is clear that v_j is reachable from v_i. Of course, to determine reachability, we only need to know the existence of a path. In any case, the matrix B_n furnishes the required information about the reachability of any node of the graph from any other node.

Definition 7.15: Let $G = (V, E)$ be a simple digraph in which $\#V = n$ and the nodes of G are assumed to be ordered. The $n \times n$ matrix P whose elements are given by

$$p_{ij} = \begin{cases} 1, & \text{if there exists a path from } v_i \text{ to } v_j \\ 0, & \text{otherwise} \end{cases}$$

is called the *path matrix (reachability matrix)* of the graph G.

The main diagonal entry $p_{ii} = 1$ if and only if there is a path from v_i to itself.

Note that the path matrix only shows the presence or absence of at least one path between a pair of points and also the presence or absence of a cycle at any node. It does not, however, show all the paths that may exist. In this sense, a path matrix does not give complete information about a graph, as does the adjacency matrix. The path matrix is important in its own right.

The path matrix can be calculated from the matrix B_n by choosing $p_{ij} = 1$ if the element in the ith row and jth column of B_n is nonzero and $p_{ij} = 0$ otherwise. We shall apply this method of calculating the path matrix to our sample problem, whose graph is given in Figure 7.26. The adjacency matrix A and its powers A^2, A^3, A^4, and A^5 have already been calculated. We thus have B_5 and the path matrix P given by

$$B_5 = \begin{pmatrix} 1 & 7 & 6 & 2 & 2 \\ 0 & 2 & 3 & 0 & 0 \\ 0 & 3 & 2 & 0 & 0 \\ 2 & 10 & 9 & 1 & 2 \\ 2 & 7 & 6 & 2 & 1 \end{pmatrix}, \qquad P = \begin{pmatrix} 1 & 1 & 1 & 1 & 1 \\ 0 & 1 & 1 & 0 & 0 \\ 0 & 1 & 1 & 0 & 0 \\ 1 & 1 & 1 & 1 & 1 \\ 1 & 1 & 1 & 1 & 1 \end{pmatrix}$$

It may be remarked that, if we are only interested in knowing the reachability of one node from another, it is sufficient to calculate B_{n-1}, because a path of length n cannot be elementary. The only difference between P calculated from B_{n-1} and P calculated from B_n is in the diagonal elements. Some authors calculate the path matrix from B_{n-1}, while others do it from B_n.

The method of calculating the path matrix P of a graph by calculating first A, A^2, $\ldots$, A^n and then B_n is cumbersome. We shall now describe another method based on a similar idea, but which is more efficient. Observe that we are not interested in the number of paths of any particular length from a node, say v_i, to a node v_j. This information is obtained during the course of our calculation of the powers of A, and later it is suppressed because these actual numbers are not needed. To reduce the amount of calculation involved, this unwanted information is not generated. This is achieved by using Boolean matrix operations in our calculations, which will now be defined.

The operations $\wedge$ (and) and $\vee$ (or) on Boolean values are given in Table 7.2. For any two $n \times n$ Boolean matrices A and B, the Boolean sum and Boolean product of A and B are written as $A \vee B$ and $A \odot B$, which are also Boolean matrices, say C and D. The operator $\odot$ was introduced in Section 5.3.5. The elements of C and D are given by

$$c_{ij} = a_{ij} \vee b_{ij} \quad \text{and} \quad d_{ij} = \bigvee_{k=1}^{n} (a_{ik} \wedge b_{kj}), \qquad \text{for all } i, j = 1, 2, \ldots, n$$

Note that the element d_{ij} is easily obtained by scanning the ith row of A from left to right and simultaneously the jth column of B from top to bottom. If, for any k, the kth element in the row and the kth element in the column are both 1, then $d_{ij} = 1$; otherwise, $d_{ij} = 0$.

The adjacency matrix is a Boolean matrix, and so also is the path matrix. Let us write $A \odot A = A^{(2)}$ and $A \odot A^{(r-1)} = A^{(r)}$. Comparing $A^{(2)}$ and A^2, the entry in the ith row and jth column of $A^{(2)}$ is 1 if there is at least one path of length 2 from v_i to v_j, while in

TABLE 7.2.   Boolean Operators

$\wedge$	0	1		$\vee$	0	1
0	0	0		0	0	1
1	0	1		1	1	1

A^2 the entry in the ith row and jth column shows the number of paths of length 2 from v_i to v_j. Similar remarks apply to A^3 and $A^{(3)}$ or in general A^r and $A^{(r)}$ for any positive integer r. From this description, it is clear that the path matrix P is given by

$$P = A \vee A^{(2)} \vee A^{(3)} \vee \cdots \vee A^{(n)} = \bigvee_{k=1}^{n} A^{(k)}$$

If we take the sum from $k = 1$ to $k = n - 1$, we get a matrix that may differ if at all from P in the diagonal terms only.

For our sample example of the graph given in Figure 7.26,

$$A^{(2)} = \begin{pmatrix} 0 & 1 & 1 & 0 & 1 \\ 0 & 1 & 0 & 0 & 0 \\ 0 & 0 & 1 & 0 & 0 \\ 1 & 1 & 1 & 0 & 0 \\ 0 & 0 & 1 & 1 & 0 \end{pmatrix}, \quad A^{(3)} = \begin{pmatrix} 1 & 1 & 1 & 0 & 0 \\ 0 & 0 & 1 & 0 & 0 \\ 0 & 1 & 0 & 0 & 0 \\ 0 & 1 & 1 & 1 & 0 \\ 0 & 1 & 1 & 0 & 1 \end{pmatrix}$$

$$A^{(4)} = \begin{pmatrix} 0 & 1 & 1 & 1 & 0 \\ 0 & 1 & 0 & 0 & 0 \\ 0 & 0 & 1 & 0 & 0 \\ 0 & 1 & 1 & 0 & 1 \\ 1 & 1 & 1 & 0 & 0 \end{pmatrix}, \quad A^{(5)} = \begin{pmatrix} 0 & 1 & 1 & 0 & 1 \\ 0 & 0 & 1 & 0 & 0 \\ 0 & 1 & 0 & 0 & 0 \\ 1 & 1 & 1 & 0 & 0 \\ 0 & 1 & 1 & 1 & 0 \end{pmatrix}$$

$$A \vee A^{(2)} \vee A^{(3)} \vee A^{(4)} \vee A^{(5)} = P = \begin{pmatrix} 1 & 1 & 1 & 1 & 1 \\ 0 & 1 & 1 & 0 & 0 \\ 0 & 1 & 1 & 0 & 0 \\ 1 & 1 & 1 & 1 & 1 \\ 1 & 1 & 1 & 1 & 1 \end{pmatrix}$$

The matrices A, $A^{(2)}$, $A^{(3)}$, ... can be interpreted in a different way. In a simple digraph $G = (V, E)$, $E \subseteq V \times V$, so E can be interpreted as a relation in V. The adjacency matrix A is the relation matrix of the relation E. Recall that the composite relation $E \circ E = E^2$ was defined in Section 5.3.5 as the relation such that $v_i \, E^2 \, v_j$ if there exists a v_k such that $v_i \, E \, v_k$ and $v_k \, E \, v_j$. In other words, the relation matrix of E^2 has 1 in the ith row and jth column if there is at least a path of length 2 from v_i to v_j. This shows that $A^{(2)}$ is the relation matrix of the relation E^2. Similarly, $A^{(3)}$, $A^{(4)}$, ... are the relation matrices of the relations $E \circ E \circ E = E^3$, E^4, ... in V.

Next, let E_1 and E_2 be two relations in V and A_1 and A_2 be the corresponding relation matrices. For the relations $E_1 \cup E_2$ and $E_1 \circ E_2$, the relation matrices are given by $A_1 \vee A_2$ and $A_1 \odot A_2$, respectively.

For a given relation E in V, a relation E^+, called the transitive closure of E, was defined in Section 5.4.5 as

$$E^+ = E \cup E^2 \cup \cdots$$

Clearly, the relation matrix of E^+ is given by

$$A^+ = A \vee A^{(2)} \vee A^{(3)} \vee \cdots$$

where A is the relation matrix of E. It has been shown that if the number of elements in V is n then no elementary path or cycle exceeds n in length; therefore, A^+ can be obtained by simply considering the sum up to $A^{(n)}$, for powers higher than n will not change A^+. Therefore,

$$A^+ = A \vee A^{(2)} \vee A^{(3)} \vee \cdots \vee A^{(n)} = P$$

The matrix A^+ is the same as the path matrix. The time complexity of this approach is $O(n^4)$. This result follows from the fact that the computation of each $A^{(i)}$ is $O(n^3)$, and there are $n-1$ matrix multiplications to perform. This approach is not very efficient except, possibly, for very small values of n.

We now explore a more efficient method of computing the path matrix known as Warshall's algorithm.

The goal of this approach is to generate a sequence of matrices $P^0, P^1, P^2, \ldots, P^k,$ $\ldots, P^n$ for a graph of n vertices with $P^n = P$ (the path matrix). Initially, $P^0 = A$ (the adjacency matrix).

The first iteration consists of exploring the existence of paths from any vertex to any other either directly via an edge or indirectly through the intermediate or pivot vertex, say, v_1. P^1 denotes the resulting matrix with its general element $p_{ij}^{(1)}$ obtained as follows:

$$p_{ij}^{(1)} = \begin{cases} 1, & \text{if there exists an edge from } v_i \text{ to } v_j \text{ or there is a path (of length} \\ & \text{2) from } v_i \text{ to } v_1 \text{ and } v_1 \text{ to } v_j \\ 0, & \text{otherwise} \end{cases}$$

As an example, consider the digraph in Figure 7.26. Using v_1 as a pivot, there is a path of length 2 from v_5 to v_4. This is the only new path that results from using v_1 as a pivot. The resulting matrix, P^1, appears in Figure 7.27a.

The second iteration is to explore any paths from any vertex to any other with v_1 or v_2 or both as pivots. We compute P^2 and define its general element $p_{ij}^{(2)}$ as follows:

$$p_{ij}^{(2)} = \begin{cases} 1, & \text{if there exists an edge from } v_i \text{ to } v_j \text{ or a path from } v_i \text{ to } v_j \text{ using} \\ & \text{only pivots (intermediate vertices) from } \{v_1, v_2\} \\ 0, & \text{otherwise} \end{cases}$$

In the current example, using v_2 as a pivot reveals paths v_5 to v_3, v_3 to v_3, and v_4 to v_3. Note that the last path is not new since there is an edge from v_4 to v_3. The resulting matrix P^2 is given in Figure 7.27b.

$$\begin{pmatrix} 0 & 0 & 0 & 1 & 0 \\ 0 & 0 & 1 & 0 & 0 \\ 0 & 1 & 0 & 0 & 0 \\ 0 & 1 & 1 & 0 & 1 \\ 1 & 1 & 0 & 1 & 0 \end{pmatrix}$$
Pivot set: $\{v_1\}$

(a) P^1

$$\begin{pmatrix} 0 & 0 & 0 & 1 & 0 \\ 0 & 0 & 1 & 0 & 0 \\ 0 & 1 & 1 & 0 & 0 \\ 0 & 1 & 1 & 0 & 1 \\ 1 & 1 & 1 & 1 & 0 \end{pmatrix}$$
Pivot set: $\{v_1, v_2\}$

(b) P^2

$$\begin{pmatrix} 0 & 0 & 0 & 1 & 0 \\ 0 & 1 & 1 & 0 & 0 \\ 0 & 1 & 1 & 0 & 0 \\ 0 & 1 & 1 & 0 & 1 \\ 1 & 1 & 1 & 1 & 0 \end{pmatrix}$$
Pivot set: $\{v_1, v_2, v_3\}$

(c) P^3

$$\begin{pmatrix} 0 & 1 & 1 & 1 & 1 \\ 0 & 1 & 1 & 0 & 0 \\ 0 & 1 & 1 & 0 & 0 \\ 0 & 1 & 1 & 0 & 1 \\ 1 & 1 & 1 & 1 & 1 \end{pmatrix}$$
Pivot set: $\{v_1, v_2, v_3, v_4\}$

(d) P^4

$$\begin{pmatrix} 1 & 1 & 1 & 1 & 1 \\ 0 & 1 & 1 & 0 & 0 \\ 0 & 1 & 1 & 0 & 0 \\ 1 & 1 & 1 & 1 & 1 \\ 1 & 1 & 1 & 1 & 1 \end{pmatrix}$$
Pivot set: $\{v_1, v_2, v_3, v_4, v_5\}$

(e) P^5

Figure 7.27. Trace of successive iterations using Warshall's approach

In general, during the kth iteration, pivots used are from the set $\{v_1, v_2, \ldots, v_k\}$. The result of the kth iteration is to compute P^k, where

$$p_{ij}^{(k)} = \begin{cases} 1, & \text{if there exists an edge from } v_i \text{ to } v_j \text{ or a path from } v_i \text{ to } v_j \text{ using} \\ & \text{only pivots from } \{v_1, v_2, \ldots, v_k\} \\ 0, & \text{otherwise} \end{cases}$$

Observe that we can compute $p_{ij}^{(k)}$ from the previous iteration as follows:

$$p_{ij}^{(k)} := p_{ij}^{(k-1)} \lor (p_{ik}^{(k-1)} \land p_{kj}^{(k-1)})$$

The only way that the value of $p_{ij}^{(k)}$ can change from 0 is to find a path through the pivot v_k, as shown in Figure 7.28. That is, there is a path from v_i to v_k and a path from v_k to v_j.

The completion of computing the path matrix for our current example produces P^3 to P^5 during the third to fifth iterations, respectively. These results are given in Figures 7.27c to 7.27e. Therefore, the path matrix $P = P^5$. In general, for a graph of n vertices, $P = P^n$.

An implementation of the previous strategy, which is due to Warshall, is given in Figure 7.29. Note that the integers 1 through n are used instead of the vertex set $\{v_1, v_2, \ldots, v_n\}$.

To show that this procedure produces the required matrix, note that step 1 produces a matrix in which $P[i, j] = 1$ if there is a path of length 1 from v_i to v_j. Assume that for a fixed k the intermediate matrix $P[i, j]$ produced by the procedure is such that the element in the ith row and jth column is 1 iff there is a path from v_i to v_j that may pass through the nodes $v_1, v_2, \ldots, v_k$, or an edge from v_i to v_j. During the next iteration, we find that $P[i, j] = 1$ either if $P[i, j] = 1$ earlier or if there is a path from v_i to v_j that traverses through v_{k+1}. This means that $P[i, j] = 1$ iff there is a path from v_i to v_j that may pass through the nodes $v_1, v_2, \ldots, v_{k+1}$ or an edge from v_i to v_j, which completes the proof.

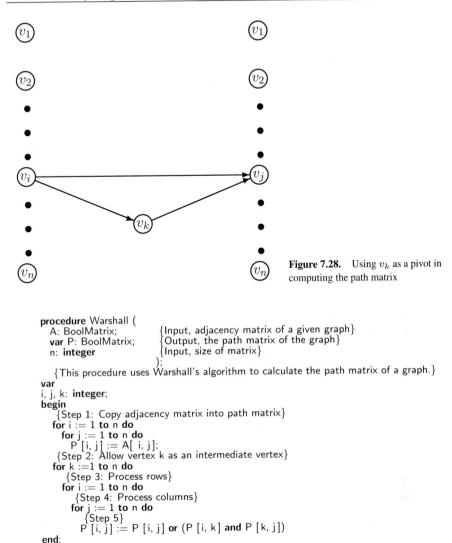

Figure 7.28.   Using v_k as a pivot in computing the path matrix

```
procedure Warshall (
   A: BoolMatrix;          {Input, adjacency matrix of a given graph}
   var P: BoolMatrix;      {Output, the path matrix of the graph}
   n: integer              {Input, size of matrix}
                           );
   {This procedure uses Warshall's algorithm to calculate the path matrix of a graph.}
var
i, j, k: integer;
begin
   {Step 1: Copy adjacency matrix into path matrix}
   for i := 1 to n do
      for j := 1 to n do
         P [i, j] := A[ i, j];
   {Step 2: Allow vertex k as an intermediate vertex}
   for k :=1 to n do
      {Step 3: Process rows}
      for i := 1 to n do
         {Step 4: Process columns}
         for j := 1 to n do
            {Step 5}
            P [i, j] := P [i, j] or (P [i, k] and P [k, j])
end;
```

Figure 7.29.   Procedure for computing the path matrix of a graph

Warshall's algorithm is easy to analyze. The number of times that the assignment statement is performed is $O(n^3)$. This algorithm is more efficient than the brute-force approach discussed earlier, whose complexity is $O(n^4)$.

Now that we know how to determine the existence of paths and since everyone wants to go from some place to another as quickly as possible, let us try to find shortest paths (minimum weighted path lengths). Warshall's algorithm to compute the path matrix can be modified to obtain a matrix that contains the shortest paths between all pairs of vertices provided that all weights are nonnegative. For this purpose, we use a modified adjacency matrix of the graph called Dist. Replace all those elements not on the main diagonal in the adjacency matrix that are zero by ∞ or some very large integer INF, which shows that there is no edge between the nodes in question. The corresponding entry of an edge in the modified matrix is the weight of that edge. The distance from any node to itself, without traversing any edges, is assumed to be 0, and its corresponding entry in the matrix is 0.

As was done in Warshall's algorithm, we use some vertex, say k, as a pivot or intermediate vertex between all pairs of vertices. For a selected pair of vertices (e.g., i and j), the current shortest path from vertices i to k, Dist [i, k], is added to the current shortest path from vertices k to j. If this sum is less than the current shortest path between vertices i and j, then the shorter path through intermediate vertex k is used. That is,

> **if** Dist [i, k] + Dist [k, j] < Dist [i, j]
> **then** Dist [i, j] := Dist [i, k] + Dist [k, j]

By repeating this process for all other pairs of vertices in the graph, one can obtain for each pair the minimum weighted path length through vertex k. By trying all vertices, in turn, as pivot elements, the desired minimum weighted path length matrix is obtained.

As well, we generate additional information that can be used to determine the actual minimum weighted path, that is, the vertices that lie on each minimum weighted path.

The procedure in Figure 7.30, which is called Floyd's algorithm, is very similar to that given in Figure 7.29. Observe that we have introduced a second matrix, Path, which is used to record intermediate node information. This information can be used to generate the nodes that lie in each minimum path (see Problems 7.4). All elements in Path are initially 0. When the procedure completes its task, Path [i, j] contains the node number (or iteration number) that caused a change to Dist [i, j]. If Path [i, j] = 0, then the shortest path is a direct path along the edge connecting nodes i and j; otherwise, if Path [i, j] = k, the shortest path from i to j passes through node k. This allows one to reconstruct the entire path. Of course, if Dist [i, k] < INF and Dist [k, j] < INF, then the path is from i to k and from k to j. If Dist [i, k] = INF, then one can find a path from i to k recursively, and if Dist [k, j] = INF, one can find a path from k to j recursively. By examining recursively Path [i, k] and Path [k, j], other intermediate node information on the shortest path can be generated.

We now give a partial trace of this procedure using the simple weighted digraph and its initial Dist and Path matrices given in Figure 7.31a.

The first iteration uses node 1 as a pivot. The first pair of nodes of interest is when i = 2 and j = 4. Since Dist [2, 4] has a value of INF, it is changed to Dist [2, 1] + Dist [1, 4] = 3 + 1 = 4. Also, Path [2, 4] is set to 1, indicating that node 1 is an intermediate node in the path from node 2 to node 4. Similarly, Dist [3, 4] is set to Dist [3, 1] + Dist [1, 4] = 5 + 1 = 6 with Path [3, 4] = 1. The revised matrices for Dist and Path at the end of the first iteration are given in Figure 7.31b. The revised matrices at the end of the second through the fourth iterations appear in Figures 7.31c to e, respectively.

The information in the Path matrix can be used to generate the nodes on each minimum path. For example, assume that we want to find the nodes that lie on the minimum path from node 1 to node 3. The element Path [1, 3] = 4 indicates that node 4 is an intermediate between nodes 1 and 3. Because Path [1, 4] = 0, there is an edge from node 1 to node 4.

```
procedure MinWeightPaths (
    var Dist, Path : IntegerMatrix;
    n: integer              );
    {Given a Dist matrix where Dist[i, i] = 0, and Dist [i,j] is set to the nonnegative weight of the edge
    between nodes i and j, or INF (a large constant) otherwise; this procedure determines the minimum
    path length for each pair of nodes. Also, a second array Path is used to generate the intermediate
    node information which can be used to generate the nodes that are on each minimum path}
    var
    k, i, j : integer;
begin
    {Step 1: Initialize Path matrix}
    for i := 1 to n do
        for j := 1 to n do
            Path [i, j] := 0;
    {Step 2: Compute minimum weighted path lengths}
    for k := 1 to n do
        for i := 1 to n do
            for j := 1 to n do
                if Dist [i, k] + Dist [k, j] < Dist [i, j]
                    then begin
                            Dist [i, j] := Dist [i, k] + Dist [k, j];
                            Path [i, j] := k
                    end
end; {MinWeightPaths}
```

Figure 7.30.   Procedure to generate minimum weighted path information in a graph

Since Path $[4, 3] = 2$, node 2 is an intermediate between node 4 and node 3. Finally, Path $[4, 2]$ and Path $[2, 3]$ are both 0, indicating that there are no more intermediate nodes. Therefore, the minimum path is the sequence of directed edges: $\langle (1, 4), (4, 2), (2, 3) \rangle$.

The timing of the procedure for determining the minimum weighted path lengths in a graph of n nodes, based on its adjacency matrix representation, is $O(n^3)$.

Many other properties of a graph can be determined by using the adjacency matrix and the path matrix of a graph. As an example, we shall show how the path matrix can be used to obtain the strong component containing any particular node of a graph.

Let v_i be any node of a simple digraph G and P be its path matrix. Notice that if v_j is reachable from v_i then $p_{ij} = 1$; also, if v_i is reachable from v_j then $p_{ji} = 1$. If $P = [p_{ij}]$ and if $P' = [p_{ji}]$, one can form $P * P' = [p_{ij} \cdot p_{ji}]$. The elements of $P * P'$ are 1 if i and j are mutually reachable and 0 otherwise. Therefore, the element in the ith row and jth column of $P * P'$ is 1 iff v_i and v_j are mutually reachable. This is true for all j. Hence the ith row of the matrix $P * P'$, which is obtained by the element-wise product of the elements, gives the strong component containing v_i.

We shall end this subsection by showing how the path matrix of a digraph can be used to determine whether certain procedures in a program are recursive. This approach can also be used to detect deadlock in the operating system example given earlier.

In some programming languages, a programmer must explicitly state that a procedure is recursive. In other languages that do not require any such specification, it is possible to use concepts from graph theory to determine which procedures are recursive. A recursive procedure is not necessarily one that invokes itself directly. If procedure p_1 invokes p_2, p_2 invokes $p_3, \ldots, p_{n-1}$ invokes p_n, p_n invokes p_1, then the procedure is also called recursive. To detect recursive procedures, one uses a call graph (see Section 7.1).

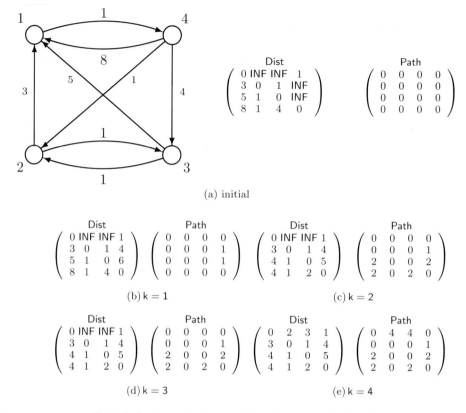

Figure 7.31. Trace of minimum path lengths in a weighted digraph

Let $P = \{p_1, p_2, \ldots, p_n\}$ be the set of procedures in a program. A directed or call graph consists of nodes representing elements of P. If procedure p_i invokes p_j, then there is an edge from p_i to p_j. Figure 7.32 shows a directed graph and its adjacency matrix representing the calls made by the set of procedures $P = \{p_1, p_2, \ldots, p_5\}$.

A procedure p_i is recursive if there exists a cycle involving p_i in the graph. Such cycles can be detected from the diagonal elements of the path matrix of the graph. Thus p_i is recursive if and only if the entry in the ith row and the ith column of the path matrix is 1. The path matrix can be obtained by using Warshall's algorithm. The path matrix is given by

$$\begin{pmatrix} 0 & 1 & 0 & 1 & 1 \\ 0 & 1 & 0 & 1 & 1 \\ 1 & 1 & 0 & 1 & 1 \\ 0 & 1 & 0 & 1 & 1 \\ 0 & 1 & 0 & 1 & 1 \end{pmatrix}$$

which shows that the procedures p_2, p_4, and p_5 are recursive.

procedure p1;
  begin
    ...
    p2;
    ...
  end{p1};

procedure p2;
  begin
    ...
    p4;
    ...
  end{p2};

procedure p3;
  begin
    ...
    p1;
    ...
  end{p3};

procedure p4;
  begin
    ...
    p5;
    ...
  end{p4};

procedure p5;
  begin
    ...
    p2;
    ...
  end{p5};

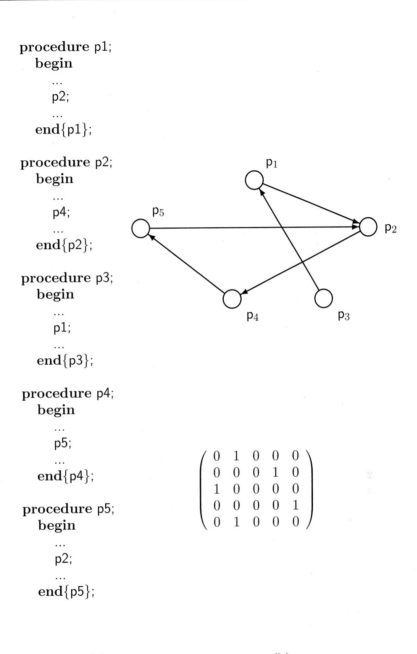

$$\begin{pmatrix} 0 & 1 & 0 & 0 & 0 \\ 0 & 0 & 0 & 1 & 0 \\ 1 & 0 & 0 & 0 & 0 \\ 0 & 0 & 0 & 0 & 1 \\ 0 & 1 & 0 & 0 & 0 \end{pmatrix}$$

(a)   (b)

Figure 7.32.   Procedure calls among p_1, p_2, p_3, p_4, and p_5

A matrix representation of a graph is usually preferred when the graph is *dense*, that is, when the number of edges is close to the maximum possible number, which is n^2 for a graph of n nodes (assuming that loops are allowed). Another method of representing graphs, used for sparse graphs, is explored in Section 7.5.2.

Problems 7.4

1. Obtain the adjacency matrix A of the digraph given in Figure 7.33. Find the elementary paths of lengths 1 and 2 from v_1 to v_4. Show that there is also a simple path of length 4 from v_1 to v_4. Verify the results by calculating A^2, A^3, and A^4.

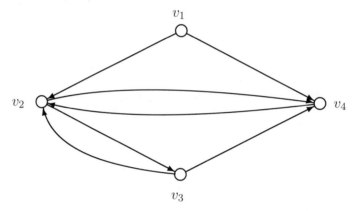

Figure 7.33

2. For any $n \times n$ Boolean matrix A, show that

$$(I \vee A)^{(2)} = (I \vee A) \odot (I \vee A) = I \vee A \vee A^{(2)}$$

where I is the $n \times n$ identity matrix and $A^{(2)} = A \wedge A$. Show also that for any positive integer r

$$(I \vee A)^{(r)} = I \vee A \vee A^{(2)} \vee \cdots \vee A^{(r)}$$

3. Given the adjacency matrix A of the digraph in Figure 7.31, obtain its path matrix.

4. For a simple digraph $G = (V, E)$ whose adjacency matrix is denoted by A, its *distance matrix* is given by

$$
\begin{aligned}
d_{ij} &= \infty, & &\text{if } (v_i, v_j) \notin E \\
d_{ii} &= 0 & &\text{for all } i = 1, 2, \ldots \\
d_{ij} &= k & &\text{where } k \text{ is the smallest integer for which } a_{ij}^{(k)} \neq 0
\end{aligned}
$$

Determine the distance matrix of the digraph given in Figure 7.33. What does $d_{ij} = 1$ mean?

5. Show that a digraph G is strongly connected if all the entries of the distance matrix except the diagonal are nonzero. How will you obtain the path matrix from a distance matrix? How will you modify the diagonal entries?

6. Suppose that we are given an adjacency matrix representation of a graph and we wish to find the following matrix:

$$
C[i, j] = \begin{cases} 1, & \text{if there is a path from } i \text{ to } j \text{ subject to the restriction that only} \\ & \text{vertices 1, 3, 7, and 9 can appear on the path between vertices} \\ & i \text{ and } j \\ 0, & \text{if no such path exists} \end{cases}
$$

Assume that the vertices are labeled $1, 2, \ldots, n$.
   (a) Formulate an algorithm for computing the desired information.
   (b) Perform a timing analysis of the algorithm.

7. Use Warshall's algorithm on the graph given in Figure 7.33 and output the path matrix, P, after each iteration on the loop index k.

8. Obtain the distance matrix for the graph in Figure 7.34b by using procedure MinWeightPaths (Figure 7.30). Give a trace of your result in the same form as that given in Figure 7.31.

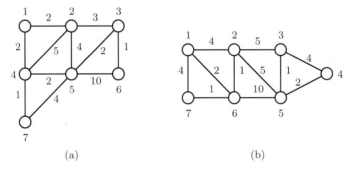

(a)   (b)

Figure 7.34

9. Repeat Problem 8 for Figure 7.34a.

10. In procedure MinWeightPaths, a matrix Path was used to record intermediate node information about the minimum weighted paths between pairs of vertices in a graph. Formulate an algorithm that, when given the matrix Path and the number of vertices in the graph (n), generates a sequence of nodes that lies on a minimum path for each pair of vertices. Assume that the vertices in the graph are labeled by the enumerated type 1..n.

11. For each of the weighted graphs in Figure 7.35, use procedure MinWeightPaths to compute the Dist matrix. Give the initial Dist matrix for the digraph and the Dist matrix after each k has been used as a pivot element.

12. In applications such as those dealing with network scheduling the maximum weighted path lengths in a digraph are of interest. In such applications the resulting digraphs are always acyclic. One approach to generating the desired information is to alter the procedure MinWeightPaths discussed earlier in this section. The maximum weighted path information can be stored in a matrix MaxDist. Explore this possibility and in particular:
   (a) Discuss your approach for representing the graph in a matrix form suitable for the computation of the matrix MaxDist.
   (b) Using the representation of a graph determined in part (a) and the number of nodes in the graph (n) as inputs, formulate an algorithm for generating the maximum weighted path information that is to be stored in MaxDist.

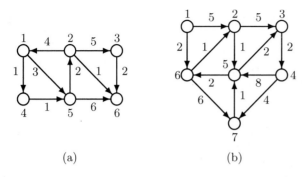

(a)   (b)   **Figure 7.35**

(c)   Trace the algorithm obtained in part (b) for the graph in Figure 7.43 on page 404. First show
the initial matrix MaxDist. Then give the snapshot of the MaxDist after each iteration of
using node k as a pivot element.

7.5 TRAVERSING GRAPHS REPRESENTED AS ADJACENCY LISTS

7.5.1 Introduction

In the previous section we explored determining the existence of paths between pairs of
vertices in a graph. An adjacency matrix representation of a graph was used for this
purpose.

A more general problem involves searching a graph for a particular vertex or node
with some specific property or with some associated specific data. Furthermore, in certain
applications it is required to traverse (i.e., visit or process each node in) a graph in some
particular order. This section focuses on two general methods for searching or traversing
a graph that arise in a variety of applications: breadth-first search and depth-first search.
Before going into the details of these general traversal methods, we explore an alternative
(and sometimes more suitable) representation of a graph to the adjacency matrix approach
taken in the previous section.

7.5.2 Adjacency Lists Representation of Graphs

In general, the best storage representation for a graph depends on the nature of the data.
Furthermore, the choice of a suitable representation is affected by other factors, such as
number of nodes, the average number of edges leaving a node, whether a graph is directed,
the frequency of insertions and/or deletions to be performed, and so on.

The use of an adjacency matrix to represent a graph as described in Section 7.4 has a
number of drawbacks. This representation makes it difficult to store additional information
about the graph. If information about a node is to be included, it would have to be represented
by an additional storage structure. Weighted graphs could have their edge values stored
in place of a bit value, but such a configuration would force us to modify the Warshall
algorithm described in the previous section. The most severe problem with using a matrix
to represent a graph is its static nature. To use this representation, we find that the number of
nodes must be known beforehand to set up the storage array. Also, an insertion or deletion of
a node requires changing the dimensions of the array, which is both difficult and inefficient

with large graphs. This approach is not very suitable for a graph that has a large number of nodes or that has many nodes that are connected to only a few edges; that is, the graph is sparse.

A graph can be represented in many different ways, the most appropriate depending on the application. In the following section we will use an adjacency list. For a graph $G = (V, E)$, an adjacency list is formed for each element x of V. Hence, the list for x contains all nodes y such that (x, y) is an element of E.

The adjacency list representation of the graph in Figure 7.36a, where the adjacency lists are given in part (b), could be as shown in part (c). Observe that each adjacency list in Figure 7.36(b) is in alphabetical order. This need not be the case in general. However, in the discussion to follow, as a matter of convenience we assume such an order. Undirected graphs can also be stored using this data structure; however, each edge will be represented twice, once in each direction, as shown in Figure 7.37. Observe that the storage representation of the graph consists of a node table directory, and that we have an edge list, associated with each entry in this directory. A typical node directory entry consists of a node number, the data associated with it, and a pointer field that gives the address of the list of edges for this node. Each list of edges, stored as a linked list, has an entry that can contain the weight of the edge (optional) and the node number at which the particular edge terminates. For a completely dynamic representation, the node table directory could be replaced by a linked list, in which the terminating node number in the edge list is changed to a pointer to the appropriate node in the linked list table directory. This representation would simplify the insertion and deletion of nodes (see Problem 5).

The declarations for a node directory table and associated adjacency lists can be represented in Pascal by an array of record structures and linked lists, respectively, as follows. We assume that the data associated with each node is of type NameType.

```
type
   GraphType = array [Node] of
           record
                   Data : NameType;
                   Head : List
           end;
```

Here Node is a subrange of the form 1..n, where n is the number of nodes, and List is an adjacency (linked) list defined by

```
type
   List = ↑ EdgeNode;
   EdgeNode = record
                      NodeIndex : Node;
                      Next : List
              end;
```

For example, the declarations for the graph in Figure 7.36(a) are as follows:

```
type Node = 1..6;
   NameType = char;
var Graph : GraphType;
```

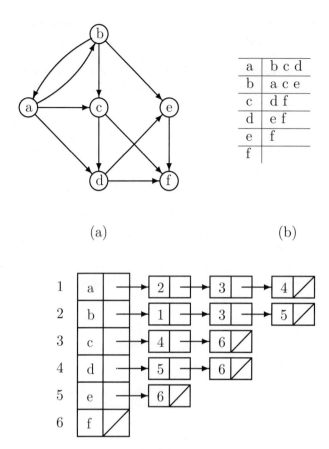

a	b c d
b	a c e
c	d f
d	e f
e	f
f	

(a)   (b)

(c)

Figure 7.36.   Adjacency list representation
of a directed graph

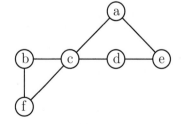

a	c e
b	c f
c	a b d f
d	c e
e	a d
f	b c

(a)   (b)

Figure 7.37.   Adjacency list representation
of an undirected graph

The adjacency list representation of a graph is usually preferred when the graph is sparse; that is, for each node there are not many edges incident to it. On the other hand, an adjacency matrix representation may be more suitable if the graph is dense.

The particular representation chosen may depend on the types (and frequencies) of the operations in algorithms on graphs. To demonstrate the effect of the choice of a graph representation on the efficiency of an algorithm, consider the following problem. Suppose that we wish to determine whether a graph contains at least one edge. If the graph is stored as an adjacency matrix, the matrix must be searched until a 1 bit is found. In the worst case there are n^2 elements to be examined, where n is the number of nodes. If an adjacency list representation is used, each node in the node table directory must be searched to determine if a linked list is present, requiring in the worst case n comparisons.

7.5.3 Breadth-first Search

There are two methods of traversing a graph: breadth-first search (BFS) and depth-first search (DFS). This section is concerned with BFS. The emphasis is on simple connected graphs given in adjacency list representation, as discussed in the previous section.

Breadth-first search can be used to find the shortest distance between some starting node and the remaining nodes of the graph. This shortest distance is the minimum number of edges traversed in order to travel from the start node to the specific node being examined. Starting at a node s, this distance is calculated by examining all incident edges to node s and then moving on to an adjacent node w and repeating the process. The traversal continues until all nodes in the graph have been examined.

Using the BFS strategy just described on the graph in Figure 7.38, the indicated traversal results, assuming that node v_1 is the start position and each edge is assigned a value of 1. The shortest distance from the start node is to be calculated and associated with each node. All nodes adjacent to the current node are numbered before the search is continued. This ensures that every node will be examined at least once (if the graph is connected).

In a breadth-first search, each node is visited or processed in some sense depending on a particular application. We start searching the graph at a specified node. The search is then expanded to the vertices in the graph that are nearest to the starting node before visiting any others. These vertices are related in some way to the specified node and are part of a cluster that depends on a particular application. Initially, we visit each node that is adjacent to the starting node. Then we visit all the nodes that are a distance of 2 from the starting node. This process is repeated until all possible nodes are visited. This searching approach, however, can lead to problems. A previously visited node is not to be revisited. This situation can be avoided by marking a node when it is visited. If a node has been previously marked (i.e., it has been reached or visited), it is never revisited. The node directory table described in the previous section can be altered so as to have an extra Boolean field Reach to keep track of visited nodes. Then we add a field called Dist to store the minimum distance information.

A breadth-first search of the graph given in Figure 7.38(a) with the adjacency lists ordered as shown in Figure 7.38(b) starting at node v_1 results in the following sequence of visited nodes:

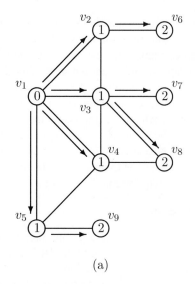

(a)

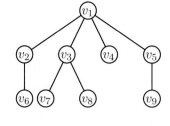

v_1	v_2 v_3 v_4 v_5
v_2	v_1 v_3 v_6
v_3	v_1 v_2 v_4 v_7 v_8
v_4	v_1 v_3 v_5 v_8
v_5	v_1 v_4 v_9
v_6	v_2
v_7	v_3
v_8	v_3 v_4
v_9	v_5

(b)   (c) Shortest path BFS tree

Figure 7.38

$$v_1, \ v_2, \ v_3, \ v_4, \ v_5, \ v_6, \ v_7, \ v_8, \ v_9$$

The starting node v_1 is visited first. Then the nodes that are adjacent to v_1 are considered next. Each adjacent node is a distance of 1 from v_1. Because the nodes in the adjacency list for v_1 are ordered as shown in Figure 7.38(b), the nodes adjacent to v_1 are visited in the following order:

$$v_2, \ v_3, \ v_4, \ v_5$$

We then proceed to visit the nodes that are a distance of 2 from v_1. The visiting of these nodes is facilitated by having stored all adjacent nodes to v_1 in a first-in, first-out queue structure. Since v_2 was the first adjacent node to v_1 to be visited (and saved in the queue), we explore the nodes adjacent to v_2 that have not been visited. The only such node is v_6. The process is repeated on v_3, the second node that is adjacent to v_1, which results in the visits of nodes v_7 and v_8. We next explore the unvisited nodes of v_4. There are none. Finally, the only unvisited node adjacent to v_5 is v_9. This last visit completes the breadth-first search (traversal) of the graph. The BFS strategy results in the traversal indicated by the arrows.

A first-in, first-out queue structure is a convenient way to keep track of nodes whose neighbors may not have been visited. Such a queue structure is appropriate since vertices are visited in the order in which they occur in an adjacency list. Therefore, a vertex that precedes another in an adjacency list will be placed in the queue before the other, that is, on a first-come, first-served basis. Given a start node s, a general algorithm for the BFS approach is as follows:

1. Mark all vertices in the graph as not visited (or reached).

2. Mark and visit s.

3. Set the Dist for s to zero.

4. Put s in queue.

5. While the queue is not empty:

> Remove the front element from the queue
>> and call it the current node.

> For each neighbor of the current node

>> If the neighbor is not marked,

>> then Visit and mark the neighbor.

>>> Update the Dist value of neighbor.

>>> Put neighbor in the queue.

Note that a visited vertex has an associated mark value of true.

The following declarations for a directed graph are an update of those given in the previous section with a field (Reach) to record whether a node has been visited, a subrange type for Node for a graph (as in Figure 7.38) of nine nodes, and a previously constructed storage representation for the given graph:

```
type Node : 1..9;
GraphType = array [Node] of
            record
                Reach : boolean;
                Data : NameType;
                Dist : integer;
                Head : List
            end;
List = ↑EdgeNode;
EdgeNode = record
                NodeIndex : Node;
                Next : List
            end;
var Graph : GraphType;
```

Figure 7.39 is a Pascal procedure for a breadth-first search of the graph. This procedure assumes the existence of a queue structure of type Queue. The procedure QInitialize initializes the queue to be empty. The procedures QInsert and QDelete are queue routines for inserting and deleting elements into and from a queue, respectively. Observe that each element that is deleted from the queue is copied in the variable Current. QEmpty is a routine for testing whether the queue is empty.

Let us examine the timing analysis of procedure BFS. In our analysis, n and m denote the number of nodes and the number of edges in the graph, respectively. Steps 1 to 4 are executed once. The "then part" of the if-statement in step 9 is performed $n - 1$ times. This follows from the fact that the "then part" is executed only when the Reach value of a node is false. Note that the Reach value is set to true in the "then part." Hence, one node from step 4 and $n - 1$ nodes from step 9 are inserted into the queue for a total of n nodes. Consequently, step 5 is repeated $n + 1$ times (counting the time that the queue is empty). The adjacency lists contain a total of $2m$ edges, since all edges in all the adjacency lists are examined. It then follows that step 8 and the assignment statement in step 10 are each performed $2m$ times. Consequently, the time analysis for the procedure is $O(n + m)$.

The BFS approach will visit every node regardless of the start node if the simple graph is connected. If the graph is not connected, procedure BFS will not visit each node. Procedure BFS can be modified to determine whether a graph is connected. This modification is left as an exercise.

During a breadth-first search of a graph, a shortest path (or spanning) tree is followed. For example, the shortest path tree of the example graph in Figure 7.38(a) appears in Figure 7.38(c) where v_1 is the root node. The vertices in the tree are searched in increasing level-number order and within each level from left to right.

7.5.4 Depth-first Search

Suppose that a person is placed in some interconnected system of caves or a maze at some given intersection (or node), and this person is asked to find his or her way out to some recognizable node. Several approaches could be used in this search. One approach that would probably not be used is a breadth-first search. Intuitively, a strategy that starts at some designated node in the cave and then visits each adjacent node by following the cave does not appear to be that promising.

A much more likely strategy is to begin at the start point and follow, say, the rightmost cave from there to the next intersection. When a person arrives at this intersection, a rightmost branch is again taken and so on until the desired node is found. If this rightmost branching search process is not successful because no new intersections are discovered, the individual retraces the path taken to the most recent previous intersection and then takes the next rightmost branch, if any, forward. If all branches at this node have been explored, the individual retraces to the next previous intersection, and so on. The retracing may very well bring a person back to the starting point, and the next rightmost branch is then explored. The forward moves, and then sometimes backward moves, are essentially a depth-first search method that we now describe and study in detail.

A depth-first search (DFS) of an arbitrary graph can be used to perform a traversal of a general graph. As each new node is encountered, it is marked to prevent the revisiting of

```
procedure BFS (Start : Node);
   {Given a start node, this procedure performs a breadth-first search on a simple connected undirected
   graph}
   var NodePtr : List;  {Pointer to a node in an adjacency list}
      Neighbor, Current : Node;
   Queue : array [1..MaxVert] of Node;   {The queue structure}
   begin
      {Step1: Mark all vertices as not reached}
      for Current := 1 to n do
         Graph [Current].Reach := false;
      {Step 2: Mark and visit start node}
      Graph [Start]. Reach := true;
      write (Graph [Start]. Data);
      {Step 3: Set the start node distance to 0}
      Graph [Start]. Dist := 0;
      {Step 4: Initialize and place start node in queue}
      QInitialize (Queue);
      QInsert (Queue, Start);
      {Step 5: Traverse graph until all remaining nodes are visited}
      while not QEmpty (Queue) do
         begin
            {Step 6: Remove front node from queue and call it the current node}
            QDelete (Queue, Current);
            {Step 7: Obtain first neighbor of current node}
            NodePtr := Graph [Current]. Head;
            {Step 8: Find all neighbors of current node}
            while NodePtr <> nil do
               begin
                  Neighbor := NodePtr ↑. NodeIndex;
                  {Step 9: If the neighbor is unvisited; mark and visit it.}
                  if not Graph [Neighbor]. Reach
                     then begin
                             Graph [Neighbor]. Reach := true;
                             write (Graph[Neighbor]. Data);
                             Graph [Neighbor]. Dist :=
                                    Graph [Current]. Dist+ 1;
                             QInsert (Queue, Neighbor)
                          end;
                  {Step 10: Advance to next neighbor}
                  NodePtr := NodePtr ↑ .Next
               end
         end
   end;
```

Figure 7.39.   Procedure to perform a breadth-first search

that node. Recall that such a marking was also used in the breadth-first search. The DFS
strategy is as follows: A node s is picked as a start node and marked. An unmarked adjacent
node to s is now selected and marked, becoming the new start node, possibly leaving the
original start node with unexplored edges for the present. The search continues in the graph
until the current path ends at a node with outdegree zero or at a node with all adjacent nodes
already marked. Then the search returns to the last node that still has unmarked adjacent
nodes and continues marking in a recursive manner until all nodes are marked.

As was done in the case of a breadth-first search, we must also mark nodes in a depth-first search. If this were not done, a graph that contains cycles would lead to infinite loops! As was the case in a BFS, marking a node prevents a node from being revisited. Compared with a BFS strategy, which goes for breadth, this search strategy goes as far, or deep, as possible from the start node.

A general algorithm for a depth-first search consists of a single main routine that calls a recursive procedure as follows:

Main
      1. Mark all vertices in the graph as not marked.
      2. Invoke DFS (s) for some start node s.
Procedure DFS (s)
      1. Mark and visit s
      2. For each neighbor w of s
            If the neighbor w is not marked
            then invoke DFS (w).

A depth-first search of the graph given in Figure 7.38 appears in Figure 7.40. Starting at node v_1 results in the following sequence of visited nodes:

$$v_1, \ v_2, \ v_3, \ v_4, \ v_5, \ v_9, \ v_8, \ v_7, \ v_6$$

We now give a trace of the result. The starting vertex v_1 is marked and visited. The first node in the adjacency list for v_1 is v_2. We mark and visit this node. Now the first node in the adjacency list for v_2 is v_1 and this node has already been visited. Consequently, the next adjacent node to be marked and visited is v_3. The first two nodes in the adjacency list for v_3 (i.e., v_1 and v_2) have already been visited. Therefore, v_4 is the next node to be marked and visited in the search. Again, since the first two nodes in the adjacency list for v_4 have been visited, the next node to be marked and visited is v_5. The next unvisited node in the adjacency list for v_5 is v_9. We then mark and visit v_9. At this point there are no new adjacent nodes to v_9 to be explored. We return to node v_5 and find that all its adjacent nodes have been visited. We then return to node v_4 and find that node v_8 has not been visited. We next mark and visit v_8. Since there are no more unmarked nodes that are adjacent to v_4, we return to node v_3, which results in the marking and visiting of v_7. Because v_8 has already been visited, a return to v_2 occurs, which results in the marking and visiting of node v_6. Finally, a return to v_1 occurs and the search ends. The graph traversal is shown in Figure 7.40, where the arrows show a trace of the nodes visited and the number at each node gives its visit sequence number. For example, v_1 and v_6 were the first and last nodes to be visited, respectively.

Since adjacent nodes are needed during the traversal, the most efficient representation again is an adjacency list. The same data structure as presented in the previous subsection will be used. The procedure to perform the depth-first search of a graph is recursive, as is shown in Figure 7.41.

The main routine to invoke procedure DFS is as follows:

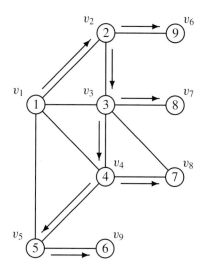

Figure 7.40.   Depth-first search traversal of a simple connected graph

procedure DFS (Current : Node);
  {Given the node directory table and associated adjacency list structure for a simple graph as described earlier, and a current node, this recursive procedure performs a depth-first search traversal of the graph or the part of the graph not already reached. Note that the Reach fields must be set to false prior to invoking this procedure.}
var
  NodePtr : List;
  Neighbor : Node;
begin
  {Step 1: Mark and visit present node}
  Graph [Current]. Reach := **true**;
  write (Graph [Current]. Data);
  {Step 2: Obtain pointer to first neighbor of present node}
  NodePtr := Graph [Current]. Head;
  {Step 3: Repeat for all neighbors of present node}
  while NodePtr <> **nil do**
    begin
      {Step 4: Obtain index of neighbor}
      Neighbor := NodePtr ↑. NodeIndex;
      {Step 5: If neighbor has not been marked then explore it}
      if not Graph [Neighbor]. Reach
        then DFS (Neighbor);
      {Step 6: Obtain pointer to next neighbor}
      NodePtr := NodePtr ↑. Next
    end
end;

Figure 7.41.   Procedure for performing a depth-first traversal on a graph

```
{Step 1: Set all reach fields to false}
    for i := 1 to n do Graph[i]. Reach := false;
{Step 2: Invoke DFS with some start node s}
    DFS (s)
```

The timing analysis for this procedure is similar to that obtained for procedure BFS. In the worst-case analysis, at most $n - 1$ recursive calls are made since a recursive call is performed only once for each unmarked node. The time complexity of step 3 (and step 6) is $O(m)$, where m denotes the number of edges in the graph. Therefore, the worst-case time complexity is $O(n + m)$.

We now examine briefly the use of the depth-first search approach to finding cycles in graphs. There are applications in which cycles are not allowed (e.g., the deadlocks discussed earlier). Also, a tree is a special kind of graph in which cycles are not allowed (see Section 7.6). Finally, in scheduling networks, which were mentioned in Section 7.1, where activities correspond to directed edges and nodes correspond to events or points in time, cycles of activities are forbidden. Since each activity usually takes some amount of time to complete, a cycle of activities, from a time point of view, does not make sense. The reason for this is that a cycle starting at some node s, by performing a sequence of activities that terminates at s, would require some time. After performing these activities, you cannot return to a previous point in time. Another more general application involves the use of a cycle detector in attempting to linearize or topologically sort an acyclic graph (see Section 7.7.3). The purpose of such a sort is to order the events in a linear manner, that is, first, second, third, and so on, with the restriction that an event cannot precede other events that must first take place.

Although cycles can exist in both directed and undirected graphs, we concentrate here on using a depth-first search strategy to find a cycle containing a specified vertex v_0 in an undirected graph. If, during the search, an edge is found from the current vertex p to v_0, then a cycle has been found. The cycle consists of the search path from v_0 to p followed by the edge $\{p, v_0\}$. If no such edge is found, then no cycle exists for v_0.

A procedure for the desired task is given in Figure 7.42. The routine is very similar to procedure DFS and is invoked by

$$\text{CycleDetector } (v_0, v_0)$$

Observe that this procedure only detects a cycle that involves vertex v_0 and that the Reach fields must be initialized to false prior to its invocation. In general, a graph can contain many cycles. The detection of all cycles in a graph is left as an exercise. Another important aspect of detecting cycles is to determine (and display) the vertices that lie on a cycle. Procedure CycleDetector can be altered to generate this additional information.

7.5.5 Dijkstra's Algorithm for Finding Minimum Paths

Suppose that a trip by car is being planned from some city to another, for example, from Boston to Los Angeles. Using a road map, there are several ways to arrive at an answer. Floyd's algorithm, which is based on the matrix representation of a graph, could be used here. Since this problem involves a weighted graph in which the weight of each edge is the

```
function CycleDetector (var Current : Node;
                            Fixed : Node): boolean;
  {Given a node directory table and associated adjacency list structure for a simple graph as described
  earlier, a specified fixed node Fixed and a current node, Current, this routine searches for a non-zero
  length path from Current to Fixed. As the search starts at Fixed, this will test for a cycle containing
  Fixed.}
  var
    NodePtr : List;
    Neighbor : Node;
    Cycle : boolean;
  begin
    Cycle := false;
    Graph [Current] .Reach := true;
    NodePtr := Graph [Current] .Head;
    while (NodePtr <> nil ) and (not Cycle) do
      begin
        Neighbor := NodePtr ↑ .NodeIndex;
        if not Graph [Neighbor] .Reach
        then Cycle := CycleDetector (Neighbor, Fixed)
        else {Check for cycle}
            if Neighbor = Fixed then Cycle := true;
        NodePtr := NodePtr ↑ .Next
      end;
    CycleDetector := Cycle
  end; {CycleDetector}
```

Figure 7.42.   Function for detecting cycles in a graph

distance between two cities, the breadth-first search and depth-first search are not directly usable, because these approaches assume that every edge in a graph has a weight of 1.

This subsection examines an algorithm, attributed to Dijkstra, to find the minimum weighted path length from one vertex to another. Assume that a given graph contains n nodes $v_1, v_2, \ldots, v_n$ and that it is required to find the shortest weighted path length from v_1 to v_n. Although we want the shortest path from one node to another, we find it useful (and efficient) to generate the shortest path lengths from v_1 to all other nodes in the graph. A distance field Dist is associated with each node in the graph. Initially, the start node v_1 is selected and its shortest distance to itself is set to 0. The approach is to find the node that is closest to v_1. Next, we search for the next closest vertex to v_1, and so on. Eventually, the node v_n will be the next closest from v_1, and at this point the shortest path length has been found. As each successive node that is closest to v_1 is found, it is placed in a set S and ignored in the sense that its minimum distance from v_1 has been found. At any given stage, Dist[v_j] is the minimum distance of a path from v_1 to v_j for those vertices that are in S.

As an example of this approach, consider the weighted digraph in Figure 7.43. In finding the shortest distances from v_1 to other nodes in the graph, the current distance vector Dist is updated each time a new element is selected and placed in S. During the kth iteration, let S_k denote the next closest set of selected vertices and u_k the next closest vertex. The recomputed shortest distances for all vertices not in S_k that are adjacent to u_k are accomplished by the following program fragment:

for i := 1 **to** n
   if (v_i is adjacent to u_k) **and** (**not** (v_i in S_k))
     then Dist [v_i] := min (Dist [v_i], Dist [u_k] + Weight [u_k, v_i])

At each iteration, before finding the next closest node, the current distance vector will contain the shortest distance from v_1 to each node in the graph, provided that all intermediate nodes on each path are selected from S. Observe that the current distance of u_k from v_1 remains the same (i.e., it is the shortest distance from v_1). This observation also holds for each vertex not adjacent to u_k (see Problems 7.5). If there is no path from v_1 to a certain node with this property, the current distance of the latter node is set to INF. Initially, we set Dist[v_1] = 0, Dist [v_2] = 3, Dist [v_3] = 6, and Dist[v_i] = INF for all other nodes, and S = $\{v_1\}$. This situation is shown in the first row of Figure 7.44.

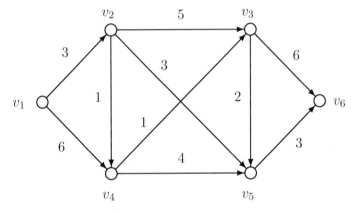

Figure 7.43.   Weighted digraph used to illustrate Dijkstra's algorithm

   In the first iteration we select the next closest node to the start node v_1. The closest node to v_1 is v_2 and we therefore place v_2 in S. We next compute the shortest distance of all nodes not in S that are adjacent to v_2. The recomputation of path lengths through v_2 to nodes v_3, v_4, and v_5 yields values of 8, 4, and 6, respectively. This completes the first iteration, which is summarized in the second row of Figure 7.44. The updated distances are in parentheses.

   During the second iteration we search for the second closest node to v_1. Three nodes are adjacent to v_2: v_3, v_4, and v_5. Therefore, we choose node v_4 as the second closest node to v_1 and place it in S. The updated distances for the nodes that are not in S and are adjacent to v_4 are recomputed. The recomputation of the distances to v_3 and v_5 yields values of 5 and 8, respectively. The results are summarized in the third row of Figure 7.44.

   In the third iteration, nodes v_3 and v_5 are adjacent to v_4, with v_3 being the closest. Therefore, the next node to be selected is v_3. Nodes v_5 and v_6 are adjacent to v_3 and the recomputation of their distances yields values of 7 and 11, respectively. The distance of 11 is the current minimum distance to v_6, however, the distance of 7 is not the minimum distance to v_5, and therefore the minimum distance of 6 is maintained. The fourth row of Figure 7.44 summarizes the results.

Iteration	v_1	v_2	Dist v_3	v_4	v_5	v_6	u	S
0	0	3	INF	6	INF	INF		$\{v_1\}$
1	0	③	INF(8)	6(4)	INF(6)	INF	v_2	$\{v_1, v_2\}$
2	0	3	8(5)	④	6 (8)	INF	v_4	$\{v_1, v_2, v_4\}$
3	0	3	⑤	4	6 (7)	INF(11)	v_3	$\{v_1, v_2, v_4, v_3\}$
4	0	3	5	4	⑥	11(9)	v_5	$\{v_1, v_2, v_4, v_3, v_5\}$
5	0	3	5	4	6	⑨	v_6	$\{v_1, v_2, v_4, v_3, v_5, v_6\}$

Figure 7.44.   Trace of Dijkstra's algorithm

The next two iterations select v_5 and v_6 in turn. The last iteration yields a shortest distance of 9 from v_1 to v_6.

The previous approach is formalized in the procedure given in Figure 7.45. Observe that some statements in the procedure are not syntactically valid Pascal statements. The exact form of these statements depends on how the graph structure is represented, that is, by a weighted adjacency matrix or a node table directory with adjacency lists. We will return to this point shortly.

Let us briefly examine the worst-case timing analysis of Dijkstra's algorithm. Step 5 may need to be executed for as many as $n - 1$ nodes since the initial selected vertex is Start. Now consider the execution of the if-statements within steps 6 and 8, which are labels for the two inner loops in the main loop of step 5. If an adjacency matrix representation is used, then the worst case can require $n - 1$ executions of the if-statements. Therefore, the worst-case timing analysis of the algorithm is $O(n^2)$. However, in the average case, u may equal Find in the repeat ... until loop considerably earlier than having to perform the maximum $n - 1$ iterations. On the other hand, if adjacency lists are used to represent the graph and the number of edges in the graph is much less than n^2 (i.e., the graph is sparse), say, $m = O(n)$ where m denotes the number of edges, then, assuming that the weight matrix is stored, the algorithm might be considerably better than $O(n^2)$.

The current algorithm finds the shortest distance from one node to another. An approach similar to that used in Floyd's algorithm given earlier can be followed to find nodes on the shortest path. To that end, we can save at each iteration the index of the predecessor node on the shortest path to the next closest node that is placed in S. This information can be obtained by modifying step 8 in Figure 7.45. When a decrease in the current distance occurs because there is an edge from node u to node i, we simply record node u as the predecessor. Eventually, when node Final enters S, we can work backward to the specified node Start using the predecessor node information. Figure 7.46 contains the

```
procedure Dijkstra (Start, Final : Node;
                    n : integer;
                    var Dist : DistType);
```
{Given a weighted graph and its representation which is considered to be global to this procedure, this routine computes the shortest distance from the initial node, Start, to the destination node, Final. Node is assumed to be a subrange type 1..n, for simplicity. If a current distance from the initial node does not exist, it is set to INF. The weight of an edge from v_i to v_j is denoted by Weight[v_i, v_j]. The actual representation depends on how the graph is represented. The function min returns the smallest argument.}
```
var
   i, u : Node;
   MinDist : integer;
   S: set of Node;
begin
   {Step 1: Set distance field of each node to infinity}
   for i := 1 to n do
      Dist [i] := INF;
   {Step 2: Select initial node as start node}
   Dist [Start] := 0;
   S := [Start];
   {Step 3: Select current node to be the start node}
   u := Start;
   {Step 4: Compute shortest distances to all nodes adjacent to start node}
   for i := 1 to n do
      if not (i in S)
      then Dist [i] := min (Dist [i], Dist [u] + Weight [u, i])
         {Step 5: Select next closest node (u) to start node and recompute shortest distances of
         nodes not in S through node u}
         repeat
            {Step 6: Select next closest node to start node}
            MinDist := INF;
            for i := 1 to n do
               if not (i in S) and (Dist [i] < MinDist)
               then begin
                       MinDist := Dist [i];
                       u := i
                    end;
            {Step 7: Add next closest node to set of closest nodes }
            S := S + [u];
            {Step 8: Recompute shortest distances for nodes not in S through u}
            for i := 1 to n do
               if not (i in S)
               then Dist [i] := min (Dist [i], Dist [u] + Weight [u, i]);
         until u = Final
end;
```

Figure 7.45. Procedure for Dijkstra's algorithm

predecessor node information for the current example. For example, in the first iteration the shortest path to v_2 contains the predecessor node v_1, and at the end of the third iteration the shortest path to v_3 has v_4 as the predecessor node. Let us now determine from Figure 7.46 the nodes on the path from v_1 to v_6. The predecessor of the destination node v_6 on the shortest path is v_5. On the shortest path, v_2 is the predecessor to v_5. The retracing activity terminates when v_1 is the predecessor to v_2. At this point the nodes on the shortest path are given by the sequence $\langle v_1, v_2, v_5, v_6 \rangle$.

Recall that Floyd's algorithm, which is based on an adjacency matrix representation of a graph, is used to obtain all shortest distances and paths in a graph. Dijkstra's approach can also be used for this purpose by invoking procedure Dijkstra for each node in the given

Iterations	Predecessor node on shortest path to:				
	v_2	v_3	v_4	v_5	v_6
0					
1	v_1		v_1		
2		v_2	v_2	v_2	
3		v_4			
4					
5					v_5

Figure 7.46.   Predecessor node information for shortest paths

graph. If the graph is sparse, this approach, from a timing analysis viewpoint, may be superior to Floyd's algorithm. On the other hand, for dense graphs both approaches are $O(n^3)$, but Floyd's algorithm will probably be faster.

Problems 7.5

1. For each graph represented by the following adjacency lists, list the vertices in the order that they are reached in a breadth-first search and a depth-first search of the graph starting at vertex 1.

   (a)

Vertex	:	Adjacency list
1	:	2, 5, 7, 4
2	:	3, 6, 1, 4
3	:	2, 4
4	:	1, 3, 2
5	:	6, 1
6	:	2, 5
7	:	1

   (b)

Vertex	:	Adjacency list
1	:	2, 7, 8, 4
2	:	3, 7, 6, 1
3	:	2, 7, 4
4	:	1, 7, 3
5	:	8, 6
6	:	2, 9, 8, 5
7	:	1, 3, 4, 2
8	:	5, 1, 6
9	:	6

   (c)

Vertex	:	Adjacency list
1	:	9, 5, 2
2	:	8, 5, 3, 4
3	:	9, 2, 6, 7
4	:	8, 2
5	:	1, 2
6	:	3
7	:	8, 3
8	:	4, 7, 2
9	:	3, 1

   (d)

Vertex	:	Adjacency list
1	:	4, 6, 9
2	:	12, 10, 5
3	:	8, 7
4	:	6, 11
5	:	
6	:	2, 8
7	:	9, 3
8	:	9
9	:	1, 8
10	:	
11	:	6, 1
12	:	5

2. Augment the node directory table of a graph to include an additional field, DFN, that is to be maintained. The DFN number of the first node to be visited is 1, the second, 2, and so on, and

the last is n, assuming that the graph is connected. Modify procedure DFS to compute the DFN number of each vertex in a connected graph.

3. Using either a breadth-first or depth-first search approach, formulate an algorithm to determine whether a given graph is connected. (Recall that a graph is connected if there is a path between any two distinct nodes in the graph.)

4. Using a breadth-first search approach, formulate an algorithm to traverse an undirected graph that is not connected. The result of performing such an algorithm on a graph is to traverse each strong component in the graph.

5. Repeat Problem 4 using a depth-first search approach.

6. Modify procedure CycleDetector so that the labels of the nodes that lie in a cycle are output.

7. Explore an algorithm that generates all cycles in a given graph.

8. Formulate an algorithm to test whether a given graph is acyclic.

9. You are given an *undirected* graph stored by means of adjacency lists, such that each vertex is labeled left or right. Suppose that label(v) yields left or right for each vertex v. In the diagram of such a graph, as given in Figure 7.47, all vertices labeled left are placed on the left and all vertices labeled right are placed on the right. A path in such a graph is called *alternating* if the vertices on the path alternate between being left and right. For example,

1–5–8–6–4–7 is alternating

2–3–9 is alternating

8–6–1–5–8 is an alternating cycle

5–6 is not alternating

1–5–8–6–9–3 is not alternating

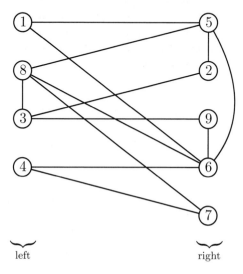

left   right   **Figure 7.47**

Given two vertices x and y, use the depth-first search approach to develop an algorithm to determine if there is an alternating path from x to y. If such a path exists, then the algorithm should output the length of one such path.

10. Given the graph in Figure 7.34(a) where all edges are assumed to have a weight of 1, perform a breadth-first search of it using procedure BFS starting at node 1. Give a trace of the search in the form given in Figure 7.38. Repeat the exercise for each of the following starting vertices: 5, 3, and 7.

11. Repeat Problem 10 for Figure 7.34(b).

12. For each weighted digraph in Figure 7.35, give a trace of procedure Dijkstra (as in Figure 7.44) starting at vertex 1.

13. Show in Dijkstra's approach that the distances computed from the start node to each node that has been placed in the next closest set of nodes S are the shortest distances.

14. Using the predecessor node information for Dijkstra's approach as given in Figure 7.46, formulate an algorithm for generating the sequence of nodes on the minimum weighted path from vertex 1 to vertex n.

15. Formulate an algorithm based on Dijkstra's approach to generate all minimum weighted paths for a given graph.

16. Illustrate, by using an example, that Dijkstra's approach may not work if edges are assigned negative weights.

7.6 TREES AND SPANNING TREES

7.6.1 Introduction

One of the most important special kinds of graph is a tree. A special kind of tree called a *rooted tree* has already been introduced in an informal manner in Section 1.3 in conjunction with the parse tree of a proposition and also more formally when dealing with recursion in Section 3.3.2. Trees are used in many other application areas. For example, trees are used in computer science to organize information so that operations on this information can be performed in an efficient manner. Also, a complicated process can often be decomposed and represented as a tree structure. Furthermore, trees arise in networks that are modeled with graphs. In a communication network, for example, it may be required that any pair of nodes in the network be connected at the least possible cost. The solution to this problem involves building another kind of tree called a spanning tree.

In this section we first introduce free trees. We next introduce the notion of a spanning tree in conjunction with a graph. The emphasis of the discussion is on various approaches to generating spanning trees. Finally, we focus on algorithms for generating a minimum spanning tree for a weighted graph.

7.6.2 Free Trees

There are two broad classes of trees: free trees and rooted trees. A free tree is a special kind of undirected graph. A rooted tree is a special case of a directed graph. Rooted trees, which are important in computer science for organizing, searching, and sorting data, were discussed in Section 3.3.2.

Definition 7.16:   A *free tree (tree)* is an undirected simple graph that is connected and acyclic. Recall that an acyclic graph is a graph that has no cycles.

A free tree does not contain cycles, and each of its pairs of vertices are connected. Examples of free trees are given in Figure 7.48. A tree must have at least one vertex.

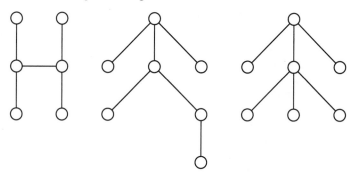

Figure 7.48.   Examples of free trees

Theorem 7.5.   A free tree containing n vertices has $n - 1$ edges.

Proof. We formulate an inductive proof. Let the base case be $n = 1$. In this case the tree has no edges. The inductive hypothesis assumes that a tree of k vertices has $k - 1$ edges. The induction step involves showing that a tree of $k + 1$ vertices contains k edges. Observe that a tree that contains at least one edge must have at least two vertices with a degree of 1. This follows from the fact that the largest acyclic path, say, $v_1, v_2, \ldots, v_m$ with $v_1 \neq v_m$, has v_1 and v_m as end points, which means that their degree is 1. Therefore, if we remove one such vertex and its incident edge from the tree of $k + 1$ vertices, we have a tree of k vertices. By the inductive hypothesis, this tree contains $k - 1$ edges. Consequently, this shows the induction step. ∎

7.6.3 Spanning Trees

An important problem associated with a network that is represented by a graph is to obtain a spanning tree for the graph. Such a spanning tree is to contain all the vertices in the graph and some edges from the graph, which ensures connectivity.

Definition 7.17:   A *spanning tree* of a connected undirected graph $G = (V, E)$ is a free tree with vertex set V that is a subgraph of G; that is, a spanning tree is connected, acyclic, and has all of V as its vertex set and some of E as its edge set.

There are many approaches to generating a spanning tree for a given graph. One approach is to remove edges belonging to cycles one at a time until no cycles remain in the graph. If only edges in cycles are removed, then the graph will remain connected, and this is essential for the generation of a spanning tree.

As an example of this approach, consider the graph in Figure 7.49(a). We can begin by deleting edge $\{v_2, v_5\}$. This removes one simple cycle. The resulting subgraph is still connected. The next steps remove edges $\{v_2, v_6\}$, $\{v_3, v_6\}$, $\{v_3, v_7\}$, $\{v_4, v_7\}$, and $\{v_6, v_7\}$ in turn. This sequence of edge removals yields the subgraph in Figure 7.49(b).

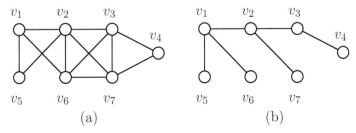

Figure 7.49.   Obtaining a spanning tree for a graph

Observe that many spanning trees could be generated for the given graph. Examples of other spanning trees for the graph in Figure 7.49(a) appear in Figure 7.50.

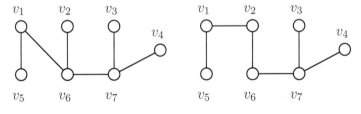

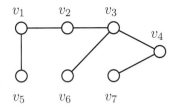

Figure 7.50.   Other spanning trees for the graph in Figure 7.49(a)

The current approach to generating a spanning tree may result in the removal of many edges in a dense graph. Since any spanning tree for a graph of n vertices always contains $n - 1$ edges, many more edges may have to be removed than kept.

An alternative (and possibly more efficient) approach to generating a spanning tree is to choose a sequence of $n - 1$ edges, one at a time, so that at each step the current subgraph is acyclic. The breadth-first search and depth-first search use this approach. Before formulating an algorithm based on this approach, however, we explore the generation of a spanning tree by using the breadth-first search and depth-first search procedures developed earlier in Section 7.5.

Let us consider a BFS of the graph given in Figure 7.49(a). Assume that an adjacency list representation is used for the graph and it is

$$v_1 : v_2, v_5, v_6$$
$$v_2 : v_1, v_3, v_5, v_6, v_7$$
$$v_3 : v_2, v_4, v_6, v_7$$
$$v_4 : v_3, v_7$$
$$v_5 : v_1, v_2$$
$$v_6 : v_1, v_2, v_3, v_7$$
$$v_7 : v_2, v_3, v_4, v_6$$

A breadth-first search of the graph starting at vertex v_1 yields a directed tree rooted at v_1. If we ignore the root node and the direction of the arcs in the directed tree, we obtain a free tree for the graph. An elaboration on how to build the tree during the search appears as an exercise. For the example graph, starting at v_1, a BFS approach includes arcs to vertices v_2, v_5, and v_6. Since these are the only vertices that are incident to v_1, we next proceed to include arcs to the vertices that are adjacent to v_2, that is, arcs to vertices v_3 and v_7. Continuing in this manner results in the rooted tree in Figure 7.51(a). The underlying free tree of this rooted tree appears in Figure 7.51(b).

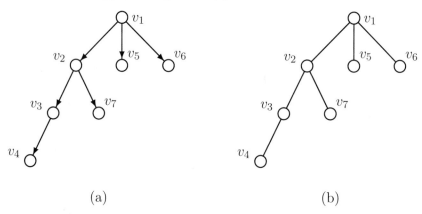

(a)   (b)

Figure 7.51. Obtaining a spanning tree using a breadth-first search

A depth-first search of a graph can also yield a spanning tree. The example graph starting at vertex v_1 results in the rooted tree in Figure 7.52(a). Its corresponding free tree appears in Figure 7.52(b).

We now consider a more general approach to generating a spanning tree by choosing a sequence of edges, one at a time, so that at each step the current subgraph is acyclic. Figure 7.53 contains a program skeleton for generating a spanning tree for a given graph. Nodes represents the selected vertices at each iteration step. Similarly, Edges represents

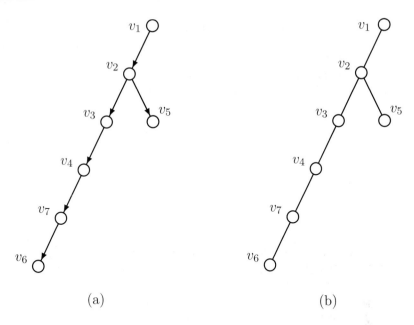

Figure 7.52.   Obtaining a spanning tree using a depth-first search

```
procedure BuildSpanTree (u: Node;
                    var Nodes : set of Node;
                    var Edges : set of EdgeType);
   {Given a starting vertex u and a connected graph G = (V, E) which is assumed to be global to
   this routine, this procedure generates a spanning tree for G.  The spanning tree is denoted by
   T = (Nodes, Edges) where Nodes contains the vertices and Edges are edges selected from E.}
var
   Rest: set of Node;
   j, k : Node;
begin
   {Step 1: Initialize}
   Nodes := [u];
   Edges := [];
   Rest := V − [u];
   {Step 2: Generate spanning tree}
   while Rest <> [] do
      begin
         Choose an edge [j, k] in E
            such that j in Nodes and k in Rest
         Nodes := Nodes + [k];
         Edges := Edges + [[j,k]];
         Rest := Rest − [k]
      end;
end;   {BuildSpanTree}
```

Figure 7.53.   Program skeleton for obtaining a spanning tree of a graph

the selected edges at each iteration. Rest denotes vertices that have not yet been selected. Note that Nodes and Rest are disjoint sets.

Suppose again that the given graph is Figure 7.49(a) and the starting vertex is v_1. Initially, Nodes contains vertex v_1 and no edges have been selected. Starting at v_1, we want to choose an edge that joins v_1 and some other vertex in Rest, which at this point is the set $\{v_2, v_3, v_4, v_5, v_6, v_7\}$. Several choices are possible at this point: edges $\{v_1, v_2\}$, $\{v_1, v_5\}$, and $\{v_1, v_6\}$. In this trace we make the choice based on the numerical increasing order of the vertex index until all edges that preserve the acyclicity of the subgraph have been selected. Therefore, we choose v_2 and the edge $\{v_1, v_2\}$, resulting in the changes Nodes = $\{v_1, v_2\}$ and Edges = $\{\{v_1, v_2\}\}$. It should be pointed out that, in practice, a random order of selection should probably be used. We next select v_5 and edge $\{v_1, v_5\}$ and then v_6 and edge $\{v_1, v_6\}$. At this point no further edge selection is possible from vertex v_1, and Nodes = $\{v_1, v_2, v_5, v_6\}$ and Edges = $\{\{v_1, v_2\}, \{v_1, v_5\}, \{v_1, v_6\}\}$. We next focus on vertex v_2. The edges that could be selected are $\{v_2, v_3\}$ and $\{v_2, v_7\}$. Note that the selection of either edge $\{v_2, v_5\}$ or edge $\{v_2, v_6\}$ would cause a cycle in the subgraph being constructed. We select vertex v_3 and edge $\{v_2, v_3\}$ and then v_7 and $\{v_2, v_7\}$. At this point, Nodes = $\{v_1, v_2, v_3, v_5, v_6, v_7\}$ and Edges = $\{\{v_1, v_2\}, \{v_1, v_5\}, \{v_1, v_6\}, \{v_2, v_3\}, \{v_2, v_7\}\}$. Finally, we focus on vertex v_3. Only one choice is possible, and the edge $\{v_3, v_4\}$ must be chosen. The process terminates with Nodes = $\{v_1, v_2, v_3, v_4, v_5, v_6, v_7\}$ and Edges = $\{\{v_1, v_2\}, \{v_1, v_5\}, \{v_1, v_6\}, \{v_2, v_3\}, \{v_2, v_7\}, \{v_3, v_4\}\}$. A trace of the construction phase appears in Figure 7.54.

Let us now examine the timing analysis of procedure BuildSpanTree. The while-loop in step 2 depends on how the choices are made and how the list of edges is organized. The optimum organization results in each vertex and edge being handled once. In such a case the performance of the procedure is $O(n + m)$. For comparison purposes, recall that both DFS and BFS approaches to generating a spanning tree are $O(n + m)$. Actually, these searching approaches are special cases of BuildSpanTree that use special selections of the next step.

Another question that arises considers what happens when the procedure BuildSpanTree has, as input, a graph that is not connected. In this case, no spanning tree exists for the graph. However, for a graph that is not connected, we can partition the graph into its connected components. Recall from Section 7.2 that a connected *component* is a maximal connected subgraph, that is, a subgraph that is not a subgraph of any other subgraph in the given graph. A spanning tree can be found for each component in the graph. The collection of spanning trees for the graph is called a spanning *forest*.

An outline of an approach to generating such a forest based on a breadth-first search strategy is as follows:

1. For each vertex s in V set its Reach field to false
2. Set ComponentCtr to 0
3. For each vertex s in V

    If not Reach[s]

    then BFS(s)

    Increase ComponentCtr by 1

Note that we have assumed a BFS approach and the previous BFS routine would need to be enhanced to return a spanning tree each time the routine is invoked. Also, this routine

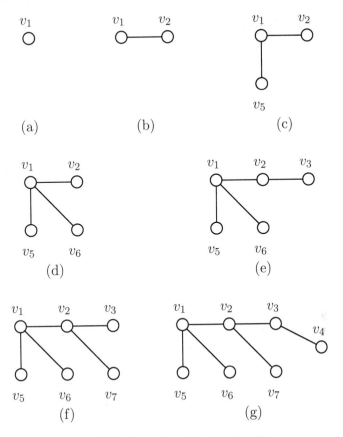

Figure 7.54.   Trace of procedure BuildSpanTree

marks the Reach field of each visited vertex. For a connected graph, ComponentCtr has a value of 1, while for a graph of n vertices with no edges (a null graph) ComponentCtr has a value of n.

As an example, consider the graph shown in Figure 7.55(a), which has three components: (V_1, E_1), (V_2, E_2), (V_3, E_3), where

$$V_1 = \{v_1, v_2, v_6, v_7\}$$
$$V_2 = \{v_3, v_8, v_9\}$$
$$V_3 = \{v_4, v_5, v_{10}, v_{11}\}$$
$$E_1 = \{\{v_1, v_2\}, \{v_2, v_7\}, \{v_1, v_6\}, \{v_6, v_7\}\}$$
$$E_2 = \{\{v_3, v_8\}, \{v_3, v_9\}, \{v_8, v_9\}\}$$
$$E_3 = \{\{v_4, v_5\}, \{v_4, v_{10}\}, \{v_5, v_{10}\}, \{v_5, v_{11}\}, \{v_{10}, v_{11}\}\}$$

Each of these components has a spanning tree as shown in Figure 7.55(b).

The previous algorithm invokes the BFS routine 3 times. The first invocation begins with the start vertex v_1. In addition to vertex v_1, the routine also marks vertices v_2, v_6, and

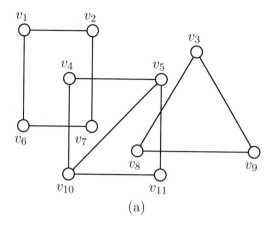

(a)

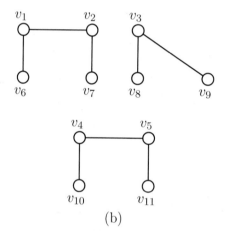

(b)

Figure 7.55.   An unconnected graph and an associated forest of spanning trees

v_7. The second invocation starts with vertex v_3 and also reaches vertices v_8 and v_9. Finally, the third invocation of the routine marks the remaining vertices in the graph.

The focus in this subsection has been on finding a spanning tree for a connected graph with all edges having a weight of 1. The next subsection examines the more general problem of finding a minimum spanning tree for a weighted connected graph.

7.6.4 Minimum Spanning Trees

Several applications can be modeled by graph structures. In many cases, each edge has an associated weight (label). For example, in an airline application the nodes are cities and the weighted edges may denote the cost of flying an airplane between pairs of cities. Obtaining a minimum spanning tree in this application that minimizes the airline's flying cost is a problem from an airline's point of view. In a computer network application, the nodes may be the computer centers in the network and the weighted edges the distances or lease line costs between certain pairs of nodes. An important problem in such a network is to obtain a spanning tree so that the sum of the weights of the edges in the tree is a minimum. Such a minimum spanning tree represents a situation in which communication costs are being minimized. This notion leads us to the next definition.

Definition 7.18:   A spanning tree of a weighted, connected undirected graph in which the sum of the weights of its edges is a minimum is called a *minimum spanning tree*.

For example, Figure 7.56(a) repeats the graph in Figure 7.1, which represents the distances between the major cities in Western Canada given in kilometers. Its corresponding minimum spanning tree appears in Figure 7.56(b). For a traveler who wishes to visit all cities once, this spanning tree minimizes the distance that must be traveled.

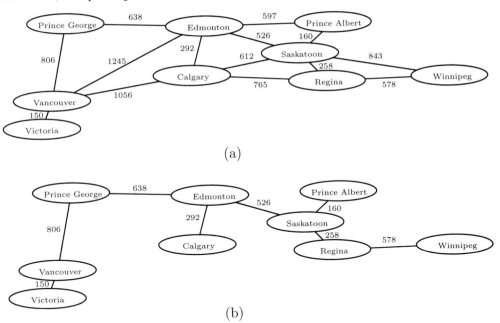

Figure 7.56.   Connected undirected graph and its minimum spanning tree

The first approach to generating a minimum spanning tree is based on procedure BuildSpanTree discussed earlier with edge weights being taken into account. Initially, the algorithm starting at a designated vertex chooses an edge with minimum weight and considers this edge and its associated vertices as part of the desired tree. We then iterate, looking for an edge with minimum weight not yet selected that has one of its nodes in the tree while the other node is not. The process terminates when $n - 1$ edges have been selected from a graph of n nodes to form a minimum spanning tree. This approach is called *Prim's method*, and an algorithm skeleton is shown in Figure 7.57.

```
procedure Prim (s : Node);
    {Given a weighted connected undirected graph G = (V, E) which is assumed to be global to this
    routine and some starting vertex s, this procedure generates a minimum spanning tree for G. The
    minimum spanning tree is denoted by T = (Nodes, Edges) where Nodes contains the vertices and
    Edges are edges selected from E.}
    var
        Nodes, Rest: set of Node;
        Edges: EdgeType;
        k, w : Node;
    begin
        {Step 1: Initialize}
        Choose a minimum weight edge, say, {s, w}, from E.
        Nodes := [s, w];
        Rest := V − Nodes;
        Edges := [[s, w]];
        {Step 2: Generate a minimum spanning tree}
        while Rest <> [] do
            begin
                Choose a smallest weight edge [k, w] in E
                    such that (k in Nodes) and (w in Rest);
                Edges := Edges + [[k,w]];
                Nodes := Nodes + [w];
                Rest := Rest − [w]
            end
    end; {Prim}
```

Figure 7.57.   Prim's method for finding a minimum spanning tree

Consider the example graph given in Figure 7.58(a). Assume that the algorithm begins at, say, starting vertex v_1. The edge with the lowest weight incident to this vertex is $\{v_1, v_2\}$, which has a weight of 1. We then enter the loop in step 2. During the first iteration, there are four edge choices: $\{v_1, v_6\}$, $\{v_2, v_3\}$, $\{v_2, v_6\}$, and $\{v_2, v_7\}$. The edge $\{v_2, v_3\}$ is chosen because its weight is the least of the four possibilities. In the second iteration, edge $\{v_1, v_6\}$ is chosen. The process continues until the seven edges forming the tree are selected. A trace of the tree construction appears in Figure 7.58(b) through (h). The minimum spanning tree has a weight of 23.

The procedure in Figure 7.57 generates a unique minimum spanning tree for a graph with distinct edge weights. If some edges in a graph, however, have the same edge weight, then there may be more than one minimum spanning tree.

The efficiency of Prim's method is examined briefly. Step 1 is $O(m)$ since a minimum weight edge must be chosen. The loop in step 2 is repeated $n - 2$ times to obtain a spanning tree of $n - 1$ edges. At each iteration in this loop, a smallest weight edge is chosen. Depending on how the information is organized, it can be shown that Prim's approach is $O(n^2)$. The details can be found in references [1] and [25] at the end of this book.

We next explore a second approach, called *Kruskal's method*, for generating a minimum spanning tree. First, the edges of a given graph are sorted in nondecreasing order of their weights. The second step examines each edge in order of nondecreasing weight. We choose to include an edge into the minimum spanning tree being constructed only if that edge does not introduce cycles. An outline of Kruskal's method is as follows:

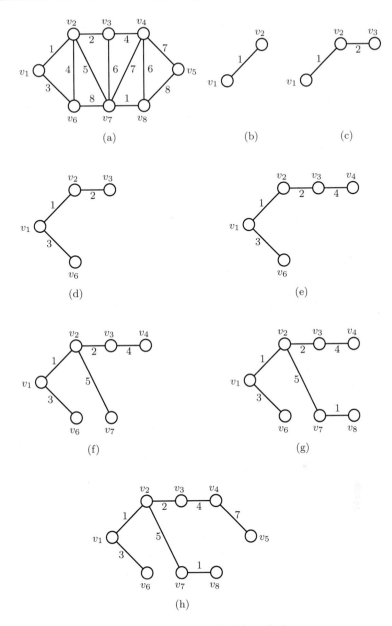

Figure 7.58.   Trace of Prim's method

1. Sort the set of weighted edges in the given graph in nondecreasing order. Assume that the output of this step is $e_1, e_2, \ldots, e_m$, where m denotes the number of edges in the graph.

2. Edges := $\{\}$

3. For i := 1 to m do

   if Edges $\cup$ e_i is acyclic

      then Edges := Edges $\cup$ e_i

Upon termination, this algorithm has determined the set of edges in a minimum spanning tree for the given graph.

   As an example, consider the graph shown in Figure 7.59(a). Figure 7.59(b) shows a possible ordering of the edges when they have been sorted in nondecreasing order. Figure 7.59(c) contains a minimum spanning tree for the graph. Observe that, had edge e_5 been selected, edges e_1, e_3, and e_5 would have formed a cycle. For the same reason, edges e_8, e_{11}, e_{12}, and e_{13} were not selected.

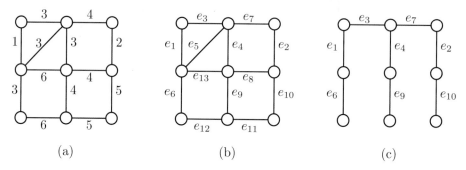

Figure 7.59. Obtaining a minimum spanning tree by Kruskal's method

   Let us examine briefly the efficiency of Kruskal's approach. The sorting of the edges in step 1 in nondecreasing order can be performed using well-known sorting techniques that are $O(m \log_2 m)$. Step 3 requires a test to determine if e_i belongs to a cycle. With the careful representation of the edges, it can be shown that this step is also $nO(m \log_2 m)$. The details can be found in references [1] and [25] at the end of this book.

Problems 7.6

1. Draw all distinct free trees having two, three, and four vertices.

2. Find a spanning tree for each graph in Figure 7.60 by successively removing edges in simple cycles.

3. Obtain a spanning tree for each graph in Figure 7.60 by using a breadth-first search approach to traversal.

4. Repeat Problem 3 using a depth-first search approach.

5. Give a trace (as in Figure 7.54) of procedure BuildSpanTree for each graph in Figure 7.60.

6. If a graph is not connected, we can obtain a spanning tree for each of its components. That is, there is a forest of spanning trees or simply a spanning forest. Formulate an algorithm, based on a depth-first search, to obtain a spanning forest for an unconnected graph.

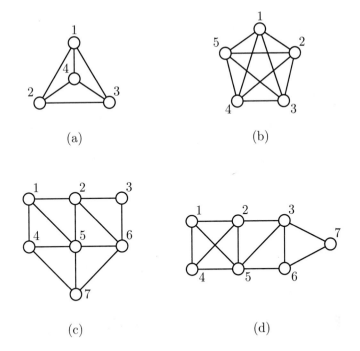

Figure 7.60

7. Apply Prim's method starting at vertex 1 to obtain a minimum spanning tree for each graph in Figure 7.61. For each graph, give a trace (as in Figure 7.58) of the process.

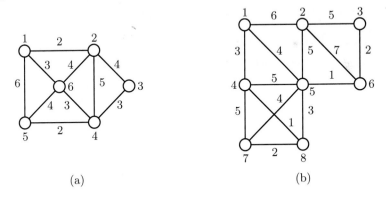

Figure 7.61

8. Use Kruskal's approach to obtain a minimum spanning tree for each graph in Figure 7.61. For each graph, label the edges from e_1 to e_m, where m denotes the number of edges, after the edges are sorted. Show a spanning tree with the edge labels displayed in the tree.

9. In this section you were introduced to Prim's method for obtaining a minimum spanning tree.
   (a) Modify Prim's method to obtain a minimum spanning forest for a graph that is not connected.

(b) Use the algorithm in part (a) to generate a minimum spanning forest for the graph in Figure 7.62.

10. Show that the graph given in Figure 7.63 has more than one minimum spanning tree.

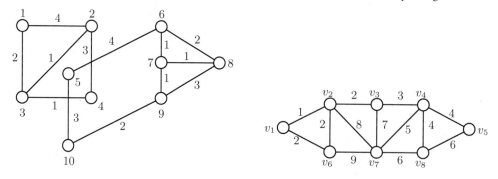

Figure 7.62   Figure 7.63

7.7 SCHEDULING NETWORKS

7.7.1 Introduction

In this section we present two applications in which graph structures are extensively used. The first application, which was introduced in Section 7.1, deals with the PERT project scheduling technique. The second focuses on performing a topological sort on a directed acyclic graph. In the context of network analysis, directed edges and nodes correspond to activities and events, respectively. The purpose of a topological sort is to order the events in a linear manner called a linear order.

7.7.2 A Project Management Model

Recall that Example 7.5 dealt, for scheduling purposes, with the modeling of a project by a directed graph in which each vertex or node represented an event (a point in time) and each directed edge an activity or task to be performed. The weight or numerical value associated with an activity is the time taken to perform that activity. The directed graph has one starting event and one terminating event, called the source and sink, respectively. One primary goal of modeling a project in this way is to obtain the critical activities on the critical path or the critical paths of the digraph. The length of any critical path is the longest path from the source to the sink. In this section we formalize and extend the notions introduced in Example 7.5.

Two basic approaches to scheduling, planning, and controlling projects appeared in the late 1950s: PERT (Program Evaluation and Review Technique) and CPM (Critical Path Method). Although there initially were major differences between the two methods, subsequent development of these methods has blurred these differences. Consequently, we do not distinguish between PERT and CPM in the following discussion.

The construction of a scheduling network is constrained by the following rules:

- Each activity must be represented by a single directed edge (i.e., the activity is unique).

- At most one activity can originate at one node and terminate at another (i.e., the graph must be simple).

The second rule may require the use of *dummy activities*. A dummy activity does not require any time to complete, but it is required because the graph must be simple, and certain logical relationships need to be modeled. As an example of the latter, consider the dummy activity D_1 in Figure 7.9. Since concurrent activities a_1 and a_2 must be completed before activities a_3, a_4, and a_8 can proceed, a dummy activity (denoted by a broken directed edge) with a time of 0 is placed from event 2 to event 3. Similarly, a dummy activity D_2 is created as a prerequisite to performing activity a_{16}, which states that a house design must be selected and a lot purchased before a house can be advertised. D_3, D_4, D_5, D_6, and D_7 represent the remaining dummy activities in the example network. In general, it can be shown that multigraphs can be avoided by using dummy activities.

The main objectives in project management are to determine the critical path and to produce a schedule that shows the start and finish times for each activity in the network. The following concepts are useful in producing such a schedule.

We start with the following definition for each event:

Definition 7.19:   The *earliest starting time*, TE_j, for an event j, is the earliest time when the event can occur such that all incoming activities to that event have been completed.

The TE_j value for each event is generated by performing a forward pass of the network from the source to the sink. In terms of the network, this corresponds to the assignment of time values to each node (event) in such a manner that the value assigned to an event is the length of time to complete the activities along the *longest path* leading into that event. That is, we assign to an event the value that is the maximum, over all incoming edges, of the weight of an edge plus the time associated with that edge's originating node (event).

By definition, the value of 0 is assigned to the source. For the current example, the earliest start time for event 2 is the time (in days) to complete the activities on the longest path from event 1 ending in event 2. There is only one such path involving the completion of activity a_2. Therefore, $TE_2 = 5$. For event 3, there are two possible paths from the source. The first path requires the completion of activity a_1, for a time of 2 days. The second path, which is the longest, requires the completion of activity a_2 and dummy activity D_1, for a total of 5 days. Consequently, the earliest time at which event 3 can begin is 5 days; that is, $TE_3 = 5$. The earliest start times for the remaining events can be obtained in a similar manner.

　　　In general, the earliest starting time for an event is given by a maximum length from the source to that event. To avoid enumerating all possible paths from the source to some event j, one can use an approach that is represented as follows:

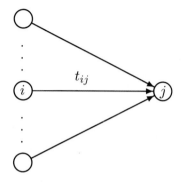

where there may be several incoming activities to event j. A typical activity from some event i to event j has an estimated duration of t_{ij} units of time. Therefore,

$$TE_1 = 0$$
$$TE_j = \max_i\{TE_i + t_{ij}\}, \qquad j \neq 1$$

for all possible activities entering event j. Using this approach, the earliest starting time for each event can be calculated in an efficient manner. For the current example,

$$TE_2 = \max\{TE_1 + t_{12}\} = \max\{0 + 5\} = 5$$
$$TE_3 = \max\{TE_1 + t_{13}, TE_2 + t_{23}\}$$
$$= \max\{0 + 2, 5 + 0\} = \max\{2, 5\} = 5$$
$$TE_4 = \max\{TE_3 + t_{34}\} = \max\{5 + 14\} = 19$$
$$TE_5 = \max\{TE_3 + t_{35}, TE_4 + t_{45}\}$$
$$= \max\{5 + 1, 19 + 2\} = \max\{6, 21\} = 21$$

Continuing the forward pass of the network yields the earliest starting times for all events, as shown in the second column of Table 7.3. Observe that the earliest completion time for the project is 66 days.

　　　The next definition associates a latest completion time with an event.

Definition 7.20:　The *latest completion time*, TL_i, is the latest time when an event can occur without delaying the completion of the project.

TABLE 7.3.
TE_i and TL_i for the Construction Project

Event	TE_i	TL_i	S_i
1	0	0	0
2	5	5	0
3	5	5	0
4	19	19	0
5	21	21	0
6	8	52	44
7	8	28	20
8	22	55	33
9	33	33	0
10	36	38	2
11	38	38	0
12	45	45	0
13	47	47	0
14	47	58	11
15	52	52	0
16	55	55	0
17	60	61	1
18	57	61	4
19	61	61	0
20	61	61	0
21	62	62	0
22	66	66	0
23	66	66	0

The TL_i value for each event is computed by making a backward pass from the sink to the source. In the current example, the latest completion time for the activity entering event 22 such that the time to complete the project is not delayed beyond 66 days is $(66 - 0)$; that is, $TL_{22} = 66$. The latest completion time for event 21 is $(66 - 4) = 62$, since 4 days are required to perform activity a_{28}.

The *latest completion time* associated with each event is the latest time an activity can be completed without causing a delay in the earliest completion date of the project (i.e., they are the latest completion times associated with the activities that do not cause the TE value of the sink node to be increased). The latest completion time, TL, value is the largest value that will still allow every activity starting at that node to be completed without an overall time increase. In terms of the graph, we assign to a node the value that is the minimum, over all outgoing edges, of the edge's destination node. The TL value of the sink equals its TE value.

The approach to compute TL_i can be represented as follows:

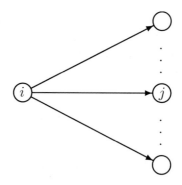

where there may be several outgoing activities from event i. In general, the latest completion time for an event can be computed as follows:

$$TL_n = TE_n, \qquad \text{where } n \text{ is the sink}$$
$$TL_i = \min_j\{TL_j - t_{ij}\}, \qquad i \neq n$$

One has to take the minimum because, by meeting the earliest deadline, one also meets the later ones. In the current example,

$$TL_{21} = \min\{TL_{23} - t_{21,23}, TL_{22} - t_{21,22}\}$$
$$= \min\{66 - 2, 66 - 4\} = \min\{64, 62\} = 62$$

The TL values for the other events can be obtained as shown in Table 7.3.

Another important notion when dealing with scheduling networks is the following:

Definition 7.21:   The *slack time* of an event i is defined by $S_i = TL_i - TE_i$, that is, the maximum time that all activities beginning at event i could be delayed without delaying the entire earliest completion time of the entire project.

In the current example the slack value for event 6 is $S_6 = TL_6 - TE_6 = 52 - 8 = 44$. This value indicates that the start of event 6 could be delayed by up to 44 days without causing a delay in the earliest completion time of the construction project. The slack value of each event appears in the last column of Table 7.3. The notion of slack time (sometimes called float time) can also be associated with an activity.

Definition 7.22:   The *total slack (float) time* for an activity (i, j), originating at event i and terminating at event j, is defined as $TS_{ij} = TL_j - TE_i - t_{ij}$.

That is, the total slack of an activity is the maximum time to perform that activity minus the time to complete the activity. The maximum time allowed to perform an activity (i, j) is the difference between its earliest starting time and its latest completion time. For example, the total slack for activity a_1 is

$$
\begin{aligned}
TS_{1,3} &= TL_3 - TE_1 - t_{13} \\
&= 5 - 0 - 2 \\
&= 3 \text{ days}
\end{aligned}
$$

which indicates that this activity can be delayed (or float) by as much as 3 days without causing a delay in the completion of the project. As another example, the total slack time for activity a_{28} is

$$
\begin{aligned}
TS_{21,22} &= TL_{22} - TE_{21} - t_{21,22} \\
&= 66 - 62 - 4 \\
&= 0
\end{aligned}
$$

For this activity, however, no delay is possible if the project is to be completed in 66 days. The completions of the TS values are summarized in Table 7.4. Notice the use of subscripting involving events i and j and the use of TE_i and TL_j values from Table 7.3.

The reader may have noticed that there are several activities in the network that have a total slack time of 0. These activities on a path from the source to the sink are said to be critical. This special path is very important in project scheduling and is defined as follows.

> **Definition 7.23:**   A *critical path* from the source to the sink for a network is a path containing those activities that have a zero total slack.

Therefore, a delay in starting any critical activity on the critical path will delay the completion of the project. Consequently, the critical path is the longest path in the network, and the sum of the times for completing the critical activities is the shortest time to complete the project. In the current example the critical path is the following sequence of activities:

$$a_2 \; D_1 \; a_4 \; a_5 \; a_{10} \; a_{11} \; a_{14} \; a_{17} \; a_{18} \; a_{19} \; a_{23} \; D_6 \; a_{26} \; a_{28} \; D_7$$

with a shortest completion time of 66 days. Recall that the dummy activities D_1, D_6, and D_7 each have a zero completion time. The critical path details are presented in Table 7.5.

We are now able to formulate an algorithm for computing the earliest completion time of a project and its associated critical path (including the critical activities). The total slack time for each noncritical activity is also generated. The network is assumed to have a source of 1 and a sink of n. Although it is possible to have more than one critical path from the source to the sink of a given graph, this situation does not appear to occur often in practice. In any case, the following algorithm determines a single critical path.

TABLE 7.4. Time Calculations for a Construction Project

Activity	Arc	t_{ij}	TE_i	TL_j	TS_{ij}
a_1	$(1,3)$	2	0	5	3
a_2	$(1,2)$	5	0	5	0
a_3	$(3,5)$	1	5	21	15
a_4	$(3,4)$	14	5	19	0
a_5	$(4,5)$	2	19	21	0
a_6	$(5,8)$	1	21	55	33
a_7	$(8,21)$	7	22	62	33
a_8	$(3,7)$	3	5	28	20
a_9	$(7,11)$	10	8	38	20
a_{10}	$(5,9)$	12	21	33	0
a_{11}	$(9,11)$	5	33	38	0
a_{12}	$(9,10)$	3	33	38	2
a_{13}	$(9,12)$	4	33	45	8
a_{14}	$(11,12)$	7	38	45	0
a_{15}	$(2,6)$	3	5	52	44
a_{16}	$(6,23)$	14	8	66	44
a_{17}	$(12,13)$	2	45	47	0
a_{18}	$(13,15)$	5	47	52	0
a_{19}	$(15,16)$	3	52	55	0
a_{20}	$(16,20)$	3	55	61	3
a_{21}	$(16,17)$	5	55	61	1
a_{22}	$(16,18)$	2	55	61	4
a_{23}	$(16,19)$	6	55	61	0
a_{24}	$(12,14)$	2	45	58	11
a_{25}	$(14,20)$	3	47	61	11
a_{26}	$(20,21)$	1	61	62	0
a_{27}	$(21,23)$	2	62	66	2
a_{28}	$(21,22)$	4	62	66	0

TABLE 7.5. Critical Path

a_2	D_1	a_4	a_5	a_{10}	a_{11}	a_{14}	a_{17}	a_{18}	a_{19}	a_{23}	D_6	a_{26}	a_{28}	D_7
5	0	14	2	12	5	7	2	5	3	6	0	1	4	0
5	5	19	21	33	38	45	47	52	55	61	61	62	66	66

1. [Perform a forward pass to compute TE values]
   $\text{TE}_1 := 0$
   For $j := 2$ to n
      $\text{TE}_j := \max_i \{\text{TE}_i + \text{t}_{ij}\}$ for all possible activities entering event j
2. [Perform a backward pass to compute TL values]
   $\text{TL}_n := \text{TE}_n$

For i := 1 to n −1
   $TL_i := \min_j\{TL_j - t_{ij}\}$ for all possible activities leaving event i.

3. [Generate a critical path]
   m := 0
   For i := 1 to n
      If $TE_i = TL_i$
      then m := m + 1
         CriticalNodes [m] := i

   For k := 1 to m − 1
      i := CriticalNodes[k]
      j := CriticalNodes[k + 1]
      Display (a_{ij})
   Write ('The earliest completion time is', TE_n)
4. [Compute total slack of each activity]
   For each activity (edge) (i, j)
      $TS_{ij} := TL_j - TE_i - t_{ij}$

Most steps in the algorithm have been discussed earlier. Step 3, however, requires some elaboration. The vector CriticalNodes contains the events that lie on a critical path, that is, those events with the same TE and TL values. These critical events are then used to determine the critical activities on that critical path. A routine Display is assumed to output an activity. Finally, in step 3 the earliest completion time of the project is the output.

Problems 7.7.2

1. Make all the following changes to the construction example explored in this section. Show how these changes affect the scheduling network in Figure 7.9 and note any changes to the critical path. Make all the changes simultaneously, not one at time.

   (a) The installation of cupboards (a_{20}) must be performed before the installation of flooring (a_{21}) and the installation of light fixtures (a_{22}).

   (b) The basement must be fully constructed (a_7) before construction can be started on the house (beginning at activity a_{10}).

   (c) The owner of the construction firm has several relatives who are subcontractors so, although bids are accepted, the subcontractors are predetermined. This allows the selection of subcontractors (a_5) to be done even before bids are accepted (a_4).

   (d) Assume that the exterior siding installation (a_{24}) and exterior painting (a_{25}) can be performed in parallel.

2. Make the following additions to the activities associated with the house construction example. Show how these changes affect the scheduling network, and note any changes to the critical path.

   (a) The house has a suspended ceiling installed. This activity takes 1 day and must be performed after the interior painting has been done.

   (b) Assume that the rafters for the roof have to be made by the firm building the house. This activity would have to appear before framing the roof and would have to appear after the house design was finalized. Assume that this activity takes 2 days.

   (c) Assume that the houses have decks put on them after they are placed on the sites. Building the deck and painting it would take 4 days. This activity would happen after the houses are placed on the sites, but before the sites are landscaped.

3. Formulate a set of activities and precedence relations for getting to the university in the morning. Activities such as the following would be reasonable candidates:

- Get up
- Wake up
- Make breakfast
- Eat breakfast
- Take a shower
- Choose clothes
- Dress
- Bring books and due assignments
- Get to university

Represent the activities and precedence relations by a scheduling network.

4. Given the activities (and their durations) and the precedence relations in Table 7.6, obtain the following:

   (a) The TE, TL, and S values for each event (see Table 7.3)

   (b) The total slack for each activity

   (c) A critical path for the scheduling network

5. Repeat Problem 4 for Table 7.7.

TABLE 7.6

Activity	Precedence Relations	Duration
a	—	15
b	—	8
c	—	6
d	—	7
e	d	12
f	e	8
g	a, f	10
h	g	9
i	b, h	9
j	b, h	8
k	i	11
l	g	8
m	c, k	9
n	k, l	12

TABLE 7.7

Activity	Precedence Relations	Duration
a	—	8
b	a	4
c	b	14
d	a	10
e	a	9
f	c, d	18
g	d	25
h	d, e	27
i	f, g, h	6

7.7.3 Topological Sorting

The purpose of a topological sort is to order the events in a linear manner, that is, first, second, third, and so on, with the restriction that an event cannot precede other events that must first take place. The earliest use of topological sorting with computers was in conjunction with network analysis, for example, with techniques such as PERT (see previous section).

Recall from Section 5.4.8 that a partial ordering or partial order is a relation on a set that is reflexive, antisymmetric, and transitive. Let $(A, \preceq)$ denote a partially ordered set. Then two elements of A, say a_1 and a_2, are comparable if either $a_1 \preceq a_2$ or $a_2 \preceq a_1$. As mentioned in Section 5.4.8 a set $(A, \preceq)$ is a linearly ordered set, or a linear order, if all elements of A are comparable. A topological sort arranges the elements of A in conformance with $\preceq$, as indicated by the following definition:

Definition 7.24:   A *topological sort* of $(A, \preceq)$ is a linear arrangement of the elements of A such that x precedes y if $x \preceq y$.

As an example of the results of performing a topological sort, consider the partial order $\langle A, \preceq \rangle$, where

$$A = \{1, 2, 3, 4, 5, 6, 7\}$$
$$\preceq = \{(2, 1), (2, 6), (3, 6), (4, 2), (5, 3), (5, 6), (7, 5)\}$$

and its associated digraph in Figure 7.64(a). One possible order for A is

a_1	a_2	a_3	a_4	a_5	a_6	a_7
7	5	3	4	2	1	6

Figure 7.64(b) shows this linear order. The interpretation of $a_i \preceq a_j$ in this diagram is that a_i precedes a_j; that is, there is an arc going from a_i to a_j, for $i < j$. Note that other linear orderings could be defined for this example.

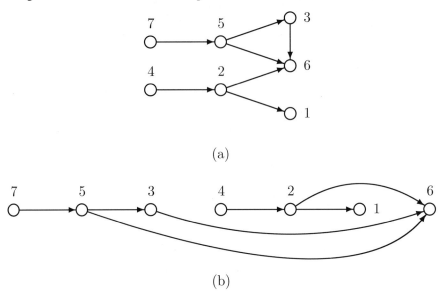

(a)

(b)

Figure 7.64.   Example of a topological sort

Another application of topological sorting is the ordering of the definitions of terms in a book. Suppose that we have a number of terms $t_1, t_2, \ldots, t_n$ that are related by pairs (t_i, t_j) such that t_i is used in the definition of t_j. Then we wish to order the terms so that each term t_i appears before each term t_j that uses t_i directly or indirectly in its definition. It is possible to have circular definitions, for example, $(t_i, t_j), (t_j, t_k)$, and (t_k, t_i). In this case, a complete linear ordering of the terms is impossible. Consider the examples in Figure 7.65; the relationships between the definitions of record, file, field, key, and transaction (as found in two different books) are shown. An edge is directed from t_i to t_j if t_i is used in the definition of t_j. In Figure 7.65(a), a linear ordering cannot be defined. For Figure 7.65(b), a topological sort could define the linear ordering

Record—Field—File—Key—-Transaction

An alternative ordering is

Record—File—Field—Key—Transaction

In Figure 7.65(a), even though we cannot successfully apply a topological sort, the attempt to do so can be useful in detecting the circular definitions. Only acyclic digraphs can be topologically sorted.

Figure 7.66 contains a suitable representation of the digraph of Figure 7.65(a) for performing a topological sort. The column labeled $PredCtr$ gives the number of immediate predecessors for each node. For example, Key has a predecessor count of 3.

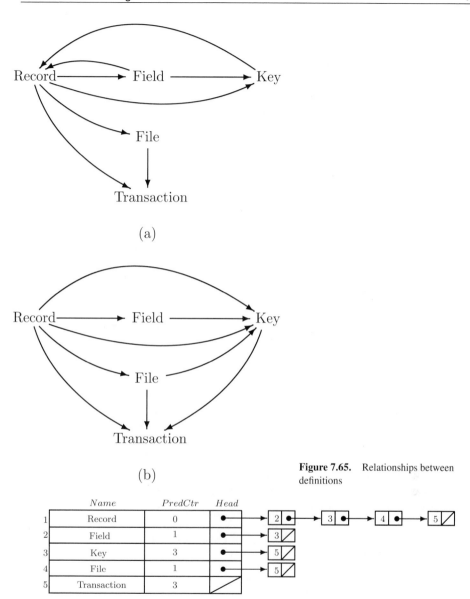

(a)

(b)

Figure 7.65.   Relationships between definitions

Figure 7.66.   Representation of a digraph for topological sorting

An algorithm for performing a topological sort is quite simple now that the structure for representing a directed graph is available. The node descriptors are to be printed in a linear order by the sort. If a node's predecessor count is zero, then it can be printed. After this, the predecessor count of each of the node's successors can be decremented by 1. Each time a node's predecessor count reaches 0, it is eligible to be printed. Since more than one node may have a zero predecessor count simultaneously, a control mechanism is required to keep track of these nodes. We use a queue structure for this purpose.

A general algorithm for performing a topological sort follows:

1. Insert all nodes without predecessors at the rear of the queue.
2. While the queue is not empty, do the following steps:

   2.1 Output the descriptor of the node at the front of the queue and delete this node from the queue.

   2.2 Decrease the predecessor count of each node in the adjacency list of the deleted node by 1.

   2.3 If the predecessor count of a node reaches 0, then add the node to the rear of the queue.

Step 2.1 can be modified to count the number of nodes that are output. If the count is less than the number of nodes in the graph, some nodes are never output, and this indicates the presence of cycles.

Another approach to performing a topological sort is to modify the depth-first search approach introduced earlier. Given an acyclic digraph $G = (V, E)$ represented as adjacency lists, the following algorithm skeleton performs a topological sort in *reverse order*.

```
procedure ReverseTopological (G)
begin
        for each v ∈ V
                Reach[v] := false;
        for each v ∈ V
                if not Reach[v] then DFS(v)
    end;


    procedure DFS (v)
    begin
        Reach[v] := true;
        for each neighbor w of v
                if not Reach[w] then DFS(w);
        write(v.Name)
    end;
```

We leave the justification that this procedure works to the reader. The output from this algorithm on the digraph of Figure 7.65(a) starting at the node with descriptor Record is

$$\text{Transaction—Key—File—Field—Record}$$

which is the reverse of the desired topological ordering.

Although we have only used small examples throughout the discussion, it should be emphasized that acyclic digraphs having thousands of nodes have been sorted topologically.

In scheduling networks with one source and one sink that have an acyclic representation, the modified depth-first search approach will traverse the entire digraph starting at its source. If the digraph, however, is not acyclic (indicated by some nodes not being output), then the approach used in Section 7.5.4 for detecting cycles can be used to detect and display the cycles.

Problems 7.7.3

1. Obtain a linear order for the digraph in Figure 7.67.

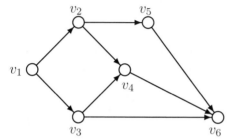

Figure 7.67

2. Obtain a linear order for the digraph with the following adjacency list representation:

1	2, 3
2	4, 5
3	5
4	6, 7
5	7
6	8
7	8
8	

3. Using the digraph of Problem 2 as input, give a trace of the modified depth-first approach for generating a reverse topological order.

4. In compiler construction an expression such as $(x - y) * (w + z) - (x - y) * (w - z)$ can be represented by the directed acyclic graph of Figure 7.68. Obtain a topological order for this graph.

5. (a) Show that every finite nonempty partially ordered set has a minimal element.
   (b) Using the result of Part (a), formulate an algorithm to obtain a topological sort of the given partially ordered set.
   (c) Using the digraph of Problem 2, give a trace of the algorithm obtained in Part (b).

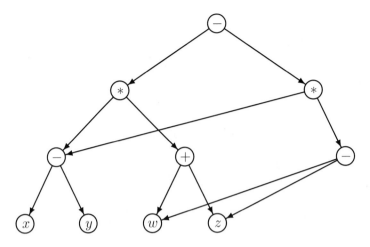

Figure 7.68

PROBLEMS: CHAPTER 7

1. When a depth-first search is done on an *undirected* graph, the edges can be classified as tree edges (edges to unreached vertices) and back edges (edges to vertices with a smaller depth first number other than the parent, where the depth first numbering of the vertices indicates the order in which they were reached). When a depth-first search is done on a *directed* graph, the edges can be classified as tree edges (edges to unreached vertices), back edges, forward edges, and cross edges. With respect to the depth-first tree formed during a depth-first search, a back edge is an edge from a vertex to an ancestor of the vertex (closer to the root), and a cross edge is an edge from one branch of the tree to a previously scanned branch of the tree. In the graph given in Figure 7.69, we have the following:

$$\text{back edges: } (e, c), (g, b), (c, b)$$
$$\text{forward edges: } (a, d), (b, f)$$
$$\text{cross edges: } (g, e), (f, e)$$

Note that in a directed graph (b, c) and (c, b) are considered distinct edges so that one can be a tree edge and the other a back edge. Give an efficient algorithm based on depth-first search in order to classify the edges of a directed graph when started at vertex s. Note: Assume that you are only given the adjacency lists of the graph to be analyzed, the vertex set and the vertex s.

2. When doing a depth-first search of an undirected graph, each back edge defines a cycle. The cycle consists of the back edge and the path in the tree between the ends of the back edge (see Figure 7.70). You are to give an algorithm based on depth-first search that will print out one cycle of the above form that has the *shortest length*. You should assume that the input consists of the adjacency lists of an undirected graph. Remember that in an undirected graph the edge from a vertex back to its parent is not considered a back edge (because the same edge is the tree edge from the parent to the vertex). In the depth-first search tree given in Figure 7.70, both $d\,c\,b\,d$ and $i\,h\,a\,i$ qualify as shortest cycles. Note: This does not necessarily produce the shortest

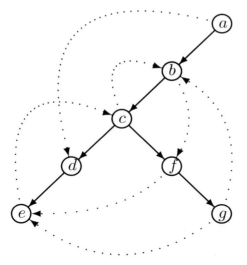

Figure 7.69

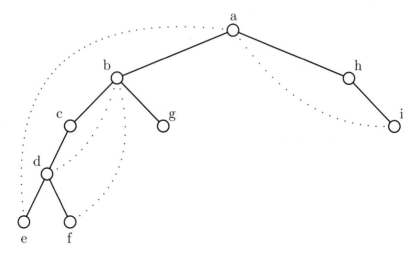

Figure 7.70

cycle in the graph, but for the purpose of this question we are only interested in one shortest cycle of the above form.

3. A breadth-first search of an undirected graph can be used to find a shortest-path tree. This tree contains all the vertices and some of the edges of the graph. Each edge of the graph that is not in the breadth-first search tree must go from one branch of the tree to another branch of the tree. Such nontree edges are called *cross edges*. In Figure 7.71 the cross edges are marked as broken lines. Thus, using a breadth-first search, the edges of a graph can be classified as either tree edges or cross edges.

Given two vertices in a tree, their *nearest common ancestor* is a vertex in the tree that is an ancestor of both the vertices and is also the closest such vertex. In Figure 7.71 the nearest common ancestor is given for the vertices of each cross edge. You are to develop an algorithm that does a breadth-first search of an undirected graph starting at vertex s. When the search detects a cross edge, the algorithm should determine the nearest common ancestor of the two vertices of the edge and print out the cross edge and its nearest common ancestor.

You should assume that the graph is stored in the form of adjacency lists. Note that if vertex b in the adjacency list of vertex a is used to form a tree edge, then vertex a in the adjacency list of vertex b is just a duplicate representation of the same edge (and does not cause the formation of a cross edge). Also, if vertex d in c's adjacency list is used to form a cross edge, vertex c in d's adjacency list is just a duplicate representation of the same cross edge.

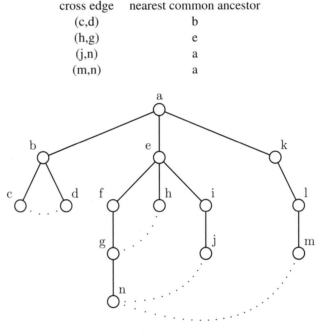

cross edge	nearest common ancestor
(c,d)	b
(h,g)	e
(j,n)	a
(m,n)	a

Figure 7.71

4. Give a function (a detailed algorithm or Pascal program) that returns a Boolean value that indicates whether a graph is in fact a tree (without a root). For example, the graph in Figure 7.72 with 10 vertices is not a tree, but part of it formed by vertices 4, 5, 6, 9, and 10 is in fact a tree. Note that in order to be a tree a graph cannot have any cycles and must be connected, that is, there must be a path between every two vertices. The graph of Figure 7.72 satisfies neither of the criteria.

Assume that the graph is undirected with n vertices and that adjacency lists are used to store its edge set. You may assume that the specification of the graph, that is, n and the adjacency

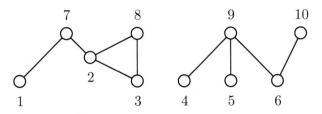

Figure 7.72.   Undirected graph that is not a tree

lists are global to the function. The algorithm for your function should be based on either *breadth-first search* or *depth-first search*, your choice, but specify which you are using.

5. Formulate an algorithm based on a depth-first search to construct the DFS (depth-first search) tree T starting at vertex v_0. For example, the adjacency lists for an undirected graph in Figure 7.73(a) on page 440 have the associated DFS tree in Figure 7.73(b), which is obtained from a depth-first search starting at vertex 1. Your algorithm is to generate the following arrays to represent the DFS tree T.

Parent[u]: where $u \neq v_0$ denotes the parent of u in T

Dist[u]: stores the distance of vertex u from vertex v_0 in T

NumDescend[u]: stores the number of descendants of u in T

The arrays for the example DFS tree are given in Figure 7.73(c).

6. Repeat Problem 5 for a breadth-first search approach.

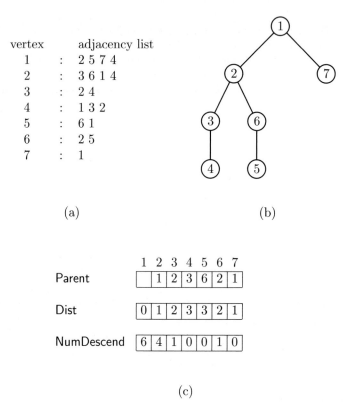

vertex   adjacency list
  1   :   2 5 7 4
  2   :   3 6 1 4
  3   :   2 4
  4   :   1 3 2
  5   :   6 1
  6   :   2 5
  7   :   1

(a)   (b)

              1 2 3 4 5 6 7
Parent | | 1 | 2 | 3 | 6 | 2 | 1 |

Dist | 0 | 1 | 2 | 3 | 3 | 2 | 1 |

NumDescend | 6 | 4 | 1 | 0 | 0 | 1 | 0 |

(c)

Figure 7.73

8

The Formal Specification of Requirements in Z

In the last decade there has been an increasing emphasis on using formal methods in developing hardware and software systems. These methods are often mathematically based and provide support environments that systems developers (such as analysts, designers, and coders) can use to specify, design, code, verify, and maintain computer systems in a systematic, rather than informal, manner.

A formal method, which is used to specify customer requirements, is often called a formal specification technique. In such a context a formal method has an associated formal specification language. Customer requirements written in such a formal language specify what a system does, not how it does it. Formally stated specifications can be analyzed for syntactical correctness, ambiguity, inconsistency, and some forms of completeness.

This chapter introduces the formal specification language Z. Most of the mathematical foundations on which Z is based have appeared in earlier chapters.

8.1 INTRODUCTION

System developers are in the business of designing solutions to problems and implementing these as computer software. This process has both a scientific aspect and an engineering aspect and is thus often referred to as *software engineering*.

Traditionally, science concerns itself with the discovery of basic principles, laws, and relationships, while engineering deals with the practical application of these. Software engineering attempts to combine sound engineering principles of design, management, testing, and so forth, with the concepts of computer science in the production of computer software.

To a person with experience only in small programming problems, the ideas of software engineering may seem unnecessary. In programming practice, however, two facts dominate. First, practical software projects are typically anything but small. Generally, these involve many tens of thousands (or hundreds of thousands) of lines of code, as well as teams of many programmers. As the size of a project grows, the development cost grows more than linearly. This is due largely to the rapidly increasing complexity of large projects and to the basic human inability to manage this complexity. Cost and time overruns are the inevitable result. This situation is often referred to as the *software crisis* or software bottleneck.

The second dominant fact is summarized in the term *software evolution*. Software changes constantly to accommodate new user requirements, to fix discovered bugs, and to fit new host systems. In many practical situations, the maintenance time for a piece of software vastly exceeds the development time under the presumption that it is easier to modify a piece of working software rather than to build something totally new.

There is now a recognized need for the creation of tools and support environments for the development of software to meet the software crisis in the development of new computer systems. Commercial-style computer-aided software engineering (CASE) environments have emerged in the last 10 years that support development efforts in the distinct phases that constitutes what is called the *software development life cycle*.

This chapter first examines the distinct phases that make up an example of a software development life cycle. The next section argues for the need for a formal notation to facilitate the capture of system requirements. The main focus of the chapter is on a tool that is used in the elicitation, capture, and analysis of customer requirements for a system that is to be developed. The Z language is a formal language, based on logic, sets, relations, and functions, that is used to specify, for example, customer requirements for a system. Because a Z program is analyzed by a compiler, a program can be checked for correct syntax, contradictions, partial completeness, and the like.

8.2 SOFTWARE LIFE CYCLE

A large-scale software project spans a considerable period of time. A number of distinct phases can be identified in the development of a computer system for some given application. Together these phases make up what is known as the *software life cycle*. Several models have been proposed to represent the software life cycle. In this chapter, the *waterfall model* is used to represent the development process. Because of its simplicity, it is used here instead of more recently proposed models.

While the actual terminology may differ, most authors identify six key phases in the software life cycle (as shown in Figure 8.1).

1. *Requirements analysis and specification*: The requirements are analyzed and specified.
2. *System (architectural) design*: A high-level design of the system is generated from the requirements specification.
3. *Detailed design*: The details of the design in the previous step are obtained.

4. *Implementation*: The detailed design is programmed (or coded) in a particular programming language.

5. *Testing*: The implemented system is tested to see that it meets the specified requirements.

6. *Operation and maintenance*: The system is installed and used. Errors found are repaired, and system enhancements and/or improvements are performed.

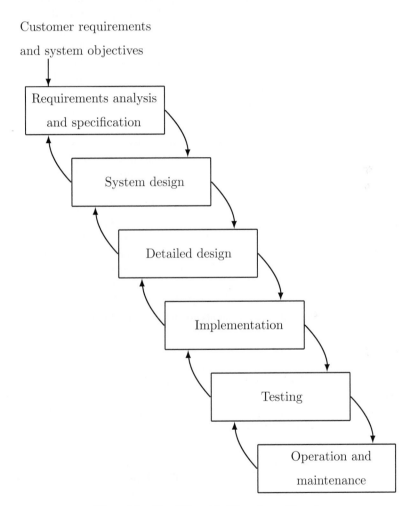

Figure 8.1.   Waterfall model of the software life cycle

While a software project can be described in terms of these six phases, the actual development process itself is iterative, with both feedforward and feedback components. Each phase feeds something forward, upon which subsequent phases are based, but also feeds information back to earlier phases. Implementation, for example, reveals design flaws, and testing reveals implementation errors. Each phase has an input and an output that must be checked carefully before being passed on.

At various times during the software development process, the evolving software product is described by artifacts (or documents) that specify the software system. A customer's statement of requirements and system objectives is a customer's view of what the software is to do. The requirements specifications document is a developer's view of what the system is to do. The architectural design gives a high-level view of the system architecture that satisfies the requirements specification. The detailed design gives a complete and detailed view of the system architecture. Finally, the result of the implementation process is a set of programs that is a realization of the system. Unfortunately, in many cases, each of these artifacts is expressed in different notations. During the development process, these artifacts must be checked for errors and modified if necessary. This activity is known as verification and validation.

Verification is concerned with checking that the input to a phase is properly transformed to the output of the phase. Since the input and output of a phase are usually represented by different notations, the verification activity is tedious and difficult. *Validation* is concerned with checking that the output artifact of each phase matches the customer requirements and system objectives (i.e., the developer is building the right system for the customer).

The first phase, *requirements analysis and specification*, refers to the period during which the requirements of the system desired, that is, its functional characteristics and operational details, are specified. The inputs to this phase are the (often rather loosely) stated needs for the software in the form of customer requirements and system objectives. Typically, a requirements document is the output of this phase, a set of precisely stated properties or constraints that the final product must satisfy. This is not a design, but rather precedes the design, specifying *what* the system should do without specifying *how* it is to do it. The existence of a requirements document provides something against which a design (the next phase in the life cycle) can be validated.

As in any of the phases, it is important that errors not be allowed to move into subsequent phases. An error in requirements, such as a misstated function, leads to a faulty design and an implementation that does not do what is required. If this is allowed to proceed undetected, say, until the testing phase, the cost of repairing this error (including redesign and reimplementation) can be substantial. It has been observed that the cost of fixing such an error grows exponentially with the phase in which it is detected. For example, the cost of repairing an error detected in the maintenance phase may be one hundred times greater than that of repairing the error if found in the requirements analysis phase. This has been referred to as the "price of procrastination" with respect to error detection. Requirements are most effectively validated by reviewing the requirements document with the client for whom the software is being developed.

The second phase, *architectural or system design*, is predominately creative. While some would argue that creativity is inherent and cannot be taught or improved, it can certainly be enhanced by the use of good procedures and tools. The input to this phase is a validated requirements specifications document. The output is a design expressed in some appropriate form, such as a directed acyclic graph of modules or components that work together, in general, to perform several functions or to achieve several objectives. A function of a system is the work, activity, or task that a system performs. An objective of a system is the purpose or goal that the system serves in carrying out its function(s).

The third phase, *detailed design*, is concerned with the transformation of the architectural design into a detailed form from which programmers can generate programs. Each module or component in the architectural design is fleshed out with the necessary details, including the particular algorithms that are required for the module to achieve its function(s). The result of this phase is usually some form of graphical notation of the system augmented with pseudocode. Validation in both design phases is important. Each requirement in the requirements document must have a corresponding design fragment to meet it. Informal reviews involve the entire design team, management, and even the customer.

The fourth phase, *implementation*, is the actual coding of the design developed in the third phase. The lure of this phase is strong, and many a foolhardy programmer has been drawn to it before the groundwork in the first three phases has been adequately laid out. Each module in this phase consists of variables, types, and program units (such as functions and procedures in some languages and abstract data types in others).

In the fifth phase, *testing*, the concern is with demonstrating the correctness of the programs for the system. Invariably, some testing is performed as part of the previous four phases as well. Individual modules are first unit tested; that is, each module is tested in isolation. After the modules are unit tested, they are gradually integrated and tested.

The final phase of the systems development, *operation and maintenance*, is an ongoing phase in the lifetime of a computer system. Operation refers to the time when the system is delivered to the customers and becomes functional within an organization.

The importance of maintenance in the real world, however, cannot be overemphasized, because the cost of maintaining a widely used program can match or exceed the cost of developing it. Unlike hardware maintenance, software maintenance deals not with the repair of deteriorated components, but with repair of design defects, which may include the provision of added functions to meet new needs. The ability of programmers to produce *new* programs is clearly affected by the amount of time that they spend maintaining *old* ones. The necessity of maintenance must be recognized and steps taken to reduce its time consumption.

The total cost of a software product is a function of the time involved and the number of people working on the project over its entire lifetime. The breakdown of the software life cycle into constituent phases provides for a finer analysis of this cost. It has been observed repeatedly that these phases contribute unequally to the cost of a project. For example, the maintenance phase may contribute as much as all the development phases (phases 1 through 5) combined. It is the job of the software engineer to keep *total* cost as low as possible. This is done by apportioning time judiciously among all the phases. Inadequate time spent in one phase, such as testing, leads to problems in subsequent phases (here, maintenance) and increases total cost. It has been claimed that much of the maintenance effort in actual projects is due not to coding (or implementation) errors, but to changes or errors in requirements and poor design.

8.3 NEED FOR FORMAL SPECIFICATIONS

Several problems occur during the development of information systems. Some of these arise from inadequacies of the notations that are used to describe the software product at each stage of the development life cycle. Many of these notations include natural language as a vehicle for describing the different artifacts. The notations that are most dependent on natural language are those that are used upstream in the life cycle (i.e., in its early phases of development). Also, in many development approaches to producing software, a different notation is used for each phase of the life cycle, and because of these very "visible seams" between phases, usually interface errors result.

Developers have attempted to avoid some of the problems associated with developing software through exploring the use of mathematical notations based on logic, sets, functions, and relations for describing the software product. Mathematics is also well suited to represent different levels of abstraction. A specification can be given initially at a high level and then successively refined at lower levels much in the same way allowed by most current programming languages. Such approaches to assist in software development have been called *formal methods*.

This section first focuses on the weakness of natural language as a notation for analyzing and specifying requirements. Second, some of the advantages and disadvantages of using formal specification methods in the early phases of software development are briefly described. Formal specification methods are considered to be a subset of formal methods. Formal methods is used generically to also include those methods associated with formal validation and verification.

It should be stressed that formal specification approaches are not meant to replace natural language descriptions completely. Rather, formal approaches use natural language much in the same way that computer languages contain comments in their statement repertoire. Comments are used for annotations and explanations.

Recall from the previous section that customer statement of requirements consists of various types of statements that describe a system from the customers' viewpoints. Requirements at this stage include the following types of statements:

Functional requirements: those that detail *what* a system is to do.

Nonfunctional requirements: those that contain directives (both design and implementation) that constrain the software developer in the downstream phases of the life cycle. An example of such a constraint is to implement a system in C++.

Goals: requirements that may guide the software developer when choices exist (e.g., minimum time, minimum space).

In practice, a customer statement of requirements given in natural language will usually be ambiguous, inconsistent, incomplete, and randomly mixed with different levels of abstraction (i.e., not given in a structured form). Furthermore a natural language description tends to contain redundancy. Also, such a description will often contain several forward references to features or elements of the system not defined until later.

The principal goal of a system analyst is to analyze the customer statement of requirements and strive to obtain a complete, consistent, unambiguous, nonverbose, and modular version of these requirements. There is close analyst–customer interaction during the preparation of a requirements specification document that describes the desired system.

This chapter will concentrate on specifying requirements using an approach based on mathematics. More specifically, the formal specification language, Z, is introduced in the next section to achieve this goal.

The use of formal specification techniques such as those provided by using the Z language has not been without controversy. Some advantages of using a formal approach include the following:

- The use of a formal notation (based on mathematics) produces or provides the precise statement of what a system is to do. It can be a convenient and structured form for discussing and specifying the requirements of a system.
- Since a specification language is a formal computer language, specifications can be checked for syntactic correctness, consistency, and contradiction by support tools.
- Certain kinds of partial completeness can be checked. Of course, the total completeness of a specification can never be checked. Since environmental factors and customer omissions could have occurred at the customer requirement stage, a support environment such as that provided for the Z language cannot check for that kind of total completeness.
- Finding errors in the requirements specification document early will save substantial time and effort as compared to finding these errors in the later phases of the life cycle.
- Formal specifications can be used to perform verification and validation throughout the phases of the life cycle.

Some disadvantages of using a formal specification approach include the following:

- Customers cannot understand a formal specification document.
- There is a major technology transfer problem (many practitioners do not know much about discrete mathematics; the emphasis in most computer science curricula is still on the integral and differential calculus).
- Lack of integrated support environments for analysts and designers using formal techniques.

This section has identified where in the life cycle formal specification techniques can be used. The next section illustrates how this can be done by using a popular formal specification language.

8.4 INTRODUCTION TO Z

8.4.1 Introduction

Z is a formal specification language based on logic, sets, relations, and functions for stating what a system should do and in what order it should be done without stating how it should be done. Z, therefore, is considered to be a declarative language, as compared to a procedural or imperative one such as Pascal. Recall that Prolog is also considered to be a declarative language. In specifying a system in Z, issues concerning efficiency and implementability of the system are not of importance at the specification stage of software development.

Z has become one of the most popular specification languages in recent years. It was initiated by Jean-Raymond Abrial and subsequently developed by a team at Oxford University.

Section 8.4.2 gives a brief overview of the Z language elements. Section 8.4.3 introduces the notions of types and declarations. Z is a strongly typed language. Section 8.4.4 uses the basic notions of logic and sets to model the requirements of a system. The main Z construct is the schema. A schema is similar to a subprogram, and its primary purpose is to draw to the attention of the reader important aspects of the software system being modeled. The schema also contains statements about the relationships that exist among the various things that have been identified. The remaining three subsections focus on relations, functions, and sequences, respectively. These data structures are increasingly more powerful and less abstract as vehicles for modeling system requirements.

8.4.2 Alphabet and Lexical Elements

The Z notation is used to construct specification documents that contain natural language text interspaced with sections of Z code. The alphabet of symbols used in Z is made up of characters that are available on most computer keyboards and other characters that are not. Several symbols are mathematical symbols that have already been discussed in the text, while others are new and are only used in Z programs. A Z program has two forms: textual and graphical. Both forms are discussed in this chapter, but the graphical form is emphasized.

Usually, Z programs cannot be generated directly from conventional typewriting equipment. Consequently, several tools have been constructed to allow the typesetting of Z text. In this book, Latex (equipped with a specialized style) has been used for this purpose.

In addition to the set of mathematical and other special-purpose symbols used in Z, other lexical elements in the language include the set of integers, some punctuation symbols such as the colon(:), comma (,), and semicolon (;), and names or programming identifiers. Identifiers are constructed from letters (in either lower or uppercase), digits, and the break (_) character. Upper- and lowercase letters are considered to be distinct symbols. Identifiers can be of any length, but their first character must be a letter. The following are examples of distinct and valid identifiers:

- *AddStudent*
- *addstudent*
- *addStudent*
- *Add_Student*
- *add_student*

Identifiers can also begin with a special prefix symbol, either Δ or Ξ. Also, identifiers can end with *decorations*, the use of which will be discussed later. The most frequently used decorations are the question mark (?), the exclamation mark (!), and the prime ('). It should be noted that the Z language is still under development, and the notation used in this text may vary slightly from notations used in other texts.

8.4.3 Types and Declarations

Z is a strongly typed language and is similar in that respect to languages such as Pascal, Ada, and Miranda. As discussed in Section 5.2.5, types are sets. The use of types (sets) in a language offers some advantages, which include avoiding certain mathematical paradoxes and allowing type checks on Z programs to be done at compile time by the Z compiler. Z contains three kinds of types:

1. *Built in*: consisting of the set $\mathbb{Z}$ of integers.

2. *Basic* or *given*: depending on the sets used in the various applications being modeled.

3. *Free type*: enumerations, which are similar to those available in Pascal and Modula 2.

The integer set $\mathbb{Z}$ can be restricted to the natural numbers, represented by $\mathbb{N}$, and the positive integers, represented by $\mathbb{N}_1$. Real numbers and Z's character set are not considered to be built-in types. Real numbers can be considered a basic type. Z's character set could be represented as a free type.

The basic types in a specification are stated in a somewhat abstract manner. Such a statement is not concerned with the details of the elements of the basic types. For example, in a course registration system, a specification may refer to the set of all possible students, regardless of how student records might be represented in the software product that eventually would result. The set of students could be called *Student* and written in brackets ([and]) as

$[Student]$ the set of all uniquely identifiable students

As another example, consider the modeling of a library application. A set of basic types might include the following:

$[Book]$ the set of all possible books
$[Person]$ the set of all uniquely identifiable people
$[Author]$ the set of all uniquely identifiable authors
$[Subject]$ the set of all possible subject areas

These four types can be specified in the single equivalent form

$[Book, Person, Author, Subject]$

The following are examples of free types:

$$
\begin{aligned}
Reply \quad &::= \text{no} \mid \text{yes} \\
Transaction\_Type &::= \text{add} \mid \text{delete} \mid \text{change} \\
Student\_Type \quad &::= \text{undergraduate} \mid \text{graduate}
\end{aligned}
$$

As in other typed languages, all identifiers representing variables in a Z specification must be declared. For example,

$$patron : Person$$

declares the variable *patron* to be of type *Person*. The example

$$x, y, z : \mathbb{Z}$$

declares three integer variables. This form of declaration in Z is similar to that used in procedural languages such as Pascal.

This section has been a brief introduction to types and declarations. Their use in Z specifications is illustrated in the next section.

Problems 8.4.3

1. Given the declarations

$$i, j : \mathbb{N};\ b : Book;\ a : Author$$
$$on\_loan : \mathbb{P}\ Book;\ scientific\_authors : \mathbb{P}\ Author$$

where $\mathbb{P}$ denotes the powerset symbol, determine the type of each of the following:

(a) i (e) $\{on\_loan\} \cup \{b\}$

(b) $i * j$ (f) $i + j$

(c) $\{i, j\}$ (g) $\{scientific\_authors\} \cap \{a\}$

(d) $\{on\_loan\}$

2. Given the declarations of Problem 1, determine which of the following expressions are acceptable and why they are acceptable in a strong typing sense:

(a) $i = j$ (d) $\{a, b\}$

(b) $a \in scientific\_authors$ (e) $on\_loan \subseteq Book$

(c) $\{scientific\_authors \subset on\_loan\}$ (f) $i \in scientific\_authors$

3. Consider an application involving the recording of passengers on a bus. The seats on a bus are numbered from 1 to *bus_capacity*. A bus is driven by a licensed bus driver. Determine the types for this application.

4. Consider the file system within an operating system at a university computing center. The purpose of the file system is to maintain various files belonging to users of the computing center. There are three kinds of users: student, faculty, and system staff. System staff users have special powers, such as creating new accounts and deleting old ones. Determine the types for this system.

8.4.4 Specifying a System with Logic and Sets

A Z specification of a system can contain three kinds of entities: states, events, and observations. A *state* is given by the values associated with a mathematical structure that models a system. A structure, for example, can be a set, a relation, a function, or a sequence. The emphasis in this section is on sets. An *event* is an occurrence, for example, an operation, that is of interest to an analyst who is modeling a system. In a course registration system, adding a student to a class and deleting a student from that class are examples of events. An *observation* is associated with the examination of the value of some variable before or after an event has occurred. For example, in the specification of a stack structure, looking at its top element or inquiring as to whether the stack is empty are examples of observations.

In this section the mathematical notions of logic and sets introduced earlier in the text are used to model a system in Z. The set notation used in Z is given in Table 8.1. Most of the symbols have already been introduced. The table contains the notation for a set in the most general form of

$$\{\text{declaration} \mid \text{constraint} \bullet \text{expression}\}$$

In practice, some parts of this general form may be omitted. For example, the set

$$\{n : \mathbb{N} \mid n > 9 \bullet (n * (n * n))\}$$

specifies an infinite set of cubes of all natural numbers greater than or equal to 10. The notation m. .n denotes the set of integers between m and n inclusive.

To illustrate the Z notation, consider a simple application involving keeping track of students enrolled in a class. A class has a maximum enrollment limit. A student can enroll in a class if the class is not full. Also, a student can drop a class. For this application, there is one basic type:

$$[Student] \text{ the set of all uniquely identifiable students}$$

Also, the maximum class size is given by the following declaration:

$$class\_limit : \mathbb{N}$$

The state of the system is the set of students enrolled in the class at any given time. Let *enrolled* denote the set of students currently taking a class.

$$enrolled : \mathbb{P} \ Student$$

One kind of property that a Z specification contains is a collection of predicates that always holds regardless of the events that may occur. These are called *invariants*. In the current example the invariant is that the number of students enrolled in a class must never exceed the class limit; therefore,

$$\#enrolled \leq class\_limit$$

No operation can violate this invariant property. Identifying the invariant properties that exist among the components of a system is a key step in creating a formal specification of that system.

The initial state of the system is specified by

$$enrolled = \varnothing$$

TABLE 8.1.   Summary of Z Notation for Sets and Logic

Notation	Meaning
$\mathbb{Z}$	the set of integers
$\mathbb{N}$	the set of natural numbers
$\mathbb{N}_1$	the set of positive integers
$e \in S$	e is an element of S
$e \notin S$	e is not an element of S
$A \subseteq B$	A is a subset of B
$A \subset B$	A is a proper subset of B
$\{\}, \varnothing$	the empty set
$\{e_1, e_2, \ldots, e_m\}$	the set containing m elements $e_1, e_2, \ldots, e_m$
$A \cup B$	the union of sets A and B
$A \cap B$	the intersection of sets A and B
$A \setminus B$	the elements of A that are not in B
$=, \neq, <, \leq, >, \geq$	the relational operators
$\mathbb{P}S$	the powerset of S
$\#S$	the cardinality or number of elements in S
$m \mathinner{..} n$	the set of integers between m and n inclusive
$P \vee Q$	the logical ORing of P and Q
$P \wedge Q$	the logical ANDing of P and Q
$\neg P$	the negation of P
$\{D \mid P \bullet x\}$	the set of x from declaration D such that predicate P holds
$P \Rightarrow Q$	P implies Q
$P \Leftrightarrow Q$	P is equivalent to Q
$x_1 = x_2$	equal terms
$x_1 \neq x_2$	$\neg(x_1 = x_2)$
$x_1 \leftrightarrow x_2$	relation, in which x_1 and x_2 are sets
$x_1 \rightarrow x_2$	total function, in which x_1 and x_2 are sets
$x_1 \nrightarrow x_2$	partial function, in which x_1 and x_2 are sets

Observe that the initial state satisfies the invariant property since

$$\#enrolled = 0 \leq class\_limit$$

The events or operations in this example include allowing a student to pick up a class. This event changes the value of *enrolled*. The value of *enrolled* after an operation is denoted by *enrolled'*, which is pronounced "enrolled prime." Performing an operation must not result in the violation of an invariant property. To prevent such a violation requires that certain preconditions hold before a student is allowed to pick up a class. In this case

$$\#enrolled < class\_limit$$

would be such a precondition. Also, the student who wishes to enroll in a class must not already be enrolled in that class. The following Z fragment summarizes the specification of picking up a class:

$$\left. \begin{array}{l} s? : Student \\ enrolled' : \mathbb{P}\ Student \end{array} \right\} \text{declarations}$$

$$\left. \begin{array}{l} s? \notin enrolled \\ \#enrolled < class\_limit \end{array} \right\} \text{preconditions}$$

$$enrolled' = enrolled \cup \{s?\} \ \} \text{ update of state}$$

In this Z fragment the ? symbol after s indicates that s is an input variable. Provided that the preconditions hold, the new state is the set of students in the class after a student has picked up the class. The specification of the behavior of the system when the preconditions are not satisfied is left until later.

It is also necessary to have an operation to allow a student to drop a class. The effect on *enrolled* is

$$enrolled' = enrolled \setminus \{s?\}$$

The precondition for this operation is that the student $s?$ must be in the class; that is,

$$s? \in enrolled$$

Observations in the form of inquiry operations also need modeling. For example, the following Z fragment models an inquiry for the number of students in the class:

$$number\_in\_class! : \mathbb{N}$$
$$number\_in\_class! = \#enrolled$$
$$enrolled' = enrolled$$

The ! symbol at the end of the identifier denotes that it is an output variable. Note that it is required to state explicitly that *enrolled* does not change after the inquiry operation. This is indicated by having the next state equal to the previous state.

Another inquiry operation is to determine whether a given student is in the class. This operation is specified as follows:

$$Reply ::= \text{no}|\ \text{yes}$$
$$response! : Reply$$
$$(s? \in enrolled \land response! = \text{yes})$$
$$\lor$$
$$(s? \notin enrolled \land response! = \text{no})$$
$$enrolled' = enrolled$$

A Z document usually consists of the following sections:

- An introduction describing the problem to be modeled in narrative form
- The types used in the specification (global declarations)
- An initial state

- Operations and inquiries
- Error handling
- Final versions of operations and inquiries including error handling

Discussion of the error handling parts of a Z document has been delayed until the next section.

The Z specification for the trivial class registration system just discussed is simple, short, and straightforward. For more complicated applications, however, it is necessary to have language features that permit the handling of complexity (i.e., "divide and conquer"). The next section introduces such a language feature.

Problems 8.4.4

1. Using predicates, precisely specify the following sets:
   (a) The set of natural numbers that are greater than 10 and less than 50
   (b) The set of natural numbers whose cubes are greater than 20
   (c) The set of pairs of natural numbers that have a sum of 200
   (d) The set of natural numbers that are odd numbers in the range 1 to 100
   (e) The set of pairs of natural numbers for which the first element is greater than 20 and the second element is the cube of the first element

2. This problem concerns keeping track of passengers that are on a bus. The bus has a fixed number of seats and each passenger on the bus must have a seat. For simplicity, it is assumed that the seats are not numbered and passengers are seated in the order of their arrivals (i.e., first-come, first-serve). For this system:
   (a) State its invariant properties.
   (b) Define an initial state.
   (c) Define an operation for a passenger to board the bus.
   (d) Define an operation for a passenger to leave the bus.
   (e) Formulate an inquiry operation for determining the number of passengers on the bus.
   (f) Formulate an inquiry operation to determine whether a person is on the bus.

3. A computer system maintains computer accounts of its users. There are three kinds of users: student, faculty, and system. System users have special powers, such as creating new accounts and deleting old accounts. For this system, it is assumed that each user has one account and only one user can log on at one time. For this system:
   (a) Determine its types.
   (b) State its invariant properties.
   (c) Define an initial state.
   (d) Define operations for a user to log on and log off the system.
   (e) Define an operation to add a new user.
   (f) Define an inquiry operation to determine whether a person is a user.
   (g) Define an operation that determines how many users have computer accounts.

8.4.5 Schemas

Schemas permit the decomposition of specifications (at a higher level of abstraction) into more detailed (and less abstract) forms, much like procedures and functions do in conventional computer language such as FORTRAN and Pascal. This section focuses on the Z schema. Another reason for using schemas is to clearly separate the natural language passages in a Z document from its formal components written in Z.

A schema is a unit of mathematical text that describes some part of a system and has the following general two-dimensional graphical form:

```
┌─ Schema_Name ────────
│  declaration part or signature
├───────────────────
│  predicate part
└───────────────────
```

A schema contains a schema name, a declaration part (or signature), and a predicate part. The schema name identifies the schema. The declaration part specifies the local declarations to the schema. These declarations can refer to other schemas as a way of importing observations into the schema being defined. The predicate part specifies requirements about the values of the variables. The predicate part can be omitted. In this case the schema's function is to modularize declarations that are used in other schemas. It is also possible to omit the schema name. Such a schema is called an *anonymous schema*. This type of schema permits global declarations that can be used in other schemas without having to formally state their importation.

Another form of a schema is its textual form:

$$Schema\_Name \; \hat{=} \; [\text{declaration part} \mid \text{predicate part}]$$

The symbol $\hat{=}$ means "is defined as" and the symbols [and] are schema brackets. It is sometimes convenient to write a schema in its textual form if the schema contains short declaration and predicate parts.

An example of global declarations for the class recording system discussed in the previous section is specified in the following syntax definition for *Reply* and anonymous schema:

$$Reply ::= \text{no} \mid \text{yes}$$

```
│  class_limit : ℕ
├──────────────
```

These declarations can be used in any other schemas. A schema for the class recording system follows:

```
┌─ Class ─────────────
│  enrolled : ℙ Student
├──────────────────
│  #enrolled ≤ class_limit
└──────────────────
```

Each line in the declaration part is assumed to be terminated by a semicolon. Also, all lines in the predicate part are assumed to be logically anded together.

Schemas can be combined or manipulated by using several kinds of operators. Specifically, the logical connectives $\wedge$, $\vee$, $\Rightarrow$, and $\Leftrightarrow$ can be used to combine schemas. Other operations will be illustrated later in this section.

A schema for the initial state is

$$
\begin{array}{|l}
\hline
\;\textit{Initial} \underline{} \\
\;\textit{Class} \\
\hline
\;\textit{enrolled} = \varnothing \\
\hline
\end{array}
$$

Note that the declaration part refers to the schema *Class* and therefore *enrolled* becomes available to the schema *Initial*.

As described in the previous section, the specification of a state transformation due to some operation requires the representation of the before and after states. Recall that in Z the after state is represented by decorating a variable with a prime. *Class* in the current example is the name of the schema that represents a before state. The after state representation of *Class*, denoted as $Class'$, is

$$
\begin{array}{|l}
\hline
\;\textit{Class}' \underline{} \\
\;\textit{enrolled}' : \mathbb{P} \; \textit{Student} \\
\hline
\;\#\textit{enrolled}' \le \textit{class\_limit} \\
\hline
\end{array}
$$

Usually a Δ (delta) schema is obtained by combining the before and after specification of a state. For example, a delta schema for the current example is

$$
\begin{array}{|l}
\hline
\;\Delta\textit{Class} \underline{} \\
\;\textit{Class} \\
\;\textit{Class}' \\
\hline
\end{array}
$$

or, alternatively, using the delta notation,

$$\Delta Class \mathrel{\hat=} Class \wedge Class'$$

which can be exploded to the following graphical form:

$$
\begin{array}{|l}
\hline
\;\Delta\textit{Class} \underline{} \\
\;\textit{enrolled} : \mathbb{P} \; \textit{Student} \\
\;\textit{enrolled}' : \mathbb{P} \; \textit{Student} \\
\hline
\;\#\textit{enrolled} \le \textit{class\_limit} \\
\;\#\textit{enrolled}' \le \textit{class\_limit} \\
\hline
\end{array}
$$

This form is called the standard form of using a Δ schema.

Using this notation, the solution of the class registration example can be further modularized. The operation of adding a student to a class is specified by the following schema:

$$
\begin{array}{|l}
\hline Add\_Class_o \\
\hline \Delta Class \\
s? : Student \\
\hline s? \notin enrolled \\
\#enrolled < class\_limit \\
enrolled' = enrolled \cup \{s?\} \\
\hline
\end{array}
$$

Note that the schema name has a subscript zero as its last symbol.

In a similar fashion, the following schema specifies the operation of a student dropping a class:

$$
\begin{array}{|l}
\hline Drop\_Class_o \\
\hline \Delta Class \\
s? : Student \\
\hline s? \in enrolled \\
enrolled' = enrolled \setminus \{s?\} \\
\hline
\end{array}
$$

Another type of schema is the Xi (Ξ) schema. This schema is used in the specification of operations (inquiries) that do not change the state of a structure. In the current example the $\Xi Class$ schema is defined as

$$
\begin{array}{|l}
\hline \Xi Class \\
\hline enrolled : \mathbb{P}\ Student \\
enrolled' : \mathbb{P}\ Student \\
\hline \#enrolled \leq class\_limit \\
\#enrolled' \leq class\_limit \\
enrolled' = enrolled \\
\hline
\end{array}
$$

Using the Ξ notation, schemas for the inquiry operations of determining the size of a class and whether a given student is enrolled in a class are as follows:

$$
\begin{array}{|l}
\hline Class\_Size \\
\hline \Xi Class \\
number\_in\_class! : \mathbb{N} \\
\hline number\_in\_class! = \#enrolled \\
\hline
\end{array}
$$

```
┌─ Student_In_Class ─────────────────
│ ΞClass
│ response! : Reply
│ s? : Student
├────────────────────────────────────
│ (s? ∈ enrolled ∧ response! = yes)
│
│ ∨
│
│ (s? ∉ enrolled ∧ response! = no)
└────────────────────────────────────
```

The previous discussion has ignored what happens when the preconditions do not hold. Z schemas can be formulated to deal with such errors. An enumeration of possible error messages is contained in the following declarations:

$$Possible\_Messages ::= \text{ok} \mid \text{already\_in\_class} \mid \text{full} \mid \text{not\_in\_class} \mid \text{two\_errors}$$

An error message schema, which uses this declaration, to output an ok response is as follows:

```
┌─ Ok_Message ───────────────
│ output! : Possible_Messages
├────────────────────────────
│ output! = ok
└────────────────────────────
```

This schema outputs the message "ok".

A schema to signal possible errors when adding a student to a class follows:

```
┌─ Add_Class_Error ──────────────────────────────────
│ ΞClass
│ s? : Student
│ output! : Possible_Messages
├────────────────────────────────────────────────────
│ (s? ∉ enrolled ∧ #enrolled = class_limit ∧ output! = full)
│ ∨ (s? ∈ enrolled ∧ output! = already_in_class)
│ ∨ (s? ∈ enrolled ∧ #enrolled = class_limit ∧ output! =
│        two_errors)
└────────────────────────────────────────────────────
```

Note that if an attempt is made to add a student, who is already in a class, to that class which is already full, this results in the generation of the response "two_errors".

The complete schema, which includes error handling, is

$$Add\_Class \cong (Add\_Class_o \wedge Ok\_Message) \vee Add\_Class\_Error$$

Handling errors when a student drops a class results in the following schema:

$$
\begin{array}{|l}
\hline
\;Drop\_Class\_Error \underline{\hspace{3cm}} \\
\;\Xi Class \\
\;s? : Student \\
\;output! : Possible\_Messages \\
\hline
\;(s? \notin enrolled \wedge output! = \text{not\_in\_class}) \\
\hline
\end{array}
$$

This schema can be combined to yield another that handles errors:

$$Drop\_Class \mathrel{\widehat{=}} (Drop\_Class_o \wedge Ok\_Message) \vee Drop\_Class\_Error$$

This section has introduced the schema as a vehicle for handling complexity; however, modeling a system usually requires more complex structures than just sets. The next section introduces more powerful structures for this purpose.

Problems 8.4.5

1. Formulate a schema called $Square\_Root\_Spec$ to output the square root, $root!$, of an input variable, $x?$, where $x?$ is a nonnegative real number.

2. Consider the following schemas:

$$
\begin{array}{|l}
\hline
\;List \underline{\hspace{1.5cm}} \\
\;LA : \mathbb{P}\ Name \\
\;max : \mathbb{N}_1 \\
\hline
\;\#LA \leq max \\
\hline
\end{array}
$$

and

$$
\begin{array}{|l}
\hline
\;Calc \underline{\hspace{1.5cm}} \\
\;m? : \mathbb{N} \\
\;n! : \mathbb{N} \\
\hline
\;n! = 2 * m? \\
\hline
\end{array}
$$

   (a) List all variables in the schema $\Delta List$.
   (b) List all predicates in the schema $List \wedge Calc$.
   (c) What is the input of schema $Calc$, and what is its output?

3. Use schemas to solve Problem 2 in Problems 8.4.4 including the handling of errors.

4. Use schemas to solve Problem 3 in Problems 8.4.4 including the handling of errors.

5. This problem deals with the management of bank accounts. Assume that each customer bank account is identified by a unique account number. Using schemas, prepare a Z document that includes types used, global declarations, initial state, operations (such as add a new account and delete an account), inquiries, error handling, and the final versions of operations and inquiries including error handling.

8.4.6 Relations

The set structures that have been encountered so far are essentially one dimensional, since they are limited to one basic type. In more complex modeling situations, there are needs to relate sets to each other. As discussed in Section 5.3.2, a relation that is based on the Cartesian product is such a structure.

A relation type defines all possible relations drawn from X and Y and is written as $X \leftrightarrow Y$. $X \leftrightarrow Y$ is the same as $\mathbb{P}(X \times Y)$.

Relations are illustrated in this section by specifying a simple university parking lot system. Assume that such a system involves the assignment of parking stalls to university faculty. Some faculty members can be assigned more than one parking stall. It is also possible for more than one faculty to share one parking stall. The basic types in this application are

$$[Person,\ Parking\_Stall]$$

where

$$[Person] \text{ is the set of uniquely identifiable individuals}$$

and

$$[Parking\_Stall] \text{ is the set of uniquely identifiable parking stalls}$$

The many–many relationship between the sets $Parking\_Stall$ and $Person$ is specified by

$$Parking\_Stall \leftrightarrow Person$$

A defining schema for this system is

$$
\begin{array}{|l}
\ Parking \underline{\hspace{4cm}} \\
\ faculty : \mathbb{P}\ Person \\
\ assigned\_to : Parking\_Stall \leftrightarrow Person \\
\ available, occupied : \mathbb{P}\ Parking\_Stall \\
\hline
\ \ \text{ran } assigned\_to \subseteq faculty \\
\ available \cap occupied = \varnothing \\
\end{array}
$$

In this schema $faculty$ is a subset of people. The identifier $assigned\_to$ is the name of the relation that relates parking stalls to faculty members. There are two invariants in the schema. The first states that the range of the relation $assigned\_to$ must be a subset of $faculty$. Also, the intersection of the stalls that have not been assigned ($available$) and those that have ($occupied$) must always be empty.

The initial state for the system is given by the following schema:

$$
\begin{array}{|l}
\ Initial \underline{\hspace{2cm}} \\
\ Parking \\
\hline
\ faculty = \varnothing \\
\ assigned\_to = \varnothing \\
\ available = \varnothing \\
\ occupied = \varnothing \\
\end{array}
$$

Observe that the class invariants are satisfied by the schema.

The following schema specifies the operation of assigning a parking stall to a faculty member:

$$
\begin{array}{|l}
\underline{\quad Assign\_A\_Stall_o \rule{5cm}{0pt}} \\
\Delta Parking \\
name? : Person \\
s? : Parking\_Stall \\
\hline
name? \in faculty \\[4pt]
s? \in available \\[4pt]
s? \mapsto name? \notin assigned\_to \\[4pt]
available' = available \setminus \{s?\} \\[4pt]
occupied' = occupied \cup \{s?\} \\[4pt]
faculty' = faculty \\[4pt]
assigned\_to' = assigned\_to \cup \{s? \mapsto name?\}
\end{array}
$$

In Z, $s? \mapsto name?$ is called a *maplet* and stands for the ordered pair whose first element is $s?$ and second element is $name?$. The preconditions to the operation require that the person being assigned a stall be a faculty member who does not already own that stall. Also, the stall denoted by $s?$ should be available. The specified faculty member could already have another stall.

An error condition can arise in assigning a stall if the person being given the stall is not a faculty member. The following schema handles this kind of error:

$$
\begin{array}{|l}
\underline{\quad Not\_Faculty \rule{3cm}{0pt}} \\
\Xi Parking \\
name? : Person \\
message! : Possible\_Messages \\
\hline
name? \notin faculty \\[4pt]
message! = \text{not\_on\_faculty}
\end{array}
$$

where the enumeration $Possible\_Messages$ is defined as

$$
\begin{aligned}
Possible\_Messages ::= \ & \text{okay} \mid \text{not\_on\_faculty} \mid \\
& \text{stall\_already\_assigned} \mid \\
& \text{unknown\_faculty\_name} \mid \\
& \text{unknown\_parking\_stall}
\end{aligned}
$$

Another possible error arises when the given parking stall has already been assigned to the specified faculty member. The following schema details this kind of error:

```
┌─ Stall_Already_Assigned ────────
│ ΞParking
│ name? : Person
│ s? : Parking_Stall
│ message! : Possible_Messages
├───────────────────────────────
│ s? ↦ name? ∈ assigned_to
│ message! = stall_already_assigned
└───────────────────────────────
```

Usually, if no errors have occurred, then the system should report that the operation has been completed successfully. This is handled by the following schema:

```
┌─ OK ──────────────────────
│ message! : Possible_Messages
├──────────────────────────
│ message! = okay
└──────────────────────────
```

Therefore, the schema for assigning a stall to a person with error handling is

$$Assign\_A\_Stall \mathrel{\widehat{=}} (Assign\_A\_Stall_0 \wedge OK) \vee$$
$$Not\_Faculty \vee Stall\_Already\_Assigned$$

A schema for canceling a parking stall that is not shared is as follows:

```
┌─ Free_A_Stall_0 ──────────────────────────────
│ ΔParking
│ name? : Person
│ s? : Parking_Stall
├───────────────────────────────────────────
│ name? ∈ faculty
│ s? ∈ occupied
│ s? ↦ name? ∈ assigned_to ∧ #({s} ◁ assigned_to) = 1
│ available' = available ∪ {s?}
│ occupied' = occupied \ {s?}
│ faculty' = faculty
│ assigned_to' = assigned_to \ {s? ↦ name?}
└───────────────────────────────────────────
```

where the operator $\triangleleft$ restricts the domain of the relation $assigned\_to$ to $s?$. Again there are error handling situations that can arise, but their details are not given here.

There are several inquiry operations that can be specified. The operation $Find\_Stalls$ has a person's name as input and the set of stall numbers assigned to this person as output. The schema for this operation is

$$\begin{array}{|l}
\_Find\_Stalls_0 \underline{\hspace{4cm}} \\
\Xi Parking \\
name? : Person \\
numbers! : \mathbb{P}\ Parking\_Stall \\
\hline
name? \in \mathrm{ran}\ assigned\_to \\
numbers! = assigned\_to^{-1}\ (\!|\{name?\}|\!) \\
\end{array}$$

The precondition for this schema is that the given name belong to the range of the relation *assigned_to*. The relational image of the set *numbers!* in the relation *assigned_to* is denoted by $assigned\_to\ (\!|numbers!|\!)$, where the symbols $(\!|$ and $|\!)$ are the relational image brackets. In this example, the relational image is the set of faculty members who currently have parking stalls. The stalls assigned to a particular faculty member are given by the inverse of the relation. Recall that the inverse of the relation *assigned_to* is denoted by $assigned\_to^{-1}$. In Z both $assigned\_to^{-1}$ and $assigned\_t\tilde{o}$ (refer to Section 5.3.4) are admissible.

An error situation occurs when the given identifier *name?* does not have an assigned parking stall. The following schema handles this case:

$$\begin{array}{|l}
\_Unknown\_Name \underline{\hspace{3cm}} \\
\Xi Parking \\
name? : Person \\
message! : Possible\_Messages \\
\hline
name? \notin \mathrm{ran}\ assigned\_to \\
message! = \text{unknown\_faculty\_name} \\
\end{array}$$

The complete schema for *Find_Stalls* is

$$Find\_Stalls \cong (Find\_Stalls_o \wedge OK) \vee Unknown\_Name$$

Another inquiry operation involves finding the names of faculty members that share a particular parking stall.

$$\begin{array}{|l}
\_Find\_Faculty\_Names_o \underline{\hspace{3cm}} \\
\Xi Parking \\
names! : \mathbb{P}\ Person \\
number? : Parking\_Stall \\
\hline
number? \in \mathrm{dom}\ assigned\_to \\
names! = assigned\_to\ (\!|\{number?\}|\!) \\
\end{array}$$

The desired set of faculty names is obtained by using the relational image of *assigned_to*.

To specify the total operation, an error schema dealing with an unknown or invalid stall number is required. Such a schema is the following:

$\underline{\quad Unknown\_Stall\_Number \quad\rule{1.5cm}{0.4pt}}$
$\Xi Parking$
$number? : Parking\_Stall$
$message! : Possible\_Messages$

$\rule{6cm}{0.4pt}$

$number? \notin \text{dom } assigned\_to$

$message! = \text{unknown\_parking\_stall}$

The total schema for the inquiry is

$$Find\_Faculty\_Names \mathrel{\widehat{=}} (Find\_Faculty\_Names_o \wedge OK)$$
$$\vee\ Unknown\_Stall\_Number$$

In concluding this example specification, there are schemas required for creating a new stall and retiring an existing stall. Also, the university hires new faculty members and existing members leave the university.

Schemas for creating a new stall and a faculty member leaving the university follow:

$\underline{\quad Create\_New\_Stall_o \quad\rule{1.5cm}{0.4pt}}$
$\Delta Parking$
$s? : Parking\_Stall$

$\rule{6cm}{0.4pt}$

$available' = available \cup \{s?\}$

$\underline{\quad Faculty\_Leaving_o \quad\rule{1.5cm}{0.4pt}}$
$\Delta Parking$
$name? : Person$

$\rule{6cm}{0.4pt}$

$name? \in faculty$

$faculty' = faculty \setminus \{name?\}$

The operation for faculty leaving is

$$Faculty\_Leaving \mathrel{\widehat{=}} (Faculty\_Leaving_o \wedge OK) \vee Not\_Faculty$$

The remaining schema can be obtained in an analogous manner.

Relations are used frequently in database applications. The most popular database systems are based on relations. This important area of computer science is dealt with in Chapter 12.

Problems 8.4.6

1. For the university parking lot system, formulate an error schema for the case when an attempt is made to assign an invalid (nonexistent) stall to a faculty member.

2. For the university parking lot system, formulate schemas for the retirement of an existing parking stall and the hiring of a new faculty member.

3. This problem concerns the modeling of a university committee application. Faculty members may sit on university committees dealing with budget, academic affairs, research, and other issues. The basic types in this application are:

> $[Faculty]$ the set of all uniquely identifiable faculty members
> $[Committee]$ the set of all uniquely identifiable committees

The relation $member\_of$ specifies which faculty members sit on these committees.

Using schemas, prepare a Z document that includes types used, global declarations, initial state, operations (such as adding another faculty member to a given committee, deleting a certain faculty member from a designated committee, and creating a new committee), inquiries (such as listing all committees that a particular faculty member sits on and listing all faculty members who sit on a specific committee), and error handling.

4. Consider a simplified Medicare application involving the treatment of patients by doctors. The basic types are:

> $[Person]$ the set of all uniquely identifiable people in the Medicare system

The relation *treats* specifies the association between doctors and their patients; that is, a particular doctor treats a specific patient. A doctor may treat several patients and a patient may be treated by more than one doctor.

Using schemas, prepare a Z document that includes types used, global declarations, initial state, operations (such as a new treatment of a particular patient by a doctor, adding a new doctor to the system, and deleting a patient from the system), inquiries (such as generating all patients treated by a specific doctor and generating all doctors who have treated a specific patient), and error handling.

5. Consider a bibliographical application that has the following basic types:

> $[Author]$ the set of uniquely identifiable authors
> $[Title]$ the set of all uniquely identifiable references
> $[Publisher]$ the set of all uniquely identifiable publishers

There are two relations in this application. The first relation, *writes*, associates authors and titles. The second relation, *published_by*, associates titles and publishers. An author may have written several titles and several authors may co-author a specific title. Also, several titles may be published by the same publisher.

Using schemas, prepare a Z document that includes types used, global declarations, initial state, operations (such as adding a new entry to each of the *writes* and *published_by* relations), simple queries (such as listing titles written by a specific author, listing all co-authors of a given author, and generating all titles published by a certain publisher), more complex queries (such as finding all publishers associated with a given author and listing all authors associated with a given publisher), and error handling. Note that the more complex queries require the composition of relations. The composition of two relations, such as *writes* and *published_by*, is represented in Z as $writes \, \S \, published\_by$.

8.4.7 Functions

Recall from Section 6.1.2 that a partial function is a special kind of relation in which each element in its domain can be associated with *at most one element* in its range.

A Z function that maps, for example, a Library of Congress number to a book is represented as

$$assigned\_to : ISBN \nrightarrow Book$$

The domain of the function is a set of uniquely identifiable Library of Congress numbers.

Recall from Section 6.1.2 that a partial function f in Z from X to Y is represented as

$$f : X \nrightarrow Y$$

This means that there are possibly some values in X for which the function is not defined. Given a value $x \in X$, the value of the function is represented as fx. Note that in several programming languages fx is represented as $f(x)$. This is also represented as the maplet $x \mapsto y$.

A total function is represented as

$$f : X \rightarrow Y$$

In a total function, f is defined for every value of $x \in X$. Although in Z there are *different* arrow symbols for representing "onto," "one–one," "one–one onto," and total functions, in this section only the arrow symbol ($\nrightarrow$) is used. These different arrow symbols are very similar and therefore confusing. Any special properties of functions (e.g., one–one onto) will be explicitly stated.

The use of a function in Z is introduced by considering a simple example. A manufacturer wants to monitor its short-term debts incurred from purchases from its suppliers. The basic types in this system are

$$[Firm] \text{ set of uniquely identifiable firms}$$
$$[Dollars] \text{ set of real numbers}$$

The following schema introduces the main structures:

$$
\begin{array}{|l}
\hline
\;Accounts\_Payable \underline{\hspace{3cm}} \\
\;\; suppliers : \mathbb{P} \; Firm \\
\;\; amounts\_due : Firm \nrightarrow Dollars \\
\hline
\;\; suppliers \subseteq \text{dom } amounts\_due \\
\hline
\end{array}
$$

Given a supplier's unique identification, the function *amounts_due* generates the amount owed to the supplier.

The schema that describes the initial state of the system is

$$
\begin{array}{|l}
\hline
\_\,Initial _____ \\
\quad Accounts\_Payable \\
\hline
suppliers = \varnothing \\
amounts\_due = \varnothing \\
\hline
\end{array}
$$

Consider the operation of updating a supplier's account by a purchase transaction. The specification for this operation is the following:

$$
\begin{array}{|l}
\hline
\_\,Purchase\_Transaction _____ \\
\Delta\,Accounts\_Payable \\
supplier\_id? : Firm \\
purchase\_amount? : Dollars \\
\hline
supplier\_id? \in suppliers \\
amounts\_due' = amounts\_due \\
\qquad \oplus \{supplier\_id? \mapsto ((amounts\_due\ supplier\_id?) \\
\qquad +purchase\_amount?)\} \\
\hline
\end{array}
$$

The only precondition in the schema is that the transaction must involve an existing supplier. The update of the supplier's account involves the notion of functional overriding. In the purchase operation, the amount owed to the supplier is to be updated. The overriding operator $\oplus$ in this case specifies that all ordered pairs in the function *amounts_due* be retained, except for the ordered pair that must be updated. The replacement maplet appears to the right of $\oplus$. This ordered pair becomes part of the new function representing the new state *amounts_due'*. That is, the amount owed to the supplier *supplier_id?* is updated by the current purchase amount. The term *amounts_due supplier_id?* gives the old amount owed to *supplier_id?*. Observe that errors have not been dealt with in the schema *Purchase_Transaction*.

Another operation involves making a payment to a supplier. The following schema specifies this operation:

$$
\begin{array}{|l}
\hline
\_\,Payment\_Transaction _____ \\
\Delta\,Accounts\_Payable \\
supplier\_id? : Firm \\
payment? : Dollars \\
\hline
supplier\_id? \in suppliers \\
amounts\_due' = amounts\_due \\
\qquad \oplus \{supplier\_id? \mapsto ((amounts\_due\ supplier\_id?) \\
\qquad -payment?)\} \\
\hline
\end{array}
$$

Again, error conditions have not been addressed here.

A schema for adding a supplier to the system follows:

$$
\begin{array}{l}
\rule{6cm}{0.4pt}\; Add\_Supplier \rule{2cm}{0.4pt} \\
\Delta\ Accounts\_Payable \\
supplier\_id?: Firm \\
initial\_purchase?: Dollars \\
\rule{8cm}{0.4pt} \\
supplier\_id? \notin suppliers \\
amounts\_due' = amounts\_due \\
\qquad\qquad \cup \{supplier\_id? \mapsto initial\_purchase?\}
\end{array}
$$

If the supplier is new, then a new maplet is added to the function *amounts_due*. Again, no error handling is addressed in the schema.

A number of inquiry operations in this application arise. As an example, consider an inquiry as to how much is owed to a specified supplier. The following schema describes this inquiry in Z:

$$
\begin{array}{l}
\rule{4cm}{0.4pt}\; Find\_Amount \rule{3cm}{0.4pt} \\
\Xi\ Accounts\_Payable \\
supplier\_id?: Firm \\
amount\_owed!: Dollars \\
\rule{7cm}{0.4pt} \\
supplier\_id? \in suppliers \\
amount\_owed! = amounts\_due\ supplier\_id?
\end{array}
$$

Up to this point in the chapter, examples have involved the use of binary relations. This section concludes with an example that illustrates the use of a function that involves ternary relations. A function of two variables is a special kind of ternary relation. A real estate company manages the rental of suites to tenants for fixed periods of time, in most instances for one-year periods. However, lease periods can be for shorter or longer periods. At any given time a suite may only be leased to one person. A rental system is required to handle leases. Over time the company may take on new suites and give up other suites (i.e., certain suites may be taken out of circulation). An outline of a formal specification of this system is now given. The error conditions will not be addressed in the presentation. The given types of this system include

[*Person*] the set of all people

[*Suite*] the set of all rental units (apartments) currently managed by the company

[*Lease_Period*] the set of all time intervals for which an apartment can be leased

The state for the rental system is as follows:

```
┌─ Rental_System ──────────────────────────────
│ apartments : ℙ Suite
│ tenants : ℙ Person
│ leased : Suite ⇸ Lease_Period ⇸ Person
├──────────────────────────────────────────────
│ dom leased ⊆ apartments
│ (∀ ru : Suite | ru ∈ apartments •
│       ran (leased ru) ⊆ tenants)
└──────────────────────────────────────────────
```

The set *tenants* consists of individuals some of whom may not be assigned a suite. A schema for adding a tenant will be given shortly. The mapping *leased* relates the rental of a particular suite for a fixed period of time to a certain tenant. That is, $Suite \nrightarrow Lease\_Period \nrightarrow Person$ [or $Suite \nrightarrow (LeasePeriod \nrightarrow Person)$] denotes a function whose elements are triples. The second precondition states that leases are only offered to tenants that are on record. The term

$$\text{ran } (leased\ ru)$$

refers, by definition, to the tenant who has leased a specific suite *ru*. Note that *leased ru* is a function.

The initial state for the system is

```
┌─ Initial_Rental_System ──────
│ Rental_System
├──────────────────────────────
│ apartments = ∅
│ tenants = ∅
│ leased = ∅
└──────────────────────────────
```

This schema asserts that there are no apartments being managed, no tenants, and no suites that are leased.

Several operations are possible in the rental system. An important operation involves the rental of a suite for some period of time to a particular tenant. The following schema is a specification of that operation:

```
┌─ Rent_A_Suite ────────────────────────
│ ΔRental_System
│ s? : Suite
│ p? : Lease_Period
│ t? : Person
├────────────────────────────────────────
│ s? ∈ apartments
│ t? ∈ tenants
│ p? ∉ dom (leased s?)
│ leased' = leased ⊕ {s? ↦ {p? ↦ t?}}
│ apartments' = apartments
│ tenants' = tenants
└────────────────────────────────────────
```

The first precondition states that the suite must be one managed by the company. The second precondition requires that a tenant be known. The final precondition asserts that the given suite must not already be leased for the desired period. If all three preconditions hold, then the suite is leased. The new state for *lease* is updated using a function overriding. The maplet $s? \mapsto \{p? \mapsto t?\}$ denotes a triple in which the suite $s?$ with a lease period $p?$ is assigned to a tenant $t?$. This notation is similar to that introduced earlier in Section 5.1.2. There is no change in the state of *apartments* and *tenants*. As stated earlier, error conditions will not be addressed in this section.

In a similar fashion the following schema frees a suite so that it becomes available for leasing:

$$
\begin{array}{|l}
\hline
\_Free\_A\_Suite \underline{\hspace{3cm}} \\
\Delta Rental\_System \\
s? : Suite \\
p? : Lease\_Period \\
t? : Person \\
\hline
s? \in apartments \\
t? \in tenants \\
p? \in \mathrm{dom}\,(leased\ s?) \\
leased' = leased \setminus \{s? \mapsto \{p? \mapsto t?\}\} \\
apartments' = apartments \\
tenants' = tenants \\
\hline
\end{array}
$$

Another operation involves the addition of a new tenant. The following schema specifies this operation:

$$
\begin{array}{|l}
\hline
\_Add\_Tenant \underline{\hspace{2cm}} \\
\Delta Rental\_System \\
h? : Person \\
\hline
h? \notin tenants \\
tenants' = tenants \cup \{h?\} \\
apartments' = apartments \\
leased' = leased \\
\hline
\end{array}
$$

The operation adds a new tenant to the list of people who are leasing from the company. Note that an apartment is not allocated to a tenant in this operation. This operation is only performed if the given person is not already a tenant.

From time to time a new suite is added to the set of suites already being managed by the company. A schema for this operation is

```
 ___Add_New_Suite_____
| ΔRental_System
| s? : Suite
|_____
| s? ∉ apartments
| apartments' = apartments ∪ {s?}
| tenants' = tenants
| leased' = leased
|_____
```

Provided that the indicated suite is not being currently managed by the company, the given suite is added to the set of managed apartments.

As a final operation, consider a query to determine which suites are available for lease during a given time period. A schema for such a query is

```
 ___Available_Suites_____
| ΞRental_System
| p? : Lease_Period
| available! : ℙ Suite
|_____
| available! = {s : Suite | s ∈ apartments ∧
|     p? ∉ dom (leased s) • s}
|_____
```

This operation examines the mapping *leased* to determine which suites that are being managed by the company are not rented during the indicated time period. More specifically, the term

$$p? \notin \text{dom} \ (leased \ s)$$

holds if suite s is not leased during period p?.

Problems 8.4.7

1. For the accounts payable system maintained by a manufacturer, formulate an error schema for each of the following schemas:
   - Payment_Transaction
   - Purchase_Transaction
   - Add_Supplier

2. For the rental system maintained by a real estate company, formulate error schemas to deal with various error conditions.

3. A vehicle wrecking company carries various used motor vehicle parts. A computer system is used to maintain a parts inventory. An inventory level, which indicates the quantity on hand, is maintained for each vehicle part in the inventory. This level is changed as parts are purchased from or replaced in the inventory. Over time, new parts (including their levels) are added to the inventory or parts are discontinued, provided that their inventory level is zero. Each vehicle

part is identified by a unique part number. A function maps the part number of a vehicle part into an integer that denotes the inventory level for that part.

Using schemas, prepare a Z document that includes types used, global declarations, initial state, operations (such as removing a specified quantity of a designated part from inventory, adding a specified quantity of a designated part already in inventory, introducing a new part, with an initial inventory level, to the inventory, and deleting a discontinued part), inquiries (such as obtaining the inventory level for a specific part), and error handling.

4. A dental office maintains an information system about dental patients. A patient number is used to uniquely identify a patient. A function maps a patient number into associated dental patient information. A second function maps a patient number into a patient's name and address.

Using schemas, formulate a Z document that includes types used, global declarations, initial state, operations (such as deleting a patient, adding a new patient, and adding the particulars of a new dental visit of a patient), inquiries (such as finding the name and address of a specified patient and displaying dental information kept on an individual statement), and error handling.

5. Formulate a Z document for a banking application where a bank customer has a checking account identified by a unique account number. Operations such as opening an account, closing an account, making a deposit, and cashing a check are to be handled. Also, an inquiry such as displaying the balance_on_hand in a particular checking account is to be dealt with. Finally, error handling is to be addressed.

8.4.8 Sequences

The modeling of certain applications requires that some orderings on data elements be specified. Frequently, there is a need for allowing the presence of duplicate elements. Also, there are situations where values in a set must be distinguished by position. A structure that satisfies these needs is a sequence. Recall that sequences have already been discussed with respect to recurrence relations (see Section 6.4) and were introduced in Section 5.2.3. This section, however, focuses on sequences from a requirements perspective. In particular, the support of sequences in Z is emphasized.

8.4.8.1. Basic Notions.   In Z, a sequence x of elements drawn from a set of type T is declared:

$$x : \text{ seq } T$$

For example, let the following set define the days of the week.

$$Days\_of\_Week = \{sun, mon, tue, wed, thu, fri, sat\}$$

The following are examples of sequences drawn from this set:

$$Week\_Day = \langle mon, tue, wed, thu, fri \rangle$$
$$Week\_End = \langle sun, sat \rangle$$
$$Days = \langle mon, mon, fri, sat, sat, fri, mon \rangle$$

In each sequence, the elements are listed in order and separated by commas. The elements are enclosed by the corner brackets, $\langle$ and $\rangle$.

Another example of a sequence declaration drawn from natural numbers is

$$x : \text{ seq } \mathbb{N}$$

The following is an example of a sequence:

$$x = \langle 2, 3, 7, 3, 2, 2, 7 \rangle$$

In Z, the declaration of a sequence by

$$x : \text{ seq } T$$

is equivalent to the following function:

$$x : \mathbb{N}_1 \twoheadrightarrow T$$

where

$$\text{dom } x = 1 \mathinner{\ldotp\ldotp} \#x$$

and $\#x$ denotes the length of the sequence x.   Using this approach, the sequence $\langle mon, tue, wed, thu, fri \rangle$ of five elements can be represented by the function

$$\{1 \mapsto mon, 2 \mapsto tue, 3 \mapsto wed, 4 \mapsto thu, 5 \mapsto fri\}$$

As another example, the sequence $\langle 2, 3, 7, 3, 2, 2, 7 \rangle$ of seven elements is equivalent to the mapping

$$\{1 \mapsto 2, 2 \mapsto 3, 3 \mapsto 7, 4 \mapsto 3, 5 \mapsto 2, 6 \mapsto 2, 7 \mapsto 7\}$$

The sequence $\langle \rangle$ denotes the null or empty sequence of length zero.

8.4.8.2. Operations on Sequences.   In this section, several operations on sequences are defined and illustrated.

The *selection* operation: The first operation is that of selecting an element in a sequence by its position. For example,

$$\langle mon, tue, wed, thu, fri \rangle\, 4 = thu$$

selects the fourth element in the sequence. In general, the position value specified must be between 1 and the length of the sequence.

The *concatenation* operation: Another important operation deals with the concatenation (catenation) of sequences. Given the sequences $\langle x_1, x_2, \ldots, x_n \rangle$ and $\langle y_1, y_2, \ldots, y_m \rangle$ whose lengths are n and m, respectively, their concatenation is represented as

$$\langle x_1, x_2, \ldots, x_n \rangle \frown \langle y_1, y_2, \ldots, y_m \rangle = \langle x_1, x_2, \ldots, x_n, y_1, y_2, \ldots, y_m \rangle$$

where $\frown$ denotes the concatenation operator. The length of the resulting sequence is $n + m$.

The *head* operation: The *head* of a given sequence is the first element of that sequence. For example,

$$\text{head } \langle mon, tue, wed, thu, fri \rangle = mon$$

The head function is undefined for the empty sequence.

The *last* operation: The *last* of a given sequence is the last element of that sequence. For example,

$$\text{last } \langle mon, tue, wed, thu, fri \rangle = fri$$

Again, the function is undefined for the empty sequence.

The *tail* operation: Given a nonempty sequence, the *tail* of a sequence is all the elements of that sequence except the first. For example,

$$\text{tail } \langle mon, tue, wed, thu, fri \rangle = \langle tue, wed, thu, fri \rangle$$

The *front* operation: Given a nonempty sequence, the *front* of a sequence is all the elements of that sequence except the last. For example,

$$\text{front } \langle mon, tue, wed, thu, fri \rangle = \langle mon, tue, wed, thu \rangle$$

The *filtering* ($\upharpoonright$) operation: Given a sequence $x = \langle x_1, x_2, \ldots, x_n \rangle$ and a filtering set $y = \{y_1, y_2, \ldots, y_m\}$, the filtering operation

$$\langle x_1, x_2, \ldots, x_n \rangle \upharpoonright \{y_1, y_2, \ldots, y_m\}$$

yields as a result the sequence of elements from x that are in the set y. For example,

$$\langle 2, 3, 7, 3, 2, 2, 7 \rangle \upharpoonright \{2, 3\}$$

yields the sequence $\langle 2, 3, 3, 2, 2 \rangle$.

The *rev* operation: Given a sequence $x = \langle x_1, x_2, \ldots, x_n \rangle$, the reverse of that sequence is a sequence containing the same elements of x, but in reverse order. For example,

$$\text{rev } \langle mon, tue, wed, thu, fri \rangle = \langle fri, thu, wed, tue, mon \rangle$$

Observe that a sequence has been defined as a function that is a special kind of relation. Therefore, the operators that have already been introduced for relations also hold for sequences. In particular, the *dom* and *ran* of a sequence are valid operators.

The *squash* operation: The restriction of a sequence usually does not yield a sequence because the resulting domain is not a set of contiguous natural numbers from 1. The *squash* operation transforms a relation into a sequence. For example,

$$\text{squash}(2 \ldots 3 \lhd \langle mon, tue, wed, thu, fri \rangle) = \langle tue, wed \rangle$$

That is, it yields the sequence of elements from position 2 to position 3 of the given sequence.

Other operations on sequences are available in Z, but they are not covered here.

8.4.8.3. Queue: An Example of a Sequence.

A simple queue structure is a well-known example of a linear data structure in which elements are inserted at one end (the rear) and deletions are performed at the opposite end (the front). In this section a specification of a simple queue is given. This data structure behaves in a first-in, first-out manner.

Assume that $[T]$ represents an arbitrary type. Schemas using this notation are called *generic schemas*. The state of the queue structure is given by

$$
\begin{array}{|l}
\hline
\_Queue[T]\_\!\!\_ \\
q:\ \mathrm{seq}\ T \\
\hline
\end{array}
$$

That is, a queue is represented as a sequence (possibly empty) of elements of type T.

The initial state of the queue structure is

$$
\begin{array}{|l}
\hline
\_Q\_Init\_\!\!\_ \\
Queue \\
\hline
q = \langle\rangle \\
\hline
\end{array}
$$

The queue is initialized to the empty sequence.

An insert operation places a new element at the rear of an existing structure. The length of the resulting sequence increases by 1 after an insert operation. The following schema specifies the insert operation:

$$
\begin{array}{|l}
\hline
\_Q\_Insert[T]\_\!\!\_ \\
\Delta Queue \\
new?:T \\
\hline
q' = q \frown \langle new?\rangle \\
\hline
\end{array}
$$

Similarly, a deletion operation removes the front (first) element from the given structure. The schema for this operation is

$$
\begin{array}{|l}
\hline
\_Q\_Delete[T]\_\!\!\_ \\
\Delta Queue \\
x!:T \\
\hline
q \neq \langle\rangle \\
x! =\ \mathrm{head}\ q \\
q' =\ \mathrm{tail}\ q \\
\hline
\end{array}
$$

The precondition for this operation is that the queue must be nonempty. The output variable $x!$ contains the front element of the queue after the operation is performed. The state of the structure is changed by removing the first element of the given structure. The statement sequence

$$
\begin{aligned}
x! &=\ \mathrm{head}\ q \\
q' &=\ \mathrm{tail}\ q
\end{aligned}
$$

can be replaced by the equivalent statement

$$q = \langle x! \rangle \frown q'$$

The length of the structure is realized in the following schema:

```
┌─ Q_Length[T] ──────
│ ΞQueue
│ size! : ℕ
├─────────────────────
│ size! = #q
└─────────────────────
```

The cardinality operator # also applies to sequences.

The following schema determines whether a given structure is empty:

```
┌─ Q_Empty ──────────────
│ ΞQueue
│ x! : OK
├─────────────────────────
│ ((q = ⟨⟩) ∧ (x! = yes))
│ ∨
│ ((q ≠ ⟨⟩) ∧ (x! = no))
└─────────────────────────
```

where

$$OK ::= \text{no} \mid \text{yes}$$

In this schema, "yes" denotes an empty queue.

8.4.8.4. A Sequential File: Another Example of a Sequence.

A sequential file is a sequence of records that are stored one after the other on an external storage device such as a magnetic tape or a magnetic disk. The operations performed on a sequential file may differ slightly, depending on the storage device used. For example, a file on magnetic tape can be either an input or an output file, but not both at the same time. A sequential file on disk can be used strictly for input, strictly for output, or for update. Update means that, as records are read, the record most recently read can be rewritten on the *same* file, if so desired. In this section the update operation is not addressed.

In a sequential file, to access the nth record requires that the previous $n - 1$ records be first read in order. A specification of a sequential file, which allows a file to read records in input mode and write records in output mode, is now given.

A schema for the state of the sequential file is

```
┌─ Sequential_File[T] ─────────────────────────────────
│ file : seq T
│ unread : seq T
│ mode : File_Mode
├───────────────────────────────────────────────────────
│ ∃ already_read : seq T • already_read ⌢ unread = file
└───────────────────────────────────────────────────────
```

where $[T]$ represents a record type and *unread* represents that part of the file not yet read. Also,

$$File\_Mode ::= \text{input} \mid \text{output}$$

The following is a schema for opening a file for input:

```
┌─ Open_For_Input ─────────────
│ ΔSequential_File
├──────────────────────────────
│ mode′ = input
│ unread′ = file
│ file′ = file
└──────────────────────────────
```

The file mode is set to input and no records have been read at this point. This operation does not alter the file.

A schema that specifies a read operation is

```
┌─ Read[T] ─────────────────────
│ ΔSequential_File
│ x! : T
├───────────────────────────────
│ mode = input
│ unread ≠ ⟨⟩
│ ⟨x!⟩ ⌢ unread′ = unread
│ mode′ = mode
│ file′ = file
└───────────────────────────────
```

The file mode must be input and the file must contain records not yet read. The record that is returned is the first in that part of the file yet to be read. Note that a read operation does not change the file contents or its mode.

Opening a file for output is described in the following schema:

```
┌─ Open_For_Output ─────────────
│ ΔSequential_File
├───────────────────────────────
│ mode′ = output
│ file′ = ⟨⟩
└───────────────────────────────
```

The file mode is set to output and the file contains no records.

A schema for a write operation follows:

$$
\begin{array}{|l}
\_Write[T] _____ \\
\Delta Sequential\_File \\
x? : T \\
\hline
mode = \text{output} \\
file' = file \frown \langle x? \rangle \\
mode' = mode \\
\end{array}
$$

As a precondition, the file mode must be output before the next record is written to the file. The file mode remains unchanged.

The following schema checks for an end-of-file condition:

$$
\begin{array}{|l}
\_End\_Of\_File _____ \\
\Xi Sequential\_File \\
x! : OK \\
\hline
((unread = \langle \rangle) \wedge (x! = \text{yes})) \\
\vee \\
((unread \neq \langle \rangle) \wedge (x! = \text{no})) \\
\end{array}
$$

If all records have been read, then the end of the file has been reached.

Problems 8.4.8

1. Another well-known example of a linear data structure that can be represented by a sequence is a stack. In a stack structure, elements are inserted and deleted from the same end, called its top. This structure exhibits a last-in, first-out behavior because the last element to be inserted onto the stack is the first to be removed.

   As was done for the queue structure, formulate a Z document, where $[T]$ is an arbitrary type, for the following:
   - A create operation for an empty stack
   - An insert operation that pushes an element on top of the stack
   - A delete operation that pops (or removes) the top element in a nonempty stack
   - An inquiry that returns the top element in a nonempty stack
   - An *is_empty* inquiry that returns true if the stack is empty or false, otherwise

   Your document should include error handling.

2. A simple queue has been described as behaving in a first-in, first-out manner in the sense that each deletion removes the oldest remaining item in the structure. A *double-ended queue* (or deque) is a sequence in which insertions and deletions can be made to or from either end of the structure. A deque is more general than a stack or a simple queue. There are two variations of a deque, the input-restricted deque and the output-restricted deque. An input-restricted deque allows insertions at only one end, while an output-restricted deque permits deletions from only one end. Formulate a Z document for each of the following:

- A deque
- An input-restricted deque
- An output-restricted deque

The operations should include insert (possibly at the rear or front, if applicable) and delete (again from the front or rear, if applicable). Queries consist of examining the front and rear elements and reporting whether a deque is empty.

3. Consider a travel bus (coach) with a certain fixed number of seats for which the passenger list is kept as a sequence of names with no duplicates. The person with the first name in the sequence is assigned the first seat, the person with the second name the second seat, and so on.

Using schemas, formulate a Z document that includes types used, initial state, operations (such as boarding and leaving the bus), inquiries (such as the number of passengers on board and whether a particular passenger is on the bus), and error handling.

PROBLEMS: CHAPTER 8

1. Given the schema

$$
\begin{array}{|l}
\hline
\quad NewSchema \underline{\qquad} \\
\quad LA : Names \rightarrow \mathbb{N}_1 \\
\quad LB : Names \\
\hline
\end{array}
$$

answer the following questions:

(a) In the structure LA, is it possible that a name can be associated with several numbers? State why or why not.

(b) What variables are generated by $\Xi NewSchema$? What conditions, if any, are generated by $\Xi NewSchema$?

2. Consider a simplified student registration system for some university that involves students, courses, and faculty. Two relations can be defined for this system. The first relation, *takes*, associates students enrolling in courses. The second relation, *given_by*, associates courses taught by faculty members.

Using schemas, prepare a Z document that includes types used, global declarations, initial state, operations (such as a student picking up a new course, a student dropping a course, and a course assigned to a faculty member), queries (such as the list of all courses taken by a student, the list of all students in all courses, the list of faculty members that teach a particular student, and the list of all students taking a specific course), and error handling.

3. A system that involves actors and actresses, movies, and directors is to be specified. An actor or actress may *star_in* a movie. Also, a movie is *directed_by* a director. Both of these relationships are many-to-many; that is, a movie can have more than one star, and an actor or actress can star in several movies. Similarly, a director can direct several movies or can codirect a movie with other directors.

Using schemas, prepare a Z document that includes types used, global declarations, initial state, operations (such as adding a tuple to each of the relations *star_in* and *directed_by*), queries (such as finding the movies that a specified actor or actress has starred in, listing all movies directed by a specified director, generating the directors that have directed a given movie star, listing each movie in which the director has also been the movie's star, and finding the movie stars in a given movie), and error handling.

4. Consider a small library system with the following transactions:
- Check out a book to a particular patron
- Return a book by a patron
- Generate a list of books written by a particular author
- Generate a list of books on loan to a particular patron
- Determine which patron last checked out a book

The library satisfies the following constraints:
- Each book in the library is either available or on loan, but not both at the same time.
- A patron cannot check out more than some fixed number of books at one time.

The basic types are the following:

$$[Book] \text{ the set of all uniquely identifiable books}$$
$$[Person] \text{ the set of all uniquely identifiable people}$$
$$[Author] \text{ the set of all possible authors}$$

There is a relation that maps a book into one or more authors.

Using schemas, formulate a Z document for this library system with error handling and suitable error messages and reports.

5. A direct file contains a set of records stored typically on some secondary storage memory device such as a magnetic disk. A particular record is accessed in any order by specifying the unique key of that record. The basic types for the file systems are as follows:

$$[Key] \text{ the set of all distinct values used to identify records}$$
$$[Info] \text{ the set of remaining items in a record besides the key}$$

Given a particular key value, we can use this value to find the record's associated information items.

Using schemas, formulate a Z document that includes the following:
- Read a record with a specified key
- Given a new key, insert a new record in the file
- Given a specific key, update the record with the key in the file
- Given a specific key, delete the record from the file that contains that key

Your Z document should include error handling.

9

Program Correctness Proofs

Program construction is based on rational arguments, and it is therefore reasonable to apply formal proof techniques to show that a program works correctly. It is therefore not surprising that there is a fair amount of research investigating how programs can be formally proved to be correct. The resulting formal techniques are not yet widely used in practice, but they have had great influence on the way people think about programs, and some of the concepts developed for doing correctness proofs have been implemented in modern computer languages. Moreover, the authors of this book feel that proving programs correct can pay its way by reducing the number of errors, which in turn reduces the effort needed for debugging. In fact, one author of this book has proved the correctness of many programs he used in his research, and he found that this not only helped uncover errors, but that it also forced him to formulate the objectives of his efforts more concisely, which in turn lead to clearer and better structured programs.

Basic to correctness proofs are *preconditions* and *postconditions*. Preconditions indicate what conditions the input data must satisfy such that the program can accomplish its mission. Postconditions, on the other hand, indicate what output states are acceptable as correct solutions of the problem at hand. A program calculating a square root, for instance, has the precondition that the argument of the square root must be nonnegative and the postcondition that the result returned be the desired square root. The precondition of a binary search is that the file to be searched must be sorted, and the postcondition expresses the fact that the program will return the position of the item in question.

We use the word *code* or *pieces of code* to denote any part of a program, from a single statement up to the entire program. To show that the different pieces of code work correctly, one must associate preconditions and postconditions with each piece of code. One must then show that the preconditions and postconditions of all the pieces of code are compatible with each other and with the preconditions and postconditions of the entire program. In fact, when the program is executed, the postconditions of each piece of code must imply the preconditions of the next piece. This should be established, no matter how large the pieces of code happen to be. In fact, these considerations are even more important when constructing and proving large programs than for small ones: in large programs, there is a much greater likelihood of committing errors, and it is more difficult to locate the errors. It is for this reason that an increasing number of programming languages have facilities to deal with preconditions and postconditions.

When doing program correctness proofs, one needs a rule of inference to deal with each type of executable statement. In this chapter, we only consider a subset of Pascal, and we introduce a rule of inference for each type of statement within this subset. Specifically, there will be a rule for the assignment statement, a rule for the if-statement, and a rule for the while statement. In programs containing a while statement, one must not only prove that the program is correct if and when it terminates, but one must also prove that the program actually terminates. All these issues will be addressed in this chapter.

9.1 PRELIMINARY CONCEPTS

9.1.1 Introduction

Following Section 3.4, we divide programs into pieces of code, which are sequences of statements that convert the initial state into a final state. The initial state is the state before the piece of code is executed, and the final state is the state after the code is executed. The state of the system, in turn, is given by the values of variables as they appear in the declarations.

Every logical statement regarding a state is called an *assertion*. The two most important assertions are the *preconditions* and the *postconditions*. A precondition is an assertion regarding the initial state of a piece of code, and a postcondition is an assertion about its final state. Typically, the postconditions of a piece of code describe what has to be accomplished, and the preconditions give the conditions that must be satisfied such that the code will end with the stipulated postconditions. The piece of code is considered correct if all states satisfying the preconditions will lead to states satisfying the postconditions once the code has been executed. These issues will now be discussed. Note, however, that this discussion does not address the issue of whether the preconditions and postconditions actually reflect what the programmer had in mind. Since the design, too, is increasingly formalized (see Chapter 8), it is often possible to obtain such rigorously defined preconditions and postconditions.

9.1.2 Programs and Codes

In this chapter, it is assumed that the language for writing code is a subset of Pascal. This subset contains the following:

1. All arithmetic operators and all library functions of Pascal
2. Pascal blocks, starting with the keyword begin and ending with the keyword end

3. The if-then and the if-then-else constructs

4. The while loop construct

We do not use declarations. In all cases, the type of the variables should be obvious from the context. Furthermore, we do not use input–output statements. The input may occur before the first statement of the code under consideration, and the output may occur after the last statement.

As mentioned already in Section 3.4, programs work with *program states*, and these program states are changed by *pieces of code*. The program state is described by the set of variables used within the program. Formally, if the variables are x_1, x_2, ..., x_n, and if they are of type X_1, X_2, ..., X_n, then the state is an element of the Cartesian product $X_1 \times X_2 \times \cdots \times X_n$. In words, every possible configuration of the values x_1, x_2, ..., x_n describes a state.

The states are changed by *pieces of code*. As indicated in Section 3.4, pieces of codes are functions that map the *initial state*, which is the state at the start, into a *final state*, which is the state at the time of completion of the piece of code. To become a function, the statements comprising the piece of code must be self-contained. For instance, a Pascal block, starting with begin and ending with end, is a piece of code. If there is a begin but no end, then the sequence of statements is no longer self-contained, and it no longer defines a function. Of course, every single statement is self-contained, and it is therefore a piece of code on its own.

Several pieces of code can be *concatenated*, which merely means that they are executed in sequence. Specifically, if C_1 and C_2 are two pieces of code, their concatenation is denoted by $C_1; C_2$. Concatenation is really function composition, except that in function composition the order of the operands is reversed. In other words, when C_1 and C_2 are interpreted as functions, then $C_1; C_2$ corresponds to $C_2 \circ C_1$. The semicolon is thus a composition operator in a relational sense.

The concatenation of two or more pieces of code constitutes again a piece of code. This means that the following sequence of statements forms a piece of code.

$$h := a; \ a := b; \ b := h \tag{9.1}$$

The effect of this code obviously is to interchange the values of a and b.

9.1.3 Assertions

Basic to correctness proofs are *assertions*.

Definition 9.1:   Any statement regarding a program state is called an *assertion*.

To indicate that a statement A is an assertion, one often encloses it in braces, such as $\{A\}$. If the braces are confusing, they are of course omitted.

Assertions relating to the initial state and the final state of a program are of particular importance. We therefore define the following:

Definition 9.2:   If C is a piece of code, then any assertion $\{P\}$ is called a *precondition* of C if $\{P\}$ only involves the initial state. Any assertion $\{Q\}$ is called a *postcondition* if $\{Q\}$ only involves the final state. If C has $\{P\}$ as a precondition and $\{Q\}$ as a postcondition, one writes $\{P\}C\{Q\}$. The triple $\{P\}C\{Q\}$ is called a *Hoare* triple.

As an example of a precondition and a postcondition, consider the code consisting of the single statement x := 1/y. A precondition of this statement is $\{y \neq 0\}$, and a postcondition is $\{x = 1/y\}$. Sometimes there is no precondition. For instance, the statement a := b has $\{a = b\}$ as a postcondition, regardless of its initial state. For this case, we introduce the empty assertion $\{\ \}$, which can be read as "true for all possible states." The following Hoare triple contains the empty assertion as a precondition.

$$\{\ \}\, \text{a} := \text{b}\, \{a = b\}$$

In most pieces of code, the final state obviously depends on the initial state. This implies that preconditions and postconditions are not independent. For instance, if one stipulates $\{a < b\}$ as a precondition of the code given by (9.1), then the postcondition becomes $\{b < a\}$. If, on the other hand, the precondition is changed to $\{a > b\}$, then the postcondition becomes $\{b > a\}$. To capture effects like this, two methods will be used. The first method to make the distinction between the initial and the final state involves subscripts. Specifically, the subscript α is used to denote the initial value of a variable, and the subscript ω is used to denote the final value of a variable. For instance, in the code given by (9.1), the initial values of the variables a and b are a_α and b_α, and their final values are a_ω and b_ω. In this notation, one can express the effect of (9.1) as

$$(a_\omega = b_\alpha) \wedge (b_\omega = a_\alpha)$$

Since all identifiers of this statement are associated with a state at a particular time, the placement of this assertion within the code becomes irrelevant. This assertion can even be used outside any code, say for such purposes as specification. This is exactly what is done in specification languages such as Z (see Chapter 8).

A different method to handle the difference between the initial and the final states is by means of *shadow variables*. Shadow variables are variables that do not appear in the code and that are introduced to store initial values of certain storage locations. For instance, one can use the symbols A and B to denote the initial values of a and b, respectively. The definition of the shadow variables is done in the preconditions. In the code given by (9.1), this would be done as follows:

$$\{a = A,\, b = B\}\, \text{h} := \text{a};\, \text{a} := \text{b};\ \text{b} := \text{h}\, \{b = A,\, a = B\}$$

The method of denoting the initial state by a subscript is convenient when deriving new rules, and we will use it extensively in this context. Otherwise, shadow variables tend to lead to clearer proofs, and they are preferred when working with established rules.

9.1.4 Correctness

If $\{P\}C\{Q\}$ is a code with precondition $\{P\}$ and postcondition $\{Q\}$, then $\{P\}C\{Q\}$ is correct if every possible initial state satisfying P results in a final state satisfying Q. Actually, as indicated in Section 3.4, one must distinguish between *partial* and *total* correctness. These terms are defined as follows:

> **Definition 9.3:**   If C is some code with precondition $\{P\}$ and postcondition $\{Q\}$, then $\{P\}C\{Q\}$ is said to be *partially correct* if the final state of C satisfies $\{Q\}$, provided the initial state satisfies $\{P\}$. C is also partially correct if there is no final state due to the fact that the program does not terminate. If $\{P\}C\{Q\}$ is partially correct and if C terminates, then $\{P\}C\{Q\}$ is said to be *totally correct*.

In other words, a code is not totally correct if there are some initial states satisfying P that lead to an infinite loop. Such programs may be partially correct, however, because for partial correctness one only requires that they satisfy the postconditions if these programs terminate. Loopless code always terminates, and partial correctness implies total correctness. Hence, the distinction between partial and total correctness is only essential for code containing loops or recursive procedures.

Consider now the correctness of a single statement. In many cases, the postcondition for a given precondition follows immediately from the definition of the statement in question. For instance, it is clear that $\{\ \}\,\mathsf{x} := 3\,\{x = 3\}$ is correct: the statement $\mathsf{x} := 3$ is exactly the statement that demands that the postcondition $x = 3$ be satisfied. Similarly, as pointed out earlier,

$$\{\ \}\,\mathsf{a} := \mathsf{b}\,\{a = b\}$$

A more precise description of the postcondition is used in the following expression:

$$\{\ \}\,\mathsf{a} := \mathsf{b}\,\{a = b,\ b = b_\alpha\}$$

This is essentially the definition of the assignment statement. As our last example, we present the following statement:

$$\{\ \}\ \textbf{if}\ \mathsf{x} > \mathsf{y}\ \textbf{then}\ \mathsf{m} := \mathsf{x}\ \textbf{else}\ \mathsf{m} := \mathsf{y}$$
$$\{((x > y) \Rightarrow (m = x_\alpha)) \wedge (\neg(x > y) \Rightarrow (m = y_\alpha))\}$$

All these examples are intuitively clear. We will derive rules that formalize this intuition, and we will prove that these rules are sound.

As was indicated previously, preconditions and postconditions are important for doing correctness proofs. In fact, the very definition of the word "correct" is based on preconditions and postconditions. However, the importance of these conditions is yet more general. They help one to clarify the objectives of the code under investigation. Moreover, by inserting preconditions and postconditions into pieces of code within a program, testing can be done in a more systematic way: if the preconditions of some code are satisfied, yet its postconditions are not, then the error must have occurred within the code in question. This allows one to narrow down the search for errors. This is one reason that preconditions and postconditions are supported by a number of programming languages, in particular by Eiffel.

Problems 9.1

1. Consider the following code:

$$x := x * x; \ y := y * y; \ z := x + y$$

   Suppose that the initial state is $x = 3$, $y = 4$, and $z = 0$. Find the state after each statement, and find the final state.

2. Use the subscripts α and ω to describe the effect of the following pieces of code:
   (a) i := j; j := i
   (b) i := 2 * i + j
   (c) sum := sum + a

3. Which of the following Hoare triples are correct and which are not? State why or why not.
   (a) $\{i > 10\}$ i := i + 3 $\{i > 20\}$
   (b) $\{i > 10\}$ i := i + 3 $\{i > 10\}$
   (c) $\{a < b\}$ a := b $\{a < b\}$
   (d) $\{a = 3, b = 4\}$ a := a + b $\{a \geq 7\}$

9.2 GENERAL RULES INVOLVING PRECONDITIONS AND POSTCONDITIONS

9.2.1 Introduction

In this section, we deal with *precondition strengthening* and *postcondition weakening*. We also discuss rules that can strengthen or weaken preconditions and postconditions simultaneously. The following definition explains what is meant by *strengthening* and *weakening* assertions.

> **Definition 9.4:**   If R and S are two assertions, then R is said to be *stronger* than S if $R \Rightarrow S$. If R is stronger than S, then S is said to be *weaker* than R.

For instance, $i < 0$ is stronger than $i < 1$, because whenever $i < 0$ then it is certain that $i < 1$. If an assertion R is stronger than an assertion S, then all states satisfying R also satisfy S; but there may be some states that satisfy S, but not R. Consequently, there are typically fewer states that satisfy R. For instance, there are fewer states satisfying $i < 0$ than $i < 1$. Stronger means, in this sense, more selective or more specific. Weaker, on the other hand, means more frequent or more general. The weakest assertion is of course the assertion $\{\}$, which holds for every state. This assertion can be identified with the logical constant T. The strongest assertion is F, because there is no state satisfying F.

9.2.2 Precondition Strengthening

If a piece of code is correct under precondition $\{P\}$, then it remains correct if $\{P\}$ is strengthened. For instance, if a code is correct for precondition $i > 0$, it remains correct for precondition $i > 1$, which is stronger. This leads to the following rule:

Precondition Strengthening: Suppose that $\{P\}C\{Q\}$ is correct and $P_1 \Rightarrow P$ has been proved. In this case, one is allowed to conclude that $\{P_1\}C\{Q\}$ is correct. This leads to the following rule of inference:

$$P_1 \Rightarrow P$$
$$\frac{\{P\}C\{Q\}}{\{P_1\}C\{Q\}}$$

Before proving precondition strengthening, we illustrate it in the context of an example. Suppose that the following Hoare triple has been proved.

$$\{y \neq 0\}\, \mathsf{x} := 1/\mathsf{y}\, \{x = 1/y\} \tag{9.2}$$

In this case, the following is also correct.

$$\{y = 4\}\, \mathsf{x} := 1/\mathsf{y}\, \{x = 1/y\}$$

The reason for this is simple. Clearly, every state that satisfies $y = 4$ also satisfies $y \neq 0$. If the initial state satisfies $y \neq 0$, then the final state satisfies $x = 1/y$ because of (9.2). It follows that every initial state satisfying $y = 4$ will lead to a final state satisfying $x = 1/y$. The general proof of precondition strengthening is quite similar. If $P_1 \Rightarrow P$ is true, then every state for which P_1 is true also satisfies P. If each initial state that satisfies P leads to a final state that satisfies Q, then each state in which P_1 holds leads to a final state in which Q is true. This, in turn, means that $\{P_1\}C\{Q\}$ holds, as indicated by the rule in question.

Example 9.1   Use the Hoare triple

$$\{i < 4\}\, \mathsf{i} := \mathsf{i} + 1\, \{i < 5\}$$

to prove that

$$\{i < 3\}\, \mathsf{i} := \mathsf{i} + 1\, \{i < 5\}$$

Solution   Informally, all one needs to show is that $(i < 3) \Rightarrow (i < 4)$, which is obviously true. To make the proof formal, we use the pattern given previously.

$$(i < 3) \Rightarrow (i < 4)$$
$$\frac{\{i < 4\}\, \mathsf{i} := \mathsf{i} + 1\, \{i < 5\}}{\{i < 3\}\, \mathsf{i} := \mathsf{i} + 1\, \{i < 5\}}$$

Note how $i < 3$, $i < 4$, and $\mathsf{i} := \mathsf{i} + 1$ match P_1, P, and C, respectively.   ■

Example 9.2   Assume that the following Hoare triple holds.

$$\{x \ge 0\}\, \mathsf{y} := \mathsf{sqrt(x)}\, \{y^2 = x\}$$

Prove that

$$\{x \ge 1\}\, \mathsf{y} := \mathsf{sqrt(x)}\, \{y^2 = x\}$$

Solution   Obviously, if $x \ge 1$, then $x \ge 0$, which proves that the Hoare triple in question is correct.   ■

The empty assertion $\{\,\}$ holds for every state, which makes it the weakest possible precondition. The empty assertion may therefore be strengthened to yield any precondition $\{P\}$ whatsoever. For instance, $\{\,\}\, \mathsf{a} := \mathsf{b}\, \{a = b\}$ can be used to justify $\{P\}\, \mathsf{a} := \mathsf{b}\, \{a = b\}$ for every imaginable condition P, such as $P = (a < b)$, $P = (a > 0)$, or any other condition that comes to mind. Other things being equal, this makes the empty precondition the ideal precondition, because once a piece of code is proved to be true with the empty precondition it is also true for any other precondition. More generally, it is always advantageous to try to formulate the weakest precondition that assures that a given postcondition is met. Any stronger precondition will automatically be satisfied because of the principle of precondition strengthening. Moreover, other things being equal, programs should be written such that they are as versatile as possible. This means that they should cover as many initial states as possible or that the precondition should be as weak as possible. The program with the weakest precondition is the most general program.

9.2.3 Postcondition Weakening

Earlier, we considered the two Hoare triples $\{\,\}\, \mathsf{a} := \mathsf{b}\, \{a = b\}$ and $\{\,\}\, \mathsf{a} := \mathsf{b}\, \{a = b, b = b_\alpha\}$, where b_α is the value of b immediately before $\mathsf{a} := \mathsf{b}$ is executed. It is clear that the postcondition of the second Hoare triple is more specific, or stronger: there are fewer states that satisfy $a = b$, $b = b_\alpha$ than there are states that satisfy $a = b$. Generally, one is allowed to replace more specific postconditions by less specific or weaker postconditions, as indicated by the following rule:

Postcondition Weakening: The principle of *postcondition weakening* allows one to conclude that $\{P\}C\{Q_1\}$ once $\{P\}C\{Q\}$ and $Q \Rightarrow Q_1$ are established. Formally, this can be expressed as follows:

$$\{P\}C\{Q\}$$
$$\frac{Q \Rightarrow Q_1}{\{P\}C\{Q_1\}}$$

The proof of postcondition weakening is as follows.   According to definition, $\{P\}C\{Q\}$ means that the final state of C satisfies Q, provided the initial state satisfies P. If, in addition to this, $Q \Rightarrow Q_1$, then the final state must also satisfy Q_1, which translates into $\{P\}C\{Q_1\}$.

Example 9.3   The following Hoare triple is given.

$$\{\,\} \mathsf{max} := \mathsf{b} \{max = b\}$$

Prove that

$$\{\,\} \mathsf{max} := \mathsf{b} \{max \geq b\}$$

Solution   Clearly, $max = b$ implies that $max \geq b$, and, according to the principle of postcondition weakening, one finds

$$\{\,\} \mathsf{max} := \mathsf{b} \{max = b\}$$
$$\frac{(max = b) \Rightarrow (max \geq b)}{\{\,\} \mathsf{max} := \mathsf{b} \{max \geq b\}}$$
∎

Example 9.4   The following Hoare triple is given.

$$\{a > 4\} \mathsf{b} := \mathsf{a} + \mathbf{1} \{b > 5, a > 4\}$$

Prove that

$$\{a > 4\} \mathsf{b} := \mathsf{a} + \mathbf{1} \{b > 5\}$$

Solution   Because of the rule of simplification, $(b > 5) \wedge (a > 4)$ implies that $b > 5$. One now applies postcondition weakening as follows:

$$\{a > 4\} \mathsf{b} := \mathsf{a} + \mathbf{1} \{b > 5, a > 4\}$$
$$\frac{(b > 5) \wedge (a > 4) \Rightarrow (b > 5)}{\{a > 4\} \mathsf{b} := \mathsf{a} + \mathbf{1} \{b > 5\}}$$
∎

If possible, one should try to find postconditions that are as specific or as strong as possible, especially if a piece of code is used in several places. If, in a particular application, a weaker postcondition is acceptable, one can always use postcondition weakening.

9.2.4 Conjunction and Disjunction Rules

The following rule allows one to strengthen the precondition and the postcondition simultaneously.

Conjunction Rule: If C is a piece of code and if one has established that $\{P_1\}C\{Q_1\}$ and $\{P_2\}C\{Q_2\}$, one is allowed to conclude that $\{P_1 \wedge P_2\}C\{Q_1 \wedge Q_2\}$. Formally, this can be expressed as follows:

$$\frac{\begin{array}{c} \{P_1\}C\{Q_1\} \\ \{P_2\}C\{Q_2\} \end{array}}{\{P_1 \wedge P_2\}C\{Q_1 \wedge Q_2\}}$$

To prove the conjunction rule, note that if the initial state satisfies P_1 then the final state satisfies Q_1, and if the initial state satisfies P_2, then the final state satisfies Q_2. Consequently, if the initial state satisfies both P_1 and P_2, then the final state must satisfy both Q_1 and Q_2.

The empty precondition represents T, and $\text{T} \wedge P \equiv P$. This means that the conjunction rule implies that

$$\frac{\begin{array}{c} \{\}C\{Q_1\} \\ \{P_2\}C\{Q_2\} \end{array}}{\{P_2\}C\{Q_1 \wedge Q_2\}}$$

Example 9.5   Use the Hoare triples

$$\{\,\}\,i := i + 1 \,\{i_\omega = i_\alpha + 1\}$$
$$\{i_\alpha > 0\}\,i := i + 1 \,\{i_\alpha > 0\}$$

to prove that

$$\{i > 0\}\,i := i + 1 \,\{i > 1\}$$

Solution   According to the conjunction rule, one has

$$\frac{\begin{array}{c} \{\,\}\,i := i + 1 \,\{i_\omega = i_\alpha + 1\} \\ \{i_\alpha > 0\}\,i := i + 1 \,\{i_\alpha > 0\} \end{array}}{\{i_\alpha > 0\}\,i := i + 1 \,\{(i_\omega = i_\alpha + 1) \wedge (i_\alpha > 0)\}}$$

Now

$$(i_\omega = i_\alpha + 1) \wedge (i_\alpha > 0)$$
$$\Leftrightarrow (i_\alpha = i_\omega - 1) \wedge (i_\alpha > 0) \qquad \text{arithmetic}$$
$$\Leftrightarrow (i_\alpha = i_\omega - 1) \wedge (i_\omega - 1 > 0) \qquad \text{substitution}$$
$$\Rightarrow (i_\omega - 1 > 0) \qquad \text{simplification law}$$
$$\Leftrightarrow (i_\omega > 1)$$

Hence, postcondition weakening yields

$$\{i_\alpha > 0\}\, \mathsf{i} := \mathsf{i} + 1 \,\{i_\omega > 1\}$$

All subscripts now refer to the present state, that is, i_α in the precondition refers to the initial state, and i_ω in the postcondition refers to the final state. One may therefore drop the subscripts to find

$$\{i > 0\}\, \mathsf{i} := \mathsf{i} + 1 \,\{i > 1\} \qquad\qquad \blacksquare$$

Example 9.6   Given the Hoare triple

$$\{\,\}\, \mathsf{h} := \mathsf{a};\ \mathsf{a} := \mathsf{b};\ \mathsf{b} := \mathsf{h}\, \{a = b_\alpha,\, b = a_\alpha\}$$

prove that

$$\{a > b\}\, \mathsf{h} := \mathsf{a};\ \mathsf{a} := \mathsf{b};\ \mathsf{b} := \mathsf{h}\, \{b > a\}$$

Solution   One obviously has

$$\{a_\alpha > b_\alpha\}\, \mathsf{h} := \mathsf{a};\ \mathsf{a} := \mathsf{b};\ \mathsf{b} := \mathsf{h}\, \{a_\alpha > b_\alpha\}$$

The conjunction rule can now be applied as follows:

$$\{\,\}\, \mathsf{h} := \mathsf{a};\ \mathsf{a} := \mathsf{b};\ \mathsf{b} := \mathsf{h}\, \{a = b_\alpha,\, b = a_\alpha\}$$
$$\{a_\alpha > b_\alpha\}\, \mathsf{h} := \mathsf{a};\ \mathsf{a} := \mathsf{b};\ \mathsf{b} := \mathsf{h}\, \{a_\alpha > b_\alpha\}$$

$$\{a_\alpha > b_\alpha\}\, \mathsf{h} := \mathsf{a};\ \mathsf{a} := \mathsf{b};\ \mathsf{b} := \mathsf{h}\, \{a = b_\alpha,\, b = a_\alpha,\, a_\alpha > b_\alpha\}$$

The postcondition implies that $b > a$, which completes the proof. $\blacksquare$

Instead of a conjunction, one may also use a disjunction, as indicated by the following rule:

Disjunction Rule: If C is a piece of code and if one has established $\{P_1\}C\{Q_1\}$ and $\{P_2\}C\{Q_2\}$, one is allowed to conclude that $\{P_1 \vee P_2\}C\{Q_1 \vee Q_2\}$. Formally, this can be expressed as follows:

$$\{P_1\}C\{Q_1\}$$
$$\{P_2\}C\{Q_2\}$$

$$\{P_1 \vee P_2\}C\{Q_1 \vee Q_2\}$$

If either P_1 or P_2 is empty, then $\{P_1 \vee P_2\}$ becomes true, and this assertion reduces to the empty assertion. The proof of the disjunction rule is similar to the conjunction rule, and it is left as an exercise.

Problems 9.2

1. Suppose that $\{sum > 1\}$ sum := sum + 4 $\{sum > 5\}$ is correct. Which of the following codes can be proved correct also? Indicate the rules and the implications used.
   (a) $\{sum > 2\}$ sum := sum + 4 $\{sum > 5\}$
   (b) $\{sum \geq 1\}$ sum := sum + 4 $\{sum > 5\}$
   (c) $\{sum > 0\}$ sum := sum + 4 $\{sum > 5\}$
   (d) $\{sum > 1\}$ sum := sum + 4 $\{sum > 6\}$
   (e) $\{sum > 1\}$ sum := sum + 4 $\{sum > 4\}$
   (f) $\{sum > 1\}$ sum := sum + 4 $\{sum > 5 \vee sum < -2\}$
   (g) $\{sum > 1 \wedge i > 0\}$ sum := sum + 4 $\{sum > 5\}$

2. Consider the assertion $\{i > 2, j > 3, k = i * j\}$. Are you allowed to replace this assertion by $\{k > 6\}$ if the assertion is (a) a precondition or (b) a postcondition? State why or why not.

3. Given that

$$\{x < y\}C_1\{u < v\}$$

which of the following expressions can be proved, and, if so, by what rule (precondition strengthening or postcondition weakening)?
   (a) $\{x \leq y\}C_1\{u < v\}$
   (b) $\{x \leq y - 2\}C_1\{u < v\}$
   (c) $\{x < y\}C_1\{u \leq v\}$
   (d) $\{x < y\}C_1\{u \leq v - 2\}$
   (e) $\{x \leq y\}C_1\{u \leq v\}$

4. If $\{P\}C\{Q\}$ holds, which of the following Hoare triples can be proved? State whether the precondition is strengthened or weakened, and do the same for the postcondition.
   (a) $\{P \vee D\}C\{Q\}$
   (b) $\{P \wedge Q\}C\{Q\}$
   (c) $\{P\}C\{Q \vee D\}$
   (d) $\{P\}C\{Q \vee P\}$
   (e) $\{P\}C\{Q \wedge P\}$

5. Prove the disjunction rule.

6. Using the conjunction rule, prove that $\{i > 2\}$ i := 2 * i $\{i > 4\}$.

7. The following Hoare triples are given:

$$\{j > 1\}\, i := i + 2;\ j := j + 3\,\{j > 4\}$$
$$\{i > 2\}\, i := i + 2;\ j := j + 3\,\{i > 4\}$$

Show that these triples imply that

$$\{j > i,\, i > 2\}\, i := i + 2;\ j := j + 3\,\{j > 4, i > 4\}$$

Which rule must be used for the proof?

9.3 CORRECTNESS PROOFS IN LOOPLESS CODE

9.3.1 Introduction

The *assignment statement*, also called *assignment*, is the most frequent statement in almost every program, and it will be discussed in Section 9.3.2. Section 9.3.3 then shows how to concatenate a number of assignment statements. The rules for concatenation hold generally. In fact, these rules are useful whenever several pieces of code, no mater how large or small, must be concatenated. Finally, Section 9.3.4 discusses the if-statement.

To deal with assignment statements, we introduce the *assignment rule*. This rule is not as general as one would like. It requires that no two variables with different names share the same storage location. This excludes, for instance, the use of pointers. More importantly, this also excludes the use of subscripted variables. To see this, note that the two variables $a[i]$ and $a[j]$ have different names, yet if $i = j$, they share the same storage location. If $i = j$, then $a[j]$ is just another name for, or an *alias* of, $a[i]$. The fact that two variables are aliases of one another is called *aliasing*. We postpone the discussion of rules dealing with aliasing until later.

To prove the correctness of pieces of code containing several statements, one uses the postcondition of the entire code as a starting point. This is quite natural. First, one asks what results have to be obtained and what conditions these results must satisfy, and this is expressed by the postcondition. The precondition is then derived from the postcondition. For this reason, one proves code in the opposite order in which it is executed. The postcondition of the last statement is identical with the postcondition of the entire code. One then derives the precondition of the last statement, which becomes the postcondition of the previous statement. This allows one to find the precondition of the second to last statement, which becomes the postcondition of the third to last statement. In this way, one continues until one finds the precondition of the first statement, which becomes the precondition of the entire code. This idea is elaborated in Section 9.3.3, and it will be used throughout the chapter.

9.3.2 Assignment Statements

Assignment statements are statements of the form $V := E$, where V is a variable and E is an expression. Here $:=$ is the assignment operator, V will be called the *left-hand side,* and E the *right-hand side* of the assignment. The effect of the assignment operator is that the right-hand side is evaluated using the initial values of the variables, and it is assigned to the left-hand side to obtain the final state. Hence, if E_α is the value of the right-hand side if evaluated using the initial state, then, by definition, the postcondition of the assignment statement $V := E$ becomes

$$\{\,\} V := E \{V = E_\alpha\} \tag{9.3}$$

This assumes, of course, that E can be evaluated for all possible states. If $E = 1/y$, for instance, this is not true. In this case, one must add a term to the precondition to indicate that $y \neq 0$. Equation (9.3) ignores this possibility. We return to this issue later.

It is inconvenient to track the values of the initial variables beyond the precondition. Since the initial values of a variable is marked by α, this means that all instances of variables subscripted with α should be removed from the postcondition of an assignment statement. In order to show how this is done, suppose that one wants to find a precondition P such that $\{P\}\, \mathsf{i} := 2 * \mathsf{i}\, \{i < 6\}$ is correct. According to (9.3), one has

$$\{\,\}\, \mathsf{i} := 2 * \mathsf{i}\, \{i = 2 * i_\alpha,\, i < 6\}$$

Clearly,

$$(i = 2 * i_\alpha) \wedge (i < 6) \Rightarrow (2 * i_\alpha < 6) \wedge (i < 6)$$

Hence, postcondition weakening yields

$$\{\,\}\, \mathsf{i} := 2 * \mathsf{i}\, \{2 * i_\alpha < 6,\, i < 6\}$$

The term $2 * i_\alpha < 6$ can be converted into a precondition, which leads to the following result:

$$\{2 * i < 6\}\, \mathsf{i} := 2 * \mathsf{i}\, \{i < 6\}$$

Hence, P is $2 * i < 6$ or $i < 3$. Note that we found the precondition in effect by replacing the i in the original postcondition $\{i < 6\}$ by $2 * i$, the right-hand side of the assignment. The following rule generalizes this method.

Assignment Rule: Let E be an expression and let V be an unsubscripted variable. Furthermore, if C is a statement of the form $V := E$ with postcondition $\{Q\}$, then the precondition of C can be found by replacing all instances of V in Q by E. If Q_E^V is the expression thus obtained, one has the following:

$$\{Q_E^V\}\, V := E\, \{Q\}$$

To prove the assignment rule, suppose that the postcondition of $V := E$ is Q. Because of (9.3), one can add the term $V = E_\alpha$ to the postcondition, which yields the new postcondition $\{Q \wedge (V = E_\alpha)\}$. Since V and E_α are equal, one can be substituted for the other, and

$$Q \wedge (V = E_\alpha) \equiv Q_{E_\alpha}^V \wedge (V = E_\alpha)$$

$Q_{E_\alpha}^V$ does not contain V because all instances of V have been substituted by E_α. The assignment does not change any variable except V. Consequently, all variables of $Q_{E_\alpha}^V$ refer to the initial state; that is, $Q_{E_\alpha}^V$ can be converted into a precondition. Once this is done, the subscript α may be dropped. This completes the proof of the assignment rule.

Example 9.7 Consider the statement $\mathsf{j} := \mathsf{i} + 1$. Suppose that the postcondition of this statement is $j > 0$. Find the precondition.

Solution According to the assignment rule, one has to replace j in the postcondition $\{j > 0\}$ by $i + 1$, which yields P as $i + 1 > 0$. Consequently,

$$\{i + 1 > 0\}\, \mathsf{j} := \mathsf{i} + 1\, \{j > 0\} \qquad \blacksquare$$

Example 9.8   Let C be the statement y := x * x, and let $y > 1$ be the postcondition of C. Find the precondition.

Solution   To find the required precondition, replace y in $y > 1$ by x^2. This yields $x^2 > 1$.
∎

In the case of an assignment statement of the form $V := E$, where E does not contain the variable V, (9.3) can be applied directly. In particular, in statements such as a := b, one easily finds from (9.3) that

$$\{\,\} \, \text{a} := \text{b} \, \{a = b\}$$

Even in this case, however, it is often better to use the assignment rule. To do this, one starts with the postcondition $\{a = b\}$, and one proves that the corresponding precondition is always true, which allows one to reduce it to the empty precondition. This is done as follows: According to the assignment rule, one finds the precondition by replacing a in $a = b$ by b. This yields

$$\{b = b\} \, \text{a} := \text{b} \, \{a = b\}$$

However, $b = b$ is always true. Hence, one may replace $\{b = b\}$ by $\{\,\}$, and the result is the same as before.

If the precondition is given and the postcondition is to be determined, then the assignment rule is somewhat inconvenient. In this case, it is better to work directly with (9.3). The following example shows how.

Example 9.9   Given the statement x := x * x, which has the precondition $x > 2$, find the postcondition.

Solution   The precondition must first be combined with the postcondition given in (9.3), which can be achieved by writing the precondition $x > 2$ as $x_\alpha > 2$. This yields

$$\{x > 2\} \, \text{x} := \text{x} * \text{x} \, \{x_\alpha > 2, \, x = x_\alpha^2\}$$

We now remove x_α from the postcondition. To find x_α, we solve the equation $x = x_\alpha^2$, which yields $x_\alpha = +\sqrt{x}$ or $x_\alpha = -\sqrt{x}$. Since $x_\alpha > 2$, the negative root can be discarded, and $x_\alpha > 2$ becomes $\sqrt{x} > 2$ or $x > 4$. Hence

$$\{x > 2\} \, \text{x} := \text{x} * \text{x} \, \{x > 4\}$$

This result may be checked by the assignment rule. If x in $x > 4$ is replaced by x^2, one obtains $x^2 > 4$, which is equivalent to $(x > 2) \vee (x < -2)$. Since $x > 2$ logically implies that $(x > 2) \vee (x < -2)$, the preceding result follows because of precondition strengthening.
∎

The previous example demonstrates that it is in general much easier to find the precondition for a prescribed postcondition than to find the postcondition for a prescribed precondition. To find the precondition may involve finding an inverse function, and this is sometimes impossible. Fortunately, a proof typically starts with the postcondition, and the objective is to find the precondition.

Some arithmetic operations are undefined for some states. In particular, if E is an expression, $1/E$ is undefined if E evaluates to zero. Similarly, if sqrt(x) is the square root function of x, then x must not be negative. In such cases, the condition that makes the evaluation of E possible must be added to the precondition.

Example 9.10 Find the precondition of the statement x := 1/x, given that the postcondition is $x \geq 0$.

Solution By the assignment rule, one has $1/x \geq 0$, and because of the division, one has $x \neq 0$. Together, these two conditions imply that $x > 0$. In summary,

$$\{x > 0\}\, \text{x} := 1/\text{x}\, \{x \geq 0\}$$

∎

9.3.3 Concatenation of Code

Two pieces of code that are separated by a semicolon are said to be *concatenated*. Concatenation means that the pieces of code are executed in sequence, such that the final state of the first piece of code becomes the initial state of the second piece of code. Consequently, if $C_1; C_2$ is the concatenation of C_1 with C_2, then the final state of C_1 becomes the initial state of C_2, and any assertion that holds for the final state of C_1 must be equally true for the initial state of C_2. This leads to the following rule:

Concatenation Rule: Let C_1 and C_2 be two pieces of code, and let $C_1; C_2$ be their concatenation. If $\{P\}C_1\{R\}$ and $\{R\}C_2\{Q\}$ are both correct, one is allowed to conclude that $\{P\}C_1; C_2\{Q\}$. Hence,

$$\frac{\{P\}C_1\{R\}}{\{R\}C_2\{Q\}}$$
$$\overline{\{P\}C_1; C_2\{Q\}}$$

The concatenation rule is applied intuitively by all programmers. For instance, a file update is faster if the update file is sorted. Hence, it is reasonable to write an update program that requires as a precondition that the update file be sorted. To satisfy this precondition, one uses a sort program to sort the update file first. If the sort is identified with C_1 and the update program with C_2, then the postcondition of C_1 must satisfy the precondition of C_2. The next example illustrates the concatenation rule.

Example 9.11 Prove that the following code is correct.

$$\{\,\}\, \text{c} := \text{a} + \text{b};\ \text{c} := \text{c}/2\, \{c = (a + b)/2\}$$

Solution   One has

$$
\begin{array}{c}
\{\,\}\,\mathsf{c} := \mathsf{a} + \mathsf{b}\,\{c/2 = (a+b)/2\} \\[4pt]
\{c/2 = (a+b)/2\}\,\mathsf{c} := \mathsf{c}/2\,\{c = (a+b)/2\} \\[2pt]
\hline \\[-6pt]
\{\,\}\,\mathsf{c} := \mathsf{a} + \mathsf{b};\ \mathsf{c} := \mathsf{c}/2\,\{c = (a+b)/2\}
\end{array}
$$

To understand this proof, note that the postcondition of the second statement is given and that its precondition can be found by the assignment rule. Specifically, by replacing c in $\{c = (a+b)/2\}$ by $c/2$, one obtains $\{c/2 = (a+b)/2\}$, which is simultaneously the precondition of the second statement and the postcondition of the first statement. To find the precondition of the first statement, one uses the assignment rule once more: one replaces c by the $a+b$, the left-hand side of the assignment statement, and one obtains $\{(a+b)/2 = (a+b)/2\}$. This obviously reduces to the empty assertion. In the proof, one could have simplified the assertion $\{c/2 = (a+b)/2\}$ to $\{c = a+b\}$.   ∎

In the example, the last statement was dealt with first and the first statement last. This is the typical way that correctness proofs proceed. The postcondition typically defines the desired end result of the computation, and this result must be clearly stated. The postcondition is therefore usually given, and it can be used to find the precondition of the last statement. In addition, it is easier to work with the assignment rule if the postcondition is given and the precondition must be found. The precondition of the last statement is obviously the postcondition of the preceding statement, which means that, after the proof of the last statement, one can proceed with the proof of the second to last statement.

The concatenation rule can be generalized as follows:

Modified Concatenation Rule: Let C_1 and C_2 be two pieces of code, and let $C_1; C_2$ be their concatenation. If $\{P\}C_1\{R\}$ and $\{S\}C_2\{Q\}$ are both correct, and if $R \Rightarrow S$, then one is allowed to conclude that $\{P\}C_1; C_2\{Q\}$. Hence,

$$
\begin{array}{c}
\{P\}C_1\{R\} \\
\{S\}C_2\{Q\} \\
R \Rightarrow S \\
\hline
\{P\}C_1; C_2\{Q\}
\end{array}
$$

This rule is correct because $\{R\}C_2\{Q\}$ can be obtained from $\{S\}C_2\{Q\}$ by precondition strengthening.

Both the original and the modified concatenation rule can be generalized to more than two pieces of code. To express this, we use a three-column format for correctness proofs. The first column contains the precondition for the statement appearing in the second column, and the third column gives its postcondition. It is automatically understood that the postcondition of each statement implies the precondition of the next statement. If the precondition of the entire code is strengthened, a line is added at the top of the proof, with the strengthened precondition appearing in column 3. Similarly, if the postcondition is weakened, a line is added at the bottom of the proof, with the weakened postcondition appearing in column 1.

Example 9.12   The following code has no precondition, and its postcondition is $s = 1 + r + r^2$. Prove that the code is correct.

$$s := 1;$$
$$s := s + r;$$
$$s := s + r * r$$

Solution   The solution is given by the following schema

Precondition	Statement	Postcondition
		$\{\}$
$\{1 = 1\}$	$s := 1;$	$\{s = 1\}$
$\{s + r = 1 + r\}$	$s := s + r;$	$\{s = 1 + r\}$
$\{s + r^2 = 1 + r + r^2\}$	$s := s + r * r$	$\{s = 1 + r + r^2\}$

To obtain this table, one starts with the last line, where one enters the postcondition of the entire code. Starting with the last statement, one recursively finds the precondition of each statement, which becomes the postcondition of the previous statement. To do this, one uses the assignment rule. In detail, this works as follows. The postcondition of the last line is $\{s = 1 + r + r^2\}$. According to the assignment rule, s must be replaced by the $s + r * r$, the right-hand side of the assignment statement in question. This yields $\{s + r^2 = 1 + r + r^2\}$, as indicated in column 1 of the last line. One can simplify this expression to $\{s = 1 + r\}$, which provides the postcondition of the previous line. The assignment statement of this line has $s + r$ as its right-hand side, which means that one must replace s in $\{s = 1 + r\}$ by $s + r$, which yields the precondition $\{s + r = 1 + r\}$, or $\{s = 1\}$. This becomes the postcondition of the first statement, and since the right-hand side is now 1, s must be replaced by 1, which yields $\{1 = 1\}$, or $\{\ \}$. This completes the proof. One can write the result as follows

$$\{\}$$
$$s := 1;$$
$$s := s + r;$$
$$s := s + r * r$$
$$\{s = 1 + r + r^2\}$$ ■

It is sometimes convenient to insert all assertions on a line by themselves between statements. This is particularly convenient if the precondition of the previous statement is identical to the postcondition of the following statement. This method is demonstrated by the following example:

Example 9.13   Prove that

$$\{b = B, a = A\}$$
h := a; a := b; b := h
$$\{a = B, b = A\}$$

Solution   The proof of the program fragment is as follows:

$$\{b = B,\ a = A\}$$
h := a;
$$\{b = B, h = A\}$$
a := b;
$$\{a = B, h = A\}$$
b := h
$$\{a = B, b = A\}$$

As before, one starts with the postcondition of the entire code, and one recursively finds the preconditions of all statements, starting with the last statement.   ■

Any assertion that is at the same time a precondition and a postcondition of a piece of code is called an *invariant*. Invariants are particularly important for any correctness proof that is based on complete induction. This will be discussed in Sections 9.4.2 and 9.4.3, where we analyze loops. The proof that an assertion is an invariant is not different from the proofs given before. The precondition and the postcondition just happen to be the same assertion.

Example 9.14   Show that $\{r = 2^i\}$ is an invariant of i := i + 1; r := r * 2.

Solution   To prove that $\{r = 2^i\}$ is an invariant, one must show that

$$\{r = 2^i\}\ i := i + 1;\ r := r * 2\ \{r = 2^i\}$$

This is done as follows:

Precondition	Statement	Postcondition
		$\{r = 2^i\}$
$\{r = 2^{i+1-1}\}$	i := i + 1;	$\{r = 2^{i-1}\}$
$\{2 * r = 2^i\}$	r := r * 2	$\{r = 2^i\}$

The proof is conducted in the same fashion as the proof of Example 9.12. One starts with the last line, which has the invariant as a postcondition. The r in this postcondition is replaced by $2 * r$, which yields $\{2 * r = 2^i\}$ or, equivalently, $\{r = 2^{i-1}\}$. This is the postcondition of the previous line. Since the present statement has the right-hand side $i + 1$, one must replace i by $i + 1$ in $\{r = 2^{i-1}\}$, which yields $\{r = 2^{i+1-1}\}$ or $\{r = 2^i\}$. Hence, $\{r = 2^i\}$ is both the precondition and the postcondition of the two-line code; that is, $\{r = 2^i\}$ is an invariant.

■

9.3.4 The If-Statement

If C_1 and C_2 are two pieces of code and if B is some condition, then the statement

if B then C_1 else C_2

has the following interpretation. If B is true, then C_1 is executed, and if B is false, C_2 is executed. Suppose now that one needs to prove that an if-statement with precondition $\{P\}$ and postcondition $\{Q\}$ is correct. Clearly, if the initial state satisfies B in addition to $\{P\}$, then C_1 is executed, and the proof amounts to a demonstration that $\{P \wedge B\}C_1\{Q\}$ is correct. If, on the other hand, the initial state satisfies $\neg B$, then C_2 is executed, and it must be shown that $\{P \wedge \neg B\}C_2\{Q\}$ is correct. This yields the following rule.

Rule for Conditions with Else Clause: If C_1 and C_2 are two pieces of code and if B is a logical expression, then one has

$$\frac{\{P \wedge B\}C_1\{Q\}}{\{P \wedge \neg B\}C_2\{Q\}}$$

$$\{P\} \text{ \textbf{if} } B \text{ \textbf{then} } C_1 \text{ \textbf{else} } C_2\{Q\}$$

Example 9.15   Prove

$$\{\,\} \text{ \textbf{if} } a > b \text{ \textbf{then} } m := a \text{ \textbf{else} } m := b \,\{(m \geq a) \wedge (m \geq b)\} \tag{9.4}$$

Solution   According to the rule for conditions, one must prove that

$$\{a > b\}\, m := a \,\{(m \geq a) \wedge (m \geq b)\} \tag{9.5}$$

and

$$\{\neg(a > b)\}\, m := b \,\{(m \geq a) \wedge (m \geq b)\} \tag{9.6}$$

First, consider (9.5). The assignment rule allows one to replace m in the postcondition $(m \geq a) \wedge (m \geq b)$ by a, which yields the precondition $\{(a \geq a) \wedge (a \geq b)\}$, or $\{a \geq b\}$. Since $a > b$ implies that $a \geq b$, (9.5) follows by precondition strengthening. To prove (9.6), use the assignment rule once more: replace m in the postcondition by b, which yields the precondition $(b \geq a) \wedge (b \geq b)$, which is equivalent to $b \geq a$ and to $\neg(a > b)$. This proves (9.6). According to the rule for conditions, (9.5) and (9.6) together prove (9.4).   ■

The else clause of the if-statement is optional. Hence, one must consider statements of the form

$$\{P\} \text{ if } B \text{ then } C \{Q\}$$

In this case, C is done only if B is true, and nothing is done if B is false. This means that one must prove that $\{P \wedge B\}C\{Q\}$ and $(P \wedge \neg B) \Rightarrow Q$.

Rule for Conditions: Let B be a condition, and let C be a piece of code. In this case, the following rule applies:

$$\{P \wedge B\}C\{Q\}$$
$$\frac{(P \wedge \neg B) \Rightarrow Q}{\{P\} \text{ if } B \text{ then } C\{Q\}}$$

Example 9.16   Prove that, for any possible precondition, the following program fragment has $\{max \geq a\}$ as a postcondition.

$$\text{if } max < a \text{ then } max := a$$

Solution   According to the preceding rule, one must prove that

$$\{max < a\} \text{ max} := \text{a } \{max \geq a\}$$

and

$$\neg(max < a) \Rightarrow (max \geq a)$$

The second of these two statements is immediately obvious. To prove the first statement, the assignment rule is applied, and max is replaced by a, which yields $\{a \geq a\}$, and this reduces to $\{\ \}$. Since $\{max < a\}$ implies $\{\ \}$, the result follows by precondition strengthening. Hence,

$$\{max < a\}\text{max} := \text{a}\{max \geq a\}$$
$$\frac{\neg(max < a) \Rightarrow (max \geq a)}{\{\ \} \text{ if max} < a \text{ then max} := \text{a } \{max \geq a\}}$$

■

When using the if-statement, it is relatively easy to find the postcondition for a given precondition, but it is relatively difficult to do the reverse. Unfortunately, the reverse is what is usually required; that is, one usually wants to find the precondition corresponding to a given postcondition. In this case, the following rule, which is proved later, is often more convenient.

$$\{B\}C_1\{B\}$$
$$\{\neg B\}C_2\{\neg B\}$$
$$\{P\}C_1\{Q_1\}$$
$$\{P\}C_2\{Q_2\}$$

$$\overline{\{P\}\ \textbf{if}\ B\ \textbf{then}\ C_1\ \textbf{else}\ C_2\{(B \wedge Q_1) \vee (\neg B \wedge Q_2)\}}$$

Example 9.17 Write and prove the correctness of an if-statement that satisfies the postcondition

$$\{((\text{income} < 6000) \wedge (\text{tax} = 0)) \vee (\neg(\text{income} < 6000) \\ \wedge (\text{tax} = 0.4 * \text{income}))\}$$

Solution The program that accomplishes the task is

$$\textbf{if}\ (\text{income} < 6000)\ \textbf{then}\ \text{tax} := 0 \\ \textbf{else}\ \text{tax} := 0.4 * \text{income}$$

The proof that this statement works is as follows:

$$\{income < 6000\}\ \text{tax} := 0\ \{income < 6000\}$$
$$\{\neg(income < 6000)\}\ \text{tax} := 0.4 * \text{income}\ \{\neg(income < 6000)\}$$
$$\{\ \}\ \text{tax} := 0\ \{tax = 0\}$$
$$\{\ \}\ \text{tax} := 0.4 * \text{income}\ \{tax = 0.4 * income\}$$

$$\overline{\{\ \}\ \textbf{if}\ \text{income} < 6000\ \textbf{then}\ \text{tax} := 0\ \ \textbf{else}\ \text{tax} := 0.4 * \text{income}}$$
$$\{(income < 6000 \wedge tax = 0) \vee (\neg(income < 6000) \wedge tax = 0.4 * income\} \qquad \blacksquare$$

To prove the derived rule for the if-statement, note that $\{B\}C_1\{B\}$ and $\{P\}C_1\{Q_1\}$ allow one to derive $\{B \wedge P\}C_1\{B \wedge Q_1\}$ because of the conjunction rule. Next, one uses postcondition weakening to show that

$$\{B \wedge P\}C_1\{(B \wedge Q_1) \vee (\neg B \wedge Q_2)\}$$

A similar argument leads to

$$\{\neg B \wedge P\}C_1\{(B \wedge Q_1) \vee (\neg B \wedge Q_2)\}$$

The proof now follows directly from the rule for the conditional.

Note that there is a different way to write this postcondition, as indicated by the following logical equivalence:

$$(B \wedge Q_1) \vee (\neg B \wedge Q_2) \equiv (B \Rightarrow Q_1) \wedge (\neg B \Rightarrow Q_2)$$

For instance, the postcondition of Example 9.17 could have been written as

$$\{((income < 6000) \Rightarrow (tax = 0)) \wedge (\neg(income < 6000) \Rightarrow (tax = 0.4 * income))\}$$

The easiest way to prove this equivalence is by writing the truth tables of the expressions involved.

Problems 9.3

1. The following code has the postcondition $\{x + b * y = u, c * y - 2 * x = 0\}$. Find the precondition.

$$y := 2/(2 * b + c); \ x := 1 - b * y$$

2. Let A and B be the initial values of a and b, respectively. Write a piece of code that has $\{a = A + B, b = A - B\}$ as postcondition, and prove that the code is correct.

3. Prove that the following statement has the postcondition $\{x \geq 0, x^2 = a^2\}$.

$$\textbf{if } a > 0 \textbf{ then } x := a \ \textbf{ else } x := -a$$

4. The following program fragment has $sum = j * (j - 1)/2$ as a precondition and as a postcondition. Prove that this is true.

$$sum := sum + j; \ j := j + 1$$

5. Prove that $\{i * j + 2 * j + 3 * i = 0\} \ j := j + 3; \ i := i + 2 \ \{i * j = 6\}$.

9.4 LOOPS AND ARRAYS

9.4.1 Introduction

Sections 9.4.2 and 9.4.3 introduce rules for proving while loops. Loops frequently make use of arrays, and we therefore discuss arrays in Section 9.4.4. Whereas all partially correct programs without loops and without recursion will always terminate, the same is not true for programs with loops or recursion. We therefore show in Section 9.4.5 how to prove termination. We should mention, however, that there is no method that allows one to decide in every case whether a particular program will ever come to an end. This is the famous halting problem, a problem that has been shown to be unsolvable (see Section 3.5.4).

9.4.2 A Preliminary While Rule

In this section, we discuss how to prove the correctness of code containing a Pascal while statement. When doing this, we are only concerned with partial correctness. In other words, we prove that if the code ends its final state satisfies the postcondition. Termination proofs will be postponed until Section 9.4.5. In a while loop, a certain code is repeated. This makes sense if certain things stay the same from iteration to iteration. Features that stay the same are captured by an invariant, the *loop invariant*. A loop is only entered if the *entry condition* is satisfied at the beginning of the loop. The negation of the entry condition is the *exit condition*. Once the exit condition is satisfied, the loop terminates. The *loop variant* is an expression that measures the progress made toward satisfying the exit condition.

To illustrate the main concepts, consider the code given in Figure 9.1. The entry condition of this loop is $j <> n$. If this expression is false at the beginning of the loop, the loop is terminated. For this reason, one calls the negation of the entry condition the exit condition. In our case, the exit condition is $\neg(j <> n)$, or $j = n$. At the end of the piece of code, $sum = n * a$, and this condition is the postcondition. During each iteration, progress is made toward this postcondition. At the end of each iteration, $sum = j * a$, and this assertion turns out to be the loop invariant. The loop variant is $n - j$. The loop variant decreases with each iteration, and as soon as it reaches zero, the loop terminates. At the exit of the loop, $j = n$, and $j * a = n * a$. Hence, after termination, $sum = n * a$, which is the postcondition, as indicated earlier.

```
sum := 0;
j := 0;
while j <> n do
    begin
        sum := sum + a;
        j := j + 1
    end
```

Figure 9.1.   Code with loop

We first present a preliminary rule for proving the correctness of while loops. A generalization of this rule is given in Section 9.4.3.

Preliminary While Rule: If there is an invariant I of a code C such that $\{I\}C\{I\}$ holds, then the following rule applies:

$$\frac{\{I\}C\{I\}}{\{I\} \textbf{ while } D \textbf{ do } C \ \{\neg D \wedge I\}}$$

Here D is the *entry condition* and $\neg D$ is the *exit condition*.

To prove this rule, we first use induction to show that I is true at the end of each iteration. We have the following:

Inductive base   If I is the precondition of the while loop, then I is true at the beginning of the first iteration. Furthermore, if the loop consists of the code C with $\{I\}C\{I\}$, then I must also be true at the end of the first iteration.

Inductive hypothesis   The inductive hypothesis is that I is true at the end of iteration n or that the loop has been terminated.

Inductive step   If I is true at the end of iteration n and if the loop has not been terminated, then I is also true at the beginning of iteration $n+1$. Because $\{I\}C\{I\}$ holds, this means that I is also true at the end of iteration $n + 1$.

The loop does not end until the entry condition D is false at the end of an iteration. Since I holds at the end of each iteration, including the iteration causing the termination, both I and $\neg D$ must hold when the loop has terminated. This proves the soundness of the preliminary while rule.

To show how the while rule works, we reconsider the introductory example once more.

Example 9.18   Consider the code given in Figure 9.1. Use the while rule to show that the code is correct with respect to the precondition $\{\ \}$ and the postcondition $\{sum = n * a\}$.

Solution   The loop invariant is $sum = j * a$. According to the while rule, we first show that the body of the loop has the invariant as precondition and postcondition. This is done in the usual way, starting with the last statement of the loop.

Precondition	Statement	Postcondition
		$\{sum = j * a\}$
$\{sum + a = (j + 1) * a\}$	sum := sum + a;	$\{sum = (j + 1) * a\}$
$\{sum = (j + 1) * a\}$	j := j + 1	$\{sum = j * a\}$

This establishes $\{I\}C\{I\}$, where I is $sum = j * a$ and C consists of the two statements sum := sum + a and j := j + 1. Next one proves that sum := 0 and j := 0 have the invariant as a postcondition. Using the assignment axiom twice, first for j := 0 and then for sum := 0, one has the following:

Precondition	Statement	Postcondition
		$\{\ \}$
$\{0 = 0\}$	sum := 0;	$\{sum = 0\}$
$\{sum = 0 * a\}$	j := 0	$\{sum = j * a\}$

Hence, before starting the while, the loop invariant $sum = a * j$ holds. One now has

$$\{sum = j * a\}\ \text{sum} := \text{sum} + \text{a};\ \ j := j + 1\ \{sum = j * a\}$$

$$\{sum = j * a\}\ \textbf{while}\ j <> n\ \textbf{do}$$
$$\textbf{begin}$$
$$\text{sum} := \text{sum} + \text{a};\ j := j + 1$$
$$\textbf{end}\ \{\neg(j \neq n) \wedge (sum = j * a)\}$$

To complete the correctness proof, one must still show that $\neg(j \neq n)$ and $sum = j * a$ together imply that $sum = j * n$. This, however, is trivial. Clearly,

$$\neg(j \neq n) \wedge (sum = j * a) \equiv (j = n) \wedge (sum = n * a)$$

Because of postcondition weakening, one is allowed to drop the term $j = n$, which completes the proof.   ∎

It is often difficult to find the loop invariant. Typically, the loop invariant is a precursor of the postcondition, which means that it must somehow be similar to the postcondition. In addition to that, some progress toward the postcondition must be made in each iteration, which means that the loop invariant must contain variables that are changed inside the loop. To illustrate how this works, consider the loop invariant of the code given in Figure 9.1, which is $sum = j * a$. This loop invariant $sum = j * a$ is similar to the postcondition $sum = n * a$, except that the variable j is replaced by n. Of course, j increases from iteration to iteration, indicating some progress toward the postcondition, whereas n is fixed. The loop invariant may be more complicated than the postcondition. Typically, the loop invariant contains all variables that change from iteration to iteration.

The concept of a loop invariant is also useful for program construction. In program construction, one typically has an objective indicating what has to be accomplished by the end of each loop. This goal can be formalized, and this formalization yields the loop invariant. Hence, correctness proofs and program constructions go hand in hand. In particular, if the objective to be reached at the end of each loop is clearly spelled out, it is much easier to find the loop invariant. In this sense, the loop invariant is only a formalization of the programmer's objectives.

Loop invariants frequently deal with sums, and it is therefore important that the manipulations involving sums be fully understood. As described in Section 3.2.3, all sums have a variable, called the *index*, that runs from its lower bound to its upper bound. The index can be treated like the bound variable of a quantifier. In particular, the name of the index variable is irrelevant as long as it is not identical with some other variable appearing in the expression in question. For instance, $\sum_{k=1}^{i} k^2$ and $\sum_{j=1}^{i} j^2$ denote the same sum. Moreover, the index is strictly local to the sum in the sense that any variable of the program having the same name as the index refers to a different variable. For instance, if the program contains the statement $i := i + 1$, and if this statement has the postcondition $\sum_{i=1}^{n} i$, then the i in the statement is distinct from the i used as the index for the sum. To avoid problems arising from having the same name for two different variables, one should change all summation indexes such that they are distinct from the variables used in the program. For instance, assume that one has to find the precondition P in the code

$$\{P\}\, \mathsf{i} := \mathsf{i} + 1\, \{sum = \sum_{i=1}^{n} i^2\} \tag{9.7}$$

In this case, one should write

$$\{P\}\, \mathsf{i} := \mathsf{i} + 1\, \{sum = \sum_{k=1}^{n} k^2\}$$

After these initial remarks, an example involving a program with sums is presented.

Example 9.19   In the following code, find the loop variant, the loop invariant, and the postcondition. The program fragment is supposed to find the sum of squares from 1 to n.

```
i := 1;
sum := 0;
while not (i = n + 1) do
    begin
        sum := sum + i * i;
        i := i + 1;
    end
```

Solution   Since the code is supposed to calculate $\sum_{i=1}^{n} i^2$ and since this sum is obviously accumulated by the variable sum, the postcondition becomes

$$sum = \sum_{i=1}^{n} i^2$$

Since the index variable i is also used in the statement i := i + 1, we substitute j for i in order to avoid clashes:

$$sum = \sum_{j=1}^{n} j^2 \tag{9.8}$$

To find the loop invariant, note that the variables i and sum are the only variables that change inside the loop. Since sum accumulates our result, bringing us closer and closer to the postcondition, we try the following tentative solution

$$sum = \sum_{j=1}^{i} j^2$$

This is not correct because the exit condition is $i = n + 1$, which leads to the postcondition

$$sum = \sum_{j=1}^{n+1} j^2$$

and this is wrong. The problem is easily fixed, however. Reduce the upper bound of the tentative loop invariant by 1, to find the correct invariant as

$$sum = \sum_{j=1}^{i-1} j^2$$

The procedure used here can be applied in many situations. A tentative loop invariant is found first, which is later adjusted to yield the correct postcondition.

The proof is given in Figure 9.2, and a similar setup as for the solution of Example 9.12 is used. A last line is added indicating that the exit condition, together with the loop invariant, implies the postcondition given in (9.8). In the proof, the loop invariant appears at both the end of the loop (after i := i + 1) and at the beginning of the loop (after **begin**). Once the loop

Precondition	Statement	Postcondition
		$\{\,\}$
$\{0 = \sum_{j=1}^{1-1} j^2\}$	i := 1;	$\{0 = \sum_{j=1}^{i-1} j^2\}$
$\{0 = \sum_{j=1}^{i-1} j^2\}$	sum := 0;	$\{sum = \sum_{j=1}^{i-1} j^2\}$
	while not (i $=$ n $+$ 1) **do**	
	begin	$\{sum = \sum_{j=1}^{i-1} j^2\}$
$\{sum + i^2 = \sum_{j=1}^{i} j^2\}$	sum := sum $+$ i $*$ i;	$\{sum = \sum_{j=1}^{i} j^2\}$
$\{sum = \sum_{j=1}^{i+1-1} j^2\}$	i := i $+$ 1	$\{sum = \sum_{j=1}^{i-1} j^2\}$
	end	

$$(i = n + 1) \wedge (sum = \textstyle\sum_{j=1}^{i-1} j^2) \Rightarrow (sum = \sum_{j=1}^{n} j^2)$$

Figure 9.2.   Proof of Example 9.19

invariant is written at the end of the last statement of the loop, the proof proceeds in the normal way, starting with the last statement and ending with the first one. Each statement, in turn, is proved by the assignment rule. One identifies all instances where the variable on the left of the assignment appears in the postcondition and replaces all these instances by the right-hand side of the assignment statement. This yields the precondition of the statement in question, which is then simplified and used as the postcondition of the previous statement. For instance, the postcondition of i := i $+$ 1 is $sum = \sum_{j=1}^{i-1} j^2$. Replacing i by $i + 1$ in this expression yields the precondition $sum = \sum_{j=1}^{i+1-1} j^2$. This expression simplifies to $sum = \sum_{j=1}^{i} j^2$, which becomes the postcondition of the preceding statement. The preceding statement is sum := sum $+$ i $*$ i, and one now must replace sum in $sum = \sum_{j=1}^{i} j^2$ by $sum + i * i$, which yields $sum + i^2 = \sum_{j=1}^{i} j^2$. Since $\sum_{j=1}^{i} = 1^2 + 2^2 + \cdots + i^2$, one can drop the term i^2 from both sides of the equation, which yields $sum = \sum_{j=1}^{i-1} j^2$. This, however, is exactly the loop invariant, and the loop is proved. To complete the proof, one still must show that the loop invariant is satisfied before entering the loop for the first time. To do this, one uses this loop invariant as postcondition of the statement preceding the loop, and one continues the proof in the same fashion as before, proving each statement in turn, starting with the last statement. Specifically, the statement sum := 0 requires one to replace all instances of sum in $sum = \sum_{j=1}^{i-1} j^2$ by 0, which yields $0 = \sum_{j=1}^{i-1} j^2$. This assertion is the postcondition of the previous statement $i = 1$, and the assignment now yields $0 = \sum_{j=1}^{1-1} j^2$. This sum has an upper summation bound of 0, and since the lower summation bound is 1, the sum contains no terms; it is zero. The precondition of the entire code is therefore $\{0 = 0\}$, which reduces to the empty precondition. The proof of the program is now complete. The variable sum will store the sum of the first n squares no matter what the initial state happens to be.   ∎

9.4.3 The General While Rule

Whenever the loop of a while statement is executed, the entry condition D must be true before the execution of the first statement of the loop is performed. This leads to the following while rule, which is a generalization of the preliminary while rule given previously.

The While Rule: If C is a code such that $\{D \wedge I\}C\{I\}$ holds, then one is allowed to make the following inference:

$$\frac{\{D \wedge I\}C\{I\}}{\{I\} \text{ while } D \text{ do } C \ \{\neg D \wedge I\}}$$

The proof of the general while rule is similar to proof of the preliminary while rule, and it is left as an exercise. The general while rule must be used for loops where the exit condition is an inequality rather than an equality. The following example demonstrates this.

Example 9.20   Prove that the following program segment calculates $\sum_{j=0}^{n} r^j$ and stores the result in the variable t.

```
f := 1; t := 0; i := 0;
while ( i <= n) do
    begin
        t := t + f; f := f * r; i := i + 1
    end
```

Solution   The entire proof is given in Figure 9.3. We now discuss the individual steps in detail. The program must end with $t = \sum_{j=0}^{n} r^j$, and this expression becomes the postcondition. Each variable that changes inside the loop should be considered for inclusion in the loop invariant. In addition, the loop invariant must contain the precursor of the postcondition. What is new now is that one also needs a term in the loop invariant that keeps track of the progress made toward satisfying the exit condition. Taking all these factors into account, one finds that

$$I = \left(t = \sum_{j=0}^{i-1} r^j\right) \wedge (f = r^i) \wedge (i \leq n+1) \tag{9.9}$$

Here the first term is the precursor of the postcondition, the second term reflects that $f = r^i$, and the third term is used to record the progress made toward satisfying the exit condition. As soon as $i = n + 1$, the exit condition holds. The entire invariant can be found after the statement i := i + 1 in Figure 9.3. The last line of this figure shows that the invariant, together with the exit condition, implies the postcondition. The implication is proved as follows. The last two terms of the antecedent are logically equivalent to $i = n + 1$. If $i = n + 1$, the first term of the antecedent can be written as $t = \sum_{j=0}^{n} r^j$. The implication now follows from the rule of simplification.

The term $i \leq n+1$ of the invariant does not hold at the end of the loop unless the entry condition $i \leq n$ is true at the beginning of the loop. This means that one must use the general while rule, because only the general while rule makes use of the entry condition. One must form $I \wedge D$, which instantiates to

$$I \wedge D = \left((t = \sum_{j=0}^{i-1} r^j\right) \wedge (f = r^i) \wedge (i \leq n+1) \wedge (i \leq n)) \tag{9.10}$$

Precondition	Statement	Postcondition
		$\{n \geq -1\}$
$\{1 = 1, n \geq -1\}$	f := 1;	$\{f = 1, n \geq -1\}$
$\{0 = 0, f = 1, n \geq -1\}$	t := 0;	$\{t = 0, f = 1, n \geq -1\}$
$\{t = \sum_{j=0}^{-1} r^j,$		
$f = r^0, 0 \leq n+1\}$	i := 0;	$\{t = \sum_{j=0}^{i-1} r^j,$
		$f = r^i, i \leq n+1\}$
	while (i <= n) **do**	
	begin	$\{t = \sum_{j=0}^{i-1} r^i, f = r^i,$
		$i \leq n+1, i \leq n\}$
$\{t + f = \sum_{j=0}^{i} r^j$		
$f = r^i, i \leq n\}$	t := t + f;	$\{t = \sum_{j=0}^{i} r^j$
		$f = r^i, i \leq n\}$
$\{t = \sum_{j=0}^{i} r^j$		
$f * r = r^{i+1}, i \leq n\}$	f := f * r;	$\{t = \sum_{j=0}^{i} r^j,$
		$f = r^{i+1}, i \leq n\}$
$\{t = \sum_{j=0}^{i+1-1} r^j, f = r^{i+1}$		
$i+1 \leq n+1\}$	i := i + 1	$\{t = \sum_{j=0}^{i-1} r^j$
		$f = r^i, i \leq n+1\}$
	end	

$$\left(t = \sum_{j=0}^{i-1} r^j\right) \wedge (f = r^i) \wedge (i \leq n+1) \wedge \neg(i \leq n) \Rightarrow \left(t = \sum_{j=0}^{n} r^j\right)$$

Figure 9.3.   Proof of a piece of code

In Figure 9.3, assertion (9.10) appears at the beginning of the loop, right after the keyword begin. Assertion (9.10) is logically equivalent to the assertion preceding the statement t := t + f. The proof of the loop itself is standard. The loop invariant appears again as the postcondition of the statement preceding the keyword **while**. This postcondition allows one to work backward in order to find the precondition of the entire program, which is $\{n \geq -1\}$. In case $n < -1$, a separate proof is needed which is then combined with the previous proof by the disjunction rule. If $n < -1$, the loop is not entered, and $t = 0$ after the loop. This satisfies the postcondition since for $n < 0$, $\sum_{j=0}^{n} r^j = 0$.   ■

9.4.4 Arrays

The assignment rule for finding the precondition of $V := E$ is not sound if the assignment changes any variable other than V. In particular, if V shares a common storage location with a variable named W, then the assignment not only changes V, but also W. Hence, if the storage location of V has an alias besides V, then the assignment rule is no longer sound. Aliasing is unavoidable when dealing with arrays. For instance, if $V = a[i]$, then $a[j]$ is an alias for $a[i]$ if $i = j$. In this case, $a[i] := E$ not only changes $a[i]$, but also $a[j]$. The following example shows this in more detail.

Example 9.21   Show that the following Hoare triplet is not sound.

$$\{\,\} \, \mathsf{a[j]} := \mathsf{a[i]} + 1 \, \{a[j] = a[i] + 1\} \tag{9.11}$$

Solution   If $i = j$ in the initial state, then the postcondition becomes $\{i = j, \, a[j] = a[i]+1\}$, which implies that $a[i] = a[i] + 1$. Since this is impossible, the application of the assignment rule cannot be sound in cases involving subscripted variables. ∎

The assignment rule is sound only if $a[i]$ and $a[j]$ refer to different storage locations; that is, these two variables must not be aliases for each other. Note that if we knew which variable name is an alias for which other variable name the problem would not be difficult to resolve. In this case, one would treat the two variable names as identical. Unfortunately, when working with arrays, the aliases change dynamically, and only a detailed study, which distinguishes between all possible alias combinations, will lead to a sound correctness proof. In (9.11), in particular, one must distinguish between the case $i \neq j$, in which case the postcondition is correct, and $i = j$, in which case the postcondition is false. At the end of the proof, all possible cases must then be combined by means of the disjunction rule introduced in Section 9.2.4.

Example 9.22   Find the precondition $\{P\}$ in the following Hoare triple:

$$\{P\} \, \mathsf{a[i]} := \mathsf{b} \, \{a[j] = 2 * a[i]\}$$

Solution   For $i \neq j$, the assignment rule can be applied directly and yields the precondition $a[j] = 2 * b$. Hence,

$$\{i \neq j, \, a[j] = 2 * b\} \, \mathsf{a[i]} := \mathsf{b} \, \{a[j] = 2 * a[i]\}$$

If $i = j$, then $a[i]$ and $a[j]$ must be considered as the same variable, and any substitution of $a[i]$ must also be applied to $a[j]$. The precondition obtained by using the assignment rule therefore becomes $b = 2 * b$ or, equivalently, $b = 0$. By adding this to the condition $i = j$, one obtains

$$\{i = j, \, b = 0\} \, \mathsf{a[i]} := \mathsf{b} \, \{a[j] = 2 * a[i]\}$$

Combining these results by means of the disjunction rule yields

$$\{((i \neq j) \wedge (a[j] = 2 * b)) \vee ((i = j) \wedge (b = 0))\}$$
$$\mathsf{a[i]} := \mathsf{b}$$
$$\{a[j] = 2 * a[i]\}$$

Here the preconditions are given at the top and the postconditions are given at the bottom. ∎

Example 9.23   Prove that the following code is correct.

$$\{a[i] = A[i], \, a[j] = A[j]\}$$
$$\mathsf{h} := \mathsf{a[i]}; \ \mathsf{a[i]} := \mathsf{a[j]}; \ \mathsf{a[j]} := \mathsf{h}$$
$$\{a[i] = A[j], \, a[j] = A[i]\}$$

Note that the initial values of the variables are capitalized.

Solution   We must conduct two proofs, one for the case $i = j$ and the other for $i \neq j$. These proofs are now given, with the assertions inserted between the statements on an extra line. One finds for $i \neq j$ that

$$\{i \neq j,\, a[j] = A[j],\, a[i] = A[i]\}$$

$$\mathsf{h := a[j];}$$

$$\{i \neq j,\, h = A[j],\, a[i] = A[i]\}$$

$$\mathsf{a[j] := a[i];}$$

$$\{i \neq j,\, h = A[j],\, a[j] = A[i]\}$$

$$\mathsf{a[i] := h}$$

$$\{i \neq j,\, a[i] = A[j],\, a[j] = A[i]\}$$

Similarly, one has for $i = j$

$$\{i = j,\, a[j] = A[j],\, a[j] = A[i]\}$$

$$\mathsf{h := a[j];}$$

$$\{i = j,\, h = A[j],\, h = A[i]\}$$

$$\mathsf{a[j] := a[i];}$$

$$\{i = j,\, h = A[j],\, h = A[i]\}$$

$$\mathsf{a[i] := h;}$$

$$\{i = j,\, a[i] = A[j],\, a[j] = A[i]\}$$

Now

$$((i = j) \wedge (a[j] = A[j]) \wedge (a[j] = A[i]))$$
$$\vee\ ((i \neq j) \wedge (a[j] = A[j]) \wedge (a[i] = A[i]))$$
$$\equiv (a[j] = A[j]) \wedge (a[i] = A[i])$$

Because of the disjunction rule, one therefore has

$$\{a[j] = A[j],\, a[i] = A[i]\}$$
$$\mathsf{h := a[i];\ \ a[i] := a[j];\, a[j] := h}$$
$$\{a[i] = A[j],\, a[j] = A[i]\}$$ ∎

The distinction between different cases in an assignment statement becomes unwieldy if the postcondition has many variables that can alias the left-hand side of the assignment statement. If V in $V := E$ is a subscripted variable and if this statement contains a postcondition with n variables from the same array as V, then 2^n cases must be distinguished. For instance, if $V = a[i]$ and if the postcondition contains the two variables $a[j]$ and $a[k]$ besides $a[i]$, then the following $2^2 = 4$ cases arise.

$$i = j, \qquad i = k$$
$$i = j, \qquad i \neq k$$
$$i \neq j, \qquad i = k$$
$$i \neq j, \qquad i \neq k$$

Hence, the number of cases to be considered increases exponentially with n, which makes the handling of postconditions containing many subscripted variables impractical for high values of n. Fortunately, in many practical cases, n is either small or shortcuts can be found. In particular, the postcondition can often be used to exclude the majority of cases. The next example shows how this works.

Example 9.24   Find the precondition P that makes the following Hoare triple correct:

$$\{P\}\, \mathsf{a}[i] := 2 * \mathsf{b}\, \{j \leq i,\, k < i,\, a[i] + a[j-1] + a[k] > b\}$$

Solution   When considering the postcondition, one immediately sees that neither $j - 1$ nor k can be equal to i, and aliasing can be excluded. The assignment rule may therefore be applied directly. This yields the following precondition:

$$\{j \leq i,\, k < i,\, 2 * b + a[j-1] + a[k] > b\}$$

or

$$\{j \leq i,\, k < i,\, b + a[j-1] + a[k] > 0\} \qquad\qquad \blacksquare$$

In all cases, it was assumed that the assignment statement does not change the value of any subscript. If subscripts may contain subscripts, this possibility cannot be excluded. For instance, consider the statement a[a[2]] := 1. One would believe that this statement has the postcondition $a[a[2]] = 1$. This is not the case. To see this, the reader may want to run the following Pascal segment:

```
a[1] := 2;
a[2] := 2;
a[a[2]] := 1;
write ('a[a[2]] = ', a[a[2]]);
```

The output of this segment is

$$\mathsf{a}[\mathsf{a}[2]] = 2$$

The reader may check this. Hence, the postcondition of a[a[2]] := 1 cannot be $a[a[2]] = 1$! To avoid such misleading cases, we disallow nested subscripts.

To perform a certain operation for each element of an array, one typically uses a loop. The loop invariant then indicates for which element in the array the operation in question has already been carried out. For instance, suppose that it is desired to copy an array a into an array b. The array a extends from 1 to n. The loop invariant of the loop copying the array must now contain a statement to the effect that all elements up to and including the loop index, say i, have already been copied. This can be expressed as

$$\bigwedge_{j=1}^{i} \left(a[j] = b[j] \right)$$

If the upper bound i is below the lower bound, then the conjunction evaluates to true.

Example 9.25   Show that the following program fragment has the postcondition $\bigwedge_{j=1}^{n}\left(a[j] = b[j]\right)$.

```
i := 0;
while (i <> n) do
    begin
        i := i + 1;
        a[i] := b[i]
    end
```

Solution   The loop invariant in this case indicates how many of the $a[j]$ have already been assigned the value of $b[j]$ and are therefore equal to $b[j]$. This leads to the following expression:

$$\bigwedge_{j=1}^{i}\left(a[j] = b[j]\right)$$

The proof is now as follows:

Precondition	Statement	Postcondition
$\{\,\}$	i := 0;	$\{\bigwedge_{j=1}^{i}(a[i] = b[i])\}$
	while (i $<>$ n) **do**	
	begin	$\{\bigwedge_{j=1}^{i}(a[j] = b[j])\}$
$\{\bigwedge_{j=1}^{i+1-1}(a[j] = b[j])\}$	i := i + 1;	$\{\bigwedge_{j=1}^{i-1}(a[j] = b[j])\}$
$\{\bigwedge_{j=1}^{i-1}(a[j] = b[j])\}$	a[i] := b[i]	$\{\bigwedge_{j=1}^{i}(a[j] = b[j])\}$
	end	
$\{\bigwedge_{j=1}^{i}(a[j] = b[j]) \wedge (i = n) \Rightarrow \bigwedge_{j=1}^{n}(a[j] = b[j])$		

This proof follows the same principles discussed earlier. Only the following line merits a special discussion.

$$\{\bigwedge_{j=1}^{i-1}(a[j] = b[j])\}\; a[i] := b[i]\; \{\bigwedge_{j=1}^{i}(a[j] = b[j])\}$$

To see how the precondition of this statement was obtained, note that

$$\bigwedge_{j=1}^{i}\left(a[j] = b[j]\right) \equiv \left(a[i] = b[i]\right) \wedge \bigwedge_{j=1}^{i-1}\left(a[j] = b[j]\right)$$

Now, none of the j in $\bigwedge_{j=1}^{i-1}(a[j] = b[j])$ can be equal to i, and the assignment rule can be applied without the necessity to distinguish between different cases. This yields

$$\left(b[j] = b[j]\right) \wedge \bigwedge_{j=1}^{i-1}\left(a[j] = b[j]\right) \equiv \bigwedge_{j=1}^{i-1}\left(a[j] = b[j]\right)$$

This explains the precondition of the statement a[i] := b[i].   ∎

9.4.5 Program Termination

A program is partially correct if it can be proved that, if the program terminates, it terminates satisfying its postcondition. If, in addition to partial correctness, one can prove that the program does indeed terminate, then the program is *totally correct*. Normally, the proof for total correctness is divided into two parts, a partial correctness proof and a proof for termination. In this section, we prove termination.

In the subset of Pascal considered here, only programs that contain while loops can exhibit nontermination. The termination of a while loop depends only on the exit condition and on the variables contained in the exit condition. We can assume that at least one variable in the exit condition changes its value inside the loop. If this is not so, then the exit condition must either be always true, and the loop is never entered, or the exit condition must always be false, in which case the loop never terminates. Hence, at the minimum, one variable in the exit condition must be affected by the statements inside the loop. There may, of course, be several such variables. The vector of all variables that occur in the exit condition and that can change their values inside the loop will be called the *variant vector*. Typically, each iteration changes the variant vector. Successive iterations therefore produce a sequence of variant vectors. If every sequence consisting of the variant vectors reaches the exit condition after a finite number of iterations, then the loop must obviously terminate. This observation forms the basis of termination proofs.

Example 9.26   Consider the following program segment:

```
i := 0;
f := 1;
while i <> n do
    begin
        i := i + 1; f := f * r
    end
```

Prove that the program fragment terminates for $n \geq 0$.

Solution   The only variable in the entry condition that changes inside the loop is i. Hence, the variant vector only consists of i. The value of i describes the sequence $0, 1, \ldots, n - 1$. As soon as $i = n$, the entry condition fails and the loop terminates. Hence, the sequence is finite. However, for $n < 0$, the sequence described by i is $0, 1, 2, \ldots$, and this sequence is infinite.   ■

Frequently, there is a strict partial order among the variant vectors that may allow one to simplify proofs for termination. To make the explanation easier, we call the sequence of variant vectors the *variant sequence*. Moreover, we assume that for each pair A, B of variant vectors there is a strict partial order $\prec$. If $A \prec B$, we say that A is less than B. The loop terminates if the sequence of variant vectors decreases in the sense of $\prec$ and if every decreasing sequence is finite (see Sections 3.3 and 5.4.8 for details of decreasing sequences). The proof for termination is then reduced to a proof that all elements of the variant sequence belong to a well-founded set and that the variant sequence is decreasing.

Example 9.27   Write a program fragment that determines the greatest common factor of 2 given positive integers $x1$ and $y1$, and prove that the program fragment terminates.

Solution   To solve the problem, we reduce the problem to a smaller one, and we show that the smaller problem has the same solution as the original problem. As the problem decreases in size, we should find a case that has a trivial solution. The problem of finding the greatest common factor of x and y can be reduced as follows: If a is a factor of both x and y, then a is also a factor of $x - y$, as is easily verified. Since this is true for any a, it must also be true for the largest common factor a of x and y. Since we can always arrange that y is smaller than x by interchanging the two numbers, this means that the size of the problem can always be reduced. It turns out that this reduction eventually leads to $x = 0$. Since zero has every number as a factor, the greatest common factor of x and y is now y. The following program segment implements this idea, and it now must be proved that this piece of code terminates.

```
x := x1;  y := y1;
while ( x > 0) do
    begin
        if ( x < y ) then
            begin
                s := y;
                y := x;
                x := s
            end
        else x := x − y;
    end
```

The loop variant is x, and this variant decreases by at least 1 in each iteration. The sequence of these values must be finite, which means that the program fragment terminates. ∎

Problems 9.4

1. Prove the while rule in its general form. Why must D be true at the beginning of the loop?
2. Consider a function with two arguments used in a program. State why aliasing could be a problem in this case.
3. Prove the partial correctness of the following program fragment:

```
sum := 0; j := 1;
while (j <> c) do
    begin
        sum := sum + j;  j := j + 1
    end
    {sum = c * (c − 1)/2}
```

   [Hint: Use $sum = (j − 1) * j/2$ as the loop invariant.]
4. Prove the correctness of the following Hoare triple:

$$\{a[i] \geq 0\}\, a[i] := a[i] + a[j]\, \{a[i] \geq a[j]\}$$

5. Prove the following program segment:

```
{n ≥ 0}
i := 1;
while i <= n do
    begin
        a[i] := b[i];  i := i + 1
    end
    {⋀ⁿᵢ₌₁(a[i] = b[i])}
```

6. The following program fragment calculates $\sum_{i=1}^{n} i!$. Prove that this is correct.

```
i := 1; sum := 0; f := 1;
while i <> n + 1 do
    begin
        sum := sum + f; i := i + 1; f := f * i
    end
```

PROBLEMS: CHAPTER 9

1. For each of the following pieces of code, find an appropriate postcondition.
   (a) $\{i < 10\}$ i := 2 * i + 1
   (b) $\{i > 0\}$ i := i - 1
   (c) $\{i > j\}$ i := i + 1; j := j + 1
   (d) $\{\,\}$ i := 3; j := 2 * i
2. For each of the following pieces of code, find appropriate preconditions.
   (a) i := 3 * k $\{i > 6\}$
   (b) a := b * c $\{a = 1\}$
   (c) b := c - 2; a := a/b
3. Prove correctness for the following code. State all rules used.

$$\{y > 0\}\, xa := x + y;\; xb := x - y\, \{xa > xb\}$$

4. Prove the following code, indicating all rules used.

$$\{\,\}\; \textbf{if } x < 0 \textbf{ then } x := -x\, \{x \geq 0\}$$

5. Prove the following program segment:

```
max := a[1]; i := 1;
while (i <> n + 1) do
    begin
        if (a[i] >= max) then max := a[i];
        i := i + 1
    end
```
$$\{\textstyle\bigwedge_{i=1}^{n}(max \geq a[i])\}$$

6. Prove partial correctness for the following code:

```
max := a[1]; i := 1;
while (i < n) do
    begin
        i := i + 1;
        if (a[i] > max) then max := a[i];
    end
```
$$\{\textstyle\bigwedge_{i=1}^{n}(max \geq a[i]),\, \bigvee_{i=1}^{n}(max = a[i])\}$$

7. Prove partial correctness for the following code:

```
i := 0; j := n;
while (i < n) do
    begin
        i := i + 1;  a[i] := b[j]; j := j − 1
    end
{⋀ⁿᵢ₌₁(a[i] = b[n + 1 − i])}
            + 1
```

8. Prove partial correctness for the following code:

```
{⋀ⁿᵢ₌₁(a[i] = A[i])}
i := 1; s := 0;
while (i <= n) do
    begin
        s := s + a[i];  a[i] := s;  i := i + 1
    end
{⋀ⁿᵢ₌₁(a[i] = ∑ⁱₖ₌₁ A[k])}
```

9. Given $0 \le i \le n, n \ge 0$, prove that the following program segment evaluates $n!/(i!*(n-i)!)$.

```
k := 0;  fact := 1;
while (k <> n) do
    begin
        k := k + 1;  fact := fact * k;
        if (k <= i)
            then afact := fact;
        if (k <= n − i)
            then bfact := fact
    end;
bcof := fact/(afact * bfact)
```

Hint: Use the following terms in your loop invariant:

$$\{((i > k) \vee (afact = i!)), (n - i > k) \vee (bfact = (n - i)!), fact = k!, n \ge 0, i \le n\}$$

10. Prove the termination of the program segment given in Problem 5. Which condition do you have to impose in order to do the proof?

11. Prove the termination of the program segment given in Problem 6. Use the notion of a well-founded set in your proof.

10

Grammars, Languages, and Parsing

To use computers efficiently, good programming languages are essential, and such languages require good programming paradigms. New paradigms can change the way we think about programming. To see this, only think about structured programming or, more recently, object-oriented programming. Such new paradigms are constantly being developed, and programmer productivity suffers significantly if these paradigms are not supported by the language used by the programmer. This alone requires an ongoing effort to create new languages. Add to this novel computer architectures, such as parallelism, and the great variety of users, having different expertise and working in different application areas, and it becomes obvious that a great variety of languages is needed. It should therefore come as no surprise that to this date several hundred programming languages have been proposed and implemented. To support the ongoing demand for new languages, efficient language and compiler design tools are needed.

Programming language designers have made extensive uses of a finite device called a grammar in their design efforts. A *grammar* is a vehicle for precisely describing a language because a formal set of grammatical rules is used. Furthermore, a grammar gives structure to the sentences of a language.

A set of grammatical rules in a grammar is specified as a finite relation. By mechanically analyzing these rules, certain *parsing relations* can be defined and obtained from these rules. The resulting parsing relations are used to perform syntactic analysis, that is, to syntactically validate, say, a given computer program.

Syntactic analysis frequently takes the form of constructing a parse tree (recall Section 1.3.3) for a given computer program or logical expression. Two popular approaches have been used in the construction of such a parse tree. First, the tree can be constructed by starting at its root and proceeding downward toward its leaves. This approach is known as top-down parsing. A second parsing approach to constructing a parse tree consists of starting at its leaves and moving upward toward its root. This approach is called bottom-up parsing.

This chapter begins with a formal description of context-free grammars. This discussion contains basic terminology, which is used throughout the rest of the chapter. The next section introduces the class of LL(1) grammars and its associated top-down parsing strategy.

10.1 LANGUAGES AND GRAMMARS

10.1.1 Introduction

The basic machine instructions of a digital computer are very primitive compared with the complex operations that must be performed in various disciplines, such as engineering, commerce, and mathematics. Although a complex procedure can be programmed in machine language, it is desirable to use a high-level language that contains instructions similar to those required in a particular application. For example, in a payroll application, one wants to manipulate employee records in a master file, generate complex reports, and perform rather simple arithmetic operations on certain data. A language such as COBOL, which has high-level commands that manipulate records and generate reports, is a definite asset to a programmer.

While high-level programming languages reduce much of the drudgery of machine language programming, they also introduce new problems. A program (compiler) that converts a program to some object language such as machine language must be written. Also, programming languages must be precisely defined. Sometimes it is the existence of a particular compiler that finally provides the precise definition of a language (not a very satisfactory situation for either the programmer or the compiler writer!). The specification of a programming language involves the definition of the following:

1. The set of symbols (or alphabet) that can be used to construct correct programs
2. The set of all correct programs
3. The "meaning" of all correct programs

In this section we shall be concerned with the first two items in the specification of programming languages.

First, however, we introduce some terminology associated with grammars and formal languages. For this purpose, let V denote a nonempty set of symbols. Such a set V is called an *alphabet*. An alphabet need not be finite or even countable. However, for our purpose here we shall assume V to be finite. For example, we may have

$$V_1 = \{0, 1\}$$
$$V_2 = \{a, b, \ldots, z\}$$
$$V_3 = \{a, 1, *, +, A\}$$

$$V_4 = \{\text{cat, book, dog}, x\}$$
$$V_5 = \{a, \ldots, z, 0, 1, \ldots, 9, ', .\}$$

In alphabet V_4 we assume that each word is a symbol and is indivisible. An element of an alphabet is called a *letter*, a *character*, or a *symbol*. A *string* over an alphabet is a sequence of symbols from the alphabet. A string is also called a *sequence*, a *word*, or a *sentence*, depending on its nature. A string consisting of m symbols ($m > 0$) is called a string of *length* m. For example, let $V = \{a, b\}$; then aa, bb, ab, and ba are all possible strings of length 2. If we admit an empty string, that is, a string of length 0 ($m = 0$), which is usually denoted by ε, then we can have strings of length m for $m \geq 0$. The set of strings over an alphabet V is generally denoted by V^* and the set of nonempty strings by $V^+ = V^* - \{\varepsilon\}$. We shall assume that the strings in V^* are of finite length (see Section 5.2.3).

A language L can be considered a subset of V_T^*, where V_T is the alphabet associated with L. The language consisting of V_T^* is not particularly interesting since it is too large. Our definition of a language L is a set of strings or sentences over some finite alphabet V_T so that $L \subseteq V_T^*$.

How can a language be represented? A language consists of a finite or an infinite set of sentences. Finite languages can be specified by exhaustively enumerating all their sentences. However, for infinite languages such an enumeration is not possible. On the other hand, any device that specifies a language should be finite. One method for specification that satisfies this requirement uses a generative device called a *grammar*. A grammar consists of a finite set of rules or productions that specifies the syntax of the language. These productions are often recursive (see Chapter 4). In addition, a grammar imposes structure on the sentences of a language. The study of grammars constitutes an important subarea of computer science called formal languages. This area emerged in the mid-1950s as a result of the efforts of Noam Chomsky, who gave a mathematical model of a grammar in connection with his study of natural languages. In 1960, the concept of a grammar became important to programmers because the syntax of ALGOL 60 was described by a grammar. Today, grammars are used widely to describe many programming languages.

In this section we are concerned with a grammar as a mathematical system for defining languages and as a device for giving some useful structure to sentences in a language. The problem of syntactic analysis will be discussed briefly.

10.1.2 Discussion of Grammars

A grammar imposes a structure on the sentences of a language. For a sentence in English, such a structure is described in terms of subject, predicate, phrase, noun, and so on. On the other hand, for a program the structure is given in terms of procedures, statements, expressions, and the like. In any case, it may be desirable to describe all such structures and to obtain a set of all the correct or admissible sentences in a language. For example, we may have a set of correct sentences in English or a set of valid Pascal programs. The grammatical structure of a language helps us to determine whether a particular sentence belongs to the set of correct sentences. The grammatical structure of a sentence is generally studied by analyzing the various parts of a sentence and their relationships to one another; this analysis is called *parsing*.

Consider the sentence "a monkey ate the banana." Its structure, or parse, is shown in Figure 10.1. This diagram of a parse displays the syntax of a sentence in the form of a syntax tree or parse tree. Each node in the diagram represents a phrase of the syntax. The words, such as "the" and "monkey," are the basic symbols, or primitives, of the language.

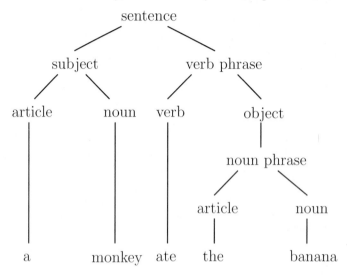

Figure 10.1.   A parse tree for an English sentence

The syntax of a small subset of the English language can be described by using the symbols

S: sentence V: verb O: object A: article N: noun

SP: subject phrase VP: verb phrase NP: noun phrase

These symbols are used to formulate the following rules:

$$S \rightarrow SP \; VP \qquad N \rightarrow \text{tree}$$
$$SP \rightarrow A \; N \qquad VP \rightarrow V \; O$$
$$A \rightarrow a \qquad V \rightarrow \text{ate}$$
$$A \rightarrow \text{the} \qquad V \rightarrow \text{climbs}$$
$$N \rightarrow \text{monkey} \qquad O \rightarrow NP$$
$$N \rightarrow \text{banana} \qquad NP \rightarrow A \; N$$

These rules state that a sentence is composed of a *subject phrase* followed by a *verb phrase*; the subject phrase is composed of an *article* followed by a *noun*; a verb phrase is composed of a *verb* followed by an *object*; and so on.

The structure of a language is discussed by using symbols such as sentence, verb, subject phrase, and verb phrase, which represent *syntactic classes* of elements. Each *syntactic class* consists of a number of alternative structures, and each structure consists of a sequence of items that are either primitives (of the language) or syntactic classes. These alternative structures are called *productions* or *rules of syntax*. For example, the production $S \rightarrow SP\ VP$ defines a sentence to be composed of a subject phrase followed by a verb phrase. The symbol $\rightarrow$ separates the syntactic class sentence from its definition.

The syntactic class and the arrow symbol along with the interpretation of a production enable us to describe a language. As discussed in Section 1.4.2, a system or language that describes another language is known as a *metalanguage*. The metalanguage used to teach German at most Canadian universities is English, while the metalanguage used to teach English is English. The diagram of a sentence describes its *syntax*, but not its meaning or *semantics*. Currently, we are mainly concerned with the syntax of a language, and the device that we have just defined to give the syntactic definition of the language is called a *grammar*.

Using the grammatical rules for our example, we can either *produce* (*generate*) or *derive* a sentence in the language. A computer programmer is concerned with using the productions (grammatical rules) to produce syntactically correct programs. A compiler of a language, on the other hand, is faced with the problem of determining whether a given sentence (source program) is syntactically correct, based on the given grammatical rules. If the syntax is correct, then it produces object code.

Consider the problem of trying to generate or produce the sentence "a monkey ate the banana" from the set of productions given. It is accomplished by starting first with syntactic class symbol S and looking for a production that has S to the left of the arrow. There is only one such production:

$$S \rightarrow SP\ VP$$

We have replaced the class S by its only possible composition. We then take the string

$$SP\ VP$$

and look for a production whose left-hand side is SP and then replace it with the right-hand side of that production. The application of the only production possible produces the string

$$A\ N\ VP$$

We next look for a production whose left part is A, and two such productions are found. By selecting the production $A \rightarrow a$ and upon substituting its right-hand side in the string $A\ N\ VP$, we obtain the string

$$a\ N\ VP$$

This process is continued until we arrive at the correct sentence. At this point, the sentence contains only primitive or terminal elements of the language (no syntactic classes). A complete derivation or generation of the sentence "a monkey ate the banana" is as follows:

$$S \Rightarrow SP\ VP$$
$$\Rightarrow A\ N\ VP$$
$$\Rightarrow a\ N\ VP$$
$$\Rightarrow a\ monkey\ VP$$

$\Rightarrow$ a monkey $V\ O$

$\Rightarrow$ a monkey ate O

$\Rightarrow$ a monkey ate NP

$\Rightarrow$ a monkey ate $A\ N$

$\Rightarrow$ a monkey ate the N

$\Rightarrow$ a monkey ate the banana

Here the symbol $\Rightarrow$ denotes that the string on the right-hand side of the symbol can be obtained by applying one rewriting rule to the previous string.

The rules for the example language can produce a number of sentences.

- the monkey ate the banana
- the monkey climbs a tree
- the monkey climbs the tree
- the banana ate the monkey

The last of these sentences, although grammatically correct, does not really make sense. This situation is often allowed in the specification of languages. There are many valid FORTRAN and Pascal programs that do not make sense. It is easier to define languages if certain sentences of questionable validity are allowed by the rewriting rules.

The set of sentences that can be generated by the example rules is finite. Any interesting language usually consists of an infinite set of sentences. As a matter of fact, the importance of a finite device such as a grammar is that it permits the study of the structure of a language consisting of an infinite set of sentences.

Let the symbols L, D, and I denote the classes L: letter, D: digit, and I: identifier. The productions that follow are recursive and produce an infinite set of names, because the syntactic class I is present on both the left and the right sides of certain productions.

$$I \rightarrow L \qquad D \rightarrow 0$$
$$I \rightarrow ID \qquad D \rightarrow 1$$
$$I \rightarrow IL \qquad \vdots$$
$$L \rightarrow a \qquad D \rightarrow 9$$
$$L \rightarrow b$$
$$\vdots$$
$$L \rightarrow z$$

It is easily seen that the class I defines an infinite set of strings or names in which each name consists of a letter followed by any number of letters or digits. This set is a consequence of using recursion in the definition of the productions $I \rightarrow ID$ and $I \rightarrow IL$. In fact, recursion is fundamental to the definition of an infinite language by the use of a grammar (see also Section 3.3).

10.1.3 Formal Definition of a Language

Let us now formalize the idea of a grammar and how it is used. For this purpose, let V_T be a finite nonempty set of symbols called the *alphabet*. The symbols in V_T are called *terminal symbols*. The *metalanguage* that is used to generate strings in the language is assumed to contain a set of syntactic classes or variables called *nonterminal symbols*. The set of nonterminal symbols is denoted by V_N, and the elements of V_N are used to define the syntax (structure) of the language. Furthermore, the sets V_N and V_T are assumed to be disjoint. The set $V_N \cup V_T$ consisting of nonterminal and terminal symbols is called the *vocabulary* of the language. We shall use capital letters such as $A, B, C, \ldots, X, Y, Z$ to denote nonterminal symbols, while $S_1, S_2, \ldots$ represent the elements of the vocabulary. The strings of terminal symbols are denoted by lowercase letters $x, y, z, \ldots$, while strings of symbols over the vocabulary are given by $\alpha, \beta, \gamma, \ldots$. The length of a string α will be denoted by #α.

Definition 10.1: A *context-free grammar* is defined by a 4-tuple $G = (V_N, V_T, S, \Phi)$, where V_T and V_N are sets of terminal and nonterminal (syntactic class) symbols, respectively. S, a distinguished element of V_N and therefore of the vocabulary, is called the starting symbol. Φ is a finite relation from V_N to $(V_T \cup V_N)^+$. In general, an element $\langle \alpha, \beta \rangle$ is written as $\alpha \rightarrow \beta$ and is called a *production rule* or a *rewriting rule*. $V_N = V_N \cup V_T$ is called the vocabulary of grammar.

Note that the above definition does not allow the use of empty rules (ϵ-rules). An empty rule is one with its right part as the empty string. Although most context-free parsers can handle empty rules, we do not further discuss them in the interest of simplicity.

For our example given earlier, dealing with variable names, we may write the grammar as $G_1 = (V_N, V_T, S, \Phi)$ in which

$$V_N = \{I, L, D\}$$
$$V_T = \{a, b, c, d, e, f, g, h, i, j, k, l, m, n, o, p, q, r, s, t, u, v, w, x, y, z,$$
$$0, 1, 2, 3, 4, 5, 6, 7, 8, 9\}$$
$$S = I$$
$$\Phi = \{I \rightarrow L, I \rightarrow IL, I \rightarrow ID, L \rightarrow a, L \rightarrow b, \ldots, L \rightarrow z,$$
$$D \rightarrow 0, D \rightarrow 1, \ldots, D \rightarrow 9\}$$

Definition 10.2: Let $G = (V_N, V_T, S, \Phi)$ be a grammar. For $\sigma, \psi \in (V_N \cup V_T)^+$, σ is said to be a *direct derivative* of ψ, written as $\psi \Rightarrow \sigma$, if there are strings ϕ_1 and ϕ_2 (including possibly empty strings) such that $\psi = \phi_1 B \phi_2$, $\sigma = \phi_1 \beta \phi_2$, $B \in V_N$, and $B \rightarrow \beta$ is a production of G.

If $\psi \Rightarrow \sigma$, we may also say that ψ directly produces σ or σ directly reduces to ψ. For grammar G_1 of our example, we have listed in Table 10.1 some illustrations of direct derivations.

TABLE 10.1.   Direct Derivatives in G_1

ψ	σ	Rule Used	ϕ_1	ϕ_2
I	L	$I \to L$	ε	ε
Ib	Lb	$I \to L$	ε	b
Lb	ab	$L \to a$	ε	b
LD	$L1$	$D \to 1$	L	ε
LD	aD	$L \to a$	ε	D

These concepts can now be extended to produce a string σ not necessarily directly, but in a number of steps from a string ψ.

Definition 10.3:   Let $G = (V_N, V_T, S, \Phi)$ be a grammar. The string ψ produces σ (σ reduces to ψ or σ is the derivation of ψ), written as $\psi \overset{+}{\Rightarrow} \sigma$, if there are strings $\phi_1, \phi_2, \ldots, \phi_n (n > 0)$ such that $\psi = \phi_0 \Rightarrow \phi_1, \phi_1 \Rightarrow \phi_2, \ldots, \phi_{n-1} \Rightarrow \phi_n$ and $\phi_n = \sigma$. The relation $\overset{+}{\Rightarrow}$ is the transitive closure of the relation $\Rightarrow$. If we let $n = 0$, then we can define the reflexive transitive closure of $\Rightarrow$ as

$$\psi \overset{*}{\Rightarrow} \sigma \equiv \psi \overset{+}{\Rightarrow} \sigma \text{ or } \psi = \sigma$$

Returning to the grammar G_1, we show that the string $abc12$ is derived from I by following the derivation sequence:

$$I \Rightarrow ID$$
$$\Rightarrow IDD$$
$$\Rightarrow ILDD$$
$$\Rightarrow ILLDD$$
$$\Rightarrow LLLDD$$
$$\Rightarrow aLLDD$$
$$\Rightarrow abLDD$$
$$\Rightarrow abcDD$$
$$\Rightarrow abc1D$$
$$\Rightarrow abc12$$

Note that as long as we have a nonterminal character in the string we can produce a new string from it. On the other hand, if a string contains only terminal symbols, then the derivation is complete, and we cannot produce any further strings from it.

Definition 10.4:   A *sentential form* is any derivative of the unique nonterminal symbol S. The language L generated by a grammar G is the set of all sentential forms whose symbols are terminal; that is,

$$L(G) = \{\sigma | S \overset{*}{\Rightarrow} \sigma \text{ and } \sigma \in V_T^+ \}$$

Therefore, the language is merely a subset of the set of all terminal strings over V_T.

We shall now give a number of examples of grammars.

Example 10.1   Let $G_2 = (\{E, T, F\}, \{a, +, *, (,)\}, E, \Phi)$, where Φ consists of the productions

$$E \rightarrow E + T$$
$$E \rightarrow T$$
$$T \rightarrow T * F$$
$$T \rightarrow F$$
$$F \rightarrow (E)$$
$$F \rightarrow a$$

The variables E, T, and F, here represent the names *expression*, *term*, and *factor*, respectively, commonly used in conjunction with arithmetic expressions. A derivation for the expression $a * a + a$ is

$$E \Rightarrow E + T$$
$$\Rightarrow T + T$$
$$\Rightarrow T * F + T$$
$$\Rightarrow F * F + T$$
$$\Rightarrow a * F + T$$
$$\Rightarrow a * a + T$$
$$\Rightarrow a * a + F$$
$$\Rightarrow a * a + a$$

Example 10.2   The set of completely parenthesized logical expressions is generated by the grammar

$$G_3 = (\{L\}, \{a, \neg, \wedge, \vee, \Rightarrow, \Leftrightarrow\}, L, \Phi)$$

where Φ is the set of productions

$$L \rightarrow a$$
$$L \rightarrow t$$
$$L \rightarrow f$$
$$L \rightarrow (\neg L)$$
$$L \rightarrow (L \wedge L)$$
$$L \rightarrow (L \vee L)$$
$$L \rightarrow (L \Rightarrow L)$$
$$L \rightarrow (L \Leftrightarrow L)$$

Note that the set of terminals in this grammar consists of a, which represents an atomic expression, t (true), f (false), and $\neg, \wedge, \vee, \Rightarrow$, and $\Leftrightarrow$, which are the logical connectives.

Example 10.3   The language $L(G_4) = \{a^n b a^n | n \geq 1\}$ is generated by the grammar

$$G_4 = (\{S, C\}, \{a, b\}, S, \Phi)$$

where Φ is the set of productions

$$S \rightarrow aCa$$
$$C \rightarrow aCa$$
$$C \rightarrow b$$

A derivation for $a^2 b a^2$ consists of the following steps:

$$S \Rightarrow aCa$$
$$\Rightarrow aaCaa$$
$$\Rightarrow aabaa$$

Example 10.4   The language $L(G_5) = \{a^n b a^m | n, m \geq 1\}$ is generated by the grammar

$$G_5 = (\{S, A, B, C\}, \{a, b\}, S, \Phi)$$

where the set of productions is

$$S \rightarrow aS$$
$$S \rightarrow aB$$
$$B \rightarrow bC$$
$$C \rightarrow aC$$
$$C \rightarrow a$$

The sentence $a^2 b a^3$ has the following derivation:

$$S \Rightarrow aS$$
$$\Rightarrow aaB$$
$$\Rightarrow aabC$$
$$\Rightarrow aabaC$$
$$\Rightarrow aabaaC$$
$$\Rightarrow aabaaa$$

With such grammars, the rewriting variable in a sentential form is rewritten regardless of the other symbols in its vicinity or context. It has led to the term *context free* for grammars consisting of productions whose left-hand side consists of a single class symbol. Context-free grammars do not have the power to represent even significant parts of the English language since context dependency is often required in order to properly analyze the structure of a sentence. Context-free grammars are not capable of specifying (or determining) that a certain variable was declared when it is used in some expression in a subsequent statement of a source program. However, these grammars can specify most of the syntax for computer or artificial languages since these are, by and large, simple in structure. Context-free grammars are said to generate context-free languages.

We conclude this subsection by introducing a different metalanguage from the one that was previously used. The metavariables or syntactic classes will be enclosed by the symbols ⟨and⟩. Using this terminology, the symbol ⟨sentence⟩ is a symbol of V_N, and the symbol "sentence" is an element of V_T. In this way, no confusion or ambiguity arises when attempting to distinguish the two symbols. This metalanguage is known as *Backus–Naur form* (*BNF*). The BNF metalanguage has been used extensively in the formal definition of many programming languages. A popular language described using BNF is ALGOL. For example, the definition of an identifier in BNF is given as

$$\langle \text{identifier} \rangle \ ::= \ \langle \text{letter} \rangle \mid \langle \text{identifier} \rangle \ \langle \text{digit} \rangle \mid \langle \text{identifier} \rangle \ \langle \text{letter} \rangle$$
$$\langle \text{letter} \rangle \ ::= a \mid b \mid c \mid \cdots \mid y \mid z$$
$$\langle \text{digit} \rangle \ ::= 0 \mid 1 \mid 2 \mid \cdots \mid 8 \mid 9$$

The symbol ::= replaces the symbol → in the grammar notation, and | is used to separate different right-hand sides of productions corresponding to the same left-hand side. The symbol ::= is interpreted as "is defined as" and | as "or." BNF gives a much more compact description of a language than could be achieved with the previous metalanguage.

10.1.4 Notions of Syntax Analysis

In the following pages we briefly discuss the problem of syntax analysis or parsing. The *parse* of a sentence is the construction of a derivation for that sentence; that is, a sequence of productions used in generating a given sentence from the starting symbol is required.

An important aid to understanding the syntax of a sentence is a *syntax tree* (again recall Section 1.3.3). The structural relationships between the parts of a sentence are easily seen from its syntax tree. Consider the grammar described earlier for the set of valid identifiers or variable names. The derivation of the identifier $a1$ is

$$\langle \text{identifier} \rangle \ \Rightarrow \ \langle \text{identifier} \rangle \ \langle \text{digit} \rangle \ \Rightarrow \ \langle \text{letter} \rangle \ \langle \text{digit} \rangle \ \Rightarrow a \ \langle \text{digit} \rangle \ \Rightarrow a1$$

Let us now illustrate how to construct a syntax tree corresponding to this derivation. This process is shown as a sequence of diagrams in Figure 10.2, where each diagram corresponds to a sentential form in the derivation of the sentence. The syntax tree has a distinguished point called its root, which is labeled by the starting symbol ⟨identifier⟩ of the grammar. From the root we draw two downward branches (see Figure 10.2b) corresponding to the rewriting of ⟨identifier⟩ by ⟨identifier⟩ ⟨digit⟩. The symbol ⟨identifier⟩ in the sentential form ⟨identifier⟩ ⟨digit⟩ is then rewritten as ⟨letter⟩ by using the production ⟨identifier⟩ ::= ⟨letter⟩ (see Figure 10.2c). This process continues for each production applied until Figure 10.2e is obtained.

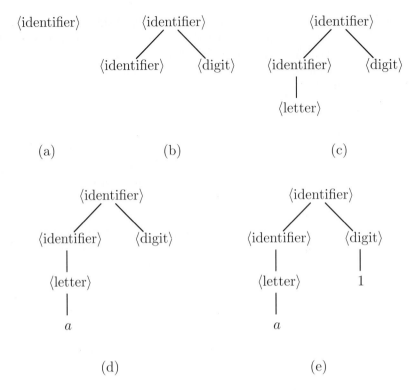

Figure 10.2. Syntax tree for sentence $a1$

Given a sentence in the language, the construction of a parse can be described pictorially as in Figure 10.3, where the root and leaves (which represent the terminal symbols in the sentence) of the tree are known and the rest of the syntax tree must be found. There are a number of ways by which this construction can be accomplished. First, an attempt to construct the tree can be initiated by starting at the root and proceeding downward toward the leaves. This method is called a *top-down parse*. Alternatively, the completion of the tree can be attempted by starting at the leaves and moving upward toward the root. This method is called a *bottom-up parse*. The top-down and bottom-up approaches can be combined to yield other possibilities.

Let us briefly discuss top-down parsing. Consider the identifier $c2$ generated by the BNF grammar of the previous subsection. The first step is to construct the direct derivation ⟨identifier⟩ ⇒ ⟨identifier⟩ ⟨digit⟩. At each successive step, the leftmost variable A of the current sentential form $\phi_1 A \phi_2$ is replaced by the right part of a production $A ::= \psi$ to obtain the next sentential form. This process is shown for the identifier $c2$ by the five trees in Figure 10.4.

We have very conveniently chosen the rules that generate the given identifier. If the first step had been the construction of the direct derivation ⟨identifier⟩ ⇒ ⟨identifier⟩ ⟨letter⟩, then we would have eventually produced the sentential form c ⟨letter⟩, where it would have been impossible to obtain $c2$. At this point, a new alternative would have to be tried by

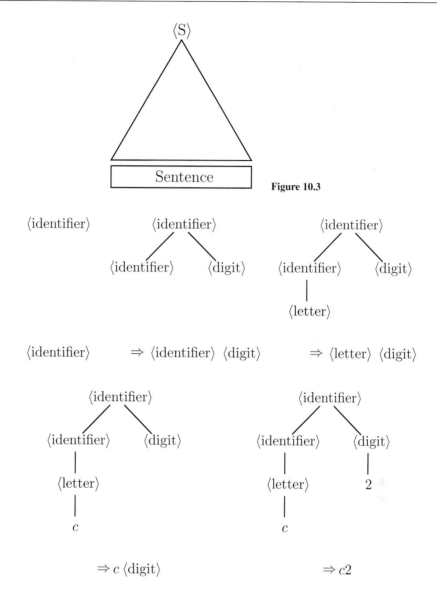

Figure 10.3

$\langle \text{identifier} \rangle \Rightarrow \langle \text{identifier} \rangle \langle \text{digit} \rangle \Rightarrow \langle \text{letter} \rangle \langle \text{digit} \rangle$

$\Rightarrow c \langle \text{digit} \rangle \qquad\qquad \Rightarrow c2$

Figure 10.4.   Trace of a top-down parse

restarting the procedure and choosing the rule $\langle \text{identifier} \rangle ::= \langle \text{identifier} \rangle \langle \text{digit} \rangle$. We examine top-down parsing in more detail in Section 10.2.

A bottom-up parsing technique begins with a given string and tries to reduce it to the starting symbol of the grammar. The first step in parsing the identifier $c2$ is to reduce c to $\langle \text{letter} \rangle$, resulting in the sentential form $\langle \text{letter} \rangle 2$. The direct derivation $\langle \text{letter} \rangle 2 \Rightarrow c2$ has now been constructed, as shown in Figure 10.5d. The next step is to reduce $\langle \text{letter} \rangle$ to $\langle \text{identifier} \rangle$, as represented by Figure 10.5c. The process continues until the entire syntax tree of Figure 10.5a is constructed. Note that it is possible to construct other derivations, but the resulting syntax tree is the same.

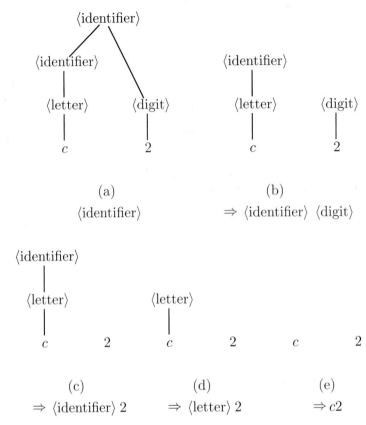

Figure 10.5.   Trace of a bottom-up parse

Example

10.5   The following grammar generates the set of logical expressions

$$G_6 = (V_N, V_T, \langle \text{logical wff} \rangle, \Phi)$$

where

$$V_N = \{ \langle \text{logical wff} \rangle, \langle \text{if then} \rangle, \langle \text{disjunction} \rangle,$$
$$\langle \text{conjunction} \rangle, \langle \text{negation} \rangle, \langle \text{primary} \rangle \}$$
$$V_T = \{ a, t, f, \neg, \vee, \wedge, \Rightarrow, \Leftrightarrow \}$$

and the set of productions is

$$\langle \text{logical wff} \rangle ::= \langle \text{if then} \rangle \mid \langle \text{logical wff} \rangle \Leftrightarrow \langle \text{if then} \rangle$$
$$\langle \text{if then} \rangle ::= \langle \text{disjunction} \rangle \mid \langle \text{if then} \rangle \Rightarrow \langle \text{disjunction} \rangle$$
$$\langle \text{disjunction} \rangle ::= \langle \text{conjunction} \rangle \mid \langle \text{disjunction} \rangle \vee \langle \text{conjunction} \rangle$$

$$\langle \text{conjunction} \rangle ::= \langle \text{negation} \rangle \mid \langle \text{conjunction} \rangle \wedge \langle \text{negation} \rangle$$
$$\langle \text{negation} \rangle ::= \langle \text{primary} \rangle \mid \neg \langle \text{negation} \rangle$$
$$\langle \text{primary} \rangle ::= a|t|f|(\ \langle \text{logical wff} \rangle\ )$$

Unlike grammar G_3 for completely parenthesized logical expressions given earlier, this grammar "enforces" the normal precedence associated with the logical connectives; that is, parentheses have the highest precedence and $\Leftrightarrow$ the lowest. Observe that the grammar contains a distinct nonterminal for each level of logical connective (or operator) precedence. Also, the logical expression need not be completely parenthesized. The precedence levels and the ability to write partially parenthesized logical expressions are implicitly contained in the structure of the grammar. Observe that, the deeper the position of an operator in an associated parse tree for some given input, the higher is the evaluation precedence for this operator; and conversely. Thus parentheses have the highest precedence because they will be deep in a parse tree. In an unparenthesized expression, the operator $\Leftrightarrow$ has the lowest evaluation precedence. The reader is encouraged to generate a derivation (and its associated parse tree) for the logical expression

$$((P_1 \wedge P_2) \Rightarrow (P_1 \vee \neg P_3))$$

where the terminal symbol "a" can represent P_1, P_2, P_3, and P_4, and verify that, from a precedence viewpoint, the syntax tree represents the expression properly.

We now turn to a more general discussion of language translation. A compiler for a certain language is concerned with several tasks: determining whether a sentence belongs to the language, constructing a syntax tree for the sentence, and generating object code for the given sentence if its syntax and semantics are valid. This process can be represented by Figure 10.6.

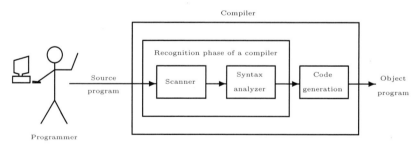

Figure 10.6. Block diagram of a compiler

The source program is input to a scanner whose purpose is to separate the incoming text into pieces such as constants, variable names, key words (such as **for, if,** and the like, in Pascal) and operators. This type of analysis is quite simple to perform. Usually, the scanner constructs tables that contain variable names, constants, and labels.

The scanner feeds the syntax analyzer, whose task is essentially to construct a syntax tree for the given source program. The syntax analyzer is much more complicated than the scanner. The output of the syntax analyzer is fed to the code-generation block, which uses the syntax tree for the sentence and other things (which are not specified for simplicity) to generate object code for that source program.

Consider the following scanner problem. It is required to construct a scanner that will split a sentence into a number of parts. These parts are strings such as identifiers, literal constants, adding operators, multiplying operators, and assignment operators. The syntax of these primitive classes is now given. An identifier is described by the following rules:

$$\langle \text{identifer} \rangle \ ::= \ \langle \text{identifier} \rangle \ \langle \text{letter} \rangle \ | \ \langle \text{identifier} \rangle \ \langle \text{digit} \rangle \ | \ \langle \text{letter} \rangle$$
$$\langle \text{letter} \rangle \ ::= \ A|B|C| \cdots |Y|Z$$
$$\langle \text{digit} \rangle \ ::= \ 0|1| \cdots |8|9$$

Literal constants are described by the rules

$$\langle \text{literal constant} \rangle \ ::= \ \langle \text{digit string} \rangle \ | \ \langle \text{digit string} \rangle \ . \ \langle \text{digit string} \rangle$$
$$| \ \langle \text{digit string} \rangle \ .$$
$$\langle \text{digit string} \rangle \ ::= \ \langle \text{digit} \rangle \ | \ \langle \text{digit string} \rangle \ \langle \text{digit} \rangle$$

The different classes of operators are described by the productions

$$\langle \text{adding operator} \rangle \ ::= \ + \ | -$$
$$\langle \text{multiplying operator} \rangle \ ::= \ * \ | \ /$$
$$\langle \text{assignment operator} \rangle \ ::= \ :=$$

The syntactic classes $\langle \text{identifier} \rangle$, $\langle \text{literal constant} \rangle$, $\langle \text{adding operator} \rangle$, $\langle \text{multiplying operator} \rangle$, and $\langle \text{assignment operator} \rangle$ in a line of text are to be singled out. For example, an input statement such as

$$\text{Answer} := 15.5 * G + H * F/12 - Q$$

would break down to the following table:

Answer	$\langle \text{identifier} \rangle$
:=	$\langle \text{assignment operator} \rangle$
15.5	$\langle \text{literal constant} \rangle$
*	$\langle \text{multiplying operator} \rangle$
G	$\langle \text{identifier} \rangle$
+	$\langle \text{adding operator} \rangle$
H	$\langle \text{identifier} \rangle$
*	$\langle \text{multiplying operator} \rangle$
F	$\langle \text{identifier} \rangle$
/	$\langle \text{multiplying operator} \rangle$
12	$\langle \text{literal constant} \rangle$
−	$\langle \text{adding operator} \rangle$
Q	$\langle \text{identifier} \rangle$

10.1.5 Ambiguous Grammars

An important question that arises in formal languages is whether a sentence generated by a grammar has a unique syntax tree.

Example 10.6   Consider the simple grammar that has the following productions:

$$\langle expr \rangle := i \mid \langle expr \rangle + \langle expr \rangle \mid \langle expr \rangle * \langle expr \rangle$$
$$\mid \langle expr \rangle \uparrow \langle expr \rangle \mid (\langle expr \rangle)$$

where i and $\uparrow$ denote an identifier and the exponential operator, respectively. Show that some expressions have more than one syntax tree.

Solution   Let us find a derivation for the sentence $i + i * i \uparrow i$. One such derivation is

$$\begin{aligned}
\langle expr \rangle &\Rightarrow \underline{\langle expr \rangle} + \langle expr \rangle \\
&\Rightarrow \langle expr \rangle + \langle expr \rangle * \underline{\langle expr \rangle} \\
&\Rightarrow \langle expr \rangle + \langle expr \rangle * \langle expr \rangle \uparrow \langle expr \rangle \\
&\overset{+}{\Rightarrow} i + i * i \uparrow i
\end{aligned}$$

where each underlined nonterminal denotes that it is being rewritten in the next step of the derivation. Another possible derivation is

$$\begin{aligned}
\langle expr \rangle &\Rightarrow \underline{\langle expr \rangle} \uparrow \langle expr \rangle \\
&\Rightarrow \langle expr \rangle + \underline{\langle expr \rangle} \uparrow \langle expr \rangle \\
&\Rightarrow \langle expr \rangle + \langle expr \rangle * \langle expr \rangle \uparrow \langle expr \rangle \\
&\overset{+}{\Rightarrow} i + i * i \uparrow i
\end{aligned}$$

Both of these possibilities are diagrammed in Figure 10.7. It is clear that the two syntax trees are different. That is, we have two different parses for the same sentence. Figure 10.7a shows the correct interpretation, assuming that exponentiation has precedence over multiplication, which, in turn, has precedence over addition. In Figure 10.7b, however, the multiplication is done first, addition second, and $\uparrow$ last. This interpretation is clearly not the one that we associate with ordinary mathematics.   ∎

The existence of more than one parse for some sentence in a language can cause a compiler to generate a different set of instructions (object code) for different parses. Usually, this phenomenon is intolerable. If a compiler is to perform valid translations of sentences in a language, then that language must be unambiguously defined. This concept leads to the following definition:

Definition 10.5:   A sentence generated by a grammar is *ambiguous* if there exists more than one syntax tree for it. A grammar is ambiguous if it generates at least one ambiguous sentence; otherwise, it is *unambiguous*.

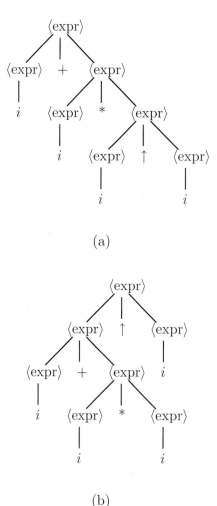

(a)

(b)

Figure 10.7.   Two distinct syntax trees for the sentence $i + i * i \uparrow i$

It should be noted that we called the grammar ambiguous and not the language that it generates. Many grammars can generate the same language; some are ambiguous and some are not. However, there are certain languages for which no unambiguous grammars can be found. Such languages are said to be *inherently ambiguous*. For example, the language $\{x^i y^j z^k \mid i = j \text{ and } j = k\}$ is an inherently ambiguous context-free language.

The question that naturally arises at this point is whether there exists an algorithm that can accept any context-free grammar and decide whether, in some finite time, it is ambiguous. The answer is no! This problem, in general, is undecidable. However, a simple set of sufficient conditions can be developed such that, when they are applied to a grammar and are found to hold, the grammar is guaranteed to be unambiguous. We wish to point out that these conditions are sufficient but not necessary. In other words, even if a grammar does not satisfy the conditions, it may still be unambiguous. We will formulate such a set of conditions in Section 10.2.3.

It should be noted that the grammar in the previous example can be rewritten so that the new version is unambiguous. Such an unambiguous grammar contains the following productions:

$$\langle expr \rangle ::= \langle term \rangle \mid \langle expr \rangle + \langle term \rangle$$
$$\langle term \rangle ::= \langle factor \rangle \mid \langle term \rangle * \langle factor \rangle$$
$$\langle factor \rangle ::= \langle primary \rangle \mid \langle primary \rangle \uparrow \langle factor \rangle$$
$$\langle primary \rangle ::= i \mid (\langle expr \rangle)$$

In this version $\uparrow$ has precedence over $*$, which in turn has precedence over $+$. Of course, parentheses have the highest precedence. Note that we have made $\uparrow$ right associative with a right-recursive rule, while $*$ and $+$ are left associative.

As a second example of ambiguity, we consider the "dangling else" problem associated with a conditional statement in certain procedural programming languages.

Example 10.7   Consider the following set of productions:

$$\langle if\ stat \rangle ::= \text{if } b \text{ then } \langle stat \rangle \mid$$
$$\text{if } b \text{ then } \langle stat \rangle \text{ else } \langle stat \rangle$$
$$\langle stat \rangle ::= s \mid \langle if\ stat \rangle$$

where b denotes a Boolean result and s is a single statement, such as an assignment statement. Show that this grammar is ambiguous.

Solution   Consider the following nested statement:

$$\text{if } b \text{ then if } b \text{ then } s \text{ else } s$$

We now show that this statement is ambiguous since it is not known if the *else* is associated with the first or second *then*. As a first derivation we have

$$\langle if\ stat \rangle \Rightarrow \text{if } b \text{ then } \langle stat \rangle$$
$$\Rightarrow \text{if } b \text{ then } \langle if\ stat \rangle$$
$$\Rightarrow \text{if } b \text{ then if } b \text{ then } \langle stat \rangle \text{ else } \langle stat \rangle$$
$$\Rightarrow \text{if } b \text{ then if } b \text{ then } s \text{ else } \langle stat \rangle$$
$$\Rightarrow \text{if } b \text{ then if } b \text{ then } s \text{ else } s$$

As a second derivation, consider the following:

$$\langle if\ stat \rangle \Rightarrow \text{if } b \text{ then } \langle stat \rangle \text{ else } \langle stat \rangle$$
$$\Rightarrow \text{if } b \text{ then } \langle if\ stat \rangle \text{ else } \langle stat \rangle$$
$$\Rightarrow \text{if } b \text{ then if } b \text{ then } \langle stat \rangle \text{ else } \langle stat \rangle$$
$$\Rightarrow \text{if } b \text{ then if } b \text{ then } s \text{ else } \langle stat \rangle$$
$$\Rightarrow \text{if } b \text{ then if } b \text{ then } s \text{ else } s$$

It is clear that the syntax trees given in Figure 10.8 are different. In Figure 10.8a the *else* is associated with the *then* to its immediate left, while in Figure 10.8b the *else* is associated with the first *then*. In any case, these two distinct parse trees have different meanings.   ∎

The grammar in the preceding example can be rewritten so that the new version is unambiguous. This is left as an exercise. In some programming languages the "dangling

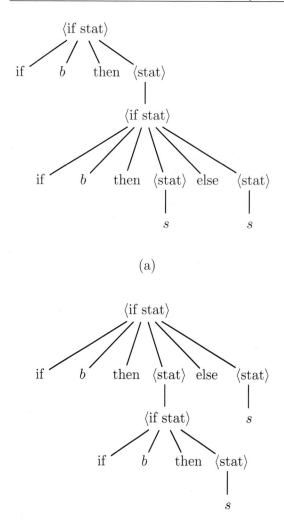

(a)

(b)

Figure 10.8.   Ambiguous syntax trees for two conditional statements

else" problem has been resolved by introducing an additional keyword (or delimiter) such as fi at the end of a conditional statement. Using this approach, an unambiguous grammar for the conditional statement consists of the following productions:

$$\langle \text{if stat} \rangle ::= \text{if } b \text{ then } \langle \text{stat} \rangle \text{ fi } |$$
$$\text{if } b \text{ then } \langle \text{stat} \rangle \text{ else } \langle \text{stat} \rangle \text{ fi}$$
$$\langle \text{stat} \rangle ::= s \mid \langle \text{if stat} \rangle$$

The example conditional statement is then rewritten as

if b
then if b then s else s fi fi

or

if b
then if b then s fi
else s fi

depending on which interpretation is desired. In either case, each of the two statements has a unique parse tree, as shown in Figure 10.9.

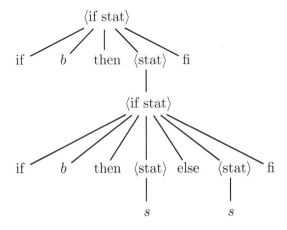

(a) if b then if b then s else s fi fi

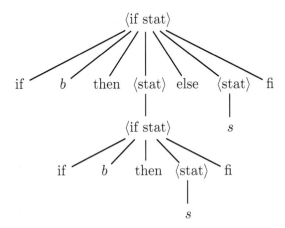

(b) if b then if b then s fi else s fi

Figure 10.9. Unambiguous syntax trees for two conditional statements

Problems 10.1.5

1. Rewrite the grammar in Example 10.7 so that the new version is unambiguous.

2. Suppose we want to implement a DCC compiler for the DDC (decimal digit calculator) language which performs arithmetic operations on integer arguments. The BNF grammar description below was written to describe the DDC language syntactically. Unfortunately, the grammar is ambiguous.

$$
\begin{array}{lll}
\langle \text{DDC expr} \rangle & ::= & \langle \text{DDC term} \rangle \\
& & | \ \langle \text{DDC expr} \rangle \ \langle \text{op1} \rangle \ \langle \text{DDC expr} \rangle \\
\langle \text{DDC term} \rangle & ::= & \langle \text{decimal arg} \rangle \\
& & | \ \langle \text{DDC term} \rangle \ \langle \text{op2} \rangle \ \langle \text{decimal arg} \rangle \\
\langle \text{decimal arg} \rangle & ::= & \langle \text{digit} \rangle \\
& & | \ \langle \text{decimal arg} \rangle \ \langle \text{digit} \rangle \\
\langle \text{digit} \rangle & ::= & 0 \mid 1 \mid 2 \mid 3 \mid 4 \mid 5 \mid 6 \mid 7 \mid 8 \mid 9 \\
\langle \text{op1} \rangle & ::= & + \mid - \\
\langle \text{op2} \rangle & ::= & * \mid / \\
\end{array}
$$

   (a) Demonstrate that the grammar is, indeed, ambiguous.
   (b) Correct the grammar so that it is unambiguous.
   (c) According to your grammar, what is the value of $3 * 6/2 + 4$?
   (d) If we change the BNF description of $\langle \text{op1} \rangle$ and $\langle \text{op2} \rangle$ to read

$$
\begin{array}{lll}
\langle \text{op1} \rangle & ::= & * \mid / \\
\langle \text{op2} \rangle & ::= & + \mid - \\
\end{array}
$$

   what is the value of the expression $3 * 6/2 + 4$ in the language described by your corrected grammar?

10.1.6 Reduced Grammars

Given a context-free grammar, there are a number of questions concerning its simplicity which can be asked. Some of the questions are the following:

1. Are rules of the form $A \rightarrow A$ absent from the grammar?

2. Does each nonterminal produce some part of a terminal sentence?

3. Beginning with the starting symbol of the grammar, does each symbol of the grammar occur in some sentential form?

If the answer to each of these questions is yes, then a grammar is said to be *reduced*. Note that other definitions of reduced grammars have been used in the literature. This subsection is concerned with obtaining from a grammar its reduced equivalent. By equivalent we mean that both grammars generate the same language.

Let us now consider the problem of obtaining the reduced equivalent of a given context-free grammar. Rules of the form $A \rightarrow A$ can be removed from the grammar. Such productions are not only useless, but they make the grammar ambiguous. Since such a rule can be applied any number of times, there are many possible distinct syntax trees for some sentence in the language.

Consider the following grammar:

$$S \to aAa \quad B \to abb$$
$$A \to Sb \quad B \to aC$$
$$A \to bBB \quad C \to aCA$$

On examining this grammar, we find that it is impossible to produce any terminal sentences from the nonterminal C. That is, $\{x | C \overset{*}{\Rightarrow} x \text{ and } x \in \{a, b\}^+\}$ is empty. Consequently, any rule that contains C can be deleted. Making these deletions yields the following grammar:

$$S \to aAa \quad A \to bBB$$
$$A \to Sb \quad B \to abb$$

If a nonterminal symbol generates at least one terminal string, such a symbol is said to be an *active nonterminal*. Given a context-free grammar, we would like to formulate an algorithm that will determine the set of active symbols. Initially, each symbol in the set $\{A | A \to \alpha \text{ and } \alpha \in V_T^+\}$ is active. We can obtain, in a recursive manner, other active symbols from the following observation: If all nonterminal symbols in the right part of a rule are active, then the symbol in its left part is also active. Observe that all terminal symbols are considered to be active.

For the grammar just given, the initial set of active symbols is $\{B\}$. Since the symbol B is active, so is A. S then becomes active because the symbols a and A are active. At this point, no additional symbols (such as C) are found to be active; so the process halts. The set $\{A, B, S\}$ represents the active symbols of the sample grammar. This computational process is incorporated in the following informal algorithm, where the symbol := denotes the assignment operator.

Algorithm ACTIVE   Given a context-free grammar $G = (V_N, V_T, S, \Phi)$, this algorithm determines the active set of nonterminals V_N' in G. The rules of the grammar are then scanned for selection. A rule is selected if all its nonterminals are active. The resulting set of rules is denoted by Φ'. The number of nonterminals in G is denoted by m. The sets W_k for $1 \leq k \leq m$ represent sets of nonterminals.

1. [Compute initial set of active nonterminals]

    $W_1 := \{A | (A \to \alpha) \in \Phi \text{ and } \alpha \in V_T^+\}$

2. [Calculate successive sets of active nonterminals]

    Repeat for $k = 2, 3, \ldots, m$

    $W_k := W_{k-1} \cup \{A | (A \to \alpha) \in \Phi \text{ and } \alpha \in (V_T \cup W_{k-1})^+\}$

    If $W_k = W_{k-1}$ or $k = m$

    then $V_N' := W_k$

    $\Phi' := \{(A \to \alpha) \in \Phi | A, \alpha \in (V_T \cup V_N')^+\}$

    Exit

Note that $W_k \supseteq W_{k-1}$ for all k. Also, since the grammar contains m nonterminals, there is no need to compute any set beyond W_m. When the equality of two successive W sets is detected, the desired set of active nonterminals has been obtained, and the sets V_N' and Φ' can then be generated. The reader should verify that L(G) = L(G'), where $G' = (V_N', V_T, S, \Phi')$. That is, both grammars generate exactly the same language.

For the sample grammar given earlier, the sequence of active nonterminal sets is $W_1 = \{B\}$, $W_2 = \{B, A\}$, $W_3 = \{B, A, S\}$, and $W_4 = W_3$.

Another subclass of undesirable symbols is those that cannot occur in any sentential form derivable from the starting symbol of the grammar. Alternatively, we are only interested in those symbols that belong to the set $\{A \mid S \overset{*}{\Rightarrow} \phi_1 A \phi_2 \text{ where } A \in V_T \cup V_N\}$. A symbol that occurs in some sentential form derivable from the starting symbol is called a *reachable symbol*; otherwise, it is called an *unreachable symbol*.

The set of reachable symbols can be easily obtained by first noting that the starting symbol of the grammar is reachable. Second, if the nonterminal in the left part of a production is reachable, then so are all the symbols in its right part. This fact can be used repeatedly until all reachable symbols are found. As an example, consider the grammar

$$
\begin{aligned}
S &\to aSb & A &\to aAc \\
S &\to bAB & C &\to aSbS \\
S &\to a & C &\to aba \\
B &\to d
\end{aligned}
$$

The initial set of reachable symbols is $\{S\}$. Since S is reachable, then from productions $S \to a$, $S \to aSb$, and $S \to bAB$ it directly follows that the new set of reachable symbols is $\{S, A, B, a, b\}$. Now that A and B are reachable, it follows from productions $B \to d$ and $A \to aAc$ that c and d are reachable. Note that C is unreachable. Consequently, the set of reachable symbols is $\{S, A, B, a, b, c, d\}$. The rules with C as their left part can therefore be deleted from the grammar. The reduced grammar then becomes

$$
\begin{aligned}
S &\to aSb & A &\to aAc \\
S &\to bAB & B &\to d \\
S &\to a
\end{aligned}
$$

A more formal algorithm that is similar in structure to Algorithm ACTIVE follows:

Algorithm REACHABLE   Given a context-free grammar $G' = (V_N', V_T', S, \Phi')$, this algorithm determines the set of reachable symbols in the vocabulary $V_N' \cup V_T'$ according to the informal method just described. The sets W_k for $0 \le k \le m$, where m denotes the number of symbols in V_N', represent a family of sets. The new reachable sets of terminal and nonterminal symbols are represented by V_T'' and V_N'', respectively. The set of remaining productions is denoted by Φ''.

1. [Compute initial set of reachable symbols]
   $W_0 := \{S\}$

2. [Calculate successive sets of reachable symbols]

Repeat for $k = 1, 2, \ldots, m$

$$W_k := W_{k-1} \cup \{A \in V_N' \cup V_T' | (B \to \phi_1 A \phi_2) \in \Phi' \text{ where } B \in W_{k-1}\}$$

If $W_k = W_{k-1}$ or $k = m$

then $V_N'' := W_k \cap V_N'$

$V_T'' := W_k \cap V_T'$

$\Phi'' := \{(A \to \alpha) \in \Phi' | A, \alpha \in W_k^+\}$

Exit

This algorithm is similar to Algorithm ACTIVE in the sense that $W_k \supseteq W_{k-1}$ and the stopping criterion is the same as that used in the previous algorithm. Note that $L(G') = L(G'')$, where $G'' = (V_N'', V_T'', S, \Phi'')$.

A nonterminal is called *useful* if it is both active and reachable. Otherwise, a nonterminal is said to be *useless*. A grammar is *reduced* if all *useless* symbols and associated rules have been eliminated and all rules of the form $A \to A$ have been removed.

A reduced grammar is obtained by applying Algorithms ACTIVE and REACHABLE to its unreduced equivalent and removing rules of the form $A \to A$. It should be noted, however, that the order of application of these algorithms is important. First, we should eliminate from the original grammar those nonterminals (and associated rules) that are inactive. Second, the resulting grammar from the first step can then be used as input to Algorithm REACHABLE. With this order of application, some symbols may become unreachable only after an inactive nonterminal and its associated rules have been removed, but not conversely. As an example of this phenomenon, consider the following grammar, which is meant to (but does not) generate the language $\{a^m b^n a^m b^k ccc \cup ccc | m, n, k \geq 1\}$:

$$S \to ccc \qquad B \to aBa$$
$$S \to Abccc \qquad B \to AC$$
$$A \to Ab \qquad C \to Cb$$
$$A \to aBa \qquad C \to b$$

By first applying Algorithm ACTIVE, we find that the set of inactive nonterminals is $\{A, B\}$. Upon removing those productions involving these inactive symbols, we obtain the grammar

$$S \to ccc, \qquad C \to Cb, \qquad C \to b$$

With the application of Algorithm REACHABLE, the symbol C has now become unreachable. The resulting reduced grammar contains the single rule $S \to ccc$. It should be emphasized that previously the application of Algorithm ACTIVE C was reachable, but it then became unreachable after it was determined that A and B were inactive. Had we applied the two algorithms in reverse order, a different grammar would have been obtained. This grammar obviously would not have been truly reduced.

Reduced grammars are used in Section 10.2.3, which deals with deterministic top-down parsing.

Problems 10.1.6

1. Show that Algorithm ACTIVE does compute the active set of nonterminals in a given grammar.
2. Give a trace of Algorithm ACTIVE for the following input grammar:

$$
\begin{array}{ll}
S \rightarrow aAbA & A \rightarrow bBC \\
S \rightarrow aba & A \rightarrow a \\
A \rightarrow aAb & B \rightarrow aBc
\end{array}
$$

3. In Algorithm ACTIVE, show for each integer $n > 1$ that there exists a grammar with n variables such that $W_k \neq W_{k+1}$ for $1 \leq k \leq n - 1$.
4. Reduce the following grammar:

$$
\begin{array}{l}
S \rightarrow bS \mid ba \mid A \\
A \rightarrow a \mid Bb \\
B \rightarrow C \mid ba \mid aCC \\
C \rightarrow CCa \mid b
\end{array}
$$

5. Reduce the following grammar:

$$
\begin{array}{l}
S \rightarrow E + T \\
E \rightarrow E \mid Z + F \mid T \\
F \rightarrow F \mid FP \mid P \\
P \rightarrow G \\
G \rightarrow G \mid GG \mid F \\
T \rightarrow T * i \mid i \\
Q \rightarrow E \mid E + F \mid T \mid Z \\
Z \rightarrow i
\end{array}
$$

Problems 10.1

1. How many different English sentences can the example grammar on page 522 generate?
2. Consider the following grammar with the set of terminal symbols $\{a, b\}$:

$$
\begin{array}{llll}
S \rightarrow a & S \rightarrow Sa & S \rightarrow b & S \rightarrow bS
\end{array}
$$

Describe (in a closed form) the set of strings generated by the grammar.

3. Write grammars for the following languages:
   (a) The set of nonnegative odd integers
   (b) The set of nonnegative even integers with no leading zeros permitted
   (c) The set of all odd integers

4. Give a grammar that generates

$$L = \{w|w \text{ consists of an equal number of } a\text{'s and } b\text{'s}\}$$

5. Give a context-free grammar that generates

$$L = \{w|w \text{ contains twice as many 0s as 1s}\}$$

6. Construct a grammar that will generate all strings of 0s and 1s having both an odd number of 0s and an odd number of 1s.

7. Obtain a grammar for the language

$$L = \{0^i 1^j | i \neq j \text{ and } i, j > 0\}$$

8. **(a)** Obtain a context-free grammar for generating the set of parenthesized arithmetic expressions consisting of the binary operators $+$, $-$, $*$, $/$, and $\uparrow$ (exponentiation) and the single variable i. Assume all operations are left-associative except for $\uparrow$, which is right-associative. Assume the usual precedence rules in algebra.

   (b) Construct a parse tree for each of the following expressions:
   - $i + i * i$
   - $(i + i)/i$
   - $i + i * i \uparrow i$
   - $i \uparrow i \uparrow i + i$

9. **(a)** Obtain a context-free grammar for all nonnegative real numbers including those in floating-point form. The form of a floating-point number consists of a nonnegative real number followed by the letter "E," an optional sign ($+$ or $-$) and one or two decimal digits.

   (b) Construct a parse tree for each of the following:
   - 15.4E5
   - 2.0E-03
   - 0.5E+21

10.2 TOP-DOWN PARSING

10.2.1 Introduction

In the previous section the notions of top-down parsing and bottom-up parsing were introduced. Recall that top-down parsing was characterized as a parsing method that, beginning with the goal symbol of the grammar, attempted to produce a string of terminal symbols that was identical to a given input string. This matching process proceeded by successively applying the productions of the grammar to produce substrings from nonterminals. In this section we will examine methods of performing top-down parsing.

The types of grammars that we will be dealing with fall into the class of context-free grammars. Empty productions are not permitted.

This section is divided into two subsections. In the first subsection we will deal with an overview of a general method of top-down parsing. The second subsection will discuss deterministic top-down parsing.

10.2.2 General Top-down Parsing Strategy

Recall that top-down parsing moves from the goal symbol to a string of terminal symbols. In the terminology of trees, this is moving from the root of the tree to a set of the leaves in the syntax tree for a given input string. In using a general approach, we are willing to attempt to create a syntax tree by following branches until the correct set of terminals is reached. This approach is similar to the depth-first approach discussed in Section 7.5.4. In the worst possible case, that of trying to parse a string that is not in the language, all possible combinations are attempted before the failure to parse is recognized.

Top-down parsing in this manner is a brute-force method of parsing. In general terms, this method operates as follows:

1. Given a particular nonterminal that is to be expanded, the first production for this nonterminal is applied.

2. Then, within this newly expanded string, the next (leftmost) nonterminal is selected for expansion and its first production is applied.

3. This process (step 2) of applying productions is repeated for all subsequent nonterminals that are selected until such time as the process cannot or should not be continued. This termination (if it ever occurs) may be due to two causes. First, no more nonterminals may be present, in which case the string has been successfully parsed. Second, it may result from an incorrect expansion, which would be indicated by the production of a substring of terminals that does not match the appropriate segment of the input string. In the case of such an incorrect expansion, the process is "backed up" by undoing the most recently applied production. Instead of using the particular expansion that caused the error, the next production of this nonterminal is used as the next expansion, and then the process of production application continues as before.

   If, on the other hand, no further productions are available to replace the production that caused the error, this error-causing expansion is replaced by the nonterminal itself, and the process is backed up again to undo the next most recently applied production. This backing up continues either until we are able to resume normal application of productions to selected nonterminals or until we have backed up to the goal symbol and there are no further productions to be tried. In the latter case, the given string must be unparsable because it is not part of the language determined by this particular grammar.

Example 10.8   As an example of this general top-down technique, let us consider the simple grammar

$$S \rightarrow aSe \mid B \qquad B \rightarrow bBe \mid c$$

where S is the goal or start symbol.   The language generated by this grammar is $\{a^m b^n c e^{m+n} \mid m \geq 0,\ n \geq 0\}$. Obtain a parse for the string $abcee$.

Solution   Figure 10.10 illustrates the working of this general parsing technique by showing the sequence of syntax trees generated during the parse of the string $abcee$.

Initially, we start with the tree of Figure 10.10a, which merely contains the goal symbol. We next select the first production for S, thus yielding Figure 10.10b. At this point we have matched the symbol a in the string to be parsed. We now choose the first production for S and obtain Figure 10.10c. Note, however, that we have a mismatch between the second symbol b of the input string and the second symbol a in the sentential form $aaSee$. At this point in the parse, we must back up. The previous production application for S must be deleted and replaced with its next choice. The result of performing this operation is to transform Figure 10.10c into Figure 10.10d. We next choose the first production for B and obtain Figure 10.10e with the leftmost two characters of the given string being matched. Again choosing the first production for B transforms Figure 10.10e into Figure 10.10f. When the third symbol c of the string is compared with the next symbol in the current sentential form; however, a mismatch occurs. The previously chosen production for B must be deleted. We then apply the second production for B, which gives Figure 10.10g. At this point in the parsing process the third input symbol, c, has been matched. The remaining input symbols are then matched with the remaining symbols in the sentential form of Figure 10.10g, thereby resulting in a successful parse. ∎

The top-down parsing method just described will not work on every context-free grammar. We now examine a particular subclass of the context-free grammars that causes this method to fail.

Given a context-free grammar $G = \{V_N, V_T, S, \Phi\}$, a nonterminal X is said to be *left recursive* if $X \overset{+}{\Rightarrow} X\alpha$ for some string $\alpha \in V^*$. If a grammar G contains at least one left-recursive nonterminal, G is said to be *left recursive*.

A grammar that has one or more nonterminals that are left recursive can be presented with strings to parse that will cause our informal top-down parsing method to enter an infinite loop of production applications. As an example of this phenomenon, consider the grammar

$$S \to aAc \qquad A \to Ab \,|\, b$$

which generates the language $\{ab^n c \,|\, n > 0\}$. Figure 10.11 illustrates a trace of this situation for the attempted parse of the string abc.

Of course, this particular infinite loop can be avoided by reversing the order of the productions for A to $A \to b \,|\, Ab$ or by rewriting the first production for A so that the rule becomes $A \to bA$ (thus eliminating the left recursion). Not all left recursions, however, can be handled in such an easy manner.

10.2.3 Deterministic Top-down Parsing with LL(1) Grammars

In this subsection, top-down parsing with no backup is discussed. The information required during a no-backup parse is first discussed informally and then formally characterized in terms of LL(1) grammars. This class of grammars can be parsed in a deterministic manner by looking at the next input symbol in the input string to be parsed. The discussion will proceed as follows: First the class of simple LL(1) grammars is introduced. This class of grammars is very easily parsed. The next step is to permit a more general form of production.

Parsing algorithms are given for each subclass of grammars just discussed. The parsing tables required in these algorithms can be obtained directly from the rules of the

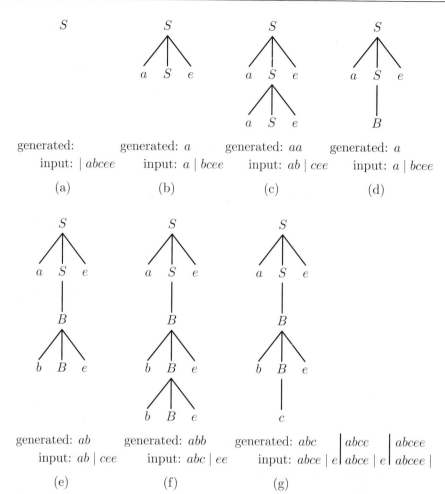

Figure 10.10.   Trace of a general top-down parse for string *abcee* (*Note*: The symbol | denotes the extent of the scanning process from left to right in the input string)

grammars. The use of bit matrices (see Section 5.3.5) can facilitate the generation of such parsing tables.

LL(1) grammars have been used in defining programming languages for some 25 years. The parsing algorithms for this class of grammars are efficient. Actually, the performance of these algorithms is a linear function of the input string length.

The main idea of this section deals with the notion of a *parser generator system*. A block diagram of such a system and how it fits into the front end of a compiler appears in Figure 10.12. The parser generator has as its main input a context-free grammar, which should be an LL(1) grammar. If the grammar is LL(1), then the system generates a parsing table from that input grammar. The parsing table, which is unique for the input grammar, is used as input to the LL(1) parser. The LL(1) parser is said to be table driven. If, on

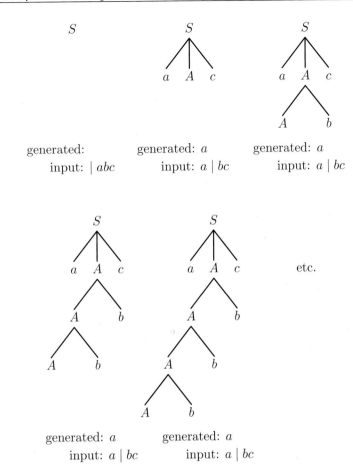

generated: generated: a generated: a

input: $|\ abc$ input: $a\ |\ bc$ input: $a\ |\ bc$

generated: a generated: a

input: $a\ |\ bc$ input: $a\ |\ bc$

Figure 10.11.   Nonterminating sequence of syntax trees generated from a left-recursive grammar in a top-down parse

the other hand, the grammar is not LL(1), then diagnostics as to why the grammar is not LL(1) are provided to the compiler writer. Note that the parser generator system generates a parsing table once by the LL(1) table constructor module.

The LL(1) parser is a universal parser in that, given a parsing table for an input grammar that describes, say, Pascal and a (Pascal) source program, the parser will generate a parse tree (or its equivalent) if the source program is syntactically valid or syntax errors in the case of a syntactically invalid source program.

The demonstration that a grammar is LL(1) has other benefits as well. For example, an LL(1) grammar is guaranteed to be unambiguous. Also, an LL(1) grammar cannot contain any left-recursive rules.

10.2.3.1. Notions of Deterministic Parsing.   In the situation where no backup is allowed, our problem becomes one of determining which production is to be

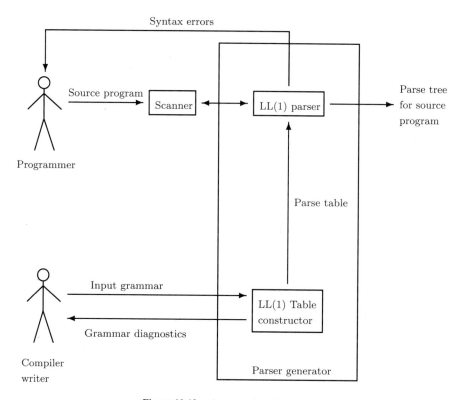

Figure 10.12. A parser generator system

applied, given that we know which nonterminal we are attempting to expand and which head of the input string has been successfully parsed to this point. We are looking for a *left canonical parse*, that is, a parse in which the leftmost nonterminal in the current sentential form is always chosen for expansion.

Another way of stating this requirement is that we must always make the correct choice among alternative productions for a particular nonterminal so that we never undo a particular expansion. Either the parse moves directly from left to right across the input string and concludes successfully, or it reaches a point where no production can correctly be applied, in which case the input string is rejected as being invalid.

Figure 10.13 illustrates this situation in terms of a generalized syntax diagram. At the point illustrated, the head $t_1 t_2 \ldots t_i$ of the input string $t_1 t_2 \ldots t_m$ has been matched, and the nonterminal A is to be expanded. We must now correctly choose which production to use to expand A, the leftmost nonterminal in the current sentential form $t_1 t_2 \ldots t_i A \alpha_1 \ldots \alpha_n$.

A large class of left-parsable grammars, the LL(1) grammars, allows this determination of the correct production to apply. Given the input head (prefix) parsed to this point and the leftmost nonterminal, the correct production can be applied provided that we also know the next symbol of the unparsed input substring, say the symbol t_{i+1}. We next turn to a discussion of simple LL(1) grammars.

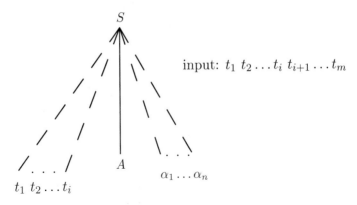

input: $t_1\ t_2 \ldots t_i\ t_{i+1} \ldots t_m$

Figure 10.13.   Left canonical parse

10.2.3.2. Simple LL(1) Grammars.

In this section we examine a class of grammars that can be parsed by simply looking at the next symbol in the unparsed input string in order to decide which production is to be applied. The parsing method described here is deterministic in the sense that no backup is required. We first describe in an informal manner a parser for this class of grammars. We then examine how to construct a parsing table from a given grammar in this class.

Definition 10.6:   A *simple LL(1) grammar* is a context-free grammar without ε-rules such that for every $A \in V_N$ the alternates for A each begin with a different terminal symbol. More formally, a context-free grammar is called a *simple LL(1) grammar* or an *s-grammar* if all its rules are of the form

$$A \rightarrow a_1\alpha_1 \,|\, a_2\alpha_2 \,|\, \cdots \,|\, a_m\alpha_m,$$
$$a_i \neq a_j \text{ for } i \neq j \text{ and } a_i \in V_T \text{ for } 1 \leq i \leq m$$

The language generated by an *s-grammar* is called an *s-language*.

As an example, consider the grammar whose productions are

1.	$S \rightarrow aA$	5.	$A \rightarrow aA$
2.	$S \rightarrow b$	6.	$A \rightarrow b$
3.	$S \rightarrow cB$	7.	$B \rightarrow cB$
4.	$S \rightarrow d$	8.	$B \rightarrow d$

Clearly, this grammar is a simple LL(1) grammar. It is convenient to have some end-marker symbol at the end of all input strings to be parsed. We use the symbol # to denote the end of the input string. Because # is not in the terminal vocabulary V_T of the grammar, it is impossible for some malicious person to insert the # in the middle of the input string.

Conveniently, the grammar is augmented with an extra production

$$0. \quad S' \rightarrow S\#$$

which facilitates the construction of certain relations and tables later in this section.

Let us construct the parse of the string $aaab\#$ according to the preceding grammar. The leftmost canonical derivation for this input string is as follows:

$$S' \underset{L}{\Rightarrow} S\# \underset{L}{\Rightarrow} aA\# \underset{L}{\Rightarrow} aaA\# \underset{L}{\Rightarrow} aaaA\# \underset{L}{\Rightarrow} aaab\#$$

where $\underset{L}{\Rightarrow}$ denotes a left-canonical step (i.e., the leftmost nonterminal is rewritten). The parse tree for this derivation is given in Figure 10.14.

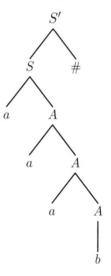

Figure 10.14.   Parse tree for the string $aaab\#$

We now describe an informal method for constructing, in a top-down manner without backup, the parse tree of Figure 10.14. This construction process will then be formalized. In a simple LL(1) grammar parser, one stack is required. Its purpose is to record a portion of the current sentential form. If the current sentential form is $t_1 t_2 \ldots t_i A\alpha_1 \ldots \alpha_n$, where the substring $t_1 t_2 \ldots t_i$ denotes the prefix of the given input string that is matched so far, then the portion of the current sentential form that is retained in the stack is the string $A\alpha_1 \ldots \alpha_n$. This string, in the case of a valid input string, will produce the remaining unmatched input symbols; that is, for the input string $t_1 t_2 \ldots t_m$, $A\alpha_1\alpha_2 \ldots \alpha_n \overset{+}{\Rightarrow} t_{i+1} \ldots t_m$. In addition to the stack, an output tape is used to record the production numbers in the order in which they were applied. The symbol # is used as the stack bottom symbol. Initially, the stack is set to the contents

$$S\#$$

with the leftmost symbol denoting the top symbol in the stack. From this initial configuration, a rule must be selected on the basis of the top stack symbol and the current input character being examined. With a stack symbol of S and the current input symbol a, rule 1 must be chosen if we are to avoid backup. So S is replaced by the right part of rule 1, thus giving a stack contents of $aA\#$. Now, since the stack symbol is a terminal that matches the current input symbol, the stack is popped and the input cursor is advanced. The current stack contents are

$$A\#$$

The second symbol a in the original input string then becomes the new current symbol. Rule 5 is chosen. Note that in each of these two cases the choice is made in a deterministic manner. With this rule application the stack contents change to

$$aA\#$$

The stack symbol is a, which matches the second input symbol. This condition causes the stack to be popped. The input cursor is then advanced, with a becoming the new current input symbol. The stack contents are

$$A\#$$

Thus, rule 5 is again chosen and the stack contents become

$$aA\#$$

The stack symbol is a, which matches the third input symbol. This condition again causes the stack to be popped. Based on the fact that the stack symbol is A and the current input symbol is b, rule 6 can be chosen in a deterministic manner as the production to apply at this point. The stack symbol is replaced by the right part of this rule, thus yielding a stack contents of

$$b\#$$

The stack and the current input symbol now match and the stack is popped. At this point the stack contents are

$$\#$$

and the current input symbol is also $\#$. This situation signals the end of the parsing process for the given input string. A trace of this parse is given in Table 10.2.

We can now, in a more formal manner, formulate a parsing strategy for the process just described. The heart of this strategy is a parsing function that is represented in a table. Given the symbol at the top of the stack and the current input symbol, the table yields as a value either the production to apply or else an indication of how to continue or terminate. This table is defined as follows:

$$ACTION : \{V \cup \{\#\} \times \{V_T \cup \{\#\}\} \to \{(\beta, i), pop, accept, error\}$$

where $\#$ marks the bottom of the stack and the end of the input string and where (β, i) is an ordered pair such that β is the right part of the production number i.

In short, if A is the symbol on top of the stack and a is the current input symbol, then $ACTION(A, a)$ is defined as follows:

$$ACTION(A, a) = \begin{cases} pop & \text{if } A = a \text{ for } a \in V_T \\ accept & \text{if } A = \# \text{ and } a = \# \\ (a\alpha, i) & \text{if } A \to a\alpha \text{ is the } i\text{th production} \\ error & \text{otherwise} \end{cases}$$

TABLE 10.2.

Trace of a Top-down Parse for the String $aaab\#$

Unused Input String	Stack Contents	Rules Used
$aaab\#$	$S\#$	ε
$aaab\#$	$aA\#$	1
$aab\#$	$A\#$	1
$aab\#$	$aA\#$	15
$ab\#$	$A\#$	15
$ab\#$	$aA\#$	155
$b\#$	$A\#$	155
$b\#$	$b\#$	1556
$\#$	$\#$	1556

The parsing function $ACTION$ for the current example grammar is given in Table 10.3.

TABLE 10.3. Parsing Function for a Simple LL(1) Grammar

Stack	Current Input Symbol				
Symbol	a	b	c	d	$\#$
S	$(aA, 1)$	$(b, 2)$	$(cB, 3)$	$(d, 4)$	
A	$(aA, 5)$	$(b, 6)$			
B			$(cB, 7)$	$(d, 8)$	
a	pop				
b		pop			
c			pop		
d				pop	
$\#$					$accept$

Blank entries are all *error* entries.

In summary, we have introduced a class of grammars that is very easy and efficient to parse. The restrictions placed on the rules of a simple LL(1) grammar, however, are severe. It is highly desirable, for practical reasons, to expand the class of grammars that can be handled by such a straightforward approach. This problem is examined in the next subsection.

Problems 10.2.3.2

1. For the sample grammar used in this section, give a trace (as in Table 10.2) of the parse for each of the following input strings:
   (a) $cd\#$ **(b)** $cccd\#$

2. Formulate a parsing table for the simple LL(1) grammar whose rules are

$$
\begin{aligned}
0.&\quad S' \rightarrow S\# \\
1.&\quad S \rightarrow aS \\
2.&\quad S \rightarrow bA \\
3.&\quad A \rightarrow d \\
4.&\quad A \rightarrow ccA
\end{aligned}
$$

Using this parsing table, give a trace of the parse for each of the following input strings:
(**a**) $aabccd\#$
(**b**) $bccd\#$
(**c**) $abccccd\#$

3. Obtain a simple LL(1) grammar for the language $\{ab^n ab^{n+1} \mid n \geq 0\}$.

4. Obtain a simple LL(1) grammar for the language $\{1^n a0^n \mid n > 0\} \cup \{a1^m a0^{2m} \mid m > 0\}$.

10.2.3.3. LL(1) Grammars without Empty Rules.

In this section we pursue the discussion of top-down deterministic parsers that require only one look-ahead symbol. Our aim is to generalize the simple LL(1) grammars discussed in the previous subsection. This generalization will eliminate some of the restrictions that are imposed on these simple grammars. Specifically, the condition that the leftmost symbol in the right part of a production must be a terminal symbol is removed. One restriction that still remains, however, is that empty rules are not allowed. Furthermore, it should be noted that LL(1) grammars without empty rules have the same generative power as s-grammars, which were discussed in the previous subsection. That is, given an LL(1) grammar without empty rules we can obtain an equivalent s-grammar that generates the same language.

Throughout the following discussion, it is assumed that all useless symbols (recall Section 10.1.6) and left-recursive rules have been eliminated from the grammar. We will show later that left-recursive grammars are not LL(1).

Let us attempt to parse, in a deterministic manner that is based on a single look-ahead character, the input string $adccd\#$ generated by the following grammar:

$$
\begin{aligned}
0.&\quad S' \rightarrow S\#   &\quad 4.&\quad B \rightarrow aA \\
1.&\quad S \rightarrow BA   &\quad 5.&\quad B \rightarrow bS \\
2.&\quad A \rightarrow BS   &\quad 6.&\quad B \rightarrow c \\
3.&\quad A \rightarrow d
\end{aligned}
$$

The leftmost derivation for the given string is

$$
\begin{aligned}
S' &\underset{L}{\Rightarrow} S\# \underset{L}{\Rightarrow} BA\# \underset{L}{\Rightarrow} aAA\# \underset{L}{\Rightarrow} adA\# \underset{L}{\Rightarrow} adBS\# \\
&\underset{L}{\Rightarrow} adcS\# \underset{L}{\Rightarrow} adcBA\# \underset{L}{\Rightarrow} adccA\# \underset{L}{\Rightarrow} adccd\#
\end{aligned}
$$

and the associated syntax tree is given in Figure 10.15. Note that this grammar is not in its present form a simple LL(1) grammar. Productions 0, 1, and 2 violate the simple LL(1) condition.

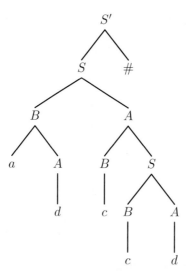

Figure 10.15.   Parse tree for the string $adccd\#$

Using the approach taken in the previous subsection, let us now attempt to parse the given string based on a one-character look-ahead. As before, the stack is initialized to the contents

$$S\#$$

with the leftmost symbol denoting the top symbol in the stack. From this initial configuration, a rule must be selected on the basis of the top stack symbol (S) and the current input character being examined (a). For the simple LL(1) grammar case, we merely had to select the rule whose left part and the leftmost symbol in its right part were the stack symbol and the current input symbol, respectively. In this instance, however, such an approach is inadequate, since the only rule that has S as its left part contains a nonterminal (B) as the leftmost symbol in its right part. If B can eventually leftmost-produce the symbol a, then we are assured that by selecting rule 1 we are on the right path, and backup beyond this point will never be necessary. More formally, $B \stackrel{+}{\Rightarrow} a \ldots$, where the notation $\ldots$ denotes a string (possibly empty) of characters selected from the vocabulary. The replacement of the stack symbol S by the string BA yields a new stack contents of

$$BA\#$$

We now must select a second production based on the knowledge that the stack symbol is B and the current input symbol is a. Three rules are possible: rules 4, 5, and 6. The decision in this case, however, is clear: rule 4 must be selected. This choice results in the following stack contents:

$$aAA\#$$

Now that the stack symbol matches the current input symbol, the stack is popped and the input cursor is advanced. As a result of these operations, the new stack symbol and current input symbol become the symbols A and d, respectively, thus yielding a stack contents of

$$AA\#$$

At this point in the parse, we observe that the symbol d is leftmost derivable from A; that is, $A \Rightarrow d$. Consequently, we select rule 3 as the next rule to apply in the parsing process. This choice results in the stack being changed to

$$dA\#$$

As a result, the stack and current input symbols match. We can now pop the stack and advance the input cursor, thus obtaining the stack contents

$$A\#$$

and the new input symbol c.

Two rules for A are possible, rules 2 and 3. Because of the current input symbol, we cannot choose rule 3, since this choice will leftmost-produce symbol d. Consequently, we must choose rule 2. The new stack contents now become

$$BS\#$$

The parsing process can be continued until the complete parse is obtained. A trace of this parse is given in Table 10.4.

TABLE 10.4.   Trace of the Parse of $adccd\#$

Unused Input String	Stack	Output Tape
$adccd\#$	$S\#$	ε
$adccd\#$	$BA\#$	1
$adccd\#$	$aAA\#$	14
$dccd\#$	$AA\#$	14
$dccd\#$	$dA\#$	143
$ccd\#$	$A\#$	143
$ccd\#$	$BS\#$	1432
$ccd\#$	$cS\#$	14326
$cd\#$	$S\#$	14326
$cd\#$	$BA\#$	143261
$cd\#$	$cA\#$	1432616
$d\#$	$A\#$	1432616
$d\#$	$d\#$	14326163
$\#$	$\#$	14326163

The sample grammar under consideration seems to be LL(1). Let us now introduce certain notions that will be used to formalize the definition of an LL(1) grammar without empty rules. In the previous discussion, we introduced the notion of leftmost derivable. Given some string $\alpha \in V^+$, the set of terminal symbols that is leftmost derivable from α is given by the equation

$$\mathrm{FIRST}(\alpha) = \{w \mid \alpha \overset{*}{\Rightarrow} w \ldots \text{ and } w \in V_T\}$$

For example, in the current example grammar the set

$$\text{FIRST}(B) = \{w \mid B \overset{*}{\Rightarrow} w \ldots \text{ and } w \in V_T\} = \{a, b, c\}$$

and

$$\text{FIRST}(BA) = \{w \mid BA \overset{*}{\Rightarrow} w \ldots \text{ and } w \in V_T\} = \{a, b, c\}$$

Again referring to our example grammar, we note that there are three rules with a left part of B: the rules $B \to aA$, $B \to bS$, and $B \to c$. The FIRST set of each right part is given by

$$\text{FIRST}(aA) = \{a\}, \text{ FIRST}(bS) = \{b\}, \text{ FIRST}(c) = \{c\}$$

These FIRST sets are disjoint. There are also two rules with a left part of A whose FIRST sets are also disjoint:

$$\text{FIRST}(BS) = \text{FIRST}(BA) = \{a, b, c\}, \text{ FIRST}(d) = \{d\}$$

We are now in a position to generalize these observations.

Definition 10.7: A grammar without empty rules is an *LL(1) grammar* if for all rules of the form $A \to \alpha_1 \mid \alpha_2 \mid \ldots \mid \alpha_n$ the sets $\text{FIRST}(\alpha_1)$, $\text{FIRST}(\alpha_2)$, $\ldots$ and $\text{FIRST}(\alpha_n)$ are pairwise disjoint; that is,

$$\text{FIRST}(\alpha_i) \cap \text{FIRST}(\alpha_j) = \emptyset \text{ for } i \neq j$$

By applying this definition to the sample grammar, we obtain the following:

1. Rules $A \to BS \mid d$:

$$\text{FIRST}(BS) \cap \text{FIRST}(d) = \{a, b, c\} \cap \{d\} = \emptyset$$

2. Rules $B \to aA \mid bS \mid c$:

$$\text{FIRST}(aA) \cap \text{FIRST}(bS) = \{a\} \cap \{b\} = \emptyset$$
$$\text{FIRST}(aA) \cap \text{FIRST}(c) = \{a\} \cap \{c\} = \emptyset$$
$$\text{FIRST}(bS) \cap \text{FIRST}(c) = \{b\} \cap \{c\} = \emptyset$$

Therefore, this grammar is LL(1).

Using the approach of the previous subsection, we can formulate in a formal manner a parsing algorithm for the process just described. The parsing function associated with this parsing algorithm is defined as follows:

$$\text{ACTION} : \{V \cup \{\#\}\} \times \{V_T \cup \{\#\}\} \to \{(\beta, i), pop, accept, error\}$$

where $\#$ marks the bottom of the stack and the end of the input string and where (β, i) is an ordered pair such that β is the right part of production number i.

If A is the top symbol on the stack and a is the current input symbol, then ACTION is defined as follows:

$$\text{ACTION}(A, a) = \begin{cases} pop & \text{if } A = a \text{ for } a \in V_T \\ accept & \text{if } A = \# \text{ and } a = \# \\ (\beta, i) & \text{if } a \in \text{FIRST}(\beta) \\ & \text{and } A \to \beta \text{ is the } i\text{th production} \\ error & \text{otherwise} \end{cases}$$

The parsing function of our current grammar is given in Table 10.5.

TABLE 10.5.   Parsing Function (ACTION) for an LL(1) Grammar

Stack	Current Input Symbol				
Symbol	a	b	c	d	$\#$
S	$(BA, 1)$	$(BA, 1)$	$(BA, 1)$		
A	$(BS, 2)$	$(BS, 2)$	$(BS, 2)$	$(d, 3)$	
B	$(aA, 4)$	$(bS, 5)$	$(c, 6)$		
a	pop				
b		pop			
c			pop		
d				pop	
$\#$					$accept$

Blank entries are all *error* entries.

We have defined the FIRST relation, and it is clear that this relation is important in computing the parsing function associated with a particular grammar. Since the mechanical evaluation of FIRST may be cumbersome, we redefine this relation in terms of another relation over a finite set in the following manner: Let F be a relation over the vocabulary such that

$$U F X \text{ iff there exists a production } U \to X \ldots$$

The transitive closure F^+(recall Section 5.4.6) of relation F holds between U and X iff there exists some sequence of rules (at least one) such that

$$U \to A_1 \ldots, A_1 \to A_2 \ldots, \ldots, A_n \to X \ldots$$

It is clear that $U F^+ X$ iff $U \overset{+}{\Rightarrow} X \ldots$. This relation is easily computed by using Warshall's algorithm, as shown in Section 7.4.

Our aim is to compute the FIRST(α) set for $\alpha = x_1 x_2 \ldots x_n$. Now FIRST($\alpha$) = FIRST($x_1$), since we are assuming no empty rules except for the case where $S' \to \varepsilon$ and S' does not occur in the right part of any rule in the grammar. The FIRST sets for terminals are trivial in form. Therefore, in order to compute the FIRST sets, we need only compute those associated with nonterminals. These FIRST sets are given by

$$\text{For each } A \in V_N, \text{FIRST}(A) = \{x \mid x \in V_T \text{ and } AF^+x\}$$

The F, F^+, and FIRST relations for the sample grammar are given in Table 10.6. Note that in this table F entries are marked by F. F^+ entries are marked by F or F^+. Contributors to a FIRST set by nonterminals are underlined. They are the F^+ entries in $V_N \times V_T$.

TABLE 10.6.
F and F^+ Matrices and FIRST Sets for the Sample Grammar

	S'	S	A	B	a	b	c	d
S'		F		F^+	$\underline{F^+}$	$\underline{F^+}$	$\underline{F^+}$	
S				F	$\underline{F^+}$	$\underline{F^+}$	$\underline{F^+}$	
A				F	$\underline{F^+}$	$\underline{F^+}$	$\underline{F^+}$	F
B					$\underline{F}$	$\underline{F}$	$\underline{F}$	
a								
b								
c								
d								

We now culminate the previous discussion with the presentation of a Pascal procedure for parsing strings according to the top-down approach based on one-symbol look-ahead. It is assumed that the grammar is reduced (see Section 10.1.6) and all left recursion has been eliminated. It is also assumed that the table of active operations is global to the procedure and that it has been set up appropriately.

The procedure LL1Parser presented in Figure 10.16 uses a built-in routine called SUB. This routine has three parameters, say I, P, L, and returns a substring of string I starting at position P which is L characters long. The parameter L is optional, and if omitted, the routine will return the substring of I consisting of the character at position P, and all other characters after position P as ordered in string I. This routine is used in LL1Parser for examining the leftmost symbol in the sentential form and for examining the current symbol in the string being parsed.

It should be noted that in order to present a procedure that is both clear to the reader and manageable in size, the syntax used for the procedure is not true Pascal, and therefore the procedure is not compilable code. This alleviates much of the excessive code needed to deal with the variable record structures that are stored in the active table. This table can contain entries of the form accept, pop, ' ', or (β, i), where $\beta \in V^+$ and i is the number of the production rule used to produce β. In the procedure, all these variants must be accounted for, and they are in the order accept, ' ', pop, (β, i). If none of the first three is found, it is assumed that the entry is an ordered pair, and the assignment statement

$$(\text{RHS}, \text{Prod\#}) := \text{ACTION}(\text{Top}, \text{Cur})$$

```
procedure LL1PARSER (InString : string;  {Input String}
                     Start : char);
{Given the parsing function ACTION for a particular LL(1) grammar G whose V_N and V_T sets contain
m and n elements, respectively, this procedure parses strings to determine whether or not these strings
belong to L(G). The current sentential form is represented by a string variable, Sent, where Top
contains the leftmost symbol in the sentential form.  The variable Cur contains the current input
symbol. For convenience, we assume each terminal and nonterminal symbol is a string of length 1.
The list of production numbers for the parse is represented by the string variable Parse. The starting
symbol of the unaugmented grammar is the string variable Start.}
var
   Sent, Parse, CurString, RHS : string;
   Top, Cur : char;
   P, Prod# : integer;   {Index of InString}
   Done : boolean;
begin
   {Step1: Initialize}
   CurString := InString + '#';   {'+' denotes concatenation}
   P := 1; Sent := Start + '#';  Done := false;
   Parse := '';    {The empty string}
   {Step2: Attempt to perform the required parse}
   while not Done do
      begin
      Top := SUB (Sent, 1, 1);  { selects leftmost symbol in Sent}
      Cur := SUB (CurString, P, 1);
      if (ACTION (Top, Cur) = accept) or (ACTION (Top, Cur) = ' ') then
         Done := true
      if ACTION (Top, Cur) = accept then
         write ('SUCCESSFUL PARSE');
      else if ACTION (Top, Cur) = ' ' then
         write ('UNSUCCESSFUL PARSE');
         else if ACTION (Top, Cur) = pop then
            begin
               Sent := SUB (Sent, 2);{pop first character from Sent}
               P := P + 1 {move input cursor forward}
            end
            else
               begin
                  (RHS, Prod#) := ACTION(Top, Cur);
                  Sent := RHS + SUB (Sent, 2);    {Top is replaced by RHS}
                  Parse := Parse + ':' + Prod# {record production number}
               end
      end
end;
```

Figure 10.16.   A procedure for parsing LL(1) grammars

is used to extract the production information from the table so that it can be used to continue the parsing of the string.

From the discussion dealing with the construction of the parser, it may seem that an LL(1) grammar must be unambiguous. The proof of this point is left as an exercise.

Another property of a grammar that satisfies the LL(1) condition is that such a grammar cannot contain left-recursive nonterminals. Let us give an informal proof of this fact. If X_0 is a left-recursive nonterminal, then there exists a sequence of production applications $X_{i-1} \Rightarrow X_i \alpha_i$ for $1 \leq i \leq m$ such that $X_m = X_0$. Clearly, all the X_i symbols are left-recursive nonterminals. At least one of these, however, must have another production if terminal strings are to be produced (the grammar is assumed to be reduced). Let this particular nonterminal be X and let its productions be

$$X \to \alpha_1 \mid \alpha_2 \mid \ldots \mid \alpha_n \qquad (n \geq 2)$$

Now let us further assume without loss of generality that α_1 is the right part of X that leads to the left-recursion problem. Thus we have

$$X \Rightarrow \alpha_1 \overset{*}{\Rightarrow} X\beta \Rightarrow \alpha_2\beta, \qquad \text{where } \beta \in V^*$$

Since $\alpha_1 \overset{*}{\Rightarrow} \alpha_2\beta$, it is obvious that the LL(1) condition is not satisfied because the FIRST sets associated with α_1 and α_2 are not disjoint.

Problems 10.2

1. Obtain a simple LL(1) grammar that generates the same language as the example grammar used in this subsection.

2. For the LL(1) grammar whose rules are

0.	$S' \to S\#$	5.	$J \to, EJ$
1.	$S \to LB$	6.	$J \to)$
2.	$B \to; S; L$	7.	$E \to a$
3.	$B \to := L$	8.	$E \to L$
4.	$L \to (EJ$		

   give a trace (as in Table 10.4) of the parse for each of the following strings:
   (a) $(a, a) := (a, a)\#$
   (b) $((a, a), a); (a) := (a); (a)\#$

3. Obtain the parsing table (as in Table 10.5) for the grammar given in Problem 2.

4. Obtain the F and F^+ matrices (as in Table 10.6) for the example grammar of Problem 2. Also obtain the FIRST sets associated with this grammar.

5. Verify that the following grammar is LL(1).

0.	$S' \to S\#$	4.	$A \to c$
1.	$S \to ABe$	5.	$B \to AS$
2.	$A \to dB$	6.	$B \to b$
3.	$A \to aS$		

   Give a trace of the parse for the input string $adbbebe\#$.

6. The following BNF describes an unsigned number in ALGOL 60.

   $$\langle \text{unsigned integer} \rangle ::= \langle \text{digit} \rangle \mid \langle \text{unsigned integer} \rangle \langle \text{digit} \rangle$$
   $$\langle \text{digit} \rangle ::= 0 \mid 1 \mid 2 \mid 3 \mid 4 \mid 5 \mid 6 \mid 7 \mid 8 \mid 9$$

   (a) Obtain an LL(1) grammar that generates the same language.
   (b) Obtain the parsing function for the grammar of part (a).

PROBLEMS: CHAPTER 10

1. Show that an LL(1) grammar is unambiguous.

2. Obtain a context-free grammar that generates the language

$$L = \{a^n 0 a^i b^n \mid i, n \geq 0\} \cup \{0 a^n 1 a^i c^n \mid n \geq 0\}$$

3. Are the grammars with the following rules LL(1)? Show your work.
   (a) $S \rightarrow abSa \mid aaAb \mid b$
       $A \rightarrow baAb \mid b$
   (b) $S \rightarrow i; \mid i(L); \mid i : S$
       $L \rightarrow i \mid i, L$
   (c) $S \rightarrow aSe \mid B$
       $B \rightarrow bBe \mid C$
       $C \rightarrow cCe \mid d$

4. What language is generated by each of the following grammars?
   (a) $S \rightarrow 0S1 \mid 01$
   (b) $S \rightarrow +SS \mid -SS \mid a$
   (c) $S \rightarrow a \mid S + S \mid SS \mid S^* \mid (S)$

5. Which grammar(s) of Problem 4 is (are) ambiguous? Show why!

6. Obtain a context-free grammar that generates binary strings that have values divisible by 3.

7. Find a context-free grammar that generates the following language:

$$\{1^n 0^m \mid n > m > 0\}$$

8. Find a grammar that generates Pascal-like procedure declarations. A typical example of such a declaration is the following:

   function subr (**var** arg1 : **string**; x : **integer**;
                   y : **real**) : **integer**;

11

Derivations

In this chapter, we discuss two types of logical derivations: natural derivation and resolution theorem proving. Natural derivations are similar to the derivations introduced in Sections 1.6 and 2.3. They start with a set of premises from which the conclusion is obtained by using a sequence of sound arguments. Resolution theorem proving is a method for doing derivations in which all expressions are disjunctions. Hence, before starting a proof, all premises must be converted into sets of disjunctions. There is only one rule of inference in resolution theorem proving, namely, resolution. Resolution theorem proving is used as the main engine of many automatic reasoning programs. Moreover, resolution theorem proving is closely related to Prolog. Natural derivation and resolution theorem proving are first discussed in the context of propositional calculus. Then a number of results about predicate calculus is discussed. In particular, we extend duality to predicate calculus and we introduce *prenex normal forms*. These concepts are essential for understanding natural derivation and resolution methods in the presence of predicates. Moreover, we describe canonical derivations, which provide fundamental concepts used by both natural derivation and resolution theorem proving.

11.1 DERIVATIONS IN PROPOSITIONAL CALCULUS

11.1.1 Introduction

The following sections are devoted to natural derivations and to resolution theorem proving in the context of propositional calculus. Section 11.1.2 gives all rules of natural derivation not using the deduction theorem. The rules involving the deduction theorem are discussed in Section 11.1.3. Section 11.1.4 describes resolution theorem proving.

11.1.2 Basics of Natural Derivation

Like any other system for doing derivations, natural derivation makes use of a number of *rules of inference* (see Section 1.6.5). These rules are given in Table 11.1. There are two rules for each logical connective, one used for introducing the connective and the other for eliminating it. This is suggested by the names of the rules, which consist of two symbols. The first symbol indicates which connective is involved, and the second one is an I or E, depending on whether the connective is introduced or eliminated. For example, the rule $\wedge I$ introduces the connective $\wedge$, whereas $\wedge E$ eliminates it. In all rules, the symbols A and B represent arbitrary expressions. A number of rules rely on the deduction theorem (see Section 1.6.6). These rules are implemented by means of subderivations, and their discussion is postponed to the next section. In addition to the rules of Table 11.1, there is rule R, which allows the repetition of any line that has appeared earlier in the derivation.

TABLE 11.1.   Rules of Inference for Natural Deduction

Introduce Connective		Eliminate Connective	
Name	Rule	Name	Rule
$\neg I$	Deduction theorem	$\neg E$	$\neg\neg A \vdash A$
$\Rightarrow I$	Deduction theorem	$\Rightarrow E$	$A, A \Rightarrow B \vdash B$
$\wedge I$	$A, B \vdash A \wedge B$	$\wedge E$	$A \wedge B \vdash A$
			$A \wedge B \vdash B$
$\vee I$	$A \vdash A \vee B$	$\vee E$	Deduction theorem
	$A \vdash B \vee A$		
$\Leftrightarrow I$	$A \Rightarrow B,$	$\Leftrightarrow E$	$A \Leftrightarrow B \vdash A \Rightarrow B$
	$B \Rightarrow A \vdash A \Leftrightarrow B$		$A \Leftrightarrow B \vdash B \Rightarrow A$

All rules not involving the deduction theorem will be called *simple rules*. All simple rules have *rule premises*. The premises are given as schemas to the left of the symbol $\vdash$. If all rule premises can be matched with lines that have already appeared in the derivation, then one is allowed to add the corresponding conclusion to the derivation. The use of some rules of inference is now demonstrated by an example. The format used is similar to the format one typically finds in textbooks on this subject. In particular, all premises are listed first, separated from the remainder of the derivation by a horizontal line. To justify a new line, the rule is given to the right, together with the lines that are used as rule premises.

Example 11.1   Prove $P \wedge Q \vdash (P \vee R) \wedge (Q \vee R)$.

Solution   The derivation is given in Table 11.2.   ∎

11.1.3 Implementation of the Deduction Theorem

Three rules in Table 11.1 make use of the deduction theorem: $\neg I$, $\Rightarrow I$, and $\vee E$. In all cases, one has to create a subordinate derivation or, in the case of $\vee E$, two subordinate

TABLE 11.2.
A Derivation Using Natural Deduction

1	$P \wedge Q$	
2	P	$\wedge E, 1$
3	Q	$\wedge E, 1$
4	$P \vee R$	$\vee I, 2$
5	$Q \vee R$	$\vee I, 3$
6	$(P \vee R) \wedge (Q \vee R)$	$\wedge I, 4, 5$

derivations. Subordinate derivations, or subderivations as they are called for short, may be nested; that is, a subderivation may make use of other subderivations. The derivation creating the subderivation is the *calling derivation*. All subderivations introduce an additional assumption, which is dropped, or *discharged*, once the subderivation is complete. The $\Rightarrow I$ rule is a direct formalization of the deduction theorem. It is illustrated in Table 11.3, which proves the hypothetical syllogism (see also Section 1.6.6). Generally, rule $\Rightarrow I$ is used to derive lines of the form $A \Rightarrow B$. The subderivation must have A as an assumption, and it must end with B. As soon as the conclusion B is reached, the assumption A is discharged, and the line $A \Rightarrow B$ is added to the calling derivation. This line is justified by writing $\Rightarrow I$, followed by the line numbers of the subderivation. The details can be seen from Table 11.3, which shows the derivation for the hypothetical syllogism. The subderivation is separated from the calling derivation by a horizontal line. Since $P \Rightarrow R$ is to be proved, the subderivation introduces the assumption P. The subderivation also has access to all the lines of the calling derivation. The conclusion of the subderivation is R. Once R has been derived, P is discharged, that is, it can no longer be used, and $P \Rightarrow R$ is added to the calling derivation. In the example of Table 11.3, the assumption P is discharged in line 8. To indicate how this result was obtained, one writes $\Rightarrow I$, 3–7, indicating that lines 3 through 7 were used in the derivation of line 8.

TABLE 11.3.
Derivation of the Hypothetical Syllogism in Natural Deduction

1	$P \Rightarrow Q$	
2	$Q \Rightarrow R$	
3	P	
4	$P \Rightarrow Q$	R, 1
5	Q	$\Rightarrow E, 3, 4$
6	$Q \Rightarrow R$	R, 2
7	R	$\Rightarrow E, 5, 6$
8	$P \Rightarrow R$	$\Rightarrow I, 3\text{–}7$

If an expression of the form $\neg A$ is to be derived, one uses a subderivation in which the assumption A is introduced. If the subderivation results in a contradiction, the line $\neg A$ is added to the calling derivation. On the same line, one writes $\neg I$, followed by the line numbers used to do the subderivation. An example of the rule $\neg I$ is the derivation of the modus tollens given in Table 11.4. The subderivation is on lines 3 through 7 and the conclusion is on line 8.

TABLE 11.4.
Derivation of the Modus Tollens in Natural Deduction

1	$P \Rightarrow Q$	
2	$\neg Q$	
3	P	
4	$P \Rightarrow Q$	R, 1
5	Q	$\Rightarrow E, 3, 4$
6	$\neg Q$	R, 2
7	$Q \wedge \neg Q$	$\wedge I, 5, 6$
8	$\neg P$	$\neg I, 3\text{--}7$

A very simple, but very important result is $P \wedge \neg P \vdash Q$. This is the inconsistency law, which was discussed in Section 1.6.5. Because of the inconsistency law, a single falsehood in the premises allows one to prove everything. Table 11.5 derives the inconsistency law.

TABLE 11.5.
Contradiction Implies Everything

1	P	
2	$\neg P$	
3	$\neg Q$	
4	P	R, 1
5	$\neg P$	R, 2
6	$P \wedge \neg P$	$\wedge I, 4, 5$
7	$\neg \neg Q$	$\neg I, 3\text{--}6$
8	Q	$\neg E, 7$

The most complicated rule of inference is $\vee E$. This rule is used if one has to prove some expression C and if one is given an expression of the form $A \vee B$. In this case, two subderivations are created, one with A as an assumption, and the other with B as an assumption. To establish C, one must show that both subderivations have C as their conclusion. As with all rules of inference based on the deduction theorem, the assumptions A and B are dropped after the subderivations are complete. Table 11.6 applies this rule to prove the disjunctive syllogism. To do this, one has to show that, given the premises $P \vee Q$ and $\neg Q$, one can derive P. The derivation contains two subderivations, one containing the assumption P and the conclusion P and the other containing the assumption Q and the conclusion P. According to $\vee E$, these two subderivations, together with the premise $P \vee Q$, allow one to conclude P. This is indicated in line 10 of the derivation. The rule

$\lor E$ is justified by line 1, which is $P \lor Q$, lines 3 and 4, which give the first subderivation, and lines 5 through 9, which give the second subderivation. The subderivation starting in line 5 with the assumption Q has in turn a subderivation, consisting of lines 6 and 7, which prove $\neg\neg P$ by means of the rule $\neg I$.

TABLE 11.6.   Derivation for the Disjunctive Syllogism

1	$P \lor Q$	
2	$\neg Q$	
3	P	
4	P	R, 3
5	Q	
6	$\neg P$	
7	$Q \land \neg Q$	$\land I$, 2, 5
8	$\neg\neg P$	$\neg I$, 6–7
9	P	$\neg E$, 8
10	P	$\lor E$, 1, 3–4, 5–9

It is possible to change the rules of inference used in natural derivation such that $\Rightarrow I$ is the only rule of inference using subderivations. To do this, one replaces $\neg I$ and $\lor E$ by

$$A \Rightarrow B \land \neg B \vdash \neg A \tag{11.1}$$

$$A \lor B, \quad A \Rightarrow C, \quad B \Rightarrow C \vdash C \tag{11.2}$$

The syllogism given by (11.1) was already given in Table 1.25. This rule is called the inconsistency law. The syllogism given by (11.2) is called the *dilemma*. To obtain the effect of the $\neg I$, one uses $\Rightarrow I$ to prove $A \Rightarrow (B \land \neg B)$ and applies the contradiction law. Similarly, instead of the $\lor E$, one proves $A \Rightarrow C$ and $B \Rightarrow C$ by $\Rightarrow I$, and then one uses the dilemma to find C.

Incidentally, the dilemma was known already to the Greeks. It means that one has two alternatives and that both alternatives lead to the same conclusion. Often, the conclusion under each alternative is unattractive, which explains why the word is now frequently used to express the choice between two unpleasant alternatives.

11.1.4 Resolution

Resolution theorem proving is a method for doing derivations that has the following features:

1. The only expressions allowed in resolution theorem proving are disjunctions of literals. A disjunction of literals is called a *clause*. Hence, all expressions involved in resolution theorem proving must be clauses.

2. Resolution follows the refutation principle; that is, it shows that the negation of the conclusion is inconsistent with the premises.

3. There is essentially only one rule of inference, *resolution*.

In a *refutation system*, one proves that $A_1, A_2, \ldots, A_n$ and $\neg C$ cannot all be true. In other words, one shows that the statements

$$A_1, A_2, \ldots, A_n, \ \neg C$$

are inconsistent. This is shown by proving that for some proposition P both P and $\neg P$ can be derived.

Resolution works with clauses, that is, all A_i, as well as $\neg C$, must be disjunctions of literals. This does not restrict the applicability of resolution theorem proving because any set of expressions can be converted into a set of clauses, as will be shown by examples. The first example involves the modus ponens, which can be expressed as

$$P, P \Rightarrow Q \models Q$$

The negation of the conclusion is now added to the premises, and the following set of statements is obtained.

$$P, P \Rightarrow Q, \neg Q$$

It must be shown that it is impossible to find any assignment that can satisfy all these statements; that is, it must be shown that the statements are inconsistent. To convert all these expressions into disjunctions, we write $\neg P \vee Q$ instead of $P \Rightarrow Q$. Once this is done, the following set of clauses is obtained.

$$P, \neg P \vee Q, \neg Q \tag{11.3}$$

Note that a single literal can be interpreted as a disjunction with a single term. Hence, a clause may contain only one literal. Such clauses are called *unit clauses*. For example, P and $\neg Q$ are unit clauses. The empty clause is a disjunction with no literal, and, as shown in Section 1.5.6, such a clause is always false. Deriving the empty clause is in this sense equivalent to deriving F.

As a second example to show how to convert a logical implication into clauses, consider the hypothetical syllogism

$$P \Rightarrow Q, Q \Rightarrow R \models P \Rightarrow R$$

This yields the following set of expressions, which must be shown to be inconsistent.

$$P \Rightarrow Q, \ Q \Rightarrow R, \ \neg(P \Rightarrow R)$$

All these statements must be converted into clauses. For the expressions $P \Rightarrow Q$ and $Q \Rightarrow R$, this causes no problem. The expression $\neg(P \Rightarrow R)$, in turn, is first written as $\neg(\neg P \vee R)$. After we apply De Morgan's law, this yields $P \wedge \neg R$. Any conjunction of clauses may be broken down into separate clauses. In fact, each term of the conjunction becomes a clause on its own. Hence, $P \wedge \neg R$ results in the two clauses P and $\neg R$, and one has

$$\neg P \vee Q, \neg Q \vee R, P, \neg R \tag{11.4}$$

In general, one can convert any expression into one or more clauses. To do this, one first converts all expressions into a conjunction of disjunctions; that is, one converts the expressions into conjunctive normal forms. This can always be done, as was shown in Section 1.5.7. Each term of the conjunction is then made into a clause of its own.

Example 11.2   Convert $P \Rightarrow (Q \wedge R)$ into clauses.

> **Solution**   We first eliminate the $\Rightarrow$ by writing $\neg P \vee (Q \wedge R)$, and we then apply the distributive law to obtain
>
> $$P \Rightarrow (Q \wedge R) \equiv (\neg P \vee Q) \wedge (\neg P \vee R)$$
>
> This yields the two clauses $\neg P \vee Q$ and $\neg P \vee R$.   ∎

We now come to the rule of inference, that is, resolution.

> **Definition 11.1:**   Two clauses can be *resolved* if and only if they contain two complementary literals. In this case, they give rise to a new clause, called the *resolvent*. If the complementary literals are P and $\neg P$, one says the resolution is *on* P. The clauses giving rise to the resolvent are called *parent clauses*. The resolvent on P is the disjunction of all literals of the parent clauses, except that P and $\neg P$ are omitted from the resolvent.

Example 11.3   Find the resolvent of $P \vee \neg Q \vee R$ and $\neg S \vee Q$.

> **Solution**   The two clauses $P \vee \neg Q \vee R$ and $\neg S \vee Q$ can be resolved on Q, because Q is negative in the first clause and positive in the second. The resolvent is the disjunction of $P \vee R$ with $\neg S$, which yields $P \vee R \vee \neg S$.   ∎

Any clause corresponds to a set of literals, that is, the literals contained within the clause. For instance, the clause $P \vee \neg Q \vee R$ corresponds to the set $\{P, \neg Q, R\}$, and $\neg S \vee Q$ corresponds to the set $\{\neg S, Q\}$. In fact, since the order in which literals appear in a disjunction is irrelevant and since the same is true for the multiplicity in which the terms occur, the set associated with a clause completely determines the clause. For this reason, one frequently treats clauses as sets, which allows one to speak about the union of two clauses. If this is done, one can write the resolvent of two clauses A and B on P as follows.

$$C = A \cup B - \{P, \neg P\}$$

In words, the resolvent is the union of all literals of A and B, except that the two literals involving P are omitted.

The resolvent is logically implied by its parent clauses, which makes it a sound rule of inference. To see this, let P be a propositional variable, and let A and B be (possibly empty) clauses. One has

$$P \vee A, \neg P \vee B \models A \vee B$$

This is valid for the following reason. If P is false, then A must be true, because otherwise $P \vee A$ is false. Similarly, if P is true, then B must be true, because otherwise $\neg P \vee B$ is false. Since P must be true or false, either A or B must be true, and the result follows. Of course, $A \vee B$ is the resolvent of the parent clauses $P \vee A$ and $\neg P \vee B$ on P, which proves the soundness of resolution.

Resolution is particularly effective if one of the parent clauses is a unit clause, that is, a clause that contains only one literal. An example is the resolution of $\neg P \vee Q \vee R$ and $\neg R$. The resolvent of these two clauses is $\neg P \vee Q$. The resolution of $P \vee Q$ with $\neg P$ yields Q, which agrees with the disjunctive syllogism. Also, the two clauses P and $\neg P$ have the empty clause as resolvent, which is correct, since P and $\neg P$ are contradictory and therefore false like the empty clause.

Let us now return to the modus ponens and the hypothetical syllogism, as given in (11.3) and (11.4). To prove the modus ponens, one uses the following derivations:

1. P Premise

2. $\neg P \vee Q$ Premise

3. $\neg Q$ Negation of conclusion

4. Q Resolvent of 1, 2

5. F   Resolvent of 3, 4

For the hypothetical syllogism, one has

1. $\neg P \vee Q$ Premise

2. $\neg Q \vee R$ Premise

3. P Derived from negation of conclusion

4. $\neg R$ Derived from negation of conclusion

5. Q Resolvent of 1, 3

6. $\neg Q$ Resolvent of 2, 4

7. F   Resolvent of 5, 6

We now consider once more the derivation given in Figure 1.3. In this case, the problem was to derive P_4 from $P_1 \Rightarrow P_2$, $\neg P_2$, $\neg P_1 \Rightarrow P_3 \vee P_4$, $P_3 \Rightarrow P_5$, $P_6 \Rightarrow \neg P_5$ and P_6. We convert these expressions to clauses, to which we add $\neg P_4$, the negation of the conclusion. This yields the first seven lines of Table 11.7. The remaining lines are obtained by resolution. The process culminates in the empty clause, which is reached in line 13.

When doing resolution automatically, one has to decide in which order to resolve the clauses. This order can greatly affect the time needed to find a contradiction. In Table 11.7, *unit resolution* is used. This means that all resolutions involve at least one unit clause. Moreover, preference is given to clauses that have not been used yet. Within these constraints, the resolution is done in sequence, starting from the top. Following these rules, $\neg P_1 \vee P_2$ and $\neg P_2$ are resolved first, which yields $\neg P_1$. At this stage, the preference for the two clauses involved in the resolution is lowered. Next, clause 5 is resolved against the unit

TABLE 11.7. Derivation of P_4 by Resolution

1.	$\neg P_1 \lor P_2$	Premise
2.	$\neg P_2$	Premise
3.	$P_1 \lor P_3 \lor P_4$	Premise
4.	$\neg P_3 \lor P_5$	Premise
5.	$\neg P_6 \lor \neg P_5$	Premise
6.	P_6	Premise
7.	$\neg P_4$	Negation of conclusion
8.	$\neg P_1$	Resolvent of 1, 2
9.	$\neg P_5$	Resolvent of 5, 6
10.	$P_1 \lor P_3$	Resolvent of 3, 7
11.	$\neg P_3$	Resolvent of 4, 9
12.	P_3	Resolvent of 8, 10
13.	F	Resolvent of 11, 12

clause 6, which yields $\neg P_5$. This continues until a contradiction is found. This happens in line 13.

Unit resolution is not the only strategy. Another strategy is the *set of support strategy*. The set of support strategy works as follows. One partitions all clauses into two sets, the *set of support* and the *auxiliary set*. The auxiliary set is formed in such a way that the expressions in the auxiliary set are not contradictory. For instance, the premises are usually not inconsistent. The inconsistency only arises after one adds the negation of the conclusion. Accordingly, one often uses the premises as the initial auxiliary set and the negation of the conclusion as the initial set of support. Since one cannot derive any contradiction by resolving clauses within the auxiliary set, one avoids such resolutions. Stated positively, each resolution takes at least one clause from the set of support. The resolvent is then added to the set of support. We now apply the set of support strategy to the example given in Table 11.7. Initially, the set of support is given by $\neg P_4$, the negation of the conclusion. One then does all possible resolutions involving $\neg P_4$, then all possible resolutions involving the resulting resolvents, and so on. If the initial 7 clauses are omitted, this yields the following derivation:

8.	$P_1 \lor P_3$	Resolvent of 7, 3
9.	$P_2 \lor P_3$	Resolvent of 1, 8
10.	P_3	Resolvent of 2, 9
11.	P_5	Resolvent of 4, 10
12.	$\neg P_6$	Resolvent of 5, 11
13.	F	Resolvent of 6, 12

The set of support strategy is complete, but unit resolution is not. This is demonstrated by the following example. The premises are $Q \lor R$, $Q \lor \neg R$, and $\neg Q \lor R$, and the conclusion is $Q \land R$. In this case there is no unit clause, which makes unit resolution impossible. The set of support strategy, however, yields the derivation given in Table 11.8. Unit clause resolution can be modified in the sense that if there are no unit clauses one is allowed to resolve on other clauses. In this way, a *unit preference strategy* is obtained. In fact, the derivation of Table 11.8 can be viewed as an example of a unit preference derivation.

TABLE 11.8.
Example for Which Unit Resolution Is Impossible

1.	$Q \vee R$	Premise
2.	$Q \vee \neg R$	Premise
3.	$\neg Q \vee R$	Premise
4.	$\neg Q \vee \neg R$	Negation of conclusion
5.	$\neg R$	Resolvent of 2, 4
6.	Q	Resolvent of 1, 5
7.	R	Resolvent of 3, 6
8.	F	Resolvent of 5, 7

One additional rule is sometimes applied in resolution theorem proving, and this is factoring. The introduction of factoring allows one to eliminate certain inefficient resolutions while maintaining completeness. To show what is meant, consider lines 1 and 4 of Table 11.8. There are two possible resolvents of these two lines, $Q \vee \neg Q$ and $R \vee \neg R$. Both of these resolvents are always true. Clauses that are always true are normally useless, and there is no point generating them. A strategy that forbids this type of resolution is incomplete, however. There are exceptional cases for which one must resolve two clauses that share more than one pair of opposing literals. To see this, consider the two clauses $P \vee P$ and $\neg P \vee \neg P$. If one allows the resolution of these two clauses, one can use the following derivation to prove inconsistency.

1. $P \vee P$ Premise

2. $\neg P \vee \neg P$ Premise

3. $P \vee \neg P$ Resolvent of 1, 2

4. P Resolvent of 1, 3

5. $\neg P$ Resolvent of 2, 4

6. F   Resolvent of 4, 5

Most systems for doing derivations do not allow resolutions of clauses containing more than one pair of opposing literals. The only place where such a resolution is meaningful is if a clause contains the same literal more than once. It is obviously easier to remove this literal right away, and this rule is called *factoring*. For instance, if one applies factoring to $P \vee Q \vee P$, one obtains $P \vee Q$. Factoring, together with resolution that is restricted to clauses with exactly one pair of opposing literals, is complete.

Problems 11.1

1. Use natural derivation to prove

   (a) $P \vee Q \vdash \neg P \Rightarrow Q$

   (b) $(P \wedge Q) \Rightarrow R \vdash P \Rightarrow (Q \Rightarrow R)$

   (c) $\vdash P \vee \neg P$

2. Use natural derivation to prove

   (a) $Q \vdash P \Rightarrow Q$

   (b) $\neg P \vdash P \Rightarrow Q$

   (c) $P, \neg Q \vdash \neg(P \Rightarrow Q)$

3. Use natural derivation to prove

   (a) $P \vee (Q \wedge R) \vdash P \vee Q$

   (b) $P \vee Q \vdash P \vee (Q \vee S)$

   (c) $\neg P \vee \neg Q \vdash \neg(P \wedge Q)$

4. Suppose that you have the premise $P \Leftrightarrow Q$. Using the rule $\Leftrightarrow E$ only, what can you conclude from this premise?

5. Derive

   (a) $\neg P, \neg Q \vdash \neg(P \vee Q)$

   (b) $\neg P \vdash \neg(P \wedge Q)$

   (c) $P, \neg Q \vdash \neg(P \Rightarrow Q)$

6. The following is a list of parent clauses. Find their resolvents.

   (a) $P \vee Q \vee \neg R, \neg Q \vee S$

   (b) $\neg P \vee \neg Q \vee \neg R, R$

   (c) $\neg P \vee \neg Q \vee \neg R, R \vee \neg S$

7. Given that $P \Rightarrow Q, Q \Rightarrow \neg R$, and $\neg P \Rightarrow \neg R$ are the premises and $\neg R$ is the conclusion, convert the premises and the negation of the conclusion into clauses and derive a contradiction through resolution.

8. The premises $P_1 \Rightarrow Q_1$ and $P_2 \Rightarrow Q_2$ entail $P_1 \wedge P_2 \Rightarrow Q_1 \wedge Q_2$.

   (a) Prove this by natural derivation.

   (b) Prove this by resolution.

9. The premises $P_1 \Rightarrow Q_1$ and $P_2 \Rightarrow Q_2$ entail $P_1 \vee P_2 \Rightarrow Q_1 \vee Q_2$.

   (a) Prove this by natural derivation.

   (b) Prove this by resolution.

11.2 SOME RESULTS FROM PREDICATE CALCULUS

11.2.1 Introduction

The following sections present complementation and the prenex normal form, two concepts important for proving theorems in predicate calculus.

11.2.2 Complements

Negated quantifiers impede the freedom with which atomic formulas can be moved within expressions and, for this reason, one would like to remove negations in front of quantifiers. For this purpose, the following equations can be used. See Table 2.4, Section 2.4.2.

$$\neg \exists x A \Leftrightarrow \forall x \neg A$$

$$\neg \forall x A \Leftrightarrow \exists x \neg A$$

Here A can be any expression. To remove all negations, one often has to use these equivalences more than once. In this case, it is advantageous to use complementation instead. In Section 1.5.9, it was shown how to form complements in propositional calculus and how to use them for the purpose of negation. It was stated that the complement of an expression can be obtained by replacing all $\wedge$ by $\vee$, and vice versa, and by replacing every literal by its complement. These rules remain valid in predicate calculus. In addition to that, $\exists$ and $\forall$ are dual pairs; that is, when forming the complement, all $\forall$ must be replaced by $\exists$, and vice versa. This follows from the discussion of Section 2.2.2, equations (2.2) and (2.3), which relate the universal quantifier to conjunctions and the existential quantifier to disjunctions. For instance, the complement of $\forall x \exists y \forall z P(x, y, z)$ is $\exists x \forall y \exists z \neg P(x, y, z)$. Since the complement is always the negation of the primal, this means

$$\neg \forall x \exists y \forall z P(x, y, z) \equiv \exists x \forall y \exists z \neg P(x, y, z)$$

Complementation can also be used to negate schemas. In fact, there is no need to distinguish between identifiers that represent expressions and propositional variables. For instance, to negate $\forall x \forall y \neg \forall z P(x, y, z)$, one can form the complement of $\forall x \forall y \neg A$, where $A = \forall z P(x, y, z)$. The result is $\exists x \exists y A$. Hence,

$$\neg \forall x \forall y \neg \forall z P(x, y, z) \equiv \exists x \exists y \forall z P(x, y, z)$$

Complementation is very convenient if one wants to move all negations inward, creating in this fashion expressions where all negations are part of some literal. Before applying complementation, one must remove all $\Rightarrow$ and $\Leftrightarrow$.

Example 11.4   In the following expression, move all negations inward.

$$\neg \exists x ((P(x) \wedge Q(x)) \Rightarrow \forall y R(x, y))$$

Solution   To solve this problem by complementation, one must first remove the implication and obtain

$$\neg \exists x (\neg (P(x) \wedge Q(x)) \vee \forall y R(x, y))$$

We now form the complement of this expression, except that we do not change the expression under the scope of the inner negation. We merely remove the inner $\neg$. This yields

$$\forall x ((P(x) \wedge Q(x)) \wedge \exists y \neg R(x, y)) \qquad \blacksquare$$

Example 11.5   In the following expression, move all negations inward.

$$\neg(\forall x P(x) \wedge \neg \forall x Q(x) \wedge \exists x R(x))$$

Solution   In this case, it is best to treat $\neg \forall x Q(x)$ as an expression on its own, which means that we are really dealing with the complement of

$$\neg(\forall x P(x) \wedge \neg A \wedge \exists x R(x))$$

When doing this, one finds

$$\exists x \neg P(x) \vee A \vee \forall x \neg R(x) \equiv \exists x \neg P(x) \vee \forall x Q(x) \vee \forall x \neg R(x) \qquad \blacksquare$$

11.2.3 Prenex Normal Forms

It is often more convenient to deal with expressions in which all quantifiers have been moved to the front of the expression. These types of expressions are said to be in *prenex normal form*. Generally, one defines the following:

Definition 11.2:   An expression is in *prenex normal form* if there is no quantifier in any scope of the logical connectives $\neg$, $\wedge$, $\vee$, $\Rightarrow$, and $\Leftrightarrow$.

Example 11.6   Which of the following logical expressions are in prenex normal form?

$$\forall x P(x) \vee \forall x Q(x)$$
$$\forall x \forall y \neg (P(x) \Rightarrow Q(y))$$
$$\forall x \exists y R(x, y)$$
$$R(x, y)$$
$$\neg \forall x R(x, y)$$

Solution   The first expression is the disjunction of two expressions, both of which contain quantifiers. Hence, it is not in prenex normal form. The second expression contains the connectives $\neg$ and $\Rightarrow$. The scope of $\neg$ is $P(x) \Rightarrow Q(y)$, and the two scopes of $\Rightarrow$ are $P(x)$ and $Q(y)$. Neither of the two scopes contains a quantifier, which means that the expression is in prenex normal form. The expression $\forall x \exists y R(x, y)$ is also in prenex normal form, and so is $R(x, y)$. Both expressions have no logical connectives. Consequently, they have no connectives that have quantifiers in their scope. The last expression has the connective $\neg$, which has a universally quantified expression in its scope, and it is therefore not in prenex normal form. $\qquad \blacksquare$

Any expression can be converted into prenex normal form. To do this, the following steps are needed.

1. Eliminate all occurrences of $\Rightarrow$ and $\Leftrightarrow$ from the expression in question.
2. Move all negations inward such that, in the end, negations only appear as part of literals.

3. Standardize all variables apart.

4. The prenex normal form can now be obtained by moving all quantifiers to the front of the expression.

Example 11.7   Find the prenex normal form of

$$\forall x(\exists y R(x, y) \wedge \forall y \neg S(x, y) \Rightarrow \neg(\exists y R(x, y) \wedge P))$$

Solution   According to step 1, we must eliminate the $\Rightarrow$, which yields

$$\forall x(\neg(\exists y R(x, y) \wedge \forall y \neg S(x, y)) \vee \neg(\exists y R(x, y) \wedge P))$$

We now move all negations in, which yields

$$\forall x(\forall y \neg R(x, y) \vee \exists y S(x, y) \vee \forall y \neg R(x, y) \vee \neg P)$$

Next, all quantifiers are standardized apart.

$$\forall x(\forall y_1 \neg R(x, y_1) \vee \exists y_2 S(x, y_2) \vee \forall y_3 \neg R(x, y_3) \vee \neg P)$$

We can now move all quantifiers to the front, which yields

$$\forall x \forall y_1 \exists y_2 \forall y_3 (\neg R(x, y_1) \vee S(x, y_2) \vee \neg R(x, y_3) \vee \neg P) \qquad \blacksquare$$

Problems 11.2

1. Use complementation to negate the following expressions:
   (a) $\exists x \forall y (P(x) \wedge \neg Q(y)) \wedge \forall z \exists y R(x, y)$
   (b) $\forall x \forall y \exists z ((P(x, y) \vee P(x, z)) \wedge \neg P(y, z))$

2. Find the prenex normal form of the following expressions:
   (a) $\forall x \exists y P(x, y) \wedge \exists y \forall x P(x, y)$
   (b) $\forall x \forall y (P(x, y) \Rightarrow \exists z Q(z)) \vee \exists x \forall y R(x, y, z)$.

3. In the following expressions, remove all $\Rightarrow$ and $\Leftrightarrow$, move all negations in, and convert to prenex normal form.
   (a) $\neg \forall x \exists y \, Q(x, y) \vee (\forall x \forall y (R(x, y) \Leftrightarrow Q(x, y)))$
   (b) $\neg \forall x \exists y \forall z (R(x, y, z) \Rightarrow Q(x, y, z))$
   (c) $\neg \forall x (\exists y P(x, y) \vee \neg \exists z P(z, x))$

11.3 DERIVATIONS IN PREDICATE CALCULUS

11.3.1 Introduction

The following sections deal with derivations in predicate calculus. We first discuss *canonical derivation*. In a canonical derivation, the negation of the conclusion is added to the premises, and a contradiction is derived. This is done by systematically generating all finite models.

We then give a short discussion of the treatment of quantifiers in natural derivations. After that, we generalize resolution theorem proving to expressions involving predicates. In fact, we interpret resolution theorem proving as a refinement of canonical derivations in the sense that only expressions that have the potential to allow one to derive a contradiction are generated. This is done through unification. Moreover, resolution theorem proving makes use of a shorthand notation that avoids quantifiers.

11.3.2 Canonical Derivations

In any finite universe of discourse, one can decide whether a set of formulas has a model. Generally, canonical derivation is a method to generate such universes. Obviously, one wants to keep the universe of discourse at each stage as small as possible and add individuals only if and when required. We now elaborate on this general idea. Initially, we assume that the expressions in question do not contain any constants or free variables. To put it positively, all variables must be bound. This assumption will be relaxed later. Moreover, all expressions must first be converted into prenex normal form. This can always be done as shown in Section 11.2.3. Canonical derivations are rather inefficient, but they give the foundations of more efficient methods, such as resolution.

We now discuss how to generate a universe. Generally, the universe of discourse is obtained in stages. Initially, one has no individuals; that is, the preliminary universe U_0 is empty. One then adds as few individuals as possible to the universe, which results in a new preliminary domain U_1. At this stage, one can conduct all possible instantiations that have not been done yet and that do not add any individuals to the universe. After this is done, all instantiations that require one to create new individuals are done, which leads to a new preliminary domain U_2. All instantiations not creating new individuals are now done, followed by all instantiations that do. In this way, one continues.

In this procedure, one must avoid all instantiations that may later lead to contradictions. For instance, suppose that one wants to find a model for a set of expressions, the first two of which are $\exists x P(x)$ and $\exists x Q(x)$. One could now introduce an individual, say a_1, and set $P(a_1)$ and $Q(a_1)$ both to T. This would indeed satisfy $\exists x P(x)$ and $\exists x Q(x)$, but the existence of the individual that satisfies both $P(x)$ and $Q(x)$ may lead to a contradiction later. For instance, the chosen interpretation no longer allows one to satisfy $\forall x(\neg P(x) \lor \neg Q(x))$, as one can verify. On the other hand, $P(a_1)$ true, $P(a_2)$ false, $Q(a_2)$ true, and $Q(a_1)$ false are a model for $\exists x P(x)$, $\exists x Q(x)$, and $\forall x(\neg P(x) \lor \neg Q(x))$. The problem with the earlier interpretation was that the same individual was selected to satisfy both $\exists x P(x)$ and $\exists x Q(x)$, and this led to a contradiction later. More generally, if the same individual is used to instantiate two different existential quantifiers, then this may prevent one from finding a model even if one exists. Hence, if the objective is to prove that there is no model for a set of expressions, one must *never* use the same individual in two different existential instantiations. For each existential instantiation, a new individual must be added to the universe.

For every preliminary universe U_i, $i \geq 0$, one can do all possible universal instantiations. For instance, if $U_1 = \{a_1, a_2\}$ is the present preliminary universe, then $\forall x(\neg P(x) \lor \neg Q(x))$ allows the instantiations $\neg P(a_1) \lor \neg Q(a_1)$ and $\neg P(a_2) \lor \neg Q(a_2)$. No new individuals are generated in universal instantiations. However, universal instantiations may generate new existential quantifiers. For instance, if the present universe is $U_1 = \{a_1, a_2\}$ as before, then $\forall x \exists y P(x, y)$ leads to the instantiations $\exists y P(a_1, y)$ and $\exists y P(a_2, y)$. These are two different expressions, and each of them has to be instantiated with a different variable.

We are now ready to describe what we mean by a canonical derivation. A canonical derivation is a method that generates sets U_1, U_2, ..., of individuals and that generates interpretations according to the following procedure:

1. Set $U_0 = \{\}$, and set $i = 0$.

2. Repeat steps 3 through 5.

3. Do all existential instantiations.

4. Add all individuals created by existential instantiations to U_i, and call the new set U_{i+1}.

5. Increase i by 1.

In principle, a canonical derivation will never end. In this text, however, we will always stop the canonical derivation as soon as a contradiction can be derived, which shows that the set of statements in question is inconsistent, or as soon as there are no existential quantifiers left to instantiate. We now demonstrate the concepts discussed so far by an example.

Example 11.8   Show that $\forall x(P \lor Q(x))$ logically implies $P \lor \forall x Q(x)$. Here P does not contain x free.

Solution   Instead of the logical implication, it will be proved that the following statements are inconsistent.

$$\forall x(P \lor Q(x)), \neg(P \lor \forall x Q(x)) \tag{11.5}$$

We convert these statements into prenex normal form. To do this, we first have to find the complement of $P \lor \forall x Q(x)$, which is $\neg P \land \exists x \neg Q(x)$. This statement is equivalent to the two statements $\neg P$ and $\exists x \neg Q(x)$. Hence, one has

$$\forall x(P \lor Q(x)), \neg P, \exists x \neg Q(x) \tag{11.6}$$

To start a canonical derivation, we set $U_0 = \{\}$, as indicated by the previous procedure. Next all existential instantiations are done. There is only one here, $\exists x \neg Q(x)$, which yields $\neg Q(a)$. We now add this individual to U_0, finding in this fashion $U_1 = \{a\}$. We now do all universal instantiations. There is only one; $\forall x(P \lor Q(x))$ is instantiated to $P \lor Q(a)$. Since there are no further existential instantiations, we stop at this point. The derivation has yielded the following expressions:

$$\forall x(P \lor Q(x)), \neg P, \exists x \neg Q(x), P \lor Q(a), \neg P, \neg Q(a)$$

We now search for a contradiction. Clearly, $\neg P$ and $P \lor Q(a)$ imply $Q(a)$, and $Q(a)$ and $\neg Q(a)$ are contradictory. Hence, the set of statements is inconsistent. According to the refutation principle, we can therefore conclude that

$$\forall x(P \lor Q(x)) \models P \lor \forall x Q(x)$$

Like other derivations, one can express canonical derivations in the form of a list. One obtains

1. $\forall x(P \vee Q(x))$ Given

2. $\neg P$ Given

3. $\exists x \neg Q(x)$ Given

4. $\neg Q(a)$ EI of line 3 with $x := a$

5. $P \vee Q(a)$ UI of line 1 with $x := a$

6. $Q(a)$ Disjunctive syllogism, lines 2 and 5

Lines 4 and 6 are contradictory ∎

Canonical derivation, as specified here, only works if there is at least one expression that starts with an existential quantifier. This, however, is no real restriction. Since the final universe of discourse must contain at least one element, if $U(x)$ is true if x belongs to the universe, then $\exists x U(x)$ must always be true. If this expression is added to the statements in question, an initial individual can always be created.

Example 11.9   Prove that $\forall x P(x)$ and $\forall x \neg P(x)$ are two inconsistent statements.

Solution   We add $\exists x U(x)$ to the statements, which allows us to construct the following derivation:

1. $\forall x P(x)$ Given

2. $\forall x \neg P(x)$ Given

3. $\exists x U(x)$ The universe has an individual

4. $U(a)$ EI of line 3 with $x := a$

5. $P(a)$ UI of line 1 with $x := a$

6. $\neg P(a)$ UI of line 2 with $x := a$

Lines 5 and 6 are contradictory ∎

Note that in each iteration of a canonical derivation, all existential quantifications are done first. This means that in an expression like $\exists x \exists y P(x, y)$ one instantiates both existential quantifiers. Here we do this in one step, which yields, if a_n and a_{n+1} are two new individuals, $P(a_n, a_{n+1})$. However, in an expression like $\exists x \forall y \exists z A$, where A is any expression, one only instantiates the first existential quantifier. At this point, one has to wait until the universal quantifiers have their turn. The inner $\exists$ can only be done after all universal instantiations have been completed, and this only happens when the present iteration is complete. A similar philosophy is used in the case of universal quantifiers. There, too, instantiation is delayed to the next iteration as soon as an existential quantifier is encountered. Otherwise, all consecutive quantifiers are instantiated simultaneously. For instance, if the preliminary universe consists of the individuals a and b, if the formula $\forall x \forall y P(x, y)$ is to be instantiated, and if this formula has already been instantiated for $x = y = a$, then the following new instantiations arise.

$$P(a, b), \qquad P(b, a), \qquad P(b, b)$$

However, no universal quantifier is instantiated unless all existential quantifiers preceding it are instantiated. For instance, if the universe is $\{a, b\}$, then $\forall x \exists y \forall z P(x, y, z)$ allows the instantiations

$$\exists y \forall z P(a, y, z), \qquad \exists y \forall z P(b, y, z)$$

If the formula $\forall x \exists y \forall z P(x, y, z)$ has already been instantiated for $x = a$, only the second instantiation is added to the derivation.

We now provide a nontrivial example. We prove that a relation $R : A \leftrightarrow A$ with the domain A is reflexive if it is transitive and symmetric. To do the proof, one must formulate the theorem in terms of logic. We define $R(x, y)$ to be true if x is related to y. As a universe of discourse, we use A, and since A is also the domain, we have

$$\forall x \exists y R(x, y) \tag{11.7}$$

The relation is transitive, which means that

$$\forall x \forall y \forall z (R(x, y) \wedge R(y, z) \Rightarrow R(x, z)) \tag{11.8}$$

Finally, the relation is symmetric; that is,

$$\forall x \forall y (R(x, y) \Rightarrow R(y, x)) \tag{11.9}$$

The claim is that equations (11.7)–(11.9) imply reflexivity; that is,

$$\forall x R(x, x) \tag{11.10}$$

To prove the theorem by refutation, negate the conclusion given by (11.10), which yields

$$\exists x \neg R(x, x) \tag{11.11}$$

The following derivation shows that the set of statements given by (11.7)–(11.9) and (11.11) is inconsistent.

1. $\forall x \exists y R(x, y)$ (11.7)
2. $\forall x \forall y \forall z (R(x, y) \wedge R(y, z)$
       $\Rightarrow R(x, z))$ (11.8)
3. $\forall x \forall y (R(x, y) \Rightarrow R(y, x))$ (11.9)
4. $\exists x \neg R(x, x)$ (11.11)
5. $\neg R(a, a)$ EI of line 4 with $x := a$
6. $\exists y R(a, y)$ UI of line 1 with $x := a$
7. $R(a, a) \wedge R(a, a) \Rightarrow R(a, a)$ UI of line 2 with $x := a, y := a, z := a$
8. $R(a, a) \Rightarrow R(a, a)$ UI of line 3 with $x := a, y := a$
9. $R(a, b)$ EI of line 6 with $y := b$
10. $\exists y R(b, y)$ UI of line 1 with $x := b$

11. $R(a,a) \land R(a,b) \Rightarrow R(a,b)$ UI of line 2 with $x := a, y := a, z := b$

12. $R(a,b) \land R(b,a) \Rightarrow R(a,a)$ UI of line 2 with $x := a, y := b, z := a$

13. $R(a,b) \land R(b,b) \Rightarrow R(a,b)$ UI of line 2 with $x := a, y := b, z := b$

14. $R(b,a) \land R(a,a) \Rightarrow R(b,a)$ UI of line 2 with $x := b, y := a, z := a$

15. $R(b,a) \land R(a,b) \Rightarrow R(b,b)$ UI of line 2 with $x := b, y := a, z := b$

16. $R(b,b) \land R(b,a) \Rightarrow R(b,a)$ UI of line 2 with $x := b, y := b, z := a$

17. $R(b,b) \land R(b,b) \Rightarrow R(b,b)$ UI of line 2 with $x := b, y := b, z := b$

18. $R(a,b) \Rightarrow R(b,a)$ UI of line 3 with $x := a, y := b$

19. $R(b,a) \Rightarrow R(a,b)$ UI of line 3 with $x := b, y := a$

20. $R(b,b) \Rightarrow R(b,b)$ UI of line 3 with $x := b, y := b$

21. $R(b,a)$ Lines 9, 18, modus ponens

22. $R(a,b) \land R(b,a)$ Lines 9, 21, combination

23. $R(a,a)$ Lines 22, 12, modus ponens

           Lines 5 and 23 are contradictory

In line 5, the individual a is generated through existential instantiation, which yields $U_1 = \{a\}$. Using a, one can instantiate lines 1, 2, and 3, which yields lines 6, 7, and 8. The next iteration starts with the existential instantiation of line 6, which yields $R(a,b)$ in line 9 and which adds the new individual b, resulting in $U_2 = \{a, b\}$. This leads to a number of new universal instantiations of lines 1 through 3. These instantiations are given in lines 10 through 20. From these lines, a contradiction can be derived, which means that the conclusion given in (11.10) logically follows from (11.7)–(11.9).

So far, we have excluded the case where the initial expressions contain constants or free variables. If there are constants, then U_0 is the set of these constants. Variables are of two kinds. Frequently, if one has a free variable within an expression, then one assumes that it is universally quantified. As an example, under this interpretation, $\{P(x), \neg P(y)\}$ really means that $\{\forall x P(x), \forall y \neg P(y)\}$. In other cases, free variables are used to establish certain connections between statements, as is the case in the following premises:

$$P(x), \; P(x) \Rightarrow Q(x) \; \vdash \; Q(x)$$

In this case, the variable must be treated like a constant. No matter how a free variable is to be interpreted, it must be either quantified or else treated like a constant before one starts a canonical derivation.

11.3.3 Quantifiers in Natural Deduction

There are four rules in natural deduction to deal with quantifiers: $\forall I$, $\forall E$, $\exists I$, and $\exists E$. These four rules correspond to universal generalization, universal instantiation, existential generalization, and existential instantiation, respectively. These rules were already discussed extensively in Section 2.3. Moreover, existential and universal instantiations were used extensively in canonical derivations (see Section 11.3.2). Our discussion is therefore kept short. The reader may get a feeling of how derivations work by considering the example given in Table 11.9, which shows that $\forall x \neg P(x)$ logically implies $\neg \exists x P(x)$. It can be proved that every conclusion derivable by canonical derivation is also derivable by natural derivation. The proof essentially proceeds as follows: First, one shows that it is possible in natural derivation to derive the prenex normal form of any expression. One gives the expression as a premise and obtains its normal form as a conclusion. Once the expression is in normal form, one can follow step by step the canonical derivation. In fact, existential and universal instantiations are available in both canonical derivation and natural derivation. The only difference is that in natural derivation universal instantiation is called $\forall E$, and existential instantiation is called $\exists E$. This, however, is only notation.

TABLE 11.9.
Derivation in Natural Deduction Using Quantifiers

1	$\forall x \neg P(x)$	
2	$\exists x P(x)$	
3	$\quad P(a)$	$\exists E, 2$
4	$\quad \neg P(a)$	$\forall E, 2$
5	$\quad P(a) \wedge \neg P(a)$	$\wedge I, 3, 4$
6	$\neg \exists x P(x)$	$\neg I, 3\text{--}5$

11.3.4 Replacing Quantifiers by Functions and Free Variables

Functions are relations, and like any other relation they give rise to predicates. For instance, to express $f(x) = x^2$, we can introduce square(x, y), which is true of $y = x^2$. Similarly, $x + y$ is a function of x and y, and it can be represented by the predicate plus(x, y, z), which is assigned true whenever $x + y = z$. However, expressing functions by means of predicates is sometimes inconvenient, and explicit methods to deal with functions are often preferable.

Functions can be used as arguments of functions. For instance, $f(f(x))$ is a legitimate function, and if $f(x) = x^2$, $f(f(x))$ is x^4. Or, to use a different example, let $g(x, y) = x + y$. Then $g(x, g(g(y, x), z))$ is a function. We can write it as

$$g(x, g(g(y, x), z)) = x + g(g(y, x), z) = x + g(y, x) + z = x + y + x + z$$

One can use functions to create an efficient strategy for finding contradictions. This strategy can be combined with resolution, and it is much more efficient than canonical derivation. To see this, consider statements of the form

$$\forall x_1 \forall x_2 \ldots \forall x_n \exists y A$$

Here A is an arbitrary statement, possibly involving quantifiers. In canonical derivations, there is a different $\exists y A$ for each combination of $x_1, x_2, \ldots, x_n$. Each $\exists y A$ generates exactly one individual for each n-tuple $x_1, x_2, \ldots, x_n$. As a consequence, the individual generated is a function of $x_1, x_2, \ldots, x_n$, which can be expressed by using the term $f(x_1, x_2, \ldots, x_n)$

to denote any of these newly created individuals. This function is called a *Skolem function*. Skolem functions allow one to remove all existential quantifiers. One writes

$$\forall x_1 \forall x_2 \ldots \forall x_n A^y_{f(x_1, x_2, \ldots, x_n)} \tag{11.12}$$

For instance, in $\forall x \exists y P(y, x)$, one has a different y for each x, which means that $\exists y P(y, x)$ leads to $P(g(x), x)$. Here $g(x)$ is the individual that is created for each x by the existential instantiation of $\exists y P(y, x)$. Note that the expression obtained by using a Skolem function is not in general logically equivalent to the original expression. It is only equivalent in the universes generated by canonical derivations or similar constructions where only one individual is created by each existential instantiation. In contrast to this, it is very well possible that in the intended interpretation there is more than one individual arising for an existential quantifier. For instance, the intended interpretation of $\forall x \exists y P(y, x)$ may mean that "for every x, there is a y, such that y is the parent of x." Canonical derivation creates only one parent for each x, and this parent is now denoted by $g(x)$. In the intended interpretation, most people have two parents, which means that $g(x)$ cannot be a function. The purpose of canonical derivation is to derive a model or to show that none exists. Toward this end, it is irrelevant how many individuals satisfy A in an expression of the form $\exists x A$, as long as there is at least one individual.

It may be instructive to see how Skolem functions relate to canonical derivations. Suppose that a preliminary universe of discourse consists of the individuals a, b, and c. One can then instantiate the statement $\forall x \exists y P(y, x)$ with $x := a$, $x := b$, and $x := c$, which yields

$$\exists y P(y, a), \qquad \exists y P(y, b), \qquad \exists y P(y, c)$$

Each of these existential statements can be instantiated, and a new individual must be created when this is done. If d, e, and f are the new individuals, this yields

$$P(d, a), \qquad P(e, b), \qquad P(f, c)$$

Clearly, d is parent of a, e is parent of b, and f is parent of c. Hence, for each individual, one parent is introduced, which makes the parent a function of the individual. When using the Skolem function $g(x)$, this means that $d = g(a)$, $e = g(b)$, and $f = g(c)$.

It is convenient to consider constants as functions with zero arguments. If this is done, (11.12) remains valid if an existential quantifier is not preceded by any universal quantifier at all, in which case $n = 0$. After all existential quantifiers are eliminated through the use of Skolem functions, it is customary to drop all universal quantifiers also. This means that all variables are implicitly universally quantified.

Example 11.10   Eliminate all quantifiers from the following two expressions:

$$\exists x \forall y \forall z \exists u P(x, y, z, u)$$

$$\forall y \exists x \forall z \exists u P(x, y, z, u)$$

Solution   The first expression contains two existential quantifiers, one having x as its bound variable and the other having u as its bound variable. Since there is no universal quantifier preceding $\exists x$, x can be replaced by a Skolem function with zero arguments, or a constant. We use a for this purpose. There are two universal quantifiers preceding $\exists u$, which means that the Skolem function corresponding to u must have two arguments, y and z. We use $g(y, z)$ for this Skolem function. We now drop the universal quantifiers to obtain

$$P(a, y, z, g(y, z))$$

In the second expression, one introduces the Skolem functions $f(y)$ and $g(y, z)$, which correspond, respectively, to the bound variables x and u. The variables corresponding to the universal quantifiers remain. After dropping the universal quantifiers, this yields

$$P(f(y), y, z, g(x, y)) \qquad \blacksquare$$

11.3.5 Resolution in Predicate Calculus

In propositional calculus, it is impossible that a set of expressions be inconsistent unless the same variable occurs more than once. For instance, there is no way to derive a contradiction from the two expressions $P \wedge Q \vee R$ and $\neg S$. The two expressions do not share variables, and the truth of the first has no bearing on the truth of the second. Similarly, in predicate calculus, one cannot derive a contradiction unless the expressions in question share atomic formulas that are either equal or can be made equal through unification. Any search for a contradiction can therefore be restricted to expressions that can be unified. Unification is therefore basic to efficient refutation methods. It turns out that unification can be applied much more efficiently if the expressions in question contain Skolem functions instead of existential quantifiers.

In this section, unification is combined with resolution to obtain an efficient refutation method. The method for doing this is called *resolution theorem proving*. In resolution theorem proving, all quantifiers are eliminated. The existential quantifiers are replaced by Skolem functions, and the universal quantifiers are dropped. The resulting quantifier-free expressions are converted to clausal form; that is, they are written as a set of disjunctions (see Section 11.1.4).

Resolution theorem proving is based on literals. There are two types of literals. A *positive literal* is merely an atomic formula. A *negative literal* is the negation of an atomic formula. For instance, $P(x, y)$ and $Q(z)$ are positive literals, whereas $\neg P(x, y)$ and $\neg Q(x)$ are negative literals. A pair of literals is said to be *complementary* if one is the negation of the other. For instance, $Q(z)$ and $\neg Q(z)$ are complementary literals, but $Q(z)$ and $\neg Q(y)$ are not. Resolution can only be applied to expressions that contain complementary literals. The idea is now to create complementary literals by means of unification and determine the resolvent. This is demonstrated by the following example. Note that free variables occurring in different clauses are distinct. Each clause is really universally quantified, except that the quantifiers are dropped.

Example 11.11   Find the resolvent of the following two clauses:

$$G(a, x, y) \vee H(y, x) \vee \neg D(z)$$

$$\neg G(x, c, y) \vee H(f(x), b) \vee E(a)$$

Here a, b, and c are constants, and x, y, and z are variables.

Solution To obtain two complementary literals, we unify $G(a, x, y)$ in the first clause with $G(x, c, y)$ in the second. We can do this as follows: Since x, y, and z in the first clause are (implicitly) universally quantified, we can replace these variables by any term. In particular, we can set $x := c$, which yields

$$G(a, c, y) \lor H(y, c) \lor D(z) \tag{11.13}$$

Similarly, one can replace the free variables of the second clause by any term. We set $x := a$ and obtain

$$\neg G(a, c, y) \lor H(f(a), b) \lor E(a) \tag{11.14}$$

Note that the variables x in the two expressions have to be considered as distinct variables, which allowed us to instantiate x in the first expression to c and x in the second expression to a. This also means that y in the first expression is a different variable than y in the second, and it is therefore necessary to explicitly unify these two variables, even though they have the same name. Once this is done, the resolvent of (11.13) and (11.14) can be found readily as

$$H(y, c) \lor D(z) \lor H(f(a), b) \lor E(a) \qquad\blacksquare$$

We now provide a number of examples to demonstrate resolution theorem proving. In the first example, we prove that everybody has a grandparent, given that everybody has a parent. To express this problem in logic, let $P(x, y)$ represent "x is parent of y." The premise can now be stated as $\forall x \exists y P(y, x)$. From this, we conclude that there is a parent of a parent, which can be expressed as $\forall x \exists y \exists z (P(z, y) \land P(y, x))$. We must thus prove that

$$\forall x \exists y P(y, x) \models \forall x \exists y \exists z (P(z, y) \land P(y, x)) \tag{11.15}$$

We add the negation or, equivalently, the complement of the conclusion to the premise, which yields the following two statements:

$$\forall x \exists y P(y, x), \exists x \forall y \forall z (\neg P(z, y) \lor \neg P(y, x))$$

To show that these two statements are contradictory, we eliminate the existential quantifiers to obtain

$$\forall x P(f(x), x), \forall y \forall z (\neg P(z, y) \lor \neg P(y, a))$$

After dropping the universal quantifiers, this yields

$$P(f(x), x), \neg P(z, y) \lor \neg P(y, a)$$

Resolution can now be used to find a contradiction as follows:

1. $P(f(x), x)$ Given
2. $\neg P(z, y) \lor \neg P(y, a)$ Given
3. $P(f(a), a)$ UI of line 1 with $x := a$
4. $\neg P(z, f(a)) \lor \neg P(f(a), a)$ UI of line 2 with $y := f(a)$
5. $\neg P(z, f(a))$ Resolve 3 and 4

6. $P(f(f(a)), f(a))$ UI of line 1 with $x := f(a)$

7. $\neg P(f(f(a)), f(a))$ UI of line 5 with $z := f(f(a))$

Contradiction: lines 6 and 7

Note how the derivation proceeds. After the first two lines, there is only one reasonable way to apply resolution, and this is by unifying $P(f(x), x)$ in line 1 against $\neg P(y, a)$ in line 2. The instantiations to do the unifications are given in lines 3 and 4. The actual resolution is then carried out in line 5. The resolvent $\neg P(z, f(a))$ is then unified with line 1, which is done in lines 6 and 7. These two lines are contradictory, and the proof is complete.

All expressions occurring in resolution theorem proving occur in some form or other in the corresponding canonical derivation. However, making use of unification avoids all lines that contain no literals that can be unified, and which therefore can never be used to derive a contradiction. To show the correspondence between resolution theorem proving and canonical derivation, we give all lines corresponding to the preceding derivation as they would appear in a canonical derivation. We omit, however, the lines that are not used.

1. $\forall x \exists y P(y, x)$ Given

2. $\exists x \forall y \forall z (\neg P(z, y) \lor \neg P(y, x))$ Given

3. $\forall y \forall z (\neg P(z, y) \lor \neg P(y, a))$ EI of line 2 with $x := a$

4. $\exists y P(y, a)$ UI of line 1 with $x := a$

5. $P(b, a)$ EI of line 4 with $y := b$

6. $\exists y P(y, b)$ UI of line 1 with $x := b$

7. $P(c, b)$ EI of line 6 with $y := c$

8. $\neg P(c, b) \lor \neg P(b, a)$ UI of line 3 with $y := b, z := c$

9. $\neg P(c, b)$ Resolvent of 5 and 8

Contradiction: lines 7 and 9

In resolution theorem proving, it is very important to follow an appropriate strategy. The applicable strategies were already discussed in Section 11.1.4. In particular, we mentioned *unit clause resolution* and the *set of support strategy*. These strategies remain important in the context of predicate calculus. To show this, consider again the proof that, if A is the domain of a relation $R : A \leftrightarrow A$, then R is reflexive if it is symmetric and transitive. We show the proof by demonstrating through resolution that equations (11.7)–(11.9) and (11.11) are inconsistent. We use the set of support strategy, with the negation of the conclusion as the only statement in the set of support. If this strategy leaves more than one choice, we prefer resolution against unit clauses, provided the unit clause contains a constant. Specifically, we apply all such unit clauses in turn until none can be applied any more before trying the next clause. Any resolvent that is tautologically true, such as $R(a, a) \lor \neg R(a, a)$, is avoided. Whenever a new unit clause is found, it is compared to other unit clauses to see whether a contradiction can be derived. In the following proof, the first four lines of the derivation are obtained from (11.7)–(11.9) and (11.11). All other lines are found through resolution.

1. $\neg R(a, a)$ From (11.11)
2. $R(x, f(x))$ From (11.7)
3. $\neg R(x, y) \vee \neg R(y, z) \vee R(x, z)$ From (11.8)
4. $\neg R(x, y) \vee R(y, z)$ From (11.9)
5. $\neg R(a, y) \vee \neg R(y, a) \vee R(a, a)$ Line 3 with $x := a, z := a$
6. $\neg R(a, y) \vee \neg R(y, a)$ Resolve lines 1 and 5
7. $R(a, f(a))$ Line 2 with $x := a$
8. $\neg R(a, f(a)) \vee R(f(a), a)$ Line 6 with $y := f(a)$
9. $\neg R(f(a), a)$ Resolve lines 7 and 8
10. $\neg R(a, f(a)) \vee R(f(a), a)$ Line 4 with $x := a, y := f(a), z := a$
11. $\neg R(a, f(a))$ Resolve lines 9 and 10
12. $R(a, f(a))$ Line 2 with $x := a$

Contradiction: lines 11 and 12

We now look at a simple argument from daily life. Suppose that we know that Mary is either the wife of Paul or the wife of Al, but we do not know which. If we are informed that Jane is the wife of Al, we would immediately conclude that Mary is the wife of Paul. If the situation is formulated in terms of logic, one can use the predicate wife(x, y) to express the fact that x is the wife of y. The desired conclusion is now wife(Mary, Paul). If this conclusion is negated and added to the premises, the following clauses result.

wife(Mary, Paul) $\vee$ wife(Mary, Al),   wife(Jane, Al), $\neg$ wife(Mary, Paul)

It is impossible to obtain a contradiction from these clauses. In fact, if both Jane and Mary could be the wife of Al, then all clauses could be satisfied. The point is, of course, that each husband can have only one wife. This fact must be included explicitly among the premises. In logic, we express this fact by the following clause:

$$\text{wife}(x, \text{ Al}) \Rightarrow E(x, \text{ Jane})$$

Here $E(x, \text{Jane})$ stands for $x = \text{Jane}$. We now still need the fact that Mary and Jane are different persons, which can be expressed as

$$\neg E(\text{Mary, Jane})$$

We now have all the necessary premises:

1. wife(Mary, Paul) $\vee$ wife(Mary, Al)
2. wife(Jane, Al)
3. $\neg$ wife(Mary, Paul)
4. $\neg$ wife$(x, \text{Al}) \vee E(x, \text{Jane})$
5. $\neg E(\text{Mary, Jane})$

At this point, one can complete the derivation, using, for instance, the set of support strategy. The initial set of support is $\neg$ wife(Mary, Paul). In the derivation, the steps of unification and resolution are combined. One obtains the following:

6. wife (Mary, Al)   Resolution, lines 1, 3

7. E(Mary, Jane)   Resolution, lines 4, 6

　　　　　　　　　　Contradiction: lines 5 and 7

In this example, new facts were introduced as the argument progressed. This is rather typical, even when conducting mathematical arguments. One starts out with some basic assumptions and with a desired conclusion. One then looks for facts or theorems that are helpful to make progress toward the desired conclusion. Occasionally, one may introduce a counterexample, which shows that the premises considered so far do not yet support the theorem to be proved.

　　　We have not presented any example involving factoring yet, and, indeed, factoring is a rule of inference that is used only rarely. To show how factoring works, consider the clause $P(a, f(a)) \vee P(x, y)$. By setting x to a and y to $f(a)$, this can be converted to $P(a, f(a)) \vee P(a, f(a))$, which in turn yields $P(a, f(a))$. We do not further discuss factoring.

　　　As mentioned in Section 4.5.2, Prolog is based on clauses and, in fact, Prolog works by resolution. The clauses available in Prolog are restricted to *Horn clauses*, that is, clauses that contain only one nonnegated literal. One can say that Prolog uses a set of support strategy in which the query is the original set of support: all resolutions involve the query or its descendants. Still, Prolog is not designed as a theorem prover. Especially, it does not give lower priority to clauses that have already been used, and this sometimes causes infinite loops, which may prevent finding a contradiction even if one exists.

Problems 11.3

1. Use natural derivation to show that $\forall x(P \Rightarrow Q(x)) \vdash P \Rightarrow \forall x Q(x)$.

2. Use natural derivation to derive $U \wedge \exists x P(x)$ from $\exists x(U \wedge P(x))$.

3. Use natural derivation to prove that $\exists x M(x)$, $\forall x(M(x) \Rightarrow C(x, a))$, and $\forall x(\exists y C(x, y) \Rightarrow F(x))$ logically imply $\exists x F(x)$.

4. Use canonical derivation to show that the three statements $\exists x(P(x) \vee Q(x))$, $\forall x \neg P(x)$, and $\forall x \neg Q(x)$ cannot be simultaneously satisfied.

5. Use Skolem functions to remove all quantifiers from the following expressions:
   (a) $\forall x \exists y(P(x, y) \Rightarrow Q(x, y))$
   (b) $\forall x \exists y \forall z P(y, z, x)$
   (c) $\forall x \forall y \exists z P(z, x, z, y)$

6. Find the resolvents of the following parent clauses. In all cases, a, b, and c are constants, and x, y, and z are variables.
   (a) $P(g(x), x, x) \vee C(b, x)$ and $\neg P(y, a, z) \vee R(z, y, g(x))$
   (b) $P(f(x), y) \vee R(y, f(y))$ and $\neg R(a, x)$
   (c) $P(x, g(x, y), y) \vee Q(x, y)$ and $\neg P(a, x, b)$

7. Use unit resolution to show that the clauses $P(x)$, $\neg P(x) \vee Q(x)$, $\neg Q(a) \vee R(b)$, and $\neg R(x)$ have no model. Here a and b are constants.

PROBLEMS: CHAPTER 11

1. Show that there is no model simultaneously satisfying the following two expressions:

$$\forall x P(x) \vee \forall x Q(x), \quad \exists x (\neg P(x) \wedge \neg Q(x))$$

2. Give a formal derivation of $\vdash P \vee (P \wedge Q) \Leftrightarrow P$. Use the rules of natural derivation.

3. A patient is administered three different tests. Test A will give a positive result if and only if either virus X or virus Y is present. Test B will give a positive result if and only if virus Y or Z is present. If test C is positive, virus Y can be excluded. The patient reacts positively to all three tests. Prove that he has virus X and virus Z but not virus Y. Carefully formulate your propositions, and use natural derivation to obtain the conclusion. (Hint: Consult Table 11.6.)

4. Prove, using natural derivation, that under the following premises everyone will have a good time. There will be a party this weekend at Mike's place. If the weather is nice, the party will be in the backyard. If it rains, the party will be in the basement. Either the weather is nice or it rains. If the party is outside, people will have a barbecue, and they will have a good time. If the party is in the basement, they will listen to music, and they will have a good time.

5. Use natural derivation to prove
   (a) $P \Rightarrow Q, P \Rightarrow \neg Q \vdash \neg P$
   (b) $P \Rightarrow Q, \neg P \Rightarrow Q \vdash Q$

   Hint: Ideas for the proof of part (b) can be obtained from Table 11.4, as well as from the solution of part (a) of this problem.

6. In Chapter 9, use was made of the following result.

$$(B \wedge Q_1) \vee (\neg B \wedge Q_2) \equiv (B \Rightarrow Q_1) \wedge (\neg B \Rightarrow Q_2)$$

   Negate this statement, convert it into clausal form, and derive a contradiction.

7. Solve Problem 5 by resolution.

8. Solve Problem 3 by resolution.

9. Solve Problem 4 by resolution.

10. Convert the following argument into predicate calculus, and prove its correctness by natural derivation.

    If there had been a question on completeness on the exam, then nobody would have passed. Some students passed. Consequently, there cannot have been a question on completeness on the exam.

11. Use natural derivation to prove that $U \vee \forall x P(x) \vdash \forall x (U \vee P(x))$.

12. Use natural derivation to prove that $U \wedge \exists x P(x) \vdash \exists x (U \wedge P(x))$.

13. Put the following formula into clausal form. Use Skolem functions.

$$(\forall x (P(x) \Rightarrow \exists y R(y))) \vee \neg (\exists x (Q(x) \vee \forall z P(x, z))) \wedge \exists v S(v)$$

14. A certain object is either red or black, but not both. Formulate this in clausal form. Assume, in addition, that object a is red. Use resolution to prove that object a is not black.

15. $R(x, y)$ is reflexive, symmetric, and transitive. Use resolution to prove that $R(y, x) \land R(z, x) \Rightarrow (R(y, u) \Rightarrow R(z, u))$ is true for all x, y, z, and u.

12

An Overview of Relational
Database Systems

Effective decision making in an organization requires the capture and maintenance of appropriate, accurate, and reliable data. The nature of the data depends on various facets of an organization's operations. For example, at a university, various classes of entities deal with students, professors, and courses. Appropriate data models are used to represent entities and their associated relationships. Such related data in an organization are often stored in a database. In this chapter we focus on a popular data model called the relational data model, which is part of a relational database system. The mathematical underpinnings of the relational data model are a variation and extension of a relation that was introduced in Section 5.3. The relational data model was proposed in the early 1970s by E. F. Codd for describing the structure of data and data manipulation operations. Although early implementations of the model were inefficient, the relational model has become very popular because more efficient implementations have been realized, in part by cost–performance improvements in hardware systems. One major reason for this popularity is due to the data model's simplicity. It is based on data being viewed as tables—a familiar conceptual representation used in everyday life that makes the model easy to learn. Also, the data model has a strong theoretical foundation, being based on the mathematical relation.

This chapter begins with the introduction of basic concepts in database systems. The next section focuses on the relational data model. In the next three sections we examine three approaches for stating queries in relational databases. The first approach introduces a relational algebra with an associated suite of eight operators. These operators manipulate tables to produce tables. The second approach, which is based on predicate logic, formulates a relational calculus for specifying queries. The third approach gives an overview of the Structured Query Language (SQL), a programming language which is based on both relational algebra and relational calculus.

12.1 BASIC CONCEPTS

12.1.1 Introduction

We first introduce a hierarchy of information structures for organizing data. Then, a simple student registration relational database is presented to demonstrate basic database concepts. The section concludes with the presentation of a high-level overview of a database architecture.

12.1.2 Definitions and Concepts

In data modeling we conceptualize an *entity* or *object* as a person, place, thing, or event. For example, in a university environment, entities of interest are students, professors, courses, departments, colleges, staff, and so on. Each entity can be characterized by a set of *attributes* or *properties*. For example, when modeling a student entity, attributes such as student number, name, year of study, major, phone number, courses taken, and course average are of interest. Sometimes the term *record* consisting of a collection of items is used to represent more concretely a particular entity. An example of a record about a passenger on an airline flight consists of different attributes, such as passenger's name, address, seat number, and menu restrictions.

Records of the same type can be grouped together in files. A record item or attribute that uniquely identifies a record in a file is called a *key*. In the airline example, the key could be either the passenger's name or seat number. Note that it could be possible to have two passengers with the same name. Consequently, a seat number is a better choice of a key because it would guarantee uniqueness. The records in a file are commonly ordered according to this key.

Thus far we have observed a hierarchy of information structures in which attributes or items are composed to form records and records are composed to form files. Files can be composed to form a set of files. If the set of files is used by the application programs of a certain application area or within a certain enterprise, and if these files exhibit certain associations or relationships among the records of the files, then such a collection of files is often referred to as a *database*. Figure 12.1 shows an information structure hierarchy as it pertains to file-processing applications.

Attributes, records, files, and databases are abstract terms in the sense that they have been introduced without any indication of how they can be realized or stored physically on an external device, such as a hard or floppy disk drive with a magnetic disk.

12.1.3 Introductory Example of a Relational Database

Recall that a database is a collection of computerized files. This collection can be considered to be a repository that resides in an electronic filing cabinet. A database system manages this repository and performs operations such as the following:

- Inserting, deleting, and updating data in existing files
- Retrieval of data from existing files
- Summarizing data in existing files

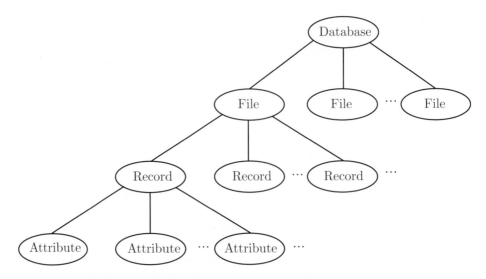

Figure 12.1.   Information structure hierarchy for information processing

- Creating new files
- Deleting old files

    A data model such as the relational model discussed in this chapter is a conceptual vehicle for modeling what is represented in the database. A conceptual model needs to deal with three types of relationships to describe associations among data. These relationship types are one-to-one, one-to-many, and many-to-many. In a one-to-one relationship, one entity is related to exactly one other entity, and vice versa. For example, a department head administers only one department, and one department has only one head. In a one-to-many relationship, one entity is related to many others, such as a professor teaching several courses or an author writing many books. In a many-to-many relationship, many entities can be related to many others, such as in a student registration system at a university where a particular course can be taken by many students and a particular student can enroll in several courses.

    Consider, as an example, a very simplified student registration system for some university. Table 12.1 contains typical information that would be kept about students. We model a student as an entity with five attributes or properties, as shown in the table. Each record in the table is considered to be a logical record. Each student is uniquely identified by a student number, which is considered to be a *primary key*. We have listed the students in the STUDENT file in increasing student number order. It should be noted that many more attributes about students would be maintained in a real system.

    Table 12.2 illustrates information about seven courses available to students. Again, each course entity has five attributes, with the course number being considered to be unique for the COURSE file.

    Table 12.3 represents relationships between students and courses by the relationship ENROLL. Observe that the table contains redundancies in the sense that some courses are

TABLE 12.1.   STUDENT

student#	student_name	major	cumulative_avg	address
278123	Susan Graham	Computer Science	81.5	721 Primrose Rd
314182	Paul Harvey	Computer Science	74.5	718 Park St
392044	Tim Horton	Mathematics	69.0	4233 Rushmore Rd
412181	Henry Silva	Electrical Engineering	78.6	1280 Taylor St
673040	Ann White	Engineering Physics	80.2	512 Avenue P
821688	Verna Friesen	Computer Science	71.4	126 Cumberland St

TABLE 12.2.   COURSE

course_num	course_name	department	room	time
CS215	Computer Systems	Computer Science	A146	TTh: 0830
CS220	Logical Design	Computer Science	A143	MWF: 1330
CS250	Data Structures	Computer Science	E1B71	MWF: 1530
CS260	Discrete Structures	Computer Science	E1B71	TTh: 1300
MA241	Probability	Mathematics	A143	TTh: 1030
MA242	Statistics	Mathematics	A146	MWF: 0930
MA266	Linear Algebra	Mathematics	A134	TTh: 1000

TABLE 12.3.   ENROLL

course_num	student#	mark	course_num	student#	mark
CS215	278123	82	CS250	673040	79
CS215	314182	61	CS260	314182	41
CS215	821688	58	CS260	392044	71
CS220	278123	73	CS260	821688	83
CS220	314182	94	MA241	392044	64
CS220	821688	80	MA241	821688	77
CS250	392044	46	MA242	392044	68
CS250	412181	65	MA266	392044	89

listed more than once. The same observation holds for student number entries. However, in the relational model to be described later in the chapter, the ENROLL relation is a bridge or linking table that associates students and courses. Observe that this linking is characterized by an attribute that records the grade assigned to a certain student taking a particular course.

Let us examine some operations that can be performed on this simple database. Students could be added to or deleted from the STUDENT file. Similarly, new courses could be added to and old courses deleted from the COURSE file. Insertions and deletions can also be performed on the ENROLL table. A new row entry would indicate a student picking up another course. The deletion of a row, on the other hand, would indicate that a student has withdrawn from a course. Also a particular grade of a particular student in some course could be changed.

Databases are used by end users for retrieval purposes. In such cases, end users formulate queries using some form of query language. The queries that we now give are somewhat informal. However, the queries resemble those given more formally later in the chapter. The following syntax is used to formulate queries:

> **select** ⟨columns⟩
> **from** ⟨table names⟩
> [**where** ⟨conditions⟩];

where ⟨columns⟩ denotes a list of column (attribute) names separated by commas from the specified ⟨table names⟩. ⟨Table names⟩ is a list of table names also separated by commas, and the optional where clause specifies the restrictions or constraints on the specified search.

As a first example, the execution of the query

> select student_name, address
> from STUDENT;

will result in the following output:

```
 _ _ _ _ _ _ _ _ _ _ _ _ _ _ _ _ _ _
 student_name    address
 _ _ _ _ _ _ _ _ _ _ _ _ _ _ _ _ _ _

 Susan Graham    721 Primrose Rd
 Paul Harvey     718 Park St
 Tim Horton      4233 Rushmore Rd
 Henry Silva     1280 Taylor St
 Ann White       512 Avenue P
 Verna Friesen   126 Cumberland St
 _ _ _ _ _ _ _ _ _ _ _ _ _ _ _ _ _ _ _ _
```

As a second example, the next query specifies a search for generating the names and averages of students with averages greater than 75.

> select student_name, cumulative_avg
> from STUDENT
>   where cumulative_avg > 75;

This query will generate the following output:

```
 _ _ _ _ _ _ _ _ _ _ _ _ _ _ _ _ _
 student_name    cumulative_avg
 _ _ _ _ _ _ _ _ _ _ _ _ _ _ _ _ _

 Susan Graham    81.5
 Henry Silva     78.6
 Ann White       80.2
 _ _ _ _ _ _ _ _ _ _ _ _ _ _ _ _ _
```

The next query specifies the names of students who are majoring in Computer Science or Mathematics:

```
        select student_name
          from STUDENT
            where major = 'Computer Science'
              or major = 'Mathematics';
```

The output generated by the query is

```
            - - - - - - -
            student_name
            - - - - - - -
            Susan Graham
            Paul Harvey
            Tim Horton
            Verna Friesen
            - - - - - - -
```

An important and powerful feature of a relational database is the ability to combine tables on a common attribute. For example, the following query combines tables STUDENT and ENROLL on the attribute student#.

```
        select STUDENT.student_name, STUDENT.major
          from STUDENT, ENROLL
            where ENROLL.course_num = 'CS260' and
              (STUDENT.student# = ENROLL.student#);
```

The query specifies the retrieval of name and major of each student who is enrolled in the discrete structures course CS260. Note that this query uses name qualification; for example, STUDENT.student# in the where clause refers to the student number in the STUDENT relation. The output generated by the query is

```
        - - - - - - - - - - - - - - - - -
        student_name    major
        - - - - - - - - - - - - - - - - -
        Paul Harvey     Computer Science
        Tim Horton      Mathematics
        Verna Friesen   Computer Science
        - - - - - - - - - - - - - - - - -
```

The informal queries given here are compiled into a set of relational operators that are part of the relational database model. These operators are discussed later in the chapter.

12.1.4 Overview of a Database System

It is important for an algorithm to be able to generate relevant information efficiently in order to carry out business. Data management, which is concerned with data collection, storage, and retrieval, is a crucial activity in any organization.

A database was defined as a collection of files used by application programs for some particular enterprise such that these files exhibit certain associations or relationships at the record level. A database system can be viewed as an automatic record-keeping system that records, manipulates, summarizes, and produces on demand information for a variety of users and purposes. Figure 12.2 gives a high-level context diagram of a database system. The Database Management System (DBMS) interfaces with users and manages the database contents. A database system contains four main components; users, data, software, and hardware. We now consider briefly each of these components.

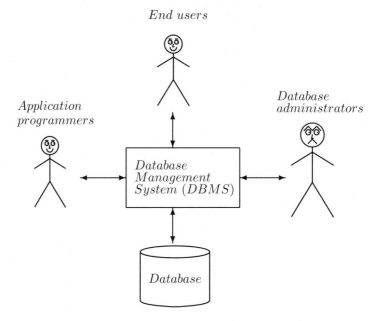

Figure 12.2.   A high-level architectural diagram of a database system

There are different kinds of users in a database system—end users, application programmers, and database administrators. The simplest type of end user is a naive user, such as an automated teller user, who may not even be aware of the presence of a database system. A more sophisticated kind of end user is an online user, such as a bank teller, who is aware of the database system and has a limited amount of expertise based on her or his permitted interaction within the system. The next more sophisticated kind of user is an application programmer who focuses on developing application programs and user interfaces for end users. These programs are written in some high-level programming language such as COBOL. The most sophisticated kind of user is the database administrator (DBA), who defines the database structure and looks after the data in the database. More specifically, a DBA's duties include deciding on the storage structure of the database, providing liaison with users, defining authorization checks, defining backup and recovery strategies, and monitoring system performance and user behavior.

Database systems can be single-user or multiuser systems. Many larger computer systems, however, are multiuser systems. In these systems, data are integrated and shared. The purpose of data integration is to unify data and reduce redundancy. Shared data are made available to several users. Another aspect of multiuser systems is concurrency control. The system must be able to handle situations where several users want to access simultaneously, say, a record in the "Albert file."

The most sophisticated software in a database system is associated with the database management system. A primary purpose for this software is to shield users from the physical database stored on secondary-storage devices such as magnetic disks. Also, hardware-level details dealing with the interaction with users are handled by the DBMS software. Software is also written by application programmers to query and manipulate the data in the database. A programming language such as the Structured Query Language (SQL), which is discussed later in the chapter, can be used for this purpose.

Finally, the hardware component of a database system includes secondary-storage devices, communication links and devices, processors, and display devices.

An important function of the DBMS software is to support the data model used in the database system, for example, the relational model based on relations in a relational database system. A data model includes the following:

1. A data structure, for example, a relation
2. Languages for the definition and manipulation of the data structure
3. Constraints for security, integrity, and concurrency control

Each of these aspects for the relational data model is discussed in more detail later in the chapter.

Database systems can be characterized by an important property called *data independence*. Data independence can be described as a condition in which the data and the application programs are independent in the sense that either may be changed without changing the other. Hence, for example, application programs can be left unaffected by changes made to the data and the way they are organized.

The term data independence is now demonstrated in the context of a manual filing system. Suppose that we ask the secretary to retrieve information in the "Albert file." Any changes to the location of the file (i.e., changing the file from a drawer A to drawer B), to the internal numbering of the file, or to the number of subfiles created from the main Albert file should not seriously affect the ability to retrieve the desired information. Therefore, the request for information is to a certain degree independent of how that information is stored or organized. However, if data independence is carried to an extreme (e.g., we remove Albert as an index in our filing system and subsume the file's information in a number of other files), then retrieval becomes difficult and time consuming. The degree of data independence that is achievable in a database system depends in part on the database management approach adopted.

Besides providing some degree of data independence, integrated database systems should offer a centralized control operation. This form of operation aids in the following:

1. Reducing the amount of redundancy in the stored data
2. Promoting data integrity and avoiding problems of data inconsistency, that is, inconsistencies due to changing one instance of a fact, but leaving other instances of the same fact unchanged
3. Enhancing the sharing of data between users
4. Providing more uniform and effective controls for the security and privacy of user data

In the remainder of the chapter we present the relational data model, describe how information can be created and manipulated within this model, and discuss some of its advantages and disadvantages.

12.2 RELATIONAL DATA MODEL

12.2.1 Introduction

In a database application, a data model is a mechanism that provides an abstraction capability. The abstraction process facilitates the highlighting of details that are pertinent to a particular application, while ignoring details that are not relevant to the application. Data modeling is concerned with identifying and representing entities of interest and their relationships in the database. Several data models have been developed over the past quarter-century. These models differ in the way in which relationships among entities are represented. Three traditional data models have emerged: the hierarchical model, the network model, and the relational model. Although we are concerned mainly with the relational model, we briefly describe the other two models.

The hierarchical data model is used for structuring the data as a hierarchy of record types in a hierarchical database. A tree structure is used to represent this type of hierarchy in which record types denote the nodes in the tree structure and the arcs between nodes denote relationships between parents and children.

The network data model is based on a graph structure, which from Chapter 7 we know to be more general than a tree structure. In a hierarchical database there can be only one type of relationship, which is a parent–child relationship. In a network database, however, there can be more than one relationship between pairs of nodes; that is, the specification of many-to-many relationships is allowed.

This section first presents the relational data structure based on a table view. Then we introduce formally the concept of a database relation and its associated schema. We next discuss briefly the representation of relations in the relational database model. An important facet of the relational database model concerns integrity rules. The section concludes with the presentation of two kinds of integrity rules.

12.2.2 Overview of the Relational Structure

Recall that a data model includes the following parts:

- A data structure, for example, a relation or table in the relational model
- A language for the definition of the data structure
- A language for the manipulation of the data, for example, a relational algebra in the relational model
- Constraints for security and integrity

Special terms are associated with each of these parts. Several of these terms are associated with the data structure—the focus of the current section. Table 12.4 illustrates the terminology required when using the relation STUDENT given earlier.

TABLE 12.4.   Terminology Used in a Relation

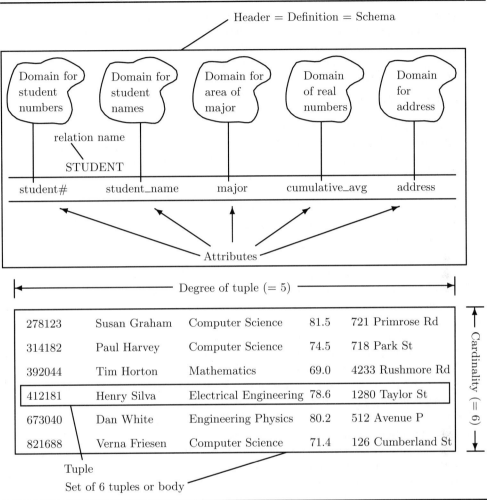

A *relation* corresponds informally to a table. The definition of a relation is specified by a schema, which is essentially a template (or type) declaration for a family of possible concrete (instantiated) tables. A *schema* includes the following:

- Table name
- Set of attribute names
- Domain for each corresponding attribute

The domains for the attributes need not be distinct; that is, attributes can share domains. It should be stressed, however, that most database systems do not *formally* support the definition of domains. Each row or record in the body of a table is a *tuple*. The number of columns (attributes) in a table is the *degree* of the relation. In the current example, the table contains 5-tuples. A *domain* for an attribute is essentially a data type or set of values from which attribute values are selected. The *cardinality* of a relation is simply the number of tuples in the relation or the number of rows in the table. Observe that the cardinality of a relation is a time-varying value because tuples (or records) can be added or deleted from the relation over time.

The notion of a primary key was introduced earlier. It is an attribute (or combination of attributes) that uniquely identifies a record or tuple in a relation. Consequently, a relation or table cannot contain duplicate records or tuples. Although the entries or records in a table seem to be listed in a top-down manner, no such order is implied. From a mathematical viewpoint, tuples in a relation form a set that is not ordered.

Recall from Section 5.2.2 that the order of the components of a tuple is significant. The tuples in a given table are ordered for presentation purposes only. However, such an order is not required, since columns in a table have labels. Because of this labeling, we can easily permutate the columns (i.e., components) of a tuple without causing any ambiguity.

12.2.3 Relations and Their Schemas

An entity (or object) has certain characteristics or properties. The properties that we select to characterize an entity, within the context of a particular application, are called its *attributes*. Attribute values are chosen from a domain. A *domain* D is a set of values of the same data type. All values in a domain are homogeneous. A domain can be atomic (simple) or composite. An *atomic domain* has values that are not decomposable without losing some meaning within some given application context. Examples of atomic domains include the set of integers, reals, and Booleans. A *composite domain* contains nonatomic values. For example, the domain for the *address* attribute in the STUDENT relation is an example of a composite domain that specifies the street number and street name. Domain attributes defined on the same domain are said to be comparable because the values for these attributes are selected from the same set.

Suppose that we have an entity characterized by n attributes $a_1, a_2, \ldots, a_n$ whose values are chosen from corresponding domains $D_1, D_2, \ldots, D_n$. Then this entity can be represented by an n-tuple (see Section 5.2.2) $(a_1, a_2, \ldots, a_n)$, which is a member of the Cartesian product set $D_1 \times D_2 \times \cdots \times D_n$. From a mathematical perspective, the order of the components in an n-tuple is significant. In the context of relational databases, however, we can relax the ordering requirement by associating an attribute name with each component in the tuple. We can view a tuple as a mapping from attribute names to the values in the corresponding domains of the attributes. More formally, let the attribute names be denoted by $A_1, A_2, \ldots, A_n$. Then an *unordered labeled tuple* can be represented by the set

$$\{(A_1 : a_1), (A_2 : a_2), \ldots, (A_n : a_n)\}$$

where the value of each a_i is selected from the corresponding domain D_i.

We can now define a relation, in the context of a relational database, on a set of domains $D_1, D_2, \ldots, D_n$, which need not be distinct. A relation is made up of two parts:

a header and a body. A *header* contains a fixed set of attribute names. Each attribute name is associated with a particular domain. The header is sometimes referred to as *relational schema* or *schema*. The body of the relation consists of a set of n-tuples. Each n-tuple or row in the table is n attribute-value pairs

$$\{(A_1 : a_{i1}), (A_2 : a_{i2}), \dots, (A_n : a_{in})\}, \qquad 1 \leq i \leq m$$

where each a_{ij} is a value selected from domain D_j, which is associated with attribute name A_j. The value m is the cardinality of the relation, and it may vary with time.

We denote a schema by $\mathcal{R}(A_1, A_2, \dots, A_n)$, where each attribute A_i is associated with domain D_i, $1 \leq i \leq n$. An instantiated relation of the schema $\mathcal{R}$ is a subset of $D_1 \times D_2 \times \cdots \times D_n$. $\mathcal{R}$ can also be considered as a relationship type or class that can be instantiated. A *database schema* is essentially a set of relational schemas.

The following are the relation schemas for the STUDENT and COURSE relations in the student registration example without the explicit specification of the domains for their attributes.

STUDENT(*student#, student_name, major, cumulative_avg, address*)
COURSE(*course_num, course_name, department, room, time*)

12.2.4 Representing Relations in the Relational Model

Because each tuple in a relation has a unique key, relationships among entities can be specified with relative ease. With respect to the relational model, the important task in database design is to choose suitable tables to represent the data. If a table has entities with too many attributes, the storing and updating of data may become inefficient because data redundancy will usually occur. Alternatively, choosing tables with too few attributes may make it impossible to link tables by relationships such that interesting facts can be derived.

In the student registration example, the relation ENROLL (see Table 12.3) represents a relationship that relates entities in the STUDENT and COURSE tables. We informally denote the relationship between the entities, say, Course and Student by

ENROLL(*Course, Student, Mark*)

The Course and Student entities are represented by records. These records are identified by their primary keys *course_num* and *student#*, respectively, yielding the relation

ENROLL(*course_num, student#, mark*)

where *mark* is an attribute of the relationship that denotes the grade obtained by a specific student in a particular course.

The attributes *course_num* and *student#* are called *foreign keys*. ENROLL can be viewed as an *associative relation* in that it associates a student to a course. From this form of relationship, new facts that are not evident from the individual participants of the relationship can be derived. Any foreign key value used in an associative relation must have a corresponding tuple in another relation. For example, ENROLL contains the value CS260 for attribute *course_num*, and there is a tuple in relation COURSE with this value as a primary key. Any foreign key value that occurs in an associative relation *must* have a corresponding tuple with that value in some other relation. Let us assume, for example, that ENROLL had a tuple with a value of CS429 for the attribute *course_num*. Since the relation COURSE does not have a tuple for this course, such a situation would not be allowed to occur.

12.2.5 Integrity Rules

Various kinds of integrity rules exist in the relational model. The data model provides for data integrity. An example of data integrity for an attribute, say *salary*, in a personnel-payroll application would be to check to ensure that the value for that item falls within a specified range, for example, $15,000 and $100,000. Another kind of integrity ensures that, if a particular entity is referenced, then that entity must exist. For example, in a university registration system one is not allowed to enroll in a nonexistent course.

The relational model also provides for two types of general integrity rules. The first type, *entity integrity*, is concerned with values for primary keys. An attribute may have a *null value*. A null value for an attribute may denote a situation where the value for that attribute is not known or does not apply to a given entity instance. Any attribute in a primary key for a certain entity *must always* have a nonnull value; otherwise, that particular entity could not be uniquely identified. For example, in the student registration system, each student must have a valid student number. A student without a valid number cannot be uniquely identified. Similarly, each course offered at a university must have a unique course number.

The entity integrity rule can be stated as follows: If an attribute A in a relational schema is part of a primary key, then the value for A in any entity (tuple) instance must have a nonnull value.

We have already encountered another kind of integrity rule dealing with associative relations. The foreign keys *course_num* and *student#* in the relation ENROLL refer to primary keys in the relations COURSE and STUDENT, respectively. As another example, Table 12.5 contains relations CUSTOMER and SALESPERSON, whose primary keys are *cust#* and *rep#*, respectively. An entity in CUSTOMER has *rep#* as a foreign key. If a particular *rep#* occurs in a tuple in CUSTOMER, then there must exist a unique tuple with that *rep#* in SALESPERSON. It is permitted, however, for the *rep#* value of a certain tuple in CUSTOMER to be undefined or *null*. In such a case there is no reference to any sales representative in relation SALESPERSON. This type of integrity is called *referential integrity*. More formally, if a particular relation has a foreign key consisting of one or more attributes that form a primary key in another relation (possibly the same), then either the foreign key value for a particular entity in the first relation must have a corresponding entity with that key value in the other relation or the foreign key value must be *null*.

Relations can be manipulated in various ways. Several relational operators have been created for this purpose. The next section focuses on such a suite of operators.

TABLE 12.5.   CUSTOMER and SALESPERSON Relations

CUSTOMER

cust#	cus_name	cust_phone	rep#
0750	John Brown	9317641	171
1011	Al Ens	2424041	185
1317	Jane Smith	2492022	NULL
1420	Judy White	6522351	201

SALESPERSON

rep#	rep_name	rep_sales_ytd
171	Elaine Smith	180951.25
185	Kent Boehm	101747.87
201	Susan Hanson	239566.40

12.3 RELATIONAL ALGEBRA

12.3.1 Introduction

An important part of the relational model deals with the manipulation of relations. In his original design of the relational model, Codd devised a relational algebra for this purpose. Like most algebras, the relational algebra has the important property of closure. The result of a relational operation always yields a relation. His original algebra consisted of eight high-level operators that manipulate relations. In this section we focus our attention on these relational operators. The algebra also contains an assignment operator whose purpose is to assign the result of a relational expression (i.e., a relation) to a named relation.

Codd's original set of eight relational operators contained both unary operators and binary operators. The first four operators deal with the union, intersection, difference, and Cartesian product of relations. These binary operators are very similar to the basic set-orientated operations introduced in Section 5.1 except that the basic relational operator have relations as operands. Section 12.3.2 examines these modified set operators so that relations can be manipulated. Section 12.3.3 introduces and illustrates four additional relational operations: projection, restriction (also sometimes called select), join, and divide.

12.3.2 Basic Operations

Since relations are sets, all set operators discussed in Section 5.1 are available for relations. This idea was used already in Section 5.3. For a relational operation to be meaningful, the two operands (relations) must have the same kinds of tuples; that is, their degrees must be the same. In addition to that, the header part of the two relational schemas for the two operands must be identical (or at least compatible). That is, both relations must be *union compatible*. This implies the following:

1. The same set of attribute names.
2. Attributes with the same names defined in the same (or in a compatible) domain.

We now define and illustrate each of the first three relational operations.

• Union ($\cup$)

Given the union-compatible relations P and Q, their relational union, $P \cup Q$, is a relation with the same header as P (or Q) and a body consisting of the tuples that belong to either P or Q or both. Duplicate tuples are eliminated.

• Intersection ($\cap$)

Given the union-compatible relations P and Q, their relational intersection, $P \cap Q$, is a relation with the same header as P (or Q) and a body consisting of the tuples belonging to both P and Q.

• Difference ($-$)

Given the union-compatible relations P and Q, their relational difference, $P - Q$, is a relation with the same header as P (or Q) and a body consisting of the tuples belonging to P and not to Q.

Example 12.1   Let relations P and Q be as shown in Table 12.6a. Then $P \cup Q$, $P \cap Q$, and $P - Q$ are as given in Tables 12.6b, c, and d, respectively.

Note that in the example given the relation headers for P and Q are identical. If, however, the attribute names of corresponding attributes in the two relations were not identical, the naming of the attributes in the resulting relation would have to be resolved. An operator sometimes referred to as a *rename* operator has been created for just this purpose. This operator changes the name of an attribute from one relation so that the new name is the same as the corresponding attribute name in the other.

• Product ($\times$)

The product operation is an extended version of the Cartesian product discussed in Section 5.2.2. Recall that the Cartesian product of two sets $A \times B$ is the set of all ordered pairs such that the first elements in each pair is selected from the first set and the second element is selected from the second set. In the case of the product of two relations, however, the desired result is a set of labeled tuples as opposed to ordered tuples. More formally, let $\{(A1 : a1), (A2 : a2), \ldots, (An : an)\}$ and $\{(B1 : b1), (B2 : b2), \ldots, (Bm : bm)\}$ be labeled unordered tuples in relations P and Q, respectively. Then the labeled unordered tuple in $P \times Q$ is

$$\{(A1 : a1), (A2 : a2), \ldots, (An : an), (B1 : b1), (B2 : b2), \ldots, (Bm : bm)\}$$

The degree of this tuple is $m + n$. A problem arises in forming the product of two relations when the relations have common attribute names. In such a case, some attributes must be renamed such that the headers of the two relations become disjoint or *product compatible*.

The product operation is not that important in practice. Basically, it does not really produce any new information. It is included in the relational algebra for conceptual reasons. However, it plays an important role in the join operation, which is to be discussed later.

TABLE 12.6

		P					*Q*		
emp#	name	dept	college		emp#	name	dept	college	
1015	Smith	HIST	Arts		651	Kaplan	HIST	Science	
1711	Carter	MATH	Science		1711	Carter	MATH	Science	
2560	Andrews	CHEM	Science		2735	Jones	PED	Education	
3673	Angel	ACCT	Commerce		3673	Angel	ACCT	Commerce	

(a)

		$P \cup Q$					$P \cap Q$	
emp#	name	dept	college		emp#	name	dept	college
651	Kaplan	HIST	Science		1711	Carter	MATH	Science
1015	Smith	HIST	Arts		3673	Angel	ACCT	Commerce
1711	Carter	MATH	Science					
2560	Andrews	CHEM	Science			(c)		
2735	Jones	PED	Education					
3673	Angel	ACCT	Commerce			$P - Q$		

emp#	name	dept	college
1015	Smith	HIST	Arts
2560	Andrews	CHEM	Science

(b)

(d)

We observe that the relational operators ∪, ∩, and × are all associative and commutative. The difference operator, however, is neither associative nor commutative.

12.3.3 Additional Relational Operations

In his initial work, Codd supplemented the basic relational operations just described by four additional operators: projection, restriction, join, and division. We now examine each of these operations.

● Projection

The projection operation selects specified attributes from a designated relation. Recall that attributes correspond to columns in a relation. The projection operation can also be viewed as a vertical subsetting operation. More formally, the projection of a relation R on attributes $A1, A2, \ldots, Ai$ is written as

$$R[A1, A2, \ldots, Ai]$$

and denotes an extraction of a relation from R consisting of the header of each specified attribute and the column values for each associated column in the relation. The cardinality of the resulting relation may be smaller than that of R since duplicate tuples may be removed.

Example 12.2   In the context of Table 12.6a the expression

$$P[dept, college]$$

specifies the following relation:

dept	college
HIST	Arts
MATH	Science
CHEM	Science
ACCT	Commerce

Example 12.3   The expression

$$P[college]$$

results in the following relation:

college
Arts
Science
Commerce

Observe that the duplicate entry for the *Science* tuple has been removed.

• Restriction

This operation horizontally subsets a specified relation by selecting a subset of its tuples. One form of this operation is

$$R[A \; \theta \; OPER]$$

where A is an attribute name, θ is a comparison operator ($=, \neq, <, \leq, >, \geq$), and $OPER$ denotes either an attribute name or a constant.

Example 12.4   The expression

$$P[college = \text{'Science'}]$$

specifies the following relation:

emp#	name	dept	college
1711	Carter	MATH	Science
2560	Andrews	CHEM	Science

A second form of restriction expression is

$$P[B]$$

where B is some logical (propositional calculus) expression involving the logical operators AND, OR, and NOT. In addition to these logical operators, attribute names and constants are used to form logical expressions.

Example 12.5   The expression

$$P[dept = \text{'HIST'} \ OR \ college = \text{'Commerce'}]$$

specifies the tuples that have a *dept* attribute value of 'HIST' or a *college* attribute value of 'Commerce' and results in the following relation:

exp#	name	dept	college
1015	Smith	HIST	Arts
3673	Angel	ACCT	Commerce

The restriction operation can be combined with a projection operation to form a nested expression such as

$$(P[college = \text{'Arts'} \ OR \ college = \text{'Science'}])[name, college]$$

where the ordinary parentheses indicate the order of evaluation. Conceptually, this expression is equivalent to first doing a restriction operation and following it by a projection operation. These two steps can be expressed as follows:

$$R1 := P[college = \text{'Arts'} \ OR \ college = \text{'Science'}]$$
$$R2 := R1[name, college]$$

where the operator := denotes a relational assignment operator. Recall that the result of performing a relational algebra operation is always a relation. Keeping this fact in mind, the evaluation of the steps on Table 12.6a results in the following relations:

R1

emp#	name	dept	college
1015	Smith	HIST	Arts
1711	Carter	MATH	Science
2560	Andrews	CHEM	Science

R2

name	college
Smith	Arts
Carter	Science
Andrews	Science

Although the explicit labeling of the relation in each of the preceding two steps is perhaps more understandable and readable, we normally prefer to use a nested expression form that is more compact and avoids the naming of intermediate relations whose permanent storage is not required.

• Join

The join operation is very important in databases because it allows the combination of two relations, which share some common attributes, to produce a single new relation as its result. The join operation has a number of variants. One variant, called the *natural join*, is widely conceded to be the most important operation in databases.

Consider the relations SALESPERSON and SALARY in Table 12.7a. Suppose that we want to find the salary of each salesperson by name. One approach to answering this query is to first generate the extended Cartesian product of the two given relations, as shown in Table 12.7b. Note that there are two attributes in the product with the name $emp\#$. We shall use name qualification to resolve this ambiguity. Taking the restriction of the product relation such that SALESPERSON.$emp\#$ = SALARY.$emp\#$ yields the relation in Table 12.7c. Finally, a projection of this relation to discard the duplicate $emp\#$ attribute in the third column yields the desired answer to the query. The result in Table 12.7d is called the natural join of the relations SALESPERSON and SALARY. Two relations whose natural join is required usually have one or more corresponding attributes. In the current example, each relation contains the common attribute $emp\#$. The natural join is expressed as

$$\text{SALESPERSON[SALESPERSON}.emp\# = \text{SALARY}.emp\#]\text{SALARY}$$

where SALESPERSON.$emp\#$ and SALARY.$emp\#$ denote the employee number in relations SALESPERSON and SALARY, respectively. Note that in a natural join operation duplicates of corresponding attributes are discarded. We now formalize the notion of natural join.

Let P and Q be relations of degree m and n, respectively. Let the relations P and Q have the headers

$$(A_1, A_2, \ldots, A_{m-r}, B_1, B_2, \ldots, B_r)$$

and

$$(B_1, B_2, \ldots, B_r, C_1, C_2, \ldots, C_{n-r})$$

respectively, with r attributes in common, where $r \leq m$ and $r \leq n$. Assume that t_1 and t_2 are tuples of the form

$$t_1 = (a_1, a_2, \ldots, a_{m-r}, b_1, b_2, \ldots, b_r)$$

and

$$t_2 = (b_1, b_2, \ldots, b_r, c_1, c_2, \ldots, c_{n-r})$$

in relations P and Q, respectively. Also, let $b = (b_1, b_2, \ldots, b_r)$ be a composite attribute. Then the natural join $P[P.b_i = Q.b_i]Q$, $1 \leq i \leq r$, is defined as

$$P[P.b = Q.b]Q = \{t = (a_1, a_2, \ldots, a_{m-r}, b_1, b_2, \ldots, b_r, c_1, c_2, \ldots, c_{n-r})$$
$$|t_1 \in P \text{ and } t_2 \in Q\}$$

TABLE 12.7

SALESPERSON		SALARY	
emp#	emp_name	emp#	commission
1519	Jones	1519	1050
1675	Boehm	1675	775
2520	Brown	2520	1125

(a)

emp#	emp_name	emp#	commission
1519	Jones	1519	1050
1519	Jones	1675	775
1519	Jones	2520	1125
1675	Boehm	1519	1050
1675	Boehm	1675	775
1675	Boehm	2520	1125
2520	Brown	1519	1050
2520	Brown	1675	775
2520	Brown	2520	1125

(b) SALESPERSON $\times$ SALARY

emp#	emp_name	emp#	commission
1519	Jones	1519	1050
1675	Boehm	1675	775
2520	Brown	2520	1125

(c) SALESPERSON $\times$ SALARY
    [SALESPERSON.$emp\#$ = SALARY.$emp\#$]

emp#	emp_name	commission
1519	Jones	1050
1625	Boehm	775
2520	Brown	1125

(d) Natural join of SALESPERSON and SALARY

The term $P.b = Q.b$ corresponds to the sequence of predicates $P.b_i = Q.b_i$, $1 \leq i \leq r$; that is, the corresponding common attributes in both relations must have the same values. Observe that the degree of the output relation is $m + n - r$. As a shorthand, we denote the natural join by $P * Q$.

Example 12.6 Obtain the natural join of the relations FLIGHTS and DEPARTURES given in Table 12.8a. In terms of the previous definition, $m = 3$, $n = 3$, and $r = 2$. That is, the two relations are both of degree 3 and have two attributes in common. The natural join, which requires that the corresponding flight and gate numbers be equal, appears in Table 12.8.

TABLE 12.8

FLIGHTS				DEPARTURES		
airline	flight_number	gate		flight_number	gate	departure_time
AC	157	89		111	76	07:30
AC	924	46		157	89	06:00
CP	111	76		612	32	08:00
CP	612	32		700	51	09:30
TWA	700	51		924	46	10:00

(a)

airline	flight_number	gate	departure
AC	157	89	06:00
AC	924	46	10:00
CP	111	76	07:30
CP	612	32	08:00
TWA	700	51	09:30

(b) Natural join of FLIGHTS and DEPARTURES

The natural join operation is both associative and commutative. The proofs of these natural join properties are left to the problems.

We can manipulate further the result of performing a natural join operation. The evaluation of the relational expression

$$((\text{FLIGHTS} * \text{DEPARTURES})[gate \leq 76 \ AND \ gate \geq 46])[airline]$$

first performs a restriction on the natural join by selecting those tuples whose departure gate number is between 46 and 76 inclusive and then projects this intermediate relation on the *airline* attribute, yielding

```
- - - -
airline
- - - -
AC
CP
TWA
- - - -
```

The natural join operation can be viewed as both a horizontal and vertical subset of the extended Cartesian product.

Other types of join operators are possible. For example, the *theta join* operation is used when we need to join two relations on the basis of some predicate other than equality. The composite predicate can be of the form

$$P.b_i \ \theta_i \ Q.b_i, \ 1 \leq i \leq r$$

where θ_i can be any comparison operator from the set $\{=, \neq, <, >, \leq, \geq\}$. Unlike a natural join operation, however, the result of performing a theta join operation *does not* eliminate the common attributes in the given input relations. That is, a tuple in the resulting relation is of the form

$$(a_1, a_2, \ldots, a_{m-r}, b_1, b_2, \ldots, b_r, b_1, b_2, \ldots, b_r, c_1, c_2, \ldots, c_{n-r})$$

That is, the output relation is of degree $m + n$ instead of $m + n - r$ as is the case for the natural join. If the two relations have no common attributes, then the theta join and the natural join operations both yield the Cartesian product.

• Divide

The division operation in relational algebra is similar to the divide operator in normal algebra on numbers. The *division* of P by Q is denoted by P/Q. As in normal algebra, there is a dividend and a divisor, which are relations, and loosely speaking the dividend may contain several times the divisor, where times represents the quotient. Before defining the relational operator for division, let us first consider a simple example.

Table 12.9 contains two relations P and Q, where y is the only attribute in Q. The corresponding attribute y in P assumed to be defined on the same domain. For each tuple in $(P/Q) \times Q$, its extended Cartesian product, with the tuples of Q, must be in P. For the relation Q in the table, the tuples (x_1, y_1), (x_1, y_2), and (x_1, y_3) are in P. The tuples (x_3, y_1), (x_3, y_2), and (x_3, y_3) also belong to P. Therefore, the product $(P/Q) \times Q \subset P$. The remainder of the division operation is the unmatched part of P consisting of the tuples (x_2, y_1), (x_2, y_2), (x_4, y_2), and (x_5, y_3). In essence, the result of a division operation is the set of entities in relation P that has all the properties specified by relation Q, and the remainder is $P - (P/Q) \times Q$. Table 12.9 also contains the results of dividing P by Q' and P by Q''.

More formally, let P and Q be relations that have the headers

$$(A_1, A_2, \ldots, A_n, B_1, B_2, \ldots, B_m)$$

and

$$(B_1, B_2, \ldots, B_m)$$

respectively. That is, the attributes $B_1, B_2, \ldots, B_m$ are common to both relations. One further assumes that corresponding attributes in the relations have the same name and are defined on the same domain. Then the result of dividing P by Q is a relation with header $(A_1, A_2, \ldots, A_n)$ and whose body contains the tuples

$$P/Q = \{t = (a_1, a_2, \ldots, a_n) \mid \text{ there is a tuple}$$
$$(a_1, a_2, \ldots, a_n, b_1, b_2, \ldots, b_m) \in P \text{ such that}$$
$$\text{for all tuples } (b_1, b_2, \ldots, b_m) \in Q\}$$

In this section we have defined a relational algebra consisting of eight operations on relations. These operations are essentially those introduced by Codd in his original work: union, intersection, difference, extended Cartesian product, projection, restriction (sometimes also called selection), join, and division. Five of these operations, union, difference,

TABLE 12.9.   Division Operation Examples

P		Q	P/Q
x	y	y	x
x_1	y_1	y_1	x_1
x_1	y_2	y_2	x_3
x_1	y_3	y_3	
x_2	y_1		
x_2	y_2		
x_3	y_1		
x_3	y_2		
x_3	y_3		
x_4	y_2		
x_5	y_3		

Q'	P/Q'
y	x
y_1	x_1
y_2	x_2
	x_3

Q''	P/Q''
y	x
y_3	x_1
	x_3
	x_5

extended Cartesian product, projection, and restriction, are considered to be *primitive* or *minimal* since none of these can be obtained from others. The remaining three operations, intersection, join, and divide, are not primitive since they can be expressed in terms of the other five primitive operations. For example, the intersection operator can be expressed as

$$P \cap Q = P - (P - Q)$$

Also, we have shown informally that a join operation is a projection of a restriction of an extended Cartesian product. The expression of the division operation in terms of the other primitive operations is left as a problem. Because of their convenience, all eight operators are supported directly.

12.3.4 Examples

We further illustrate the use of relational algebra in the formulation of queries. The example queries are based on the sample database given in Table 12.10. This database contains three relations that a university committee application would require. The FACULTY relation contains the employee numbers and names of faculty members who perform committee work. The MEMBER_OF relation contains the specifics of who sits on which committees. Finally, the relation COMMITTEE contains the specifics of each university committee: committee code, committee name, and current chairperson.

The next example illustrates the restrict and project operations.

TABLE 12.10

FACULTY

emp#	fac_name
1023	Brown
1111	Smith
1373	Cook
1447	Green
1529	Zoerb
1919	Yang

COMMITTEE

com#	com_name	chair
COM101	Appeals	Brown
COM202	Planning	Yang
COM303	Budget	Cook
COM404	Awards	Cook
COM505	Disciplinary	Green
COM606	Research	Smith

MEMBER_OF

com#	emp#
COM101	1023
COM606	1023
COM202	1919
COM303	1529
COM404	1373
COM505	1447
COM202	1023
COM606	1111
COM303	1023
COM404	1023
COM505	1023
COM505	1529
COM404	1447
COM101	1529
COM202	1373
COM202	1447
COM303	1111
COM303	1373

Example 12.7   Consider the following query: "Obtain the employee numbers of faculty sitting on committee COM202."

Solution   To evaluate this query, we first restrict the tuples in relation MEMBER_OF such that the value of the $com\#$ attribute is COM202 and then project this intermediate result on the attribute $emp\#$ to get the desired answer. The query

$$(\text{MEMBER\_OF}[com\# = \text{'COM202'}])[emp\#]$$

thus yields the following response:

```
    _ _ _ _
    emp#
    _ _ _ _
    1919
    1023
    1373
    1447
    _ _ _ _
```

■

The next example illustrates the join operation.

Example 12.8   Consider the following query: "Generate the details of each faculty member sitting on committee COM303."

Solution The first part of this query is very similar to the previous example. The second part, however, consists of the natural join of the intermediate result from the first part with the FACULTY relation to generate the faculty information of the members sitting on the committee COM303. The query

$$\text{FACULTY} * ((\text{MEMBER\_OF}[com\# = \text{'COM303'}])[emp\#])$$

yields the following response:

emp#	fac_name
1529	Zoerb
1023	Brown
1111	Smith
1373	Cook

The next example requires the use of three relations.

Example 12.9 Consider the following query: "Generate the details of the faculty members who sit on the Research committee."

Solution This query can be answered by using two joins. The first step is to find the $com\#$ of the committee named 'Research'. This step requires a restriction of the relation COMMITTEE, followed by a projection on the attribute $com\#$. We then join the result of this projection with the relation MEMBER_OF involving COM606. Finally, we project this second intermediate on $emp\#$ and then join it with the FACULTY relation to get the required answer. The query

$$\text{FACULTY} * ((\text{MEMBER\_OF} * ((\text{COMMITTEE}$$
$$[com\_name = \text{'Research'}])[com\#]))[emp\#])$$

generates the following response:

emp#	fac_name
1023	Brown
1111	Smith

The next two examples illustrate the divide operation.

Example 12.10 Obtain an answer to the following query: "Generate details of faculty sitting on both committees COM202 and COM303."

Solution The first step in evaluating this query is to obtain the unary relation consisting of the desired committee members. This unary relation is obtained by selecting the tuples of MEMBER_OF, where $com\#$ is either COM202 or COM303, and then projecting the result on $com\#$. We then divide the relation MEMBER_OF by the result of the first step to obtain another unary relation consisting of the faculty members who are sitting on both committees. Finally, a join of FACULTY with this second unary relation yields the desired result. The relational algebra query

$$\text{FACULTY} * (\text{MEMBER\_OF}/(\text{MEMBER\_OF}$$
$$[com\# = \text{'COM202'} \ OR \ com\# = \text{'COM303'}])[com\#])$$

generates the following desired result:

```
_ _ _ _ _ _ _ _ _
emp#    fac_name
_ _ _ _ _ _ _ _ _
1023    Brown
1373    Cook
_ _ _ _ _ _ _ _ _                                    ■
```

Example 12.11   Assume the following query: "Obtain the employee numbers of faculty who sit at least on the same committees as faculty member 1373."

Solution   We first determine all the committee numbers that faculty member 1373 sits on. We then divide the relation MEMBER_OF by this unary relation to obtain the result that includes member 1373. Finally, we obtain the answer by deleting member 1373 with a difference operation. The relational algebra query

$$(\text{MEMBER\_OF}/((\text{MEMBER\_OF}[emp\# = 1373])[com\#])) - \{1373\}$$

yields the result

```
_ _ _ _
emp#
_ _ _ _
1023
_ _ _ _                                              ■
```

The next example illustrates the difference operation.

Example 12.12   Formulate a relational expression for the following query: "Obtain the employee numbers of faculty who do not sit on committee COM606."

Solution   We first determine those faculty members who are members of committee COM606. We next determine all faculty who serve on at least one of the committees. Finally, from these two sets, we generate their difference to arrive at the set of faculty members who are not members of committee COM606. The relational algebra query:

$$(\text{MEMBER\_OF}[emp\#]) -$$
$$((\text{MEMBER\_OF}[com\# = 'COM606'])[emp\#])$$

generates the following answer:

```
_ _ _ _
emp#
_ _ _ _
1373
1447
1529
1919
_ _ _ _                                              ■
```

The final example illustrates the use of a join operation and a divide operation.

Example 12.13   Obtain a relational expression for the following query: "Obtain faculty information on those members who sit on all committees."

Solution   The first step is to compile a list of committees from the COMMITTEE relation by a projection on $com\#$. Then we divide the relation MEMBER_OF by the unary relation obtained in the first step. Finally, we obtain the desired faculty information by joining FACULTY with the result obtained in the second step. The relational expression for the specified query

$$\text{FACULTY} * (\text{MEMBER\_OF}/(\text{COMMITTEE}[com\#]))$$

yields the following answer:

```
 _ _ _ _ _ _ _ _ _
 emp#    fac_name
 _ _ _ _ _ _ _ _ _
 1023    Brown
 _ _ _ _ _ _ _ _ _
```
■

From the previous examples, it is clear that the formulation of a relational expression for a given query is very much a procedural process. The expression must be constructed as a sequence of steps much as when writing some program in a procedural language, such as Pascal or C. The next section explores an alternative declarative or nonprocedural approach to formalizing a query.

Problems 12.3

1. Using the relations P and Q of Table 12.11, evaluate the following relational expressions:

   (a) $P[a, c]$ (d) $(P/Q)[b]$

   (b) $P[c = c_1]$ (e) $(P/Q)[b = b_1 \ OR \ b = b_2]$

   (c) $P * Q$

TABLE 12.11

P			Q	
a	b	c	b	d
a_1	b_1	c_1	b_1	d_1
a_1	b_1	c_2	b_1	d_2
a_1	b_2	c_1	b_1	d_3
a_1	b_3	c_1	b_1	d_4
a_2	b_1	c_1	b_2	d_1
a_2	b_2	c_2	b_2	d_2
a_2	b_3	c_2	b_3	d_1
a_3	b_1	c_1		
a_3	b_2	c_2		
a_3	b_3	c_1		

2. Express the division operation in terms of the five primitive relational operators (i.e., union, difference, extended Cartesian product, projection, and restriction).

3. Using the database relations of Table 12.10, formulate the following queries in relational algebra:
   (a) List the details of the committees each faculty member sits on.
   (b) List the names of committees that faculty member 1023 sits on.
   (c) List all faculty members that sit on committees with faculty member 1023.
   (d) List all committee names on which faculty member 1529 does not sit.
   (e) List the faculty details of faculty who sit on committees not assigned to faculty member 1447.
   (f) Obtain the faculty details of faculty who chair the Budget and Awards committees.
   (g) Obtain the names of faculty who sit on the Appeals committee but not on the Planning committee.

4. How many different projections are there for a relation with n attributes?

5. Show that the five primitive relational operators form a minimal set.

6. Given the database in Table 12.12, formulate the following queries in the relational algebra:

TABLE 12.12

SP		
supplier#	part#	qty
S1	P1	3
S1	P2	1
S2	P5	4
S2	P6	2
S3	P1	3
S3	P2	4
S3	P3	2
S3	P6	2
S4	P4	6
S5	P4	2
S5	P5	5
S5	P6	4
S6	P1	3
S6	P2	1
S6	P3	2

P		
part#	pname	psize
P1	Nut	$\frac{3}{4}$
P2	Bolt	$\frac{3}{4} - 2$
P3	Bolt	$\frac{3}{4} - 1$
P4	Screw	$\frac{1}{4} - 1$
P5	Spring	2
P6	Sprocket	4

S			
supplier#	sname	city	delvry_time
S1	Black	New York	2
S2	Lee	Toronto	1
S3	Waters	Chicago	1
S4	Dyck	St. Louis	3
S5	Jones	Montreal	1
S6	Whyte	Los Angeles	2

   (a) Find the supplier numbers for the suppliers who supply part $P3$.
   (b) Find the part numbers for parts supplied by supplier $S2$.
   (c) For each supplier, find part numbers and supplier's cities from which the parts may be obtained.
   (d) Find the supplier numbers for suppliers who are from Toronto or Montreal and who supply a part of size $\frac{3}{4}$.
   (e) Find the supplier names for all suppliers who supply springs.
   (f) Find the cities where a supplier has a delivery time of 2 or 3.

7. Show that the natural join operation is both associative and commutative.

12.4 RELATIONAL CALCULUS

12.4.1 Introduction

In this section we examine an alternative approach to specifying a query from the relational algebra approach discussed in the previous section. The approach, which is based on first-order predicate calculus, focuses on specifying the query in a nonprocedural way such that the query states what is required—not how it is to be obtained or executed. This approach has led to the creation of a special language called the *relational calculus*.

Relational calculus, which deals with calculating with relations, is a query system in which queries are specified as variables and formulas on these variables. Relational calculus is divided into two parts:

1. *Tuple calculus*, in which queries are specified as tuple calculus expressions of the form

$$\{T \mid F(T)\}$$

   where T is a set of tuples and F is a logical expression such that $F(T)$ is true

2. *Domain calculus*, in which queries are specified as

$$\{V \mid F(V)\}$$

   where V denotes a set of domain (or attribute) variables and F is a logical expression such that $F(V)$ is true.

The remainder of this section first elaborates on the relational calculus. Second, several examples of queries in relational calculus form are given using the university committee database introduced earlier.

12.4.2 Tuple Calculus

In tuple calculus, a tuple variable ranges over some relation. As such, all tuples of a relation are scanned and checked to determine if the tuple expression (i.e., predicate) is satisfied. In other words, tuple calculus is implicitly tuple iterative.

As an example of a tuple calculus query, suppose that we require the airline flights in the FLIGHTS relation in Table 12.8 that are between gates 51 and 89 inclusive. The tuple (first-order predicate) calculus expression query

$$\{f \mid f \in \text{FLIGHTS} \land$$
$$\exists x(x \in \text{FLIGHTS} \land (x.gate \geq 51 \land x.gate \leq 89)$$
$$\land(f.flight\_number = x.flight\_number))\}$$

specifies the result:

airline	flight_number	gate
AC	157	89
CP	111	76
TWA	700	51

The tuple calculus expression specifies the selection of those tuples whose gate numbers are within the desired range. The tuple variables f and x range over the relation FLIGHTS.

Recall from Section 1.3.3 that propositional calculus expressions are defined by well-formed formulas (wffs). We can use a similar approach to define precisely a tuple calculus language that includes the existential and universal quantifiers discussed in Chapter 2. Well-formed formulas in tuple calculus are constructed from atoms. An atom can be one of the following forms:

- $t \in P$, where P is a relation and t is a tuple variable
- $a\,\theta\,b$ or $a\,\theta\,k$, where θ denotes a comparison operator, that is, belongs to the set $\{=, \neq, <, \leq, >, \geq\}$, a and b are qualified domain-compatible attribute variables, and k denotes a domain-compatible constant.

The following are examples of atoms in the context of the university committee database of Table 12.10:

$$f \in \text{FACULTY}$$
$$c \in \text{COMMITTEE}$$
$$m \in \text{MEMBER\_OF}$$
$$f.emp\# = m.emp\#$$
$$m.com\# = \text{'COM101'}$$
$$f.fac\_name = \text{'Smith'}$$

Formulas are built from the atoms, using formation rules similar to the ones mentioned earlier (see Sections 1.3.2 and 2.1.5). As indicated in Section 1.3.3, the formulas can be written more clearly when use is made of precedence rules, because this allows one to omit a number of parentheses. The following are examples of wffs in the context of the university committee example:

$$m \in \text{MEMBER\_OF} \wedge m.com\# = \text{'COM202'}$$
$$\exists m(m \in \text{MEMBER\_OF} \wedge f \in \text{FACULTY}$$
$$\wedge((m.com\# = \text{'COM202'}) \vee (m.com\# = \text{'COM303'}))$$
$$\wedge(f.emp\# = m.emp\#))$$
(In this example m is a bound variable, and f is a free variable)
$$\exists f(f \in \text{FACULTY} \wedge \exists m(m \in \text{MEMBER\_OF} \wedge c \in \text{COMMITTEE}$$
$$\wedge(c.com\_name = \text{'Awards'})$$
$$\wedge(m.com\# = c.com\#)$$
$$\wedge(f.emp\# = m.emp\#)))$$

Instead of having a variable that represents an entire tuple, we can use a variable to represent a domain (column) in a relation. Using variables to represent components of a tuple is called *domain calculus*. Both tuple calculus and domain calculus use the same set of operators in forming wffs. Any wff in tuple calculus can be converted to an equivalent wff in domain calculus by replacing each tuple variable by m domain variables, where m is the degree of the associated relation. Wffs for domain calculus can be defined recursively, as was done for wffs in the tuple calculus. However, we do not give such a definition here. Rather, the definition is left as a problem.

12.4.3 Examples

Using the relational calculus introduced in the previous section, we now formulate the seven queries given in Section 12.3.4.

Example 12.14   Consider the following query: "Obtain the employee numbers of faculty sitting on committee COM202."

Solution   The formulation of this query in relational calculus form involves iterating on each tuple of the relation MEMBER_OF and selecting the employee number of a tuple whose committee number is COM202. This query in relational calculus form is

$$\{t.emp\# \mid (t \in \text{MEMBER\_OF} \land \exists m(m \in \text{MEMBER\_OF}$$
$$\land(m.com\# = \text{'COM202'})$$
$$\land(t.emp\# = m.emp\#)))\}$$

where t is a free variable over the relation MEMBER_OF. Also note that, since we are only interested in faculty members, we project t on $emp\#$.   ∎

Example 12.15   As an extension of the previous example, consider the following query: "Generate the details of each faculty member sitting on committee COM303."

Solution   The relational calculus form of this query is

$$\{f \mid (f \in \text{FACULTY} \land \exists m(m \in \text{MEMBER\_OF}$$
$$\land(m.com\# = \text{'COM303'})$$
$$\land(f.emp\# = m.emp\#)))\}$$

This query is very similar to the previous example. Note, however, that we join the relations FACULTY and MEMBER_OF in the subexpression $f.emp\# = m.emp\#$.   ∎

The next example requires the use of three relations.

Example 12.16   Consider the following query: "Generate the details of the faculty members who sit on the Research committee."

Solution   The relational calculus form of this query follows:

$$\{f \mid (f \in \text{FACULTY}$$
$$\land \exists m, c(c \in \text{COMMITTEE} \land (c.com\_name = \text{'Research'})$$
$$\land(m \in \text{MEMBER\_OF}) \land (m.com\# = c.com\#)$$
$$\land(f.emp\# = m.emp\#))\}$$

The tuple variable f ranges over the relation FACULTY and we want to select the whole tuple. We are stating that there exist tuple variables m and c on the relations MEMBER_OF and COMMITTEE, respectively, such that the following condition holds. The attribute $com\#$ of the tuple c has a value of 'Research', and the committee numbers in m and c are identical. Furthermore, the employee numbers of tuples f and m are equal. Finally, the relation $\exists m, c(F(c, m))$ is a more compact and convenient notation for $\exists m(\exists c(F(m, c))$.   ∎

Example 12.17   Obtain a relational calculus expression for the following informal query: "Generate details of faculty sitting on both committees COM202 and COM303."

Solution   In formalizing this query, we require the existence of two tuples m_1 and m_2, which belong to relation MEMBER_OF. Also, the committee value of tuples m_1 and m_2 is COM202 and COM303, respectively. The query is

$$\{f \mid f \in \text{FACULTY}$$
$$\wedge \exists m_1, m_2 (m_1 \in \text{MEMBER\_OF} \wedge m_2 \in \text{MEMBER\_OF})$$
$$\wedge (m_1.emp\# = m_2.emp\#)$$
$$\wedge (m_1.com\# = \text{'COM202'}) \wedge (m_2.com\# = \text{'COM303'})$$
$$\wedge (f.emp\# = m_1.emp\#))\}$$

Note that the $emp\#$ attribute values for both the free tuple f and the bound tuples m_1 and m_2 are identical.   ■

Example 12.18   Using the relational calculus, formalize the following query: "Obtain the employee numbers of faculty who sit at least on the same committees as faculty member 1373."

Solution   We want each employee number from tuple f from relation MEMBER_OF, such that for all tuples m_1 from the same relation has $m_1.emp\# = 1373$, there exists another tuple m_2, again from the same relation, with $m_2.emp\# \neq 1373$. Also, we require that tuples m_1 and m_2 have identical values for attribute $com\#$. Finally, tuples f and m_2 must have the same $emp\#$ value. Putting all this together yields the following query:

$$\{f.emp\# \mid (f \in \text{MEMBER\_OF}$$
$$\wedge \forall m_1 ((m_1 \in \text{MEMBER\_OF}) \wedge (m_1.emp\# = 1373)$$
$$\wedge \exists m_2 ((m_2 \in \text{MEMBER\_OF}) \wedge (m_2.emp\# \neq 1373)$$
$$\wedge (m_1.com\# = m_2.com\#)$$
$$\wedge (f.emp\# = m_2.emp\#))))\}$$

■

Example 12.19   Formulate a relation calculus expression for the following query: "Obtain the employee numbers of faculty who do not sit on committee COM606."

Solution   A relational calculus form of this query is

$$\{f.emp\# \mid f \in \text{MEMBER\_OF}$$
$$\wedge \neg \exists m ((m \in \text{MEMBER\_OF}) \wedge (m.com\# = \text{'COM606'})$$
$$\wedge (f.emp\# = m.emp\#))\}$$

Observe that we have used a negated existential qualified expression to suppress the selection of those faculty members who do not sit on committee COM606.   ■

Example 12.20   Obtain a relational calculus formulation for the following query: "Obtain faculty information on those members who sit on all committees."

Solution   A relational calculus form of this query is

$$\{f \mid f \in \text{FACULTY}$$
$$\wedge \forall c (c \in \text{COMMITTEE}$$
$$\vee \exists m (m \in \text{MEMBER\_OF}$$
$$\wedge (c.com\# = m.com\#)$$
$$\wedge (f.emp\# = m.emp\#)))\}$$

■

We have examined two formal ways of specifying queries: relational algebra and relational calculus. Although we do not pursue it here, it can be shown that both ways are equivalent. Each commercial relational database system provides some query language capability for the declaration, creation, and manipulation of relations. The next section examines one of the most popular of these languages.

Problems 12.4

1. Define wffs for the tuple calculus and the domain calculus.
2. Formulate the eight relational operators of the previous section in the relational calculus.
3. Express the queries for the university committee database given in Problem 3 of Section 12.3 in relational calculus.
4. Express the queries for the part-supplier database in Problem 6 of Section 12.3 in relational calculus.

12.5 STRUCTURED QUERY LANGUAGE

12.5.1 Introduction

In the two previous sections we formally expressed queries in both relational algebra and relational calculus forms. Although corresponding languages in commercial database systems are based on these two forms, commercial languages may vary significantly from those discussed earlier. For example, most commercial languages support numeric, string, and aggregate operators. Examples of aggregate operators include those for obtaining the sum, average, minimum, and maximum of a collection of numerical values.

In this section we give an overview of a popular commercial database language called Structured Query Language (SQL). The focus here is primarily on the query features of this language. Since our emphasis is on education, we do not give complete details on the syntax and semantics of the language. However, we do give its flavor and illustrate its power through examples.

SQL originated at IBM's San Jose Research Center in the 1970s. Although SQL is based primarily on relational calculus, its design has also been influenced by relational algebra. More specifically, SQL uses the concept of a tuple variable from tuple calculus. Also, SQL is somewhat block structured in the sense that an SQL program consists of one or more blocks, some of which may be nested within another. We wish to emphasize that block structure in SQL is quite different than the corresponding concept in procedural languages such as Pascal and Ada.

SQL has three categories of language features to facilitate:

1. *Data definition* for creating a database and its associated relations (tables)
2. *Data management* for entering, correcting, updating, and deleting data within a database

3. *Data query* for specifying queries on the database

The following three subsections deal with each category in turn. Also, in the third subsection the set of seven queries formulated in the relational algebra and relational calculus earlier are specified in SQL.

12.5.2 Data Definition

SQL contains a statement type for creating a database structure and its associated relations. In describing SQL syntax, we use a context-free grammar notation (see Chapter 10). Although there are variations in commercial database systems, a database structure is created by a create statement such as

$$\textbf{create database} \ \langle\text{database name}\rangle$$

where ⟨database name⟩ is a metasymbol that denotes the name of a database.

We can use a create statement to specify relations in a database. In such a case the statement specifies the name of a relation and the names and domains of each attribute in the relation. The create statement has the following form:

$$\textbf{create table} \ \langle\text{relation name}\rangle \ (\ \langle\text{attribute entry list}\rangle \ )$$

in which an attribute entry list is

$$\langle\text{attribute entry list}\rangle \ ::= \ \langle\text{attribute entry}\rangle \ \{ \ , \ \langle\text{attribute entry}\rangle \ \}$$

where the braces in ({ and }) denote zero or more occurrences of what is contained within the parentheses. An attribute entry, which denotes an attribute and associated domain, is specified as follows:

$$\langle\text{attribute entry}\rangle \ ::= \ \langle\text{attribute name}\rangle \ \langle\text{data type}\rangle \ [\textbf{not null}]$$

where the brackets in ([and]) are metasymbols that denote the presence of exactly one copy of what is contained within the parentheses or its absence. Its presence denotes that an attribute must always have a nonnull value. The data type denotes the basic domains that are allowed in a given commercial database system. Examples of domains include integers (regular and short), floating-point numbers, and strings (both fixed and varying lengths). Some database systems also support Boolean, calendar date, and time values. Therefore, data types include the following:

$$\langle\text{data type}\rangle \ ::= \textbf{integer} \mid \textbf{smallint} \mid \textbf{float} \mid \textbf{char}(n) \mid \dots etc.$$

where n denotes the length of a string.

As an example of the use of a create statement for creating a relation, the following statement specifies the creation of the relation FACULTY in Table 12.10.

```
create table FACULTY
    (emp#      integer not null,
     fac_name   char(20))
```

Note that the create statement for relations defines a relation schema. No tuples (or records) are actually placed in the table. The insertion of records in a relation can be realized by using an insert statement, which will be described shortly.

SQL has other types of data definition statements. The alter statement allows the addition of a new attribute (column) to an existing table. The component of each tuple in the new relation corresponding to this new attribute is set to the null value. An update statement can be used to assign the value of the new attribute in a given tuple. Another statement type is the drop statement, which is used to delete a specified relation from a database.

12.5.3 Data Management

SQL contains several types of statements to support data management. This section briefly focuses on the following three statements: insert, delete, and update.

• Insert Statement

SQL has an insert statement for adding rows (tuples) to an existing relation. One form of the insert statement is

$$\textbf{insert into} \ \langle \text{relation\ name} \rangle$$
$$\textbf{values} \ (\ \langle \text{attribute\ value\ list} \ \rangle \ )$$

where an attribute value list is defined as follows:

$$\langle \text{attribute value list} \rangle \ ::= \ \langle \text{attribute value} \rangle \ \{ \ , \ \langle \text{attribute value} \rangle \}$$

A second form of insert statement is similar to the previous statement except that it also contains a specified list of attribute names whose values are given in the attribute value list. This second form is

$$\textbf{insert into} \ \langle \text{relation name} \rangle \ (\ \langle \text{attribute\ name\ list} \rangle \ )$$
$$\textbf{values} \ (\ \langle \text{attribute\ value\ list} \rangle \ )$$

where an attribute name list has the following form:

$$\langle \text{attribute name list} \rangle \ ::= \ \langle \text{attribute name} \rangle \ \{ \ , \ \langle \text{attribute name} \rangle \ \}$$

The following statement inserts a tuple in the FACULTY relation:

$$\textbf{insert into} \ \text{FACULTY}$$
$$\textbf{values} \ (1023, \ \text{'Brown'})$$

• Delete Statement

SQL has a delete statement for specifying the deletion of one or more tuples from a specified relation. The form of the delete statement is

$$\textbf{delete from} \ \langle \text{relation\ name} \rangle$$
$$[\ \textbf{where} \ \langle \text{predicate} \rangle]$$

The optional where clause contains a predicate that is used to specify the tuples to be deleted from the specified relation.

The following statement deletes the tuple just inserted into the FACULTY relation:

delete from FACULTY
where emp# = 1023

In the absence of a where clause, all tuples in the specified relations are deleted.

• Update Statement

SQL provides a statement for making corrections to attributes of specified tuples. A where clause indicates which tuples are to be changed, and a set clause specifies which attributes (columns) are to receive new values. The update statement has the following form:

$\langle$update statement$\rangle$ ::= **update** $\langle$relation name$\rangle$
set ($\langle$attribute assignment list$\rangle$)
where $\langle$predicate$\rangle$

where the $\langle$attribute assignment list$\rangle$, which denotes specific columns and their associated values, is given by

$\langle$attribute assignment list$\rangle$ ::=
$\langle$attribute name$\rangle$ = $\langle$expression$\rangle$
{ , $\langle$attribute name$\rangle$ = $\langle$expression$\rangle$ }

As an example, the following statement changes the chair of committee COM303 from 'Cook' to 'Green':

update COMMITTEE
set chair = 'Green'
where com# = 'COM303'

12.5.4 Data Queries

All queries in SQL are specified with a *select statement*. This statement type is not to be confused with the restrict (also called select) operation in relational algebra. Recall that the restrict operation in Section 12.3.3 specifies the selection of a subset of tuples from a given relation.

Although a select statement in SQL can be very powerful, we present here only a basic variant of it, which has the following form:

select [**distinct** | **unique**] $\langle$attribute list$\rangle$
from $\langle$relation list$\rangle$
[**where** $\langle$predicate$\rangle$]

where the attribute list has the form:

$\langle$attribute list$\rangle$ ::= $*$ | [$\langle$relation name$\rangle$.] $\langle$attribute name$\rangle$
{ , [$\langle$relation name$\rangle$.] $\langle$attribute name$\rangle$ }

The keywords **distinct** and **unique** both specify that duplicate tuples are to be deleted. Extensions of this basic SQL variant dealing with arithmetic and aggregate operators are discussed later.

In relational algebra, recall that identical tuples were automatically deleted. If this option is not used, duplicate tuples could appear in a relation. If the uniqueness property of tuples in a relation must be maintained, then this option is used. The ⟨attribute list⟩ specifies the attributes that are to be projected from the resulting relation. A special value of the symbol ∗ for ⟨attribute list⟩ denotes that all attributes of the resulting relation are to be projected.

The from clause indicates the relation(s) that is to be used in the evaluation of the query. The ⟨relation list⟩ in this clause has the following form:

$$⟨\text{relation name entry}⟩ \ \{ \ , \ ⟨\text{relation name entry}⟩ \ \}$$

where ⟨relation name entry⟩ is defined as

$$⟨\text{relation name}⟩ \ [\ ⟨\text{tuple identifier}⟩ \ ]$$

The domain of a tuple identifier is the relation name that precedes it.

The optional where clause contains a predicate whose attributes are chosen from the relation(s) in the from clause.

The remainder of this subsection is organized as follows. First we present several simple unnested data queries. We then illustrate the use of arithmetic and aggregate operators in the formulation of queries. We next focus on more complex queries, which are nested one within another. Finally, we present the suite of queries for the university committee database in SQL.

• Basic Data Queries

We now give several simple SQL queries and relate some of these to the relational algebra discussed earlier. In the following examples we use the relations STUDENT (Table 12.1), COURSE (Table 12.2), and ENROLL (Table 12.3) from Section 12.1.

We want to emphasize that the result of an SQL query is a relation (or table). The resulting table is derived in some specified fashion from the existing relations in a given database.

The SQL queries

select ∗
from ENROLL

and

select ENROLL
from ENROLL

both project the entire relation ENROLL. The symbol ∗ is a shorthand notation for specifying all the attributes in the relation ENROLL, which is named in the from clause.

The following query, which contains a projection and a selection,

```
    select *
      from ENROLL
        where course_num = 'CS260'
```

selects all the rows in the relation ENROLL involving the course CS260 and generates the following results:

course_num	student#	mark
CS260	314182	41
CS260	392044	71
CS260	821688	83

The where clause contains a predicate that specifies the condition under which tuples are chosen.

It is also possible to project on a specified set of attributes. The query

```
    select student#, mark
      from ENROLL
        where course_num = 'CS250'
```

selects the CS250 tuples and projects these tuples on the attributes student# and mark, producing the following results:

student#	mark
392044	46
412181	65
673040	79

The predicate in a where clause can contain logical operators, as in the query

```
    select student#, mark
      from ENROLL
        where (course_num = 'CS250' and (mark >= 50))
```

which generates the student numbers and marks of students who have obtained a passing grade, as follows:

course_num	mark
412181	65
673040	79

SQL allows the use of tuple variables from the relational calculus. The following example illustrates this feature.

Example 12.21   Consider the following query: "Obtain the student numbers and marks of students with a higher grade than the student with student number 314182 enrolled in the course CS260."

Solution   The query

$$\begin{aligned}
&\textbf{select unique } s_1.\text{student\#}, s_1.\text{mark} \\
&\textbf{from } \text{ENROLL } s_1, \text{ENROLL } s_2 \\
&\textbf{where } s_1.\text{mark} > s_2.\text{mark} \\
&\textbf{and } s_2.\text{student\#} = 314182 \\
&\textbf{and } s_1.\text{course\_num} = \text{'CS260'}
\end{aligned}$$

generates the following results:

```
- - - - - - - -
student#    mark
- - - - - - - -
392044      71
821688      83
- - - - - - - -
```

Note the use of name qualification to unambiguously denote the student numbers and marks from tuples s_1 and s_2.   ∎

The following example illustrates the formulation of a join operation in SQL.

Example 12.22   Consider the following query: "Obtain the names and addresses of all students registered in course CS250."

Solution   The query

$$\begin{aligned}
&\textbf{select } \text{STUDENT.student\_name}, \text{STUDENT.address} \\
&\textbf{from } \text{STUDENT, ENROLL} \\
&\textbf{where } \text{STUDENT.student\#} = \text{ENROLL.student\#} \\
&\textbf{and } \text{ENROLL.course\_num} = \text{'CS250'}
\end{aligned}$$

joins the relations STUDENT and ENROLL on the student number attribute and produces the following results:

```
- - - - - - - - - - - - - - - - - -
student_name    address
- - - - - - - - - - - - - - - - - -
Tim Horton      4233 Rushmore Rd
Henry Silva     1280 Taylor St
Ann White       512 Avenue P
- - - - - - - - - - - - - - - - - -
```
∎

The following is an example of a query that requires the joining of three relations.

Example 12.23   Consider the following query: "Obtain a class list (containing course name, student number, student name, and mark) for students enrolled in CS260."

Solution   The query

```
select COURSE.course_name, STUDENT.student#,
       STUDENT.student_name, ENROLL.mark
   from STUDENT, COURSE, ENROLL
   where STUDENT.student# = ENROLL.student#
       and ENROLL.course_num = COURSE.course_num
       and ENROLL.course_num = 'CS260'
```

produces the following table:

```
_ _ _ _ _ _ _ _ _ _ _ _ _ _ _ _ _ _ _ _ _ _ _ _ _ _ _ _ _

course_name        student#    student_name    mark
_ _ _ _ _ _ _ _ _ _ _ _ _ _ _ _ _ _ _ _ _ _ _ _ _ _ _ _ _

Discrete Structures   314182    Paul Harvey      41
Discrete Structures   392044    Tim Horton       71
Discrete Structures   821688    Verna Friesen    83
_ _ _ _ _ _ _ _ _ _ _ _ _ _ _ _ _ _ _ _ _ _ _ _ _ _ _ _ _
```

• Arithmetic and Aggregate Operators in SQL Queries

SQL provides a suite of built-in or aggregate functions to obtain the average (**avg**), minimum (**min**), maximum (**max**), and summation (**sum**) of some collection of numeric values. Also, SQL supports the use of the four common arithmetic operators.

We again stress that only a brief overview of these features is given here. For that reason, no attempt is made to be complete as to the features' syntax and semantics.

The following query specifies that each student grade in the course CS250 from Table 12.3 is to be adjusted upward by 10%.

```
select student#, mark * 1.1
   from ENROLL
       where course_num = 'CS250'
```

The output produced by this query is a table for the Data Structures course in which each row contains a student's number and his or her adjusted grade.

The query

```
select student#, mark
   from ENROLL
       where mark > min(mark)
           and course_num = 'CS215'
```

generates the student numbers and marks of students in course CS215 who obtained more than the minimum overall grade in all courses.

SQL also supports an operation that counts the number of occurrences of some kind. For example, the query

```
select count (distinct course_num)
   from ENROLL
```

generates the number of distinct course numbers in the ENROLL relation as follows:

```
_ _ _ _ _ _ _ _ _ _

count(course_num)
_ _ _ _ _ _ _ _ _ _

7
_ _ _ _ _ _ _ _ _ _
```

Recall that the keyword distinct in the preceding query denotes that only distinct course numbers are to be counted.

As another example of the use of a **count** function, consider the following query:

select count (mark)
from ENROLL
where mark < 50

which generates the number of failures in the ENROLL relation (assuming that a failure is a grade of less than 50), as follows:

```
_ _ _ _ _ _
count(mark)
_ _ _ _ _ _
2
_ _ _ _ _ _
```

SQL also supports the specification of queries that dictate that results are to be grouped by certain categories. For example, the query

select course_num, **avg**(mark)
from ENROLL
group by course_num

specifies the average grade for each course in the ENROLL relation. Each row in the table produced consists of the course number and the average grade in that course.

An SQL query can contain an ordering clause that specifies that the table produced be ordered on one or more attributes in either ascending or descending order. The default is assumed to be ascending order. As an example, let us assume that we require a class list in increasing grade order for the students in course CS215. The query

select *
from ENROLL
where course_num = 'CS215'
order by mark

produces the following result:

course_num	student#	mark
CS215	821688	58
CS215	314182	61
CS215	278123	82

So far, we have formulated queries that consist of single blocks; that is, they are unnested. We now explore nested queries.

• Nested Data Queries

SQL queries can be nested. In such queries a select–from–where part (block) is often nested within a where clause of another part or block, that is, a query within a query.

Example 12.24   Consider the following query: "Obtain the tuple(s) in the ENROLL relation with the highest grade."

Solution   The query

$$
\left.\begin{array}{l}
\textbf{select } * \\
\quad \textbf{from ENROLL}
\end{array}\right\} \text{Outer loop}
$$
$$
\textbf{where mark } =
$$
$$
\left.\begin{array}{l}
(\textbf{select max}(\text{mark}) \\
\quad \textbf{from ENROLL})
\end{array}\right\} \text{Inner loop}
$$

produces the result

course_num	student#	mark
CS220	314182	94

The query has a block level of 2; that is, it consists of two parts: an inner loop and an outer loop. The inner loop is executed before the outer loop. In this example the inner loop first determines tuples with the greatest grade in the ENROLL relation. The result of executing this loop returns a value of 94 for the fifth tuple of Table 12.3. The outer loop then selects the only tuple with that grade and produces the desired result. Observe that if the ENROLL table had contained more than one tuple with a grade of 94 then all tuples with this value would be generated.   ■

The following nested query involves the use of three relations.

Example 12.25   Formulate a query to determine the names of the courses in which Verna Friesen is currently enrolled.

Solution   The query

```
select course_name
    from COURSE
        where COURSE.course_num in
        (select course_num
            from STUDENT, ENROLL
                where STUDENT.student# = ENROLL.student#
                and STUDENT.student_name = 'Verna Friesen')
```

produces the results

course_name
Computer Systems
Logical Design
Discrete Structures
Probability

The inner loop in this query first selects the course numbers taken by Verna Friesen. The outer loop then uses these course numbers to project the course names associated with these numbers. The keyword **in** is used to restrict the courses selected in the outer loop to those returned from the inner loop.　　　　　　　　　　　　　　　　　　■

SQL supports several set operators. Although we do not wish to discuss all these operators, we do describe a few of them here. The union, intersect, and minus set operators correspond to set union, set intersection, and set difference, respectively.

The divide operator in relational algebra has no direct counterpart in SQL. This operation can be realized in SQL by using the not exists set operator. The set operator exists is similar to the existential quantifier $\exists$ in predicate calculus. The exists operator, which can be used in a nested query, tests for the existence of data.

For example, in the query

```
select student_name, address
    from STUDENT
        where exists (select *
            from ENROLL
            where STUDENT.student# = ENROLL.student#
                and mark > 79)
```

produces the following results:

```
- - - - - - - - - - - - - - - - - -
student_name     address
- - - - - - - - - - - - - - - - - -
Susan Graham     721 Primrose Rd
Paul Harvey      718 Park St
Verna Friesen    126 Cumberland St
Tim Horton       4233 Rushmore Rd
- - - - - - - - - - - - - - - - - -
```

In this example, for each student tuple from the STUDENT relation, the exists clause is evaluated. If there is at least one tuple in ENROLL for that student with a mark of at least 80, then the second select statement produces a nonempty result and the exists expression evaluates to true.

The negation of an exists expression must be used in SQL to represent the divide operation in relational algebra. For example, it was shown earlier (see Section 12.3.3 and Table 12.10) that an informal query such as "Obtain the employee numbers of faculty who serve on all committees." has the following form in relational algebra

$$(\text{MEMBER\_OF}/(\text{COMMITTEE}[com\#]))[emp\#]$$

where / denotes the divide operator. The SQL equivalent for this relational expression is

```
select distinct emp#
from MEMBER_OF m₁
where not exists (select com#
            from COMMITTEE c
            where not exists (select *
                        from MEMBER_OF m₂
                        where m₂.emp# = m₁.emp#
                            and m₂.com# = c.com#))
```

which produces the following result:

$$- \ - \ - \ -$$
emp#
$$- \ - \ - \ -$$
1023
$$- \ - \ - \ -$$

The preceding query contains the set operator not exists, which is the negation of **exists**. Observe that, although the informal query and its formulation in relational algebra are reasonably straightforward, the SQL query is complex and not very readable. Instead of using the operator not exists, an SQL clause based on the universal quantifier $\forall$ would probably alleviate this situation. SQL, however, does not have such a clause.

Referring to the STUDENT relation in Table 12.1, consider the following informal query: "Obtain the student information of those students whose cumulative average is greater than the average of all students." The following SQL query will produce the desired results:

> **select** ∗
> **from** STUDENT
> **where** cumulative_avg >
> (**select avg** (cumulative_avg)
> **from** STUDENT)

• Additional SQL Examples

Using SQL, we now formulate the seven queries given in Section 12.3.4.

Example 12.26   Consider the following query: "Obtain the employee numbers of faculty sitting on committee COM202."

Solution   The formulation of this query in SQL involves iterating on each tuple in the relation MEMBER_OF and selecting the employee number of a tuple whose committee is COM202. The SQL query is

> **select** emp#
> **from** MEMBER_OF
> **where** com# = 'COM202'   ∎

Example 12.27   As an extension of the previous example, consider the following query: "Generate the details of each faculty member sitting on committee COM303."

Solution   An SQL form for this query is

> **select** FACULTY.emp#, FACULTY.fac_name
> **from** FACULTY, MEMBER_OF
> **where** (FACULTY.emp# = MEMBER_OF.emp#
> **and** MEMBER_OF.com# = 'COM303')   ∎

The next example requires the use of three relations.

Example 12.28   Consider the following query: "Generate the details of the faculty members who sit on the Research committee."

Solution   An SQL form of this query follows:

> **select** FACULTY.emp#, FACULTY.fac_name
> **from** FACULTY, MEMBER_OF, COMMITTEE
> **where** (COMMITTEE.com_name = 'Research'
>   **and** COMMITTEE.com# = MEMBER_OF.com#
>   **and** FACULTY.emp# = MEMBER_OF.emp#)   ■

Example 12.29   Obtain an SQL query for the following informal query: "Generate details of faculty on both committees COM202 and COM303."

Solution   An SQL formulation of this query is

> **select** *
> **from** FACULTY
> **where** (**select** *
>   **from** MEMBER_OF m_1, MEMBER_OF m_2
>   **where** (m_1.emp# = m_2.emp#
>   **and** (m_1.com# = 'COM202' **and**
> m_2.com# = 'COM303')
>   **and** FACULTY.emp# = m_1.emp#))   ■

Example 12.30   Using SQL, formalize the following query: "Obtain the employee numbers of faculty who sit at least on the same committees as faculty member 1373."

Solution   An SQL form of this query is

> **select** m_1.emp#
> **from** MEMBER_OF m_1
> **where** m_1.emp# <> 1373 **and**
>   (**select count** (com#)
>   **from** MEMBER_OF
>   **where** emp# = 1373)
>   =
>   (**select count** (com#)
>   **from** MEMBER_OF
>   **where** emp# = m_1.emp# **and** com# **in**
>   (**select** m_2.com#
>   **from** MEMBER_OF m_2
>   **where** m_2.emp# = 1373))

where <> denotes the comparison operator $\neq$.   ■

Example 12.31   Formulate in SQL the following query: "Obtain the employee numbers of faculty who do not sit on committee COM606."

Solution   An SQL formulation of this query is

> **select distinct** emp#
> **from** MEMBER_OF
> **where** emp# **not in** (**select** emp#
>   **from** MEMBER_OF
>   **where** com# = 'COM606')   ■

Example 12.32   Obtain an SQL formulation of the following query: "Obtain faculty information on those members who sit on all committees."

Solution   A solution to this query involves joining the relation FACULTY with the result of the divide operation developed earlier. This approach yields the following query:

```
select *
from FACULTY
where emp# in (select emp#
            from MEMBER_OF m₁
            where not exists (select com#
                        from COMMITTEE c
                        where not exists (select *
                                    from MEMBER_OF m₂
                                    where m₂.emp# = m₁.emp#
                                        and m₂.com# = c.com#)))      ■
```

Problems 12.5

1. Express in SQL each of the eight relational operators in the relational algebra.

2. Express in SQL the queries for the university committee database in Problem 3 of Section 12.3.

3. Express in SQL the queries for the part-supplier database in Problem 6 of Section 12.3.

4. All the following queries concern the parts-supplier database in Table 12.12. Formulate these queries in SQL.

   (**a**) Obtain the supplier details for all suppliers.
   (**b**) Obtain the details of U.S. suppliers.
   (**c**) Obtain part numbers for parts supplied by a supplier in Chicago.
   (**d**) Obtain part numbers for parts supplied by Canadian suppliers.
   (**e**) Get the total number of parts supplied by supplier S1.
   (**f**) Obtain the total quantity of part P6 supplied by all suppliers.
   (**g**) Obtain all part names supplied by supplier S1.
   (**h**) Obtain the part numbers of parts supplied by the New York supplier.
   (**i**) Obtain the suppliers that supply the same part.
   (**j**) Calculate the total quantity of parts supplied by supplier S2.
   (**k**) Calculate the total quantity of part P1.
   (**l**) Calculate the total quantity of each kind of part.

12.6 CONCLUDING REMARKS

In this chapter we first introduced the relational data model. We next focused on three approaches to specifying queries: relational algebra, relational calculus, and SQL. Although these approaches are essentially equivalent in the context of specification power, several issues have not been addressed. One important issue relates to the efficiency of executing a given query on an existing database. This issue has led to a new research area within the realm of database research called *query optimization*. We now comment briefly on this issue.

   Although query properties such as clarity and simplicity are always important, there are cases when a given query can be rewritten such that its execution time may change from hours to minutes. A user may specify a query in relational algebra, relational calculus, or SQL in many ways—some good and some bad from an efficiency perspective. For a query specified, say, in SQL, the SQL translator usually performs certain simplifications and optimizations on the given input query.

The SQL translator in the query system may perform certain transformations on a given query to obtain an equivalent optimized form. Such transformations include the following:

- Converting a given query to an equivalent query that is simpler
- Reducing the cardinalities of relations by performing selection operations early, and before join operations if possible

The first type of transformation makes use of logical equivalences in that a query can be formulated in many equivalent ways. Some of these queries, from a computational viewpoint, are less costly to execute than others. The second type of transformation is based on the observation that a selection operation can reduce a relation's cardinality. Therefore, whenever possible, we should perform any selection operation before a join operation.

PROBLEMS: CHAPTER 12

1. The following relations define a restaurant database:

$$\text{MENU } (dish\#,\ description,\ unit\_price)$$
$$\text{ORDER } (bill\#,\ dish\#,\ quantity)$$
$$\text{BILL } (bill\#,\ date,\ server\#,\ table\#,\ amount,\ tip)$$

where it is assumed that

- $dish\#$ uniquely identifies a menu item.
- $bill\#$ and $dish\#$ form a composite key.
- $bill\#$ uniquely identifies a bill.

Furthermore, the domains of the attributes are as follows:

$$dish\#,\ bill\#,\ quantity,\ server\#,\ table\#:\ \text{integer}$$
$$description:\ \text{string}$$
$$unit\_price,\ amount,\ tip:\ \text{real(to the penny)}$$
$$date:\ \text{yyyymmdd(e.g., } 19950223 \text{ denotes Feb. } 23, 1995)$$

Give the relational algebra, relational calculus, and SQL queries for the following informal queries:

(a) Generate the bill numbers issued on a given date.
(b) Generate the bill numbers issued during a given time period.
(c) Generate the distinct server numbers in the database.
(d) Generate an itemized bill for a given bill number.
(e) Generate the server numbers of those servers that worked on a specified date.

(f) Generate the total amount attributed to a given table number on a specified date.

(g) Generate the total revenue (excluding tips) from the restaurant on a given date.

2. For the restaurant database given in Problem 1, express in SQL the following queries:

(a) Determine the average dish price offered in the menu.

(b) Calculate the average amount of tips for each day.

(c) Compute the amount of an average bill in the database.

(d) Generate the distinct table numbers that each server has served on a given date.

(e) Generate in increasing order the number of times each dish has been ordered.

(f) Generate for each dish the total amount of revenue for that dish.

(g) Generate, in increasing order, the amount attributed to each server on a given date.

(h) Generate, in increasing order, the tips given to each server on a stated day.

3. The following relations define a simple Medicare database:

$$DOCTOR\ (sin, name, specialty)$$
$$TREAT\ (sin, phn, date, fee)$$
$$PATIENT\ (phn, name, address, birth\_date)$$

where it is assumed that

- sin (social insurance number) uniquely identifies a doctor.
- phn (personal hospitalization number) uniquely identifies a patient.
- sin and phn form a composite key.

Furthermore, the domains of the attributes are as follows:

$$sin,\ phn,\ specialty : \text{integer}$$
$$name,\ address : \text{string}$$
$$date,\ birth\_date : \text{yyyymmdd}$$
$$fee : \text{real(to the penny)}$$

Give the relational algebra, relational calculus, and SQL queries for the following informal queries:

(a) Obtain the doctor details of doctors with a given specialty.

(b) Generate the names of patients treated by a certain doctor.

(c) Get the names of patients treated by a certain doctor on a given day.

(d) Obtain the total fees paid on behalf of a specified patient in a given year.

(e) Determine the fees paid to a specified doctor in a given year.

(f) Produce the sin number and name of each doctor that has treated a specified patient.

4. The following relations describe a movie database:

$$DIRECTOR\ (dir\#,\ dir\_name,\ country)$$
$$MOVIE\ (id\#,\ dir\#,\ star\#,\ movie\_name,\ type,\ classification,\ year)$$
$$MOVIE\_STAR\ (star\#,\ star\_name,\ age)$$

where it is assumed that

- $dir\#$ uniquely identifies a movie director.
- $star\#$ uniquely identifies a movie star.

Furthermore, the domains of the attributes are as follows:

$dir\#$, $id\#$, $star\#$, age, $year$: integer

dir_name, $country$, $movie\_name$, $type$, $classification$, $star\_name$: string

Give the relational algebra, relational calculus, and SQL queries for the following informal queries:

(a) Find the names of movies that a specified actor or actress has starred in.

(b) List all the movies directed by a specified director.

(c) List all movies with a drama type in a given year.

(d) Generate the names of directors for a given movie star.

(e) Determine the names of movies directed by American directors in a given year.

(f) List each movie in which a director has also been the movie's star.

(g) List the movie stars who acted in movies with a restricted classification.

(h) Find the names of movie stars who have acted in war movies in a given year.

(i) Find the names of the movie stars in a given movie.

Bibliography

1. Aho, A. V., and Ullman, J. D. 1992. *Foundations of Computer Science.* New York: Computer Science Press.

2. Aho, A. V., Hopcroft, J. E., and Ullman, J. D. 1974. *The Design and Analysis of Computer Algorithms.* Reading, Mass.: Addison–Wesley.

3. Andrews, P. B. 1986. *An Introduction to Mathematical Logic and Type Theory: To Truth through Proof.* Orlando, Fla.: Academic Press.

4. Backhouse, R. C. 1986. *Program Construction and Verification.* Upper Saddle River, N.J.: Prentice Hall.

5. Ben-Ari, M. 1993. *Mathematical Logic for Computer Science.* Upper Saddle River, N.J.: Prentice Hall.

6. Bird, R., and Wadler, P. 1988. *Introduction to Functional Programming.* Upper Saddle River, N.J.: Prentice Hall.

7. Boolos, G. S., and Jeffrey, R. C. 1987. *Computability and Logic,* 2nd ed. New York: Cambridge University Press.

8. Clocksin, W. F., and Mellish, S. 1987. *Programming in Prolog,* 3rd ed. New York: Springer–Verlag.

9. Date, C. J. 1990. *Database Systems,* Volume 1, 5th ed. Reading, Mass.: Addison–Wesley.

10. Desai, B. C. 1990. *An Introduction to Database Systems.* St. Paul, Minn.: West.

11. Diller, A. 1990. *Z: An Introduction to Formal Methods.* New York: Wiley.

12. Dodd, T. 1990. *An Advanced Logic Programming Language. Prolog_2 User Guide.* Oxford, England: Blackwell Scientific Publications.

13. Dowsing, R. D., Rayward-Smith, V. J., and Walter, C. D. 1986. *A First Course in Formal Logic and Its Applications in Computer Science.* Oxford, England: Blackwell Scientific Publications.

14. Enderton, H. B. 1972. *A Mathematical Introduction to Logic.* New York: Academic Press.

15. Enderton, H. B. 1977. *Elements of Set Theory.* New York: Academic Press.

16. Fejer, P. A., and Simovici, D. A. 1990. *Mathematical Foundations of Computer Science, Vol. 1: Sets, Relations and Induction.* New York: Springer–Verlag.

17. Francez, N. 1992. *Program Verification.* Reading, Mass.: Addison–Wesley.

18. Fitting, M. 1990. *First-order Logic and Automated Theorem Proving.* New York: Springer–Verlag.

19. Gallier, J. H. 1988. *Logic for Computer Science.* New York: Wiley.

20. Galton, A. 1990. *Logic for Information Technology.* Chinchester, England: Wiley.

21. Gries, D., and Schneider, F. B. 1994. *A Logical Approach to Discrete Math.* New York: Springer–Verlag.

22. Gordon, M. J. C. 1988. *Programming Language Theory and Its Implementation.* Upper Saddle River, N.J.: Prentice Hall.

23. Hindley, J. R., and Seldon, J. P. 1996. *Introduction to Combinators and λ-Calculus.* New York: Cambridge University Press.

24. Hodges, W. 1988. *Logic—An Introduction for Elementary Logic.* London: Penguin Books.

25. Horowitz, E., and Sahni, S. 1978. *Fundamentals of Computer Algorithms.* New York: Computer Science Press.

26. Hunter, G. 1973. *Metalogic: An Introduction to the Metatheory of Standard First Order Logic.* Berkeley, Calif.: University of California Press.

27. Imperato, M. 1991. *An Introduction to Z.* Lund, Sweden: Chartwell–Bratt.

28. Ince, D. C. 1988. *An Introduction to Discrete Mathematics and Formal System Specification.* New York: Oxford University Press.

29. Leblanc, H., and Wisdom, W. A. 1976. *Deductive Logic,* 2nd ed. Boston: Allyn and Bacon.

30. Lightfoot, D. 1991. *Formal Specification Using Z.* New York: Macmillan.

31. Mannila, H., and Räihä, K.-J. 1992. *The Design of Relational Databases.* Reading, Mass.: Addison–Wesley.

32. Maurer, S. B., and Ralston, A. 1991. *Discrete Algorithmic Mathematics.* Reading, Mass.: Addison–Wesley.

33. McArthur, R. P. 1991. *From Logic to Computing.* Belmont, Calif.: Wadsworth.

34. Mendelson, E. 1987. *Introduction to Mathematical Logic.* Monterey, Calif.: Wadsworth & Brooks.

35. Ozkarahan, E. 1990. *Database Management: Concepts, Design, and Practice.* Upper Saddle River, N. J.: Prentice Hall.

36. Pospesel, H. 1976. *Predicate Logic.* Upper Saddle River, N.J.: Prentice Hall.

37. Pospesel, H. 1984. *Propositional Logic,* 2nd ed. Upper Saddle River, N.J.: Prentice Hall.

38. Quintus Computer System, Ltd. 1988. *Quintus Prolog Reference Manual.* Mountain View, Calif.: Quintus.

39. Reeves, S., and Clarke, M. 1990. *Logic for Computer Science.* Reading, Mass.: Addison–Wesley.

40. Research Software Ltd. 1989. *Miranda System Manual.* Canterbury, England: Research Software.

41. Rosen, K. H. 1991. *Discrete Mathematics and Its Applications,* 2nd ed. New York: McGraw–Hill.

42. Ross, K. A., and Wright, C. R. B. 1992. *Discrete Mathematics,* 3rd ed. Upper Saddle River, N.J.: Prentice Hall.

43. Spivey, J. M. 1991. *The Z Notation: A Reference Manual.* Upper Saddle River, N.J.: Prentice Hall.

44. Sterling, L., and Shapiro, E. 1986. *The Art of Prolog.* Cambridge, Mass.: MIT Press.

45. Tremblay, J. P., and Manohar, R. 1975. *Discrete Mathematical Structures with Applications to Computer Science.* New York: McGraw–Hill.

46. Tremblay, J. P., and Sorenson, P. G. 1985. *The Theory and Practice of Compiler Writing.* New York: McGraw-Hill.

47. Wordsworth, J. B. 1992. *Software Development with Z.* Reading, Mass.: Addison–Wesley.

48. Wos, L. 1984. *Automated Reasoning.* Upper Saddle River, N.J.: Prentice Hall.

Solutions to Even-numbered Problems

Solutions 1.1

2. **(a)** is not a proposition, but **(b)**, **(c)**, and **(d)** are.

4. **(a)** F   **(b)** T   **(c)** F

Solutions 1.2

2. **(a)** P: Jim is in the barn. Q: Jack is in the barn. $P \Rightarrow Q$

(b) P: The getaway car was red. Q: The getaway car was brown. $P \vee Q$

(c) P: The news is good. $\neg P$

(d) P: You will be on time. Q: You hurry. $P \Rightarrow Q$

(e) P: He will come. Q: He has time. $P \Leftarrow Q$

(f) P: She was there. Q: She has heard it. $P \Rightarrow Q$

4.

P	$P \wedge P$	$P \vee P$	$P \wedge \text{T}$	$P \wedge \text{F}$
T	T	T	T	F
F	F	F	F	F

Solutions 1.3

2. **(a)** $((P \wedge Q) \wedge R) \Rightarrow P$ **(c)** $(\neg(P_1 \wedge P_2)) \Rightarrow ((\neg Q) \vee P_1)$

(b) $((P \wedge R) \vee Q) \Leftrightarrow \neg R$ **(d)** $(P \Rightarrow Q) \Leftrightarrow ((\neg Q) \Rightarrow \neg P)$

4. $\neg\neg P$ and $P \vee Q$ are not literals; P and $\neg P_2$ are.

6. **(a)** T   **(b)** T   **(c)** T

8. **(a)**

P	Q	$(\neg P \vee \neg Q)$	$\neg(\neg P \vee \neg Q)$
T	T	F	T
T	F	T	F
F	T	T	F
F	F	T	F

(b)

P	Q	$(\neg P \wedge \neg Q)$	$\neg(\neg P \wedge \neg Q)$
T	T	F	T
T	F	F	T
F	T	F	T
F	F	T	F

(c)

P	Q	$(P \vee Q)$	$P \wedge (P \vee Q)$
T	T	T	T
T	F	T	T
F	T	T	F
F	F	F	F

(d)

P	Q	$(Q \wedge P)$	$P \wedge (Q \wedge P)$
T	T	T	T
T	F	F	F
F	T	F	F
F	F	F	F

(e) Let $A = (\neg P \wedge (\neg Q \wedge R)) \vee (Q \wedge R) \vee (P \wedge R)$

P	Q	R	$(\neg Q \wedge R)$	$(\neg P \wedge (\neg Q \wedge R))$	$(Q \wedge R)$	$(P \wedge R)$	A
T	T	T	F	F	T	T	T
T	T	F	F	F	F	F	F
T	F	T	T	F	F	T	T
T	F	F	F	F	F	F	F
F	T	T	F	F	T	F	T
F	T	F	F	F	F	F	F
F	F	T	T	T	F	F	T
F	F	F	F	F	F	F	F

(f) Let $A = (P \wedge Q) \vee (\neg P \wedge Q) \vee (P \wedge \neg Q) \vee (\neg P \wedge \neg Q)$

P	Q	$(P \wedge Q)$	$(\neg P \wedge Q)$	$(P \wedge \neg Q)$	$(\neg P \wedge \neg Q)$	A
T	T	T	F	F	F	T
T	F	F	F	T	F	T
F	T	F	T	F	F	T
F	F	F	F	F	T	T

Solutions 1.4

2. Convert $P \vee \neg P$ into the schema $A \vee \neg A$.
    (a) Let $A = P \Rightarrow Q$ **(b)** Let $A = \neg P$ **(c)** Let $A = ((P \wedge S) \vee Q)$

4. A material implication may be true or false. A material implication that is always true is a logical implication.

Solutions 1.5

2. (a) $(P \wedge Q) \vee (Q \wedge R) \equiv (Q \wedge P) \vee (Q \wedge R)$ Commutative
$\equiv Q \wedge (P \vee R)$ Distributive

(b) $\neg(\neg(P \wedge Q) \vee P) \equiv (\neg\neg(P \wedge Q) \wedge \neg P)$ De Morgan
$\equiv (P \wedge Q) \wedge \neg P$ Double negation
$\equiv (Q \wedge P) \wedge \neg P$ Commutative
$\equiv Q \wedge (P \wedge \neg P)$ Associative
$\equiv Q \wedge F$ Contradiction
$\equiv F$ Domination

(c) $\neg(\neg P \vee \neg(R \vee S)) \equiv (\neg\neg P \wedge \neg\neg(R \vee S))$ De Morgan
$\equiv P \wedge (R \vee S)$ Double negation
$\equiv (P \wedge R) \vee (P \wedge S)$ Distributive

(d) $(P \vee R) \wedge (Q \vee S) \equiv ((P \vee R) \wedge Q) \vee ((P \vee R) \wedge S)$ Distributive
$\equiv (P \wedge Q) \vee (R \wedge Q) \vee (P \wedge S) \vee (R \wedge S)$ Distributive and Commutative

4. (a) $(Q \wedge R \wedge S) \vee (Q \wedge \neg R \wedge S) \equiv (Q \wedge S) \wedge (R \vee \neg R) \equiv Q \wedge S$ Distributive and
Opposing literals, followed by identity

(b) $(P \vee R) \wedge (P \vee R \vee S) \equiv P \vee R$ (by (1.14))

(c) $(P \vee (Q \wedge S)) \vee (\neg Q \wedge S) \equiv P \vee ((Q \wedge S) \vee (\neg Q \wedge S)) \equiv P \vee S$ (by (1.15))

6. (a) $P \vee \neg Q \vee (P \wedge Q) \wedge (P \vee \neg Q) \wedge \neg P \wedge Q$
$\equiv P \vee \neg Q \vee (P \wedge Q \wedge (P \vee \neg Q) \wedge \neg P \wedge Q)$ Associative
$\equiv P \vee \neg Q \vee F$ Opposing literals
$\equiv P \vee \neg Q$ Identity

(b) $(P \vee \neg Q) \wedge (\neg P \vee Q) \vee \neg(\neg(P \vee \neg R) \wedge Q)$
$\equiv ((P \vee \neg Q) \wedge (\neg P \vee Q)) \vee (P \vee \neg R \vee \neg Q)$ De Morgan
$\equiv ((P \vee \neg Q) \vee (P \vee \neg R \vee \neg Q))$
$\quad \wedge ((\neg P \vee Q) \vee (P \vee \neg R \vee \neg Q))$ Distributive
$\equiv (P \vee \neg Q \vee P \vee \neg R \vee \neg Q) \wedge F$ Opposing literals
$\equiv P \vee \neg Q \vee \neg R$ Identity; remove duplicates

(c) $\neg((P \vee Q) \wedge R) \vee Q \equiv (\neg(P \vee Q) \vee \neg R) \vee Q$ De Morgan
$\equiv (\neg P \wedge \neg Q) \vee \neg R \vee Q$ De Morgan
$\equiv ((\neg P \vee Q) \wedge (\neg Q \vee Q)) \vee \neg R$ Distributive
$\equiv ((\neg P \vee Q) \wedge T) \vee \neg R$ Excluded middle
$\equiv \neg P \vee Q \vee \neg R$ Identity

8. $(\neg P \wedge \neg Q \wedge R) \vee (P \wedge \neg Q \wedge \neg R) \vee (P \wedge Q \wedge R)$

10. Prove (1.14) $P \wedge (P \vee Q) \equiv (P \vee F) \wedge (P \vee Q) \equiv P \vee (F \wedge Q) \equiv P$
Prove (1.16) $(P \vee Q) \wedge (\neg P \vee Q) \equiv (Q \vee P) \wedge (Q \vee \neg P) \equiv Q \vee (P \wedge \neg P) \equiv Q$
Prove (1.9) Use truth table.

Solutions 1.6

2. (a)

P	Q	P	$P \Rightarrow Q$	Premises	Q	Valid
T	T	T	T	T	T	T
T	F	T	F	F	F	T
F	T	F	T	F	T	T
F	F	F	T	F	F	T

(b)

P	Q	$P \Rightarrow Q$	$\neg P \Rightarrow Q$	Premises	Q	Valid
T	T	T	T	T	T	T
T	F	F	T	F	F	T
F	T	T	T	T	T	T
F	F	T	F	F	F	T

(c)

P	Q	$P \Leftrightarrow Q$	$P \Rightarrow Q$	Valid
T	T	T	T	T
T	F	F	F	T
F	T	F	T	T
F	F	T	T	T

4. (a)

1. P	Premise
2. $P \Rightarrow (Q \vee R)$	Premise
3. $Q \vee R$	1, 2, MP
4. $(Q \vee R) \Rightarrow S$	Premise
5. S	3, 4, MP

(b)

1. $P \Rightarrow Q$	Premise
2. $Q \Rightarrow R$	Premise
3. $\neg R$	Premise
4. $\neg Q$	2, 3, MT
5. $\neg P$	1, 4, MT

(c)

1. P	Premise
2. $P \Rightarrow Q$	Premise
3. Q	1, 2, MP
4. $P \wedge Q$	1, 3, Combination

6. Theorem 1.2 follows from equivalence elimination.

CHAPTER 1 SOLUTIONS

2. $P_1 \Leftarrow ((P_2 \wedge P_3 \wedge P_4) \vee (P_5 \wedge P_6))$

4. *Note:* In the derivations below, we use the shortcuts discussed in Section 1.5.6 for simplifying conjunctions and disjunctions. The reader should have no difficulties in stating the steps needed when these shortcuts are not allowed. Also, generalized distributive laws are used.

(**a**) $(P \wedge Q \wedge R) \vee (P \wedge \neg Q) \vee (P \wedge \neg R)$
$\equiv P \wedge ((Q \wedge R) \vee \neg Q \vee \neg R)$ Generalized distributive
$\equiv P \wedge (Q \vee \neg Q \vee \neg R) \wedge (R \vee \neg Q \vee \neg R)$ Distributive

$\equiv P \wedge \mathrm{T} \wedge \mathrm{T}$ Opposing literals

$\equiv P$

(b) $(P \wedge Q) \vee (P \wedge R) \vee (P \wedge (Q \vee \neg R))$

$\equiv P \wedge (Q \vee R \vee Q \vee \neg R)$ Generalized distributive

$\equiv P \wedge \mathrm{T}$ Opposing literals

$\equiv P$ Identity

(c) $(\neg P \wedge R \wedge \neg (P \wedge \neg (P \vee Q))) \equiv \neg P \wedge R \wedge (\neg P \vee (P \vee Q))$ De Morgan

$\equiv \neg P \wedge R \wedge ((\neg P \vee P) \vee Q)$ Commutative

$\equiv \neg P \wedge R \wedge (\mathrm{T} \vee Q)$ Opposing literals

$\equiv \neg P \wedge R$ Identity

6. **(a)** $((P \wedge Q) \Rightarrow P) \equiv (\neg (P \wedge Q) \vee P)$ by (1.9)

$\equiv (\neg P \vee \neg Q \vee P)$ De Morgan

$\equiv \mathrm{T}$ Excluded middle and Domination

(b) $(\neg Q \wedge (P \Rightarrow Q)) \Rightarrow \neg P \equiv \neg (\neg Q \wedge (\neg P \vee Q)) \vee \neg P$ by (1.9)

$\equiv (Q \vee \neg (\neg P \vee Q) \vee \neg P)$ De Morgan

$\equiv (Q \vee (P \wedge \neg Q) \vee \neg P)$ De Morgan

$\equiv ((Q \vee P) \wedge (Q \vee \neg Q)) \vee \neg P$ Distributive

$\equiv Q \vee P \vee \neg P$ Excluded middle and Identity

$\equiv \mathrm{T}$ Excluded middle and Domination

(c) $((P \vee Q) \wedge \neg P) \Rightarrow Q \equiv \neg ((P \vee Q) \wedge \neg P) \vee Q$ by (1.9)

$\equiv \neg (P \vee Q) \vee P \vee Q$ De Morgan

$\equiv (\neg P \wedge \neg Q) \vee P \vee Q$ De Morgan

$\equiv ((\neg P \vee P) \wedge (\neg Q \vee P)) \vee Q$ Distributive

$\equiv (\mathrm{T} \wedge (\neg Q \vee P)) \vee Q$ Excluded middle

$\equiv \neg Q \vee P \vee Q$ Identity

$\equiv \mathrm{T}$ Opposing literals

8.

P	Q	R	$P \vee Q$	$R \Rightarrow P$	$R \wedge (R \Rightarrow P)$	Result
T	T	T	T	T	T	T
T	T	F	T	T	F	F
T	F	T	T	T	T	T
T	F	F	T	T	F	F
F	T	T	T	F	F	F
F	T	F	T	T	F	F
F	F	T	F	F	F	T
F	F	F	F	T	F	T

$(P \wedge Q \wedge R) \vee (P \wedge \neg Q \wedge R) \vee (\neg P \wedge \neg Q \wedge R) \vee (\neg P \wedge \neg Q \wedge \neg R)$

$\equiv (\neg P \vee \neg Q \vee R) \wedge (\neg P \vee Q \vee R) \wedge (P \vee \neg Q \vee \neg R) \wedge (P \vee \neg Q \vee R)$

10.

1. $P \vee Q$	Premise
2. $\neg P \Rightarrow Q$	1, DS, and deduction theorem
3. $Q \Rightarrow R$	Premise
4. $\neg P \Rightarrow R$	2, 3, HS
5. $P \Rightarrow R$	Premise
6. R	4, 5, Cs

12.

1.	Q	Premise
2.	P	Assumption
3.	Q	1, Copy
4. $P \Rightarrow Q$		2, 3, Deduction

1.	Q	Assumption
2.	$P \Rightarrow Q$	1, previous proof
3. $Q \Rightarrow (P \Rightarrow Q)$		1, 2, Deduction

Solutions 2.1

2. **(a)** $L(x, y) : x$ likes y, M, K, J : Mary, Kim, Julie
$L(M, K) \wedge L(K, J) \Rightarrow L(M, J)$

(b) $b(x) : x$ is busy, J, B : John, Bill
$b(J) \wedge \neg b(B)$

(c) $K(x, y) : x$ knows y, S, B : Mr. Smith, Ben
$K(B, S) \wedge \neg K(S, B)$

4. $E(x), (F(x)) : x$ speaks English (French)
$\forall x(E(x) \vee F(x))$

6. $P(3, y), P(y, y), P(y) \wedge \forall x Q(x), P(x) \wedge Q(2) \wedge R(x, 2)$

8. **(a)** $\forall x(\text{Lion}(x) \Rightarrow \text{predator}(x))$

(b) $\exists x(\text{Lion}(x) \wedge \text{in Africa}(x))$

(c) $\forall x(\text{Roars}(x) \Rightarrow \text{Lion}(x))$

(d) $\exists x(\text{Lion}(x) \wedge \exists y(\text{Zebra}(y) \wedge \text{Eats}(x, y)))$

(e) $\exists x(\text{Lion}(x) \wedge \forall y(\text{Eats}(x, y) \Rightarrow \text{Zebra}(y)))$

Solutions 2.2

2. $\exists x \neg Q(a, x)$ is T because $\neg Q(a, b)$ is T.
$\forall y Q(b, y)$ is F because $Q(b, a)$ is F.
$\forall y Q(y, y) \wedge \exists x \forall y Q(x, y)$ is F since
$\forall y Q(y, y)$ is T because $Q(a, a)$, $Q(b, b)$, and $Q(c, c)$ all are T, but
$\exists x \forall y Q(x, y)$ is F because $\forall y Q(a, y)$, $\forall y Q(b, y)$, and $\forall y Q(c, y)$ all are F.

4. The expression is valid because $P(x) \Rightarrow (P(x) \vee Q(x))$ is of the form $A \Rightarrow (A \vee B)$ that is tautologically true.

6. Assign $x = a, y = b, P(a) = F, P(b) = T, Q(a) = T, Q(b) = F$.

Solutions 2.3

2.

1. $\forall x \neg Q(x)$	Premise
2. $\forall x(P(x) \Rightarrow Q(x))$	Premise
3. $\neg Q(x)$	1, S_x^x
4. $P(x) \Rightarrow Q(x)$	2, S_x^x
5. $\neg P(x)$	3, 4, MT
6. $\forall x \neg P(x)$	5, UG

4.

1. $\forall x P(x)$	Premise
2. $P(x)$	1, S_x^x
3. $\exists x P(x)$	2, EG

6. P, M, B: Peter, Mary, Bill

1. $L(x, y) \wedge L(y, z) \Rightarrow L(x, z)$	x, y, z true variables
2. $L(P, M)$	Premise
3. $L(M, B)$	Premise
4. $L(P, M) \wedge L(M, B)$	2, 3, Combination
5. $L(P, M) \wedge L(M, B) \Rightarrow L(P, B)$	1, $S_{PMB}^{x\ y\ z}$
6. $L(P, B)$	4, 5, MP

8. Write $P(a, y_1) \wedge R(x_1, b)$ and $P(x_2, b) \wedge R(y_2, x_2)$ to obtain $P(a, b) \wedge R(x_1, a)$.

Solutions 2.4

2. $(\forall x B) \wedge A \equiv \forall x B \wedge \forall x A \quad 1$
$\equiv \forall x (B \wedge A) \quad 5$

4. $\exists x P(x) \wedge \exists x (Q(x) \wedge P(x)) \equiv \exists y P(y) \wedge \exists x (Q(x) \wedge P(x))$
$\equiv \exists y \exists x (P(y) \wedge Q(x) \wedge P(x))$

6. If anybody finds a bug, the program error can be corrected.

8. $\exists x (\exists y R(y) \vee Q(x)) \equiv \exists x (\exists y R(y) \vee \exists y Q(x)) \quad 1d$
$\equiv \exists x \exists y (R(y) \vee Q(x)) \quad 5d$
$\equiv \exists x \exists y (Q(x) \vee R(y)) \quad$ Commutative

Solutions 2.5

2. $\forall x (E(x) \Rightarrow x = \text{jim})$

4. $x \circ y = y \circ x$: because if d is the greatest common divisor of x and y, d is also the greatest common divisor of y and x.
$x \circ (y \circ z) = (x \circ y) \circ z$: They both are the greatest common divisors of x, y, and z as can be proven. There is no identity. However, since $x \circ 1 = 1 \circ x = 1$, 1 is a zero.

6. (a)

	T	F
T	T	F
F	T	T

(b) Left identity: T
$\forall x ((T \Rightarrow x) \Leftrightarrow x)$ is true for all instantiations:
$(T \Rightarrow T) \Leftrightarrow T$ and $(T \Rightarrow F) \Leftrightarrow F$.
No right identity:
$\forall x ((x \Rightarrow T) \Leftrightarrow x)$ and $\forall x ((x \Rightarrow F) \Leftrightarrow x)$ are both false.

(c) Left zero: None exists since $\forall x ((T \Rightarrow x) \Leftrightarrow T)$ and $\forall x ((F \Rightarrow x) \Leftrightarrow F)$ are both false.
Right zero: T, because $\forall x ((x \Rightarrow T) \Leftrightarrow T)$.

8. $a^{-1}baab^{-1} = ba^{-1}aab^{-1}$ Commutative
$\quad = b(a^{-1}a)ab^{-1}$ Associative
$\quad = b1ab^{-1}$ Inverse
$\quad = bab^{-1}$ Identity
$\quad = bb^{-1}a$ Commutative
$\quad = (bb^{-1})a$ Associative
$\quad = 1a$ Inverse
$\quad = a$ Identity

10.

1. $S(0) = 1$	Premise
2. $S(n+1) = S(n) + n$	Premise, n free
3. $S(1) = S(0) + 0$	2, $n := 0$
4. $S(1) = 1 + 0$	3, 1
5. $S(1) = 1$	4, Arithmetic
6. $S(2) = S(1) + 1$	2, $n := 1$
7. $S(2) = 1 + 1$	5, 6
8. $S(2) = 2$	7, Arithmetic
9. $S(3) = S(2) + 2$	2, $n := 2$
10. $S(3) = 2 + 2$	8, 9
11. $S(3) = 4$	10, Arithmetic

12. $f(f(f(x)))$ denotes the third person to the left of x. If this person is x, there must be 3 people at the table.

CHAPTER 2 SOLUTIONS

2. 3^2

4. **(a)** $W(x) : x$ has wings, $F(x) : x$ can fly
$\quad \forall x W(x) \wedge \exists x \neg F(x)$

  (b) $B(x) : x$ is a bird
$\quad \forall x(B(x) \Rightarrow W(x)) \wedge \exists x(B(x) \wedge \neg F(x))$

  (c) $H(x, y) : x$ has y
$\quad \forall x(B(x) \Rightarrow H(x, \text{wings})) \wedge \exists x(B(x) \wedge \neg F(x))$

6. **(a)** $P(x) \vee P(y)$: satisfiable because $P(a) = \mathrm{T}$, $P(a) \vee P(a) = \mathrm{T}$.
$\qquad\qquad$ not valid because $P(a) = \mathrm{F}$, $P(b) = \mathrm{F}$, $P(a) \vee P(a) = \mathrm{F}$.

  (b) $P(x) \Rightarrow (P(x) \vee Q(x))$: valid because of instantiation of tautological schema
$\qquad\qquad\qquad A \Rightarrow (A \vee B)$.
$\qquad\qquad$ satisfiable because valid.

  (c) $\forall x(P(x) \vee \neg P(x))$: valid because $P(x) \vee \neg P(x)$ is valid.
$\qquad\qquad$ satisfiable because valid.

  (d) $\exists x(P(x) \wedge \neg P(x))$: contradictory since $P(x) \wedge \neg P(x)$ is contradictory.
$\qquad\qquad$ not valid because contradictory.

  (e) $\forall x P(x)$: satisfiable: Use universe with a as its only element; let $P(a) = \mathrm{T}$.
$\qquad\qquad$ not valid: Use the same universe as before, with $P(a) = \mathrm{F}$.

8.

1. $\exists x M(x)$	Premise
2. $\forall x(M(x) \Rightarrow \exists y C(x, y))$	Premise
3. $M(a)$	1, S_a^x
4. $M(a) \Rightarrow \exists y C(a, y)$	2, S_a^x
5. $\exists y C(a, y)$	3, 4, MP
6. $\forall x(\exists y C(x, y) \Rightarrow F(x))$	Premise
7. $\exists y C(a, y) \Rightarrow F(a)$	6, S_a^x
8. $F(a)$	5, 7, MP
9. $\exists y F(y)$	8, EG

10.

1. $P(a) \wedge \forall x P(x)$	Assumption
2. $\forall x P(x)$	1, Simplification
3. $P(a) \wedge \forall x P(x) \Rightarrow \forall x P(x)$	Deduction
4. $\forall x P(x)$	Assumption
5. $P(a)$	4, S_a^x
6. $P(a) \wedge \forall x P(x)$	4, 5, Combination
7. $\forall x P(x) \Rightarrow P(a) \wedge \forall x P(x)$	Deduction
8. $(P(a) \wedge \forall x P(x)) \Leftrightarrow \forall x P(x)$	3, 7 Equivalence introduction

12. **(a)** $P(x)$ $(S(x))$: x is a professor (student)
$G_M(x)$ $(G_W(x))$: x gives an assignment on Monday (Wednesday)
$C(x)$: x complains
$F(x)$: x finishes work
$P(a) \wedge G_M(a) \wedge G_W(a) \wedge \forall x(S(x) \Rightarrow C(x)) \wedge \exists x(S(x) \wedge \neg F(x))$
(b) $B(x, y, z)$: y is between x and y
$B(a, y, b) \Leftrightarrow (a < y) \wedge (y < b)$
(c) $\forall x(\text{Person}(x) \Rightarrow \exists y(\text{day}(y) \wedge \text{fool}(x, y)))$
   $\wedge \ \exists x(\text{Person}(x) \wedge \forall y(\text{day}(y) \Rightarrow \text{fool}(x, y)))$
   $\wedge \ \neg \forall x(\text{Person}(x) \Rightarrow \forall y(\text{day}(y) \Rightarrow \text{fool}(x, y)))$
(d) $\neg \exists x(\text{free}(x) \wedge x = Lunch)$

14. **(a)** $R(x, y) \wedge R(y, z_1) \Rightarrow R(x, z_1)$ and $R(a, b) \wedge R(b, z_2) \Rightarrow R(u, a)$
$x = a, y = b, z_1 = a, z_2 = a, u = a$ which yields the following:
   $R(a, b) \wedge R(b, a) \Rightarrow R(a, a)$
(b) $G(x, y) \wedge G(f(x), f(y))$ and $G(a, b) \wedge G(z, u)$
$x = a, y = b, z = f(x), u = f(y)$ which yields the following:
   $G(a, b) \wedge G(f(a), f(b))$

16.

1. $\exists x P(x) \wedge \forall x \forall y(P(x) \wedge P(y) \Rightarrow (x = y))$	Premise
2. $P(a) \wedge \forall x \forall y(P(x) \wedge P(y) \Rightarrow (x = y))$	1, S_a^x
3. $\forall x \forall y(P(x) \wedge P(y) \Rightarrow (x = y))$	2, Simplification
4. $\forall y(P(a) \wedge P(y) \Rightarrow (a = y))$	3, S_a^x
5. $P(a) \wedge (P(y) \Rightarrow (a = y))$	4, S_y^y
6. $P(y) \Rightarrow (a = y)$	5, Simplification
7. $\forall y(P(y) \Rightarrow (a = y))$	6, UG
8. $P(a)$	2, Simplification
9. $P(a) \wedge \forall y(P(y) \Rightarrow (a = y))$	7, 8, Combination
10. $\exists x(P(x) \wedge \forall y(P(y) \Rightarrow (x = y)))$	9, EG

18. The result is not sound, as the following model indicates:
$$P(a) = T, P(b) = F, Q(a) = T, Q(b) = T$$
Then $\exists_1 x P(x)$ and $\forall x (P(x) \Rightarrow Q(x))$ are both true, but $\exists_1 x Q(x)$ is false.

20. Let $S(x)$ be x is a student and let the committee be the universe.
    (a) At most one: $\forall y (S(y) \Rightarrow (x = y))$
    (b) At least one: $\exists x S(x)$
    (c) Exactly one: $\exists x (S(x) \wedge \forall y (S(y) \Rightarrow (x = y)))$

22. $\forall x \exists_1 y (f(x) = y)$ or $\forall x \exists y ((f(x) = y) \wedge \forall z (f(x) = z \Rightarrow z = y))$

24. The universe consists of two elements, namely, a and b, with $f(a) = b$ and $f(b) = a$. Then $f(f(a)) = a$ and $f(f(b)) = b$. For example, the universe consists of 3 and -3 and $f(x) = x \times -1$.

26. $x \circ y = x + y - xy = y + x - yx = y \circ x$
$x \circ (y \circ z) = x + (y + z - yz) - x(y + z - yz)$
$(x \circ y) \circ z = (x + y - xy) + z - (x + y - xy)z$
Moreover, $x + (y + z - yz) - x(y + z - yz) = (x + y - xy) + z - (x + y - xy)z$
Identity is zero, because $x \circ 0 = x + 0 - x \times 0 = x$
Inverse : $x \circ x^{-1} = x + x^{-1} - xx^{-1} = 0$
$x^{-1}(1 - x) = -x$ or $x^{-1} = \frac{x}{x-1}$

Solutions 3.1

2. $\forall n(n \le n) \wedge \forall m \forall n((n \le m) \Rightarrow n \le s(m))$

4. There are many models that satisfy the requirements. The most trivial model is $s(n) = 0$ for all n, which produces the sequence $0, 0, 0, \ldots$. Another example is based on a universe containing only the numbers $\{0, 1, 2\}$ with $s(0) = 1$, $s(1) = 2$, and $s(2) = 0$.

6. **Inductive base**   For $n = 0$, $(0 + 2)! = 2$ is even.

    Inductive hypothesis   Assume $(n + 2)!$ is even.

    Inductive step $((n + 1) + 2)! = (n + 2)!(n + 3)$ is even.

8. **Inductive base**   For $n = 0$, $0 \times n = 0$, using (3.3).

    Inductive hypothesis   Assume $0 \times n = 0$.

    Inductive step $0 \times s(n) = 0 \times n + 0$ (3.4) with $m := 0$
                    $= 0 \times n$ (3.1) with $m := 0 \times n$
                    $= 0$ Inductive hypothesis

    Conclusion $0 \times n = 0$ for all n.

10. **Inductive base**   For $n = 4$, $2^4 \ge 4^2$ or $16 \ge 16$.

    Inductive hypothesis   Assume $2^n \ge n^2$.

    Inductive step For $n \ge 4$, $(n + 1)^2 = n^2 + 2n + 1 \le n^2 + 2n + n$
                    $= n^2 + 3n \le 2n^2$
                    Now: $2^{n+1} = 2 \times 2^n$
                    $\ge 2 \times n^2$ Inductive hypothesis
                    $\ge (n + 1)^2$

Solutions 3.2

2. **Inductive base**   For $n = m - 1$, both sides of (3.11) are 0.

   Inductive hypothesis   Assume $\sum_{i=m}^{n} (a_i b) = b \sum_{i=m}^{n} a_i$.

   Inductive step $\sum_{i=m}^{n+1} (a_i b) = (\sum_{i=m}^{n} a_i b) + a_{n+1} b$

   $$= b((\sum_{i=m}^{n} a_i) + a_{n+1})$$

   $$= b \sum_{i=m}^{n+1} a_i$$

4. **Inductive base**   For $n = 1$, $(\prod_{i=1}^{1} a_i)^2 = \prod_{i=1}^{1} a_i^2 = a_1^2$.

   Inductive hypothesis   Assume $(\prod_{i=1}^{n} a_i)^2 = \prod_{i=1}^{n} a_i^2$.

   Inductive step $(\prod_{i=1}^{n+1} a_i)^2 = (\prod_{i=1}^{n} a_i a_{n+1})^2$

   $$= (\prod_{i=1}^{n} a_i)^2 a_{n+1}^2$$

   $$= \prod_{i=1}^{n} a_i^2 a_{n+1}^2$$

   $$= \prod_{i=1}^{n+1} a_i^2$$

6. **Inductive base**   For $n = 0$, both sides are 0.

   Inductive hypothesis   Assume $\sum_{i=0}^{n} i(i+1) = n(n+1)(n+2)/3$.

   Inductive step $\sum_{i=0}^{n+1} i(i+1) = \sum_{i=0}^{n} i(i+1) + (n+1)(n+2)$

   $$= n(n+1)(n+2)/3 + (n+1)(n+2)$$

   $$= (n+1)(n+2)(n+3)/3$$

8. $\sum_{i=1}^{n} a_i - \sum_{i=1}^{n} a_{i+1} = \sum_{i=1}^{n} a_i - \sum_{i=2}^{n+1} a_i$

   $$= \sum_{i=1}^{n} a_i - (\sum_{i=1}^{n} a_i) + a_1 - a_{n+1}$$

   $$= a_1 - a_{n+1}$$

Solutions 3.3

2. Take the G-sequence $x, y, x, y, \ldots$, which is infinite.
4. Call the string $a^m b a^m$ an s-string.
   1. b is an s-string.
   2. If x is an s-string, so is axa.
   3. Nothing else is an s-string.

   Inductive base b has an odd number of characters.

   Inductive hypothesis   Assume that x is an s-string and that x has an odd number of characters.

   Inductive step   If x has an odd number of characters, so does axa. Hence, all s-strings have an odd number of characters.

6. **Inductive base**   For $n = 1$, $m^{(n-1)} = 1$ leaf.

   Inductive hypothesis   Assume trees of height n have at most $m^{(n-1)}$ leaves.

Inductive step   Trees of height $n + 1$ have at most $m^{(n-1)}$ leaves for each leaf in old tree, yielding $m \cdot m^{n-1} = m^n$ leaves.

8. **Domain well founded**   The domain is well founded provided that $n \geq 0$, but not if $n < 0$.

Inductive base   For $n = 0$, $2^0 = 1$ is returned.

Inductive hypothesis   Assume power$(n) = 2^n$.

Inductive step   power$(n + 1) = 2 *$ power$(n) = 2 * 2^n = 2^{n+1}$

Solutions 3.4

2. **type**
   ```
   Tree3Type = ↑NodeType;
   NodeType = record
     Info : InfoType;
     Ltree, Mtree, Rtree : Tree3Type;
   end;
   ```

   The function itself is the same as the function in Figure 3.7 except for two changes. The header of the function is changed to

   function copy(Tree : Tree3Type) : Tree3Type;

   The two lines underneath the comment {Copy left and right subtrees} are replaced by

   ```
   NewTree↑.Ltree := copy(Tree↑.Ltree);
   NewTree↑.Mtree := copy(Tree↑.Mtree);
   NewTree↑.Rtree := copy(Tree↑.Rtree);
   ```

4. Assume that equal(tree1↑.Info, tree2↑.Info) returns true if tree1↑.Info = tree2↑.Info.

   ```
   function compare(tree1, tree2 : TreeType) : boolean;
     begin
       if ((tree1 = nil) and (tree2 = nil))
         then compare := true
         else if (((tree1 = nil) and (tree2 <> nil)) or
                 ((tree1 <> nil) and (tree2 = nil)))
         then compare := false
         else compare := equal(tree1↑.Info, tree2↑.Info)
           and compare(tree1↑.Ltree, tree2↑.Ltree)
           and compare(tree1↑.Rtree, tree2↑.Rtree)
     end;
   ```

6. **type**
   ```
   ListType = ↑NodeType;
   NodeType = record
     Info : InfoType;
     Next : ListType;
   end;
   ```

```
function copy(List : ListType) : ListType;
  var NewList : ListType;
  begin
    if List = nil then copy := nil
      else begin
        new(NewList);
        NewList↑.Info := List↑.Info;
        NewList↑.Next := copy(List↑.Next);
        copy := NewList
      end
  end;
```

Solutions 3.5

2. $x^0 = 1$
$x^{s(y)} = x \times x^y$

4. $H(n)$, as defined, is primitive recursive.

6. Assume that $+$ and $\times$ have already been defined.
$f(0) = sum_0$ where sum_0 is the value of the variable sum before entering the for loop statement.
$f(s(n)) = f(n) + s(n) \times s(n)$

CHAPTER 3 SOLUTIONS

2. False for $n = 0, 1, 2$, and 3. True for $n = 4$.

Inductive base   For $n = 4$, $3 \times 4 + 2 \leq 16$.

Inductive hypothesis   Assume $3n + 2 \leq n^2$.

Inductive step $3(n + 1) + 2 = 3n + 2 + 3$ Arithmetic
$\leq n^2 + 3$ Inductive hypothesis
The last line is less than $(n + 1)^2 = n^2 + 2n + 1$, since for $n > 4$, $2n + 1 > 3$.

4. Inductive base   Atomic propositions need no connectives and the connectives $\Rightarrow$ and $\neg$ suffice to write all atomic propositions.

Inductive hypothesis   Let C be a logical expression, and assume that all components of C can be written using $\Rightarrow$ and $\neg$ only.

Inductive step
Let $D \equiv A$ and $E \equiv B$ where D and E are written using $\Rightarrow$ and $\neg$ only.
Case a: $C = A \vee B$ can be written as $\neg D \Rightarrow E$.
Case b: $C = A \wedge B$ can be written as $\neg(D \Rightarrow \neg E)$.
Case c: $C = A \Leftrightarrow B$ can be written as
$D \Leftrightarrow E \equiv (D \Rightarrow E) \wedge (E \Rightarrow D) \equiv \neg((D \Rightarrow E) \Rightarrow \neg(E \Rightarrow D))$.
Case d: $C = A \Rightarrow B$ can be written as $D \Rightarrow E$.
Case e: $C = \neg A$ can be written as $\neg D$.

6. **function** count(tree : TreeType) : **integer**;
      begin
        if tree = **nil then** count := 1
          else count := count(tree↑.Ltree) + count(tree↑.Rtree)
      end;

8. **type**
       Operation = (XAndY, XOrY, NotX, XThenY, XIffY);
       NodeKind = (Op, VC);
       ExpTree = ↑ TreeNode;
       TreeNode = **record**
         case NodeType : NodeKind **of**
           Op : (Lptr: ExpTree; OpType: Operation; Rptr: ExpTree);
           VC: (Value: **boolean**)
       end;

      function eval (E: ExpTree): **boolean**;
      {Given an expression that is represented by a binary tree with a root node address of E, this function returns the value (true or false) of the given expression.}
      var
        Left, Right : **boolean**;
      begin
        case E↑.NodeType **of**
         VC: eval := E↑.Value;
         Op: **case** E↑.OpType **of**
           XAndY: eval := eval(E↑.Lptr) **and** eval(E↑.RPtr);
           XOrY: eval := eval(E↑.Lptr) **or** eval(E↑.RPtr);
           NotX: eval := **not** eval(E↑.Lptr);
           XThenY: eval := **not** eval(E↑.Lptr) **or** eval(E↑.RPtr);
           XIffY:
             begin
              Left := eval(E↑.Lptr);
              Right := eval(E↑.Rptr);
              eval := (**not** Left **or** Right) **and** (**not** Right **or** Left)
             end;
          end
        end
      end;

10. **Inductive base**   For $n = 1$, $a_1 < b_1$.

    Inductive hypothesis   Assume $\sum_{i=1}^{n} a_i < \sum_{i=1}^{n} b_i$.

    Inductive step $\sum_{i=1}^{n+1} a_i = \left(\sum_{i=1}^{n} a_i \right) + a_{n+1}$

$$< \left(\sum_{i=1}^{n} b_i \right) + a_{n+1} \quad \text{Inductive hypothesis}$$

$$< \left(\sum_{i=1}^{n} b_i \right) + b_{n+1} \quad \text{since } a_{n+1} < b_{n+1}$$

$$= \sum_{i=1}^{n+1} b_i$$

12. Inductive base   All atomic expressions are contingent.

Inductive hypothesis   Let C be an expression and assume all subexpressions of C are contingent.

Inductive step   If C is of the form $\neg A$, then C is obviously contingent. If C is of the form $A \vee B$, then A and B cannot share variables because each variable appears only once. Hence, by the inductive hypothesis, there is an assignment for A which makes A true and an assignment for B which makes B true. Under this assignment, C becomes true. Similarly, by choosing an assignment making A and B both false, we find an assignment making C false. Hence, $C = A \vee B$ is contingent. A similar argument shows that $C = A \wedge B$, $C = A \Rightarrow B$, and $C = A \Leftrightarrow B$ are all contingent.

14. **(a)**
```
procedure DisplayHigh (Root : TreeType);
   begin
      if Root <> nil then
         begin
            if Root↑.Info.AmountOwing > 1000 then writeln(Root↑.Info.Name);
            DisplayHigh(Root↑.Ltree);
            DisplayHigh(Root↑.Rtree)
         end
   end;
```
(b)
```
function CopyHigh (Tree : TreeType) : TreeType;
   var
      NewTree : TreeType;
   begin
      if Tree = nil then CopyHigh := nil
         else
            begin
               if Tree↑.Info.AmountOwing > 1000 then
                  begin
                     new(NewTree);
                     NewTree↑.Info.Name := Tree↑.Info.Name;
                     NewTree↑.Info.AmountOwing := Tree↑.Info.AmountOwing;
                     CopyHigh := NewTree
                  end;
               NewTree↑.Ltree := CopyHigh(Tree↑.Ltree);
               NewTree↑.Rtree := CopyHigh(Tree↑.Rtree)
            end
   end;
```

Solutions 4.1

2.
```
times(X, 0, 0).
times(X, 1, X).
times(X, Y, Z) :- times(Y, X, Z).
```

4. likes(jane, X) :- man(X), tall(X), dark(X).
likes(joe, X) :- woman(X), tall(X), blonde(X).
likes(suzy, X) :- man(X), tall(X), blonde(X).

6. (**a**) X = abc, Cde = cde
(**b**) X = jim, Y = beth, Z = john
(**c**) X = mary, Y = beth, Z = mary

8. brotherinlaw(X, Y) :- brother(X, Z), married(Z, Y).
brotherinlaw(X, Y) :- married(X, Z), sister(Z, Y).
sisterinlaw(X, Y) :- sister(X, Z), married(Z, Y).
sisterinlaw(X, Y) :- married(X, Z), brother(Z, Y).

10. sentence(X, Y, Z) :- article(X), noun(Y), verb(Z).

Solutions 4.2

2. Charles and Elizabeth are presumably atoms and must start with a lower-case letter.

Solutions 4.3

2. id (X, Y) :- employee (name (_, X), Y).

4. convert (H, M, Result) :- Result is $60 * H + M$

6. $3 + 5$ is 8

8. howmuch(X, Y) :- cost(X, Y), nl, write('The '),
write(X), write(' costs '), write(Y).

Solutions 4.4

2. power (0, X, 1).
power (N, X, Y) :- Nless1 is $N - 1$,
power (Nless1, X, YNless1),
Y is $X * $YNless1.

4. member (3, [1, 2, 3, 4]) :- member (3, [2, 3, 4]).
member (3, [2, 3, 4]) :- member (3, [3, 4]).
member (3, [3, 4]).

6. multiply (Y, [], []).
multiply (Y, [A | As], [NewA | NewAs])
:- NewA is $Y * A$,
multiply (Y, As, NewAs).

8. howoften(X, [], 0).
howoften(X, [X | T], N) :- howoften(X, T, N2), N is $N2 + 1$.
howoften(X, [H | T], N) :- howoften(X, T, N).

Solutions 4.5

2. exclusive ([], _).
exclusive ([A | As], Bs) :- \+ member (A, Bs),
                                     exclusive (As, Bs).

4. find (X, Y) :- carpenter (X, Y).
find (_, _) :- write ('no carpenter in city'), nl.

6. common ([], B, []).
common ([A|As], B, [A|C]) :- member (A, B), !,
                                     common (As, B, C).
common ([_ | As], B, C) :- common(As, B, C).

This procedure assumes that there are no duplicates in either A or B. If there are duplicates in these lists, duplicates may exist in the common list as well.

Solutions 4.6

2. movenegin : As in text

normform (X, Y) :- movenegin (X, Z), moveconin (Z, Y).
moveconin (X, X) :- literal (X).
moveconin (A and (B or C), D or E) :-moveconin (A and B, D),
                                               moveconin (A and C, E).
moveconin ((A or B) and C, D or E) :-moveconin (A and C, D),
                                               moveconin (B and C, E).
moveconin (A and B, D and E) :- moveconin (A, D), moveconin (B, E).
moveconin (A or B, D or E) :- moveconin (A, D), moveconin (B, E).

CHAPTER 4 SOLUTIONS

2. morebooks(X, Y) :- book(Z, X), book(Z, Y), X \== Y.

4. age(X, Y, Z) :- birth(X, U), Y is Z − U.

This procedure will return the age of X in year Z, and it can also return the names of all people who were age Y in year Z. This procedure cannot, however, return the year in which X is Y years old.

6. supphas(X, Y) :- item(X, Z), supplier(Z, Y).

8. printfunction :- doX(1).
doX(21).
doX(X) :- Z is X + X * X, Y is X + 1, write(Z), nl, doX(Y).

10. fullname([], []).
fullname([Y | Ys], [[X | Y] | XsYs]) :- name(X, Y), fullname(Ys, XsYs).

12. numbered(X) :- numberedr(X, 0).
numberedr([], _).
numberedr([H | T], X) :- Y is X + 1, write(Y), write(' '), write(H), nl,
                                numberedr(T, Y).

14. bill([ ], 0).
bill([item(X, Y, Z) | T], Bill) :- bill(T, Bill2), Bill is Bill2 + Y * Z.

16. translate([ ], [ ]).
translate([H | T], [H2 | T2]) :- french(H, H2), translate(T, T2).

18. checkcon(X, Y, Z) :- X = Y; connected(X, Y);
                                connected(X, A), connected(B, Y),
                                \+ member(A, Z), \+ member(B, Z),
                                checkcon(A, B, [X | [Y | Z]]).
connection(X, Y) :- checkcon(X, Y, [ ]).

This solution uses two Prolog definitions. The definition for connection is essentially just an interface to the recursive code in checkcon. There are three cases to consider in the solution: The destination city is the same as the start city, the destination city is directly connected to the start city, or there exist two cities, say A and B, such that A is directly connected to the start, and B is directly connected to the destination, and A is connected to B.

20. derivative(A, 0) :- constant(A).
derivative(x, 1).
derivative(A+B, DA) :- derivative(A, DA), constant(B).
derivative(A+B, DB) :- derivative(B, DB), constant(A).
derivative(A+B, DA+DB) :- derivative(A, DA), derivative(B, DB).
derivative($-$A, $-$DA) :- derivative(A, DA).
derivative(A$-$B, DA) :- derivative(A, DA), constant(B).
derivative(A$-$B, $-$DB) :- derivative(B, DB), constant(A).
derivative(A$-$B, DA$-$DB) :- derivative(A, DA), derivative(B, DB).
derivative(A*B, DA*B) :- derivative(A, DA), constant(B).
derivative(A*B, A*DB) :- derivative(B, DB), constant(A).
derivative(A*B, A*DB+DA*B) :- derivative(A, DA), derivative(B, DB).
derivative(A/B, DA/B) :- derivative(A, DA), constant(B).
derivative(A/B, $-$A/(DB*DB)) :- derivative(B, DB), constant(B).
derivative(A/B, (B*DA$-$A*DB)/(B*B)) :- derivative(A, DA), derivative(B, DB).
constant(A) :- atomic(A), A \== x.
constant(A+B) :- constant(A), constant(B).
constant(A$-$B) :- constant(A), constant(B).
constant(A*B) :- constant(A), constant(B).
constant(A/B) :- constant(A), constant(B).
constant($-$A) :- constant(A).

There are several different cases that are accounted for with the derivative predicate, involving both simple and complex expressions. As well, the predicate constant is used to combine terms.

Solutions 5.1

2. **(a)** F   **(b)** T   **(c)** T   **(d)** T

4. $\#\emptyset = 0, \#\{a, b\} = 2, \#\{999\} = 1, \#\{1, 2, 3\} = 3$

6. $X \cup Y = \{1, 2, 3, 4, 5\}, \sim(X \cup Y) = \{0, 6, 7, 8, 9\},$
$\sim X = \{0, 1, 5, 6, 7, 8, 9\}, \sim Y = \{0, 3, 4, 6, 7, 8, 9\}, \sim X \cup \sim Y = \{0, 1, 3, 4, 5, 6, 7, 8, 9\},$

$Y \cap Z = \{2,5\}$, $X \cup (Y \cap Z) = \{2,3,4,5\}$,
$X \cup Y = \{1,2,3,4,5\}$, $(X \cup Y) \cap Z = \{2,5\}$

8. $\sim A = \{Saturday, Sunday\}$

10. Omitted

12. $(\sim(A \cup \sim(B \cup C) \cap \sim A \cap \sim(B \cap C) \cup A))$

$= \sim(A \cup (\sim(B \cup C) \cap \sim A \cap \sim(B \cap C)) \cup A)$ Insert parentheses

$= \sim(A \cup (\sim(B \cup C) \cap \sim A \cap \sim(B \cap C)))$ Idempotent

$= \sim A \cap \sim(\sim(B \cup C) \cap \sim A \cap \sim(B \cap C))$ De Morgan

$= \sim A \cap ((B \cup C) \cup A \cup (B \cap C))$ De Morgan

$= \sim A \cap (A \cup B \cup C \cup (B \cap C))$ Commutative

$= \sim A \cap (A \cup B \cup C)$ Absorption

$= (\sim A \cap A) \cup (\sim A \cap B) \cup (\sim A \cap C)$ Distributive

$= (\sim A \cap B) \cup (\sim A \cap C)$ Exclusion and Identity

$= \sim A \cap (B \cup C)$ Distributive

14. $A \cap A = A : x \in A \wedge x \in A \equiv x \in A$

$A \cap \emptyset = \emptyset : x \in A \wedge x \in \emptyset \equiv x \in \emptyset$

$A \cap E = A : x \in A \wedge x \in E \equiv x \in A$

$A \cup E = E : x \in A \vee x \in E \equiv x \in E$

16. If $C \subseteq A$, then every element of C is an element of A, which means $x \in C \Leftrightarrow x \in C \wedge x \in A$,
or $C = C \cap A$. Hence, $A \cap (B \cup C) = (A \cap B) \cup (A \cap C) = (A \cap B) \cup C$.

18. Omitted

20. $x \in A \wedge x \in \sim A \equiv \text{F} \equiv x \in \emptyset$

$x \in A \vee x \in \emptyset \equiv x \in A \vee \text{F} \equiv x \in A$

22. $(A \cap B) \cap (C \cap D) = A \cap (B \cap (C \cap D))$ Associative

$= A \cap ((C \cap D) \cap B)$ Commutative

$= A \cap (C \cap (D \cap B))$ Associative

$= (A \cap C) \cap (D \cap B)$ Associative

$= (A \cap C) \cap (B \cap D)$ Commutative

Solutions 5.2

2. $\#(A^2) = 16$, $\#(A^3) = 64$, $\#(2^A) = \#(\mathbb{P}A) = 2^4 = 16$

4. $A^2 = \{(3,3), (3,5), (3,7), (5,3), (5,5), (5,7), (7,3), (7,5), (7,7)\}$

$B^2 = \{(a,a), (a,b), (b,a), (b,b)\}$

$A \times B = \{(3,a), (3,b), (5,a), (5,b), (7,a), (7,b)\}$

$B \times A = \{(a,3), (a,5), (a,7), (b,3), (b,5), (b,7)\}$

6. $A \times (B \cap C) = \{(x,y) | x \in A \wedge (y \in B \wedge y \in C)\}$

$= \{(x,y) | x \in A \wedge y \in B \wedge y \in C\}$

$(A \times B) \cap (A \times C) = \{(x,y) | (x \in A \wedge y \in B) \wedge (x \in A \wedge y \in C)\}$

$= \{(x,y) | x \in A \wedge y \in B \wedge x \in A \wedge y \in C\}$

$= \{(x,y) | x \in A \wedge y \in B \wedge y \in C\}$

8. $A \cup B = \{1,2,3,4\}$

$\mathbb{P}(A \cup B) = \{\emptyset, \{1\}, \{2\}, \{3\}, \{4\}, \{1,2\}, \{1,3\}, \{1,4\}, \{2,3\}, \{2,4\}, \{3,4\},$
$\{1,2,3\}, \{1,2,4\}, \{1,3,4\}, \{2,3,4\}, \{1,2,3,4\}\}$

10. The statement is false because, for example, 2.5 is neither even nor odd. Correct: $\forall x(x : \mathbb{Z} \bullet (even(x) \vee odd(x)))$

Solutions 5.3

2. ran $S = \{0, 1, 4, 9, \ldots\}$, ran $T = \{0, 2, 4, 6, 8, \ldots\}$

4. $R \cup \breve{R}$

6. $S \circ P$

8. $R \circ S = \{(1, A), (1, B), (1, D)\}$

10. $xR^2y \equiv (x = y - 2)$
$xRy \vee xR^2y \equiv (x = y - 1 \vee x = y - 2)$

Solutions 5.4

2. Empty relation: irreflexive, symmetric, antisymmetric, transitive, strict partial order
Universal relation: reflexive, symmetric, transitive, equivalence

4. The identity relation is both symmetric and antisymmetric. More generally, any relation in which xRy is true only if $x = y$ is both symmetric and antisymmetric.

6. Reflexive: $xRx \wedge xSx \Rightarrow x(R \cap S)x$
Symmetric: $(xRy \Rightarrow yRx) \wedge (xSy \Rightarrow ySx)$
$\Rightarrow (xRy \wedge xSy) \Rightarrow (yRx \wedge ySx)$
$\equiv x(R \cap S)y \Rightarrow y(R \cap S)x$
Transitive: $(xRy \wedge yRz \Rightarrow xRz) \wedge (xSy \wedge ySz \Rightarrow xSz)$
$\Rightarrow ((xRy \wedge xSy) \wedge (yRz \wedge ySz) \Rightarrow xRz \wedge xSz)$
$\Rightarrow x(R \cap S)y \wedge y(R \cap S)z \Rightarrow x(R \cap S)z$

8. Symmetric: $x + y = 10 \Rightarrow y + x = 10$

10. $M_{R} = \begin{bmatrix} 1 & 1 & 1 \\ 0 & 1 & 1 \\ 1 & 0 & 1 \end{bmatrix}$

$M_{R^2} = \begin{bmatrix} 1 & 0 & 1 \\ 1 & 1 & 0 \\ 1 & 1 & 1 \end{bmatrix} \begin{bmatrix} 1 & 0 & 1 \\ 1 & 1 & 0 \\ 1 & 1 & 1 \end{bmatrix} = \begin{bmatrix} 1 & 1 & 1 \\ 1 & 1 & 1 \\ 1 & 1 & 1 \end{bmatrix}$

$M_{R^3} = \begin{bmatrix} 1 & 1 & 1 \\ 1 & 1 & 1 \\ 1 & 1 & 1 \end{bmatrix} \quad M_{R \circ R} = \begin{bmatrix} 1 & 1 & 1 \\ 1 & 1 & 1 \\ 1 & 1 & 1 \end{bmatrix}$

12. Reflexive: $xSx \Rightarrow x\breve{S}x$
Symmetric: $(xSy \Rightarrow ySx) \Rightarrow (y\breve{S}x \Rightarrow x\breve{S}y)$
Transitive: $(xSy \wedge ySz \Rightarrow xSz) \Rightarrow (y\breve{S}x \wedge z\breve{S}y \Rightarrow z\breve{S}x) \equiv z\breve{S}y \wedge y\breve{S}x \Rightarrow z\breve{S}x$
This holds for all x, y, and z.

14. (a) Irreflexive, antisymmetric
(b) Irreflexive, antisymmetric, transitive; strict partial order

(**c**) Irreflexive, symmetric

(**d**) Irreflexive, antisymmetric

(**e**) Irreflexive, antisymmetric

(**f**) Irreflexive, antisymmetric, transitive; strict partial order

(**g**) Reflexive, symmetric, transitive as long as half-siblings are excluded; equivalence in this case

(**h**) Reflexive, antisymmetric transitive: weak partial order

16. $\{\mathbb{N}, \leq\}$ not well founded: $1 \leq 2 \leq 3 \ldots$ is a descending sequence
   $\{\mathbb{N}, \geq\}$ well founded

18. (**a**) r (x divides x), t (if x divides y and y divides z, then x divides z)

(**b**) r, s, t (if inlaws and remarriage are excluded)

(**c**) r, s, t (if half-brothers are excluded)

(**d**) r, s (not transitive because father and mother are typically not related.)

20. **Inductive base**   For $n = 1$, $R^n = R$ and $xRy \Rightarrow xRy$.

Inductive hypothesis   Assume $xR^n y \Rightarrow xRy$.

Inductive step
   $xR^{n+1}y \equiv xRz \wedge zR^n y$
   $zR^n y \Rightarrow zRy$ by inductive hypothesis
   Also: $A \Rightarrow C$ logically implies $A \wedge B \Rightarrow C \wedge B$
   Hence: $xRz \wedge zR^n y$
       $\Rightarrow xRz \wedge zRy$ because $zR^n y \Rightarrow zRy$
       $\Rightarrow xRy$ transitivity

CHAPTER 5 SOLUTIONS

2. $((A \cup B) \cap \sim(C \cup A)) \cup ((C \cap B) \cup A) = ((A \cup B) \cap \sim C \cap \sim A) \cup (C \cap B) \cup A$
   $= (((\sim A \cap A) \cup (\sim A \cap B)) \cap \sim C) \cup (B \cap C) \cup A$
   $= (\sim A \cap B \cap \sim C) \cup (B \cap C) \cup A$
   $= ((A \cup \sim A) \cap (A \cup B) \cap (A \cup \sim C)) \cup (B \cap C)$
   $= ((A \cup B) \cap (A \cup \sim C)) \cup (B \cap C)$
   $= A \cup (B \cap \sim C) \cup (B \cap C)$
   $= A \cup B$

4. $A = \{0, 1, 2, 3\}$ $\#A = 4$

6. (**a**) $A \cap B \subseteq A$ by (5.6)
       $A \subseteq A \cup B$ (Section 5.1.5)
       $A \cap B \subseteq A \cup B$

(**b**) This follows from the definition of difference, since some elements are removed from A to form $A - B$.

(**c**) $x \in A \Rightarrow x \in C$, $x \in B \Rightarrow x \in C$
       Hence: $x \in A$ or $x \in B \Rightarrow x \in C$

8. (**a**) $A \circ A^{\smallsmile} = \{(a, a), (b, b), (c, c), (a, c), (c, a)\}$

(**b**) $(A \circ A) - A \circ A^{\smallsmile}$
       $= \{(a, a), (a, c), (b, b), (b, d), (c, a), (c, c)\} - \{(a, a), (b, b), (c, c), (a, c), (c, a)\}$
       $= \{(b, d)\}$

    (c) $(A \circ A) \cup A = \{(a,a),(a,b),(a,c),(b,a),(b,b),(b,c),(b,d),(c,a),(c,b),(c,c),(c,d)\}$

    (d) $A \circ B = \{(a,2),(b,1),(b,3),(b,2),(c,2)\}$

10. R is reflexive, symmetric, and transitive if R is an identity relation on some set A.

12. **(a)** Equivalence, weak partial order

    (b) None

    (c) Equivalence

    (d) None (equivalence requires at least $(3,3)$)

    (e) None

14. Since S is a universal relation, xSy for all x and y.

    Hence: $xRx \land xSy$ for all x and y.

16. $S \cup B$

18. Let R be on $\{0,1\}$. $R = \{(1,1)\}$

20. For all $x \in \text{dom } A$, there is a y such that xAy holds. Hence, there is a y such that $xAy \land yĂx$, which implies $x(A \circ Ă)x$. If $x \notin \text{dom } A$, then there is no such y.

Solutions 6.1

2. $\{(1,(2,3)),(2,(3,4)),(3,(1,4)),(4,(2,4))\}$ is a function with domain $\{1,2,3,4\}$ and range $\{(2,3),(3,4),(1,4),(2,4)\}$.

    $\{((1,2),3),((2,3),4),((3,3),2)\}$ is a function with domain $\{(1,2),(2,3),(3,3)\}$ and range $\{2,3,4\}$.

    $\{(1,(2,3)),(2,(3,4)),(1,(2,4))\}$ is *not* a function: 1 is associated with both $(2,3)$ and $(2,4)$.

    $\{(1,(2,3)),(2,(2,3)),(3,(2,3))\}$ is a function with domain $\{1,2,3\}$ and range $(2,3)$.

4. $(\lambda a.(\lambda b.(x+2b))3)4 = \lambda b.(x+2b)3 = x + 2 \times 3 = x + 6$

    $(\lambda x.(\lambda y.((x+y)(x-y)))2)2 = (\lambda y.(2+y)(2-y))2 = (2+2)(2-2) = 0$

    $(\lambda x.(\lambda y.(x+y))a)b = (\lambda y(b+y))a = b + a$

    $(\lambda x.4)2 = 4$

6. $f \circ g(x) = f(x/y + 2) = 3(x/y + 2)^2 + y$

    $g \circ f(x) = g(3x^2 + y) = (3x^2 + y)/y + 2$

    $f \circ h(x) = f((x+y)^2) = 3(x+y)^4 + y$

    $f \circ g \circ h(x) = f(g(h(x))) = f(g((x+y)^2)) = f((x+y)^2/y + 2) = 3((x+y)^2/y + 2)^2 + y$

8. **(a)** one-to-one, not onto   **(c)** neither

    (b) neither   **(d)** onto

10. $\#X \leq \#Y$

12. Every function $f : A \to A$ must associate a unique $y \in A$ with $x \in A$. Hence, each x must match with a y.

14. $y = x^3 - 2$ $y + 2 = x^3$ $x = \sqrt[3]{y+2}$

16. Let $f = \{(1,2),(2,3),(3,4),(4,1)\}$

    $f^2 = \{(1,3),(2,4),(3,1),(4,2)\}$

    $f^3 = \{(1,4),(2,1),(3,2),(4,3)\}$

    $f^{-1} = \{(2,1),(3,2),(4,3),(1,4)\}$

    $f \circ f^{-1} = \{(1,1),(2,2),(3,3),(4,4)\}$

18. $2, 6, 4$

20. $1 \bmod 3 = 1$ $-1 \bmod 3 = 2$
      $2 \bmod 3 = 2$ $-2 \bmod 3 = 1$
      $3 \bmod 3 = 0$ $-3 \bmod 3 = 0$
      $4 \bmod 3 = 1$ $-4 \bmod 3 = 2$

        $\vdots$ $\vdots$

22. $x < \lfloor x \rfloor + 1$
      $x^2 < \lfloor x \rfloor^2 + 2\lfloor x \rfloor + 1$
      $x^2 - \lfloor x \rfloor^2 < 2\lfloor x \rfloor + 1$

Solutions 6.2

2. $(i - 1)n_2 n_3 + (j - 1)n_3 + k = m$
     Inverse: $i = \lfloor m/(n_2 n_3) \rfloor + 1$
        $j = \lfloor [m - (i - 1)n_2 n_3]/n_3 \rfloor + 1$
        $k = m - (i - 1)n_2 n_3 - (j - 1)n_3$

4.   Use the bijection $f(x) = 10x$.

6. $(6 - 1)!$ (1 person is fixed).

8. $\frac{2^{10} - 2}{2} = 511$: The power set of $\{1, 2, \ldots, 10\}$ contains 2^{10} elements. Each element of the power set has a complement and together they divide the group of people into 2 sets. The elements of the power set include the entire set and the empty set and we deduct 2 because these two sets must be excluded. We then divide the result by 2 in order to exclude equivalent divisions. For example, the set $\{1, 2, 3, 4, 5\}$ (along with its complement $\{6, 7, 8, 9, 10\}$) and the set $\{6, 7, 8, 9, 10\}$ (along with its complement $\{1, 2, 3, 4, 5\}$) both represent the same division.

10.   The homomorphism is $t = \{(a, 1), (b, 1), (c, 2), (d, 3)\}$.

12.   The function t is not isomorphic; $t(11 + 9) \neq t(11) + t(9)$ or : $2 \neq 2 + 9$

Solutions 6.3

2.   Use (6.18) to find $n_c = 10 + 1 + 1 + 14 = 26$.

4.   For every c one can find an n_c such that $cn! < (n + 1)!$ for $n > n_c$. In fact, $n_c = c - 1$ will do.

6.   Assume: 1. copy(A, B) copies matrix A into B.
               2. mult(A, B, C) multiplies A with B, yielding C.
               3. The type matrix has been defined.

```
procedure Aton (A : matrix; var Result : matrix; n : integer);
var
   Res, Res1 : matrix;
    m : integer;
begin
   if n = 1 then copy(A, Result)
```

```
    else
      begin
        m := n div 2;
        Aton(A, Res, m);
        mult(Res, Res, Res1);
        if n - 2 * m > 0 then mult(A, Res1, Result)
          else copy(Res1, Result);
      end
end;
```

Multiplications : n even : $T(n) = T(n/2) + 1$
$\qquad\qquad$ n odd : $T(n) = T(\lfloor n/2 \rfloor) + 2$
$\qquad\qquad\qquad T(1) = 0$
$\qquad\qquad\qquad T(2) = 1$

3	$T(1) + 2 = 2$		10	$T(5) + 1 = 4$
4	$T(2) + 1 = 2$		11	$T(5) + 2 = 5$
5	$T(2) + 2 = 3$		12	$T(6) + 1 = 4$
6	$T(3) + 1 = 3$		13	$T(6) + 2 = 5$
7	$T(3) + 2 = 4$		14	$T(7) + 1 = 5$
8	$T(4) + 1 = 3$		15	$T(7) + 2 = 6$
9	$T(4) + 2 = 4$		16	$T(8) + 1 = 4$

8. Scheduling is NP-complete, and if the company can solve the problem, it would provide a solution method for all NP-problems, which is unlikely.

Solutions 6.4

2. $2^n > 0.9(5 \times 1.5^n + 2^n)$
$0.1 \times 2^n > 0.9 \times 5 \times 1.5^n$
$2^n > 45 \times 1.5^n$
$(2/1.5)^n > 45$
$n > \log 45 / \log(2/1.5) = 13.2$
Hence, $n \geq 14$

4. (a) $H_2 = 4 + 0 = 4$
$\qquad H_3 = 6 + 20 = 26$
$\qquad H_4 = 8 + 130 - 24 = 114$

$\quad$ **(b)** Particular solution:
$\qquad H_n^p = d_0 + d_1 n = 2n + 5H_{n-1} - 6H_{n-2}$
$\qquad\qquad = 2n + 5(d_0 + d_1(n-1)) - 6(d_0 + d_1(n-2))$
$\qquad d_0 + d_1 n = 2n - d_0 - d_1 n + 7d_1$
$\qquad\quad d_1 = 1$
$\qquad\quad d_0 = 3.5$
$\qquad$ Complementary equation: $H_n = 5H_{n-1} - 6H_{n-2}$. Characteristic roots, 3 and 2.
$\qquad H_n = 3.5 + n + a_1 3^n + a_2 2^n$
$\qquad 0 = 3.5 + a_1 + a_2 \quad a_1 = -(3.5 + a_2)$

$$0 = 4.5 + 3a_1 + 2a_2 \quad 0 = 4.5 - 3(3.5 + a_2) + 2a_2$$
$$a_2 = -6 \quad a_1 = 2.5$$
$$H_n = 3.5 + n + 2.5 \times 3^n - 6 \times 2^n$$

(c) $H_0 = 3.5 + 2.5 - 6 = 0$
$H_1 = 4.5 + 2.5 \times 3 - 6 \times 2 = 0$
$H_2 = 5.5 + 2.5 \times 9 - 6 \times 4 = 4$
$H_3 = 6.5 + 2.5 \times 27 - 6 \times 8 = 26$
$H_4 = 7.5 + 2.5 \times 81 - 6 \times 16 = 114$

Solutions 6.5

2. normalgrade e1 e2 e3 = (e1 + e2 + e3) / 3
   adjgrade low high1 high2 = (0.5 * low + high1 + high2) / 2.5
   avg n1 n2 = (n1 + n2) / 2
   grade e1 e2 e3 = adjgrade e1 e2 e3,   **if** e1 < 0.8 * avg e2 e3
   $\qquad\qquad\quad$ = adjgrade e2 e1 e3,   **if** e2 < 0.8 * avg e1 e3
   $\qquad\qquad\quad$ = adjgrade e3 e1 e2,   **if** e3 < 0.8 * avg e1 e2
   $\qquad\qquad\quad$ = normalgrade e1 e2 e3,   **otherwise**

4. twice (this 2) 4   yields
   twice (2 + 2 * y) 4 =(2 + 2 * y)((2 + 2 * y)4)
   $\qquad\qquad\qquad\;$ = (2 + 2 * y)(2 + 2 * 4)
   $\qquad\qquad\qquad\;$ = (2 + 2 * y)10
   $\qquad\qquad\qquad\;$ = 2 + 2 * 10 = 22

6. constructlist i a b c
   $\qquad$ = (a + b * i + c * i^2) : constructlist (i + 1) a b c,   **if** i <= 10
   $\qquad$ = [ ],   **otherwise**
   list a b c = constructlist 1 a b c

CHAPTER 6 SOLUTIONS

2. Relations 1 and 3: partial functions, total functions, onto, one-to-one, one-to-one onto. Relation 2 not a function.

4. $f(g(x)) = (x + 4)^2 - 2$
   $g(f(x)) = x^2 - 2 + 4 = x^2 + 2$
   f: neither injective nor surjective
   g: injective, surjective, bijective
   $f \circ g$: neither
   $g \circ f$: neither

6. f, g: total: then $z = f(g(x))$ gives a unique z for each x.

8. (a) There are n ways to choose $y \in Y$.   (b) $m \times n$

10. (a) $\{(a, 3), (c, 2), (b, 1)\}$: onto, one-to-one
    (b) $\{(a, 3), (c, 2), (b, 1)\}$: onto, one-to-one
    (c) $\{(c, 1), (b, 3), (a, 3)\}$: into, not one-to-one

12. **(a)** $\lambda g.(g(g\,2))(\lambda x.(x+3)) = (\lambda x.(x+3))(\lambda x.(x+3)2)$
$= (\lambda x.(x+3))(2+3) = 5+3 = 8$
(b) $\lambda h.(h\,2)(\lambda x.(x+3)) = \lambda x.(x+3)2 = 5$

14. $t(f(x_1, x_2, \ldots, x_n)) = g(t(x_1), t(x_2), \ldots, t(x_m))$

16. $(n+3)(4n^2+5) \in \Theta(n^3)$
$n^3 + 100/n \in \Theta(n^3)$
$105 + (n^3/2 + n)/(1 + 1/n) \in \Theta(n^3)$

18. **(a)** $H_n = 2H_{n-1} - 0.99H_{n-2}$
$x_1 = 1.1 \quad x_2 = 0.9$
$H_n = b_1(1.1)^n + b_2(0.9)^n$
$H_0 = 1 = b_1 + b_2$
$H_1 = 1 = b_1(1.1) + b_2(0.9)$
$b_1 = 0.5 \quad b_2 = 0.5$
$H_n = (0.5)(1.1)^n + (0.5)(0.9)^n$
(b) $H_n = b_1(1.1)^n + b_2(0.9)^n$
$H_0 = 1 = b_1 + b_2$
$H_3 = 1.5 = b_1(1.1)^3 + b_2(0.9)^3$
$1.5 = b_1(1.331) + (1 - b_1)(0.729)$
$b_1 = 1.28 \quad b_2 = -0.28$
$H_n = (1.28)(1.1)^n + (-0.28)(0.9)^n$

20. update x [ ] = x
update ((namedue, due) : listdue) ((namepaid, paid) : listpaid)
    = (namedue, due−paid) : update listdue listpaid,   **if** namedue = namepaid
    = (namedue, due) : update listdue ((namepaid, paid):listpaid), **otherwise**

22. square x = x ∗ x
distance x1 y1 x2 y2 = sqrt(square(x2 − x1) + square(y2 − y1))
lenpoly [ ] = 0
lenpoly ((x1, y1) : [ ]) = 0
lenpoly ((x1, y1) : ((x2, y2) : rest)) = distance x1 y1 x2 y2 + lenpoly ((x2, y2) : rest)

24. From the definition of homomorphism, $t(x \circ y) = t(x) \Box t(y)$ and $t(y \circ x) = t(y) \Box t(x)$.
If $\circ$ is commutative, $t(x \circ y) = t(y \circ x)$, and this implies $t(x) \Box t(y) = t(y) \Box t(x)$. Consequently, $\Box$ is commutative. The unit element of $\Box$ is $t(e)$:
$$x \circ e = x$$
$$t(x \circ e) = t(x)$$
$$t(x \circ e) = t(x) \Box t(e) = t(x)$$
Consequently, $t(e)$ is the identity element for $\Box$.

Solutions 7.2

2. To show that the digraphs given in Figure 7.15(b) and (b′) are isomorphic, we must show that there exists a mapping from the nodes in b to the nodes in b′ such that if an edge exists between two nodes in b, an edge exists between the corresponding nodes in b′. Such a mapping is given below:

$v_5 \rightarrow u_4 \quad v_2 \rightarrow u_3 \quad v_3 \rightarrow u_5 \quad v_1 \rightarrow u_2 \quad v_4 \rightarrow u_1$

Under this mapping, the edges

$$(v_1, v_4), (v_1, v_2), (v_2, v_5), (v_3, v_2), (v_3, v_4), (v_4, v_5), (v_5, v_3), (v_5, v_1)$$

are mapped into

$$(u_2, u_1), (u_2, u_3), (u_3, u_4), (u_5, u_3), (u_5, u_1), (u_1, u_4), (u_4, u_5), (u_4, u_2)$$

This is only one mapping, and, in fact, any mapping from $\{v_2, v_4\}$ to $\{u_1, u_3\}$, from $\{v_1, v_3\}$ to $\{u_2, u_5\}$, and from $\{v_5\}$ to $\{u_4\}$ will work to show that the graphs are isomorphic.

4. One possible mapping from nodes in (a) to nodes in (b) is as follows:

$$u_1 \rightarrow v_1 \quad u_3 \rightarrow v_2 \quad u_5 \rightarrow v_3 \quad u_2 \rightarrow v_4 \quad u_4 \rightarrow v_5 \quad u_6 \rightarrow v_6$$

Under this mapping, the edges

$$(u_1, u_6), (u_1, u_4), (u_1, u_2), (u_5, u_6), (u_5, u_2), (u_5, u_4), (u_3, u_2), (u_3, u_4), (u_3, u_6)$$

are mapped into

$$(v_1, v_6), (v_1, v_4), (v_1, v_4), (v_3, v_6), (v_3, v_4), (v_3, v_5), (v_2, v_4), (v_2, v_5), (v_2, v_6)$$

Since in graph (a) there exists an edge from all nodes in $\{u_1, u_3, u_5\}$ to all nodes in $\{u_2, u_4, u_6\}$, and in graph (b) there exists an edge from all nodes in $\{v_1, v_2, v_3\}$ to all nodes in $\{v_4, v_5, v_6\}$, any mapping from $\{u_1, u_3, u_5\}$ to $\{v_1, v_2, v_3\}$ and from $\{u_2, u_4, u_6\}$ to $\{v_4, v_5, v_6\}$ is valid for showing that (a) and (b) are isomorphic.

6. By definition, each node in a complete digraph is connected to every other node in the digraph. If the digraph contains n nodes, each node is therefore connected to $n - 1$ other nodes; hence, there are $n - 1$ edges originating from each node. Since there are n nodes, there must be $n(n - 1)$ edges.

Solutions 7.3

2.

Node	Indegree	Outdegree
v_1	1	1
v_2	1	2
v_3	1	2
v_4	2	2
v_5	2	0

The following are the elementary cycles contained in the graph:

$$\langle v_1, v_2, v_4, v_1 \rangle$$
$$\langle v_1, v_2, v_3, v_4, v_1 \rangle$$

If either (v_4, v_1) or (v_1, v_2) are removed the graph becomes acyclic. All other nodes in the digraph are reachable from v_1, v_2, v_3, and v_4. From v_5, however, no other node is reachable.

4. The digraph in Figure 7.24 is not strongly connected because there are no paths from node v_5 to any other node in the graph. However, the digraph is unilaterally connected as there is a path between each pair of nodes in at least one direction. The digraph in Figure 7.23 is not strongly connected as there is no edge from the node set $\{v_3, v_4\}$ to the node set $\{v_1, v_2\}$; however, the digraph is unilaterally connected.

6. The diameter in both cases is ∞ since there are nodes in the graph that cannot be reached from other nodes.

8. If a node is an isolated node, it appears alone, while if a node u is adjacent to v, then both u and v appear in its weak component. If now the node v is connected to another node, then that node will also be included in the same weak component. Hence, no node can appear in two distinct weak components.

Solutions 7.4

2. $(I \vee A) \odot (I \vee A) = I \odot I \vee I \odot A \vee A \odot I \vee A \odot A$

$$= I \vee A \vee A \vee A \odot A$$

$$= I \vee A \vee A^{(2)} \ \ (\text{since } A \vee A = A)$$

The proof for any positive integer r can be shown using mathematical induction.

Inductive Base   As shown above, $(I \vee A)^{(2)} = I \vee A \vee A^{(2)}$.

Inductive Hypothesis   Assume $(I \vee A)^{(r-1)} = I \vee A \vee \ldots \vee A^{(r-1)}$.

Inductive Step $(I \vee A)^{(r)} = (I \vee A)^{(r-1)}(I \vee A)$

$$= (I \vee A \vee \ldots \vee A^{(r-1)})(I \vee A)$$

$$= (I \vee A \vee \ldots \vee A^{(r-1)}) \vee (A \vee \ldots \vee A^{(r)})$$

4.
$$\begin{pmatrix} 0 & 1 & 2 & 1 \\ \infty & 0 & 1 & 1 \\ \infty & 1 & 0 & 1 \\ \infty & 1 & 2 & 0 \end{pmatrix}$$

$d_{ij} = 1$ means that the nodes i and j are adjacent to each other.

6. **procedure** OddPath(A : BoolMatrix; {The adjacency matrix}
                var C : IntegerMatrix; {The output matrix}
                n : **integer** {Number of nodes in the graph});
var
  i, j, k : Node;
begin
  {Step 1: Initialize C matrix.}
  for i := 1 **to** n **do**
    for j := 1 **to** n **do**
      if A[i, j] = **true**
      then C[i, j] := 1
      else C[i, j] := 0;
  {Step 2: Set pivot value to 1.}
  k := 1;
  {Step 3: Cycle through all desired pivot points.}
  while k <= 9 **do**
    begin
      {Step 4: Process pivot points 1, 3, 7, 9 but not 5.}
      if k <> 5 **then**
        for i := 1 **to** n **do**

```
    for j := 1 to n do
        {Step 5: If a new path has been found, then record it.}
        if (C[i,k] + C[k,j] > 1) and (i <> j)
        then C[i,j] := 1;
    {Step 6: Step to next pivot.}
    k := k + 2;
    end
end; {OddPath}
```

Choosing the if statement as our active operation, we can see that it is executed n times inside the j loop. This loop itself is executed n times inside the i loop, so the if statement is executed n^2 times. The outer loop is executed a constant c number of times. Therefore, $cn^2 = O(n^2)$.

8.

$Dist$
$$\begin{pmatrix} 0 & 4 & \infty & \infty & \infty & 2 & 4 \\ 4 & 0 & 5 & \infty & 5 & 1 & \infty \\ \infty & 5 & 0 & 4 & 1 & \infty & \infty \\ \infty & \infty & 4 & 0 & 2 & \infty & \infty \\ \infty & 5 & 1 & 2 & 0 & 10 & \infty \\ 2 & 1 & \infty & \infty & 10 & 0 & 1 \\ 4 & \infty & \infty & \infty & \infty & 1 & 0 \end{pmatrix}$$

$Path$
$$\begin{pmatrix} 0 & 0 & 0 & 0 & 0 & 0 & 0 \\ 0 & 0 & 0 & 0 & 0 & 0 & 0 \\ 0 & 0 & 0 & 0 & 0 & 0 & 0 \\ 0 & 0 & 0 & 0 & 0 & 0 & 0 \\ 0 & 0 & 0 & 0 & 0 & 0 & 0 \\ 0 & 0 & 0 & 0 & 0 & 0 & 0 \\ 0 & 0 & 0 & 0 & 0 & 0 & 0 \end{pmatrix}$$

$Initial$

$Dist$
$$\begin{pmatrix} 0 & 4 & \infty & \infty & \infty & 2 & 4 \\ 4 & 0 & 5 & \infty & 5 & 1 & 8 \\ \infty & 5 & 0 & 4 & 1 & \infty & \infty \\ \infty & \infty & 4 & 0 & 2 & \infty & \infty \\ \infty & 5 & 1 & 2 & 0 & 10 & \infty \\ 2 & 1 & \infty & \infty & 10 & 0 & 1 \\ 4 & 8 & \infty & \infty & \infty & 1 & 0 \end{pmatrix}$$

$Path$
$$\begin{pmatrix} 0 & 0 & 0 & 0 & 0 & 0 & 0 \\ 0 & 0 & 0 & 0 & 0 & 0 & 1 \\ 0 & 0 & 0 & 0 & 0 & 0 & 0 \\ 0 & 0 & 0 & 0 & 0 & 0 & 0 \\ 0 & 0 & 0 & 0 & 0 & 0 & 0 \\ 0 & 0 & 0 & 0 & 0 & 0 & 0 \\ 0 & 1 & 0 & 0 & 0 & 0 & 0 \end{pmatrix}$$

$k = 1$

$Dist$
$$\begin{pmatrix} 0 & 4 & 9 & \infty & 9 & 2 & 4 \\ 4 & 0 & 5 & \infty & 5 & 1 & 8 \\ 9 & 5 & 0 & 4 & 1 & 6 & 13 \\ \infty & \infty & 4 & 0 & 2 & \infty & \infty \\ 9 & 5 & 1 & 2 & 0 & 6 & 13 \\ 2 & 1 & 6 & \infty & 6 & 0 & 1 \\ 4 & 8 & 13 & \infty & 13 & 1 & 0 \end{pmatrix}$$

$Path$
$$\begin{pmatrix} 0 & 0 & 2 & 0 & 2 & 0 & 0 \\ 0 & 0 & 0 & 0 & 0 & 0 & 1 \\ 2 & 0 & 0 & 0 & 0 & 2 & 2 \\ 0 & 0 & 0 & 0 & 0 & 0 & 0 \\ 2 & 0 & 0 & 0 & 0 & 2 & 2 \\ 0 & 0 & 2 & 0 & 2 & 0 & 0 \\ 0 & 1 & 2 & 0 & 2 & 0 & 0 \end{pmatrix}$$

$k = 2$

$Dist$
$$\begin{pmatrix} 0 & 4 & 9 & 13 & 9 & 2 & 4 \\ 4 & 0 & 5 & 9 & 5 & 1 & 8 \\ 9 & 5 & 0 & 4 & 1 & 6 & 13 \\ 13 & 9 & 4 & 0 & 2 & 10 & 17 \\ 9 & 5 & 1 & 2 & 0 & 6 & 13 \\ 2 & 1 & 6 & 10 & 6 & 0 & 1 \\ 4 & 8 & 13 & 17 & 13 & 1 & 0 \end{pmatrix}$$

$Path$
$$\begin{pmatrix} 0 & 0 & 2 & 3 & 2 & 0 & 0 \\ 0 & 0 & 0 & 3 & 0 & 0 & 1 \\ 2 & 0 & 0 & 0 & 0 & 2 & 2 \\ 3 & 3 & 0 & 0 & 0 & 3 & 3 \\ 2 & 0 & 0 & 0 & 0 & 2 & 2 \\ 0 & 0 & 2 & 3 & 2 & 0 & 0 \\ 0 & 1 & 2 & 3 & 2 & 0 & 0 \end{pmatrix}$$

$k = 3$

$Dist$
$$\begin{pmatrix} 0 & 4 & 9 & 13 & 9 & 2 & 4 \\ 4 & 0 & 5 & 9 & 5 & 1 & 8 \\ 9 & 5 & 0 & 4 & 1 & 6 & 13 \\ 13 & 9 & 4 & 0 & 2 & 10 & 17 \\ 9 & 5 & 1 & 2 & 0 & 6 & 13 \\ 2 & 1 & 6 & 10 & 6 & 0 & 1 \\ 4 & 8 & 13 & 17 & 13 & 1 & 0 \end{pmatrix}$$

$Path$
$$\begin{pmatrix} 0 & 0 & 2 & 3 & 2 & 0 & 0 \\ 0 & 0 & 0 & 3 & 0 & 0 & 1 \\ 2 & 0 & 0 & 0 & 0 & 2 & 2 \\ 3 & 3 & 0 & 0 & 0 & 3 & 3 \\ 2 & 0 & 0 & 0 & 0 & 2 & 2 \\ 0 & 0 & 2 & 3 & 2 & 0 & 0 \\ 0 & 1 & 2 & 3 & 2 & 0 & 0 \end{pmatrix}$$

$k = 4$

$Dist$
$$\begin{pmatrix} 0 & 4 & 9 & 11 & 9 & 2 & 4 \\ 4 & 0 & 5 & 7 & 5 & 1 & 8 \\ 9 & 5 & 0 & 3 & 1 & 6 & 13 \\ 11 & 7 & 3 & 0 & 2 & 8 & 15 \\ 9 & 5 & 1 & 2 & 0 & 6 & 13 \\ 2 & 1 & 6 & 8 & 6 & 0 & 1 \\ 4 & 8 & 13 & 15 & 13 & 1 & 0 \end{pmatrix}$$

$Path$
$$\begin{pmatrix} 0 & 0 & 2 & 5 & 2 & 0 & 0 \\ 0 & 0 & 0 & 5 & 0 & 0 & 1 \\ 2 & 0 & 0 & 5 & 0 & 2 & 2 \\ 5 & 5 & 5 & 0 & 0 & 5 & 5 \\ 2 & 0 & 0 & 0 & 0 & 2 & 2 \\ 0 & 0 & 2 & 5 & 2 & 0 & 0 \\ 0 & 1 & 2 & 5 & 2 & 0 & 0 \end{pmatrix}$$

$k = 5$

$Dist$
$$\begin{pmatrix} 0 & 3 & 8 & 10 & 8 & 2 & 3 \\ 3 & 0 & 5 & 7 & 5 & 1 & 2 \\ 8 & 5 & 0 & 3 & 1 & 6 & 7 \\ 10 & 7 & 3 & 0 & 2 & 8 & 9 \\ 8 & 5 & 1 & 2 & 0 & 6 & 7 \\ 2 & 1 & 6 & 8 & 6 & 0 & 1 \\ 3 & 2 & 7 & 9 & 7 & 1 & 0 \end{pmatrix}$$

$Path$
$$\begin{pmatrix} 0 & 6 & 6 & 6 & 6 & 0 & 6 \\ 6 & 0 & 0 & 5 & 0 & 0 & 6 \\ 6 & 0 & 0 & 5 & 0 & 2 & 6 \\ 6 & 5 & 5 & 0 & 0 & 5 & 6 \\ 6 & 0 & 0 & 0 & 0 & 2 & 6 \\ 0 & 0 & 2 & 5 & 2 & 0 & 0 \\ 6 & 6 & 6 & 6 & 6 & 0 & 0 \end{pmatrix}$$

$k = 6$

$Dist$
$$\begin{pmatrix} 0 & 3 & 8 & 10 & 8 & 2 & 3 \\ 3 & 0 & 5 & 7 & 5 & 1 & 2 \\ 8 & 5 & 0 & 3 & 1 & 6 & 7 \\ 10 & 7 & 3 & 0 & 2 & 8 & 9 \\ 8 & 5 & 1 & 2 & 0 & 6 & 7 \\ 2 & 1 & 6 & 8 & 6 & 0 & 1 \\ 3 & 2 & 7 & 9 & 7 & 1 & 0 \end{pmatrix}$$

$Path$
$$\begin{pmatrix} 0 & 6 & 6 & 6 & 6 & 0 & 6 \\ 6 & 0 & 0 & 5 & 0 & 0 & 6 \\ 6 & 0 & 0 & 5 & 0 & 2 & 6 \\ 6 & 5 & 5 & 0 & 0 & 5 & 6 \\ 6 & 0 & 0 & 0 & 0 & 2 & 6 \\ 0 & 0 & 2 & 5 & 2 & 0 & 0 \\ 6 & 6 & 6 & 6 & 6 & 0 & 0 \end{pmatrix}$$

$k = 7$

10. This procedure works under the assumption there exists a path matrix that has been computed by the procedure MinWeightPaths.

> **procedure** GetPaths(Path : IntegerMatrix; n : Node);
> {Given a path matrix computed by the MinWeightPaths procedure, and the number of nodes in the graph, this procedure will cycle through all node pairs and output the nodes (if any) on the minimum path between each node pair.}
> > **var**
> > > i, j, current : Node
> >
> > > **procedure** GetPath (s, e : Node)
> > > {When given a start node and an end node, this procedure will output the nodes on the minimum path from s to e.}
> > > > **begin**
> > > > > {Step 1: Check for a pivot node.}
> > > > > **if** Path[s,e] <> 0 **then**
> > > > > > **begin**
> > > > > > > {Step 2: Output nodes from start to pivot.}
> > > > > > > GetPath(s, Path[s,e]);
> > > > > > > {Step 3: Output the current pivot node.}
> > > > > > > **write** (Path[s,e]:4, ', ');
> > > > > > > {Step 4: Output nodes from pivot to end.}
> > > > > > > GetPath(Path[s,e], e);
> > > > > > **end**
> > > > **end**; {GetPath}
> >
> > **begin**
> > {Step 5: Cycle through all nodes and output the path between each node pair.}
> > **for** i := 1 **to** n **do**
> > > **for** j := 1 **to** n **do**
> > > > **begin**
> > > > > **write**('The path between ', i, 'and ', j,
> > > > > > 'includes the following nodes : ');
> > > > > GetPath (i, j);
> > > > > **writeln**;
> > > > **end**
> > **end**; {GetPaths}

The nested procedure GetPath(i, j) obtains the path from i to j. The main procedure varies i and j through all pairs of nodes in order to obtain all paths in the graph. GetPath works by first traversing the path from the start node to the pivot node, then printing out the pivot node, and finally traversing the path from the pivot node to the end node. If there is a direct path from the start node to the end node, then the base case has been reached and the recursion stops.

12. (a) In order to compute the maximum distances, change the representation so that nonexistent edges are represented by $-INF$.

> (b) **procedure** MaxWeightPaths (**var** MaxDist, Path : IntegerMatrix;
> > n : **integer**);

{This procedure accepts distance matrix for a graph such that each indexed position in the matrix represents the distance between the two adjacent nodes indexing that position, and if the nodes are not adjacent, the matrix contains −INF at that position. Given such a matrix (representing an acyclic graph), this procedure will compute the maximum paths between all node pairs in the graph.}

```
var
    k, i, j : integer;
begin
    {Step 1: Initialize Path matrix.}
    for i := 1 to n do
        for j := 1 to n do
            Path [i, j] := 0;
    {Step 2: Compute maximum weighted path lengths.}
    for k := 1 to n do
        for i := 1 to n do
            for j := 1 to n do
                {Step 3: Check for greater distances not containing nonexistent edges.}
                if (MaxDist [i, k] + MaxDist [k, j] > MaxDist [i, j])
                    and (MaxDist [i, k] + MaxDist [k, j] > 0)
                then
                    begin
                        MaxDist [i, j] := MaxDist [i, k] + MaxDist [k, j];
                        Path [i, j] := k
                    end
end; {MaxWeightPaths}
```

(c)

$$
\begin{pmatrix}
0 & 3 & -\infty & 6 & -\infty & -\infty \\
-\infty & 0 & 5 & 1 & 3 & -\infty \\
-\infty & -\infty & 0 & -\infty & 2 & 6 \\
-\infty & -\infty & 1 & 0 & 4 & -\infty \\
-\infty & -\infty & -\infty & -\infty & 0 & 3 \\
-\infty & -\infty & -\infty & -\infty & -\infty & 0
\end{pmatrix}
\qquad
\begin{pmatrix}
0 & 3 & -\infty & 6 & -\infty & -\infty \\
-\infty & 0 & 5 & 1 & 3 & -\infty \\
-\infty & -\infty & 0 & -\infty & 2 & 6 \\
-\infty & -\infty & 1 & 0 & 4 & -\infty \\
-\infty & -\infty & -\infty & -\infty & 0 & 3 \\
-\infty & -\infty & -\infty & -\infty & -\infty & 0
\end{pmatrix}
$$

Initial $\qquad\qquad\qquad\qquad$ $k = 1$

$$
\begin{pmatrix}
0 & 3 & 8 & 6 & 6 & -\infty \\
-\infty & 0 & 5 & 1 & 3 & -\infty \\
-\infty & -\infty & 0 & -\infty & 2 & 6 \\
-\infty & -\infty & 1 & 0 & 4 & -\infty \\
-\infty & -\infty & -\infty & -\infty & 0 & 3 \\
-\infty & -\infty & -\infty & -\infty & -\infty & 0
\end{pmatrix}
\qquad
\begin{pmatrix}
0 & 3 & 8 & 6 & 10 & 14 \\
-\infty & 0 & 5 & 1 & 7 & 11 \\
-\infty & -\infty & 0 & -\infty & 2 & 6 \\
-\infty & -\infty & 1 & 0 & 4 & 7 \\
-\infty & -\infty & -\infty & -\infty & 0 & 3 \\
-\infty & -\infty & -\infty & -\infty & -\infty & 0
\end{pmatrix}
$$

$k = 2$ $\qquad\qquad\qquad\qquad$ $k = 3$

$$\begin{pmatrix} 0 & 3 & 8 & 6 & 10 & 14 \\ -\infty & 0 & 5 & 1 & 7 & 11 \\ -\infty & -\infty & 0 & -\infty & 2 & 6 \\ -\infty & -\infty & 1 & 0 & 4 & 7 \\ -\infty & -\infty & -\infty & -\infty & 0 & 3 \\ -\infty & -\infty & -\infty & -\infty & -\infty & 0 \end{pmatrix} \qquad \begin{pmatrix} 0 & 3 & 8 & 6 & 10 & 14 \\ -\infty & 0 & 5 & 1 & 7 & 11 \\ -\infty & -\infty & 0 & -\infty & 2 & 6 \\ -\infty & -\infty & 1 & 0 & 4 & 7 \\ -\infty & -\infty & -\infty & -\infty & 0 & 3 \\ -\infty & -\infty & -\infty & -\infty & -\infty & 0 \end{pmatrix}$$

$$k = 4 \qquad\qquad\qquad\qquad k = 5$$

$$\begin{pmatrix} 0 & 3 & 8 & 6 & 10 & 14 \\ -\infty & 0 & 5 & 1 & 7 & 11 \\ -\infty & -\infty & 0 & -\infty & 2 & 6 \\ -\infty & -\infty & 1 & 0 & 4 & 7 \\ -\infty & -\infty & -\infty & -\infty & 0 & 3 \\ -\infty & -\infty & -\infty & -\infty & -\infty & 0 \end{pmatrix}$$

$$k = 6$$

Solutions 7.5

2. To change the procedure to keep track of Depth-First numbers, only the following changes have to be made: An extra parameter is needed so that the DFSN can be passed from one procedure call to the next. In addition, this parameter needs to be a **var** parameter so that its value is preserved when the recursive procedures terminate.

var DFSN : **integer**

Two lines of code need to be added in step 1 to update the DFN fields of the nodes and to keep the DFSN number current.

```
DFSN := DFSN + 1;
Graph [Current] .DFN := DFSN;
```

4. **procedure** Traverse (Start : Node; n : **integer**; Graph : GraphType);
   {This procedure will perform a BFS on all components of an unconnected graph}
   var
      NodePtr : List; {Pointer to a node in an adjacency list}
      Neighbor, Index, Current : Node;
      Done: **boolean**;
      Queue : **array** [1..MaxVert] **of** Node; {The queue structure}
   begin
      {Step 1: Mark all vertices as not reached.}
      for Current := 1 **to** n **do**
         Graph [Current].Reach := **false**;
      Current := 1;
      Done := **false**;

```
{Step 2: Process all components.}
while not Done do
  begin
    Start := Current;
    {Step 3: Mark and visit start node.}
    Graph [Start]. Reach := true;
    write (Graph [Start]. Data);
    {Step 4: Initialize and place start node in queue.}
    QInitialize (Queue);
    QInsert (Queue, Start);
    {Step 5: Traverse graph until all remaining nodes are visited.}
    while not QEmpty (Queue) do
    begin
      {Step 6: Delete front node from queue into current.}
      QDelete (Queue, Current);
      {Step 7: Obtain first neighbor of current node.}
      NodePtr := Graph [Current]. Head;
      {Step 8: Find all neighbors of current node.}
      while NodePtr <> nil do
        begin
          Neighbor := NodePtr ↑. NodeIndex;
          {Step 9: Mark and visit unvisited neighbors.}
          if not Graph [Neighbor]. Reach then
          begin
            Graph [Neighbor]. Reach := true;
            write (Graph [Neighbor]. Data);
            QInsert (Queue, Neighbor)
          end;
          {Step 10: Advance to next neighbor.}
          NodePtr := NodePtr ↑.Next
        end
    end;
    Done := true;
    {Step 11: Find next component that has not been traversed.}
    for Index := 1 to n do
      if (Graph [ Index ] .Reach = false) and Done then
        begin
          Current := Index;
          Done := false
        end
  end;
end;
```

6. Add the following two lines to the procedure CycleDetector:

```
if Cycle
  then write (Current, ', ');
```

before the line

CycleDetector := Cycle

These two lines will cause all nodes in a cycle to be output to the screen once a cycle has been found. The nodes in the cycle cause recursive calls in the section of code that checks the neighbor nodes. When the cycle is detected, there is a sequence of these recursive calls in which Current is equal to different nodes in the cycle. Once a cycle is found, these recursive calls terminate, and each one will output its Current node before terminating, thereby outputting the cycle itself.

8. **function** TestAcyclic(G : GraphType; n : Node): **boolean**;
   var
      Acyclic : **boolean**;
      Temp, Index : Node;
   begin
      {Step 1: Initialize local variables.}
      Acyclic := **true**;
      Index := 1;
      {Step 2: Cycle through nodes to check for cycles.}
      while (Index <= n) **and** Acyclic **do**
         begin
            for Temp := 1 **to** n **do**
               Graph [temp] .Reach := **false**;
            if CycleDetector (Index, Index) = **true**
            then Acyclic := **false**;
            Index := Index + 1
         end;
      TestAcyclic := Acyclic
   end;

10. The following graphs show traces of the BFS procedure. It is assumed that the nodes are visited in ascending numeric order.

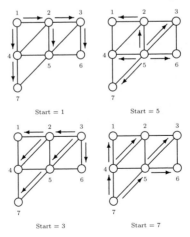

12.

Iteration	1	2	Dist 3	4	5	6	u	S
0	0	INF	INF	1	3	INF		$\{1\}$
1	0	INF	INF	(1)	3 (2)	INF	4	$\{1,4\}$
2	0	INF (4)	INF	1	(2)	INF(8)	5	$\{1,4,5\}$
3	0	(4)	INF (9)	1	2	8 (5)	2	$\{1,2,4,5\}$
4	0	4	9	1	2	(5)	6	$\{1,2,4,5,6\}$
5	0	4	(9)	1	2	5	3	$\{1,2,3,4,5,6\}$

Iteration	1	2	3	Dist 4	5	6	7	u	S
0	0	5	INF	INF	INF	2	INF		$\{1\}$
1	0	5 (3)	INF	INF	INF	(2)	INF (8)	6	$\{1,6\}$
2	0	(3)	INF (8)	INF	INF (4)	2	8	2	$\{1,2,6\}$
3	0	3	8 (6)	INF	(4)	2	8	5	$\{1,2,5,6\}$
4	0	3	(6)	INF (8)	4	2	8	3	$\{1,2,3,5,6\}$
5	0	3	6	(8)	4	2	8	4	$\{1,2,3,4,5,6\}$
6	0	3	6	8	4	2	(8)	7	$\{1,2,3,4,5,6,7\}$

14. The procedure below assumes that the predecessor node information is stored in a matrix called Pred, that 0's in the matrix represent no predecessor information, and that positive integers represent the names of predecessor nodes. It is assumed that the nodes are named according to the numeric sequence $\langle 1,2,3,\ldots,n\rangle$ and that the desired path is to node n such that $1 \leq m \leq n$. The procedure uses recursion to backtrack through the predecessors. The iterative loop in the procedure is used to find the predecessor of the current node being processed.

procedure PredPath(Pred : IntegerMatrix; Start, Finish, n: Node);
var
    Last, Index : Node;

```
begin
    {Step 1: Check to see if the recursion is done.}
    if Finish <> Start then
    begin
        {Step 2: Find the final node's predecessor.}
        for index := 1 to n do
            {Step 3: Keep track of the latest potential predecessor.}
            if Pred [Index, Finish] <> 0
            then Last := Pred [Index, Finish]
            {Step 4: Check out the predecessor node.}
            PredPath (Pred, Start, Last, n)
    end;
    write (Finish :4, ' ')
end;
```

16. Consider a graph with three nodes 1, 2, and 3. Let the edge weights be as follows: W(1, 2) = 2, W(1, 3) = 1, W(2, 3) = −5. Given this graph, Dijkstra's algorithm will say that the minimum path from node 1 to node 3 is <1, 3>, whereas it is actually <1, 2, 3>. This happens because it is assumed that the minimum weight edge leaving a node will necessarily be the cheapest route to the destination node, and therefore the path <1, 3> is picked on the first iteration.

Solutions 7.6

2. **(a)** Remove edge $\{1, 4\}$ from cycle $\langle 1, 4, 3, 1 \rangle$.
Remove edge $\{4, 3\}$ from cycle $\langle 2, 4, 3, 2 \rangle$.
Remove edge $\{2, 3\}$ from cycle $\langle 1, 2, 3, 1 \rangle$.
   (b) Remove edge $\{1, 2\}$ from cycle $\langle 1, 2, 3, 1 \rangle$.
Remove edge $\{1, 3\}$ from cycle $\langle 1, 3, 4, 1 \rangle$.
Remove edge $\{1, 4\}$ from cycle $\langle 1, 4, 5, 1 \rangle$.
Remove edge $\{2, 3\}$ from cycle $\langle 2, 3, 5, 2 \rangle$.
Remove edge $\{2, 4\}$ from cycle $\langle 2, 4, 5, 2 \rangle$.
Remove edge $\{3, 5\}$ from cycle $\langle 3, 5, 4, 3 \rangle$.
   (c) Remove edge $\{4, 7\}$ from cycle $\langle 4, 7, 5, 4 \rangle$.
Remove edge $\{4, 5\}$ from cycle $\langle 1, 4, 5, 1 \rangle$.
Remove edge $\{1, 5\}$ from cycle $\langle 1, 5, 2, 1 \rangle$.
Remove edge $\{7, 6\}$ from cycle $\langle 5, 7, 6, 5 \rangle$.
Remove edge $\{5, 6\}$ from cycle $\langle 2, 5, 6, 2 \rangle$.
Remove edge $\{2, 6\}$ from cycle $\langle 2, 6, 3, 2 \rangle$.
   (d) Remove edge $\{4, 5\}$ from cycle $\langle 4, 5, 2, 4 \rangle$.
Remove edge $\{4, 2\}$ from cycle $\langle 1, 4, 2, 1 \rangle$.
Remove edge $\{1, 5\}$ from cycle $\langle 1, 5, 2, 1 \rangle$.
Remove edge $\{5, 3\}$ from cycle $\langle 2, 5, 3, 2 \rangle$.
Remove edge $\{5, 6\}$ from cycle $\langle 2, 5, 6, 3, 2 \rangle$.
Remove edge $\{6, 7\}$ from cycle $\langle 3, 6, 7, 3 \rangle$.

The spanning trees generated by removing these edges are the following:

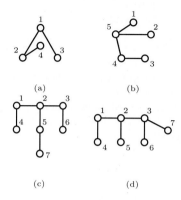

4. Assume that the DFS starts at node 1 in each graph.

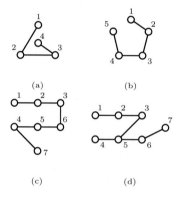

6. This algorithm uses the procedure BuildSpanTree to build spanning trees of the connected components within an unconnected graph. These spanning trees are combined to form a spanning forest, and the forest is stored in the variables Nodes and Edges.

procedure BuildSpanForest (**var** Nodes : NodeSet;
                  var Edges : EdgeSet);
{Given a starting vertex u and a graph (V, E), which is assumed to be global to this routine, this procedure generates a spanning forest for this graph. The spanning forest is denoted by (Nodes, Edges) where Nodes contains the vertices in V and Edges are edges selected from E.}
var
   Index, Vertex, w : Node;
   Done : **boolean**;
   TempNodes : **set of** Node;
   TempEdges : **set of** EdgeType;
begin
   {Step 1: Initialize Forest.}

```
    Nodes := [];
    Edges := [];
    {Step 2: Initialize variables.}
    Done := false;
    Vertex := 1;
    {Step 3: Span all components.}
    while not Done do
    begin
        {Step 4: Build Spanning tree containing Vertex.}
        BuildSpanTree (Vertex, TempNodes, TempEdges);
        {Step 5: Add tree obtained in step 4 to the tree.}
        Nodes := Nodes + TempNodes;
        Edges := Edges + TempEdges;
        Done := (V = Nodes);
        {Step 6: If some nodes are unprocessed, then pick one and span its component.}
        if not Done then
            Vertex := {w such that w in (V − Nodes)}
    end
end {BuildSpanForest}
```

8.

10. The two spanning trees shown below are distinct spanning trees of the graph given in Figure 7.63.

Solutions 7.7.2

2. **(a)** Add an edge a_{29} between nodes 16 and 23 in the PERT diagram and label the edge with a weight of 1. The critical path is not affected.

   (b) Add an edge a_{30} between nodes 2 and 5 in the PERT diagram and label the edge with a weight of 2. The critical path is not affected.

   (c) Relabel nodes 22 and 23 as 23 and 24, respectively. Add a note 22 between nodes 21 and 23. Change the source for edge a_{28} to the new node 22. Add an edge labeled a_{31}

with weight 4 between nodes 21 and 22. The augmented critical path contains activity a_{31} between activities a_26 and a_28, and takes four more days to complete.

4. The PERT diagram representing this problem is shown below.

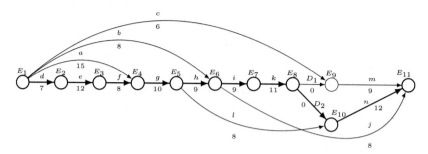

(a)

Event	Activities Completing	TE_i	Tl_i	S_i
E_1	—	0	0	0
E_2	d	7	7	0
E_3	e	19	19	0
E_4	a,f	27	27	0
E_5	g	37	37	0
E_6	b,h	46	46	0
E_7	i	55	55	0
E_8	k	66	66	0
E_9	c,D_1	66	69	3
E_{10}	l,D_2	66	66	0
E_{11}	j,m,n	78	78	0

(b)

Activity	(Start, End)	Total Slack
a	(E_1, E_4)	12
b	(E_1, E_6)	38
c	(E_1, E_9)	63
d	(E_1, E_2)	0
e	(E_2, E_3)	0
f	(E_3, E_4)	0
g	(E_4, E_5)	0
h	(E_5, E_6)	0
i	(E_6, E_7)	0
j	(E_6, E_{11})	24
k	(E_7, E_8)	0
l	(E_5, E_{10})	21
m	(E_9, E_{11})	3
n	(E_{10}, E_{11})	0

(**c**) The critical path is shown below:

Activity	d	e	f	g	h	i	k	D_2	n
t_i	7	12	8	10	9	9	11	0	12
TE_i	7	19	27	37	46	55	66	66	78

Solutions 7.7.3

2. There are many linear orders, for example

    1 3 2 5 4 7 6 8   or   1 2 3 4 5 6 7 8.

4. One linear order is

    $- * - * + z w - y x$

CHAPTER 7 SOLUTIONS

2. procedure DFSBack (Current, Start : Node;
                               DFSN : **integer**;
                               var MinLength : **integer**;
                               var Path, SmallPath : IntegerArray);
{This procedure will find the shortest cycle formed by a back edge in a depth-first traversal of a supplied graph. This procedure requires a depth-first number approach as described in Problem 7.5.2.}
var
   NodePtr : List;
   Neighbor : Node;
   TestVal, t : **integer**;
begin
   if Current = Start **then**
     begin
       DFSN := 1;
       MinLength := INF;
     end;
   {Step 1: Mark and visit present node.}
   Graph [Current]. Reach := **true**;
   write (Graph [Current]. Data);
   DFSN := DFSN + 1;
   Graph [Current] .DFN := DFSN;
   Path [DFSN] := Current;
   {Step 2: Obtain pointer to first neighbor of present node.}
   NodePtr := Graph [Current]. Head;
   {Step 3: Repeat for all neighbors of present node.}
   while NodePtr <> **nil do**
     begin
       {Step 4: Obtain index of neighbor.}

```
         Neighbor := NodePtr ↑ . NodeIndex;
         {Step 5: If neighbor has not been marked, then explore it.}
         TestVal := Graph [Current]. DFN − Graph [Neighbor]. DFN;
         if Graph [Neighbor]. Reach then
            begin
               {Step 6: Check for cross-edges, parents, and minimal cycles.}
               if (Graph [Neighbor]. DFN > −1)
                  and (TestVal > 1) and (TestVal < MinLength) then
                     begin
                        MinLength := TestVal;
                        for t := 1 to MinLength
                        SmallPath [t] := Path [Graph [Neighbor].DFN − 1 + t];
                     end
            end
         else
            DFSBack (Neighbor, Start, DFSN, MinLength, Path, SmallPath);
         {Step 7: Obtain pointer to next neighbor.}
         NodePtr := NodePtr ↑ . Next
      end;
   Graph [Current]. DFN := −1;
end;
```

Every time a previously visited neighbor is found, this represents a back-edge, and therefore a cycle of the desired type. If the cycle is a minimum length one, then the cycle is saved. The minimum cycle is printed out when the root node has no unexplored neighbors. This procedure will reject cross-edges as the reached node will have a DFN of −1. As well, this procedure will reject cycles to parents by requiring the difference in DFN to be greater than 1.

4. **function** DFSCheck (Current, Parent : Node;
                          var Tree : **boolean**;
                          var Unreached : **integer**) : **boolean**;

```
   var
      NodePtr : List;
      Neighbor : Node;
   begin
      {Step 1: Mark and visit present node.}
      Graph [Current]. Reach := true;
      {Step 2: Obtain pointer to first neighbor of present node.}
      NodePtr := Graph [Current]. Head;
      {Step 3: Mark another node as reached.}
      Unreached := Unreached − 1;
      Tree := true
      {Step 4: Repeat for all neighbors of present node.}
      while NodePtr <> nil and Tree do
         begin
            {Step 5: Obtain index of neighbor.}
            Neighbor := NodePtr ↑ . NodeIndex;
```

{Step 6: If neighbor has not been marked, then explore it.}
if not Graph [Neighbor]. Reach
 then Tree := DFSCheck (Neighbor, Current, Tree, Unreached)
 else if Neighbor <> Parent
   then Tree := **false**;
{Step 7: Obtain pointer to next neighbor.}
NodePtr := NodePtr ↑ . Next
 end;
 DFSCheck := (Unreached = 0) **and** Tree
end;

This function assumes that it is called with Unreached equal to the number of nodes in the graph. If the graph is unconnected, then Unreached will never reach 0, and if there are cycles, then Tree will be set to false.

6. **procedure** ExtendedBFS (Start : Node;
         var Parent : NodeArray;
         var Dist : IntegerArray;
         var NumDescend : IntegerArray);
{Given a start node, this procedure performs a breadth-first search on a simple connected undirected graph.}
var NodePtr : AdjList; {Pointer to a node in an adjacency list}
 Neighbor, Current, Tracer : Node;
Queue : **array** [1..MaxVert] **of** Node;   {The queue structure}
begin
 {Step 1: Mark all vertices as not reached.}
 for Current := 1 **to** n **do**
  Graph [Current].Reach := **false**;
 {Step 2: Mark and visit start node.}
 Graph [Start]. Reach := **true**;
 Parent[Start] := Start;
 write (Graph [Start]. Data);
 {Step 3: Set the start node distance to 0.}
 Dist[Start] := 0;
 {Step 4: Initialize and place start node in queue.}
 QInitialize (Queue);
 QInsert (Queue, Start);
 {Step 5: Traverse graph until all remaining nodes are visited.}
 while not QEmpty (Queue) **do**
  begin
   {Step 6: Remove front node from queue and call it the current node.}
   QDelete (Queue, Current);
   {Step 7: Obtain first neighbor of current node.}
   NodePtr := Graph [Current]. Head;
   {Step 8: Find all neighbors of current node.}
   while NodePtr <> **nil do**
    begin

```
                    Neighbor := NodePtr↑. NodeIndex;
                    {Step 9: If the neighbor is unvisited, mark and visit it.}
                    if not Graph [Neighbor]. Reach
                      then begin
                              Graph [Neighbor]. Reach := true;
                              write (Graph[Neighbor]. Data);
                              Dist[Neighbor] := Dist[Current]+ 1;
                              Parent[Neighbor] := Current;
                              QInsert (Queue, Neighbor);
                              Tracer := Neighbor;
                              {Trace through ancestors and update NumDescend}
                              while Parent[Tracer] <> Tracer do
                              begin
                                 Tracer := Parent[Tracer];
                                 NumDescend[Tracer] := NumDescend[Tracer] + 1
                              end
                           end;
                    {Step 10: Advance to next neighbor}
                    NodePtr := NodePtr↑.Next
              end
        end
  end;
```

This program is an extended form of the BFS procedure that computes an array of integers corresponding to the distances of node from the start node, an array of nodes corresponding to the Parent of each node in the BFS search tree, and an array of integers corresponding to the total number of descendants that each node contains within the spanning tree. These arrays are all passed in as **var** parameters to the procedure, and the procedure updates them accordingly.

Solutions 8.4.3

2. (a) Acceptable: i and j are of the same type, so the value in j can be equal to the value in i.
   (b) Acceptable: a is a member of the set *Author*, and therefore it is also a member of the set *scientific_authors* which is the power set of *Author*.
   (c) Unacceptable: *scientific_authors* and *on_loan* are power sets of two totally different sets, and therefore they are not compatible.
   (d) Unacceptable: When constructing sets, all elements have to be of the same type.
   (e) Acceptable: *on_loan* is a power set of *Book*, therefore there is a basis for comparison, and this statement is acceptable.
   (f) Unacceptable: i is of type $\mathbb{N}$ and *scientific_authors* is of type $\mathbb{P}$ *Author*, therefore i cannot be in the set *scientific_authors*.

4. [*Person*] the set of all possible people
   [*Account_Type*] ::= faculty | student | system

Solutions 8.4.4

2. (a) $Passengers : \mathbb{P}People$
$Bus\_Capacity : \mathbb{N}$
$\#Passengers \leq Bus\_Capacity$
(b) $Passengers = \varnothing$
(c) $p? : People$
$\#Passengers < Bus\_Capacity$
$Passengers' = Passengers \cup \{p?\}$
(d) $p? : People$
$p? \in Passengers$
$Passengers' = Passengers \setminus \{p?\}$
(e) $number\_on\_bus! : \mathbb{N}$
$number\_on\_bus! = \#Passengers$
(f) $Reply ::= \text{no} \mid \text{yes}$
$response! : Reply$
$(p? \in Passengers \wedge response! = \text{yes})$
$\vee$
$(p? \notin Passengers \wedge response! = \text{no})$

Solutions 8.4.5

2. (a) LA, LA', max, max'
(b) $\#LA \leq max, n! = 2 * m?$
(c) input $\quad m?$
output $\quad n!$

4. The following types are used to specify the registration system:

[Person]	Set of all possible people using the system
$Messages$	::= invalid_username | system_busy |
	not_logged_on | not_a_system_user |
	user_already_exists | ok
$Reply$	::= no | yes
$AccountType$	::= faculty | student | system

The following description illustrates the invariant properties of the registration system. The computer system needs to keep track of existing accounts and organize them according to the type of user who owns the accounts. Active account names are stored in $StuAccounts$, $FacAccounts$, and $SysAccounts$. These sets contain student, faculty, and system accounts, respectively. The computer system must also monitor the use of the system, so that certain restrictions can be applied depending on the class of user logged into the system. The login name of the person currently logged into the system is stored in $CurrentUser$, and if no one is logged into the system, this set is empty.

As stated in the problem, each user is restricted to one account, that is, the sets $StuAccounts$, $FacAccounts$, and $SysAccounts$ are mutually exclusive, that is, their intersections are empty. $UserAccounts$ is used to keep track of valid users of the system and, therefore, includes all valid accounts of all types.

$$
\begin{array}{|l}
\underline{\ Comp\_Sys\ \rule{0pt}{0pt}} \\
StuAccounts : \mathbb{P}\,Person \\
FacAccounts : \mathbb{P}\,Person \\
SysAccounts : \mathbb{P}\,Person \\
CurrentUser : Person \\
UserAccounts : \mathbb{P}\,Person \\
\hline
StuAccounts \cap FacAccounts = \varnothing \\
StuAccounts \cap SysAccounts = \varnothing \\
FacAccounts \cap SysAccounts = \varnothing \\
UserAccounts = StuAccounts \cup FacAccounts \cup SysAccounts \\
\end{array}
$$

The following schema is used to print ok on the screen to indicate a successful operation.

$$
\begin{array}{|l}
\underline{\ OK\_Message\ \rule{0pt}{0pt}} \\
output! : Messages \\
\hline
output! = \text{ok} \\
\end{array}
$$

When the system is in its initial state, there are no student or faculty accounts; however, a single system account called root is created so that a root user can log into the system and create any other desired accounts.

$$
\begin{array}{|l}
\underline{\ Initial\ \rule{0pt}{0pt}} \\
Comp\_Sys \\
\hline
StuAccounts = \varnothing \\
FacAccounts = \varnothing \\
SysAccounts = \{\text{root}\} \\
CurrentUser = \varnothing \\
UserAccounts = \{\text{root}\} \\
\end{array}
$$

The specification for this system requires procedures for logging on and logging off. The log on procedure adds a user $u?$ as follows: If this user name is valid (i.e., it is in the set $UserAccounts$) and there is no one else logged into the system, then the current user is updated and the user is logged on.

$\underline{\quad Log\_On_o \rule{3cm}{0.4pt}}$
$\Delta Comp\_Sys$
$u? : Person$

$u? \in UserAccounts$

$CurrentUser = \varnothing$

$CurrentUser' = \{u?\}$

$StuAccounts' = StuAccounts$

$FacAccounts' = FacAccounts$

$SysAccounts' = SysAccounts$

$UserAccounts' = UserAccounts$

Note that according to the question, this system is a single-user system; therefore, we can unambiguously logoff this user without having a username as input. The system does this by setting $CurrentUser'$ to $\varnothing$.

$\underline{\quad Log\_Off_o \rule{3cm}{0.4pt}}$
$\Delta Comp\_Sys$

$\#CurrentUser > 0$

$CurrentUser' = \varnothing$

$StuAccounts' = StuAccounts$

$FacAccounts' = FacAccounts$

$SysAccounts' = SysAccounts$

$UserAccounts' = UserAccounts$

The system now needs error schemas to handle errors that can occur with the log-on and log-off processes.

$\underline{\quad Log\_On\_Error \rule{4cm}{0.4pt}}$
$\Xi Comp\_Sys$
$output! : Messages$
$u? : Person$

$(u? \notin UserAccounts \land output! = \text{invalid\_username})$

$\lor$

$((\#CurrentUser > 0) \land output! = \text{system\_busy})$

$\underline{\quad Log\_Off\_Error \rule{4cm}{0.4pt}}$
$\Xi Comp\_Sys$
$output! : Messages$

$(\#CurrentUser < 1 \land output! = \text{not\_logged\_on})$

$$Log\_On \mathrel{\widehat{=}} (Log\_On_o \wedge OK\_Message) \vee Log\_On\_Error$$
$$Log\_Off \mathrel{\widehat{=}} (Log\_Off_o \wedge OK\_Message) \vee Log\_Off\_Error$$

The computer system needs a facility to add a new user account, and this operation is specified in the following schema. Two inputs are required for this schema: the user name for the new account and the type of account. It should be noted that only system users can update the accounts, and this is specified by the first precondition. The second precondition ensures that the account name being used is not already in the database. The postconditions ensure that the appropriate account set is updated and that the other two remain the same.

```
┌─ Add_User_o ─────────────────────────────────
│ ΔComp_Sys
│ u? : Person
│ t? : AccountType
├───────────────────────────────────────────────
│ #(CurrentUser ∩ SysAccounts) > 0
│ u? ∉ UserAccounts
│ (t? = system ∧ SysAccounts' = SysAccounts ∪ {u?}
│            ∧ StuAccounts' = StuAccounts
│            ∧ FacAccounts' = FacAccounts)
│ ∨ (t? = student ∧ StuAccounts' = StuAccounts ∪ {u?}
│            ∧ FacAccounts' = FacAccounts
│            ∧ SysAccounts' = SysAccounts)
│ ∨ (t? = faculty ∧ FacAccounts' = FacAccounts ∪ {u?}
│            ∧ StuAccounts' = StuAccounts
│            ∧ SysAccounts' = SysAccounts)
│ CurrentUser' = CurrentUser
│ UserAccounts' = UserAccounts ∪ {u?}
└───────────────────────────────────────────────
```

There are two possible errors that can arise while adding a user. First, the current user may not be a system user and, therefore, the account database cannot be updated. The second error is if the user account already exists.

```
┌─ Add_User_Error ─────────────────────────────
│ ΞComp_Sys
│ output! : Messages
│ u? : Person
├───────────────────────────────────────────────
│ (u? ∈ UserAccounts ∧ output! = user_already_exists)
│ ∨
│ ((CurrentUser ∩ SysAccounts = ∅) ∧ output! = not_a_system_user)
└───────────────────────────────────────────────
```

The operation for adding a user is combined with its corresponding error schema to produce an operation schema that handles errors.

$$Add\_User \mathrel{\widehat{=}} (Add\_User_o \wedge OK\_Message) \vee Add\_User\_Error$$

The following schema specifies the inquire operation used to find out whether or not a given user name is a valid account name. The answer is returned in the output variable *Response*!.

$$
\begin{array}{|l}
\hline
\_Inquire_____ \\
\Xi Comp\_Sys \\
u? : Person \\
Response! : Reply \\
\hline
(u? \notin UserAccounts \land Response! = \text{no}) \\
\lor \\
(u? \in UserAccounts \land Response! = \text{yes}) \\
\hline
\end{array}
$$

The schema How_Many specifies an operation that returns the number of accounts in the system.

$$
\begin{array}{|l}
\hline
\_How\_Many_____ \\
\Xi Comp\_Sys \\
NumberUsers! : \mathbb{N} \\
\hline
NumberUsers! = \#UserAccounts \\
\hline
\end{array}
$$

Solutions 8.4.6

2. The first schema specifies an operation for retiring a stall from service. The input to this schema is the stall number to be retired. If this stall is available, then it is removed from the set of available stalls.

$$
\begin{array}{|l}
\hline
\_Retire\_A\_Stall_o_____ \\
\Delta Parking \\
s? : Parking\_Stall \\
\hline
s? \in available \\
available' = available \setminus \{s?\} \\
faculty' = faculty \\
assigned\_to' = assigned\_to \\
occupied' = occupied \\
\hline
\end{array}
$$

If the stall is currently being used, then it cannot be retired. The following error schema handles this case.

$$
\begin{array}{|l}
\hline
\_Retire\_Error_____ \\
\Xi Parking \\
output! : Possible\_Messages \\
s? : Parking\_Stall \\
\hline
s? \notin available \\
output! = \text{stall\_not\_available} \\
\hline
\end{array}
$$

The retiring operation complete with its error handling schema is presented below.

$$Retire\_A\_Stall \mathrel{\widehat{=}} (Retire\_A\_Stall_o \wedge OK) \vee Retire\_Error$$

Another extension is to provide a facility for adding faculty to the parking system. The input to this operation is the name of the faculty member. If this faculty member is not already in the system, then he or she is added. This operation is handled by the following schema.

```
┌─ Hire_Faculty ──────────────
│ ΔParking
│ name? : Person
├─────────────────────────────
│ name? ∉ faculty
│ faculty' = faculty ∪ {name?}
│ assigned_to' = assigned_to
│ available' = available
│ occupied' = occupied
└─────────────────────────────
```

If a faculty member is already in the system, he or she cannot be added. The following schema handles this error.

```
┌─ Hire_Faculty_Error ─────────
│ ΞParking
│ output! : Possible_Messages
│ name? : Person
├─────────────────────────────
│ name? ∈ faculty
│ output! = already_a_faculty_member
└─────────────────────────────
```

The operation for adding faculty, complete with its error handling schema, is presented below.

$$Hire\_Faculty \mathrel{\widehat{=}} (Hire\_Faculty_o \wedge OK) \vee Hire\_Faculty\_Error$$

4. The medicare system uses the following sets:

> [$Person$]   The set of all people in the medicare system
> $Possible\_Messages$ ::= unknown_doctor | unknown_patient |
>   doctor_already_in_system

The medicare system is based on two sets: $Patient$, which is the set of all patients, and $Doctor$, which is the set of all doctors. There is one relation in the medicare system, the $treats$ relation, that specifies which doctors treat which patients. The invariant properties of this system are that only doctors can treat patients and only patients can be treated.

---
*Medicare*_____

$Patients : \mathbb{P}\,Person$
$Doctors : \mathbb{P}\,Person$
$treats : Person \leftrightarrow Person$

---

dom $treats \subseteq Doctors$
ran $treats \subseteq Patients$

When the medicare system is in its initial state, there are neither doctors nor patients and, therefore, the *treats* relation between doctors and patients is also empty.

---
*Initial*_____

$Medicare$

$Patients = \varnothing$

$Doctors = \varnothing$

$treats = \varnothing$

An operation is needed to record a relationship between a doctor and a patient. The input to this schema is the name of the doctor and the name of the patient. If the doctor's name is in the set of valid doctors and if the patient's name is in the set of valid patients, then the relation between the two can be recorded in the database if it has not already been recorded.

$Add\_Treatment_o$_____

$\Delta Medicare$
$d? : Person$
$p? : Person$

$d? \in Doctors$

$p? \in Patients$

$d? \mapsto p? \notin treats$

$treats' = treats \cup \{d? \mapsto p?\}$

$Patients' = Patients$

$Doctors' = Doctors$

Another required operation in the medicare system is an operation for adding a doctor to the database. The input to this operation is the name of the doctor to be added. The specification for an *Add_Patient* operation is quite similar but is not shown here.

Add_Doctor_o_____

$\Delta Medicare$
$d? : Person$

$d? \notin Doctors$

$Doctors' = Doctors \cup \{d?\}$

$treats' = treats$

$Patients' = Patients$

An operation for deleting patients from the medicare system is required. The input to this operation is the name of the patient to be deleted. If this patient is a valid medicare patient, then he or she is removed from the set of valid medicare patients. An operation to delete a doctor from the system is similar but is not shown here. Note how the maplets $x \mapsto p?$ are removed from the *treats* relation.

$$
\begin{array}{l}
\rule{0.5cm}{0.4pt}\ Delete\_Patient_o \rule{6cm}{0.4pt} \\
\Delta Medicare \\
p? : Person \\
\hline
p? \in Patients \\
Patients' = Patients \setminus \{p?\} \\
Doctors' = Doctors \\
treats' = treats \setminus \{x : Person \mid x \mapsto p? \in treats \wedge x \in Doctors \bullet x \mapsto p?\}
\end{array}
$$

A query operation is required for listing the patients treated by a specific doctor. The input to this operation is the name of the doctor. If this name is a valid doctor's name, the list of all patients treated by this doctor is returned in the output variable *names!*.

$$
\begin{array}{l}
\rule{0.5cm}{0.4pt}\ Treating_o \rule{2cm}{0.4pt} \\
\Xi Medicare \\
d? : Person \\
names! : \mathbb{P} Person \\
\hline
d? \in Doctor \\
names! = treats(\!(\{d?\})\!)
\end{array}
$$

Another typical query of the medicare system concerns those doctors who have a specific patient as a client. This operation is specified in the following schema that requires the name of the patient as input, and it stores the names of the doctors treating that patient in the output variable *names!*.

$$
\begin{array}{l}
\rule{0.5cm}{0.4pt}\ Treated\_By_o \rule{2cm}{0.4pt} \\
\Xi Medicare \\
p? : Person \\
names! : \mathbb{P} Person \\
\hline
p? \in Patients \\
names! = treats^{-1}(\!(\{p?\})\!)
\end{array}
$$

The following schema is used to confirm a successful operation:

$$
\begin{array}{l}
\rule{0.5cm}{0.4pt}\ OK \rule{3cm}{0.4pt} \\
output! : Possible\_Messages \\
\hline
output! = ok
\end{array}
$$

If a doctor is unknown to the system in any of the above schemas, the following error schema can be used:

```
┌─ Unknown_Doctor ──────────
│ ΞMedicare
│ d? : Person
│ output! : Possible_Messages
├─────────────────────────────
│ d? ∉ Doctors
│ output! = unknown_doctor
└─────────────────────────────
```

In the case in which a doctor is being added to the system, but the doctor's name already exists in the system, the following error schema should be used:

```
┌─ Doctor_Already_Exists ──────────
│ ΞMedicare
│ d? : Person
│ output! : Possible_Messages
├──────────────────────────────────
│ d? ∈ Doctors
│ output! = doctor_already_in_system
└──────────────────────────────────
```

In the case of a patient name that is not contained in the set of all valid patients, the following error schema should be used:

```
┌─ Unknown_Patient ──────────
│ ΞMedicare
│ p? : Person
│ output! : Possible_Messages
├─────────────────────────────
│ p? ∉ Patients
│ output! = unknown_patient
└─────────────────────────────
```

If a doctor begins treatment of a patient and is treating that patient already, then the mapping from doctor to patient does not need to be added to the $treats$ relation.

```
┌─ Already_Treating ──────────────────
│ p? : Person
│ d? : Doctor
│ output! : Possible_Messages
├──────────────────────────────────────
│ d? ↦ p? ∈ treats
│ output! = doctor_already_treating_patient
└──────────────────────────────────────
```

The following are the preceding operations complete with error schemas:

$$Add\_Treatment \mathrel{\widehat{=}} (Add\_Treatment_o \wedge OK)$$
$$\vee\, Unknown\_Doctor$$
$$\vee\, Unknown\_Patient$$
$$\vee\, Already\_Treating$$
$$Add\_Doctor \quad\ \mathrel{\widehat{=}} (Add\_Doctor_o \wedge OK)$$
$$\vee\, Doctor\_Already\_Exists$$
$$Delete\_Patient \mathrel{\widehat{=}} (Delete\_Patient_o \wedge OK)$$
$$\vee\, Unknown\_Patient$$
$$Treating \quad\quad \mathrel{\widehat{=}} (Treating_o \wedge OK)$$
$$\vee\, Unknown\_Doctor$$
$$Treated\_By \quad\ \mathrel{\widehat{=}} (Treated\_By_o \wedge OK)$$
$$\vee\, Unknown\_Patient$$

Solutions 8.4.7

2. The following schemas are extensions to the specification for the apartment example presented in this chapter. The previous specification did not handle errors. One error occurs when a tenant is unknown to the system. The following schema handles this error. The input is the name of the tenant, and if this name is not in the set of valid tenants, then the error message unknown_tenant is output.

```
┌─ Unknown_Tenant ──────────
│ ΞRental_System
│ t? : Person
│ output! : Possible_Messages
├───────────────────────────
│ t? ∉ tenants
│ output! = unknown_tenant
└───────────────────────────
```

Another error schema is needed to handle the case of a desired suite that is unknown to the system. The input to this schema is the identification of the suite. If the suite is not a valid apartment, then the appropriate error message is returned.

```
┌─ Unknown_Suite ───────────
│ ΞRental_System
│ s? : Suite
│ output! : Possible_Messages
├───────────────────────────
│ s? ∉ apartments
│ output! = unknown_suite
└───────────────────────────
```

The following schema handles the case where a user wants to add a suite to the database; however, the identification for the suite already exists in the system. The input for this schema is the identification for the suite. If this identification is already in the set of valid apartments, then the appropriate error message is produced.

> ___ *Suite_Exists* _____
> $\Xi Rental\_System$
> $s? : Suite$
> $output! : Possible\_Messages$
> _____
> $s? \in apartments$
> $output! = \text{suite\_already\_exists}$

Similar to the previous situation, the following schema handles the case where a new tenant is to be added to the system, but the identification for that tenant already exists in the system.

> ___ *Tenant_Exists* _____
> $\Xi Rental\_System$
> $t? : Person$
> $output! : Possible\_Messages$
> _____
> $t? \in tenants$
> $output! = \text{tenant\_already\_exists}$

4. The following sets are required for the dental office system being specified. It is assumed that the office has a set of codes that corresponds to different pieces of dental information one can have. Each patient has a subset of these codes in his or her record.

$[Name]$	The set of all names
$[Address]$	The set of all addresses
$[Dental\_Info]$	The set of all dental information codes
$Possible\_Messages ::=$	ok | ID_number_already_in_use |
	invalid_ID_number

   The dental office system is based on the preceding sets and two relations: *has_record* and *identifies*. *has_record* is a relation between the set of natural numbers and the power set of the dental information codes. The power set of the dental information codes is used because each patient has a subset of these codes in his or her record. The *identifies* relation relates a patient's ID number to his or her personal information. Two system invariants ensure that the domains of these relations are always within the scope of valid $ID\_Numbers$.

```
┌─ Dental_Office ──────────────────────┐
│ ID_Numbers : ℙℕ                       │
│ Names : ℙName                         │
│ Addresses : ℙAddress                  │
│ has_record : ℕ → ℙDental_Info         │
│ identifies : ℕ → Name → Address       │
├───────────────────────────────────────┤
│ dom has_record ⊆ ID_Numbers           │
│ dom identifies ⊆ ID_Numbers           │
└───────────────────────────────────────┘
```

The initial state of the dental office is defined by having all of the sets and relations initialized to $\varnothing$.

```
┌─ Initial ──────────────┐
│ Dental_Office          │
├────────────────────────┤
│ ID_Numbers = ∅         │
│ Names = ∅              │
│ Addresses = ∅          │
│ has_record = ∅         │
│ identifies = ∅         │
└────────────────────────┘
```

The first operation to be specified is the $Add\_Patient$ operation. This operation requires three input variables: the patient's ID number, name, and address. If the ID number is not one already in use, the sets $Names$, $ID\_Numbers$, and $Addresses$ get updated with the new information, the relation has_record gets augmented by the mapping from the new ID number to an empty record, and the $identifies$ relation gets updated by the mapping from the new ID number to the patient's personal information.

```
┌─ Add_Patient₀ ───────────────────────────────────┐
│ ΔDental_Office                                     │
│ n? : ℕ                                             │
│ Nm? : Name                                         │
│ Ad? : Address                                      │
├────────────────────────────────────────────────────┤
│ n? ∉ ID_Numbers                                    │
│ Names' = Names ∪ {Nm?}                             │
│ ID_Numbers' = ID_Numbers ∪ {n?}                    │
│ Addresses' = Addresses ∪ {Ad?}                     │
│ has_record' = has_record ∪ {n? ↦ ∅}               │
│ identifies' = identifies ∪ {n? ↦ {Nm? ↦ Ad?}}     │
└────────────────────────────────────────────────────┘
```

Deleting a patient is a little simpler than adding one. The only input required for this schema is the ID number of the patient involved. If the ID number provided is a valid number, then the ID number is removed from the set of valid ID numbers.

$$
\begin{array}{|l}
\underline{\;Delete\_Patient_o\;}\underline{\hspace{6cm}}\\
\Delta Dental\_Office\\
n? : \mathbb{N}\\
\hline
n? \in ID\_Numbers\\
ID\_Numbers' = ID\_Numbers \setminus \{n?\}\\
Names' = Names \setminus \mathrm{dom}\,(identifies\,n?)\\
Addresses' = Addresses \setminus \mathrm{ran}\,(identifies\,n?)\\
has\_record' = has\_record \setminus \{y : \mathbb{P}Dental\_Info \mid\\
\qquad\qquad\qquad n? \mapsto y \in has\_record' \bullet n? \mapsto y\}\\
identifies' = identifies \setminus \{y : Name \rightarrow Address \mid\\
\qquad\qquad\qquad n? \mapsto y \in identifies' \bullet n? \mapsto y\}
\end{array}
$$

Dental visits need to be recorded in the dental office system. The input to the following schema is the ID number of the patient and the codes corresponding to updates on the patient's dental information as a result of an appointment. This schema will update the has_record relation so that the new dental information is recorded.

$$
\begin{array}{|l}
\underline{\;Add\_Visit_o\;}\underline{\hspace{6cm}}\\
\Delta Dental\_Office\\
n? : \mathbb{N}\\
DI? : Dental\_Info\\
\hline
n? \in ID\_Numbers\\
has\_record' = has\_record \oplus \{n? \mapsto ((has\_record\,n?) \cup \{DI?\})\}\\
Names' = Names\\
ID\_Numbers' = ID\_Numbers\\
Addresses' = Addresses\\
identifies' = identifies
\end{array}
$$

The following schema is used to extract a patient's personal record when given their ID number. If the number provided is a valid ID number, then the name and the address of the patient is stored in the output variables $Name\_is!$ and $Address\_is!$, respectively.

$$
\begin{array}{|l}
\underline{\;Personal\_Query_o\;}\underline{\hspace{4cm}}\\
\Xi Dental\_Office\\
n? : \mathbb{N}\\
Name\_is! : Name\\
Address\_is! : Address\\
\hline
n? \in ID\_Numbers\\
\{Name\_is!\} = \mathrm{dom}\,(identifies\,n?)\\
\{Address\_is!\} = \mathrm{ran}\,(identifies\,n?)
\end{array}
$$

The following schema is used to extract a patient's dental record from the system. Using the ID number as input, if the number is valid, the desired information is stored (as a set of codes) in the output variable $Dental\_Record!$.

$$
\begin{array}{|l}
\_Dental\_Query_o \rule{4cm}{0pt} \\
\Xi Dental\_Office \\
n? : \mathbb{N} \\
Dental\_Record! : \mathbb{P} Dental\_Info \\
\hline
n? \in ID\_Numbers \\
Dental\_Record! = has\_record\ n? \\
\end{array}
$$

The following schema is designed to handle error situations that arise from a schema which is trying to add a patient. If the ID number already is active, then this is an error, and this schema will output an appropriate message.

$$
\begin{array}{|l}
\_Already\_A\_Patient \rule{3.5cm}{0pt} \\
\Xi Dental\_Office \\
n? : \mathbb{N} \\
output! : Possible\_Messages \\
\hline
n? \in ID\_Numbers \\
output! = \text{ID\_number\_already\_in\_use} \\
\end{array}
$$

If an invalid ID number is provided as input, the following schema can be used to produce an appropriate error message:

$$
\begin{array}{|l}
\_Not\_A\_Patient \rule{3cm}{0pt} \\
\Xi Dental\_Office \\
n? : \mathbb{N} \\
output! : Possible\_Messages \\
\hline
n? \notin ID\_Numbers \\
output! = \text{invalid\_ID\_number} \\
\end{array}
$$

In the event that an operation is successful, it is desirable to inform the user. This next schema specifies an operation for generating a positive response.

$$
\begin{array}{|l}
\_OK\_Message \rule{3cm}{0pt} \\
output! : Possible\_Messages \\
\hline
output! = \text{ok} \\
\end{array}
$$

The following are the previous operations complete with error schemas:

$$Add\_Patient \,\widehat{=}\, (Add\_Patient_o \wedge OK\_Message) \vee Already\_A\_Patient$$
$$Delete\_Patient \,\widehat{=}\, (Delete\_Patient_o \wedge OK\_Message) \vee Not\_A\_Patient$$
$$Personal\_Query \,\widehat{=}\, (Personal\_Query_o \wedge OK\_Message) \vee Not\_A\_Patient$$
$$Dental\_Query \,\widehat{=}\, (Dental\_Query_o \wedge OK\_Message) \vee Not\_A\_Patient$$

Solutions 8.4.8

2. (a) The deques require the following sets for their specification. The set End is a set of two elements used to designate which end of the queue a specific operation is to be performed on. The set $Possible\_Messages$ contains all possible output that can be produced by these schemas.

$$
\begin{aligned}
End &\ ::= \text{f} \mid \text{r} \\
Possible\_Messages &\ ::= \text{TRUE} \mid \text{FALSE}
\end{aligned}
$$

The insertion operation for a deque requires two input parameters: one designating which end and the other for the value to be inserted.

$$
\begin{array}{l}
\underline{\ Insert[T]\ \underline{\hspace{3cm}}} \\
\Delta Queue \\
new? : T \\
e? : End \\
\hline
(e? = \text{f} \wedge q' = \langle new? \rangle \frown q) \\
\vee\, (e? = \text{r} \wedge q' = q \frown \langle new? \rangle\,)
\end{array}
$$

The deletion operation requires a parameter to specify which end to delete from and an output variable in which the deleted element can be stored.

$$
\begin{array}{l}
\underline{\ Delete[T]\ \underline{\hspace{3.5cm}}} \\
\Delta Queue \\
x! : T \\
e? : End \\
\hline
(e? = \text{f} \wedge (x! = head\ q \wedge q' = tail\ q)) \\
\vee\, (e? = \text{r} \wedge (x! = last\ q \wedge q' = front\ q))
\end{array}
$$

The $Examine$ schema is similar to the $Delete$ schema except that the element is not actually removed from the queue.

$$
\begin{array}{l}
\underline{\ Examine[T]\ \underline{\hspace{2cm}}} \\
\Xi Queue \\
e? : End \\
x! : T \\
\hline
(e? = \text{f} \wedge x! = head\ q) \\
\vee\, (e? = \text{r} \wedge x! = last\ q)
\end{array}
$$

The following schema examines whether or not there are elements in the queue. The variable $output!$ stores the result of this query and displays it to the screen.

$$
\begin{array}{|l}
\underline{\mathit{Empty}} \\
\Xi Queue \\
output! : Possible\_Messages \\
\hline
(q = \langle \rangle \wedge output! = \text{TRUE}) \\
\vee \\
(output! = \text{FALSE}) \\
\end{array}
$$

(b) The $Insert$ operation for this type of deque can only be performed at the rear of the queue, and therefore only one parameter is needed.

$$
\begin{array}{|l}
\underline{Insert[T]} \\
\Delta Queue \\
new? : T \\
\hline
q' = \langle new? \rangle \frown q \\
\end{array}
$$

Since deletions can be performed at both ends of this deque, a parameter is needed to specify which end to delete from, and therefore the $Delete$ and $Examine$ operations are specified as in (a). There is no change from the $Empty$ operation in (a).

(c) For this type of deque the $Insert$ and $Empty$ schemas are as specified in (a). Deletions can only be performed from the front of this queue.

$$
\begin{array}{|l}
\underline{Delete[T]} \\
\Delta Queue \\
x! : T \\
\hline
x! = head\ q \\
q' = tail\ q \\
\end{array}
$$

Only the front of this queue can be examined.

$$
\begin{array}{|l}
\underline{Examine[T]} \\
\Xi Queue \\
x! : T \\
\hline
x! = head\ q \\
\end{array}
$$

CHAPTER 8 SOLUTIONS

2. The following sets are used in this specification:

[*Student*] The set of all uniquely identifiable students
[*Faculty*] The set of all uniquely identifiable faculty members
[*Courses*] The set of all uniquely identifiable courses offered

The registration system uses the sets defined above along with two relations: *takes* and *given_by*. The *takes* relation specifies which students are taking which courses. The *given_by* relation specifies what courses faculty members are teaching. The four invariants ensure that only students take courses, only courses can be taken by students, only courses can be given by faculty, and only faculty can teach courses.

$$
\begin{array}{|l}
\_\, Registration\_System \,_____ \\
Students : \mathbb{P}Student \\
Faculty\_Members : \mathbb{P}Faculty \\
Courses : \mathbb{P}Course \\
takes : Student \leftrightarrow Course \\
given\_by : Course \leftrightarrow Faculty \\
\hline
\mathrm{dom}\; takes \subseteq Students \\
\mathrm{ran}\; takes \subseteq Courses \\
\mathrm{dom}\; given\_by \subseteq Courses \\
\mathrm{ran}\; given\_by \subseteq Faculty\_Members \\
\end{array}
$$

The initial state of the registration system is specified by assigning all sets no elements and setting all relations to empty.

$$
\begin{array}{|l}
\_\, Initial \,_____ \\
Registration\_System \\
\hline
Students = \varnothing \\
Faculty\_Members = \varnothing \\
Courses = \varnothing \\
takes = \varnothing \\
given\_by = \varnothing \\
\end{array}
$$

The next two schemas specify how courses are picked up and dropped by students. In order to pick up a course, the system needs the student's identification and the name of the course as input. If the student and the course are valid, and the student is not already enrolled in the course, then the *takes* relation is updated to record the enrollment.

$$
\begin{array}{l}
\rule{0.4pt}{0pt}\underline{\ Pick\_Up\_Course_o\ } \\
\ \Delta Registration\_System \\
\ s? : Student \\
\ c? : Course \\
\hline
\ s? \in Students \\
\ c? \in Courses \\
\ s? \mapsto c? \notin takes \\
\ takes' = takes \cup \{s? \mapsto c?\} \\
\ Students' = Students \\
\ Faculty\_Members' = Faculty\_Members \\
\ Courses' = Courses \\
\ given\_by' = given\_by
\end{array}
$$

When dropping a course, the student identification and course name received as input must be valid, and, in addition, the student must be enrolled in the course. If these conditions are satisfied, then the relation between the student and the course is removed from the relation *takes*.

$$
\begin{array}{l}
\rule{0.4pt}{0pt}\underline{\ Drop\_Course_o\ } \\
\ \Delta Registration\_System \\
\ s? : Student \\
\ c? : Course \\
\hline
\ s? \in Students \\
\ c? \in Courses \\
\ s? \mapsto c? \in takes \\
\ takes' = takes \setminus \{s? \mapsto c?\} \\
\ Students' = Students \\
\ Faculty\_Members' = Faculty\_Members \\
\ Courses' = Courses \\
\ given\_by' = given\_by
\end{array}
$$

Given a set of faculty members and a set of courses, an operation is required to assign faculty to the courses. The following schema accepts a faculty member and a course as input, and given that these inputs are valid, the faculty member is assigned to teach the course if he or she is not already teaching it. This update is performed by modifying the relation *given_by*.

$$
\begin{array}{l}
\underline{\quad Assign\_Course_o \quad\rule{3cm}{0.4pt}\quad} \\
\Delta Registration\_System \\
c? : Course \\
f? : Faculty \\
\hline
c? \in Courses \\
f? \in Faculty\_Members \\
c? \mapsto f? \notin given\_by \\
given\_by' = given\_by \cup \{c? \mapsto f?\} \\
Students' = Students \\
Faculty\_Members' = Faculty\_Members \\
Courses' = Courses \\
takes' = takes
\end{array}
$$

The following schema specifies the operation for obtaining all courses a student is taking. Given a valid student as input, this operation generates all courses taken by that student and stores them in the output variable $c!$.

$$
\begin{array}{l}
\underline{\quad Courses\_By\_Student_o \quad\rule{2cm}{0.4pt}\quad} \\
\Xi Registration\_System \\
s? : Student \\
c! : \mathbb{P}Course \\
\hline
s? \in Students \\
c! = takes (\!|\{s?\}|\!)
\end{array}
$$

The next schema outputs all students enrolled in any class in the registration system. This is done by outputting the relational image of the entire set of courses offered.

$$
\begin{array}{l}
\underline{\quad Students\_In\_Courses \quad\rule{2cm}{0.4pt}\quad} \\
\Xi Registration\_System \\
s! : \mathbb{P}Student \\
\hline
s! = takes^{-1}(\!|Courses|\!)
\end{array}
$$

The operation for obtaining the list of faculty members teaching a specific student is given next. The output is produced by generating the relational image of the relation $takes$ on the student, which produces a set of classes. The relational image of this class set in the relation $given\_by$ corresponds to the desired set of faculty members. This schema has a precondition that $s?$ is a valid student.

$$
\begin{array}{|l}
\hline
\quad Faculty\_Teaches_o \underline{\hspace{2cm}} \\
\hline
\Xi Registration\_System \\
f! : \mathbb{P}Faculty \\
s? : Student \\
\hline
s? \in Students \\
f! = given\_by(takes(\{s?\})) \\
\hline
\end{array}
$$

This next schema generates the set $s!$ consisting of all students taking the course $c?$. This schema has the precondition that $c?$ is a valid course.

$$
\begin{array}{|l}
\hline
\quad Taking\_Course_o \underline{\hspace{1.5cm}} \\
\hline
\Xi Registration\_System \\
c? : Course \\
s! : \mathbb{P}Student \\
\hline
c? \in Courses \\
s! = takes^{-1}(\{c?\}) \\
\hline
\end{array}
$$

If a student is not valid, this error schema can be used.

$$
\begin{array}{|l}
\hline
\quad Student\_Error \underline{\hspace{2cm}} \\
\hline
\Xi Registration\_System \\
s? : Student \\
output! : Possible\_Messages \\
\hline
s? \notin Students \\
output! = \text{invalid\_student} \\
\hline
\end{array}
$$

If a faculty member is not valid, the following error schema is used:

$$
\begin{array}{|l}
\hline
\quad Faculty\_Error \underline{\hspace{2cm}} \\
\hline
\Xi Registration\_System \\
f? : Faculty \\
output! : Possible\_Messages \\
\hline
f? \notin Faculty\_Members \\
output! = \text{invalid\_faculty} \\
\hline
\end{array}
$$

The following error schema handles the case where a course provided as input is not in the set of valid courses:

$$
\begin{array}{|l}
\hline
\quad Course\_Error \underline{\hspace{2cm}} \\
\hline
\Xi Registration\_System \\
c? : Course \\
output! : Possible\_Messages \\
\hline
c? \notin Courses \\
output! = \text{invalid\_course} \\
\hline
\end{array}
$$

If a faculty member has already been assigned to a certain class, then the following error schema should be used:

$$
\begin{array}{|l}
\underline{\;Given\_Error\;}\rule{4cm}{0pt} \\
\Xi Registration\_System \\
c? : Course \\
f? : Faculty \\
output! : Possible\_Messages \\
\hline
c? \mapsto f? \in given\_by \\
output! = \text{faculty\_member\_already\_assigned}
\end{array}
$$

If a student is trying to drop a class he or she is not in, the following error schema should be used.

$$
\begin{array}{|l}
\underline{\;Takes\_Error\;}\rule{3cm}{0pt} \\
\Xi Registration\_System \\
s? : Student \\
c? : Course \\
output! : Possible\_Messages \\
\hline
s? \mapsto c? \notin takes \\
output! = \text{student\_not\_in\_class}
\end{array}
$$

The following error schema handles the case where a student tries to enroll in a course he or she is already enrolled in:

$$
\begin{array}{|l}
\underline{\;Already\_In\_Class\;}\rule{3cm}{0pt} \\
\Xi Registration\_System \\
s? : Student \\
c? : Course \\
output! : Possible\_Messages \\
\hline
s? \mapsto c? \in takes \\
output! = \text{student\_already\_in\_class}
\end{array}
$$

The following are the previously defined operations complete with error schemas:

$$
\begin{aligned}
Pick\_Up\_Course \quad &\cong (Pick\_Up\_Course_o \wedge OK) \vee Already\_In\_Class \\
&\quad Invalid\_Student \vee Invalid\_Course \\
Drop\_Course \quad &\cong (Drop\_Course \wedge OK) \vee Already\_In\_Class \\
&\quad Invalid\_Student \vee Invalid\_Course \\
Assign\_Course \quad &\cong (Assign\_Course_o \wedge OK) \vee Given\_Error \\
&\quad Invalid\_Student \vee Invalid\_Faculty \\
Course\_By\_Student &\cong (Course\_By\_Student_o \wedge OK) \vee Student\_Error \\
Faculty\_Teaches \quad &\cong (Faculty\_Teaches_o \wedge OK) \vee Student\_Error \\
Taking\_Course \quad &\cong (Taking\_Course_o \wedge OK) \vee Invalid\_Course
\end{aligned}
$$

4. The following sets are used in the specification of this system:

[$Book$]	The set of all uniquely identifiable books
[$Author$]	The set of all uniquely identifiable authors
[$Person$]	The set of all uniquely identifiable people
$Possible\_Messages ::=$	book_not_in_system | patron_not_in_system |
	unknown_author | book_already_signed_out |
	book_not_signed_out | quota_exceeded | ok

There is one relation between the sets in this system. The relation $written\_by$ associates authors with the books they have written. The two partial functions $signed\_out$ and $last\_person$ are used to keep track of those patrons who have signed out which books, and the patron who was the last to sign out a given book. The num_books function is used to track how many books each patron has signed out at any given time. There are three system invariants for this example which stipulate that only patrons are allowed to sign out books from the library.

$$
\begin{array}{|l}
\underline{\ Library\ } \\
books : \mathbb{P}\,Book \\
authors : \mathbb{P}\,Author \\
patrons : \mathbb{P}\,Person \\
written\_by : Book \leftrightarrow Author \\
signed\_out : Book \nrightarrow Person \\
last\_person : Book \nrightarrow Person \\
quota : \mathbb{N} \\
num\_books : Person \rightarrow \mathbb{N} \\
\hline
\mathrm{ran}\ last\_person \subseteq patrons \\
\mathrm{ran}\ signed\_out \subseteq patrons \\
\mathrm{dom}\ num\_books \subseteq patrons \\
\end{array}
$$

The initial state is defined by setting all sets, relations, functions, and partial functions to empty. In addition, a system variable to limit the number of books each patron can sign out, namely, $quota$, is set to four.

$$
\begin{array}{|l}
\underline{\ Initial\ } \\
Library \\
\hline
books = \varnothing \\
authors = \varnothing \\
patrons = \varnothing \\
written\_by = \varnothing \\
signed\_out = \varnothing \\
last\_person = \varnothing \\
quota = 4 \\
num\_books = \varnothing \\
\end{array}
$$

The following two schemas specify operations for checking out and returning books from the library. The checkout operation requires a book and a patron as input. If these two inputs exist in the system, the book is not already signed out, and the patron is below his or her quota and the patron is allowed to sign out the book.

$$
\begin{array}{l}
\underline{\quad Check\_Out_o } \\
\Delta Library \\
b? : Book \\
p? : Person \\
\hline
b? \notin \text{dom } signed\_out \\[4pt]
num\_books\ p? < quota \\[4pt]
p? \in patrons \\[4pt]
b? \in books \\[4pt]
signed\_out' = signed\_out \cup \{b? \mapsto p?\} \\[4pt]
num\_books' = num\_books \oplus \{p? \mapsto (num\_books\ p? + 1)\} \\[4pt]
books' = books \\[4pt]
authors' = authors \\[4pt]
patrons' = patrons \\[4pt]
written\_by' = written\_by \\[4pt]
last\_person' = last\_person \oplus \{b? \mapsto p?\}
\end{array}
$$

When a book is returned to the library, only the $signed\_out$ and num_books operations need to be updated. This operation is specified in the next schema.

$$
\begin{array}{l}
\underline{\quad Return\_Book_o } \\
\Delta Library \\
b? : Book \\
p? : Person \\
\hline
b? \in \text{dom } signed\_out \\[4pt]
signed\_out' = signed\_out \setminus \{b? \mapsto p?\} \\[4pt]
num\_books' = num\_books \oplus \{p? \mapsto (num\_books\ p? - 1)\} \\[4pt]
books' = books \\[4pt]
authors' = authors \\[4pt]
patrons' = patrons \\[4pt]
written\_by' = written\_by \\[4pt]
last\_person' = last\_person
\end{array}
$$

The following schema specifies an operation for returning the set of authors who wrote a specific book. Provided that a valid author is given as input, the relational image of this author through the inverse of the relation $written\_by$ generates the desired set.

$$
\begin{array}{|l}
\text{\textit{Writes}}_o \underline{\hspace{4cm}} \\
\Xi Library \\
a? : Author \\
bookswritten! : \mathbb{P}Book \\
\hline
a? \in authors \\
bookswritten! = written\_by^{-1}(\!|\{a?\}|\!) \\
\end{array}
$$

The following schema provides the set of books currently on loan to a specific patron. The precondition is that the patron is a valid user of the library system.

$$
\begin{array}{|l}
\text{\textit{On\_Loan}}_o \underline{\hspace{3cm}} \\
\Xi Library \\
p? : Person \\
loaner! : \mathbb{P}Book \\
\hline
p? \in patrons \\
loaner! = signed\_out^{-1}(\!|\{p?\}|\!) \\
\end{array}
$$

The next schema specifies an operation that will produce the identification of the last patron who signed out a specific book. The precondition for this operation is that the book is a valid book in the library system.

$$
\begin{array}{|l}
\text{\textit{Last\_Person}}_o \underline{\hspace{2cm}} \\
\Xi Library \\
b? : Book \\
p! : Person \\
\hline
b? \in books \\
p! = last\_person\ b? \\
\end{array}
$$

The following schemas are used to generate messages for the user relating to the execution of the operations. The next schema is used to print an ok message when an operation is successful.

$$
\begin{array}{|l}
\text{\textit{OK}} \underline{\hspace{3cm}} \\
\Xi library \\
output! : Possible\_Messages \\
\hline
output! = \text{ok} \\
\end{array}
$$

If a book is invalid, this schema is used.

$$
\begin{array}{|l}
\text{\textit{Unknown\_Book}} \underline{\hspace{2cm}} \\
\Xi Library \\
b? : Book \\
output! : Possible\_Messages \\
\hline
b? \notin books \\
output! = \text{book\_not\_in\_system} \\
\end{array}
$$

An invalid patron provided as input should be handled by this schema.

```
┌─ Unknown_Patron ─────────
│ ΞLibrary
│ p? : Person
│ output! : Possible_Messages
├──────────────────────────
│ p? ∉ patrons
│ output! = patron_not_in_system
└
```

Any author unknown to the library system will invoke this schema.

```
┌─ Unknown_Author ─────────
│ ΞLibrary
│ a? : Author
│ output! : Possible_Messages
├──────────────────────────
│ a? ∉ authors
│ output! = unknown_author
└
```

If a patron tries to sign out a book that is already gone, the following schema is used:

```
┌─ Already_Gone ─────────────────
│ ΞLibrary
│ b? : Book
│ output! : Possible_Messages
├────────────────────────────────
│ b? ∈ dom signed_out
│ output! = book_already_signed_out
└
```

If a book is returned that was not signed out, the following schema is used:

```
┌─ Not_Gone ─────────────────
│ ΞLibrary
│ b? : Book
│ output! : possible_Messages
├────────────────────────────
│ b? ∉ dom signed_out
│ output! = book_not_signed_out
└
```

A patron trying to violate his or her book quota can be handled by the following error schema:

```
┌─ Too_Many_Books ─────────
│ ΞLibrary
│ p? : Person
│ output! : Possible_Messages
├──────────────────────────
│ num_books p? > quota − 1
│ output! = quota_exceeded
└
```

The following are the previously specified operations complete with error schemas:

$$Check\_Out \quad \widehat{=} (Check\_Out_o \wedge OK) \vee Too\_Many\_Books$$
$$\vee\ Already\_Gone \vee Unknown\_Book \vee Unknown\_Patron$$
$$Return\_Book \widehat{=} (Return\_Book_o \wedge OK) \vee Not\_Gone$$
$$Writes \qquad \widehat{=} Writes_o \vee Unknown\_Author$$
$$On\_Loan \qquad \widehat{=} On\_Loan_o \vee Unknown\_Patron$$
$$Last\_Person \ \widehat{=} LastPerson_o \vee Unknown\_Book$$

Solutions 9.1

2. **(a)** $\{\ \}$ i := j; j := i $\{i_\omega = j_\alpha, j_\omega = j_\alpha\}$
   (b) $\{\ \}$ i := 2 ∗ i + j $\{i_\omega = 2i_\alpha + j_\alpha,\ j_\omega = j_\alpha\}$
   (c) $\{\ \}$ sum := sum + a $\{sum_\omega = sum_\alpha + a_\alpha,\ a_\omega = a_\alpha\}$

Solutions 9.2

2. $\{i > 2,\ j > 3,\ k = i \ast j\} \Rightarrow \{k > 6\}$
   Hence, if $\{i > 2,\ j > 3,\ k = i \ast j\}$ is a postcondition, it may be replaced by $\{k > 6\}$ but not if it is a precondition.

4. **(a)** $P \vee D$ weakens P: $\{P \vee D\}C\{Q\}$ not proven
   (b) $P \wedge Q$ strengthens P: $\{P \wedge Q\}C\{Q\}$ proven
   (c) $Q \vee D$ weakens Q: $\{P\}C\{Q \vee D\}$ proven
   (d) $Q \vee P$ weakens Q: $\{P\}C\{Q \vee P\}$ proven
   (e) $Q \wedge P$ strengthens Q: $\{P\}C\{Q \wedge P\}$ not proven

6.
$$\{\ \}\ i := 2 \ast i\ \{i_\omega = 2i_\alpha\}$$
$$\{i_\alpha > 2\}\ i := 2 \ast i\ \{i_\alpha > 2\}$$

$$\{i_\alpha > 2\}\ i := 2 \ast i\ \{i_\omega = 2i_\alpha,\ i_\alpha > 2\}$$
$$\{i_\omega = 2i_\alpha,\ i_\alpha > 2\} \Rightarrow \{i_\omega > 4\}$$

Solutions 9.3

2.
$$\{a = A,\ b = B\}$$
$$\{a + b = A + B,\ a - b = A - B\}\ \text{h} := \text{a} \quad \{a + b = A + B,\ h - b = A - B\}$$
$$\{a + b = A + B,\ h - b = A - B\}\ \text{a} := \text{a} + \text{b}\ \{a = A + B,\ h - b = A - B\}$$
$$\{a = A + B,\ h - b = A - B\} \qquad \text{b} := \text{h} - \text{b}\ \{a = A + B,\ b = A - B\}$$

4.
$$\{sum = (j + 1)j/2 - j$$
$$= j \ast (j - 1)/2\}$$
$$\{sum + j = (j + 1)j/2\} \qquad \text{sum} := \text{sum} + \text{j};\ \{sum = (j + 1)j/2\}$$
$$\{sum = (j + 1)(j + 1 - 1)/2\}\ \text{j} := \text{j} + 1; \qquad \{sum = j \ast (j - 1)/2\}$$

Solutions 9.4

2. If the arguments are passed by reference, then more than one argument may have the same reference, which leads to aliasing. If there are n arguments, 2^n cases have to be considered.

4. case $i \neq j$:
$\qquad \{a[i] + a[j] \geq a[j]\}$ a[i] := a[i] + a[j] $\{a[i] \geq a[j]\}$
case $i = j$: replace j by i
$\qquad \{a[i] + a[i] \geq a[i]\}$ a[i] := a[i] + a[j] $\{a[i] \geq a[j]\}$

6.
$\qquad$ { }

$\{0 = \sum_{j=1}^{1-1} j!, 1 = 1!\}$ $\qquad$ i := 1; $\qquad$ $\{0 = \sum_{j=1}^{i-1} j!, 1 = i!\}$

$\{0 = \sum_{j=1}^{i-1} j!, 1 = i!\}$ $\qquad$ sum := 0; $\qquad$ $\{sum = \sum_{j=1}^{i-1} j!, 1 = i!\}$

$\{sum = \sum_{j=1}^{i-1} j!, 1 = i!\}$ $\qquad$ f := 1; $\qquad$ $\{sum = \sum_{j=1}^{i-1} j!, f = i!\}$

$\qquad\qquad$ **while** i <> n + 1 **do**

$\qquad\qquad$ **begin** $\qquad$ $\{sum = \sum_{j=1}^{i-1} j!, f = i!\}$

$\{sum + f = \sum_{j=1}^{i} j!, f = i!\}$ $\qquad$ sum := sum + f; $\qquad$ $\{sum = \sum_{j=1}^{i} j!, f = i!\}$

$\{sum = \sum_{j=1}^{i+1-1} j!,$
$\qquad f = (i+1-1)!\}$ $\qquad$ i := i + 1; $\qquad$ $\{sum = \sum_{j=1}^{i-1} j!,$
$\qquad\qquad\qquad\qquad\qquad\qquad f = (i-1)!\}$

$\{sum = \sum_{j=1}^{i-1} j!, f*i = i!\}$ $\qquad$ f := f * i $\qquad$ $\{sum = \sum_{j=1}^{i-1} j!, f = i!\}$
$\qquad\qquad$ **end**

$(sum = \sum_{j=1}^{i-1} j! \wedge f = i! \wedge i = n + 1) \Rightarrow sum = \sum_{j=1}^{n+1-1} j!$

CHAPTER 9 SOLUTIONS

2. **(a)** $\{3 * k > 6\}$ i := 3 * k $\{i > 6\}$ or $\{k > 2\}$ i := 3 * k $\{i > 6\}$
$\quad$ **(b)** $\{b * c = 1\}$ a := b * c $\{a = 1\}$
$\quad$ **(c)** Precondition of a = a / b is $b \neq 0$. Hence $\{c - 2 \neq 0\}$ b := c $-$ 2 $\{b \neq 0\}$ or $\{c \neq 2\}$
$\qquad$ b := c $-$ 2 $\{b \neq 0\}$.

4. $\{-x \geq 0\}$ x := $-$x $\{x \geq 0\}$ or $\{x \leq 0\}$ x := $-$x $\{x \geq 0\}$. The rule for conditions yields

$$\{x \leq 0\} \text{ x} := -\text{x } \{x \geq 0\}$$
$$\neg\{x \leq 0\} \Rightarrow \{x \geq 0\}$$
$$\overline{\qquad\qquad\qquad\qquad\qquad\qquad}$$
$$\{ \} \text{ if } x < 0 \text{ then } x := -x \ \{x \geq 0\}$$

6. Loop invariant:

$$\bigwedge_{j=1}^{i} (max \geq a[j]) \wedge \bigvee_{j=1}^{i} (max = a[j]) \wedge (i < n + 1)$$

First, prove

$$\{\bigwedge_{j=1}^{i-1}(max \geq a[j]) \wedge \bigvee_{j=1}^{i-1}(max = a[j])\}$$

if (a[i] > max) **then** max := a[i];

$$\{\bigwedge_{j=1}^{i}(max \geq a[j]) \wedge \bigvee_{j=1}^{i}(max = a[j])\}$$

According to the rule for the conditional, one has to conduct the proof for $a[i] > max$ and $\neg(a[i] > max)$.

Case: $a[i] > max$

$$\{max < a[i] \wedge \bigwedge_{j=1}^{i-1}(max \geq a[j])\}$$

$$\Rightarrow \{max < a[i] \wedge \bigwedge_{j=1}^{i-1}(a[i] > a[j])\}$$

$$\Rightarrow \{\bigwedge_{j=1}^{i-1}(a[i] > a[j])\}$$

max := a[i];

$$\{\bigwedge_{j=1}^{i-1}(max > a[j]) \wedge max = a[i]\}$$

$$\Rightarrow \{\bigwedge_{j=1}^{i}(max \geq a[j])\}$$

Moreover

$$\{max < a[i] \wedge \bigvee_{j=1}^{i-1}(max = a[j])\}$$

$$\Rightarrow \{max < a[i]\}$$

max := a[i];

$$\{max = a[i]\}$$

$$\Rightarrow \{\bigvee_{j=1}^{i}(max = a[j])\}$$

By the conjunction rule

$$\{max < a[i] \wedge \bigwedge_{j=1}^{i-1}(max \geq a[j]) \wedge \bigvee_{j=1}^{i-1}(max = a[j])\}$$

max := a[i];

$$\{\bigwedge_{j=1}^{i}(max \geq a[i]) \wedge \bigvee_{j=1}^{i}(max = a[j])\}$$

Case: $\neg(a[i] > max)$

$$\{max \geq a[i] \wedge \bigwedge_{j=1}^{i-1}(max \geq a[j]) \wedge \bigvee_{j=1}^{i-1}(max = a[j])\}$$

$$\Rightarrow \{\bigwedge_{j=1}^{i}(max \geq a[j]) \wedge \bigvee_{j=1}^{i}(max = a[j])\}$$

The if rule is now satisfied. The main proof is as follows:

$\{a[1] \geq a[1], a[1] = a[1]\}$ max := a[1];

$\{\bigwedge_{j=1}^{1}(max \geq a[j]),$ i := 1;

$\bigvee_{j=1}^{1}(max = a[j]),$

$1 < n + 1\}$

$\{max \geq a[1], max = a[1],$
$\qquad n > 0\}$
$\{\bigwedge_{j=1}^{i}(max \geq a[j]),$
$\bigvee_{j=1}^{i}(max = a[j]),$
$i < n + 1\}$

while (i < n) **do**
    begin

$\{\bigwedge_{j=1}^{i}(max \geq a[j]),$
$\bigvee_{j=1}^{i}(max = a[j]),$
$i < n,$
$i < n + 1\}$

$\{\bigwedge_{j=1}^{i+1-1}(max \geq a[j]),$ i := i + 1;

$\bigvee_{j=1}^{i+1-1}(max = a[j]),$

$i + 1 < n + 1\}$

$\{\bigwedge_{j=1}^{i-1}(max \geq a[j]),$
$\bigvee_{j=1}^{i-1}(max = a[j]),$
$i < n + 1\}$

see above   **if** (a[i] > max) **then** $\{\bigwedge_{j=1}^{i}(max \geq a[j]),$

        max := a[i] $\bigvee_{j=1}^{i}(max = a[j]),$

$i < n + 1\}$

    end

$$\neg(i < n) \wedge \bigwedge_{j=1}^{i}(max \geq a[j]) \wedge \bigvee_{j=1}^{i}(max = a[j]) \wedge (i < n + 1)$$

$$\Rightarrow (i \geq n) \wedge (i < n + 1) \wedge \bigwedge_{j=1}^{i}(max \geq a[j]) \wedge \bigvee_{j=1}^{i}(max = a[j])$$

$$\Rightarrow (i = n) \wedge \bigwedge_{j=1}^{i}(max \geq a[j]) \wedge \bigvee_{j=1}^{i}(max = a[j])$$

$$\Rightarrow \{\bigwedge_{j=1}^{n}(max \geq a[j]) \wedge \bigvee_{j=1}^{n}(max = a[j])\}$$

For $n \leq 0$, the condition after the end statement cannot be satisfied. This case must be excluded.

8. Loop invariant

$$\bigwedge_{j=1}^{i-1}(a[j] = \sum_{k=1}^{j} A[k]), \bigwedge_{j=i}^{n}(a[j] = A[j]), s = \sum_{k=1}^{i-1} A[k], i \leq n + 1$$

Since each statement has a rather complex precondition and postcondition, we prove the program segment statement by statement.

Proof of loop invariant:

Last statement of loop

$$\{\bigwedge_{j=1}^{i}(a[j] = \sum_{k=1}^{j} A[k]), \bigwedge_{j=i+1}^{n}(a[j] = A[j]), s = \sum_{k=1}^{i} A[k], i \leq n\}$$
$$i := i + 1$$
$$\{\bigwedge_{j=1}^{i-1}(a[j] = \sum_{k=1}^{j} A[k]), \bigwedge_{j=i}^{n}(a[j] = A[j]), s = \sum_{k=1}^{i-1} A[k], i \leq n + 1\}$$

Rules used: assignment rule, together with $i - 1 + 1 = i$. The precondition of this statement is the postcondition of the previous statement. In this condition, we isolate all $a[i]$, which yields

$$\{(a[i] = \textstyle\sum_{k=1}^{i} A[k]), \bigwedge_{j=1}^{i-1}(a[j] = \sum_{k=1}^{j} A[k]), s = \sum_{k=1}^{i} A[k], i \leq n,$$
$$\bigwedge_{j=i+1}^{n}(a[j] = A[j])\}$$

By the assignment rule for $a[i] := s$, $a[i]$ becomes s, and the term $s = \sum_{k=1}^{i} A[k]$ appears twice. Dropping this term yields the precondition

$$\{\bigwedge_{j=1}^{i-1}(a[j] = \textstyle\sum_{k=1}^{j} A[k]), s = \sum_{k=1}^{i} A[k], \bigwedge_{j=i+1}^{n}(a[j] = A[j]), i \leq n\}$$

By precondition strengthening, this becomes

$$\{\bigwedge_{j=1}^{i-1}(a[j] = \textstyle\sum_{k=1}^{j} A[k]), s = \sum_{k=1}^{i} A[k], (a[i] = A[i]), \bigwedge_{j=i+1}^{n}(a[j] = A[j]), i \leq n\}$$

This becomes the postcondition of $s := s + a[i]$, and the assignment rule yields for this statement

$$\{\bigwedge_{j=1}^{i-1}(a[j] = \textstyle\sum_{k=1}^{j} A[k]), s + a[i] = (\sum_{k=1}^{i-1} A[k]) + A[i], (a[i] = A[i]),$$
$$\bigwedge_{j=i+1}^{n}(a[j] = A[j]), i \leq n\}$$

Since $a[i] = A[i]$,

$$s + a[i] = (\textstyle\sum_{k=1}^{i-1} A[k]) + A[i] \Rightarrow s = \sum_{k=1}^{i-1} A[k],$$

and, after combining $a[i] = A[i]$ with the conjunction that follows this term, one finds

$$\{\bigwedge_{j=1}^{i-1}(a[j] = \textstyle\sum_{k=1}^{j} A[k]), (s = \sum_{k=1}^{i-1} A[k]), \{\bigwedge_{j=i}^{n}(a[j] = A[j]), i \leq n\}$$

This is logically equivalent to the conjunction of the while-condition $i \leq n$ and the loop invariant. This proves the correctness of the loop.

Still left to do is the proof that the statements $i := 1$ and $s := 0$ generate the loop invariance. This is true, provided $n \geq -1$. If $n < -1$, the program is false in the sense that the stated postcondition cannot be satisfied. For $i := 1, s := 0$, one has

$$\{\bigwedge_{j=1}^{0}(a[j] = \textstyle\sum_{k=1}^{j} A[k]), \bigwedge_{j=1}^{n}(a[j] = A[j]), 0 = \sum_{k=1}^{0} A[k], 1 \leq n + 1\}$$

This reduces to the stated precondition, together with $n \geq 0$. Finally, the loop invariant, together with $\neg(i \leq n)$, implies the postcondition

$$\{\neg(i \leq n) \wedge (i \leq n + 1) \wedge \bigwedge_{j=1}^{i-1}(a[j] = \textstyle\sum_{k=1}^{j} A[k]) \wedge \bigwedge_{j=i}^{n}(a[j]$$
$$= A[j]) \wedge s = \textstyle\sum_{k=1}^{i-1} A[k]\} \Rightarrow \{i = n + 1 \wedge \bigwedge_{j=1}^{n}(a[j] = \sum_{k=1}^{j} A[k])\}$$

10. The loop variant is $n + 1 - i$. Initially, $i = 1$, and the loop variant is therefore n. With each iteration, the loop variant decreases by 1, until it reaches 0. Consequently, as long as $n \geq 0$, the loop variant is a natural number, and every decreasing sequence in the natural numbers is finite. The set of all loop variants is well founded. However, if $n < 0$, then the program does not terminate.

Solutions 10.1.5

2. **(a)** Recall that grammar for a language is ambiguous if a sentence from the language can be parsed in more than one way. The grammar is ambiguous because there are two parse trees for $3 + 4 + 5$ as shown below:

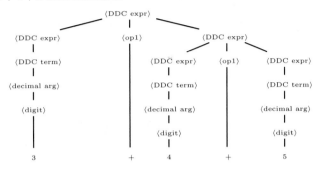

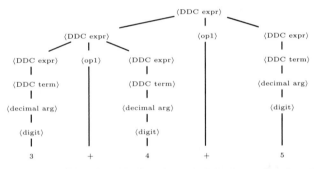

(b) To correct the grammar's ambiguity, the first production must be changed to the following:

$$\langle \text{DDC expr} \rangle ::= \langle \text{DDC term} \rangle \mid \langle \text{DDC expr} \rangle \; \langle \text{op1} \rangle \; \langle \text{DDC term} \rangle$$

(c) The expression $3 \times 6 / 2 + 4$ evaluates to 13 according to the expression tree as follows:

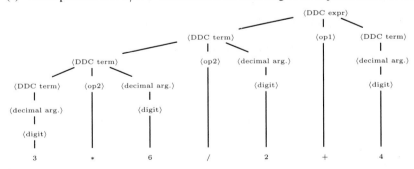

(d) When the two operator types are switched, the precedence functions are reversed (i.e., $+$ or $-$ are now done before $\times$ or $/$). The same expression now evaluates to 3 according to the new expression tree:

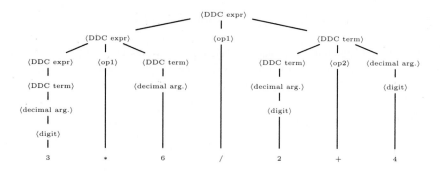

Solutions 10.1.6

2. $W_1 = \{S, A\}$
$W_2 = W_1$
$G' = (\{S, A\}, \{a, b, c\}, S, \Phi')$ where $\Phi' = \{S \to aAbA, S \to aba, A \to aAb, A \to a\}$

4. Applying algorithm ACTIVE gives the following results:

$$W_1 = \{S, A, B, C\}, W_2 = W_1$$

Therefore, all nonterminals are active and no productions from the original grammar are deleted. We now apply algorithm REDUCE with the following results:

$$W_0 = \{S\}, W_1 = \{S, A, a, b\}, W_2 = \{S, A, B, a, b\}$$
$$W_3 = \{S, A, B, C, a, b\}, W_4 = W_3$$

Again, no rules can be deleted from the original grammar.

Solutions 10.1

2. The language generated by the given grammar is

$$\{a^i | i \geq 1\} \cup \{b^i | i \geq 1\} \cup \{b^i a^j | i \geq 1 \text{ and } j \geq 1\}$$

4. $G = (V_N, V_T, S, \Phi)$ where $V_N = \{S, A, B\}$, $V_T = \{a, b\}$, and Φ consist of the following rules:

$$S \to AB \,|\, BA \,|\, SAB \,|\, SBA \,|\, ASB \,|\, BSA \,|\, ABS \,|\, BAS$$
$$A \to a$$
$$B \to b$$

6. $G = (V_N, V_T, S, \Phi)$ where $V_N = \{S, A, B, C\}$, $V_T = \{0, 1\}$, and Φ consist of the following rules:

$$S \rightarrow 1A \mid 0B$$
$$A \rightarrow 1S \mid 0C \mid 0$$
$$B \rightarrow 0S \mid 1C \mid 1$$
$$C \rightarrow 0A \mid 1B$$

8. (a) $G = (V_N, V_T, \langle \text{expr} \rangle, \Phi)$ where $V_N = \{ \langle \text{expr} \rangle \}$, $V_T = \{i, +, -, *, /, \uparrow, (,)\}$, and Φ consist of the following rules:

$$\begin{aligned}
\langle \text{expr} \rangle \;\; &::= \;\; \langle \text{term} \rangle \mid \langle \text{expr} \rangle + \langle \text{term} \rangle \\
&\mid \langle \text{expr} \rangle - \langle \text{term} \rangle \\
\langle \text{term} \rangle \;\; &::= \;\; \langle \text{factor} \rangle \mid \langle \text{term} \rangle * \langle \text{factor} \rangle \\
&\mid \langle \text{term} \rangle / \langle \text{factor} \rangle \\
\langle \text{factor} \rangle \;\; &::= \;\; \langle \text{primary} \rangle \mid \langle \text{primary} \rangle \uparrow \langle \text{factor} \rangle \\
\langle \text{primary} \rangle \;\; &::= \;\; i \mid (\langle \text{expr} \rangle)
\end{aligned}$$

Observe that left-associative operators are specified by left-recursive rules such as those for $+$ and $*$ while the right-associative operator is specified by a right-recursive rule.

(b) $i + i * i$

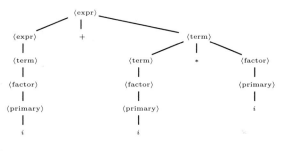

$(i + i)/i$

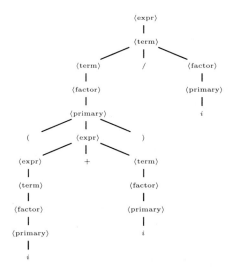

$i + i * i \uparrow i$

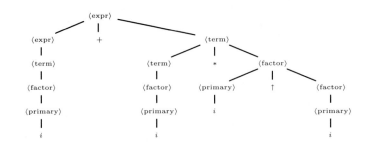

$i \uparrow i \uparrow i + i$

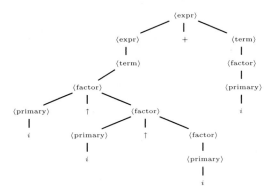

Solutions 10.2.3.2

2.

Stack	Current input symbol				
symbol	a	b	c	d	$\#$
S	$(aS, 1)$	$(bA, 2)$			
A			$(ccA, 4)$	$(d, 3)$	
a	pop				
b		pop			
c			pop		
d				pop	
$\#$					$accept$

(**a**)   Trace of the top-down parse for the string $aabccd\#$

Unused input string	Stack contents	Rules used
$aabccd\#$	$S\#$	ε
$aabccd\#$	$aS\#$	1
$abccd\#$	$S\#$	1
$abccd\#$	$aS\#$	11
$bccd\#$	$S\#$	11
$bccd\#$	$bA\#$	112
$ccd\#$	$A\#$	112
$ccd\#$	$ccA\#$	1124
$cd\#$	$cA\#$	1124
$d\#$	$A\#$	1124
$d\#$	$d\#$	11243
$\#$	$\#$	11243

(**b**)   Trace of the top-down parse for the string $bccd\#$

Unused input string	Stack contents	Rules used
$bccd\#$	$S\#$	ε
$bccd\#$	$bA\#$	2
$ccd\#$	$A\#$	2
$ccd\#$	$ccA\#$	24
$cd\#$	$cA\#$	24
$d\#$	$A\#$	24
$d\#$	$d\#$	243
$\#$	$\#$	243

(**c**)   Trace of the top-down parse for the string $abccccd\#$

Unused input string	Stack contents	Rules used
$abccccd\#$	$S\#$	ε
$abccccd\#$	$aS\#$	1
$bccccd\#$	$S\#$	1
$bccccd\#$	$bA\#$	12
$ccccd\#$	$A\#$	12
$ccccd\#$	$ccA\#$	124
$cccd\#$	$cA\#$	124
$ccd\#$	$A\#$	124
$ccd\#$	$ccA\#$	1244
$cd\#$	$cA\#$	1244
$d\#$	$A\#$	1244
$d\#$	$d\#$	12443
$\#$	$\#$	12443

4. $S \rightarrow 1A0 \mid a1C00$
       $A \rightarrow 1A0 \mid a$
       $C \rightarrow 1C00 \mid a$

Solutions 10.2

2. **(a)** $(a, a) := (a, a)\#$

Unused input string	Stack	Output Tape
$(a, a) := (a, a)\#$	$S\#$	ε
$(a, a) := (a, a)\#$	$LB\#$	1
$(a, a) := (a, a)\#$	$(EJB\#$	14
$a, a) := (a, a)\#$	$EJB\#$	14
$a, a) := (a, a)\#$	$aJB\#$	147
$, a) := (a, a)\#$	$JB\#$	147
$, a) := (a, a)\#$	$, EJB\#$	1475
$a) := (a, a)\#$	$EJB\#$	1475
$a) := (a, a)\#$	$aJB\#$	14757
$) := (a, a)\#$	$JB\#$	14757
$) := (a, a)\#$	$)B\#$	147576
$:= (a, a)\#$	$B\#$	147576
$:= (a, a)\#$	$:= L\#$	1475763
$= (a, a)\#$	$= L\#$	1475763
$(a, a)\#$	$L\#$	1475763
$(a, a)\#$	$(EJ\#$	14757634
$a, a)\#$	$EJ\#$	14757634
$a, a)\#$	$aJ\#$	147576347
$, a)\#$	$J\#$	147576347
$, a)\#$	$, EJ\#$	1475763475
$a)\#$	$EJ\#$	1475763475
$a)\#$	$aJ\#$	14757634757
$)\#$	$J\#$	14757634757
$)\#$	$)\#$	147576347576
$\#$	$\#$	147576347576

(b) Omitted due to space

4.

	S'	S	B	L	J	E	;	:	=	(	)	,	a
S'		F		F^+						F^+			
S		F								F^+			
B							F	F					
L										F			
J											F	F	
E				F						F^+			F
;													
:													
=													
(													
)													
,													
a													

The following are the first sets for the nonterminals: FIRST(S')= {(}, FIRST(S)= {(},
FIRST(B)= {;, :}, FIRST(L)= {(}, FIRST(J)= {,,)}, FIRST(E)= {(, a}

CHAPTER 10 SOLUTIONS

2. $G = (V_N, V_T, S, \Phi)$
where $V_N = \{S, A, B, C\}$, $V_T = \{a, b, c, 0, 1\}$, and Φ consist of the following rules:
$S \to A \mid 0B$
$A \to aAb \mid 0 \mid 0C$
$B \to aBc \mid 1 \mid 1C$
$C \to aC \mid a$

4. **(a)** $\{0^n 1^n \mid n \geq 1\}$
(b) Prefix Polish expressions on operators $+$ and $-$ with operands a
(c) Regular expressions on a with $+$ and $*$ denoting concatenations and closure, respectively.

6. $G = (V_N, V_T, S, \Phi)$
where $V_N = \{S\}$, $V_T = \{0, 1\}$, and Φ consist of the following rules:
$S \to A1 \mid 11 \mid 1001 \mid S0 \mid SS$
$A \to 1010A \mid 1010$

8.

$\langle \text{ProcDec} \rangle ::= $ **procedure** $\langle \text{Identifier} \rangle$; $\mid$
procedure $\langle \text{Identifier} \rangle \langle \text{Parameters} \rangle$; $\mid$
function $\langle \text{Identifier} \rangle \langle \text{Parameters} \rangle : \langle \text{IdentType} \rangle$; $\mid$
function $\langle \text{Identifier} \rangle : \langle \text{IdentType} \rangle$;

$\langle \text{Parameters} \rangle ::= (\langle \text{ParamList} \rangle)$
$\langle \text{ParamList} \rangle ::= \langle \text{ParamDec} \rangle \mid \langle \text{ParamDec} \rangle ; \langle \text{ParamList} \rangle$
$\langle \text{ParamDec} \rangle ::= \langle \text{Pnames} \rangle : \langle \text{IdentType} \rangle \mid $ **var** $\langle \text{Pnames} \rangle : \langle \text{IdentType} \rangle$
$\langle \text{Pnames} \rangle ::= \langle \text{Identifier} \rangle \mid \langle \text{Pnames} \rangle , \langle \text{Identifier} \rangle$
$\langle \text{Identifier} \rangle ::= \langle \text{Letter} \rangle \mid \langle \text{Letter} \rangle \langle \text{IdentSymbols} \rangle$
$\langle \text{IdentType} \rangle ::= $ **integer** $\mid$ **real** $\mid$ **boolean** $\mid$ **string**
$\langle \text{Letter} \rangle ::= a \mid b \mid c \mid \ldots \mid z$
$\langle \text{IdentSymbols} \rangle ::= \langle \text{Symbol} \rangle \mid \langle \text{Symbol} \rangle \langle \text{IdentSymbols} \rangle$
$\langle \text{Symbol} \rangle ::= \_ \mid a \mid .. \mid z \mid 0 \mid 1.. \mid 9$

In the parse tree that appears below, identifiers have been simplified. Rather than showing the parse of each identifier symbol by symbol, identifiers are presented as complete tokens.

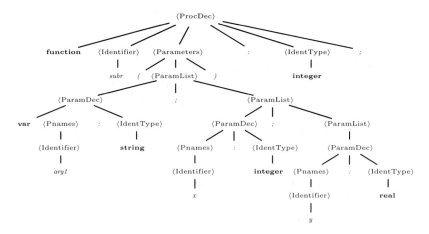

Solutions 11.1

2. (a)

```
1 │ Q
  │ ─────────
2 │   │ P
3 │   │ Q        R, 1
4 │ P ⇒ Q       ⇒I, 2-3
```

(b)

```
1 │ ¬P
  │ ──────────
2 │   │ P
  │   │ ──────────
3 │   │   │ ¬Q
  │   │   │ ──────────
4 │   │   │ ¬P         R, 1
5 │   │   │ P          R, 2
6 │   │   │ P ∧ ¬P     ∧I, 4, 5
7 │   │ ¬¬Q            ¬I, 3-5
8 │   │ Q              ¬E, 7
9 │ P ⇒ Q             ⇒I, 2-8
```

(c)

```
1 │ P
2 │ ¬Q
  │ ──────────
3 │   │ P ⇒ Q
  │   │ ──────────
4 │   │ P          R, 1
5 │   │ Q          ⇒E, 3, 4
6 │   │ ¬Q         R, 2
7 │   │ Q ∧ ¬Q     ∧E, 5, 6
8 │ ¬(P ⇒ Q)       ¬I, 3-7
```

4. $P \Rightarrow Q$
   $Q \Rightarrow P$

6. **(a)** $P \vee \neg R \vee S$ **(b)** $\neg P \vee \neg Q$ **(c)** $\neg P \vee \neg Q \vee \neg S$

8. **(a)**

1	$P_1 \Rightarrow Q_1$	
2	$P_2 \Rightarrow Q_2$	
3	$P_1 \wedge P_2$	
4	P_1	$\wedge E, 3$
5	P_2	$\wedge E, 3$
6	Q_1	$\Rightarrow E, 1, 4$
7	Q_2	$\Rightarrow E, 2, 5$
8	$Q_1 \wedge Q_2$	$\wedge I, 6, 7$
9	$P_1 \wedge P_2 \Rightarrow Q_1 \wedge Q_2$	$\Rightarrow I, 3\text{-}8$

(b) Prove that the following statements are contradictory:

$$P_1 \Rightarrow Q_1, \quad P_2 \Rightarrow Q_2, \quad \neg((P_1 \wedge P_2) \Rightarrow Q_1 \wedge Q_2)$$

Convert to clauses:

$$P_1 \Rightarrow Q_1 \equiv \neg P_1 \vee Q_1$$
$$P_2 \Rightarrow Q_2 \equiv \neg P_2 \vee Q_2$$
$$\neg((P_1 \wedge P_2) \Rightarrow Q_1 \wedge Q_2) \equiv \neg(\neg(P_1 \wedge P_2) \vee (Q_1 \wedge Q_2))$$
$$\equiv (P_1 \wedge P_2) \wedge \neg(Q_1 \wedge Q_2)$$
$$\equiv P_1 \wedge P_2 \wedge (\neg Q_1 \vee \neg Q_2)$$

We now have the following clauses:

$$\neg P_1 \vee Q_1, \neg P_2 \vee Q_2, P_1, P_2, \neg Q_1 \vee \neg Q_2$$

1. P_1	premise
2. $\neg P_1 \vee Q_1$	premise
3. Q_1	resolve lines 1, 2
4. P_2	premise
5. $\neg P_2 \vee Q_2$	premise
6. Q_2	resolve lines 4, 5
7. $\neg Q_1 \vee \neg Q_2$	premise
8. $\neg Q_1$	resolve lines 6, 7
9. F	resolve lines 8, 3

Solutions 11.2

2. **(a)** $\forall x \exists y \exists z \forall u (P(x, y) \wedge P(u, z))$
   (b) $\forall x_1 \forall y (\neg P(x_1, y) \vee \exists z Q(z)) \vee \exists u \forall v R(u, v, x)$
   $\equiv \forall x_1 \forall y \exists z \exists u \forall v (\neg P(x_1, y) \vee Q(z) \vee R(u, v, x))$

Solutions 11.3

2. 1 $\quad \exists x (U \wedge P(x))$

    2 $\quad U \wedge P(a)$ $\exists E, 1$

    3 $\quad U$ $\wedge E, 2$

    4 $\quad P(a)$ $\wedge E, 2$

    5 $\quad \exists x P(x)$ $\exists I, 4$

    6 $\quad U \wedge \exists x P(x)$ $\wedge I, 3, 5$

4. 1 $\quad \exists x (P(x) \vee Q(x))$

    2 $\quad \forall x \neg P(x)$

    3 $\quad \forall x \neg Q(x)$

    4 $\quad P(a) \vee Q(a)$ $\exists E$ of Line 1 with $x := a$

    5 $\quad \neg P(a)$ UI of Line 2 with $x := a$

    6 $\quad \neg Q(a)$ UI of Line 3 with $x := a$

    7 $\quad Q(a)$ 4, 5, DS

    8   Lines 6, 7 contradictory

6. (a) $\quad P(g(x), x, x) \vee C(b, x)$

       $\neg P(y, a, z) \vee R(z, y, g(x))$

       $P(g(a), a, a) \vee C(b, a)$ $x = a$

       $\neg P(g(a), a, a) \vee R(a, g(a), g(x))$ $y = g(a), z = a$

       $\overline{}$

       $C(b, a) \vee R(a, g(a), g(x))$

(b) $\quad P(f(x), y) \vee R(y, f(y))$

       $\neg R(a, x)$

       $P(f(x), a) \vee R(a, f(a))$ $y = a$

       $\neg R(a, f(a))$ $x = f(a)$

       $\overline{}$

       $P(f(x), a)$

(c) $\quad P(x, g(x, y), y) \vee Q(x, y)$

       $\neg P(a, x, b)$

       $P(a, g(a, b), b) \vee Q(a, b)$ $x = a, y = b$

       $\neg P(a, g(a, b), b)$ $x = g(a, b)$

       $\overline{}$

       $Q(a, b)$

CHAPTER 11 SOLUTIONS

2.
$$
\begin{array}{lll}
1 & \boxed{P \vee (P \wedge Q)} & \\
2 & \quad \boxed{P} & \\
3 & \quad \boxed{P} & R, 2 \\
4 & \quad \boxed{P \wedge Q} & \\
5 & \quad \boxed{P} & \wedge E \\
6 & \quad P & \vee E,\ 1,\ 2\text{-}3,\ 4\text{-}5 \\
7 & P \vee (P \wedge Q) \Rightarrow P & \Rightarrow I,\ 1\text{-}6 \\
8 & \quad \boxed{P} & \\
9 & \quad \boxed{P \vee (P \wedge Q)} & \vee I,\ 8 \\
10 & P \Rightarrow (P \vee (P \wedge Q)) & \Rightarrow I,\ 8\text{-}9 \\
11 & P \vee (P \wedge Q) \Leftrightarrow P & \Leftrightarrow I,\ 7,\ 10 \\
\end{array}
$$

4. Define the following:

N: The weather is nice
R: It rains
Y: The party is in the backyard
B: The party is in the basement
H: The people have a barbecue
M: The people listen to music
G: Everyone has a good time

The premises are

$$N \Rightarrow Y,\ R \Rightarrow B,\ N \vee R,\ Y \Rightarrow (H \wedge G),\ B \Rightarrow (M \wedge G)$$

The proof is as follows:

$$
\begin{array}{lll}
1 & N \Rightarrow Y & \\
2 & R \Rightarrow B & \\
3 & N \vee R & \\
4 & Y \Rightarrow (H \wedge G) & \\
5 & B \Rightarrow (M \wedge G) & \\
6 & \quad \boxed{N} & \\
7 & \quad Y & \Rightarrow E,\ 1,\ 6 \\
8 & \quad H \wedge G & \Rightarrow E,\ 4,\ 7 \\
9 & \quad G & \wedge E,\ 8 \\
10 & \quad \boxed{R} & \\
11 & \quad B & \Rightarrow E,\ 2,\ 10 \\
12 & \quad M \wedge G & \Rightarrow E,\ 5,\ 11 \\
13 & \quad G & \wedge E,\ 12 \\
14 & G & \vee E,\ 3,\ 6\text{-}9,\ 10\text{-}13 \\
\end{array}
$$

6. The conversion from the given expression to clausal form is extremely lengthy and will not be given here. The clauses are given in lines 1–5 of the following derivation:

$1\ \neg B \vee Q_1$
$2\ Q_1 \vee Q_2$

$3\ B \vee Q_2 \qquad 6\ \neg B \qquad 1, 4$

$4\ \neg B \vee \neg Q_1 \qquad 7\ B \qquad 3, 5$

$5\ B \vee \neg Q_2 \qquad 8 \qquad\qquad$ Contradiction

8. X : Patient has virus X

Y : Patient has virus Y

Z : Patient has virus Z

Because of tests, one has

1    $X \vee Y$

2    $Y \vee Z$

3    $\neg Y$

The negation of the conclusion is $\neg(X \wedge Z) \equiv \neg X \vee \neg Z$

4    $\neg X \vee \neg Z$

5   X resolve lines 1, 3

6   Z resolve lines 2, 3

7    $\neg X$ resolve lines 4, 6

8   contradiction of lines 5, 7

10. C : completeness

$P(x)$: x passed

1    $C \Rightarrow \neg \exists x P(x)$

2    $\exists x P(x)$

3   C

4        $\neg \exists x P(x)$ $\Rightarrow E,\ 1, 3$

5        $\exists x P(x)$ R, 2

6        $\exists x P(x) \wedge \neg \exists x P(x)$ $\wedge I,\ 4, 5$

7    $\neg C$ $\neg I,\ 3\text{-}6$

12.   1 $U \wedge \exists x P(x)$

2    $\exists x P(x)$ $\wedge E,\ 1$

3    $P(a)$ $\exists E,\ 2$

4   U $\wedge E,\ 1$

5    $U \wedge P(a)$ $\wedge I,\ 3, 4$

6    $\exists x(U \wedge P(x))$ $\exists I,\ 5$

14. $R(x)$: x is red

$B(x)$: x is black

1    $R(x) \vee B(x)$ Red or black

2    $\neg R(x) \vee \neg B(x)$ Not red and black

3    $R(a)$ a is red

4    $\neg B(a)$ Resolve lines 2, 3

Solutions 12.3

2. Let A be the set of attributes in P and let B be the set of attributes in Q. Then $A - B$ denotes the attributes in P which are not in Q. One has

$$P/Q = P[A - B] - (P[A - B] \times Q - P)[A - B]$$

For example, for the relations P and Q in Table 12.9, the evaluation process yields the following:

$P[x] \times Q - P$

x	y
x_2	y_3
x_4	y_1
x_4	y_3
x_5	y_1
x_5	y_2

$P[x]$

x
x_1
x_2
x_3
x_4
x_5

$P[x] \times Q$

x	y
x_1	y_1
x_2	y_1
x_3	y_1
x_4	y_1
x_5	y_1
x_1	y_2
x_2	y_2
x_3	y_2
x_4	y_2
x_5	y_2
x_1	y_3
x_2	y_3
x_3	y_3
x_4	y_3
x_5	y_3

$(P[x] \times Q - P)[x]$

x
x_2
x_4
x_5

Q

y
y_1
y_2
y_3

P/Q

x
x_1
x_3

4. $2^n - 1$

6. (a) $(SP[part\# = \text{'P3'}])[supplier\#]$

   (b) $(SP[supplier\# = \text{'S2'}])[part\#]$

   (c) $(SP * S)[supplier\#, part\#, city]$

   (d) $(((SP * P)[psize = \frac{3}{4}]) * (S[city = \text{'Toronto'} \ OR \ city = \text{'Montreal'}]))[supplier\#]$

   (e) $((S * P[pname = \text{'Spring'}]) * S)[sname]$

   (f) $(S[delvry\_time = 2 \ OR \ delvry\_time = 3])[city]$

Solutions 12.4

2. In the following table, let r be the number of attributes in P and let s be the number of attributes in Q, where P and Q are both relations. The set $1 \ldots n$ is used to designate a subset of attributes in P or Q. In the case of the join operation, this set of attributes is common to both P and Q.

Name	Relational algebra	Relational calculus
Union	$P \cup Q$	$\{f \mid f \in P \vee f \in Q\}$
Intersection	$P \cap Q$	$\{f \mid f \in P \wedge f \in Q\}$
Difference	$P - Q$	$\{f \mid f \in P \wedge \neg(f \in Q)\}$
Product	$P \times Q$	$\{(f.a_1, f.a_2, \ldots, f.a_r, g.b_1, g.b_2, \ldots, g.b_s)$ $\mid f \in P \wedge g \in Q\}$
Projection	$P[a_1, a_2, \ldots, a_n]$	$\{(f.a_1, f.a_2, \ldots, f.a_n) \mid f \in P\}$
Restriction	$P[A\ \theta\ OPER]$	$\{f \mid f \in P \wedge (f.A\ \theta\ OPER)\}$
Join	$P * Q$	$\{(f.a_1, f.a_2, \ldots, f.a_n, \ldots, f.a_r,$ $g.b_{n+1}, g.b_{n+2}, \ldots, g.b_s)$ $\mid f \in P \wedge g \in Q \wedge (f.a_1 = g.b_1)$ $\wedge (f.a_2 = g.b_2) \wedge \ldots \wedge (f.a_n = g.b_n)\}$
Divide	$P\ /\ Q$	The divide operator can be expressed in terms of the difference, product, and projection operators as was done in the second problem of Section 12.3.

4. **(a)** $\{f.supplier\# \mid f \in$ SP
  $\wedge\, \exists g(g \in$ SP
    $\wedge\, (g.part\# = \text{'P3'})$
    $\wedge\, (f.supplier\# = g.supplier\#))\}$

(b) $\{f.part\# \mid f \in$ SP
  $\wedge\, \exists g(g \in$ SP
    $\wedge\, (g.supplier\# = \text{'S2'})$
    $\wedge\, (f.part\# = g.part\#))\}$

(c) $\{(f.supplier\#, f.part\#, g.city) \mid f \in$ SP $\wedge\, g \in$ S
  $\wedge\, (f.supplier\# = g.supplier\#)\}$

(d) $\{f.supplier\# \mid f \in$ SP
  $\wedge\, \exists g, h(g \in$ P $\wedge\, h \in$ S
    $\wedge\, (g.psize = \frac{3}{4})$
    $\wedge\, (h.city = \text{'Toronto'} \vee h.city = \text{'Montreal'})$
    $\wedge\, (f.part\# = g.part\#)$
    $\wedge\, (f.supplier\# = h.supplier\#))\}$

(e) $\{f.sname \mid f \in$ S
  $\wedge\, \exists g, h(g \in$ SP $\wedge\, h \in$ P
    $\wedge\, (h.pname = \text{'Spring'})$
    $\wedge\, (g.part\# = h.part\#)$
    $\wedge\, (f.supplier\# = g.supplier\#))\}$

(f) $\{f.city \mid f \in$ S
  $\wedge\, \exists g(g \in$ S
    $\wedge\, (g.delvry\_time = 2 \vee g.delvry\_time = 3)$
    $\wedge\, (f.city = g.city))\}$

Solutions 12.5

2. **(a)** select MEMBER_OF.emp#, COMMITTEE.∗
     from MEMBER_OF, COMMITTEE
       where MEMBER_OF.com# = COMMITTEE.com#
         order by MEMBER_OF.emp#, MEMBER_OF.com#

(b) select COMMITTEE.com_name
     from COMMITTEE, MEMBER_OF
       where COMMITTEE.com# = MEMBER_OF.com#
         and MEMBER_OF.emp# = 1023

(c) select distinct emp#
     from MEMBER_OF
       where emp# <> 1023 and com# in (select com#
         from MEMBER_OF
           where emp# = 1023)

(d) select distinct com_name
     from COMMITTEE, MEMBER_OF
       where COMMITTEE.com# = MEMBER_OF.com#
         and COMMITTEE.com# not in(select com#
           from MEMBER_OF
             where emp# = 1529)

(e) select ∗
     from FACULTY
       where emp# in (select emp#
         from MEMBER_OF
           where com# not in(select com#
             from MEMBER_OF
               where emp# = 1447))

(f) select unique FACULTY.∗
     from FACULTY, COMMITTEE
       where FACULTY.fac_name = COMMITTEE.chair
         and (com_name = 'Budget' or com_name = 'Awards')

(g) select FACULTY.fac_name
     from FACULTY, MEMBER_OF, COMMITTEE
     where FACULTY.emp# = MEMBER_OF.emp#
     and MEMBER_OF.com# = COMMITTEE.com#
     and com_name = 'Appeals'
     and FACULTY.emp# not in (select emp#
         from MEMBER_OF, COMMITTEE
           where MEMBER_OF.com# = COMMITTEE.com#
             and com_name = 'Planning')

4. **(a)** select ∗
     from S

(b) select ∗
     from S
       where city = 'New York' or city = 'Chicago'
         or city = 'St. Louis' or city = 'Los Angeles'

(c) **select** part#
        from SP, S
            where SP.supplier# = S.supplier# **and** city = 'Chicago'
(d) **select distinct** part#
        from SP, S
            where SP.supplier# = S.supplier#
                and (city = 'Toronto' **or** city = 'Montreal')
(e) **select count**(part#)
        from SP
            where supplier# = 'S1'
(f) **select sum**(qty)
        from SP
            where part# = 'P6'
(g) **select** pname
        from SP, P
            where SP.part# = P.part# **and** supplier# = 'S1'
(h) **select** P.part#
        from P, SP, S
            where P.part# = SP.part#
                and SP.supplier# = S.supplier# **and** city = 'New York'
(i) **select unique** first.supplier#, second.supplier#
        from SP first, SP second
            where first.part# = second.part#
                and first.supplier# <> second.supplier#
(j) **select sum**(qty)
        from SP
            where supplier# = 'S2'
(k) **select sum**(qty)
        from SP
            where part# = 'P1'
(l) **select** part#, **sum**(qty)
        from SP
            group by part#

CHAPTER 12 SOLUTIONS

2. (a) **select avg**(unit_price)
            from MENU
    (b) **select** date, **avg**(tip)
            from BILL
                group by date
    (c) **select avg**(amount)
            from BILL
    (d) **select distinct** server#, table#
            from BILL
                where date = '19941225'
                    order by server#, table#

(e) **select** dish#, **sum**(quantity)
    from ORDER
        group by dish#
            order by **sum**(quantity)

(f) **select** MENU.dish#, **sum**(quantity $*$ unit_price)
    from MENU, ORDER
        where MENU.dish# $=$ ORDER.dish#
            group by MENU.dish#

(g) **select** server#, **sum**(amount)
    from BILL
        where date $=$ '19950609'
            group by server#
               order by **sum**(amount)

(h) **select** server#, **sum**(tip)
    from BILL
        where date $=$ '19950901'
            group by server#
               order by **sum**(tip)

4. (a) (MOVIE $*$ MOVIE_STAR[$star_name = $ 'Clint Eastwood']) [$movie\_name$]

$$\{f.movie\_name \mid f \in \text{MOVIE}$$
$$\land \exists g(g \in \text{MOVIE\_STAR}$$
$$\land (g.star\_name = \text{'Clint Eastwood'})$$
$$\land (f.star\# = g.star\#))\}$$

    select distinct MOVIE.movie_name
      from MOVIE, MOVIE_STAR
        where MOVIE.star# $=$ MOVIE_STAR.star#
         and MOVIE_STAR.star_name $=$ 'Clint Eastwood'

(b) (MOVIE $*$ DIRECTOR[$dir_name = $ 'Penny Marshall']) [$movie\_name$]

$$\{f.movie\_name \mid f \in \text{MOVIE}$$
$$\land \exists g(g \in \text{DIRECTOR}$$
$$\land (g.dir\_name = \text{'Penny Marshall'})$$
$$\land (f.dir\# = g.dir\#))\}$$

    select distinct MOVIE.movie_name
      from MOVIE, DIRECTOR
        where MOVIE.dir# $=$ DIRECTOR.dir#
         and DIRECTOR.dir_name $=$ 'Penny Marshall'

(c) (MOVIE[$type = $ 'Drama' AND $year = 1995$]) [$movie\_name$]

$$\{f.movie\_name \mid f \in \text{MOVIE}$$
$$\land \exists g(g \in \text{MOVIE}$$
$$\land (g.type = \text{'Drama'})$$

$$\wedge \, (g.year = 1995)$$
$$\wedge \, (f.movie\_name = g.movie\_name))\}$$

> **select distinct** movie_name
>   **from** MOVIE
>   **where** type = 'Drama' **and** year = 1995

(d) ((DIRECTOR * MOVIE)
      * MOVIE_STAR $[star_name = $ 'Michelle Pfeiffer'$])$ $[dir\_name]$

$\{f.dir\_name \mid f \in$ DIRECTOR
   $\wedge \, \exists g, h(g \in$ MOVIE $\wedge \, h \in$ MOVIE_STAR
     $\wedge \, (h.star_name = $ 'Michelle Pfeiffer')
     $\wedge \, (g.star\# = h.star\#)$
     $\wedge \, (f.dir\# = g.dir\#))\}$

> **select distinct** DIRECTOR.dir_name
>   **from** DIRECTOR, MOVIE, MOVIE_STAR
>   **where** DIRECTOR.dir# = MOVIE.dir#
>   **and** MOVIE.star# = MOVIE_STAR.star#
>   **and** MOVIE_STAR.star_name = 'Michelle Pfeiffer'

(e) ((DIRECTOR * MOVIE) $[country = $ 'U.S.A.' AND $year = 1995]$) $[movie\_name]$

$\{f.movie\_name \mid f \in$ MOVIE
   $\wedge \, \exists g(g \in$ DIRECTOR
     $\wedge \, (g.country = $ 'U.S.A.')
     $\wedge \, (f.dir\# = g.dir\#)$
     $\wedge \, (f.year = 1995))\}$

> **select distinct** MOVIE.movie_name
>   **from** MOVIE, DIRECTOR
>   **where** MOVIE.dir# = DIRECTOR.dir#
>   **and** DIRECTOR.country = 'U.S.A.'
>   **and** MOVIE.year = 1995

(f) (((DIRECTOR * MOVIE) * MOVIE_STAR) $[dir\_name = star\_name]$)
    $[movie\_name]$

$\{f.movie\_name \mid f \in$ MOVIE
   $\wedge \, \exists g, h(g \in$ DIRECTOR $\wedge \, h \in$ MOVIE_STAR
     $\wedge \, (g.dir\_name = h.star\_name)$
     $\wedge \, (g.dir\# = h.dir\#)$
     $\wedge \, (f.star\# = h.star\#))\}$

> **select distinct** MOVIE.movie_name
>   **from** DIRECTOR, MOVIE, MOVIE_STAR
>   **where** DIRECTOR.dir# = MOVIE.dir#

> **and** MOVIE.star# $=$ MOVIE_STAR.star#
> **and** DIRECTOR.dir_name $=$ MOVIE_STAR.star_name

(g) (MOVIE[$classification =$ 'Restricted'] $*$ MOVIE_STAR) [$star\_name$]

> $\{f.star\_name \mid f \in$ MOVIE_STAR
> $\quad \wedge \exists g(g \in$ MOVIE
> $\quad\quad \wedge \ (g.classification =$ 'Restricted')
> $\quad\quad \wedge \ (f.star\# = g.star\#))\}$

> **select distinct** MOVIE_STAR.star_name
>   **from** MOVIE_STAR, MOVIE
>   **where** MOVIE_STAR.star# $=$ MOVIE.star#
>   **and** classification $=$ 'Restricted'

(h) (MOVIE[$type =$ 'War' AND $year = 1978$] $*$ MOVIE_STAR) [$star\_name$]

> $\{f.star\_name \mid f \in$ MOVIE_STAR
> $\quad \wedge \exists g(g \in$ MOVIE
> $\quad\quad \wedge \ (g.type =$ 'War')
> $\quad\quad \wedge \ (g.year = 1978)$
> $\quad\quad \wedge \ (f.star\# = g.star\#))\}$

> **select distinct** MOVIE_STAR.star_name
>   **from** MOVIE_STAR, MOVIE
>   **where** MOVIE_STAR.star# $=$ MOVIE.star#
>   **and** type $=$ 'War'
>   **and** year $= 1978$

(i) (MOVIE[$movie\_name =$ 'Back to the Future'] $*$ MOVIE_STAR)
> [$star\_name$]

> $\{f.star\_name \mid f \in$ MOVIE_STAR
> $\quad \wedge \exists g(g \in$ MOVIE
> $\quad\quad \wedge \ (g.movie\_name =$ 'Back to the Future')
> $\quad\quad \wedge \ (f.star\# = g.star\#))\}$

> **select distinct** MOVIE_STAR.star_name
>   **from** MOVIE_STAR, MOVIE
>   **where** MOVIE_STAR.star# $=$ MOVIE.star#
>   **and** MOVIE.movie_name $=$ 'Back to the Future'

Index